The Motor Vehicle

D1354580

BELFAST INSTITUTE

3 7777 00087 9452

BELFAST INSTITUTE LIBRARIES

629·2

NEWT

✓ LS

The Motor Vehicle

Twelfth Edition

K. NEWTON

MC, BSc, ACGI, AMInstCE, MIMechE
Late Assistant Professor, Mechanical and Electrical Engineering Department,
The Royal Military College of Science

W. STEEDS

OBE, BSc, ACGI, FIMechE
Lat
The

T.

CEn
Som

BELFAST INSTITUTE LIBRARIES

DATE DUE	DATE DUE	DATE DUE
BELFAST INSTITUTE MILLFIELD LIBRARY	2 5 FEB 2003	1 4 MAR 2013
2 JAN 1999	25 MAR 2003	1 5 MAR 2013
23 MAY 2000	1 MAY 2003	4 JAN 2016
6 DEC 2000	3 MAY 2003	
0 9 MAY	2 6 OCT 2004	
11 MAR 2002	4 DEC 2004	
9.12.02	1 FEB 2005	
0 6 JAN	8 NOV 2007	
4 FEB 2003	2 6 MAY 2009	

WD 7/23

LFAST
TITUTE
LFIELD
RARY

BH

APPLICATION FOR RENEWAL MAY BE MADE
IN PERSON, BY WRITING OR BY TELEPHONE

Butterworth-Heinemann
Linacre House, Jordan Hill, Oxford OX2 8DP
A division of Reed Educational and Professional Publishing Ltd

ℛ A member of the Reed Elsevier plc group

OXFORD BOSTON JOHANNESBURG
MELBOURNE NEW DELHI SINGAPORE

First published by Iliffe & Sons 1929
Eighth edition 1966
Ninth edition 1972
Tenth edition published by Butterworths 1983
Eleventh edition 1989
Twelfth edition 1996
Paperback edition 1997

© T. K. Garrett and Exors of K. Newton and W. Steeds 1996

All rights reserved. No part of this publication may be
reproduced in any material form (including photocopying
or storing in any medium by electronic means and whether
or not transiantly or incidentally to some other use of
this publication) without the written permission of the
copyright holder except in accordance with the provisions
of the Copyright, Designs and Patents Act 1988 or under
the terms of a licence issued by the Copyright Licensing
Agency Ltd, 90 Tottenham Court Road, London, England W1P 9HE.
Applications for the copyright holder's written permission
to reproduce any part of this publication should be
addressed to the publishers

British Library Cataloguing in Publication Data
A catalogue record for this book is available from the British Library.

Library of Congress Cataloguing in Publication Data
A catalogue record for this book is available from the Library of Congress.

ISBN 0 7506 3763 3

Typeset by Vision Typesetting, Manchester
Printed and bound in Great Britain by Hartnolls Ltd, Bodmin, Cornwall.

Contents

Part 3 The Carriage Unit

Preface to the twelfth edition

Since the eleventh edition was published, the rate of technological progress in the automotive industry has been phenomenal. This situation has been primarily a result of legislation to protect the environment, though intense competition between manufacturers during the recession that started late in 1987 also contributed.

The effect has been to render a major part of what was in the eleventh edition obsolecent. Consequently, seven whole chapters have had to be almost entirely rewritten, and two new ones created. Although carburettors are obsolescent, they are still being produced for low-cost small cars, and it will be about another 10 years before they finally begin to disappear from our roads. Therefore, the chapters on carburettors and carburation have been retained, though largely rewritten and condensed. Even so, none of the essential basic information they formerly contained has been omitted. A particularly interesting addition to the chapter on engine components is a large section on the principles and practice of variable valve timing.

Since this is a work of reference, as well as a textbook, a minimal degree of duplication has been virtually inescapable because someone studying, for example, induction manifold design will want to find in the relevant chapter a brief explanation of how the fuel condenses on and evaporates from the walls. He or she certainly does not want to have to search for another chapter in which this information may have been provided for the person interested perhaps in only carburation, the properties of fuels or the processes of mixing and combustion.

One of the two additional chapters in this new edition contains a great deal of new material regarding petrol injection. Induction manifold design takes up the whole of another new chapter, which now covers induction pipe tuning and the application of the Helmholtz principle. New technology in relation to catalytic converters, the wider use of plastics fuel tanks, and the introduction of more far-reaching regulations regarding evaporative and diesel engine emissions, have together necessitated virtually completely rewriting of the chapter on emission control.

Because of the increase in the use of turbocharging, the chapter covering pressure charging in general has had to be completely rewritten. Against the background of the adoption of entirely new methods of fuel metering and mixture preparation, the chapter on fuels had become out of date and therefore, again, has been completely rewritten. This now contains a great deal of information on diesel fuel as well as petrol, and the effects of their constituents on combustion and emissions. It also includes extensive coverage of additives and alternative fuels. Yet another chapter that is almost entirely new is that on lubricants, in which large sections on additives and synthetic lubricants have been added, and which covers aspects such as the influence of lubricating oil on both exhaust and evaporative emissions.

To make room for such a huge amount of new material, something had to be cut, so Part 1, on Newton's laws, work and energy, etc., has been dropped. It was felt that this omission is unlikely to be missed by readers of a book at the level of *The Motor Vehicle*, because they obviously will already have learned these elementary fundamental principles at a relatively early stage in their education.

Units and abbreviation

Calorific value	kilojoules per kilogram	kJ/kg
	megajoules per litre	MJ/l
Specific fuel consumption	kilograms per kilowatt hour	kg/kWh
Length	millimetres, metres, kilometres	mm, m, km
Mass	kilograms, grams	kg, g
Time	seconds, minutes, hours	s, min, h
Speed	centimetres per second, metres per second	cm/s, m/s
	kilometres per hour, miles per hour	km/h, mph
Acceleration	metres-per-second per second	m/s^2
Force	newtons, kilonewtons	N, kN
Moment	newton-metres	Nm
Work	joules	J
Power	horsepower, watts, kilowatts	hp, W, kW
Pressure	newtons per square metre	N/m^2
	kilonewtons per square metre	kN/m^2
Angles	radians	rad
Angular speed	radians per second	rad/s
	radians-per-second per second	rad/s^2
	revolutions per minute	rev/min
	revolutions per second	rev/s

SI units and the old British units:

Length	1 m = 3.281 ft	1 ft = 0.3048 m
	1 km = 0.621 mile	1 mile = 1.609 km
Speed	1 m/s = 3.281 ft/s	1 ft/s = 0.305 m/s
	1 km/h = 0.621 mph	1 mph = 1.61 km/h
Acceleration	$1\ m/s^2 = 3.281\ ft/s^2$	$1\ ft/s^2 = 0.305\ m/s^2$
Mass	1 kg = 2.205 lb	1 lb = 0.454 kg
Force	$1\ N = 1\ kg\ m/s^2 = 0.225\ lbf$	1 lbf = 4.448 N

Torque	1 Nm = 0.738 lbf ft	1 lbf ft = 1.356 Nm
Pressure	1 N/m^2 = 0.000145 lbf/in^2	1 lbf/in^2 = 6.895 kN/m^2
	1 Pa = 1 N/m^2 = 0.000001 bar	
	1 bar = 14.5038 lbf/in^2	1 lbf/in^2 = 0.068947 bar
Energy, work	1 J = 0.738 ft lbf	1 ft lbf = 1.3558 J
	1 J = 0.239 calorie	1 calorie = 4.186 J
	1 kJ = 0.9478 Btu	1 Btu = 1.05506 kJ
		(1 therm = 100 000 Btu)
	1 kJ = 0.526 CHU	1 CHU = 1.9 kJ
Power	1 kW = 1.34 bhp = 1.36 PS	1 hp = 0.7457 kW
Fuel cons.	1 mpg = 0.003541/100 km	1 l/100 km = 282.48 mpg
Specific fuel	1 kg/kWh = 1.645 lb/bhp h	1 lb/bhp h = 0.6088 kg/kWh
consumption	1 litre/kWh = 1.316 pt/bhp h	1 pt/bhp h = 0.76 litre/kWh
Calorific value	1 kJ/kg = 0.4303 Btu/lb	1 Btu/lb = 2.324 kJ/kg
	1 kJ/kg = 0.239 CHU/lb	1 CHU/lb = 4.1868 kJ/kg
Standard gravity	9.80665 m/s^2 = 32.1740 ft/s^2	

Part 1

The Engine

BELFAST
INSTITUTE
MILLFIELD
LIBRARY

Chapter 1

General principles of heat engines

The petrol or oil engine, which is the source of power with which we are immediately concerned, is a form of internal combustion 'heat engine', the function of which is to convert potential heat energy contained in the fuel into mechanical work.

It is outside the scope of the present volume to go deeply into the physical laws governing this conversion, for a full study of which a work such as A. C. Walshaw's *Thermodynamics for Engineers* (Longmans-Green) should be consulted. It will not be out of place, however, to give a brief outline of the general principles.

1.1 Heat and work

A quantity of heat is conveniently measured by applying it to raise the temperature of a known quantity of pure water.

The unit of heat is defined as that quantity of heat required to raise the temperature of unit weight of water through one degree, this quantity depending, of course, on the particular unit of weight and the temperature scale employed.

The Continental European and scientific temperature scale has been the Centigrade scale, now called *Celsius* because of possible confusion with the French meaning of the word *centigrade* – one ten thousandth of a right angle. The interval between the temperatures of melting ice and boiling water (at normal pressure) is divided into one hundred, though the unsatisfactory Fahrenheit scale, which divides the foregoing interval into 180 divisions, has been the commercial standard in Britain and the USA.

It is thus necessary to define by name three different units of heat as follows—

The British Thermal Unit (Btu): The heat required to raise the temperature of 1 lb of water through 1 °F. (1 Btu = 1.05506 kJ.)
The Pound Calorie or Centigrade Heat Unit (CHU): The heat required to raise the temperature of 1 lb of water through 1 °C. (1 CHU = 1.9 kJ.)
The Kilogram Calorie: The heat required to raise the temperature of 1 kg of water through 1 °C. (1 calorie = 4.186 J.)

3

The first and second of these units are clearly in the ratio of the Fahrenheit degree to the Centigrade degree, or 5:9, while the second and third are in the ratio of the pound to the kilogram or 1:2.204. Thus the three units in the order given are in the ratio 5, 9, 19.84, or 1, 1.8, 3.97. Now we use the *joule*, which approximately equals 0.24 calories.

The *therm*, formerly used by Gas Boards, is 100 000 British thermal units (Btu).

1.2 Work

If work is done by rotating a shaft, the quantity of work is the product of the torque or turning moment applied to the shaft in newton metres, multiplied by the angle turned through measured in radians. One revolution equals 2π radians.

1.3 Joule's equivalent

Dr Joule was the first to show, in the middle of the last century, that heat and work were mutually convertible one to the other, being, in fact, different forms of energy, and that when a definite quantity of work is expanded wholly in producing heat by friction or similar means, a definite quantity of heat is produced. His experiments, confirmed and corrected by others, showed that 778 foot-pounds produce one Btu, or 1400 foot-pounds produce one CHU.

This figure is called *the mechanical equivalent of heat*, though it would perhaps be better to speak of the *thermal equivalent of work*. For though the same equivalent or rate of exchange holds for conversion in either direction, while it is comparatively simple to convert to heat by friction the whole of a quantity of work supplied, it is not possible, in a heat engine, to convert to mechanical work more than a comparatively small percentage of the total *heat* supplied. There are definite physical laws which limit this percentage – or *thermal efficiency* as it is called – to about 50% or less in the best heat engines that it is practicable to construct.

1.4 Thermal efficiency

The thermal efficiency is governed chiefly by the range of temperature through which the working fluid, be it gas or steam, passes on its way through the engine.

This range of temperature is greater in internal combustion engines than in steam engines, hence the former are inherently capable of higher thermal efficiencies, that is to say, they are capable of converting into work a higher percentage of the total heat of the fuel with which they are supplied than the latter. Even so, the physical limitations are such that the thermal efficiency of a good petrol engine is not more than about 28%. The remaining heat supplied, which is *not* converted into work, is lost in the exhaust gases and cooling water, and in radiation.

1.5 Calorific value

When unit weight of any fuel is completely burnt with oxygen (pure or diluted with nitrogen as in the air), a certain definite quantity of heat is liberated, depending on the chemical composition, that is, on the quantities of the fundamental fuels, carbon and hydrogen, which one pound of the fuel contains.

To determine how much potential heat energy is being supplied to an engine in a given time, it is necessary to know the weight of fuel supplied and its calorific value, which is the total quantity of heat liberated, when unit weight of the fuel is completely burnt.

The calorific values of carbon and hydrogen have been experimentally determined with considerable accuracy, and are usually given as—

| Carbon | 33 000 kJ/kg, or 14 200 Btu/lb |
| Hydrogen | 144 300 kJ/kg, or 62 100 Btu/lb |

The calorific value of any fuel, consisting, as all important fuels do, of a known proportion of carbon, hydrogen and incombustible impurities or diluents, may be estimated approximately on the assumption that it consists simply of a mixture of carbon, hydrogen and incombustible matter, but the state of chemical combination in the actual fuel leads to error by this method, and the only accurate and satisfactory means of determination is experimentally by the use of a suitable calorimeter.

Average petrol consists approximately of 85% carbon and 15% hydrogen by weight, the lighter fractions containing a higher percentage of hydrogen than the heavier. Refined petrol contains no measurable impurities or diluents. Its gross calorific value is about 46 000 kJ/kg, or 19 800 Btu/lb.

Liquid fuels are usually measured by volume, and therefore it is necessary to know the density before the potential heat supplied in any given case can be determined, for example—

A sample of petrol has a calorific value of 46 000 kJ/kg; its specific gravity is 0.72. How much potential heat energy is contained in 8 litres?

(1 cm^3 of water weighs 1 g.)

Weight of 8 litres = 8 × 0.72 = 5.76 kg.

Thus, the potential heat in 8 litres = 5.76 × 46 000 = 264 900 kJ.

1.6 Power

Power is the *rate* at which work is done, 1 hp being defined (by James Watt) as a rate of working of 33 000 ftlb per minute, or 550 per second. (1 hp = 745.7 W).

Problem: What is the thermal efficiency in the following case?

An engine develops 22.4 kW and consumes 10.25 litres of fuel per hour, the calorific value being 46 000 kJ/kg and the specific gravity 0.72.

Potential heat supplied per hour = 10.25 × 0.72 × 46 000 = 339 300 kJ.

Since 1 W = 1 J/s, work done per hour = 22.4 × 60 × 60 = 79 950 kJ.

Therefore the thermal efficiency is $\dfrac{79\,950}{339\,300}$ = 23.6%.

1.7 General method of conversion of heat to work

All heat engines convert heat into work by the expansion or increase in volume of a working fluid into which heat has been introduced by combustion of a fuel either external to the engine, as in a steam engine, or internally by the burning of a combustible mixture in the engine itself, a process giving rise to the phrase *internal combustion (ic) engine.*

Thus, in all so-called *static pressure engines*, as distinct from turbines, it is necessary to provide a working vessel, the volume of which is capable of variation, work being done on a moving portion of the wall by the static pressure of the working fluid as its volume increases. In general, both the pressure and temperature fall with the increase of volume.

1.8   Practical form of working vessel

In practice it has been found that for mechanical and manufacturing reasons the most satisfactory form of working chamber is a straight cylinder closed at one end and provided with a closely fitting movable plug or 'piston' on which the work is done by the pressure of the steam or gases. This arrangement is common to steam, gas, oil and petrol engines.

1.9   Rotary and reciprocating engines

The motion of the piston in the cylinder of the above arrangement is, of course, in a straight line, whereas in the majority of applications the final motion required is a rotative one.

Very many attempts have been made to devise a form of chamber and piston to give rotary motion directly, but practically all have been mechanical failures, the chief weaknesses being excessive friction and difficulty in maintaining pressure tightness. A design that achieved a limited degree of success was the NSU Wankel engine, described in Section 5.7. The universally established 'direct acting engine mechanism' with connecting rod and crank is, however, unlikely to be generally replaced in the near future. Thus in most applications the reciprocating motion of the piston must be converted to rotation of the crank by a suitable mechanism. The most important of these mechanisms are—

(1) The crank and connecting rod.
(2) The crank and cross-slide as used in the donkey pump and small steam launch engines.
(3) The 'swash-plate' or 'slant' mechanism.
(4) The 'wobble plate' or Z crank.

The second of these is not used in the applications with which we are concerned, owing to its undue weight and friction loss, and the third is, in general, confined to pumps and compressors for the conversion of rotary into reciprocating motion.

The Mitchell crankless engine, though not produced commercially, used the swash-plate in conjunction with the Mitchell thrust bearing which has eliminated the chief objection to the swash-plate, namely, excessive friction and low mechanical efficiency.

The first-mentioned mechanism is practically universal in internal combustion engines owing to its simplicity and high mechanical efficiency. We thus arrive at the fundamental parts common to all reciprocating engines having a crank and connecting rod, though the new rocking piston variant, Section 18.18, is of considerable interest.

1.10   Cylinder, piston, connecting rod and crankshaft

These fundamental parts of the conventional engine are shown in simple diagrammatic form in Fig. 1.1.

In this figure the crank is of the single web or 'overhung' type, as used in many steam engines, and certain motor cycle engines, but the double-web type, with a bearing on each side of the crank, is practically universal for internal combustion engines. This is illustrated in Fig. 1.2, which shows a cross-section through the cylinder, piston and connecting rod of the engine. A flywheel is mounted on the end of the crankshaft. The form and construction of the parts are considered later, only sufficient description being given here to enable their functions to be understood.

Cylinder. The ideal form consists of a plain cylindrical barrel in which the piston slides, the movement of the piston or 'stroke' being, in some cases, somewhat longer than the bore, but tending to equality or even less since the abandonment of the Royal Automobile Club (RAC) rating for taxation purposes. (See Section 1.19.) This is known as the *stroke : bore ratio*.

The upper end consists of a combustion or 'clearance' space in which the ignition and combustion of the charge take place. In practice it is necessary to depart from the ideal hemispherical shape in order to accommodate the valves, sparking plug, etc., and to control the process of combustion.

Piston. The usual form of piston for internal combustion engines is an inverted bucket-shape, machined to a close (but free sliding) fit in the cylinder barrel. Gas tightness is secured by means of flexible 'piston rings' fitting closely in grooves turned in the upper part of the piston.

The pressure of the gases is transmitted to the upper end of the connecting rod through the 'gudgeon pin' on which the 'small end' of the connecting rod is free to swing.

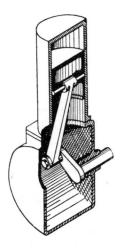

Fig. 1.1

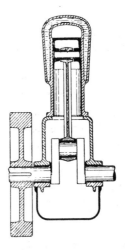

Fig. 1.2

Connecting rod. The connecting rod transmits the piston load to the crank, causing the latter to turn, thus converting the reciprocating motion of the piston into a rotary motion of the crankshaft. The lower end, or 'big end', of the connecting rod turns on the crank pin.

Crankshaft. In the great majority of internal combustion engines this is of the double-web type, the crank pin, webs and shaft being usually formed from a solid forging. The shaft turns in two or more main bearings (depending on the number and arrangement of the cylinders) mounted in the main frame or 'crankcase' of the engine.

Flywheel. At one end the crankshaft carries a heavy flywheel, the function of which is to absorb the variations in impulse transmitted to the shaft by the gas and inertia loads and to drive the pistons over the dead points and idle strokes. In motor vehicles the flywheel usually forms one member of the clutch through which the power is transmitted to the road wheels.

The foregoing are the fundamental and essential parts by which the power developed by the combustion is caused to give rotation to the crankshaft, the mechanism described being that of the *single-acting engine*, because a useful impulse is transmitted to the crankshaft while the piston moves in one direction only.

Most steam engines and a few large gas engines work on the *double-acting* principle, in which the pressure of the steam or gaseous combustion acts alternately on each side of the piston. The cylinder is then double-ended and the piston takes the form of a symmetrical disc. The force acting on the piston is transmitted through a 'piston rod' to an external 'cross-head' which carries the gudgeon pin. The piston rod passes through one end of the cylinder in a 'stuffing-box' which prevents the escape of steam or gas.

1.11 Method of working

It is now necessary to describe the sequence of operations by which the combustible charge is introduced, ignited and burned and finally discharged after it has completed its work.

There are two important 'cycles' or operations in practical use, namely, the 'four-stroke', or 'Otto' cycle as it is sometimes called (after the name of the German engineer who first applied it in practice), and the 'two-stroke', or 'Clerk' cycle, which owed its early development largely to Sir Dugald Clerk.

The cycles take their names from the number of single piston strokes which are necessary to complete a single sequence of operations, which is repeated continuously so long as the engine works.

The first named is by far the most widely adopted except for small motor cycle and motor boat engines, and for large diesels, for though it leads to greater mechanical complication in the engine, it shows higher thermal efficiency, and therefore greater economy in fuel. This cycle will therefore be described first, the two-stroke cycle being left until Chapter 7.

1.12 The four-stroke cycle

Fig. 1.3 shows in a diagrammatic manner a four-stroke engine cylinder provided with two valves of the 'mushroom' or 'poppet' type. The cylinder is shown horizontal for convenience.

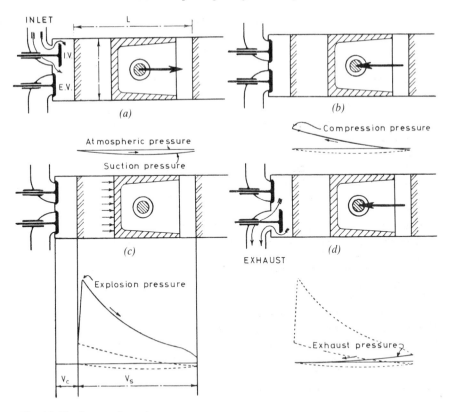

Fig. 1.3 The four-stroke cycle

The inlet valve (IV) communicates through a throttle valve with the carburettor or vaporiser, from which a combustible mixture of fuel and air is drawn. The exhaust valve (EV) communicates with the silencer through which the burnt gases are discharged to the atmosphere. These valves are opened and closed at suitable intervals by mechanisms, which will be described later.

The four strokes of the complete cycle are shown at (*a*), (*b*), (*c*) and (*d*).

Below the diagrams of the cylinder are shown the corresponding portions of what is known as the *indicator diagram*, that is to say, a diagram which shows the variation of pressure of the gases in the cylinder throughout the cycle. In practice such diagrams can be automatically recorded when the engine is running by a piece of apparatus known as an *indicator*, of which there are many types.

The four strokes of the cycle are as follows—

(*a*) *Induction stroke – exhaust valve closed: inlet valve open*
The momentum imparted to the flywheel during previous cycles or rotation by hand or by starter motor, causes the connecting rod to draw the piston outwards, setting up a partial vacuum which sucks in a new charge of combustible mixture from the carburettor. The pressure will be below atmospheric pressure by an amount which depends upon the speed of the engine and the throttle opening.

(b) Compression stroke – both valves closed
The piston returns, still driven by the momentum of the flywheel, and compresses the charge into the combustion head of the cylinder. The pressure rises to an amount which depends on the 'compression ratio', that is, the ratio of the full volume of the cylinder when the piston is at the outer end of its stroke to the volume of the clearance space when the piston is at the inner (or upper) end. In ordinary petrol engines this ratio is usually between 6 and 9 and the pressure at the end of compression is about 620.5 to 827.4 kN/m² , with full throttle opening.

$$\text{Compression ratio} = \frac{V_s + V_c}{V_c} \qquad\qquad V_s = \frac{\pi}{4} D^2 \times L$$

(c) Combustion or working stroke – both valves closed
Just before the end of the compression stroke, ignition of the charge is effected by means of an electric spark, and a rapid rise of temperature and pressure occurs inside the cylinder. Combustion is completed while the piston is practically at rest, and is followed by the expansion of the hot gases as the piston moves outwards. The pressure of the gases drives the piston forward and turns the crankshaft thus propelling the car against the external resistances and restoring to the flywheel the momentum lost during the idle strokes. The pressure falls as the volume increases.

(d) Exhaust stroke – inlet valve closed: exhaust valve open
The piston returns, again driven by the momentum of the flywheel, and discharges the spent gases through the exhaust valve. The pressure will be slightly above atmospheric pressure by an amount depending on the resistance to flow offered by the exhaust valve and silencer.

It will thus be seen that there is only one working stroke for every four piston strokes, or every two revolutions of the crankshaft, the remaining three strokes being referred to as *idle strokes*, though they form an indispensable part of the cycle. This has led engineers to search for a cycle which would reduce the proportion of idle strokes, the various forms of the two-stroke engine being the result. The correspondingly larger number of useful strokes per unit of time increases the power output relative to size of engine, but increases thermal loading.

1.13   Heat balance

It is instructive to draw up in tabular form a heat balance, arranging the figures in a manner similar to those on a financial sheet. On one side, place the figure representing the total heat input, in the form of the potential chemical energy content of the fuel supplied, assuming it is all totally burned in air. Then, on the opposite side, place the figures representing the energy output, in the form of useful work done by the engine, and all the losses such as those due to friction, heat passing out through the exhaust system, and heat dissipated in the coolant and in general radiated from the engine structure.

To draw up such a heat balance, measurements are taken of rate of mass flow and temperature of coolant and exhaust gas, radiated losses, work done and friction losses, etc. Inevitably, however, this leaves some of the heat unaccounted for. This unaccounted loss can, of course, be due to some serious

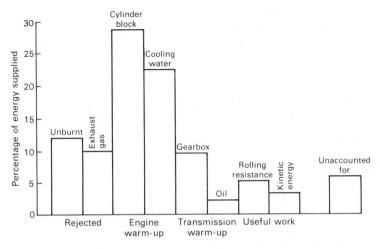

Fig. 1.4(*a*) Energy usage of a 2-litre car during the first phase (or warm-up) of the EEC 15-cycle

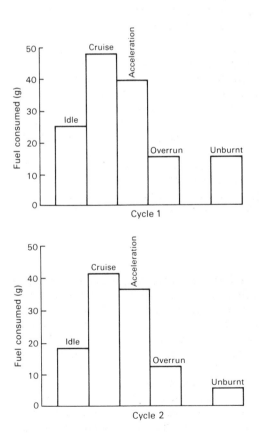

Fig. 1.4(*b*) Fuel usage of a 2-litre car during the first two stages of the EEC 15-cycle

errors of measurement, but it mostly arises mainly because the fuel has been incompletely burned.

For a diesel engine at full load, about 45% of the heat energy supplied goes to useful work on the piston, though some of this is then lost in friction. The cooling water takes away about 25% and radiation and exhaust approximately 30%. Under similar conditions in a petrol engine, approximately 32% of the total heat supplied goes to useful work on the piston, the coolant takes away about 28% and radiation and exhaust about 40%. The principal reason for the differences is that the compression ratio, and therefore expansion ratio, of the petrol engine is only about 10:1 while that of the diesel engine is around 16:1.

An extremely detailed analysis of the overall losses of energy, including those in the transmission, tyres, etc., of a saloon car, powered by a 2-litre, four-cylinder engine, operated on the EEC 15 cycle (Chapter 11), can be found in Paper 30/86 by D. J. Boam, in *Proc. Inst. Mech. Engrs*, Vol. 200, No. D1. One of the interesting conclusions in the paper was that, of the fuel energy supplied during the first phase (or warm-up) of the EEC cycle, 60% was used to warm up the engine and transmission, 12% was rejected in the form of carbon monoxide and unburned fuel, and only 8.5% went to produce useful work, Fig. 1.4 (*a*). Much of the waste was attributable to the use, during warm-up, of the strangler, or choke. In Fig. 1.4 (*b*), reproduced from the same paper, the fuel usage for the first two EEC cycles is shown.

1.14 Factors governing the mean effective pressure

The mean effective pressure depends primarily on the number of potential heat units which can be introduced into the cylinder in each charge. When the volatile liquid fuels are mixed with air in the chemically correct proportions, the potential heat units per cubic metre of mixture are almost exactly the same in all cases, being about $962 \, kcal/m^3 = 4.050 \, MJ/m^3$ at standard temperature and pressure.

The 'volumetric efficiency' represents the degree of completeness with which the cylinder is re-charged with fresh combustible mixture and varies with different engines and also with the speed.

The 'combustion efficiency' represents the degree of completeness with which the potential heat units in the charge are produced as actual heat in the cylinder. Its value depends on a variety of factors, among the more important of which are the quality of the combustible mixture, nature of fuel, quality of ignition, degree of turbulence, and temperature of cylinder walls.

Lastly, the 'thermal efficiency' governs the percentage of the actual heat units present in the cylinder which are converted into mechanical work.

In engine tests the phrase 'thermal efficiency' is taken comprehensively to include combustion efficiency as well as conversion efficiency, as in practice it is impossible to separate them.

They are further combined with the mechanical efficiency where this cannot be separately measured, as 'brake thermal efficiency'.

It can be shown theoretically that the conversion efficiency is increased with an increase in compression ratio, and this is borne out in practice, but a limit is reached owing to the liability of the high compression to lead to detonation of the charge, or *pinking* as it is popularly called. This tendency to detonation varies with different fuels, as does also the limiting compression ratio which, with low grade fuel, generally lies between 6 and $7\frac{1}{2}$. With better fuels a higher compression ratio (8 to $9\frac{1}{2}$) is possible, owing to the greater freedom from risk of detonation. (See Chapter 14.)

It thus follows that for the same volumetric efficiency, compression ratio and thermal efficiency the mean effective pressure will be practically the same for all liquid fuels. This is borne out in practice.

The thermal efficiency of an internal combustion engine of a given type does not depend very much on the *size* of the cylinders. With small cylinders, the loss of heat through the jacket may be proportionately greater, but the compression ratio may be higher.

The highest mean effective pressure obtained without supercharging, and using petrol as fuel, is about $1103.6\,\text{kN/m}^2$, but this is exceptional and very little below the theoretical maximum. A more normal figure to take in good conditions with full throttle is about $896\,\text{kN/m}^2$.

1.15 Work per minute, power and horsepower

Let p = mean effective pressure, N/m^2.
D = diameter of cylinders, m.
L = length of stroke, m.
N = revolutions per minute.
f = number of effective strokes, or combustions, per revolution per cylinder, that is, half for a four-stroke engine.
n = number of cylinders.

Then

$$\text{Force acting on one piston} = p\,\frac{\pi D^2}{4}\ \text{newtons}$$

$$\text{Work done per effective stroke} = p\,\frac{\pi D^2 L}{4}\ \text{newton-metres} = \text{joules}$$

$$\text{Work done per revolution} = p\,\frac{\pi D^2}{4}Lf\ \text{joules}$$

$$\text{Work done per minute} = p\,\frac{\pi D^2}{4}LfN\ \text{joules}$$

Since the SI unit of power is the *watt* (W), or one joule per second, the power per cylinder in SI units is—

$$\frac{p\pi D^2 LfN}{4 \times 60}\ \text{W}$$

and for the whole engine—

$$= \frac{p\pi D^2 LfnN}{4 \times 60}\ \text{W}$$

Incidentally, since 1 hp is defined as the equivalent of 550 ft lbf of work per second (Section 1.6), it can be shown that the formula for horsepower is precisely the same as that for the power output in watts, except that p, D and L

are in units of lbf/in^2, in and ft, and the bottom line of the fraction is multiplied by 550.

Following the formation of the European common market, manufacturers tended to standardise on the DIN (*Deutsche Industrie Norm* 70 020) horse-power, which came to be recognised as an SI unit. In 1995, however, the ISO (International Standards Organisation) decreed that horsepower must be determined by the ISO 1585 standard test method. This standard calls for correction factors differing from those of the DIN as follows: 25°C instead of 20°C and 99 kPa instead of 1013 bar, respectively, for atmospheric tempera-ture and pressure, and these make it numerically 3% lower than the DIN rating. The French CV (*chevaux*) and the German PS (*pferdestarke*), both meaning 'horse power', must be replaced by the SI unit, the kilowatt, 1 kW being 1.36 PS.

1.16   Piston speed and the RAC rating

The total distance travelled per minute by the piston is 2*LN*. Therefore, by multiplying by two the top and bottom of the fraction in the last equation in Section 1.15, and substituting *S* – the mean piston speed – for 2*LN*, we can express the power as a function of *S* and *p*: all the other terms are constant for any given engine. Since the maximum piston speed and bmep (see Section 1.14) tend to be limited by the factors mentioned in Section 1.19, it is not difficult, on the basis of the dimensions of an engine, to predict approximately what its maximum power output will be.

It was on these lines that the RAC horsepower rating, used for taxation purposes until just after the Second World War, was developed. When this rating was first introduced, a piston speed of 508 cm/s and an mep of 620 kN/m^2 and a mechanical efficiency of 75% were regarded as normal. Since 1 hp is defined as 33 000 lbf work per minute, by substituting these figures, and therefore English for SI metric dimensions in the formula for work done per minute, Section 1.15, and then dividing by 33 000, we get the output in horsepower. Multiplied by the efficiency factor of 0.75, this reduces to the simple equation—

$$\text{bhp} = \frac{D^2 n}{2.5} = \text{RAC hp rating}$$

For many years, therefore, the rate of taxation on a car depended on the square of the bore. However, because it restricted design, this method of rating for taxation was ultimately dropped and replaced by a flat rate. In the meantime, considerable advances had been made: mean effective pressures of 965 to 1100 kN/m^2 are regularly obtained; improved design and efficient lubrication have brought the mechanical efficiency up to 85% or more; and lastly, but most important of all, the reduction in the weight of reciprocating parts, and the proper proportioning of valves and induction passages, and the use of materials of high quality, have made possible piston speeds of over 1200 cm/s.

1.17   Indicated and brake power

The power obtained in Section 1.15 from the indicator diagram (that is, using the mep) is known as the *indicated power output* or *indicated horsepower* (ipo or ihp), and is the power developed inside the engine cylinder by the combustion of the charge.

The useful power developed at the engine shaft or clutch is less than this by the amount of power expended in overcoming the frictional resistance of the engine itself. This useful power is known as the *brake power output* or *brake horsepower* (bpo or bhp) because it can be absorbed and measured on the test bench by means of a friction or fan brake. (For further information on engine testing the reader is referred to *The Testing of Internal Combustion Engines* by Young and Pryer, EUP.)

1.18   Mechanical efficiency

The ratio of the brake horsepower to the indicated horsepower is known as the *mechanical efficiency*.

Thus—

$$\text{Mechanical efficiency} = \frac{\text{bpo}}{\text{ipo}} \quad \text{or} \quad \frac{\text{bhp}}{\text{ihp}} = \eta$$

and

bpo or bhp = Mechanical efficiency × ipo (or ihp)

For SI units, $\eta \times pD^2S \times n \times \dfrac{\pi}{4 \times 60 \times 2 \times 2}$

For BS units, divide by 550, as explained in Section 1.15.

$$= \frac{\eta pD^2S}{305.58} \quad \text{or} \quad \frac{\eta pD^2S}{168\,067}$$

$$= \frac{\eta pDSn}{305.58} \quad \text{or} \quad \frac{\eta^2Sn}{168\,067}$$

1.19   Limiting factors

Let us see to what extent these factors may be varied to give increased power.

It has been shown that the value of p depends chiefly on the compression ratio and the volumetric efficiency, and has a definite limit which cannot be exceeded without supercharging.

The diameter of the cylinder D can be increased at will, but, as is shown in Section 1.24, as D increases so does the weight per horsepower, which is a serious disadvantage in engines for traction purposes. There remain the piston speed and mechanical efficiency. The most important limitations to piston speed arise from the stresses and bearing loads due to the inertia of the reciprocating parts, and from losses due to increased velocity of the gases through the valve ports resulting in low volumetric efficiency.

A comparison of large numbers of engines of different types, but in similar categories, shows that piston speeds are sensibly constant within those categories. For example, in engines for applications where absolute reliability over very long periods is of prime importance, weight being only a secondary consideration, piston speeds are usually between about 400 and 600 cm/s, and for automobile engines, where low weight is much more important, piston speeds between about 1000 and 1400 cm/s are the general rule. In short, where the stroke is long, the revolutions per minute are low, and *vice versa*.

1.20   Characteristic speed power curves

If the mean effective pressure (mep) and the mechanical efficiency of an engine remained constant as the speed increased, then both the indicated and brake horsepower would increase in direct proportion to the speed, and the characteristic curves of the engine would be of the simple form shown in Fig. 1.5, in which the line marked 'bmep' is the product of indicated mean effective pressure (imep) and mechanical efficiency, and is known as *brake mean effective pressure* (bmep). Theoretically there would be no limit to the horsepower obtainable from the engine, as any required figure could be obtained by a proportional increase in speed. It is, of course, hardly necessary to point out that in practice a limit is imposed by the high stresses and bearing loads set up by the inertia of the reciprocating parts, which would ultimately lead to fracture or bearing seizure.

Apart from this question of mechanical failure, however, there are reasons which cause the characteristic curves to vary from the simple straight lines of Fig. 1.5, and which result in a point of maximum brake horsepower being reached at a certain speed which depends on the individual characteristics of the engine.

Characteristic curves of an early four-cylinder engine of 76.2 mm bore and 120.65 mm stroke are given in Fig. 1.6. The straight radial lines tangential to the actual power curves correspond to the power lines in Fig. 1.5, but the indicated and brake mean pressures do not, as was previously assumed, remain constant as the speed increases.

On examining these curves it will be seen first of all that the mep is not constant. It should be noted that full throttle conditions are assumed – that is, the state of affairs for maximum power at any given speed.

At low speeds the imep is less than its maximum value owing partly to carburation effects, and partly to the valve timing being designed for a moderately high speed; it reaches its maximum value at about 1800 rev/min, and thereafter decreases more and more rapidly as the speed rises. This falling off at high speeds is due almost entirely to the lower volumetric efficiency, or less complete filling of the cylinder consequent on the greater drop of pressure

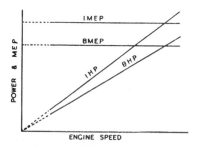

Fig. 1.5

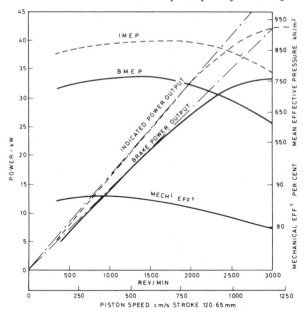

Fig. 1.6 Power curves of typical early side-valve engine, 3-in bore and $4\frac{3}{4}$-in stroke (76.2 and 120.65 mm)

absorbed in forcing the gases at high speeds through the induction passages and valve ports.

When the mep falls at the same rate as the speed rises, the horsepower remains constant, and when the mep falls still more rapidly the horsepower will actually decrease as the speed rises. This falling off is even more marked when the bmep is considered, for the mechanical efficiency decreases with increase of speed, owing to the greater friction losses. The net result is that the bhp curve departs from the ideal straight line more rapidly than does the ihp curve. The bmep peaks at about 1400 rev/min, the indicated power at 3200 and the brake power at 3000 rev/min, where 33.5 kW is developed.

Calculations of bmep P from torque, and *vice versa* are made using the following formula—

$$P = \frac{2\pi Q}{LANn} \text{ bar}$$

or

$$Q = \frac{PLANn}{2\pi} \text{ newton}$$

This applies to a two-stroke engine, which has one power stroke per revolution. Because a four-stroke engine has only one power stroke every two revolutions, we must halve the result, so we have—

$$\text{bmep} = 125.66 \times 10^{-5} \times 10^{6} \times Q/V \quad \text{kilonewton}$$

where $V = LAN$ = the piston displacement in cm^3 and 1 bar = 0.000001 N/m^2.

1.21   Torque curve

If a suitable scale is applied, the bmep curve becomes a 'torque' curve for the engine, that is, it represents the value, at different speeds, of the mean torque developed at the clutch under full throttle conditions – for there is a direct connection between the bmep and the torque, which depends only on the number and dimensions of the cylinders, that is, on the total swept volume of the engine. This relationship is arrived at as follows—

If there are n cylinders, the total work done in the cylinders per revolution is—

$$\text{Work per revolution} = p \times \frac{\pi}{4} D^2 \times L \times n \times \tfrac{1}{2} \text{ joules}$$

(see Section 1.15), if L is here the stroke in metres. Therefore the work at the clutch is—

$$\text{Brake work} = \eta \times p \times \frac{\pi}{4} D^2 L \times n \times \tfrac{1}{2} \text{ joules.}$$

But the work at the clutch is also equal to the mean torque multiplied by the angular distance moved through in radians, or $T \times 2\pi$ newton-metres per revolution if T is measured in SI units.

Therefore—

$$T \times 2\pi = \eta \times p \times \frac{\pi}{4} D^2 \times L \times n \times \tfrac{1}{2}$$

or,

$$T = \eta p \times \frac{\left(\frac{\pi}{4} D^2 \times L \times n \right)}{4\pi}$$

Now $\frac{\pi}{4} D^2 \times L \times n$ is the total stroke volume or cubic capacity of the engine, which may be denoted by V. Therefore we have—

$$T = \eta p \times \frac{V}{4\pi}$$

where ηp the bmep and $V/4\pi$ is a numerical constant for the engine, so that the bmep curve is also the torque curve if a suitable scale is applied.

In the case of the engine of Fig. 1.6, the bore and stroke are 76.2 mm and 120.65 mm respectively, and V is 2.185 litres.

Thus,

$$T = \eta p \times \frac{2.185}{4\pi}$$

and the maximum brake torque is—

$$T = 762 \times 0.1753 = 133.6 \text{ Nm.}$$

It is more usual to calculate the bmep (which gives a readier means of comparison between different engines) from the measured value of the torque obtained from a bench test.

Indicated mean pressure and mechanical efficiency are difficult to measure, and are ascertained when necessary by laboratory researches.

Mean torque, on the other hand, can be measured accurately and easily by means of the various commercial dynamometers available. The necessary equipment and procedure are in general use for routine commercial tests. It is then a simple matter to calculate from the measured torque the corresponding brake mean pressure or bmep—

$$\eta p \text{ or bmep} = T \times 4\pi/V$$

The usual form in which these power or performance curves are supplied by the makers is illustrated in Figs 1.7 and 1.8, which show torque, power, and brake specific fuel consumption curves for two Ford engines, the former for a petrol unit and the latter a diesel engine. In both instances, the tests were

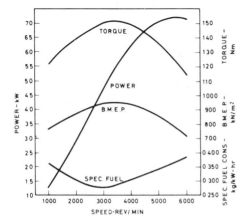

Fig. 1.7 Typical performance curves for an overhead camshaft, spark-ignition engine. High speeds are obtainable with the ohc layout

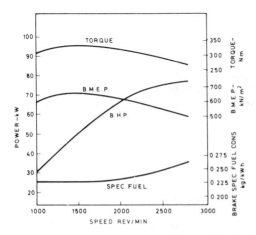

Fig. 1.8 Performance curves of a diesel engine. The fact that torque increases as speed falls off from the maximum obviates the need for excessive gear-changing

carried out in accordance with the DIN Standard 70020, which is obtainable in English from Beuth-Vertrieb GmbH, Berlin 30. The petrol unit is an overhead camshaft twin carburettor four-cylinder in-line engine with a bore and stroke of 90.8 by 86.95 mm, giving a displacement of 1.993 litres. Its compression ratio is 9.2 to 1. The diesel unit is a six-cylinder in-line engine with pushrod-actuated valve gear and having a bore and stroke of 104.8 mm by 115 mm respectively, giving a displacement of 5948 cm^3. It has a compression ratio of 16.5:1.

1.22　Effect of supercharging on bmep and power

Fig. 1.9 illustrates two aspects of supercharging and its effect on bmep (or torque) and power.

The full lines represent the performance curves of an unblown engine with a somewhat steeply falling bmep characteristic. The broken lines (a) and (b) represent two different degrees of supercharge applied to the same engine.

The curves (a) indicate a degree of progressive supercharge barely sufficient to maintain the volumetric efficiency, bmep and therefore torque, at their maximum value, through the speed range.

There would be no increase of maximum piston load or maximum torque, though there would be an appreciable increase in maximum road speed if an overspeed top gear ratio were provided – the engine speed range remaining the same.

Curves (b) show an increase of power and bmep through the whole range, due to a greater degree of supercharging. The *maximum* values of piston loads and crankshaft torque would also be increased unless modifications to compression ratio and possibly to ignition timing were made with a view to reducing peak pressures. This would have an adverse effect on specific fuel consumption, and would tend to increase waste heat disposal problems, but the former might be offset by fuel saving arising from the use of a smaller engine operating on a higher load factor under road conditions, and careful attention to exhaust valve design and directed cooling of local hot spots would minimise the latter risk.

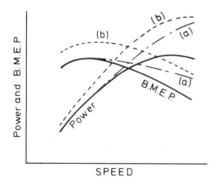

Fig. 1.9　Supercharging

1.23 Brake specific fuel consumption

When the simple term specific fuel consumption is used it normally refers to brake specific fuel consumption (bsfc), which is the fuel consumption per unit of brake horsepower. In Figs 1.7 and 1.8 the specific fuel consumption is given in terms of weight. This is more satisfactory than quoting in terms of volume, since the calorific values of fuels per unit of volume differ more widely than those per unit of weight. It can be seen that the specific fuel consumption of the diesel engine is approximately 80% that of the petrol engine, primarily due to its higher compression ratio. Costs of operation, though, depend not only on specific fuel consumption but also on rates of taxation of fuel. The curves show that the lowest specific fuel consumption of the diesel engine is attained as the fuel: air ratio approaches the ideal and at a speed at which volumetric efficiency is at the optimum. In the case of the petrol engine, however, the fuel: air ratio does not vary much, and the lowest specific fuel consumption is obtained at approximately the speed at which maximum torque is developed – optimum volumetric efficiency.

The fuel injection rate in the diesel engine is regulated so that the torque curve rises gently as the speed decreases. A point is reached at which the efficiency of combustion declines, with rich mixtures indicated by sooty exhaust. This torque characteristic is adopted in order to reduce the need for gear changing in heavy vehicles as they mount steepening inclines or are baulked by traffic. The heavy mechanical components, including the valve gear as well as connecting rod and piston assemblies of the diesel engine, and the slower combustion process, dictate slower speeds of rotation as compared with the petrol engine.

In Figs 1.7 and 1.8, the curves of specific fuel consumption are those obtained when the engine is run under maximum load over its whole speed range. However, in work such as matching turbochargers or transmission systems to engines, more information on fuel consumption is needed, and this is obtained by plotting a series of curves each at a different load, or torque, as shown in Fig. 1.10. Torque, however, bears a direct relationship to bmep and, since this is a more useful concept by means of which to make comparisons between different engines, points of constant bsfc are usually plotted against engine speed and bmep, the plots vaguely resembling the contour lines on an Ordnance Survey map.

The curves in Fig. 1.10 are those for the Perkins Phaser 180Ti, which is the turbocharged and charge-cooled version of that diesel engine in its six-cylinder form. Such a plot is sometimes referred to as a fuel consumption map. Its upper boundary is at the limit of operation above which the engine would run too roughly or stall if more heavily loaded; in other words, it is the curve of maximum torque – the left-hand boundary is the idling speed, while that on the right is set by the governor. For a petrol engine, the right-hand boundary is the limit beyond which the engine cannot draw in any more mixture to enable it to run faster at that loading.

Over much of the speed range, there are two speeds at which an engine will run at a given fuel consumption and a given torque. A skilful driver of a commercial vehicle powered by the Phaser 180Ti will operate his vehicle so far as possible over the speed range from about 1200 to 2000 rev/min, staying most of the time between 1400 and 1800 rev/min, to keep his fuel consumption

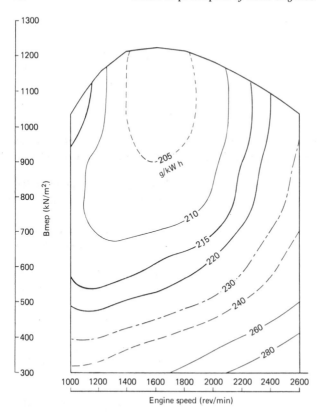

Fig. 1.10 Fuel consumption map for the Perkins Phaser 180Ti turbocharged and aftercooled diesel engine, developing 134 kW at 2600 rev/min

as low as possible. The transmission designer will provide him with gear ratios that will enable him to do so, at least for cruising and preferably over a wider range of conditions, including up- and downhill and at different laden weights.

The bearing of the shape of these curves on the choice of gear ratios is dealt with in Chapter 22, but an important difference between the petrol engine and the steam locomotive and the electric traction motor must here be pointed out.

An internal combustion engine cannot develop a maximum torque greatly in excess of that corresponding to maximum power, and at low speeds the torque fails altogether or becomes too irregular, but steam and electric prime movers are capable of giving at low speeds, or for short periods, a torque many times greater than the normal, thus enabling them to deal with gradients and high acceleration without the necessity for a gearbox to multiply the torque. This comparison is again referred to in Section 22.9.

1.24   Commercial rating

The performance curves discussed so far represent gross test-bed performance without the loss involved in driving auxiliaries such as water-pump, fan and dynamo. For commercial contract work corrected figures are supplied by manufacturers as, for example, the 'continuous' ratings given for stationary industrial engines.

Gross test-bed figures, as used in the USA, are sometimes referred to as the *SAE performance*, while Continental European makers usually quote performance as installed in the vehicle and this figure may be 10 to 15% less.

1.25   Number and diameter of cylinders

Referring again to the RAC formula (see Section 1.19), it will be seen that the power of an engine varies as the square of the cylinder diameter and directly as the numbers of cylinders.

If it is assumed that all dimensions increase in proportion to the cylinder diameter, which is approximately true, then we must say that, for a given piston speed and mean effective pressure, the power is proportional to the square of the linear dimensions. The weight will, however, vary as the cube of the linear dimensions (that is, proportionally to the volume of metal), and thus the weight increases more rapidly than the power. This is an important objection to increase of cylinder size for automobile engines.

If, on the other hand, the *number* of cylinders is increased, both the power and weight (appropriately) go up in the same proportion, and there is no increase of weight per unit power. This is one reason for multicylinder engines where limitation of weight is important, though other considerations of equal importance are the subdivision of the energy of the combustion, giving more even turning effort, with consequent saving in weight of the flywheel, and the improved balancing of the inertia effects which is obtainable.

The relationship of these variables is shown in tabular form in Fig. 1.11, in which geometrically similar engine units are assumed, all operating with the same indicator diagram. Geometrical similarity implies that the same materials are used and that all dimensions vary in exactly the same proportion with increase or decrease of cylinder size. All areas will vary as the square of the linear dimensions, and all volumes, and therefore weights, as the cube of the linear dimensions. These conditions do not hold exactly in practice, as such dimensions as crankcase, cylinder wall and water jacket thicknesses do not go up in direct proportions to the cylinder bore, while a multi-cylinder engine requires a smaller flywheel than a single-cylinder engine of the same power. The simplified fundamental relationships shown are, however, of basic importance.

It can be shown on the above assumptions that in engines of different sizes the maximum stresses and intensity of bearing loads due to inertia forces will be the same if the piston speeds are the same, and therefore if the same factor of safety against the risk of mechanical failure is to be adopted in similar engines of different size, all sizes of engine must run at the same piston speed, torque, power and weight, and gas velocities through the valves.

1.26   Power per litre

This basis of comparison is sometimes used in connection with the inherent improvement in performance of engines, but such improvement arises from increase in compression ratio giving higher brake mean pressures, the use of materials of improved quality, or by tolerating lower factors of safety or

VARIABLE	A	B	C	ENGINE
PISTON SPEED	1	1	1	(A)
STROKE	1	2	1	
REV/MIN	1	$\frac{1}{2}$	1	
BORE	1	2	1	(B)
TOTAL PISTON AREA	1	4	4	
POWER	1	4	4	
MEAN TORQUE	1	8	4	
VOLUMETRIC CAPACITY	1	8	4	
WEIGHT	1	8	4	(C)
$\frac{\text{POWER}}{\text{WEIGHT}}$	1	$\frac{1}{2}$	1	
MAX. INERTIA STRESS	1	1	1	
MEAN GAS VELOCITY	1	1	1	

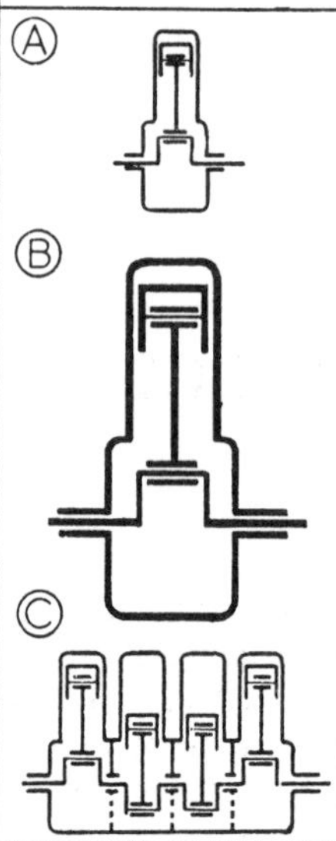

RELATIVE VALUE OF VARIABLES IN SIMILAR ENGINES [FOR SAME INDICATOR DIAGRAM]

Fig. 1.11

endurance. The comparison ceases to be a comparison of similar engines, for with similar engines the power per litre (or other convenient volume unit) may be increased merely by making the cylinder smaller in dimensions, and if the same total power is required, by increasing the number of the cylinders. Thus in Fig. 1.11 all the three engines shown develop the same power per unit of piston area at the same piston speed and for the same indicator diagram, but the power per litre of the small-cylinder engines is double that of the large cylinder, not because they are intrinsically more efficient engines but because the smaller volume is swept through more frequently.

Thus, high power per litre may not be an indication of inherently superior performance, whereas high power per unit of piston area is, since it involves high mean pressures or high piston speed or both, which are definite virtues provided that the gain is not at the expense of safety or endurance.

1.27 Considerations of balance and uniformity of torque

In the next chapter consideration is given to the best disposition of cylinders to give dynamic balance and uniformity of torque, which are factors of vital importance in ensuring smooth running.

Chapter 2

Engine balance

Any moving mass when left to itself will continue to move in a straight line with uniform speed. If a heavy mass is attached to a cord and swung round in a circle a pull, known as *centrifugal force*, will be felt in the cord. This force represents the tendency of the mass to move in a straight line. Owing to the presence of the cord the mass is compelled to move round in a circle, its tendency to move in a straight line being overcome by the pull of the cord.

Thus any mass revolving in a circle sets up an outward pull acting in the radial line through the centre of rotation and the centre of the mass. For example, the crankpin of a car engine revolves in a circle round the centre on the main bearings, and sets up a force on those bearings acting always in the direction of the crankpin. If this force is not *balanced* a vibration of the whole engine will be set up, in time with the rotation of the crankpin. This vibration will be more or less apparent according to the rigidity with which the engine is bolted to the frame.

The mathematical expression for centripetal acceleration is $\omega^2 r$, where ω is the angular speed of the mass about the centre of rotation at a radius r. If ω is expressed in radians per second (there are 2π radians in one revolution) and r is measured in metres, then the acceleration is given in metres/s^2.

The corresponding force in newtons, inwards on the mass, outwards as a reaction on the bearing, is given by—

$$F = M\omega^2 r$$

but if $2\pi N/60$ is substituted for ω, we have—

$$F = M \times \frac{4\pi^2 N^2}{60^2} \times r$$

which can be reduced to—

$$F = \frac{MrN^2}{91.1}$$

where r is expressed in metres and N in rev/min.

For example, suppose the crankpin and big end of an engine of 0.127 m stroke weigh together 1.8144 kg and the engine is turning at 3000 rev/min, then, if there were no provision for balance, the reaction on the main bearings would be—

$$F = \frac{MrN^2}{91.1}$$

$$= \frac{1.8144 \times 0.0635 \times 3000^2}{91.1}$$

$$= 11\,400 \text{ newtons}$$

due to revolving masses only.

It is, however, possible to balance the disturbing force by means of a balance 'weight' or mass, placed diametrically opposite the crankpin. (It is more correct to speak of 'mass' in connection with running or dynamic balance, since the forces are not due to 'weight' which is the attractive force of gravity.) In engines it is not possible actually to place this balance mass in the same plane as the crankpin, and it must therefore be divided into halves placed symmetrically on each side. If the balance mass is placed at the same distance as the crankpin from the crankshaft axis then it must be of the same amount; if the distance is twice as great the mass must be halved. Thus, in Fig. 2.1, $B \times r_1$, must equal $M \times r$.

Actually this force, with that due to the crank webs, would be balanced either against an opposed crank and big end, or by balance extensions to the crank webs. That part of it due to the big end of the connecting rod would represent part of the total inertia load on the big-end bearing.

2.1   Practical balancing

There is no mathematical difficulty in balancing revolving masses, and with simple disc forms or short crankshafts a static test is often sufficient. The part to be balanced is mounted on a true spindle (or its own journals) and placed on straight and carefully levelled knife edges or slips.

It will then roll until the heaviest side comes to the bottom. By attaching small counterweights until it remains indifferently in any position the error may be ascertained, and correction made by adding balance masses or removing excess metal as may be most convenient.

If the part has considerable axial length, there may be unbalanced couples which a static test will not reveal – it is not possible to ensure that Br_1 of Fig. 2.1 is in the same plane as Mr, though by a static test they may have been made equal. Such couples can be revealed and corrected only by means of a dynamic

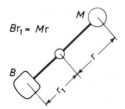

Fig. 2.1

test during which the shaft or rotor is run up to speed and the couple or moment is shown by rocking or 'pitching'. Many ingenious dynamic balancing machines have been produced and are in use to measure and locate the plane of the imbalance in order that it may be corrected.

The degree of accuracy with which the correction is made is a question of the time and cost that may be expended.

The dynamic balance of complete rotational assemblies is usually carried out with great care, and may be dependent on the selected positioning of nuts and bolts. In reassembling or refilling a fluid flywheel, for instance, care should be taken to replace plugs in the same holes from which they were removed. An article on crankshaft balancing appeared in *Automobile Engineer*, Vol. 56, No. 20.

2.2 Balance of reciprocating parts

The movement of the piston backwards and forwards in the cylinder is known as a *reciprocating movement* as opposed to the *rotative movement* of the crankshaft, flywheel, etc. The reciprocating parts of a motor engine are the piston, gudgeon pin and as much of the connecting rod as may be considered to move in a straight line with the piston (usually about one-third of the connecting rod is regarded as reciprocating, the remainder, including the big end, being considered as a revolving mass).

Now the reciprocating parts, which we will refer to simply as the *piston*, have not a uniform motion. The piston travels in one direction during the first half of a revolution and in the opposite direction during the second half. Its speed of movement in the cylinder increases during the first half of *each* stroke (that is, twice every revolution) and decreases during the second half of each stroke, the speed being greatest and *most uniform* about the middle of each stroke. To change the speed of a body requires a force whose magnitude depends on the mass of the body and the rate at which the speed is changed, that is, the acceleration. This may be realised by holding an object in the hand and moving it rapidly backwards and forwards in front of the body.

The speed of the piston is *changing* most rapidly (that is, the acceleration is greatest) at the ends of the stroke, and it follows that the force required to change the motion is greatest there also. At the middle of the stroke the speed is not changing at all, so no force is required.

The necessary force is supplied by a tension or compression in the connecting rod. If the connecting rod were to break when the piston was approaching the top of the cylinder, the engine running at a high speed, the piston would tend to fly through the top of the cylinder just as, if the cord broke, the mass referred to earlier would fly off at a tangent. Now the reaction of this force, which is required to slow the piston at the top of its stroke and to start it on its downward stroke, is transmitted through the connecting rod, big-end bearing, crankshaft and main bearings to the engine frame, and sets up a vibration. This is dealt with more fully in Sections 2.11 and 2.12.

At the two ends of the stroke the piston produces the same effect, that is, the same force, on the crankpin as if it were simply a revolving mass concentrated at the crankpin and, consequently, it may be balanced at these points by a revolving balance mass placed opposite the crankpin as in Fig. 2.2. Supposing that such a balance mass B_2 is placed on the extended crankshaft webs, of sufficient mass to balance completely the reciprocating parts. The forces set up by the movement of the piston act in a vertical direction only, that is, in the line of the stroke. They will have their greatest value at the ends of the stroke, in opposite directions, and become nothing in the middle of the stroke. Referring to Fig. 2.3, as the crankshaft revolves the centrifugal force F or the balance mass, acting always radially outwards from the shaft centre, has a decreasing effect or 'component' in the direction of the line of stroke, but an increasing one in a horizontal direction at right angles to the line of stroke. The decrease in the vertical component corresponds exactly to the decrease in the force set up by the piston. At the centre of the stroke the piston exerts no inertia force since its speed is momentarily steady, and the balance mass exerts no force in a vertical direction, since its crank is horizontal. The addition of this rotating balance mass then balances the engine completely in the vertical direction. Consider, however, the horizontal effect produced. The balance mass introduces a horizontal component force F_h (Fig. 2.3), which varies from zero when the piston is at either end of the stroke to a maximum when the piston is in the middle of the stroke when, as the crank is horizontal, $F_h = F$.

This horizontal effect is exactly equal to the original vertical effect due to the piston which it was sought to balance, and thus the only result of attempting to balance the reciprocating parts completely, by means of a revolving mass, is to transfer the disturbance from the vertical to be horizontal direction without altering its amount. In some cases this may be an advantage, but the engine is in no sense properly balanced.

In a single-cylinder engine a compromise is arrived at by adding a balance mass equivalent to a portion (usually half) of the reciprocating parts. This leaves the remainder unbalanced in a vertical direction, but the horizontal effect is only that due to the smaller balance mass. If half the piston is balanced the result is a vertical and a horizontal effect, each equal in amount to half the original unbalanced effect. If a greater balance mass is used the vertical imbalance is less, but the horizontal is greater.

It is quite impossible to balance an ordinary single-cylinder engine completely by the addition of balance masses to the revolving crankshaft.

Consider a 90° twin engine, that is, one which is provided with two equal

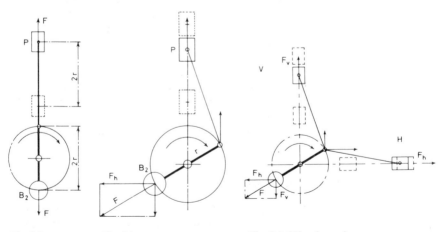

Fig. 2.2   Fig. 2.3   Fig. 2.4 90°-twin engine

cylinders having their centre lines at right angles to each other in the same plane and both connecting rods driving on to a common crankpin. Suppose that the engine is placed with the centre line of cylinder V vertical as in Fig. 2.4. The centre line of cylinder H will then be horizontal.

We will assume that the revolving parts (that is, the crankpin, etc.) are already completely balanced by the addition of a suitable balance mass opposite the crankpin, incorporated in the extended crank webs.

A further balance mass is now required for the reciprocating parts. This is placed opposite to the crankpin and is of sufficient mass to balance the *whole* of the reciprocating parts of *one* cylinder. As already explained, this will secure complete balance in the vertical direction and further, as the mass turns into the horizontal position, it will supply the increasing horizontal balancing force F_h required by the cylinder H. In other words, when one set of reciprocating parts requires the full balance mass the other requires nothing, and in intermediate positions the resultant effect of the two sets of reciprocating parts is always exactly counteracted by this single balance mass.

An engine of this type can therefore be balanced completely for what are known as the *primary forces*. The effect of a short connecting rod is to introduce the additional complications of what are known as *secondary forces*, which are dealt with in the latter portion of this chapter for the benefit of those readers with a mathematical turn of mind who may wish to study the matter further.

2.3   Other V twin engines

V twin engines in which the angle between the cylinders is not 90° but some smaller angle occupies a position intermediate between the 90° twin and the single-cylinder engine, primary balance being approached more nearly as the angle between the cylinders approaches 90°.

2.4   Horizontally-opposed twin

In engines of the flat-twin-cylinder type, the reciprocating parts are not balanced by means of revolving masses, but one set of reciprocating parts is made to balance the other set by making them operate on two diametrically-opposed crankpins. The pistons at any instant are moving in opposite directions and the inertia forces oppose and balance each other.

Owing to it being impracticable to arrange the cylinder centre lines in the same vertical plane, there is a smaller unbalanced twist or couple. This is indicated in Fig. 2.5, which is a plan view of the engine. The inertia forces F F, being equal, have no resultant force, but since they do not act along the same line they constitute a couple. The magnitude of this couple, which tends to oscillate the engine in a horizontal plane, increases as the distance d increases.

2.5   Side-by-side twin with cranks at 180°

Fig. 2.6 shows a side-by-side twin engine with cranks at 180°. The motion of the pistons is still opposed, but in addition to the fact that the couple is increased owing to the greater distance d between the cylinder centre lines, this

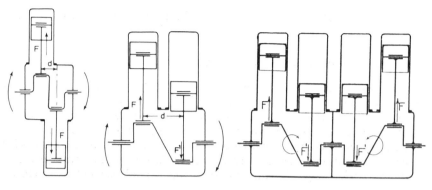

Fig. 2.5 Opposed   Fig. 2.6 Side-by-side   Fig. 2.7 (*left*) Four-cylinder in-line
twin engine   twin engine   engine

arrangement is not so good as the opposed twin because the secondary forces
due to the shortness of the connecting rod do not balance. This will be
understood after a study of the effect of the secondary forces which are dealt
with in Section 2.12. The inertia force F, due to the piston which is on its inner
dead point, is greater than the inertia force F', due to the other piston which is
at its outer dead point. With the usual ratio of connecting rod length to crank
length F is about 60% greater than F', the difference being twice the secondary
force due to one piston. Balance is, in fact, obtained only for primary forces and
not for secondary forces or for primary couples.

2.6   Four-cylinder in-line engine

Suppose, however, a second side-by-side twin, which is a 'reflection' of the first,
is arranged alongside in the same plane and driving the same crankshaft as in
Fig. 2.7. The couples due to the two pairs are now clearly acting in opposite
directions, and their effects will be opposed so that there will be no resultant
couple on the engine as a whole. The engine is then balanced for everything
except secondary forces, which in a four-cylinder engine can be dealt with only
by some device such as the Lanchester harmonic balancer, illustrated in Fig.
2.11.

It should be clearly realised that it is only the engine as a whole that is
balanced, and that the opposition of the two couples is effective only by virtue
of stresses set up in either the crankshaft or crankcase or both.

2.7   General method of balancing

This method of balancing by opposing forces and couples represents the
general method of balancing multi-cylinder engines, the cylinders and cranks
being so disposed that as far as possible or expedient the various forces and
couples, both primary and secondary, may be made to neutralise each other
throughout the engine as a whole. The best dynamic balance is not, however,
always consistent with the best distribution of the power impulses and a
compromise must therefore sometimes be made, as will be seen later. Another
consideration is that of bearing loads due to the dynamic forces, and here
again it may be desirable – as in high-speed racing cars – to tolerate some

degree of dynamic imbalance in order to reduce the load factor on a particular bearing.

2.8 Couples due to revolving masses

Referring again to Fig. 2.7, it will be appreciated that the *revolving* masses at each crank give rise to couples in just the same way as the reciprocating masses, except that here it is an 'all-round' effect instead of acting only in the vertical plane. The revolving couple is unbalanced in Fig. 2.6, while in Fig. 2.7 the two opposite couples are opposed exactly as indicated by the arrows for the reciprocating effects, and the shaft and crankcase assembly must be so stiff as to avoid whip under the combination of the two independent disturbances.

2.9 Balanced throws

The stresses and whip due to the revolving masses may be reduced by adding counterweights to the individual throws, as shown in the crankshaft illustrated in Fig. 3.15, thus eliminating the couples.

In the modern highly-rated engines, serious crankshaft whip has been eliminated only by the addition of these balance masses. They may be incorporated in the forging or be separately attached. Clearly their employment makes the crankshaft construction more expensive and in straight six- and eight-cylinder engines they may give rise to another trouble, namely *torsional vibration* of the crankshaft.

2.10 Torsional vibration

Under the combustion impulses the shaft alternately winds and unwinds to a small extent, and as with all types of strain and vibration there is a certain natural frequency of this action. The longer and more slender the shaft and the larger the crank masses incorporated in it, the lower will this natural frequency be, and it may be so low as to equal the frequency of the combustion impulses at some particular engine speed. Resonance will then occur between the forced impulses and the natural frequency of the shaft vibration giving rise to dangerous torsional strain.

Such vibrations may be damped out by the use of a vibration damper as shown in Fig. 4.4.

2.11 Secondary forces and couples

It was indicated in Section 2.2 that the motion of the piston could be regarded as the vertical component of the motion of the crankpin, and this is known as *simple harmonic motion*. If the connecting rod were infinitely long and thus always parallel to the cylinder axis, or if the crank and cross-slide mechanism referred to in Section 1.9 were used, the piston, moving in its straight line of stroke, would have this simple harmonic motion.

At any position of the piston, measured by the crank angle θ from its top dead centre as shown in Fig. 2.8, the accelerating force required for the piston would be $M\omega^2 r \cos\theta$, while the horizontal force $M\omega^2 r \sin\theta$, which would represent the horizontal component of the centrifugal force of a *revolving* mass, obviously has no existence in the context of reciprocating mass, since the

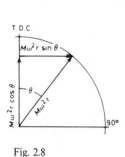

Fig. 2.8

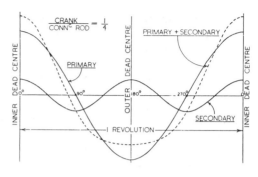

Fig. 2.9

piston has no displacement, velocity or acceleration at right angles to the axis of the cylinder. This 'primary' disturbing force is drawn in Fig. 2.9 as a full line having its maximum value when $\theta = 0°$ and $180°$. Its direction must be considered in relation to the forces on the bearings of the connecting rod, which provides the necessary accelerating force and transmits the reactions to the crankpin and main bearings.

2.12   Effect of short connecting rod

In an actual engine, owing to the shortness of the connecting rod, the motion of the piston is *not* simple harmonic, and its acceleration and the corresponding accelerating force require for their calculation other terms in the expression, of which only the 'second harmonic' is of importance in the present connection.

The total accelerating force is then—

$$F = f_1 + f_2$$
$$= M\omega^2 r \cos \theta + M\omega^2 r \cdot \frac{r}{1} \cos 2\theta$$
$$= M\omega^2 r \left(\cos \theta + \frac{r}{l} \cos 2\theta\right), \text{ where } l \text{ is the length of the connecting rod}$$

between centres.

The ratio r/l is usually about $\frac{1}{4}$, and this value has been assumed in Fig. 2.9, the lower full line representing the secondary disturbing force which, it will be seen, has twice the frequency of the primary.

At the inner dead centre the secondary is in the same direction as the primary, while at the outer dead centre they act in opposite directions.

To use a familiar term, the 'dwell' of the piston is longer at the outer dead centre than it is at the inner (or top of the stroke), owing to the fact that the swing of the connecting rod neutralises to some extent the swing of the crank.

To obtain the total disturbing force on the bearings the primary and secondary forces must be added, as shown by the broken line, but it is convenient and in many ways more enlightening to consider them separately, so that the conditions of balance in respect of primary and secondary forces, and primary and secondary couples, may be assessed.

Thus, if the four-cylinder engine of Fig. 2.7 is considered, it will be seen that when pistons 1 and 4 are at their inner dead points, 2 and 3 are at the outer position, the corresponding values of θ being 0 and 180. The primary forces and couples are balanced as indicated in Fig. 2.7, while if the direction of the secondaries is examined with the help of Fig. 2.9, it will be found that all four act together, outwards at both dead centres and inwards at mid stroke. They thus give rise to a total unbalanced force equal in magnitude – with a 4:1 connecting rod ratio – to a single primary force, and vibrating with twice the frequency.

In Fig. 2.10 are shown in diagrammatic form the conditions of balance for various arrangements of a four-cylinder engine with flat crankshaft. The reader should have no difficulty in checking the conditions of balance with the help of the curves in Fig. 2.9, if care is taken to avoid confusion as to the directions in which the primary and secondary disturbing forces act. The longer arrows represent the primary forces and the shorter the secondary.

They are drawn side by side for clarity, but actually both act along the axis of the cylinder. The double coupling lines indicate the 'arms' of the secondary couples, while the primary couples will be clear.

The following points should be noted—

The balance of primary forces and couples depends on the crankshaft arrangement. Primary forces are balanced in all the six cases, while primary couples are balanced with the 'mirror-image' crankshafts A but not with the zigzag shafts B. With the arrangements B the tendency to bend or whip the shaft due to inertia forces is less, and if a third bearing is provided, the inertia loading on this bearing is less than with the arrangements A.

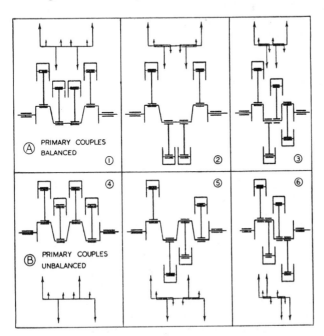

Fig. 2.10 Balancing diagrams for four cylinders

The balance of secondary forces and couples depends on cylinder disposition. No. 2 in Fig. 2.10 is the only arrangement with complete dynamic balance for primary and secondary forces and couples.

2.13　Firing intervals

It will be seen that No. 2 is also the only layout in which all four pistons are at their inner dead centres simultaneously. This makes it impossible to distribute the firing impulses at intervals of half a revolution, as is proper, because at least two pistons would be at the end of their compression strokes together, and this would result in the cylinders firing in pairs at intervals of one revolution. The smoothness of torque would then be no better than that of the 180° twin of Fig. 2.5. For this reason, arrangement No. 2 is not used in practice in spite of its perfect balance.

Arrangement No. 3 takes its place. Past examples were the Jowett *Javelin* and Lancia *Flavia*. It will be found that all arrangements except No. 2 permit of half-revolution firing intervals, and the reader may find it instructive to write down the alternative firing orders for which the camshafts might be designed.

2.14　Compactness of engine

Arrangements Nos 1 and 4 provide the most compact engine transversely to the crankshaft, while No. 6 forms the shortest engine with the simplest shaft, but has large unbalanced couples. No. 3 is a compact and favourable arrangement with perfect balance except for secondary couples, which are not a serious objection.

With suitable form of crank webs, arrangement Nos 2 and 5 can be made more compact in the axial direction than can Nos 1 and 4.

Nos 4 and 5 would be the most expensive crankshafts to manufacture, and would normally be provided with three bearings, while Nos 3 and 6 are particularly favourable to the two-bearing construction.

The most widely adopted form of the four-cylinder engine is No. 1 of Fig. 2.10 by virtue of its general compactness, symmetrical manifolding, possibilities of economic manufacture and convenience of installation in the chassis, combined with accessibility. The secondary forces remain unbalanced, however, and it is natural that engineers should have turned their attention to the problem of balancing these.

2.15　Harmonic balancer

A most ingenious device for accomplishing this object is the Lanchester harmonic balancer, which was used in early days in Lanchester, Vauxhall and Willys four-cylinder engines, and has been employed in numerous instances to cure vibration troubles in stationary and marine engines.

If suitable bob-weights are mounted on two shafts geared together to turn at the same speed in opposite directions, it is possible to produce the effect of a reciprocating mass having simple harmonic motion.

If the device is arranged to run at twice crankshaft speed, and is suitably proportioned, it may be made to balance the sum of the secondaries due to the four pistons. Such devices are generally either gear or chain driven from one end or other of the crankshaft. Another form of drive is illustrated in Fig. 2.11. C_2 and C_3 are the two inner crankpins of an ordinary four-throw shaft with three bearings.

The crank webs on each side of the centre bearing are circular in form and provided with helical gear rings which drive at double crankshaft speed the two cross-shafts carried in bearings B.

Each of these shafts carries for symmetry two bob-weights W, each of which is proportioned to develop a centrifugal force equal to the secondary inertia force of one piston, so that the four bob-weights together neutralise the combined secondary forces due to the four pistons. Since the cross-shafts revolve in opposite directions, it will be seen on referring to Fig. 4.12 that in the horizontal direction the two pairs of bob-weights neutralise each other, while they combine in the direction of the piston movement to give a reciprocating effect in opposite phase to the secondary piston disturbance. Thus complete dynamic balance of the engine is obtained.

2.16   Torsional disturbances

For balancing the torsional vibrations mentioned in Section 2.10, however, we need devices such as those described in Section 4.6. Such vibrations may be due not only to the firing impulses of the cylinders but also the reactions to them by the flywheel and other components throughout the driveline back to the driven wheels – Newton's Third Law. If a torque converter or fluid flywheel is embodied in the driveline, however, it will tend to smooth out the vibrations transmitted back from the driveline to the engine crankshaft.

Again, from Newton's Third Law, it follows that the forces originating from the pressures in the gases in the combustion chambers cannot be transmitted back as torque from the crankshaft without there being equal and opposite torsional reactions on the engine mass. These reactions arise from the side thrusts exerted through the pistons and bearings. Obviously, therefore, the value of the torque fluctuates continuously over a range of harmonics due to the variations in both the gas pressures and the reactions to them as outlined in the previous paragraph. The mean value of this torque determines the power output of the engine.

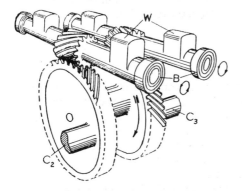

Fig. 2.11 Lanchester harmonic balancer

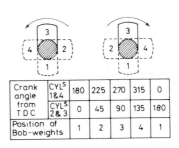

Crank angle from TDC	CYL^S 1&4	180	225	270	315	0
	CYL^S 2&3	0	45	90	135	180
Position of Bob-weights		1	2	3	4	1

Fig. 2.12

An outcome of the harmonic torsional fluctuations is a torsional vibration of the engine on its mounting, the frequencies range from the '$\frac{1}{2}$ order' or cycle frequency of the individual cylinders, to the higher orders. If the indicator diagrams of the several cylinders were precisely similar, there would be no $\frac{1}{2}$ order or half engine speed frequency, apart from camshaft effects.

'Inertia torque reaction' arising quite independently from gas pressures, may be visualised as the reaction of torque applied to the crankshaft by the flywheel to effect the piston accelerations. The connecting rod obliquities and piston side-thrusts develop the inertia torque and its reaction, which can be most noticeable under overrun conditions. Frequencies are the 1st, 2nd, 4th, 6th, etc., but only the 1st and 2nd are important.

Angular fluctuations about the longitudinal axis of the crankshaft can be balanced by contra-rotating flywheels or other contra-rotating components. This is not generally done, although Dr F. Lanchester incorporated this feature in some of his engines, *circa* 1895, and an experimental installation on a Volvo 144S car was described in the August 1970 issue of *Automobile Engineer*.

2.17   In-line engines with three cylinders

With odd numbers of cylinders, and crankpins pitched at equal intervals around the crankshaft, both primary couples and secondary forces and couples are induced in the plane of the cylinder axes. Three cylinders with a crankshaft having cranks at 120° intervals are attractive for two-stroke engines because, with the three power strokes per revolution, torque output is smooth. For four-stroke diesels, the three-cylinder layout represents a way of producing economically an engine of low power output with cylinders of the same size and spacing as a four-cylinder version and thus, at the same time, also avoiding the problems of fuel injection into very small cylinders.

In a three-cylinder engine, the primary forces are in balance because the upward inertia force on any piston at top, or inner, dead centre is balanced by the downward forces on the other two pistons. However, when No. 1 is at TDC these forces are tending to rotate the engine nose up, and when No. 3 piston is at TDC the couple acts in the opposite sense. Only when No. 2 piston is at TDC are the pitching couples in balance. When the pistons are at bottom, or outer dead centre, the directions of the couple are in each instance reversed.

Counterweights, positioned opposite the crankpins, on the end crankwebs can be used to offset these couples. However, since the centrifugal force they produce rotates about the axis of the crankshaft, while the piston inertia forces act only vertically, they introduce yawing couples.

There are two ways of overcoming or mitigating the problem. The most economical is to balance the pitching couples only partially so that the yawing couple is not too large, and to design the engine mountings to absorb both what remains of the pitching together with the yawing couples.

The second, and more costly, method is to use, in conjunction with the balance weights on the end webs, a balancer shaft rotating at crankshaft speed but in the opposite direction, to counteract the yawing couples. To keep the weights on the balancer as small as possible, the shaft should be as long as practicable and the weights on its ends. They are usually set at angles of 60° and 240° relative to the crankshaft, so that they apply maximum yawing correction couples when the effective arms of the crank balancer weights are horizontal, and zero yawing couples when they are vertical. The couples due to the secondary forces, which add to the primary forces when the pistons are at TDC and subtract at BDC, remain unbalanced, regardless of whether a balancer shaft is used.

2.18 Engines with five cylinders

The choice of five cylinders is attractive to diesel engine designers as a means of adding one more power unit to a range of engines having a high proportion of common parts and machined and assembled on common lines. Moreover, they can also be attractive to vehicle designers, because of their compactness relative to six-cylinder in-line units. In particular, they are especially suitable for buses with rear-engine installations, where they will fit more easily than a six-cylinder unit if mounted transversely to avoid encroaching too far into the space for the passengers. In any case, the power output of a six-cylinder unit is unnecessary for a bus or coach because it is not so heavily loaded as a truck of the same or similar overall dimensions.

In a five-cylinder engine, the crankpins are usually pitched 72° apart with a firing order of 1, 2, 4, 5, 3. This means that the primary forces are slightly reduced because no two pistons come together to either top or bottom dead centre. On the other hand, it also means that the intermediate main bearings are more heavily loaded than if adjacent pistons moved in precisely opposite phase. A major advantage of five cylinders, though, is that all the primary and secondary forces are in balance, which, at first sight, seems to indicate that there is nothing to choose between five and six cylinders.

This, however, is not so. Calculations show that if we take the primary force on No. 1 piston at TDC as unity, the secondary force is 0.2857, making a total of 1.2857 upwards. At the same time, the totals on the second and third pistons are 0.7207 downwards and on the fourth and fifth 0.0779 upwards. The net effect, therefore, is zero force, but the large total on No. 1 pin tends to lift the front of the engine, while those on the rear pins are tending to push the back end downwards, imposing a pitching couple on the engine. This couple is reduced progressively and then reversed by the time the fifth cylinder has rotated round to TDC. Since the pitching couples are confined to the vertical plane, a compromise solution is to put additional balance weights on the front and rear webs of the shaft to counteract the pitching couple partially, but of course this will introduce a yawing couple in the horizontal plane. The next step therefore is to design engine mountings that will accommodate the yawing motions without transmitting the vibrations to the vehicle structure.

2.19 Flexible mountings

An engine which is perfectly balanced for forces and couples in the sense described above will have no tendency to move, or to transmit vibration to the frame or foundation to which it is attached, as a result of the sources of disturbance so far enumerated.

Small imperfections of actual balance may be troublesome in extreme circumstances, but the presence of the unbalanced secondary inertia forces in four-cylinder engines not provided with harmonic balancers may involve considerable vibration troubles.

This type of simple, inexpensive engine construction is particularly popular for medium-size cars, at the expense of the admittedly superior mechanical properties of the six, which at one time appeared likely to oust the four from favour.

Resonant vibration of body panels, or 'boom', arising from the unbalanced secondaries, intensified by frameless or unitary chassis construction, has forced designers to direct their attention to methods of engine mounting to ensure that improved comfort in riding and silencing shall not be offset by fatiguing vibration effects.

2.20   Modes of vibration, natural frequency, forcing frequency and resonance

Any mass of finite dimensions, free to move in space, possesses six degrees of freedom: three of translation along, and three of rotation about three mutually perpendicular axes. These axes may be arbitrarily chosen, in relation to the crankshaft axis, which has no significance while the engine is at rest, or may be the three 'principal axes of inertia' of the engine mass, both sets passing through the mass centre of the unit. The former may have no fundamental relation to the rotational inertias of the unit, but are convenient for calculations connected with static and transient loads due to dead weight, road shocks, cornering, braking and, if the crankshaft axis is chosen, driving thrust and torque reaction due to mean torque and impulsive clutch engagement from which clutch judder may arise.

Any 'mode of motion' may be resolved into a number or all of these translations and rotations, the modes of motion becoming 'modes of vibration' when regular periodic forces and torques are applied.

Engine mountings must be designed to completely constrain the power unit against these static and vibrationary loads and disturbances, with the maximum degree of insulation of the body structure from noise and transmitted vibrations of uncomfortable frequency.

The number of modes of vibration, each with its natural frequency, of the various parts and accessories of an automobile is practically infinite, and many of these modes have frequencies liable to resonant vibration with 'forcing' vibrations emanating from the engine.

Natural frequency is determined by the mass or inertia and the stiffness (as of a spring) of the constraint. The stiffer the constraint in relation to the mass or its rotational inertia, the higher the frequency. Any component of the mounting may be called on to constrain two or more modes of motion, hence the value and simplicity of bonded rubber mountings.

Should the forcing frequency at some particular engine speed be equal to the natural frequency of the whole or part of the frame or body on which the engine is mounted, resonance will occur, and the amplitude of vibration, without damping, may theoretically become infinite.

Fig. 2.13 shows the theoretical proportion of the applied force or amplitude of movement which is transmitted from the 'exciter' to the support for different ratios of forcing frequency to natural frequency F/f, the *frequency* of the transmitted vibration being the same as the forcing frequency.

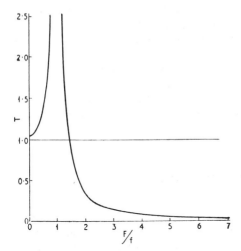

Fig. 2.13 Transmitted vibration

It will be seen that when F/f is unity, the transmitted vibration amplitude T theoretically becomes infinite. In practice inherent damping always limits the amplitude, but it is in some instances necessary to incorporate positive steps to prevent excessive movement under certain conditions.

When F/f is increased to three by reducing f, only one-eighth of the forcing amplitude or force is transmitted.

Thus for insulation of vibration a low natural frequency, i.e. a soft mounting, is required, but a compromise may be necessary with the greater stiffness likely to be needed to take the various static and transient loads enumerated above. Road shock loads, for instance, may greatly exceed the dead weight.

2.21 Principal axes of inertia

If a long, fairly regular object, such as a potato or a lump of firm plasticine, is pierced with a knitting needle in the general direction of its greatest length, an axis of rotation may be obtained about which the moment of inertia is small compared with those about axes generally at right angles to it.

There is a particular axis, passing through the mass centre, about which, owing to the general proximity of all the mass particles, the moment of inertia is a minimum for the solid. This is one of the principal axes. The two others, also passing through the mass centre, complete a trio of mutually perpendicular principal axes. Of the second and third axes, one will be the axis of maximum inertia and the other will have an intermediate value. These three principal axes, which may be described in reference to a power unit as the axes of natural roll, pitch and yaw, are the axes about which torsional oscillation can be initiated without introducing lateral or translational forces, and are the axes about which the mass, if supported in a homogeneous elastic medium (such as soft rubber or jelly) and with gravitational forces balanced, would take up component rotations when disturbed by any system of applied torques.

2.22　Importance in the design of engine mountings

These principal axes are important in the design of engine mountings, particularly the fore-and-aft 'roll' axis, which ordinarily lies at an angle of 15 to 30° with the crankshaft axis, sloping downwards from front to rear as indicated in Fig. 2.14. Ideally, the mountings should be disposed as to confine rotation to this axis, so that torsional vibrations may be constrained without introducing lateral forces. If rotation about any other axis is imposed by the mountings, such lateral forces will arise, and may require additional constraints.

It will be noticed that, assuming lateral symmetry, the principal axis intersects the crankshaft axis near the centre of the flywheel housing, so that mountings placed on the sides of the housing could deal with torsional vibrations about the principal axis as well as with direct 'bump' loads. The front mounting would be placed high, as near to the principal axis as possible, and would take the balance of the bump loads as determined by the position of the mass centre. In the Chrysler mounting shown in Fig. 2.15, all reactions are distributed between the front and rear mountings, snubbers only being provided at the flywheel housing.

Fig. 2.14 also indicates the *centre of percussion*, CP. The position of the CP is determined by the distribution of the mass relative to the rear suspension or pivot point, which in effect is the *centre of suspension*, CS. Usually it will lie near the central transverse plane of the engine, which is the plane in which the resultant of the unbalanced secondaries in a four-cylinder engine acts. If CP can be located exactly in this plane by suitable choice of CS and mass

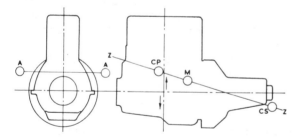

Fig. 2.14　Principal axes

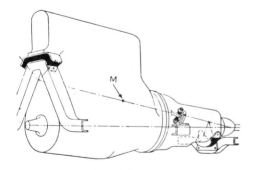

Fig. 2.15　Chrysler engine mounting

distribution, and if the front mounting can also be so placed, there will be neither pitching *moment* due to the secondary disturbing force nor reaction at the CS. The principle is similar to that applicable to a door stop, which should be positioned at the centre of percussion so that, if the door is suddenly blown open against it, the hinges are not overloaded.

In practice, it is rarely possible to arrange for exact coincidence as described above, but the central plane is a structurally desirable, though not always convenient, location for the front mountings. These are indicated at AA in Fig. 2.14, widely spaced to deal with torsional vibrations about the axis ZZ.

This percussion system appears to be an ideal means of dealing with pitching disturbances; road shocks and cornering loads would be shared by the attachments at CS and AA in a ratio determined by the position of the mass centre M. A V arrangement of links with rubber-bonded bushes, with the link centre lines meeting on the principal axis and so utilising the principle of the instantaneous centre may be used, or else a type of front V-mounting of 'compression-shear' units.

The arrangement used for the mounting of the three-cylinder Perkins diesel engine in a light van chassis is shown in Fig. 2.16.

Both vertical and horizontal primary out-of-balance couples are present in this engine, and to obtain insulation against these as well as the $1\frac{1}{2}$ order torque harmonic a suspension giving a high degree of rotational flexibility about all axes was needed. This was achieved by a V arrangement of sandwich mountings very close to the centre of gravity (or mass centre), the front rubber sandwich mounted so that its compression axis passes approximately through the centre of gravity. The degree of insulation obtained is excellent. Engine movement under shock torque reaction, and when passing through resonance on starting and stopping, is quite large, but has not proved troublesome. A pair of circular sandwich units pitched to give greater torque control may be used at the front instead of the rectangular form.

A great variety of rubber-to-steel bonded units to provide the many different constraints required has been produced by Dunlop Polymer and a few are shown in Fig. 2.17. The illustration includes early unbonded cushions, bonded double shear and compression-shear mountings, a bonded eccentric bush, rubber steel compression spring, and others.

It is possible to design units capable of resisting various combinations of

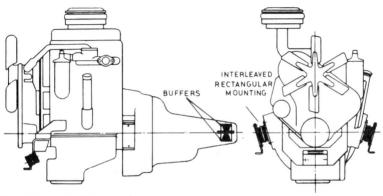

Fig. 2.16 Perkins P-3 mounting

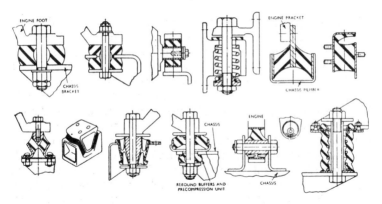

Fig. 2.17 Metalastik elastic engine mountings by Dunlop Polymer

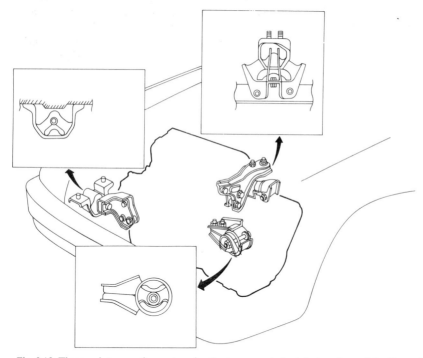

Fig. 2.18 Three-point mounting system for the transversely installed engines of the Rover 200 Series. All three are very flexible vertically but, for reacting final drive torque, their motion is limited by their metal housings. The lower mounting on the right in this illustration also limits movement of the engine parallel to the crankshaft

compression, shear and torsional loads, with appropriate variation in the elastic properties of the rubber obtained by a suitable mix.

Though no mathematical treatment has been attempted here, the reader will have realised that for quantitative analysis of inertias, modes of motion and vibration frequencies, advanced and difficult mathematics are required, combined with experimental measurement.

Similar principles apply to transversely mounted engines, but an additional, and very important, factor has to be taken into consideration. It is that, because the gearbox and final drive are combined and mounted either on or in the engine, the torque the mountings have to react is the engine torque increased by the overall ratio of the final drive and gear engaged at any particular time, instead of just that from the engine and gearbox alone. In other words, it is between about $2\frac{1}{2}$ and $3\frac{1}{2}$ times as great. In the early days of transverse engine installation, the solution was to allow the engine relatively free vertical motion, within limits, while restricting the fore-and-aft motion of the lower mountings and introducing a horizontal link, with pivot eyes at its ends, to tie the top of the crankcase to the dash. This of course enables the torque reaction to be taken between the top link and bottom mountings. Later, extra steel link was obviated by building resistance to horizontal movement into the rear high mounting, by either stiffening its rubber elements against deflection in that direction or incorporating in it stops designed to limit such movement progressively. A typical transverse engine mounting system is that of the Rover 200 range, illustrated in Fig. 2.18. For a more detailed treatise, the reader is recommended to the book *Fundamentals Balancing*, by W. Thomas, published by the Institution of Mechanical Engineers.

Chapter 3

Constructional details of the engine

From the outline of the general principles in Chapter 1, and the requirements as regards balance in Chapter 2, we now turn to the details of construction, leaving engines having six or more cylinders to Chapter 4. Sleeve valve, rotary valve, and rotary piston engine constructions will then be dealt with in Chapter 5.

The conventional layout described in Section 1.10 has become firmly established, despite attempts to develop for automotive applications others such as the swash-plate motor, widely used for hydraulic power, and the Stirling engine, the relatively large size and weight of which virtually rules it out. The gas turbine, while well established for large power units operating mainly at constant speeds, has so far defied attempts to develop it in sizes small enough and of adequate flexibility for quantity production for automotive applications.

Because the reciprocating piston type has had the benefit of so much time, effort and money spent continuously on its intensive development over the century or more since its invention, prospects are indeed remote for a successful challenge from any alternative power unit. Moreover, to justify the abandonment of the world's huge capital investment in plant and equipment for its production, the potential for gain would have to be of truly major significance. Another factor is the very large infrastructure that has been built up, again worldwide, in terms of both experience and equipment for the maintenance of such engines.

3.1 General engine parts

What may be classified as 'general' engine parts will first be described, followed by the poppet valve and its various possible positions and methods of operation. General cylinder construction, which depends on the location of the valves, will follow and, finally, descriptions of typical four-cylinder engines will be given.

3.2 The piston

The piston performs the following functions—

(1) Forms a movable gas-tight plug to confine the charge in the cylinder

(2) Transmits to the connecting rod the forces generated by combustion of the charge

(3) Forms a guide and a bearing for the small end of the connecting rod, and takes the lateral thrust due to the obliquity of that rod.

For the designer, the major problem is catering for the variation in operating temperatures – from starting from cold at sub-zero to maximum output in tropical climates, as well as the smaller, yet still large, variations encountered in any one locality. Additionally, the weight of the piston must be kept to a minimum to reduce vibration and the inertia loading on bearings, and to avoid friction and other losses entailed in accelerating the pistons in both directions. Consequently, although cast-iron pistons have been employed, to minimise the effects of differential expansion between the piston and cylinder, these are now found only in a few two-stroke engines. This is because of their superior resistance to the higher temperatures generated in that cycle, especially adjacent to the exhaust ports.

3.3 Thermal considerations

Almost all modern engines have aluminium alloy pistons. Because the aluminium alloy is of lower strength than cast iron, thicker sections have to be used so not all the advantage of the light weight of this material is realised. Moreover, because of its higher coefficient of thermal expansion, larger running clearances have to be allowed. On the other hand, the thermal conductivity of aluminium is about three times that of iron. This, together with the greater thicknesses of the sections used, enables aluminium pistons to run at temperatures about 200°C lower than cast-iron ones. Consequently, there is little or no tendency for deposition of carbon – due to thermal breakdown of lubricant – beneath the piston crown or in ring grooves. So important is this that sections thicker than necessary for carrying the mechanical loads are in many instances used to obtain a good rate of cooling by heat transfer.

The thermal flow in a piston is from the crown, out to the ring belt, whence the heat is transferred through the rings to the cylinder walls and thence to the coolant. A small proportion is transferred down to the skirt and then across the bearing surfaces – between the skirt and cylinder walls – but this is not of such great signficance, partly because of the relative remoteness of the skirt from the source of heat and partly its fairly light contact with the cylinder walls. Some heat is also taken away with the lubricant but, again, this is not a very significant proportion unless the underside of the crown is positively cooled by a jet of oil or some other system. Highly-rated engines – for example, some turbocharged diesel units – as well as engines with large diameter pistons and those operating on the two-stroke principle, may have oil-cooled pistons.

3.4 Design details

Some typical pistons for heavy-duty engines are illustrated in Figs 3.1 and 3.2. From Fig. 3.1, it can be seen that thick sections are used and there are no abrupt changes in section which could form barriers to heat flow. The aim has been not only to keep local temperatures below the level at which the mechanical properties of the material begin to fall off significantly but also to maintain fairly uniform thermal gradients to avoid thermal fatigue cracking, especially adjacent to exhaust valves.

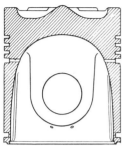

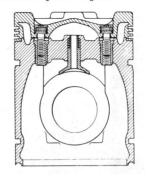

Fig. 3.1   Fig. 3.2

In Fig. 3.2 is illustrated a multi-piece oil-cooled piston, which may have either a forged steel or a Nimonic crown. Oil in this instance is taken up an axial hole in the connecting rod and, through a spring-loaded slipper pick-up, into the space between the crown and skirt portions. It drains back through holes not shown in the illustration. With this design, the two parts are secured together by set-bolts inserted from above, so the crowns can be removed for attention in service without disturbing the remainder of the reciprocating assembly.

For light duty engines it used to be common practice to employ pistons the skirts of which were split by machining a slot up one side. The object was to enable the cold clearance between skirt and cylinder bore to be reduced without risk of seizure when hot. Later, as ratings increased, T-shaped slots, as illustrated in Fig. 3.3, were used, the head of the T exercising an influence over thermal flow as well as on the resilience of the skirt under radial compressive forces. Nowadays, engines are generally too highly rated for T-slotted skirts. The modern equivalent is the piston with slots machined through the base of its bottom ring grooves. In some instances, the ends of these extend a short distance downwards, as in Fig. 3.4 which shows the Hepolite W-slot piston.

More highly-rated engines, on the other hand, have no slots, being of what is termed the *solid skirt type*. In many instances, these have steel inserts, as shown in Fig. 3.5, to help to limit the expansion, by tying the skirt in locally. Virtually all pistons are cam-ground to non-circular shapes. They are also profiled top to bottom to compensate for the differing degrees of thermal expansion due to both the thermal gradients and the variations in local thicknesses of metal needed to meet the requirements for transmitting the gas and inertia loading down from the crown to the gudgeon pin bosses. In general, the diameter on

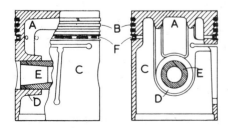

Fig. 3.3 Early light alloy piston

Fig. 3.4 In the Hepolite W-slot piston there is one slot in each face. It extends along the base of the oil control ring groove and, as shown in this illustration, its ends are swept round the gudgeon pin bosses

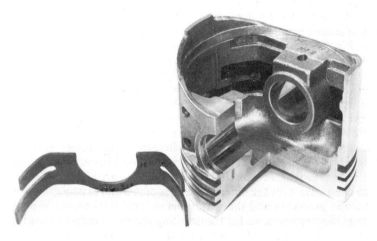

Fig. 3.5 Hepworth and Grandage Pyrostrut III piston. Its steel insert is shaped so as to exert the greatest degree of control over thermal expansion near the crown of the piston

the axis of the gudgeon pin is smaller than that on the thrust axis. This is because the region around the pin is so rigid that it could not deflect enough to accommodate tightening hot clearances. Cutting a slot in the bottom ring groove on each side, on the thrust axis, channels heat into the minor axis and gives the skirt on the thrust axis the additional flexibility required for running with tighter clearances.

Large clearances, causing piston slap when the engine is cold, are relatively harmless. Hot slap, on the other hand, can shorten fatigue life, impair the sealing of the rings and, assuming it is persistent, cause sufficient noise to induce driver fatigue. In engines with wet liners, it can also cause cavitation erosion due to resonant vibration of the liners. A remedy sometimes adopted is to offset the gudgeon pin slightly. However, this tends to be more effective the lower the engine speed and, because of inertia effects, can even be detrimental at high speeds.

If the clearances are too small, thermal expansion may cause excessive contact pressures at the rubbing surfaces between piston and cylinder bore. The result can be scuffing or seizure. These effects are usually local, hence the need for the complex profiling mentioned two paragraphs previously. For a solid skirt piston, the thermal expansion pattern is generally as follows. The crown tends to expand as a solid disc, carrying with it the rigidly attached gudgeon pin bosses. Consequently, the upper part of the skirt, being not only cooler but also less rigid, tends to be drawn in to an oval shape by the outward movement of the bosses. On the other hand, the bottom of the skirt, especially of a rigidly constructed piston for a diesel engine, may deflect in the opposite sense from the upper part. The reason for this contra deflection is not positively understood, but it is probably due to pivot action about the mid-portion.

Steel inserts, of course, as in Fig. 3.5, considerably modify the thermal expansion pattern, so the designer has to pay attention to the rigidity with which the skirt is attached to the crown. Slots machined in the base of the oil ring groove, by forming a barrier to thermal flow, significantly reduce thermal expansion of the skirt. They also of course affect the flexibility of the connection between the skirt and crown and hence the degree and mode of deflection.

The most commonly used material for pistons is an aluminium alloy containing 10 to 12.5–13% silicon, which has a coefficient of thermal expansion of $19.5 \times 10^6/°C$, as compared with 23×10^6 for aluminium and 11×10^6 for cast iron, for which reason it is called Lo-Ex alloy. It was found in the late nineteen-seventies that the addition of phosphorous, in the form of phosphor-copper, not only enhances the flow of Lo-Ex within the mould (thus improving the quality of the casting) but also refines the structure by encouraging silicon precipitation, which increases fatigue resistance. In these modern alloys the iron impurity is limited to between 0.7 and 0.5%. A hypereutectic alloy containing about 19% silicon, 1.2% nickel and 0.5 to 0.6% of each of cobalt and chromium is also available for applications in which an even lower coefficient of expansion – 17.5×10^6 – is required. It is sometimes used where cooling may not always be good, for example in portable engines for power tools.

Among the other advances that were made by AE plc in the late nineteen-seventies is the development of the *squeeze casting* technique, which is now used for the production of some heavy duty pistons. First, the molten metal is poured into a cup-shape mould, and then the die that determines the shape of the inside of the piston is lowered into the cup, to close its open end and apply pressure to the molten metal as it cools in the mould. This application of pressure has three beneficial effects. First, it increases the rate of heat transfer between the molten metal and the mould and die, thus refining the metallic structure; secondly, it compensates for the shrinkage due to cooling and solidification; and, thirdly, it prevents the liberation of dissolved gases to form bubbles, which otherwise would cause porosity.

Squeeze casting is particularly suitable for manufacture when ceramic fibre reinforcement is desired. The reinforcement comprises preformed pads. These are made by vacuum forming or pressing, during which the fibre orientation and density are controlled, and the spacing between the individual fibres is maintained by the use of complex binders. These pads are located securely in the moulds and, during the squeeze casting process, the metal is forced in, filling completely the interstices between the fibres, to which it then bonds. At low temperatures, the strengths of such composites are little more than that of the aluminium alloy matrix but, as the temperature rises to about 400 °C, and the matrix softens, the fibres progressively take over and the resultant hot strength rises steeply by about 200% relative to that of the metal alone at the same temperature.

3.5   Slipper and articulated pistons

For applications in which running temperatures are relatively constant and therefore large cold clearances are acceptable, slipper type pistons can be used. These have no skirt, the function of which is performed by two thrust pads separated from the crown and supported directly by the gudgeon pin bosses, Fig. 3.6(*a*). With this arrangement, the shape of the piston in plan view can be virtually a true circle, which improves the oil film distribution and bearing characteristics. Such a piston is sometimes termed the *cross-head type* because of the duty performed by the thrust pads. Slipper type pistons are also made in which the thrust pads are not separated from the crowns, other than perhaps by slotting in the lower ring groove.

The articulated piston is one in which the thrust pad portion, but generally in the form of a skirt, is carried independently on the gudgeon pin, Fig. 3.6(*b*). For very highly rated engines, the crown can be of iron or a ferrous alloy, while the skirt is aluminium. Because the skirt can be of virtually uniform stiffness and its running temperature is little higher than that of the lubricating oil, the clearance can be small and its bearing properties good. Moreover, by virtue of the articulation of the skirt, pivoting freely about the gudgeon pin, the lateral thrust loading tends to be well distributed over a large bearing area, while the crown remains stable and supported by its rings. Both the pistons in Fig. 3.6 are for diesel engines.

3.6   AEconoglide piston

A major advantage of the slipper and articulated types of piston is the small areas of their skill relative to those of the more conventional pistons and, in consequence, reduced oil drag between them and their cylinder bores. However, they are more complex and therefore not only costly but also the potential for weight reduction is very limited. An even simpler way to reduce the drag would be to shorten the skirt, but this would lead to piston tilting, with high contact loads on the top and bottom edges and, ultimately, deterioration of ring performance. In the early nineteen-eighties, a much better method, entailing the incorporation of three thrust pads, 0.025 mm high, on

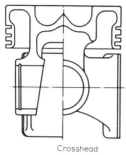

Crosshead

Fig. 3.6 (*a*)

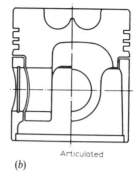

Articulated

(*b*)

the skirt, as shown in Fig. 3.7, was developed by AE plc, under the trade name AEconoglide piston.

It has three pads standing about 0.025 mm radially proud of the skirt, two near its top and one near its bottom end. During the reciprocation of the piston, therefore, none of the pads screens any of the others from their supplies of lubricating oil. Moreover, although the circumferential lengths of the two top pads add to the same as that of the single bottom pad the latter, because its loading is not so heavy as that on the top pads, is only 0.75 mm instead of 1.0 mm long axially, so the surface stresses on all three are similar. Another advantage of separating the two top pads is that it reduces the bending moments about the gudgeon pin bosses. To encourage hydrodynamic oil film formation, the faces of the pads are sloped not more than 1.5° and preferably less than 1° in the same sense as are the taper faced rings described in Section 3.10.

The benefits of the AEconoglide concept have been shown to include consistency of contact areas and stiffnesses over a wide range of temperatures and loads, and a reduction of skirt contact area of at least 75%. In consequence, reductions in friction of up to 14% and fuel consumption by up to 4% have been obtained, while power output has increased by up to 5%.

3.7   Combustion chamber in piston

Because of the need for a working clearance between the crown of the piston at TDC and the lower face of the cylinder-head casting, a conventional combustion chamber tends to take the form of a flat disc, and consequently to have a high surface:volume ratio. This, as explained at the end of Chapter 8, is liable to lead to unacceptably high proportions of unburnt hydrocarbons in the exhaust. Moreover, if the compression ratio is high, the very close tolerances necessary to ensure that it is equal in all cylinders can be difficult to maintain. This has led to the development of bowl-in-piston combustion chambers, similar to those previously found only in diesel engines.

With this arrangement, not only is a compact chamber obtained in which relatively close control of turbulence is possible, but also the maintenance of close tolerances between the flat area around the bowl and the equally flat lower face of the cylinder head is not too difficult, since both surfaces are easy to machine all over. A good example is the Austin-Morris O-Series engine, described in Section 3.65. The valve layout, for actuation by an overhead camshaft, can be extremely simple, and the gas flow is not impeded by the

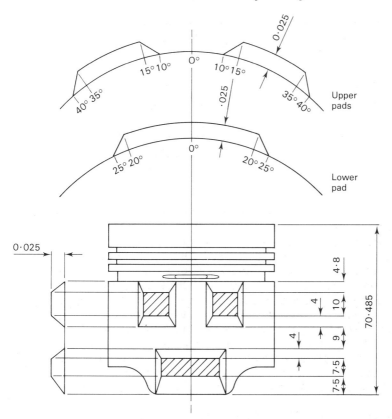

Fig. 3.7 The skirt profile of a typical AEconologide piston is dimensionally similar to that of a conventional piston except that the ovality is slightly modified. On the other hand, the thrust area, being limited to that of the pads, is 75% less

proximity of the boundary of the combustion chamber, except at TDC. With the sparking plug close to the centre, a uniform spread of combustion is obtained. Obviously the rate of heat transfer to the piston is greater than with more conventional designs, so the positioning of the rings relative to both the crown and each other is critical and the optimum must be determined by development testing.

The modern tendency is to form the combustion chamber partly in the piston crown and partly in the head. Typical examples are the Jaguar V-12 and Rover 2.3/2.6 engines described in Sections 4.17 and 4.10, respectively. The advantages relative to the total combustion chamber in piston are: heat flow problems eased and a minimum of unburnt hydrocarbons in the exhaust.

3.8   Piston rings

Piston rings tax the skills mainly of the metallurgists and production engineers. The main functions of piston rings are—

(1) To form a pressure seal, preventing blow-by of the gases, including combustion products at high temperatures.

(2) The transfer of heat from the piston to the cylinder walls.
(3) Control of the flow of oil in adequate quantity to the skirt and to the rings themselves, while preventing excessive amounts from entering the combustion chamber.

Fine-grain alloy cast iron has proved superior to any other material for this purpose. Its merits arise from its excellent heat- and wear-resistance inherent in its graphic structure. Two typical material specifications are given in Table 3.1. The medium-duty material is recommended for second compression rings, while the heavy-duty material is for top rings.

A ring of the HG10 material will not lose more than 20% of its original free gap when heated at 350°C for $6\frac{1}{2}$ hours enclosed in a sleeve the bore size of which is equal to that of the cylinder bore, and then air-cooled. One of the HG22C material, on the other hand, will lose no more than 10% in the same circumstances. Their tensile strengths (BS 4K6) are, respectively, minima of $278 \times 10^6\,\text{N/m}^2$ and $587 \times 10^6\,\text{N/m}^2$.

Piston rings are mostly cast in the open condition and then cam-turned to a profile such that, when they are closed to fit into the cylinder, their peripheries are a true circle. Final machining is done with the ring in the closed condition. An earlier method of manufacture, now discontinued, was to start with a circular ring and then open it out by peening its inner periphery.

Gaps have been cut to various shapes, some of which are shown in Fig. 3.8. Now, however, the simple square cut has superseded all others. This type does not wear a vertical ridge in the cylinder, since rings tend to rotate.

3.9 Ring sections

Cross-sectional depth is determined by the radial stiffness required, though it is necessary to ensure that the bearing areas between the ring and the sides of its groove are adequate. Thickness, although it also has an influence on radial stiffness, is primarily determined by the bearing pressures required between the outer face of the ring and the cylinder wall.

Simple rectangular-section rings, though often with modified face profiles, are mostly used. However, some others have been introduced with varying degrees of success. One is the L section, Fig. 3.9(*a*). This, when radially compressed into its groove, twists into a dished configuration owing to the

Table 3.1—CHEMICAL COMPOSITION, PER CENT

	Hepolite HG10 (medium duty)	Hepolite HG22C (heavy duty)
Total carbon	3.40 – 3.90	2.70 – 3.30
Combined carbon		0.50 – 1.10
Silicon	2.10 – 2.90	2.00 – 3.00
Sulphur	0.00 – 0.10	0.00 – 0.10
Phosphorus	0.40 – 0.80	0.00 – 0.50
Manganese	0.40 – 0.90	0.50 – 0.90
Chromium	0.00 – 0.40	0.50 – 0.85
Molybdenum	0.00 – 0.40	0.00 – 0.40
Vanadium		0.00 – 0.40

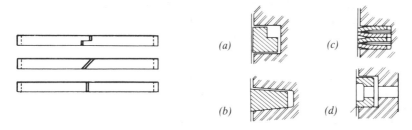

Fig. 3.8 Piston rings   Fig. 3.9

absence of material in its upper inner corner to resist the compression induced by the bending. As a result, the lower corners come more forcibly into contact with the cylinder wall and lower side of the groove. Thus both the sealing and oil control characteristics of the ring are improved, though the heat transfer may not be quite so good. At the same time, because of the relative freedom with which gas pressure can act on the top of the ring, flutter and blow-by tend to be inhibited.

The wedge section ring and groove, Fig. 3.9(*b*), was developed to obviate ring stick in compression ignition engines. It was said to be especially suitable for two-stroke engines. When it bottoms in its groove, there must be a side clearance of at least 0.0254 mm.

In Fig. 3.9(*c*), the Cords ring is illustrated. This comprises four dished and gapped alloy steel washers assembled in a manner such that the upper and lower ones are flexibly in contact with the sides of the groove. It is claimed to be especially suitable for use in worn bores.

3.10   Oil control rings

In the early days of the internal combustion engine, only one compression ring was used. Subsequently, two were found to seal better, and later even three were used. As speeds increased, greater quantities of lubricant had to be supplied and the problem of controlling it had to be faced. In the first instance, this was done simply by using a lower ring having a narrow face-width, to increase its contact pressure. Then stepped, or bevelled or taper-faced rings were employed and, ultimately, grooved and slotted rings, similar to that shown in Fig. 3.9(*d*), were adopted.

The typical oil control ring of this type, shown in the illustration, has two narrow lands bearing at a relatively high, but controlled, pressure on the cylinder walls to scrape off the oil that is surplus to requirements for lubrication. The holes, or slots, in the base of its channel section allow the oil removed to flow into the base of the groove, in which there are more holes to allow it to drain back down the cylinder to the crankcase and sump. Often the lower outer corner of the groove is chamfered and holes are drilled from the chamfer to the interior of the skirt, again for drainage. To improve the wear resistance of the narrow faces of the rings, these faces are generally chromium-plated.

Because of the high speeds and bore:stroke ratios of modern engines, a copious supply of oil is flung up the cylinder bores, so the radial pressure exerted by oil control rings has had to be intensified. Consequently, a ring comprising a helical coil spring, in compression, is often interposed between the oil control ring and the base of its groove, Fig. 3.10. Humped, or crimped, strips of spring steel exerting a radial pressure on these rings are rarely used now because they are inadequate for such high speeds and loads.

The narrow faces, or lands, of the grooved and slotted rings are difficult to chromium-plate, so three-piece oil control rings, such as that in Fig. 3.11, are now widely employed. These comprise, in effect, the two lands of the grooved and slotted ring with a spacer between them. Initially, these were of cast iron, but now steel strip is used because it is easier to produce and to plate, and is lighter. In the Hepolite SE ring illustrated, there are two rails and the expander and spacer between them are integrated to form a single component. Such a ring has the additional advantages that it is stronger and less prone to breakage than a cast-iron ring. Moreover, it can conform more easily to worn bores.

From the illustration it can be seen that the two rails are spaced apart by a rigid box section pressing and are forced radially outwards by a series of tiny cantilever springs formed in the base of the box section. The rails seat on lateral spurs, one each side of the end of each of the cantilever springs, which, if one rail lifts or sinks further than the other, act as rectangular section torsion bars to accommodate this differential deflection. Since each cantilever spring is independent of the others, failure of any one would have only a marginal effect on the performance of the ring.

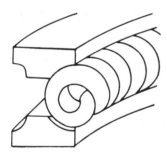

Fig. 3.10

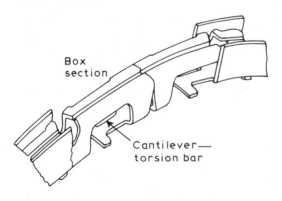

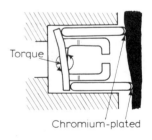

Fig. 3.11 (a)   (b)

3.11 Ring belt design

Most modern petrol, and even some diesel, engine pistons have only two compression and one oil control ring, all above the gudgeon pin. Previously, for very many years, the most common arrangement was three compression rings above the gudgeon pin and one oil control ring below. With the removal of one compression ring, space became available above the pin for the oil control ring. Locating it there also had the advantage that, at the high speeds and ratings of modern engines, the consequently increased supply of oil for lubrication of the skirt, especially for solid-skirt pistons, was desirable. Moreover, the ring grooves are now all in the part of the piston where metal sections are in any case thick and therefore less likely to be unacceptably weakened by their presence.

The top compression ring is generally chromium-plated for wear and corrosion resistance, as decribed in Section 3.12. An alternative is a sprayed molybdenum coating on the periphery of the ring to improve its scuff resistance. The outer faces of either type are generally either angled slightly – taper-faced – or lapped, to a barrel profile, to facilitate bedding in.

Spacing of rings is important, in that the lands must be thick enough to avoid their breaking up under the dynamic loading applied by the rings. The height of the land above the top ring is generally greater than the widths of the others because of the high temperatures at which it may operate which, under extreme conditions, can significantly reduce the strength in this region. For very heavy duty engines, a steel band may be bonded into the piston to carry the rings. Bonding is preferable to casting it in integrally because of the high tensile loads that would be produced at the steel-aluminium interface on cooling from the casting temperature. With bonding, the ring band can, if necessary, be shrunk on.

Alternative methods of increasing the wear resistance of the flanks of the grooves include anodising the aluminium alloy, and alloying it locally with other metals. Anodisation entails simply applying a positive charge to the component while immersing it in a suitable electrolyte. This applies a hard oxide coating about 0.05 mm thick to the surface, the hardness of which is thus increased from about 80–90 Hv to 350–450 Hv.

Local alloying, on the other hand, can reinforce the metallic structure to a depth of between 1 and 2 mm below the surface. This is a continuous process, during which the piston is rotated and an electron beam or laser is used to produce on the flank of the groove a tiny pool of molten aluminium alloy into which is fed additions of up to 30% of the alloying metal in the form of a fine rod or wire. Loss of heat to the surrounding metal is rapid, so the grain is very fine, and the bonding of the alloyed section to the parent metal is excellent. The hardness value of the parent metal is thus increased locally to about 170–190 Hv.

3.12 Cylinder bore wear and corrosion

The life of an engine between re-bores of the cylinders is determined by both abrasion and corrosion wear, the chief factors in the former being the nature of the prevailing atmospheric conditions and the efficiency of the air and oil filtration, while the latter is due to corrosive products of combustion formed during the warming-up period, and is most apparent in engines whose duty involves frequent starting from cold. Piston, piston rings and cylinder bore have to be considered together, and intensive metallurgical and engineering research is continuously devoted to the associated problems of blow-by, wear and corrosion of cylinder bores, and excessive oil consumption through pumping action due to lateral movement of the rings in their grooves.

The most widely used method of finishing cylinder bores is plateau honing. A typical machining sequence is as follows: turn to size with a diamond-tipped tool, to obtain an exceptionally clean-cut finish; coarse silicon carbide hone, plunging the tool in and out of the bore to establish two opposite-handed sets of spiral markings and thus a criss-cross pattern of extremely fine scratches; a final fine honing operation to leave only the deepest of the scratch marks, with smooth plateaux between them. The residual scratches, or minute grooves, serve to retain the lubricant, while the plateaux form the bearing surfaces to take the loading. As a result, total glazing of the bore – which would tend to prevent oil from adhering to the surface – is obviated and therefore scuffing prevented.

Chromium plating of rings or bores, and provision of 'dry' or 'wet' liners of special irons centrifugally cast, are methods of attack which are used singly or in combination, with varying degrees of success and commercial justification.

Chromium plating of bores has produced remarkable results in reducing both abrasion and corrosion wear, so reducing bore wear that life between re-bores has been extended to four or five or even more times normal experience. It is, however, costly, and running-in is a lengthy process, though assisted by the matt or slightly porous finishes used.

Application of hard chromium plating to rings rather than bore is effective in protecting both ring and bore. With the considerable difference in the hardnesses of the moving parts, abrasive particles tend to be so deeply absorbed by the softer material that they do not wear the harder.

There seems little doubt that the exceptional hardness of chromium provides protection against the lapping effect of a soft piston on the theory indicated above, while a hard plated ring has much the same effect in the reverse direction on a moderately soft liner – it is too hard either to pick up abrasive particles and act as a lap, or to be lapped by a 'loaded' bore, but has on the contrary a burnishing action on the bore.

Experience and user opinion appear to confirm the value of pre-finished chromium-plated liners in cases of heavy-duty engines in dusty atmospheres, the increased initial cost being more than balanced by reduced maintenance expenditure.

Centrifugally-cast, pre-finished liners of alloy cast iron, or made of steel tube, are very resistant to corrosion wear for applications involving frequent starting from cold, and give specially long life where abrasive conditions are adverse. These dry liners are finished to very close tolerances, and can be supplied if required to give a slip fit to aid easy renewal. They are sometimes copper-plated on their external surfaces to aid both assembly and heat flow to the water jackets.

With the pre-finished type of liner the bores in the block must be finished to a very high degree of accuracy in order to avoid distortion of the liners on assembly, the thickness of these being of the order of 2.5 mm, and not providing for re-bore.

Wet liners are made much thicker, as their necessary stiffness, reinforced by joint flanges, must be self-contained. They are usually more easily renewable than the dry type, and for two-stroke engines the necessary scavenge air ports may be readily cast in. The two types are illustrated in Figs 3.43, 3.44 and 7.10, and may be seen in various other illustrations of complete engine assemblies.

3.13   Gudgeon pin

The gudgeon pin is of case-hardened steel; usually it is hollow and the conventional way of supporting it in two internal bosses on the piston walls is shown in Fig. 3.3. Many different ways of securing the gudgeon pin against end movement have been tried and abandoned, the device now most generally used being the spring circlip. Sometimes tightness of fit is relied upon, and phosphor bronze or aluminium pads may be placed at the ends; these will not damage the cylinder should endways movement occur. An alternative arrangement is to secure the pin in the small end of the connecting rod and allow it to turn in long piston bosses without bushes, while a 'floating' pin has also been used which is free to turn in both rod and piston, end pads being fitted. When the connecting rod is free on the gudgeon pin, a phosphor bronze bush has to be provided; this bush is a tight fit in the eye of the rod. The small end bush develops very little wear and requires renewal only at long intervals.

In view of the very heavy alternating loading of the gudgeon pins of compression-ignition engines, special care is taken to avoid risk of fatigue cracks originating at the surface of the bore, by broaching or honing or rapid-traverse grinding with the object of eliminating circumferential tool marks. The external bearing surface is finished to a very high degree of accuracy to ensure correct fit in the piston and connecting rod.

3.14   Connecting rods

Most connecting rods are medium carbon steel stampings. For special applications, however, they may be forgings, and aluminium alloys or even titanium may be used. Very highly stressed rods – for example in racing car engines – are sometimes machined all over to improve fatigue strength and to reduce weight to a mnimum. Fairly recent developments are the application of malleable and SG irons and even steel castings.

As mentioned previously, the big end is usually split and has a separate cap, so that it can be assembled, together with its bearing, on to the crankpin. The cap is secured to the rod by either two or four bolts and nuts.

In the small-end eye, there is usually a force-fit phosphor-bronze or lead-bronze bush, either solid or metal-backed. The gudgeon pin generally either floats axially between circlips in the piston or is located by an interference fit in the piston bosses. Alternatively, the bearing bush and circlips are dispensed with and the pin is either a press fit in the small end or the eye is split and the pin secured by a pinch bolt.

3.15   Typical connecting rods

A typical connecting rod from an engine used in cars and trucks in the 1920s is illustrated in Fig. 3.12. At that time, big ends were secured by castellated nuts locked by split pins, though plain nuts with tab washers were employed in some engines. Alternatively, to save weight, the bearing caps can be secured by studs screwed in the end of the rod. In Fig. 3.12, A is a steel forging, B a bronze small end bush, C the big end bearing cap, and D a split big end bush of bronze lined with whitemetal. Waisting the bolts or studs to a diameter marginally less than that of the roots of the threads reduces both the rotational mass and need of additional weight for balancing it. At the same time, by obviating stress concentrations, it actually increases the fatigue strength of the shank. However, in this application, a short length in the middle of the shank is left unwaisted for positive location of the cap relative to the rod. Fillet radii between the waisted and unwaisted portions obviate local stress concentrations.

Floating bushes are sometimes employed in high-speed engines. Such a bush tends to rotate within its housing at a speed such that the values of frictional drag on its inner and outer bearing faces are equal, thus virtually halving the speed differentials, and therefore friction and wear, between the rubbing surfaces. A floating bush, however, cannot be assembled other than on to a built-up crankshaft. A example can be seen in Fig. 3.17 where, clamped firmly between the cap and rod is a steel bush A, to provide a continuous surface within which the bronze bush B can float. Oil is fed through ducts drilled in the crankshaft to lubricate the inner bearing surface of the bronze bush, and thence through holes in the bush to lubricate its outer bearing surface.

A sturdy piston and connecting rod assembly for a 10.35 litre high-speed compression ignition engine developing 97 kW at 1900 rev/min is illustrated in Fig. 3.13(*a*). The joint face between cap and the rod is inclined at 35° to the axis of the rod, to reduce the width of the assembly so that, after the bolts have been removed from below, the rod can be withdrawn upwards through the cylinder. Generally, access for the dismantling of the big end would be gained by

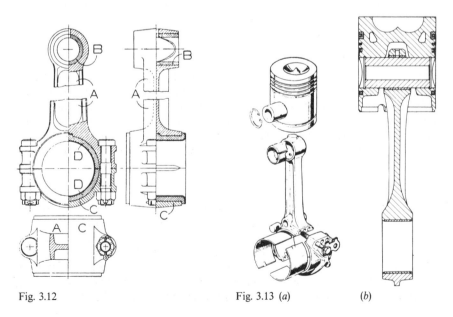

Fig. 3.12   Fig. 3.13 (*a*)   (*b*)

removing the sump but, in large engines, for marine and industrial applications, there are usually inspection covers on the side of the crankcase.

In this particular connecting rod arrangement, which was originally patented by Henry Meadows Ltd, there are four bolts, the upper pair being screwed into cylindrical nuts having diametral, instead of axial, threaded holes and which are housed in a transverse hole through the rod. This transverse hole is utilised for location during machining, as also is the extra hole between each pair of bolts. The latter can subsequently house location dowels, if required.

Conventional stud or bolt-and-nut arrangements, however, are now widely used with inclined housing joint faces, in which case the unwaisted central portions of the bolts alone will take the shear component of the loading due to the inclination of the joint faces. In some instances, the bolts have rounded heads with flats on one side to locate against shoulders rising from the edges of the spot-faces on which they seat. An alternative method of taking the shear is to machine in the joint faces either shoulders or serrations parallel to the axis of the crankpin. The shear may be at a maximum under the tension arising from the inertial loading during the induction and exhaust strokes, since this loading reduces the frictional grip due to the clamping together of the joint faces.

Other features of interest in Fig. 3.13(*a*) include the single rib around the bearing cap, the twin tab washers and the oil hole drilled axially from the big end, to take oil up to the small end. Twin ribs might have been used had the engine been more heavily loaded. On the other hand, had it been more lightly loaded, the small end would have been lubricated by splash thrown up from the big end.

For diesel engines, in which the gas loading is generally more severe than the inertia loading, chamfered small end bosses are often employed, as in the Perkins T 6.3543, Fig. 3.13(*b*). This makes the bearing areas in the critical opposite halves of the bush in the rod and the bosses in the piston as large as practical for a given cylinder bore dimension.

3.16 Bearing bushes

Except in certain applications, such as some motor-cycle engines, big end and main journal bearings are in halves – otherwise a fabricated crankshaft would have to be used so that they could be assembled on to it. One-piece cylindrical bushes are, however, used for camshaft, rocker and spiral gear bearings. In some instances these bushes are simply strips of bearing material, produced by wrapping them round a mandrel without actually joining their abutting ends.

3.17 Bearing materials

Undoubtedly whitemetal has the best bearing properties. However, its fatigue strength is limited, so other alloys have to be used for many modern highly-rated engines. Properties of bearing metals are set out in Table 3.2.

Babbitt invented whitemetal in 1839. It contained 83% tin, 11% antimony and 6% copper. The hard copper-antimony particles suspended in a matrix of soft copper-tin alloy give good wear resistance plus the ability to embed solid abrasive particles that would otherwise wear the shaft. Additionally, white-

Table 3.2—POPULAR GLACIER BEARING MATERIALS

Lining material (steel-backed)	Nominal composition					Sapphire fatigue rating**	Fatigue strength	Seizure resistance	Corrosion resistance	Embeddability and conformability	Hardness HV5	Typical usage
	Al	Cu	Pb	Sb	Sn							
Tin babbitt	–	3.5	–	7.5	89	4 500	2	10	10	9	27	Gas turbine, industrial gearbox, electric generators, slow-speed marine two-stroke engines, etc.
Lead babbitt	–	–	84	10	6	4 500	2	9	10	10	16	Large industrial machinery, petrol engines, thrust-washers, wick-lubricated fractional hp motors (lower cost than tin)
Tin-aluminium	60	–	–	–	40	8 500	4	9	10	9	27	Overlay plated in low-speed diesel cross-head, and unplated in medium-speed diesel engines and reciprocating compressors
Tin-aluminium	79	1	–	–	20	14 000	7	7	10	8	35	Overlay plated in medium- and high-speed diesel units, petrol engines
Tin-aluminium	93	1 (+1% nickel)	–	–	6	16 000	8	5*	10	6	45	Overlay plated in medium- and high-speed diesel engines
Aluminium-tin -silicon	83	2 (+4% Si)	–	–	12	14 000	8	9	10	7	60	Highly-rated petrol and high-speed diesel engines, especially with nodular iron crankshafts
Copper-lead	–	70	30	–	–	17 000	8	5*	5	7	40	Overlay plated in medium-speed diesel and petrol engines
Lead-bronze	–	73.5	25	–	1.5	18 000	9	4*	5	5	50	Overlay plated in medium-speed diesel and turbo-blowers***, petrol and high-speed diesel engines
Lead-bronze	–	73.5	22	–	4.5	18 000	9	3*	5	4	55	Overlay plated in medium- and high-speed diesel
Lead-bronze	–	80	10	–	10	18 000	10	2	5	2	100	Overlay plated in small end and rocker bushes for petrol and medium- and high-speed diesel engines
Aluminium-silicon	88.5	1	–	–	(+10.5% silicon)	18 000	9	8*	10	4	56	Overlay plated in highly-rated high-speed diesel engines

*These ratings apply to the base material. Commonly, these alloys are supplied with a layer 0.0254 mm thick of electro-deposited lead–10% tin to improve seizure resistance to a rating of 10 initially.

**Sapphire rating is the dynamic load in lbf/in² which can be withstood by an oil-lubricated bearing with minimal misalignment in a 2-in diameter sapphire test rig for more than 3×10^6 load cycles; test temperatures 110 to 130°C, depending upon load.

***Turbo-blowers have cast lead-bronze lining, Glacier GL26 (Pb 26% Sn 2%) with properties similar to lead-bronze.

metal will conform readily to inaccuracies of the shaft profile and to accommodate deflections of the shaft. Because of its low melting point, high spots in the bearing interface cause this material to soften and flow slightly to relieve excessive local pressures, instead of seizing.

Because of the increasing price of tin, there has been a tendency to use lead babbitts – which have properties similar to those of tin babbitts.

Originally, whitemetal bearings were cast in their housings. Later, they were made in the form of thick shells, sometimes in a thick bronze backing, and could therefore be more easily replaced. In either case, it was necessary first to bore them in the engine and then to scrape them manually, using prussian blue marking, to obtain a good fit.

3.18 Thin-wall bearings

In the mid-nineteen-thirties the thin-wall, or shell-type, bearing was intro-duced for cars in the UK. This had been developed originally in the USA for aero-engines and then cars. The whitemetal was applied as a very thin lining on a steel backing about 1.5 mm thick. This had two main advantages: first, the steel backing gave good support to the whitemetal, and therefore the fatigue strength of the bearing was good; and secondly, the bearings could be made with such precision that, provided that their housings were equally precisely machined, they could be assembled without the need for skilled manual fitting. Moreover, they were equally easy to replace in service.

3.19 Stronger materials

In the meantime, engine speeds, and with them gas and inertia loadings, had been increasing. Additionally, diesel engines were becoming popular for commercial vehicles. Consequently, there was a demand for stronger bearing materials. Solid bronze shells had been used, and this entailed hardening the shafts to prevent rapid wear. For heavy-duty applications, the new thin-wall bearings, with copper-lead and lead-bronze bearing materials on steel back-ings, were used. Of these materials, the latter is the stronger, but its conformability is worst. In both, the lead is held within the matrix, so that it is immediately available to smear on the bearing surface. The difference between the two is that in one case the matrix is pure copper and in the other it is the stronger copper-tin alloy. With the harder bearing materials, the crankshafts must be hardened, usually within the range 350 to 900 Vickers.

3.20 Corrosion of bearings

Unless engine oils are changed at fairly frequent intervals, copper-lead and lead-bronze bearings are liable to corrode. The lead phase is attacked by organic acids and peroxides that develop as a result of degradation of the oil at high temperatures. Weakening of the bearing structure and fatigue failure of the copper matrix ensue.

To protect these bearings from corrosion, a lead-tin overlay is almost invariably applied to their surfaces, by electroplating to a nominal thickness of 0.025 mm. Such a plating also improves both seizure-resistance and bedding-in and, provided the environment is favourable, it can last the life of the engine.

However, abrasive dirt can score the overlay and allow the corrosive elements to penetrate to the lining material. The lead-based overlay does not corrode because it is protected by a tin content, which is generally of the order of 10% – the minimum acceptable is 4%.

3.21   Aluminium-tin bearing alloys

Although aluminium-tin bearings were introduced in the mid nineteen-thirties, no more than 6% tin could be used, otherwise fatigue strength was unacceptably reduced. Because of the hardness of this alloy, the shafts had to be hardened; this problem, however, was overcome by the end of the Second World War by overlay plating with alloys of lead with tin, copper or indium. The main incentive for using aluminium is its low cost relative to that of copper. Additionally, its melting point is low enough for easy casting and application in the manufacture of bearings.

By 1950, the Glacier Metal Company Ltd, appreciating the fact that overlay plated copper-lead, lead-bronze and 6% tin-aluminium linings were expensive, intensified their efforts to develop a better material. The outcome was the introduction in 1951 of a reticular tin-aluminium alloy, containing 20% tin. This development, a joint project between Glacier and the Tin Research Institute, was a major advance.

The problem of reduction of fatigue strength was overcome by preventing the tin from remaining as grain boundary films – its natural tendency – in the aluminium. Instead the tin forms continuous films along the edges of the grains of aluminium but not across their faces, thus forming a network structure – hence the term *reticular tin*. Essentially, this development was made possible by the perfection by Glacier of a cold roll-bonding process, instead of casting, for attaching the material to a steel backing.

Increasing the tin content up to as much as 40% improves resistance to scuffing and seizure. However, additions beyond 20% reduce the mechanical strength of the alloy.

3.22   Aluminium-silicon and aluminium-tin-silicon alloys

With the increasing use of turbocharging for diesel engines, even stronger aluminium alloy bearing materials have been developed by Glacier. One is their SA78 alloy, an 11% silicon-aluminium alloy, similar to that used for pistons, but with the hard silicon particles much more finely dispersed in the aluminium matrix. This fine dispersal makes the alloy very ductile, improves its fatigue strength and bearing properties, and renders it suitable for lining on to a steel backing. The material is normally overlay-plated to improve running-in and surface properties. Protection against corrosion is unnecessary, since neither aluminium nor silicon are affected. A 1% addition of copper in solution in the aluminium matrix helps to strengthen it.

Currently, several aluminium-tin-silicon alloys are under investigation by Glacier, but at the time of writing only one automotive bearing has been developed to the production stage. It is their AS 124 which, because it needs neither corrosion protection nor enhancement of its surface properties, does not have to be overlay-plated. The figures 124 indicate 12% tin and 4% silicon. It is claimed to have excellent resistance to seizure, good strength at high temperature, and to be particularly suitable for use with nodular iron crankshafts, see Section 3.24. This alloy is continuously cast into strip and then roll-bonded to a steel backing, an aluminium foil being interposed between the two to facilitate the bonding. Subsequently, heat treatment further develops the bond strength and refines the bearing alloy. Its characteristic feature is a continuous reticular tin matrix in which are embedded the very fine particles of silicon. The copper, which serves as a solid solution hardener, is not visible in the microstructure at × 500 magnification.

3.23 The crankshaft

Crankshafts in modern engines are carried in shell type bearings similar to those of big ends, as described in Sections 3.16 to 3.19. The bearing housings at each end are generally cast integrally in the lower edges of the front and rear walls of the crankcase, as described in Section 3.59, while transverse webs inside the crankcase support the intermediate bearing housings. Their caps are below, and each is usually secured by four bolts or studs, because they have to take the full force of combustion.

A crankshaft may have as few as two main journal bearings, even in four-cylinder engines provided their rating is low. However, in highly-rated engines there is usually, in addition, one between each pair of crank throws, though some four-cylinder units have only three bearings. In the latter event, the shaft has to be very stiff, otherwise at certain speeds and loads it would whip, heavily loading the central bearing as it does so, possibly causing it to fail. This tendency can be completely eliminated, or at least reduced, by balance weights on the crank webs each side of that bearing. A major risk associated with inadequate stiffness is that bending of the shaft will transfer a high proportion of the load to the edges of the plain bearing at each end.

The larger the number of bearings to support the shaft uniformly along its length, the more slender, and therefore lighter, can it be without risk of whip due to bending. More bearings, however, will entail increased cost and, since the accuracy of their alignment is critical, the crankcase structure supporting them must be very stable under variations of temperature and load. Frictional drag of the bearings on the shaft increases as the square of the diameter but only linearly with length.

Where only two bearings, one at each end, are required, they may be of the rolling element type, though these tend to be noisy, heavy, of large diameter, more difficult to seal, and costly. Additionally, because they impose little encastré effect on the ends of the crankshaft, the latter has to be stiffer than if two plain bearings were employed. The only advantages of rolling element bearings are axial compactness and, especially if they are of the ball type, low friction. On the other hand, roller types are better for withstanding impact loading due, for example, to detonation in the cylinders or lugging at low speed.

A major factor governing the dimensions of the shaft is the torsional stiffness needed to raise its natural frequencies of vibration, together with their harmonics, above the rotational speed of the engine. For any given length of shaft both the bending and torsional stiffness depend on the diameters and overlap of the main and big end journals and the thicknesses and widths of the webs. The lengths of the journals are a function of their loadings and the strengths of the bearing materials. In the interests of compactness and stiffness, however, the aim is always at keeping the lengths of the bearings and thicknesses of webs as small as practicable, though the shorter the bearing shells, the more difficult it is to keep the lubricating oil from being squeezed out before it can spread right round their working surfaces.

The strength of the shaft depends primarily on that of the material from which it is made. Measures such as the incorporation of generous fillet radii between the webs and journals, and perhaps rolling these fillets to induce in them residual compressive stresses, can improve fatigue strength, which is also affected by heat and hardening treatments.

3.24   Crankshaft materials

Crankshafts are generally steel forgings, though high carbon, high copper, chromium silicon iron has been used and nodular, or spheroidal graphite (SG), cast iron is becoming increasingly popular. Best of the steels for crankshafts, but the most costly, are the nitrogen-hardened types, Section 3.26. Less costly, though also inferior to a significant degree, are the high carbon or alloy steels, surface hardened by the flame or induction methods, Section 3.28. Last in order of durability come the heat treated high carbon or alloy steels that have not been surface hardened, for which only the soft whitemetal bearings are suitable.

Factors favouring cast iron crankshafts are the low cost of this material, which also has a high hysteresis for damping out vibrations; shafts can be produced in more complex shapes without need for costly tooling or machining; and the larger sections required for heavily loaded shafts on account of the lower tensile strength of the material would, in any case, tend to be necessary also with steel ones, to obtain adequate stiffness. The significance of being able to produce more complex shapes is that balance weights can be cast integrally, instead of having to be bolted on during the balancing operation and, if hollow journals are called for, to reduce weight, they can be cored in the casting instead of having to be machined subsequently at a high cost. Bosses usually have to be left in the hollow sections, so that oilways from the main to the big end journals can be drilled through them.

SG irons used for crankshafts have both better tensile strength and fatigue resistance and better bearing qualities than the cast irons containing graphite in flake form. They include the following grades: 600/3, 650/2, 700/2, and 800/2 where the larger figure represents its ultimate tensile strength in N/mm^2 and the smaller one its percentage elongation. The graphite is converted to SG form by the injection of magnesium innoculants during smelting and by controlling the cooling.

Without further treatment, the safe working stresses of these grades under fatigue loading are, respectively, ± 64, 68, 72 and 80 N/mm^2. Salt bath nitriding to a depth of about 0.5 mm will increase the safe working stress by about 15% and more than double the wear resistance; induction hardening of the fillets to a depth of about 3 mm increases the working stress by about 53%, and rolling the hardened fillets at a load of between 1 and 2 tonnes for about ten revolutions increases it by approximately 80% relative to that of the untreated iron. Prior to fillet rolling, the radii are slightly undercut but the finish grinding operations are left until afterwards to ensure that the journals are truly cylindrical.

Because the strength of cast iron, in either flake or SG form, is lower than that of steel, the sections have to be larger. The critical dimension is the overlap between the main and big end journals. If the distance between the centres of adjacent main journals is taken as 1 unit and their diameter as 0.64, the proportions of a typical SG iron shaft would be approximately as follows: main journal length 0.32, crankpin journal diameter 0.52 and length 0.28, crank web thickness 0.2. If failure occurs owing to bending, the most likely fracture path is between the nearest points on adjacent fillets around the main and crankpin journals. The bending stress is given by a formula developed by MIRA from the Kerr Wilson formula:

$$\text{Stress} = 0.75 \times \text{Load on crankpin} \times \text{Main bearing span}/bt^2$$

where b and t are, respectively, the breadth and thickness of the fracture path. Some shafts have an integral collar on each web, around the main big end journals, the faces of these collars being ground to form thrust rings between which the bearings float, as in Fig. 3.80. If the webs are thick, the paths between the adjacent points on the outer peripheries of these collars might not be longer than the previously described fracture path. In a well-designed shaft the square of the length of this path should be greater than twice that between adjacent fillets, otherwise, owing to stress concentration at the edges of the collar, it might become the primary fracture path.

As regards bearing properties, cast irons are almost equal to nitrided steel, except in that they demand bearing materials of greater seizure resistance. AlSn20Cu1 alloys (aluminium with 20% tin and 1% copper) are not suitable, but AlSn10Si4Cu1 and AlSn10Si4Cu2 are. The hardness of these alloys increases with copper content. During the processing of the SG iron the graphite nodules breaking out to the surface may migrate, leaving sharp edges around the crater that had held them, while also improving the oil-retaining properties of the surface. The sharp edges, however, can adversely affect the rate of wear of the soft bearings in which they run, so the journals are first ground in a direction opposite to that in which they will be wiped by the bearing, and then lapped or polished in the other direction. With steel shafts, all grinding or lapping is done in the same direction as that in which the shaft will be wiped by the bearing.

Examples of forged crankshafts for four-cylinder engines are illustrated in Figs 3.14 and 3.15, while a cast iron crankshaft is shown in Fig. 3.16. In the three-bearing crankshaft in Fig. 3.14 the crank webs are extended to form masses to balance the revolving couples individually for each half of the shaft (see Section 2.6). Without these masses, although the shaft would still be in balance owing to its mirror symmetry, the revolving couples would load the bearings more heavily, because they would have to be reacted ultimately through engine structure. The reason for extending the webs in a fan shape is not only to increase their masses within the radius limitation imposed by the need to clear the bottom of the piston skirts during rotation of the masses past them but also in order that adjustments to the balance can be made accurately by drilling holes radially into it, from the appropriate direction over the whole range of angles.

There are no balance weights in the stiffly designed five-bearing shaft in Fig. 3.15 because a primary requirement was to keep weight to a minimum and, since this is for a compression ignition engine, the structure is in any case stiff enough to react the opposed revolving couples of each half of the shaft.

Fig. 3.16 shows how casting facilitates the production of a shaft of complex form. The purpose of the balance weights B_1 and B_2 is explained in Section 4.13, where the general arrangement of V-eight engines is described. Machining is normally confined to the journals and crankpins and the drilling of the oil ways, the balance being correct by the drilling of lightening holes and rough grinding webs as required.

Fig. 3.14 Forged crankshaft with balanced webs, by Laystall Engineering

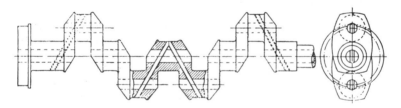

Fig. 3.15 Five-bearing shaft for ci engine

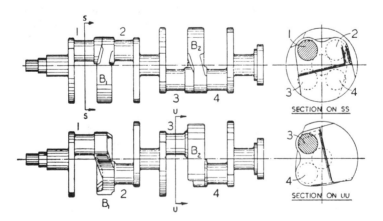

Fig. 3.16 Cast crankshaft for Ford V-eight

For a comprehensive treatise on nodular iron for crankshafts the reader is referred to a paper presented in 1954 by S. B. Bailey, *Proc. I. Mech. E.*, Vol. 168.

3.25 Built-up crankshafts

Two examples of the built-up type of crankshaft are shown in Figs 3.17 and 3.18. In Fig. 3.17 is shown one throw of a crankshaft in which the crank webs are permanently shrunk on to the journals; the case-hardened crankpins being secured in the split webs by the clamping bolts shown.

The other example, Fig. 3.18, is the crankshaft of a special racing engine. The webs of the shaft are formed of circular discs A and B, the A discs having as an integral portion the journals C while the crankpins D are integral with the B discs. The B discs are a tight fit on the journals C and are secured thereon by means of plugs E. The large ends of the latter are slightly tapered and are forced into the correspondingly tapered holes of the journals, thereby expanding the latter firmly inside the B discs. Dowel pins F fitting in holes drilled half in the journals and half in the B discs give added security against relative motion of those parts. Similar tapered plugs are used to secure the crankpins in the A discs but no dowel pins are used. When taking the shaft to pieces, plugs are screwed into the holes in the journals, thus forcing out the plugs E. To ensure correct alignment of the journals an accurately-ground rod is passed through holes H formed in the discs, during assembly of the shaft.

3.26 Surface-hardening of shafts

The term *case-hardening*, though also applicable to the nitriding and chill casting processes, is normally used for the time-honoured process of carburising the surface layer of a suitable low carbon steel to obtain a high carbon case. The carburising or carbon supplying agent may be solid, liquid, or gaseous. Subsequent quenching and heat treatment is applied with the object of producing a high degree of hardness in the case while maintaining strength and toughness in the core. Case-hardening steels are low carbon steels containing alloys which assist both the carburising process and the requirements of the core.

The depth of case is dependent on time, temperature, and composition of the steel. The time required ranges from a minimum of about a quarter of an hour in the cyanide bath to several hours in box hardening with solid carburising

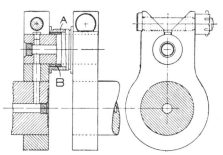

Fig. 3.17 Built-up crankshaft

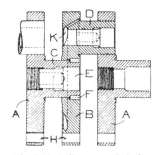

Fig. 3.18 Built-up crankshaft

agents. In complicated forms such as crankshafts, though selective hardening of pins and journals is possible, the high temperatures involved and the subsequent quenching are liable to lead to quite unmanageable distortion, since the temperature of the whole component must be raised above the critical change point.

Nitriding is a similar process in that the chemical composition of the surface layer is altered, but by the production of very hard nitrides of iron and certain alloying metals, of which aluminium and chromium are effective in producing extreme hardness in the case, while molybdenum increases the toughness and depth of penetration.

The nitriding agent is ammonia gas, which decomposes into hydrogen and nitrogen at the furnace temperature of about 500°C.

The process occupies from one to two days, and is thus a slow one compared with the rapid production methods described below. No quench is required.

The great advantage compared with carburising is the exceptional degree of hardness obtainable and the relatively low temperature necessary, this being below the change point of the parent steel. This has the double merit of reducing or preventing distortion, and permitting the normal annealing and heat treatment processes of alloy steels to be carried out beforehand, without risk of subsequent interference.

Another surface hardening process is termed *New Tufftriding*. The best results are obtained with low alloy steels containing aluminium and, perhaps, chromium, tungsten, molybdenum, vanadium or titanium. Treatment of a crankshaft generally takes about two hours, during which it is immersed in a bath of molten sodium cyanate at a temperature of 570°C.

Both nitrogen and carbon are released from the salt. Nitrogen, being more soluble than carbon in iron, diffuses deeply into the surface, forming needles of ductile iron nitride. Simultaneously, hard iron carbide particles are formed at or near the surface and act as nuclei for the precipitation of some of the diffused nitrogen, forming a tough compound zone. While this hard surface zone increases resistance to wear, galling, seizure and corrosion, the tough iron nitride needles diffused below the surface present a multitude of barriers to crack propagation and therefore increase fatigue resistance.

3.27   Chill casting

The process known as *chill casting* is a long-established one and is now being applied to the selective hardening of the cam surfaces of moulded camshafts.

By the insertion into the mould of suitably shaped iron 'chills', rapid cooling of the necessary surfaces can be effected. This results in the formation of a high proportion of combined carbon, that is of very hard carbide of iron as distinct from the free graphitic form. The hard cam surfaces can then be ground in the usual way.

Fig. 3.19 illustrates such a camshaft made by the Midland Motor Cylinder Company in their Monikrom iron. The proprietary name indicates the three important alloying elements – molybdenum, nickel and chromium.

A very valuable feature of this method of production is the incorporation of integral gear-wheel blanks, the finished gears showing, after prolonged tests, wear-resisting qualities fully comparable with the usual alternative materials.

Fig. 3.19

3.28 High-frequency induction hardening: flame hardening

Heating by the induction effects of high-frequency alternating current, of parts possessing electrical conductivity, is now used in a great variety of applications.

In the surface hardening of steel automobile parts frequencies of 2000 to 10 000 Hz are generally used for normal heavy work, and very much higher values of a 'radio' order for specially light parts requiring small penetration.

The heating is followed immediately by a quench, and the process represents the physical hardening of a suitable medium or high carbon steel in contrast to the casing processes described in Section 3.26. The general mass of the material below the surface layers remains at normal temperature and is unaffected by the operation, owing to the extreme rapidity with which the heating and quenching are accomplished.

The process requires the installation of equipment constructed to deal with large numbers of particular components, the Tocco equipment, handled in this country by the Electric Furnace Company, having reached a very high degree of specialised, high-production development.

The authors are indebted to the above company for information.

The process consists in the application to the individual crankpin or journal, or whatever part is to be hardened, of a copper muff or inductor block – split as necessary – which forms part of a high-frequency, high-current electric circuit.

There is a clearance space between the muff and the shaft into which high-pressure jets of quenching water may be introduced. Frequencies of the order of 2000 to 10 000 Hz, and currents up to 10 000 amperes are used, and the surface of the part is heated to the hardening temperature in a period of only a few seconds by the induced eddy currents, the quenching water being then introduced over a period of seven to ten seconds, the time cycle depending on the depth of penetration required.

The equipment has been developed to handle every type of automobile component made of alloy steel requiring surface hardening, and the elaborate automatic controls enable the required depth and hardness of the modified surface layer to be precisely controlled.

For rapid, precise and large-scale local hardening of suitable steels the process would seem to have no equal where the cost of equipment is justified by the large quantity of work to be dealt with.

The well-known firm of Birlec has been developing this class of apparatus in this country.

The Shorter process and equipment represent a successful attempt to introduce precision control into the use of the oxy-acetylene torch for local flame heating followed by rapid quenching.

The equipment varies in elaboration according to the nature and quantity of output, but various types of push-button apparatus have been introduced for the purpose of providing the necessary relative motion of work and torch, with the timed follow-up of the quenching sprays.

For some classes of work, particularly of the largest size where output is limited, the Shorter equipment is probably somewhat more flexible than the highly elaborate but very convenient and accurate Tocco apparatus. The Shorter process is now handled by the British Oxygen Company.

3.29   The poppet valve

A side-valve arrangement is illustrated in Fig. 3.20, an overhead valve in Fig. 3.24(*a*), and an F-head layout in Fig. 3.24(*b*). From Fig. 3.20 it can be seen that the valve head A, sliding vertically in a guide G, is held down on a conical seating by a coil spring, the washer-like retainer for which is held by a split collet C, located in an annular groove near the upper end of the valve stem.

A 45° seat angle is generally regarded as the optimum, though smaller angles down to 30° are sometimes adopted so that, by subjecting it more to compressive than shear stress, sinkage due to flow of the metal away from the seating face is avoided. This measure, however, is effective only where the compressive strength of the metal is higher than its shear strength. Flat seating valves were tried in the early days, but were never as satisfactory as the conical type.

Guides are a push or press fit in the cylinder head, for ease of replacement when worn. They are usually of cast iron, favoured because of its good bearing properties. However, other materials have been employed, and a recent development is the use of sintered powder metal. In some early engines the valves were carried simply in holes in the cylinder head, but rectification was difficult when the bores of the holes wore. Wear allows oil and air to pass through the clearance into the induction system, causing fouling of sparking

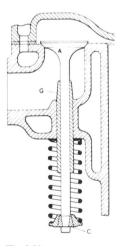

Fig. 3.20

plugs and weakening of the air : fuel ratio. Leakage of excessive quantities of oil into valve guides is likely also to cause trouble owing to carbon formation and consequent sticking of the valve, since increases of as little as 0.05 mm in the clearance in guides due to wear has been shown to increase exhaust valve temperature by as much as 50°C.

Torsion-bar and leaf springs have been used, but the coil spring is almost universal. On early engines a cotter pin passed through a slot in the valve stem was sometimes employed, instead of the split collets, for retaining the coil spring, and still is on some small industrial units. Gas-tightness is ensured by lapping the valves on their seatings, using a fine abrasive paste. Gas pressure assists the spring in keeping the valve on its seat.

Although seated by a spring, the valve is opened by any of several forms of mechanism, its lift being generally equal to about a fifth of its diameter. These mechanisms will be described later in this chapter. There are various forms of rotary and sleeve valves, but these have never been in widespread use. They are dealt with in Chapter 5.

3.30　The valve in practice

Valves are subject to both thermal and mechanical loads, the latter being applied by the springs and actuating gear. All these loads are so severe as to justify an assertion that the valves are the most heavily loaded components in an engine. Modes of potential failure include tensile elongation or fracture, either hot or cold corrosion, wear, burning, and flow of metal from the seating area, cold corrosion being caused by condensation containing acid products of combustion. The remedies are mainly in the choice of materials having adequate corrosion resistance and hot strength and hardness. Exhaust valves, in particular, operate for long periods at temperatures perhaps as high as about 800°C. Materials used include En 51 and En 52 for inlet valves and En 59 and 21-4N for exhaust valves.

A significant property is coefficient of expansion. The martensitic steels have low and the austenite materials high coefficients of expansion. Consequently, the stem of an exhaust valve of austenitic material has to be tapered to compensate for the temperature gradient along its length so that, at operating temperatures, its flanks remain as nearly parallel as possible. In many instances, a martensitic material having good wear resistance and low coefficient of expansion is used for the stem only, a head of a more costly austenitic material resistant to high temperatures being friction-welded on to it. For aircraft and some very high-performance engines, alternative materials for the valves include BAC Brightray, which is an alloy containing 20% chromium and 70–80% nickel, and the costly Nimonic alloys. In a few instances, valves with hollow stems and containing sodium, for use as a heat transfer medium when it is very hot and therefore in the liquid state, have been used. A comprehensive article on valve materials can be found in *Automobile Engineer*, March 1962.

3.31　Coated valves

For protection against corrosion, a wide variety of coatings has been applied. They include alloys containing materials such as nickel, manganese, cobalt, chromium, silicon, and, less commonly, molybdenum, tungsten, and titanium. These, however, are all costly.

A much more economical coating system has been developed and is widely employed by General Motors. This is the Aldip process for the application of an aluminium coating. After the valve seats have been finish-machined and the stems rough ground, the aluminium material, in the form of a paste, is sprayed on. The components, in a jig, are than dipped for a few seconds in a molten flux bath at 760°C and, on removal from the bath, the surplus aluminium blown off. The outcome is a smooth permanently adhering coating of aluminium on an iron-aluminium alloy underlay. No further finishing is necessary.

3.32 Corrosion and wear

Corrosion tends to cause pitting on valve seat faces, as a result of which the hot gases start to leak through, eventually burning a channel locally through which compression is lost. Among the remedies is adequate provision for cooling, to expedite the flow of heat out of the valve head, through the seats, into the coolant. Local burning and channelling can also be caused by the trapping of particles of solid products of combustion, including carbon and lead compounds, between the seating faces.

Pitting and wear can be caused, too, by local welding between the face of the valve and the material of the seat in the head. Lead deposits on the seats tend to inhibit such wear, as also can the solid lubricants in sintered seats, described in Section 3.34. When unleaded petrol was first used trouble was experienced due to what was termed valve sinkage, which rapidly took up all the valve clearance. This was caused by attrition due to local welding at the minute points of contact between the peaks of surface roughness of the seating faces and the tearing that subsequently followed as the valve opened. It was cured by better cooling, notably by the use of aluminium alloy cylinder heads and careful attention to the design and layout of the water passages around the valve seats.

In engines with cast iron heads run on liquefied petroleum gas, the same problem occurs, owing to the absence of a film of deposit on the seats. This has been overcome simply by starting and running the engine on leaded petrol until it is warm, or perhaps for the first hour of the working day. The result is the deposition of a protective film of adequate durability on the seating faces to last for the remainder of the day.

3.33 Valve rotation

Valve rotation can wipe deposits off the seat, spread lubricant, solid or liquid, around the seating faces, and distribute wear and pitting uniformly around the seat, thus ensuring that the heat flow is not impaired locally by either wear or distortion. Excessive rotation should be avoided, however, because it can actually cause wear, especially if there is a tendency to local welding.

A common method of rotating valves is to place a hard steel thimble over the end of the stem, seating its rim on the retainer washer for the spring, so that there is a small clearance between the tip of the stem and the other end of the thimble. When the cam or rocker bears down on top of the thimble it first compresses the spring, momentarily freeing the valve while the clearance between the tip of the stem and the end of the thimble closes and the valve begins to lift off its seat.

A simpler arrangement is to dispense with the thimble and have no clearance between the abutting ends of the halves of the collet, so that it does not actually grip the valve stem. At the same time, there has to be a clearance between the ends of the internal collar in the collet and the groove in which it registers in the stem, so that as soon as the valve is lifted from its seat it floats in the collet and is therefore free to rotate. With either arrangement, rotation is caused by a combination of vibration and the very slight winding and unwinding action of the coil spring as it is compressed and released. A comprehensive article on valve design can be found in the September and October issues of *Automobile Engineer*, 1954.

3.34   Seat inserts in cylinder heads

In aluminium heads seat inserts are essential, though they are also employed in some cast iron heads, especially for diesel and other heavy-duty or highly-rated engines. High-quality inserts are produced by compressing a metal powder in a ring shape mould to produce what is called a green moulding, which is then transferred into a sintering furnace. Some others, however, are Stellite faced ferrous metal; cobalt, and nickel-based alloys; or, least costly, a high-quality cast iron such as in the case of the centrifugally cast Centri-lock insert, which is an undercut, stepped ring, Fig. 3.22. The non-ferrous alloys in the Stellite range containing, in various proportions, cobalt, chromium, tungsten, and carbon, are extremely hard.

Most seat inserts, regardless of material, are first shrunk, by refrigeration, and then pressed into the head; though a few, as in Fig. 3.21, are shrunk and screwed, the radial lugs or slots for the spanner being machined off after the seats have been screwed home. Dimensional tolerances are always tight because the interference fit is critical for their remaining securely in position in service.

Potential modes of failure are loosening, corrosion, wear, and material flow at high temperatures. The requirements therefore are: a coefficient of thermal expansion compatible with that of the head material, dimensional and metallurgical stability at elevated temperatures, and resistance to loosening, corrosion, and to abrasive, adhesive, and compressive wear. Sintered powdered metal seats have the advantages of ease of production in large quantities, and material structures and compositions can be made up that are impossible

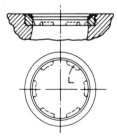

Fig. 3.21 Stellited seat

Fig. 3.22 Centri-lock seat

by the normal alloying processes. For instance, solid lubricants can be incorporated into the powder mix.

Brico produce a range of sintered powder metal inserts among which is one called the AR44. This is an 11% chromium steel, which contains 0.4% molybdenum, 6% copper, and 1% carbon for wear resistance resulting from a fine dispersion of hard alloy carbides in a matrix of martensitic chromium alloy. By heat treatment the hardness can be varied to suit the valve material used, and the ring can be machined after such treatment, whereas cast iron rings must be annealed prior to machining and then re-hardened. Good machinability is mainly attributed to the fineness of the dispersion of the carbides, though the dry lubricant additive (unspecified) contributes too. Another sintered material is the high-speed tool steel XW35, the analysis of which is Fe 62%, C 0.8%, Cr 3.5%, Mo 5%, W 5%, V 2.5%, Cu 20%. Where even better properties at high temperature are needed the XW23 grade is used, in which there is also 4% Co, but 1% less V, with 3% less Fe to make up the 100%. Cobalt, however, is particularly costly, so the latter alloy is used only for very high-performance engines.

3.35   Layout of valves and form of combustion chamber

The requirements to be met in the design of the cylinder head and location of valves are numerous and conflicting, and the search for the successful compromise has led to the designing, patenting and production of a very great number of different forms and arrangements, often with puzzling anomalies and inconsistencies in performance between different examples possessing apparently the same virtues.

The basic requirements to be met are indicated in Table 3.3. They are stated very simply, but the five main requirements given under each lettered heading, with the necessary means of provision, cover the more important factors to be reconciled.

Fig. 3.23 illustrates four conventional arrangements which are in wide general use, and Fig. 3.24 and later illustrations give examples or variations of these.

Fig. 3.23(*a*) is the side-valve construction with all valves in line, the detachable head being of the turbulence type providing for compression or squish turbulence produced as the piston closely approaches the flat portion of the cylinder head. Valve diameter and adequate valve port cooling are in conflict, as with all valves in single-line arrangements, unless the longitudinal pitch of the cylinders is increased; and volumetric efficiency is further limited by the changes in direction of the gas flow and restricted entry to the bore. Requirements A (3) and (4), in Table 3.3, are reasonably well met, though flame travel is long. Requirement B (2) is *not* met, and the high surface : volume ratio of the head, if combined with undue turbulence and over-cooled jackets, militates against economy. Requirements C and D are well met on the whole. Side-valve engines, because of their simplicity and low cost, were once very common in cars but now their use is confined mainly to industrial applications, where their very low overall height is an advantage and power output per kilogram is not so important.

Incidentally, in recent years the real significance of squish turbulence has been questioned: after all, why should the gases not move progressively from

Table 3.3—PERFORMANCE AND CONSTRUCTION

Objective	Requirements	Means of provision
A Power output Smooth running	(1) High compression ratio. (2) High volumetric efficiency. (3) Rapid and efficient combustion (4) Freedom from pinking (see Section 14.10).	Small volume of combustion chamber. Early closing of inlet valve. Large inlet valve, suitable valve timing, limited pre-heating. Short flame travel, adequate turbulence, good plug scour. Short flame travel, well-cooled 'end-gas', suitable plug position.
B Fuel economy (low specific fuel consumption or high thermal efficiency)	(1) High compression ratio. (2) Low surface:volume ratio. (3) Efficient use of weak fuel:air mixtures. (4) Adequate pre-heating.	Small volume of combustion chamber, etc. Hemispherical form of combustion chamber. Short flame travel, adequate turbulence, good plug scour. Hot jackets and induction manifold.
C Economical manufacture	(1) Economy in machining. (2) Simple castings. (3) Simplified details. (4) Ease of assembly.	Elimination of unnecessary joints and attachments. May conflict with above. Simplified design. Standardisation. Sub-, and main assemblies.
D Ease of maintenance	(1) Accessibility. (2) Easy renewal of parts. (3) Ease of decarbonising and valve grinding. (4) Limited weight of components.	Detachable head. Separate cylinder block. Renewable valve and tappet guides and cylinder liners. Unit assembly of head and valves. Sub-, and main assemblies.
E Low emissions	(1) A limited compression ratio, for NO_x. (2) Efficient use of weak air:fuel ratios, for low CO. (3) High ratio of volume:superficial area, for low HC. (4) Smooth combustion chamber, free from crevices, for low HC.	A complex series of measures needed, as described in Chapter 12.

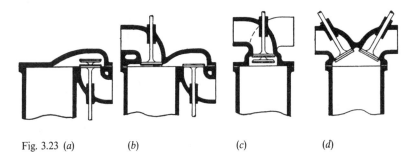

Fig. 3.23 (*a*)   (*b*)   (*c*)   (*d*)

between the approaching parallel flat faces, rather than wait there to be squirted, or squished, out at top dead centre? On the other hand, as the piston descends again, a reverse squish action certainly does occur.

Fig. 3.23(*b*) should give good volumetric efficiency as a large diameter inlet valve may be used, and the valve port gives direct access to the bore.

High compression ratio can be readily provided.

Other characteristics are similar to (*a*), though the construction is likely to be more expensive owing to the mixed direct and push-rod valve gear.

In both case combustion is initiated in a hot region and the end-gas is well cooled on a shallow flame front.

Diagram (*c*) illustrates the very widely used overhead valve (ohv) arrangement with vertical valves in single line, permitting the use of simple push-rod and rocker gear. The wider cylinder pitch at the main bearings sometimes permits of an increased length of combustion chamber, known as the 'bath-tub' type, to accommodate larger valves. The width is usually less than the bore in modern designs to provide increased compression ratio and some compression turbulence.

At (*d*) is shown the classic approximation to the ideal hemispherical head which is used in many high-performance designs. Large diameter inlet valves with free entry can readily be provided and with careful port design and possibly some degree of masking of the lower side of the inlet valve, there should be fair general turbulence though compression or squish turbulence is absent. Flame travel is short, and high compression ratio can be readily provided by a domed piston crown. However, a domed crown not only adds to the surface area but also changes the combustion chamber from hemispherical to a less than ideal hollowed-out bowl shape. Consequently, this so-called hemispherical chamber is probably more suitable for use with a piston having either a flat or slightly dished crown, when the associated loss in compression ratio is intended to be made good by turbocharging.

Fig. 3.24(*a*) shows in more detail a representative example of the type illutrated in Fig. 3.23(*c*). However, the valves are tilted over to one side to form what is sometimes termed a penthouse type combustion chamber, so that larger valves can be accommodated, the ports inclined naturally towards the manifolds, the provision of a squish shelf is easy and the sparking plug can be advantageously positioned relative to the inlet valve. Fig. 3.24(*b*) shows a form of F-head, the essential characteristic of which is an overhead inlet and side exhaust valve layout as in Fig. 3.23(*b*). With the latter layout, a semi-downdraft carburettor can be easily accommodated and there is plenty of space for a large

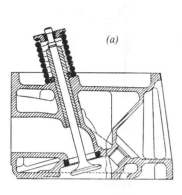

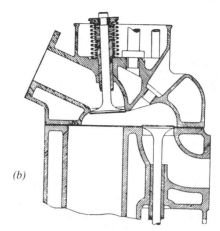

Fig. 3.24 Typical cylinder heads

inlet valve, while the exhaust ports lead the gas naturally into the manifold and thence to the downpipe.

3.36 Variable valve timing (VVT)

Since 1880, almost 800 patents on variable valve timing have been issued in the USA alone. An SAE Technical Paper No. 890764 by Dresner and Barkan, classifies all the systems into 15 basic types. In general, they can be further categorised under three main headings: variable phase control (VPC), combined valve lift and phase control (VLPC), and variable event timing (VET) systems.

Variable valve timing is generally applied in one of two ways: either the point of inlet valve closure is fixed and that of its opening varied, or both are fixed relative to each other but their timing (i.e. the inlet phase) advanced or retarded simultaneously, generally by rotating the cam relative to the shaft. The latter, termed *phase change*, has two advantages. First, the dynamic loading of the valve gear is unchanged and, secondly, the mechanisms for varying the timing are generally considerably less complex.

If the opening only is varied, retarding its timing reduces not only the overlap but also the open period. Therefore, unless the lift is reduced, the acceleration loading on the valve gear inevitably increases, causing problems at high speeds. Variation of inlet valve closure has been investigated but, at the time of writing, has not been adopted since it does not appear to offer advantages adequate to justify the further complication of the control mechanism.

3.37 Advantages of VVT

Because of the increasing stringency of legislation regarding emissions, including CO_2, interest in variable valve timing was intensifying by the early 1990s. By optimising the valve timing, volumetric efficiency and therefore power and torque can be increased. During low speed operation, the valve

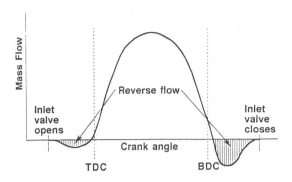

Fig. 3.25 This diagram of mass flow of air plotted against crank angle shows the effect, with fixed valve timing, of back-flow into the induction manifold at low speed

overlap period can be reduced to increase the effective expansion ratio, improve idling stability and cold starting and, especially for naturally aspirated diesel engines, reduce emissions throughout the low-speed light load range. For turbocharged engines, in particular, VVT can be utilised for recirculation of the exhaust gas. Because exhaust gas contains water vapour, sulphur and other corrosive media, the currently available EGR valves are not so reliable as might be desired.

3.38   Early inlet valve closure (EIVC)

With fixed valve timing, designers mostly aim at high nominal output, with the result that because the inlet valve is still open after BDC, Fig. 3.25, torque at the lower end of the speed range is impaired owing to back-flow over a significant proportion of the operating range. By optimising inlet valve closure over the whole range, improvements in full-load torque of over 12% and in maximum torque of about 4% can be obtained.

During low speed operation, inlet valve closure and exhaust valve opening should be relatively close to BDC. Indeed, for both stability and economy at idling, the overlap should be almost zero. Under all other conditions, larger overlaps are needed to increase the breathing potential and reduce the output of NO_x.

Going further, by utilising the valve control system to eliminate the throttle as a means of regulating the torque output, overall fuel economy can be improved by perhaps 5 to 6%. This entails adopting the Atkinson cycle (Section 3.41), defined ideally as expanding the charge down to atmospheric pressure so that all its pressure energy is utilised. In practice, however, owing to heat loss to coolant during the compression and power strokes, over-expansion can occur and lead to negative work being done on the piston. Moreover, scavenging cannot be complete without a pressure differential across the exhaust valve.

Pumping losses during idling represent 30–40% of the total power developed, and they entail a fuel consumption penalty of up to 15%. As the load is increased and the throttle opened wider, the pumping losses fall so the proportion of the power lost falls. However, since a high proportion of motoring is done at light load, the attraction of abandoning the throttle control for gasoline engines is obvious.

3.39 Problems associated with EIVC

For gasoline engines, high pressure in the inlet port is less favourable for mixture formation than are the low pressures associated with throttle control. Thus the films of fuel deposited on the walls of the ports and manifold are thicker, so large drops can be drawn intermittently into the combustion chamber, some settling on its walls. The outcome is wide variations in the air:fuel ratio, and therefore a reduction in efficiency and an increase in emissions.

If the inlet valve is closed very early, the flow into the combustion chamber occurs only when the piston speed is low. This adversely affects both the flow pattern and turbulence within the chamber, and subsequent expansion further slows both swirl and mixing. As the piston moves down to BDC, expansion can cool the gases to as low as $-15°C$, the total drop being perhaps much as $50°C$. This causes re-condensation, which continues into the early part of the compression stroke. The condensation increases with decreasing load but, as the speed falls, there is more time for heat transfer from the walls of the combustion chamber, which tends at least partly to offset the increase.

3.40 Late inlet valve closure (LIVC)

With late closure, part of the charge tends to be returned into the induction manifold as the piston passes BDC. In these circumstances, the gas exchange occurs at approximately constant pressure. Consequently, only the proportion of charge trapped in the cylinder after inlet valve closure is compressed. As the throttle is opened, therefore, closure should be progressively brought forward to the standard timing for the engine.

If the inlet valve closure is too late, the quality of combustion can be inferior. In this respect, multi-point fuel injection is at a disadvantage relative to single point or carburation unless the volumes of the primary induction pipes (branch pipes) are large enough to contain the return flow into the manifold so that there is no possibility of their being robbed, by an adjacent induction pipe, of some of the mixture they contain. At light load, any maldistribution of fuel is aggravated because back-flow into the plenum can remove during some strokes most of the fuel that has been injected and in others virtually none, so running becomes unstable.

Because variable inlet valve timing can reduce final compression and therefore combustion temperatures, it offers potential for reduction in NO_x. Moreover, by virtue of the higher induction manifold pressure, mainly during idling and slow running under light load, the pressure gradient between the inlet and exhaust ports is higher, so the residual gas content of the combustion chamber is lower, and idling stability better.

3.41 Variable valve timing and the Atkinson cycle

In general, fixed valve timing is optimised either for the speed at which maximum torque is developed or to obtain good nominal power output.

Applying the Atkinson cycle (Section 3.38) for controlling the engine solely by varying the valve timing entails opening the valves for periods that decrease as speed and load are reduced. Three benefits are thus obtained: considerable reduction or even total elimination of blow-back into the induction system, pollution of the charge by residual exhaust gas is reduced, and optimum use is made of the energy liberated by the combustion of the charge.

By increasing the duration of opening with load, the volumetric efficiency can be improved. With increasing speed too, extending the valve open periods ensures that the gas throughput obtained optimises maximum power at full throttle. As engine speeds fluctuate, the intervals, expressed in terms of crank angle, available for gas interchange differ from those in terms of time. Obviously, therefore, the independent variation of the inlet and exhaust valve event phasing has to be expresssed in terms of crank angle.

3.42 Some simple VVT mechanisms

Variable phase control (VPT) can be effected in a number of ways. One is to advance and retard the camshaft by means of a sliding muff coupling on a divided shaft, with spiral splines on the driven and straight splines on the drive interfaces, or *vice versa*. This, however, suffers the disadvantage of high frictional resistance to control operation. Another method is to install, in the belt or chain drive to the camshaft, a movable idler pulley in combination with a tensioner having a longer than usual stroke. Movement of the idler pulley towards or away from the drive, Fig. 3.26, rolls it around the half-speed wheel to advance or retard the timing while, at the same time, the tensioner compensates for the movement. In general, this appears to be the most practicable and least costly variable valve timing system.

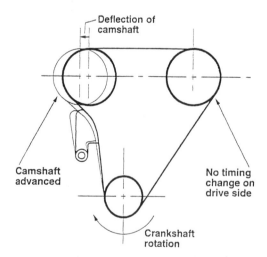

Fig. 3.26 To vary the phase of the inlet cams, the camshaft can be deflected horizontally, thus causing the half-speed wheel to roll along the belt

3.43 VPC, VLTC, VPLC and VET systems

Variable phase control, Fig. 3.27, implies varying the overlap so that low speed torque and, with it, specific fuel consumption are improved over most of the speed range. Since the duration of opening remains constant, wide-open throttle power is unaffected.

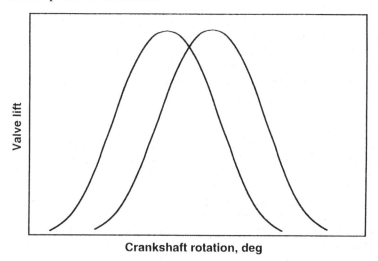

(*a*)

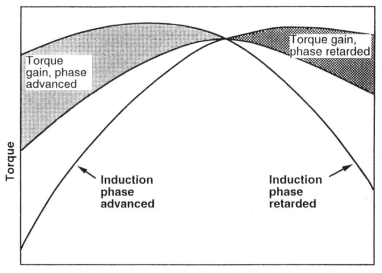

(*b*)

Fig. 3.27 With the variable phase control system (VPC), the inlet opening point can be retarded to reduce the overlap which, during idling, ensures stability and low emissions. For good low-speed torque and low fuel consumption, the inlet opening should be advanced, to increase the overlap

A combination of lift and timing control (VLTC), Fig. 3.28, can offer further performance enhancement, but is more costly than VPC. One way of combining these two is to have axially stepped cams, the variation being effected by shifting the followers from step to step, but such a mechanism is complex. Tapered cams, such as the Fiat Tittolo system combining variable lift and event timing, are an alternative but this means virtually point contact between cam and follower and, if the duration of opening is kept constant, the cam is extremely difficult to manufacture. Moreover, the axial loading introduced is about 10% of the force between the cam and follower, so a powerful controller is needed.

Honda have a commendably simple system in production, Fig. 3.29. It has three cams and rockers per pair of valves in a four-valve head. At low speed, only the outer pair of rockers actuates the valves, leaving the central one idling freely. As the speed increases, the electronic control signals a hydraulic valve to open, to allow oil pressure to move a plunger which locks all three rockers together. In this condition, since the central cam lobes are bigger than the two outer ones, the latter idle and the former actuates the valves.

Varying the valve event timing (VET) is the changing of the duration of lift while keeping the timing and magnitude of maximum lift constant, Fig. 3.30. In other words, only the opening and closing points are varied. This improves part load emissions and economy, leaving the wide-open throttle condition unchanged, and has the advantage that the ramps on the cam remain effective both as the valve begins to lift and when it re-seats. VPC and VLC can be combined (VPLC), and it is possible, as in for example the Mechadyne–Mitchell system, to combine it also with VET (VPET).

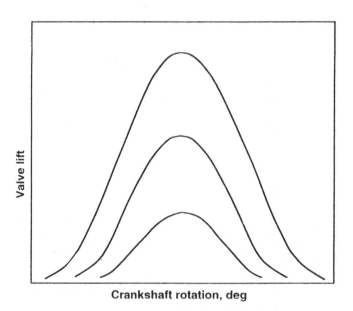

Crankshaft rotation, deg

Fig. 3.28 Stepped cams provide variable lift and timing (VLTC) with rocker-actuated valves. At the lower lifts, friction is reduced, charge swirl enhanced and fuel consumption improved. The curves represent three steps, though up to ten are practicable

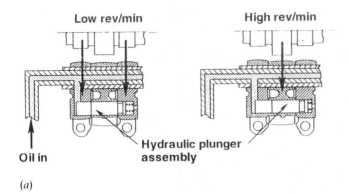

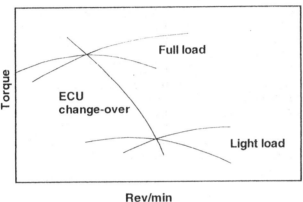

Fig.3.29 (*a*) Honda variable valve timing system. In the section on the left, the two outer rocker arms are actuated by their cams, while the central one is idling. Shown on the right is the condition when electronic control unit (ECU) has signalled a solenoid to open a valve allowing oil under pressure to push the three-piece hydraulic plunger to the right, locking all three together. In this condition, the central cam, because it is higher than the other two, actuates the valves. In the graph (*b*) the central curve shows how the ECU varies the change-over point with speed. The upper lines in the two other pairs of curves represent operation with and the lower ones without change-over

3.44 The Mechadyne–Mitchell system

The Mechadyne–Mitchell principle is applicable to almost any of the commonly used valve actuation mechanisms. Although it entails additional parts, almost all are identical for each cylinder, so the extra tooling costs are not unreasonably high. Basically, a hollow camshaft is driven by a peg on the outer end of a lever projecting from a driveshaft carried coaxially within it. This peg projects into a slot in the camshaft, Fig. 3.31. Axial location of the driveshaft is similar to that of conventional camshafts.

The arrangement is shown in greater detail in section BB of Fig. 3.32, from which it can be seen that the peg is actually a ball and slider reciprocating in a slot in a collar on the camshaft. The driveshaft is moved laterally to vary the

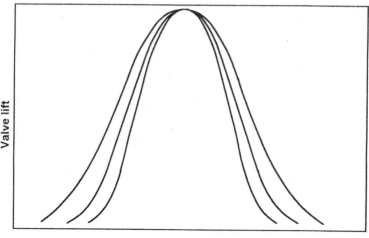

Crankshaft rotation, deg

Fig. 3.30 Curves of variable event (VET) timing, without phase change. Valve lift remains constant and the variation is effected continuously. Part-load emissions and economy are improved, and the wide-open throttle condition is unimpaired. Characteristics of a VET system with phase change are illustrated in Fig. 3.34

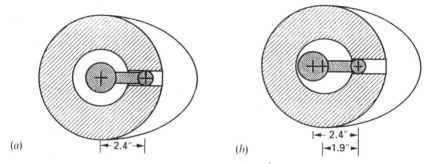

Fig. 3.31 Diagram illustrating the principle of the eccentric drive of the Mechadyne–Mitchell system, (*a*) in the co-axial and (*b*) eccentric position, in which rotation accelerates over about 25° each side of the nose of the cam, thus drawing together the valve opening and closing points, as in Fig. 3.34

drive from concentric to eccentric. When driven concentrically, as at (*a*) in Fig. 3.31, the speed of rotation of the hollow shaft and cams is constant at any given engine speed. If it is moved, say, 5 mm off centre, as at (*b*), its instantaneous speed of rotation is multiplied in the ratio $2.4:1.9 = 1.263:1$. Therefore, as the shaft rotates, the ratio reduces progressively first to $1:1$ at 90°, and then on to the inverse of $2.4:1.9$, at 180°, and finally back again to complete the 360°. As the control is moved to increase the eccentricity of the drive shaft, the duration of valve lift is progressively reduced because its opening is retarded and its closing advanced.

Appropriate phasing of each cam relative to the eccentric is rendered possible by dividing the hollow camshaft, along its length, into the same number of sections as there are cylinders. Each section is carried in its own

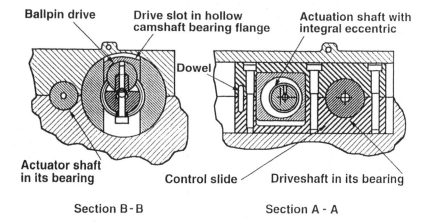

Ballpin drive

Drive slot in hollow camshaft bearing flange

Actuation shaft with integral eccentric

Dowel

Actuator shaft in its bearing

Control slide

Driveshaft in its bearing

Section B - B   Section A - A

Fig. 3.32 Sections AA and BB taken from Fig. 3.33 to show in detail the ballpin drive and actuation shaft

bearings and driven by a separate lever and peg projecting radially outwards from the one-piece drive shaft. Incidental advantages of dividing the hollow camshaft into short sections are that the ramps on the cams are always fully effective, and the short lengths of shaft are inherently very stiff both torsionally and in bending.

Radial movement of the solid driveshaft is effected by the mechanism illustrated in the plan view of the 24-valve DOHC head of a six-cylinder engine, and section BB in Figs 3.32 and 3.33. This shaft is conventionally driven by a wheel mounted on the flange at the left-hand end of the cylinder head. To one side of, and parallel to, the coaxial drive shaft and six-piece camshaft is a control shaft. This shaft is actuated by a lever on its end projecting from the right-hand end of the cylinder head assembly. In a

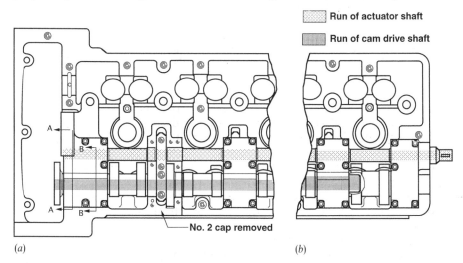

Run of actuator shaft

Run of cam drive shaft

No. 2 cap removed

(a)   (b)

Fig. 3.33 Mechadyne-Mitchell system installed on a six-cylinder engine

production version, the controller and actuator would presumably be contained within the bounds of the cylinder head assembly.

From section BB it can be seen that a mechanism comprising an eccentric in a scotch yoke slides vertically in the two-piece casting that forms not only the housing but also the driveshaft bearings and caps. Rotation of the control shaft moves the whole assembly laterally, and therefore the driveshaft into and out of concentricity with the camshaft sections. The control shaft can be rotated through only 90° which is why, when it has been actuated to bring the driveshaft into its eccentric position, as in section BB, there is no clearance above the scotch yoke.

By arranging the belt or chain drive as in Fig. 3.26, lateral movement of the driveshaft can be made to cause also a phase change. The resultant changes of the valve timing are illustrated in Fig. 3.34, which shows what the lift characteristics are with the standard timing, what they would be if only event timing were changed and what it is when both event timing and phase shift are applied. The whole system can be applied to both the inlet and exhaust valves but, for optimum cost-effectiveness, only the inlet valve timing would be varied.

The operating envelope is of course limited by the stresses superimposed on the valve train by the accelerations due to advancing and retarding the timing. However, if the eccentricity of the drive is introduced only at speeds of 2000 rev/min and below, the total stresses in the valve train need be no higher than those with fixed valve timing at 4000 rev/min. Tests with twin cylinder motorcycles have shown 32% increases in power and 43% reductions in specific fuel consumption.

3.45   Control of the Mechadyne–Mitchell system

The friction to be overcome to control the mechanism, by moving the shaft eccentrically, is not great. Consequently, either electric or hydraulic power could be used. Where hydraulic power is available, for example on power

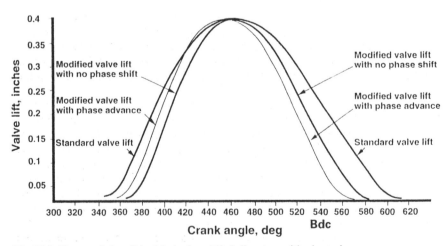

Fig. 3.34 Characteristics of the Mechadyne-Mitchell system with phase change

steered vehicles, this is more attractive, since it does not put any extra load on the battery. Alternatively, power might be taken from the engine lubrication system, though this would probably entail the introduction of a larger pump and a hydraulic accumulator, to avoid oil starvation of the engine bearings. Variations of viscosity with temperature, however, could present problems. A low-cost alternative would be an on–off solenoid control, possibly based on only engine speed sensing. The more upmarket models and commercial vehicles, however, might have a more complex continuously variable eccentricity controller, with speed, load and VET position sensing for closed loop mixture control by the ECU.

3.46 Multi-valve heads

Three valves per head, two inlet to facilitate breathing, have been used, though not widely, for many years. A typical three-valve arrangement, with a toothed belt driven single overhead camshaft, Fig. 3.35, is that of the 1342 cm^3 engine for the Rover 200 range. Although four valves per head have been virtually universal in aero engines of the reciprocating piston type since the First World War, and common in racing cars, it was not until the beginning of the nineteen-eighties that this layout began to be adopted for engines for upmarket saloon cars. The obstacle, of course, was the complexity and cost, which outweighed what in the days of low-rated engines were only slight advantages. In general, two valves are adequate for engines developing up to about 35 kW/litre but above this level of specific output four are desirable.

With four valves a greater proportion of the head area is available for porting than if only two are employed. Moreover, the sparking plug can be more easily positioned very close to the centre of the chamber, so the flame path is short. This means that not only is complete combustion easier to attain but also the ignition timing can be retarded so that the dwell of the gases at high temperature in the cylinder is reduced, with a consequent diminution of the NO_x content of the exhaust (Chapter 12). Because of the large diameters of two valves relative to the bore of the cylinder, the gas flow past the portions of their edges adjacent to the cylinder wall tends to be masked by it. With four valves, on the other hand, there need be little or no masking and moreover the interaction of the two incoming streams of gas can greatly improve mixing, again leading to more complete combustion. Furthermore, with two exhaust valves instead of one, the ratio of seat length to area exposed to the hot gases is higher and so also, therefore, is the rate of cooling by conduction through the seats.

Advantages of the straight inlet port arrangement of the Renault 1.5 litre, turbocharged V6 engine, Fig. 3.36(*a*), include not only the fact that it gives a clear downward path for the incoming gases to follow the receding piston but also swirl around a horizontal axis can be induced in the cylinder which, bearing in mind the action of the piston as it subsequently rises, can lead to exceptionally good mixing. On the other hand, the more commonly used arrangement, typified by the Saab head for their 2-litre engine, Fig. 3.36(*b*), offers a more compact installation and the prospect of introducing swirl about an axis coincident with that of the cylinder.

With pushrod actuation, four-valve heads become complex. However, the increase in complexity is less significant when applied with the valve actuation

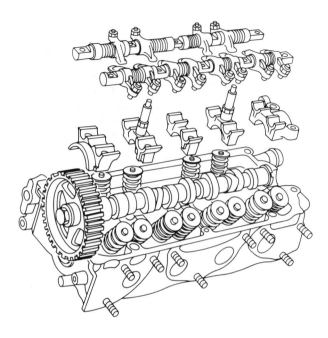

Fig. 3.35 The Rover 200 Series 1.3-litre four-cylinder engine has three valves per cylinder, in a double pent-roof combustion chamber, and a single overhead camshaft with two rocker shafts

gear on a modern twin overhead camshaft engine than with that of a pushrod-actuated unit. A compromise is the three-valve layout (two inlet, one exhaust) as used by Honda and Toyota. Particularly interesting is the use by Toyota of one large and one small diameter inlet valve, with a hinged flap in the intake to deflect all the mixture through the smaller one at low speeds, Fig. 3.37. This, by inducing high velocity flow through the valves, improves mixing of fuel and air, and thus the driveability of the car.

Kawasaki has an engine with five valves per cylinder, while several others, including Honda, are experimenting with eight valves per cylinder in engines with oval bores the major axes of which are transversely oriented. This entails the use of two sparking plugs per cylinder and two connecting rods per piston, so the engine becomes very complex. Advantages, in addition to good breathing characteristics, include the shortness of the cylinder block and, by producing engines with cylinders having identical minor axes but different major axes, a potential for manufacturing economically a range of engines having a common bore spacing but different cylinder capacities.

3.47 Cylinder head – some overall design considerations

The choice between cast iron or aluminium for the cylinder head is not simple. Aluminium has the advantages of light weight, high thermal conductivity, and ease of production to close tolerances by gravity or low-pressure diecasting.

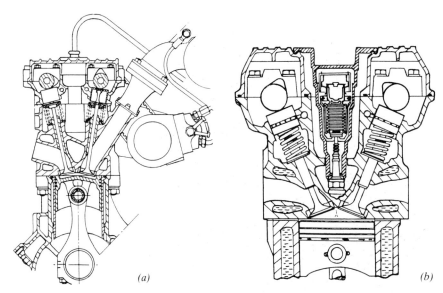

(a)

(b)

Fig. 3.36 The Renault V6 1.5-litre turbocharged engine (a) has four valves per cylinder and its straight, almost vertical, inlet ports induce swirl about a horizontal axis in the cylinder. In contrast, swirl can be induced about a vertical axis in the Saab 2-litre unit (b), with four valves per cylinder, by appropriate alignment of the inlet ports

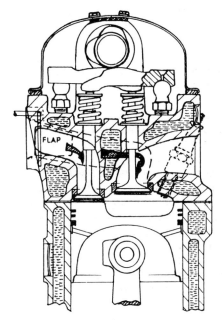

Fig. 3.37 At low speeds, a hinged flap deflects all the incoming air through the smaller of the two inlet valves in the Toyota 1.35-litre three-valve engine

On the other hand, aluminium is more expensive than iron, tooling for large quantity production is costly, porosity in the finished casting can present difficulties, aluminium is more easily damaged in service and rather more prone to gasket blow-by failure, corrosion may present problems – especially where there are copper components in the cooling system – and heat-resistant valve seat inserts are essential.

Cast iron is inherently stiffer, and therefore contains noise better, and is cheaper. On the other hand, the labour costs in making the moulds and cores are higher, and more labour may be required for removing the sand cores, and for fettling.

In a paper presented by D. A. Parker and R. H. Slee, at Symposium 86, held by AE plc, some particularly interesting comments were made on trends in engine design, Fig. 3.38. These authors pointed out that many of the overhead camshaft engines which came into vogue in the nineteen-seventies had single ohc valve gear with vertical valves and bath tub combustion chambers in cast iron heads mounted on cast iron crankcases. Screw type tappet adjusters were

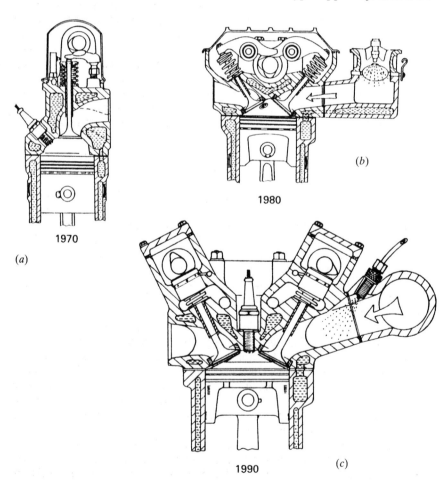

Fig. 3.38 Trend in engine design 1970–1990, as illustrated by Parker and Slee, of AE plc

used and twin valve-springs obviated a risk of the valve's dropping into the engine in the event of breakage of one, and, in any case, the installation was more compact than with a single larger spring.

By the nineteen-eighties the pistons carrying four-ring pistons had given way to three-ring types, with shorter skirts to reduce both friction and reciprocating mass. Aluminium heads became popular, partly to reduce weight, but also because their better thermal conductivity would help in the struggle to meet threatened legislation for the elimination of lead from gasoline. For the latter reason, too, sparking plug position was determined by the need to shorten flame travel, enabling the ignition to be retarded to reduce octane number sensitivity of the combustion chamber. Improvements in valve and port geometry, coupled with the introduction of electronic single-point injection, helped in the achievement of both better fuel economy and satisfying emissions regulations.

According to the two authors of the paper, in many instances, by the nineteen-nineties, not only cylinder heads but also blocks would be of aluminium alloy, while coolant volumes would be restricted to reduce weight and shorten the warm-up time. Twin overhead camshafts actuating four valves per head would provide good specific power and economy. The inherent stiffness of valve trains of this layout would enable single valve springs to be used, while self-adjusting tappets would reduce both noise and maintenance requirements. Because of its superior accuracy of metering, multi-point would replace both single-point fuel injection and carburettors for cars in all but the bottom end of the price range. Light-weight, two-ring pistons with integral hydrodynamic bearing devices for taking the thrust and minimising skirt area would reduce friction.

3.48　An interesting cylinder head design

An outstandingly good cylinder head design is that of the *Dolomite Sprint*, Fig. 3.39. Since it exemplifies the best solution to many of the problems, it will be described here. It is of aluminium and has four cylinders, four valves per cylinder, and a single overhead camshaft. The exhaust valves are inclined 16° to one side of the axis of the cylinder and the inlets 19° to the other side. Their seats are in a penthouse-type combustion chamber.

Since the engine, when installed, is tilted 45° towards the exhaust side, the inlet valve ports then slope steeply downwards. This facilitates cold starting, in the following manner. As the crankshaft is turned, any fuel remaining un-evaporated in the manifold runs down into the cylinder, where it is evaporated by the heat generated during the subsequent compression stroke. Mixing is further assisted, at TDC, by the squish effect between the flat area surrounding the slightly dished portion of the crown of the piston and the flat lower face of the casting, each side of the pairs of valves. Because of the steepness of the slope of the inlet ports, this fuel runs down positively into each cylinder in turn so that, once the first cylinder fires, the others will pick up immediately. The penalties for a slope that is either inadequate or too steep are respectively mixtures that are either too weak or too rich to fire in sequence.

A single camshaft, with eight cams, serves all the valves. Each cam actuates first an inlet and then an exhaust valve. However, whereas the inlet valve is actuated directly, through the medium of an inverted bucket tappet, the

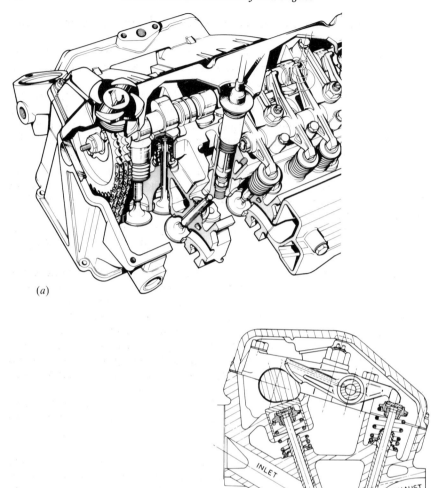

(a)

(b)

Fig. 3.39 The Triumph *Dolomite Sprint*

exhaust valve is opened by a rocker, one end of which follows the cam and the other bears on a pallet, or thick shim, seated in a recess in the top face of the valve spring retainer. A similar pallet is interposed between the exhaust valve spring retainer and its tappet.

To reduce the velocity of sliding between the rocker and cam, the pad on the end of the rocker is curved. This, however, tends to increase the velocities of both opening and closing of the valve and therefore has to be taken into account in the design of the cam profile.

The sparking plug is very close to the centre of the top of the combustion chamber, for efficient combustion. As a result, it has not been possible to make the diameter of the two inlet valves much larger than that of the exhausts but, to compensate for this, their lifts are greater – 8.712 mm as compared with 7.798 mm.

To span the distance between the camshaft and the exhaust valves, the rockers have to be long. Consequently forged En8 steel, instead of the usual cast iron, rockers are used. Because of the relatively poor bearing characteristics of steel, however, oil from a radial hole in the rocker bearing is fed down a groove on top of the rocker arm to the cam follower pad. To obviate any possibility of oil dripping down and getting into the exhaust valve guide, where it could form carbon deposits, a heat-resistant flexible seal is fitted over the upper end of the guide.

As can be seen from Fig. 3.39, the upper half-bearing for the camshaft and the semi-circular clamp for the rocker shaft are machined in a single long diecast aluminium bearing cap secured on the plane of the inclined joint face of the valve gear cover by a bolt at each end. With this arrangement, the valve actuation gear can be fitted to the cylinder head, forming a self-contained sub-assembly, with valve clearances set, all ready for mounting on the engine.

The cylinder head holding down studs on the inlet side of the head are inclined 74° relative to the cylinder gasket joint face, and this brings their upper ends out through the diecast cap, adjacent to the rocker shaft and at right angles to the inclined seating face. Consequently, when the head is tightened down, the stud pulls the clamp tightly down on to the rocker shaft so that there is absolutely no possibility of fretting fatigue between the cap and shaft. Set bolts, perpendicular to the cylinder head gasket joint face, hold down the exhaust side of the head. With this overall arrangement, the head can be removed in service without disturbing the valve gear.

The inlet valves are of Silchrome, while the exhausts are either Nimonic 80A with Stellite tips on the ends of their stems, to prevent undue wear, or of En18 with Nimonic 80A heads welded to them. All stems are chromium-plated to reduce the rate of abrasive wear in the guides.

Sparking plugs of the conical seating type are fitted, because they are screwed into bosses at the lower ends of tubular housings cored vertically in the head casting, where plug washers would be difficult both to place and to retrieve. The gap between the upper end of each of these cored housings and the valve gear cover is spanned by an aluminium tube with elastomeric seals moulded around both its ends. The lower seal is a tighter fit in the head casting than is the upper one in the valve gear cover, so that the tube will not pull away from the head when the cover is removed. Although 14-mm plugs are fitted, their hexagons are of 10-mm plug size, so that the tubular housings can be of small diameter. This, in turn, restricts as little as possible the water passages around the plug bosses.

The good thermal conductivity of the aluminium head, together with generous cooling passages around the valve seats account for the absence of valve sinkage when unleaded fuel is used. Lead in fuel is thought to act as a lubricant between the valve and its seat, so with unleaded fuel and elevated temperatures the rate of wear of less well-cooled seats can be high with the result that the valve sinks into them.

A four-valve head layout was chosen because this engine was required to have high performance. So far, no positive proof of why the four-valve layout is so efficient has been put forward. However, the central positioning of the plug probably contributes, and the scavenging on a broad front – through a pair of exhaust valves – may also help.

3.49   Cylinder block and crankcase arrangement

Until about 1925, cylinder heads and blocks were generally integral, and in highly rated aero-engines, notably the Rolls-Royce Merlin, the arrangement was still in use throughout and after the Second World War. However, with road vehicles, the need for frequent servicing and economy of manufacture led to the adoption of the separate cylinder head casting. But, with the introduction of turbocharging and the consequent development of high gas-pressures and temperatures, the designers of the Leyland 500 diesel engine, Fig. 3.40, reverted to the combined head and block casting, bolted to a separate crankcase.

The elimination of the cylinder head gasket disposes not only of the barrier to the conduction of heat, but also of the thick sections of the adjacent faces of the two separate castings, which can cause thermal distortion of the structure. In particular, with this layout, the cylinder bores and valve seats should be relatively free from distortion. The absence of cylinder head retaining studs should also help in this respect, as well as give the designer greater freedom in arranging the valve and porting layout and cooling passages. Large covers bolted on each side of the block facilitate the cleaning of sand from the cooling passages after the casting operation. However, despite all these potential advantages, the design proved extremely difficult to develop to the point of complete satisfaction in practice.

Other cylinder block and crankcase arrangements appear in Figs 3.41 to 3.45. The black sectioning represents cast iron and the cross-hatching aluminium alloy. While the figures are simplified and to a great extent diagrammatic only, each is represented by one or more examples in past or present practice.

That shown in Fig. 3.41 is the conventional arrangement used in large engines of the highest quality, where it is not essential to economise in machining and fitting. Here a monobloc cylinder casting, with overhead valves in a detachable head, is bolted to a two-part aluminium alloy crankcase split

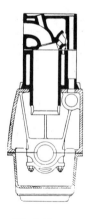

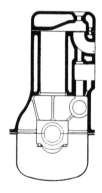

Fig. 3.40   Fig. 3.41   Fig. 3.42

on or below the crankshaft centre line. The maximum degree of accessibility is provided for valves, pistons and bearings.

Fig. 3.42 shows the side-valve arrangement of what has been the most widely used construction but which is now generally superseded by the overhead valve arrangement shown in Figs 3.43 and 3.44 with various camshaft arrangements. Modern designs simplify the cylinder block casting and improve accessibility of valves and rockers. A single monobloc casting in iron or in aluminium alloy – in which case wet cylinder liners are used – extends from the head joint face to usually well below the crankshaft centre line, forming a rigid beam structure of great depth and stiffness. Well-ribbed webs tie the crankcase walls together and carry the main and camshaft bearings. The bottom is closed by a light sump which is often ribbed for cooling purposes and may carry the main oil filters. For examination and overhaul, pistons and connecting rods must be withdrawn upwards through the cores unless removed bodily with the crankshaft.

The corresponding overhead-valve construction is illustrated in Fig. 3.43, which shows a low camshaft for push rod and rocker operation of the valves. A dry liner is indicated in this case. The valves all lie in the same longitudinal plane.

End assembly of the crankshaft is provided for in the construction of Fig. 3.44. When arranged for three bearings, the centre bearing is mounted in a circular housing of diameter exceeding the diameter of the crank webs. The old Lea–Francis design was of this construction, which provides a rigid and accurately aligned support for the crankshaft. It had high twin camshafts to operate the inclined overhead valves, which were pitched transversely.

Figs 3.45 and 7.10 show two examples of wet liner construction, one for a medium-powered petrol engine and the other for a blower-charged, poppet-exhaust two-stroke ci engine. The air gallery and scavenge ports of the latter will be noticed. Water joints are made with synthetic rubber rings (as indicated) and gas pressure joints usually by separate gasket rings to each bore. A light alloy main casting is indicated in these two cases.

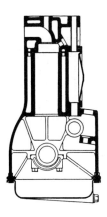
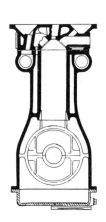
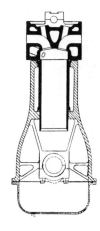

Fig. 3.43   Fig. 3.44   Fig. 3.45

3.50   The aluminium crankcase

Aluminium alloy has always been a potentially attractive material for crankcases. This is because of its light weight, good thermal conductivity, prospect for good cooling and, if aluminium pistons could be used in aluminium bores, the absence of differential expansion would enable tighter clearances to be adopted in the bores. However, until recently, this light metal has failed to gain wide acceptance because it has been necessary to fit cast iron liners, primarily owing to the unsatisfactory bearing properties of aluminium pistons in bores of the same material. This has increased the cost of production with a material that, in any case, is expensive. Moreover, there have been other problems, including the risk of electrolytic corrosion and of the head welding to the block, as well as a high noise level and, when diecasting is used because of its suitability for large quantity production, a large rejection rate owing to porosity.

Some European manufacturers, however, with access to relatively cheap supplies of aluminium, and others throughout the world who produce expensive cars, as well as Rootes – later Chrysler, now Peugeot-Talbot – have used sand-cast crankcases and cylinder blocks. The Rootes car, the Hillman *Imp*, was a special case, since weight reduction was of prime importance as the engine and transmission were to be installed at the rear. An open-top deck layout was chosen because, originally, diecasting was envisaged.

There are four iron liners, which are preheated to 204°C before they are inserted in the mould, for the aluminium to be cast around them. To key the liners in the aluminium, their outer peripheries have spiral grooves 0.331 mm deep machined around them at 3.18 mm pitch. When cast in, they are arranged in two siamesed pairs – otherwise the block would be unduly long – so water can flow transversely between only the central two cylinders and across the ends of the block.

Renault, too, have used cylinder blocks with wet cast iron liners for many years. One of their later developments has been the 1.47-litre R16 model, in which a high pressure aluminium diecast cylinder block and crankcase is employed. Although most of the walls are only 4 mm thick, the casting is well ribbed and flanged for stiffness. The liners are closely spaced in the open-top casting, and there are five main bearings – the *Imp*, being a much smaller unit, 875 cm^3, has only three bearings. An interesting feature of the Renault engine is the employment of an aluminium head too.

In 1971, General Motors announced the Chevrolet Vega 2300 engine with a diecast all-aluminium crankcase and cylinder block, Figs 3.46 and 3.47, reported in *Automobile Engineer*, August 1970. Developments that have made this possible include the Accurad method of diecasting, described in the November 1969 issue of *Automobile Engineer*. This entails control of the cooling of the metal so that areas remote from the point of injection solidify first, and a two-stage injection process, the second stage of which effectively compensates for shrinkage of the metal. Secondly, a new alloy, termed A390, has been introduced by Reynolds Metals, and this combines good fluidity in the molten condition with fine dispersion of silicon after heat treatment, which gives good bearing properties and ease of machining – large particles wear the tools and tear from the surface of the alloy. It contains 16 to 18% silicon, 4 to 5% copper and 0.45 to 0.65% magnesium.

Fig. 3.46 The diecast aluminium block of the Chevrolet Vega has 6.35 mm thick cylinder walls, and most of the other walls are 4.826 mm thick

Fig. 3.47 The open-deck design, with siamesed cylinders, was adopted for the diecast aluminium crankcase of the Vega so that the dies could be easily withdrawn

Diecasting is employed because of its suitability for large quantity production and its accuracy: the latter minimises subsequent machining and enables thin walls to be incorporated – as little as 4.826 mm, but 6.35 mm for the bores – thus economising in material. The finished block for this 2.3-litre engine weighs only 36 lb, as compared with 87 lb for the cast iron block of the comparable Chevrolet L-4 engine. However, because it is diecast and provision must therefore be made for withdrawal of metal cores, the open-top deck form, with siamesed cylinders, has to be used.

Diecast aluminium heads have not yet been used by any manufacturer in quantity production, because of the problem of withdrawal of the cores. Experiments, though, have been made with two-piece castings joined by means of adhesives or by electron beam welding. A cast iron head, however, makes up for any lack of stiffness of an aluminium block with an open-top deck and helps to contain the noise of combustion and from the overhead camshaft and valve gear. On the Vega, it is secured by ten long bolts screwed into bosses at the bases of the cylinders, so that the bores are rigidly held in compression.

The cylinder bore treatment in the Vega, to form a wear-resistant and oil-retaining surface, and to prevent scuffing, comprises exposure of the hard silicon particles by an electro-chemical etching process. In addition, the skirts of the pistons are coated with iron by a four-layer electro-deposition process, to render the bearing surfaces compatible with those of the bores. The coats, in order of application, are: zinc, copper, iron and tin. Zinc bonds well to aluminium, the copper prevents removal of the zinc by the iron plating process and the tin prevents subsequent corrosion of the iron and helps in running-in. The iron coating is 0.019 mm thick.

3.51 Camshaft drive

Whatever the type of valve used it is necessary in the four-stroke engine to drive it from a camshaft which runs at half the speed of the crankshaft, as each valve is required to function only once in two revolutions of the crankshaft. The necessary gearing for this purpose is placed, with few exceptions, at the front of the engine, that is, at the end remote from the flywheel and clutch. The camshaft or camshafts may be driven by gears, chains or toothed belts, while a few overhead camshafts have had a 'coupling rod' drive. Fig. 3.48 shows diagrammatically some typical arrangements of two-to-one drive for both low and high camshafts.

No. 1 shows the simplest possible arrangement of direct gearing for either one or two camshafts. The wheel on the camshaft has twice as many teeth as the crankshaft wheel, and therefore revolves at half the speed of the latter. Where, as is often the case, the distance between the two shafts is considerable, this arrangement requires undesirably large gear wheels, and this had led to the adoption of the arrangement shown at 2. Here an intermediate idler wheel is interposed between the crankshaft wheel and the camshaft wheel. This idler wheel may be of any convenient size, as the number of teeth in it does not affect the gear ratio. The camshaft now revolves in the same direction as the crankshaft, whereas in the former arrangement it did not.

A chain drive is shown at 3. The single chain drives the auxiliaries in addition to the camshaft, the drive thus being a triangular one, but if the chain

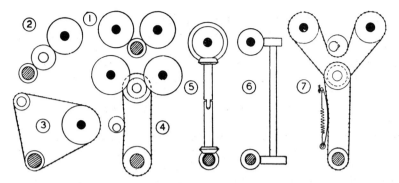

Fig. 3.48

is passed round only the crankshaft and camshaft sprockets, the shorter run thus obtained is less prone to whip.

At 4 is illustrated a combined chain and gear drive for twin high camshafts. The chain sprocket ratio is 1 : 1, and the 2 : 1 ratio is provided by the gearing.

An automatic chain tensioner of the Coventry eccentric type is indicated. The axis of rotation of the jockey chain wheel can swing eccentrically round the spindle on which it is mounted to take up slack in the chain, the desired pressure on the back of the chain being adjustable and automatically maintained by the clock-type spring, the inner end of which is secured to the mounting spindle and the outer end to the drum which carries the jockey-wheel bearing.

A later development is the Renold hydraulically-actuated tensioner, Fig. 3.49(*a*). When assembled, the cylinder casting is secured by two bolts to the front wall of the crankcase, and inserted into it is a plunger, carrying the neoprene-faced slipper together with the spring-loaded piston which forces the slipper lightly into contact with the chain. When the engine is started, oil from the pressure-lubrication system, ducted into the cylinder of the adjuster, forces the slipper firmly against the chain. The piston has a spiral ratchet-toothed slot

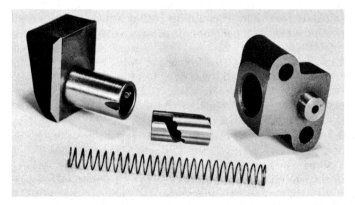

Fig. 3.49(*a*) The ratchet device on the plunger of the Renold hydraulic chain-tensioner prevents the chain from going too slack when the oil pressure is low

in its skirt, in which registers a peg projecting radially inwards in the bore of the plunger: its function is to limit the backlash, that is, movement of the slipper away from the chain.

Various means of driving overhead camshafts are illustrated at 5, 6 and 7 in Fig. 3.48. Diagrams 5 and 6 show vertical shafts driven by bevel and skew gears respectively. In either case the 2 : 1 ratio may be obtained in one or two steps as may be most convenient. In the latter case particularly, ratios 3 : 2 and 4 : 3 are advantageous.

It will be noticed that in the case of the bevel gears a tongue and slot arrangement is provided so that expansion of the cylinder block will not affect the meshing of the gears. This provision is not strictly necessary with skew gears, though the slight axial movement of the vertical gear that may place will result in a slight variation of timing between the hot and cold conditions.

No. 7 is the layout on the Jaguar in-line engine, which has stood the test of time. The 2 : 1 ratio is divided between the two stages, the tooth ratio of the four chain-wheels being 21 : 28 × 20 : 30. This tends to distribute the wear and maintains uniformity of pitch. Another advantage is that small sprockets can be used.

For the lower chain, a nylon-faced damping slipper bears on the driving strand, to prevent thrash due to torsional oscillations. On the earlier models, a spring plate tensioner bore on the slack strand, but this was subsequently superseded by the Renold hydraulic tensioner. For manual adjustment of the tension in the upper chain, the jockey sprocket is mounted on an eccentric spindle, as described in detail in *Automobile Engineer*, May 1956.

Renold also produce a commendably simple double-acting tensioner, comprising a central cast housing with twin parallel bores in which are horizontally opposed plungers, each with a shoe on its outer end. Two coil springs in compression push the plungers lightly outwards until the shoes contact the inner faces of both the taut and idle runs of the chain. At the same time, oil splashing into a pocket on top of the housing is drawn continually, by the motions of the plungers due to fluctuations in the drive, through a non-return valve to fill their bores. Movement of one plunger inwards in its bore instantly seats in the non-return valve, causing the other to move in the opposite direction so that the light contact with both runs is continuously maintained.

Another type of tensioner is that used in the Jaguar V-12 engine, Section 4.17, supplied by the Morse Chain Division of Borg-Warner Ltd. Of the four runs of chain between the sprockets – on the crankshaft, two camshafts and the jackshaft – three are controlled by damper pads and the fourth by the tensioner. Because these four runs form a strand 1.674 m long, an extremely accurate and effective tensioning is essential.

The Morse tensioner, Fig. 3.49(*b*), supports the run of chain over most of its length – its shoe is approximately 28 cm long – so the force per unit area on the shoe is small, and undue noise and wear are therefore avoided. Moreover, the shoe is a nylon moulding and therefore does not require a separate facing. Nylon, of course, has a low coefficient of friction and, in this application, fillers are added to improve both the stiffness and wear characteristics. It is also resistant to oil, fuels and temperatures of 150°C and even higher. Furthermore, its flexibility increases with temperature, so, when the engine is warm, the tensioner readily conforms to changes in shape of the chain run throughout its

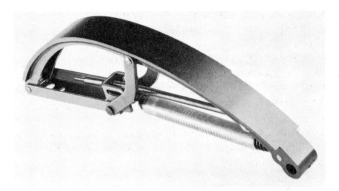

Fig. 3.49(*b*) The Morse chain-tensioner is a simple spring-loaded mechanical device

life. Finally, so that the chain cannot go slack and, possibly, ride over the sprocket, a one-way device is incorporated in the tie that holds the shoe in the bowed condition. A tendency for the chain to go slack can arise, of course, in the event of a backfire, or some other form of uneven running, or when the engine is turned backwards by hand, for timing.

From Fig. 3.49(*b*), the form that the one-way device takes can be seen. The tie comprises a steel channel section, which is attached to one end of the bowed shoe, and a plain tie-rod and tensioning spring parallel to it, which is attached to the other end, these parts being all connected, near the centre, by a small U-section bracket pivot-mounted in the channel. A pair of arms extend, one from each end of the pivot pin, and are attached mid-way between the ends of the bowed shoe, to afford additional support. The tie-rod is a clearance fit in the hole in the base of the U, and the line of action of the spring is such that it tends to cause the bracket to swing about the pivot pin. This locks the tie-rod by friction in its clearance hole so long as there is a load tending to straighten the bowed shoe, but leaves it free to slide further into the hole if that load is released altogether, so that the shoe can bow further to take up the slack in the chain. Both the bracket and the tie-rod are, of course, of case-hardened steel. The friction lock can be released in service, by the use of a key which engages in a slot in the bracket, to rotate it about its pivot until the rod is free in its hole. Around the spring is a nylon sleeve, to damp out vibrations and thus contribute to quietness and reliability.

Many manufacturers are now turning to the UniRoyal PowerGrip toothed belt drive, because of its light weight, silent operation and total absence of the following: lubrication, slip and stretch and therefore need for a tensioner. Stretch is obviated by the incorporation of a flat core of endless cables laid side by side in the plane of the pitch-line and the constituent wires of which are twisted alternately in opposite directions to avoid distortion of the finished belt. Originally, the teeth were of trapezoidal section, but now the more highly rated high torque drive (HTD) version has teeth more closely resembling those of gears, and giving transmission efficiencies of over 98%. As can be seen from Fig. 3.50, the flat outer face of such a belt can be used to drive the water pump, but double-sided belts are also available having not only toothed inner faces but also either toothed or polyvee profiles on their outer faces. Toothed belt drives are usually enclosed for both safety and cleanliness.

Fig. 3.50 The outer surface of the internally-toothed belt drive for the camshaft can be used
to drive the water pump and fan. Such a belt is wider than a chain

3.52 Camshaft brakes and compensating cams

As each cam follower passes from the front to the back of the cam at the point
of maximum lift, the effect of the valve spring changes from a resisting effort to
a driving effort, and the direction of the torque in the camshaft (as far as that
cam is concerned) is abruptly reversed. If the camshaft is long, this sudden
reversal of torque results in considerable torsional whip, while if the torque on
the camshaft as a whole (taking all the cams into consideration) reverses, then,
if there is any backlash in the camshaft drive, noise will result from the blow as
the backlash is suddenly taken up. This is avoided in some high-class engines
by fitting a lightly loaded friction brake, as shown in Fig. 3.51, which exerts a
continuous small resisting torque sufficient to prevent this torsional 'flick' or
whip. The small amount of friction necessary is often provided by the
lubricated contact of a metal or fabric disc A against the rim face of the
camshaft wheel, contact being maintained by a light compression spring or
washer B.

Another means of avoiding the trouble is the use of a compensating cam
operating a special spring-loaded tappet. The lobes of the cam are so designed
in number, shape and spacing as to compensate the fluctuations of impulse of
the main cams and thus ensure an approximately even torque on the camshaft.
A design suitable for twin overhead camshafts for a six-cylinder engine, with
spring-loaded 'lag-tappet', is shown in Fig. 3.52, and a similar application to a
four-cylinder engine has been used.

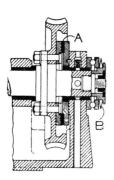

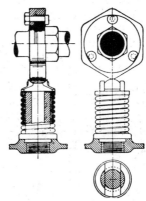

Fig. 3.51   Fig. 3.52

3.53   Camshaft bearings

Camshaft bearings do not call for any special comment. In the majority of engines they consist of plain bushed or unbushed holes in suitable bosses in the crankcase casting, the diameter being great enough to enable the shaft to be inserted. In many instances, the bearings are successively larger in diameter, from one end to the other, to facilitate insertion and withdrawal of the shaft.

Such plain bearings stand up well to the impulsive loads that arise in the operation of the cams, they are inexpensive to manufacture, and are silent in operation. In the higher class of engine, where it is desired to reduce the frictional torque to a minimum by reducing the diameter of the bearings, and also to provide adjustment for wear, the bearing bushes, usually white metal lined, are split, and provided with separate caps in a similar manner to the main bearings of crankshafts. The former arrangement is shown in Fig. 3.53, and the latter in Fig. 3.62.

3.54   Adjustment of valve timing

Clearly, with a one-piece shaft, relative variation to the inlet and exhaust valve timing or of individual cylinders is impossible, but all events can be advanced or retarded together. Where two camshafts are used, however, the exhaust and inlet valve timings, of course, can be varied independently.

Generally, the camshaft wheel is keyed or, perhaps, splined to the shaft in a

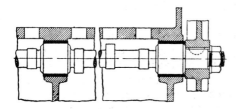

Fig. 3.53 Camshaft bearings

position such that, when certain marked teeth are meshed, the timing is most suitable for the particular engine. The smallest amount by which the timing may then be advanced or retarded is one tooth pitch, corresponding to, say, 10 to 15° of crank angle. Occasionally, finer adjustment is provided as shown in Fig. 3.54, where the cam wheel rim is formed separately from its boss, and a series of 'vernier' holes is drilled through the flanges of the two members so that they can be bolted up in relative angular positions which vary by a small fraction of a tooth pitch. While this adjustment is useful, especially during development work, it is not usually essential since, once established, the optimum timing is reproduced in each engine during manufacture.

A timing diagram forms the subject of Fig. 3.55. The five events shown are—

    (1) Inlet valve opens.
    (2) Inlet valve closes.
    (3) Spark passes.
    (4) Exhaust valve opens.
    (5) Exhaust valve closes.

These events do not take place exactly at the dead centres, but rather earlier or later as the case may be. Thus, in order to get the exhaust gases clear of the cylinder as early as possible the angle d is, on the average, about 50°, and as much as 70° in some racing engines. Very little of the useful pressure is lost since the piston is near the end of its stroke, while the total time available for the escape of the exhaust gases is very much increased.

The angle e is from 0 to 10 or 12°, a from 10° early to 15° late, while b is from 20 to 40° late, in order to take advantage of the 'ramming' or inertia effect of the rapidly moving gases in the induction pipe.

The spark occurs about 20 to 40° early when fully advanced, and approximately at TDC when fully retarded.

When the angle e is greater than the angle a there is said to be 'overlap', that is, the exhaust valve closes after the inlet valve has opened. The purpose of this is to take advantage of the momentum of the high velocity exhaust gases in producing a scavenging vacuum after the piston has reached TDC.

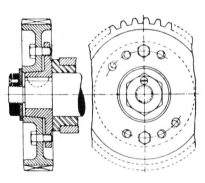

Fig. 3.54 Vernier timing adjustment

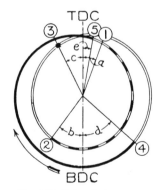

Fig. 3.55 Valve timing

3.55   Operation of the valves

The valves are operated from the camshaft in a variety of ways according to their disposition, but only in a few cases of overhead camshafts do the cam faces act directly on the valve stems. In the great majority of cases some form of cam follower or tappet is interposed between the cam and valve stem. In many cases, including some arrangements with overhead valve layouts and a few with side valves, a rocker is introduced, the leverage of which usually increases the lift of the valve in relation to the cam eccentricity. This enables the overall diameter of the camshaft to be kept small, and reduces the frictional and inertia forces.

Four typical methods of operation are shown in outline in Figs 3.56 and 3.57, namely, side valves, Fig. 3.56(*a*), overhead valves operated by push rods from a camshaft in the crankcase as in Fig. 3.56(*b*), while Fig. 3.57 shows three examples of the overhead camshaft. Direct operation of the cam on the valve stem (*a*) introduces heavy side loading, which is largely obviated in (*b*). The arrangement at (*c*), Fiat 130, is used, with various methods of tappet adjustment, on many overhead camshaft engines, to relieve the valve of all side loading.

With side valves the interposition of a tappet only is required between the cam and the valve stem. The tappet slides vertically in its guide, which is sometimes of bronze or cast iron and renewable, while the foot follows the cam profile.

In modern simplified designs, as will be seen in the illustrations showing push-rod gear, the tappet is of simple bucket form, as shown in Figs 3.56 and 3.58, moving in a plain reamed hole in the iron crankcase.

3.56   Forms of cam follower

There are four forms of tappet foot or cam follower in general use, namely—

(1) Solid curved foot (obsolescent).
(2) Mushroom or flat follower.

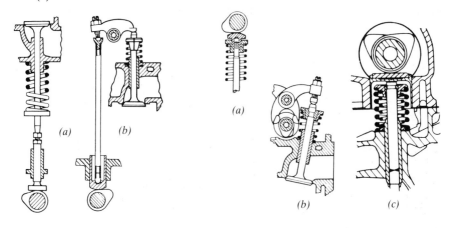

Fig. 3.56   Fig. 3.57

(3) Bucket type.
(4) Hydraulic.

These are illustrated in Fig. 3.58. A is the tappet guide, B the adjustable tappet head by means of which a suitable small clearance is maintained between the tappet and valve to allow the latter to seat properly. C shows a curved foot follower operating a push rod enclosed in an oil-tight case. E shows a flat follower with its centre line offset from the centre of the cam. This offset causes the tappet to rotate and so reduces the amount of sliding between the cam and tappet foot with consequent reduction of wear. At F is shown the 'bucket' type of follower. This is very widely used in current designs, the material being cast iron. A bucket type follower, inverted over the valve stem and spring, is used in many overhead camshaft layouts.

The rocker, or finger, type of follower G is sometimes introduced between the cam and tappet to reduce the side thrust on the latter and in some cases to increase the lift in relationship to the cam throw. This again, is used in some overhead camshaft arrangements.

D is a clamping bridge employed to hold a pair of tappet guides in position. The roller type of follower, popular at one time because of reduced friction, was liable to become eccentric through wear, with consequent variation in tappet clearance. Another example using needle roller bearings to correct this trouble may be seen in Fig. 7.10.

Theoretically the clearance should be kept at the minimum possible value that will allow the valve to seat under all conditions. With side valves, the minimum clearance usually occurs when the engine is hot, as the downward expansion of the valve stem is greater than the upward expansion of the cylinder block. With overhead valves and push rods the clearance is usually a minimum with the engine cold, as the rods tend to keep relatively cool, while the upward expansion of the cylinder block has the effect of increasing the clearance, owing to the rise of the rocker pivot. With overhead camshafts the clearance will usually be a minimum when the engine is hot. Minimum clearances are usually about 0.1 to 0.2 mm.

Hydraulic tappets (see Fig. 3.60) are of telescopic construction, the two parts being held apart by oil pressure from the engine lubrication system, thus keeping the ends of the tappet constantly in contact with the cam and pushrod, or rocker, a condition sometimes termed *zero lash*. Immediately the tappet begins to be lifted by the cam, the resultant increase in hydraulic pressure

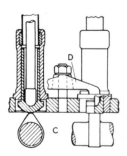
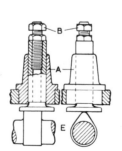
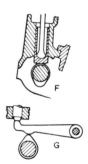

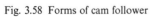
Fig. 3.58 Forms of cam follower

between the telescopic components closes a non-return ball valve, creating a hydraulic lock to maintain it at its zero lash length. When the cam releases the poppet valve, the engine oil pressure can open the non-return ball valve so that any oil leakage through the working clearances in the tappet, while it has been momentarily under heavy load, will be replaced.

3.57 Overhead valves

The push rod and rocker method of operation is illustrated in Fig. 3.59(*a*). The upper end of the push rod which is usually of tubular section for lightness is provided with a cup C in which fits the ball end of the rod B, which is screwed and lock-nutted to the rocker for purposes of adjustment. The construction in which the rounded end of the push rod bears in an inverted cup formed in the rocker is somewhat less favourable for lubrication. The rockers may be mounted on ball or needle-roller bearings, as here indicated at A, or they may be threaded on a tubular bearing bar running the whole length of the head and carried in suitable brackets. Light coil springs are usually threaded on in addition, in order to reduce noise by taking up end shake.

Except in the more simplified engines, means of adjusting the tappet clearance are provided, as shown in Figs 3.59(*a*) and (*b*), and 3.60. Fig. 3.59(*b*) is an illustration of the inverted tappet of the Vauxhall arrangement, a method of adjustment that is particularly easy to effect in service on engines of the overhead camshaft layout: alternatives are the interposition of hard shims or washers between the inverted bucket type tappet and either the cam as in Fig. 3.57(*c*), or the valve stem. With the Vauxhall system, access is gained through a hole in the side of the tappet to a grub-screw. This screw is inclined 5° 30′ relative to the flat top of the valve stem, and a flat is machined at a corresponding angle, but in the opposite sense, on its periphery in such a way that is parallel to, and seats on, the end of the valve stem. Rotation of the screw through 360° gives adjustment of 0.0762 mm.

The interesting valve spring arrangement of the old Standard *Vanguard* engine is shown in Fig. 3.61. The inner valve spring is relieved of the duty of accelerating the rocker, push rod and tappet during closure of the valve, this action being performed by the auxiliary outer spring, the spring washers being so arranged that the clearance is always concentrated at the top of the valve

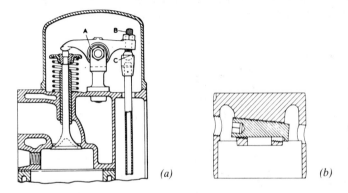

Fig. 3.59 (a) Push rod and rocker. (b) Vauxhall screw type adjuster

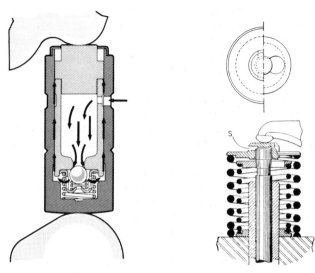

Fig. 3.60 Hydraulic tappet for Ford *Escort*   Fig. 3.61 *Vanguard* valve springs

stem. A slot S is provided for insertion of the feeler gauge during adjustment of the clearance.

The arrangement makes for reduction of noise as the clearance is taken up between only one pair of surfaces. The simple and effective provision for assembly by means of the key-hole drilling of the main spring washer may be noted.

Fig. 3.62 shows a well-thought-out and executed design, incorporating a single overhead camshaft with inclined valves and rockers. The camshaft is mounted above the rockers, which are of the third order of lever, thereby increasing the valve lift in relation to the cam throw. A further advantage of a long rocker is obtained, namely, that the 'draw' or sliding of the rocker toe on the end of the stem is less for a given valve lift, thus minimising wear. The rockers A are machined from the solid to give maximum strength with lightness, and they incorporate an oil trough in which the cam roller runs, thus ensuring excellent lubrication. The rockers are mounted on bearing spindles B carried in independent forked brackets C. The portion of the pin B on which the rocker bears has a considerable eccentricity in relation to the end portions which are clamped in the brackets, so that by rotation of the spindles in the brackets the clearance between the roller and concentric portion of the cam, which represents tappet clearance, can be adjusted. Attention may be drawn to the arrangement of light steel sleeve and soft packing around the valve stem, which prevents excess oil reaching the stem and also avoids air leakage into the induction port, a source of trouble when adjusting the carburettor for idling. A much simpler version of this finger type valve actuator can be seen in Figs 3.37 and 3.38(*a*).

3.58 Hydraulic self-adjusting tappets

Self-adjusting – in America, termed *zero-lash* – tappets have two main advantages: they compensate automatically for variations of tappet clearance as a result of both wear and differential thermal expansion of the metal components of the crankcase, cylinder head and valve actuation mechanism, and they reduce noise due to tappet clatter. The principal design aim – in addition to the obvious ones such as keeping the inertia and rates of wear on both the barrel and ends as low as possible – is the avoidance of aeration of the oil owing to the severe shaking to which it is subjected inside the tappet. In general, the remedy is to allow leakage from the top of either the tappet or its oil supply system, so that any air or vapour present vents out. At the same time a well-supported camshaft, to keep vibration as low as possible, and cam design to reduce to the minimum the instantaneous acceleration levels of the tappets help.

The Ford hydraulic tappet, Fig. 3.60, is designed for effective operation up to 7000 rev/min crankshaft speed. It comprises a hollow cylindrical body, closed as its lower end and having a thimble-shaped plunger inserted in its upper end. The top of the plunger is closed by a hardened cap, on which bears the end of the rocker, while the lower end of the body housing the plunger functions as the cam follower.

Interposed between the base of the plunger and the lower end of the body is a fairly stiff coil spring, which presses the plunger up against the rocker during the dwell periods following each closure of the valve. Between the upper end of this spring and the base of the plunger is clamped the flange around the open end of a top hat-shaped pressing that houses a non-return ball valve. The ball is seated on a port in the end of the plunger by a light coil spring interposed between it and the base of the pressing and, during actuation of the tappet, by the hydraulic pressure below it.

When the tappet is lifted by the cam, the actuation load is transmitted from the cam follower through the oil, trapped by the ball valve, between the lower end of the body and the base of the plunger, and on through the plunger itself to the rocker. The oil comes from the main pressure lubrication system. It passes from the hollow five-bearing camshaft, through a duct in the cylinder head, to an annular groove round the body of the tappet, on through a radial hole to an annular groove round the plunger, and finally through another radial hole into its centre. A second annular groove, lower down around the periphery of the body, Fig. 3.60, serves as an oil reservoir and spreader for assisting in the lubrication of the lower end of the tappet, which of course is subjected to the greatest side thrust, especially under start-up conditions when the faces of the cam and follower are relatively dry.

When the poppet valve closes, and the load is therefore released from the tappet, the spring in the base of the tappet pushes the plunger up to take up any clearance, and oil flows from the central cavity in the plunger past the non-return valve into the chamber in the base, to replace that which has leaked out during the opening of the poppet valve and to compensate for any changes in clearance that would otherwise be taking place gradually owing to thermal expansion or contraction of the metal parts. A controlled degree of leakage, between both the body and plunger and the body and the tappet housing bore, lubricates the ends of the tappet.

3.59   Alternative rocker arrangements

Fig. 3.63 shows an American construction of extreme simplicity suitable for large-scale economic manufacture. The rocker is a simple steel pressing of light

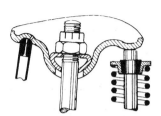

Fig. 3.63 Simplified rocker

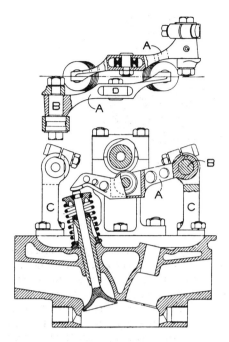

Fig. 3.62 Rocker gear of refined design for overhead camshaft

weight and great strength, taking its bearing on a spherical spacer retained by the self-locking nuts on individual mounting studs. The push rods are of light tubular construction working in close fitting guides formed by simple drillings in the upper deck of the head casting. The push rods are thus able to provide the necessary constraint to maintain transverse alignment of the rockers.

Tappet adjustment is effected by the retainer nuts. In layouts using bucket type cam followers, it may be provided by the employment of self-adjusting tappets of the type such as that in Fig. 3.60. Alternatively selective assembly by push rod length may be used.

Lubrication may be through the support studs from a longitudinal gallery, from the hydraulic tappets through the push rods, or by oil mist and splash.

Fig. 3.64 shows the typically original rocker arrangement used on the Lancia *Aurelia*, which facilitates the combination of hemispherical heads having longitudinally inclined valve stems, with a single camshaft in a V-type engine. The rocker axes are transverse, and individual bearing blocks are mounted longitudinally for each cylinder, for the pair of rockers.

The tendency to 'off-set tilt' is thus greater than in conventional arrangements, but is catered for by the bearing surface and close fit of the large diameter rocker collars.

3.60 Auxiliary drives

The following auxiliaries are common to both petrol and diesel engines, and provision for their drive must be made: oil pump, water pump, dynamo and

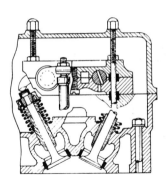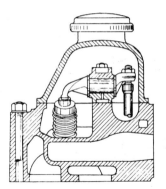

Fig. 3.64 Lancia *Aurelia* rocker

fan. For certain countries, a compressor may also be needed for air injection into the exhaust manifold, to control emissions. On petrol engines, the fan, impeller type water pump, compressor and dynamo or alternator do not call for an exact speed ratio, so some of these may have a common belt drive. Valve gear, ignition distributor and diesel injection pump require an exact ratio of 2:1 or 1:1 according to the cycle used. Such auxiliaries as exhauster and rotary blower are usually, though not necessarily, driven through longitudinal shafts by gearing or chain, while superchargers may be operated by either of these positive means or independently by engine exhaust.

In petrol engines having low camshafts, it is convenient to mount distributor and oil pump on a vertical or inclined shaft driven by 1:1 skew gearing from the camshaft, as shown in Fig. 3.68. With overhead camshafts the distributor is usually coupled directly to either its front or rear end or driven by a 1:1 spiral gear, as shown in Fig. 3.70. Magnetos, now rarely seen, were commonly driven with the water pump from a transverse shaft at a speed depending on the number of sparks provided per revolution by the magneto and required per revolution by the engine.

In most instances, the auxiliary drives are at the front of the engine. However, in this position, they are generally stationed on a crankshaft vibration antinode, which can tend to cause gear and wear and noise problems. Consequently, they are sometimes placed at the rear of the engine where they may be at or close to an antinode. The disadvantages for automotive engines are the difficulty of providing for mounting the flywheel and, in most cars, poor accessibility.

3.61   A typical all-round chain drive

A typical all-round drive employing duplex roller chain is illustrated in Fig. 3.65, which shows the auxiliary drive of an early AEC compression-ignition engine. This type of drive is popular, and has the great advantages of adaptability to any convenient position of the auxiliaries and light weight, but for the larger diesel engines gearing is preferred, despite the need for close tolerances on centre-to-centre spacing.

The crankshaft carries a spur gear G for the 1:1 drive for the oil pump, while

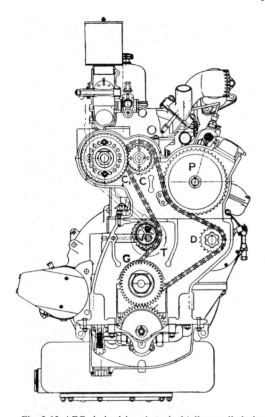

Fig. 3.65 AEC chain drive. A typical 'all-round' chain drive for a high-speed diesel engine

behind it is the twin sprocket for the chain. The $2:1$ ratio for the injection pump drive is obtained by the large sprocket P while the $2:1$ camshaft ratio is obtained by means of the spur gears C and C_1, of which the gear C is a twin gear having springs so disposed between the two parts as to impart relative rotational movement for the purpose of taking up backlash. The slight increase in pressure on the driving side of the teeth is not detrimental, and quiet running is obtained.

The gear C_1 is coupled to the camshaft by a 'vernier' coupling of the type described and illustrated in Section 3.54 and Fig. 3.54. The sprocket D drives the dynamo at $\frac{7}{6}$ engine speed. The automatic tensioner T is of the eccentric type described in Section 3.51.

3.62 All-gear and mixed drives

A typical all-gear drive arrangement is that of the Caterpillar 3406B, in-line six-cylinder, 14.6-litre, turbocharged and aftercooled diesel engine, Fig. 3.66. The wheel on the left is on the water pump, and it is driven from the twin idler cluster meshing with the crankshaft gear, the smaller diameter of the twin gears being hidden behind the larger one. Meshing with the smaller idler, and to the right of it, is the camshaft half-speed wheel which, in turn, drives two pinions: one, above and to the right, is the air compressor drive, while the second, below

Fig. 3.66 All-gear drive on the Caterpillar 3406B, in-line six-cylinder, 14.6-litre, turbocharged and aftercooled diesel engine

right, is for the fuel injection pump. In the centre, directly beneath the twin idler, is the crankshaft gear. This, because it is more heavily loaded than the others, is a pair of identical gears mounted together, that in front driving the idler and the other, the oil pump below. All these gears are hardened, have teeth set at an 18° helix angle, and their diametral pitch is 10.

Fig. 3.67 illustrates a somewhat unusual combination of chain and gear drive used on the Thornycroft KRN6 diesel engine. A three-row roller chain drives an intermediate wheel and the dynamo drive sprocket, and is tensioned by a Renold spring-loaded eccentric unit. The compound intermediate wheel is formed by bolting to the back of the driven sprocket a helical gear which, on one side, meshes with the camshaft wheel and, on the other, with the gear driving the injection pump and auxiliary compressor. Provision is made in the construction for either right- or left-hand assemblies.

Fig. 3.67 Thornycroft timing gears

3.63 Complete assembly

Some descriptions of complete engine assemblies follow in order that the relationship of the various components and the general arrangement of the engine may be clearly understood.

3.64 Vauxhall 'square' engines

The abandonment by the Treasury of the RAC horsepower formula for taxation purposes (see Section 1.19) freed design as regards stroke:bore ratio. Power developed per unit of piston area then became of relatively little significance as a fundamental criterion of performance. In the UK the first quantity-produced car engines in which full advantage was taken of the new-found freedom were the Vauxhall *Wyvern* and *Velox* units – September 1952 issue of *Automobile Engineer*.

  These engines were described as *square* – a term indicating that the bore and stroke are equal. Actually, the dimensions in this case were 79-mm bore and 76-mm stroke – slightly over-square. The *Wyvern* was a 1.5-litre four-cylinder engine and the *Velox* a 2.75-litre six-cylinder unit. A compression ratio of 6.4:1 was adopted for both.

  Fig. 3.68(*a*) shows the cross-section of the six-cylinder unit, but in general is the same for both, while (*b*) is the longitudinal section of the other. Many components were common to both. Because of the short stroke, there was considerable overlap between the journals and crankpins, giving a stiff shaft, all main journals being 53.975 mm diameter. With such a large bore, the big ends were split normal to the axes of the rods, instead of diagonally, but could nevertheless be withdrawn upwards through the cylinder.

  This is a typical example of the traditional overhead valve layout, operated by push rods and chilled cast iron tappets. The flat piston crown directs a squish, or turbulence, towards the plug, giving good scour while at the same time ensuring adequate cooling of the end-gas in the quench area – large

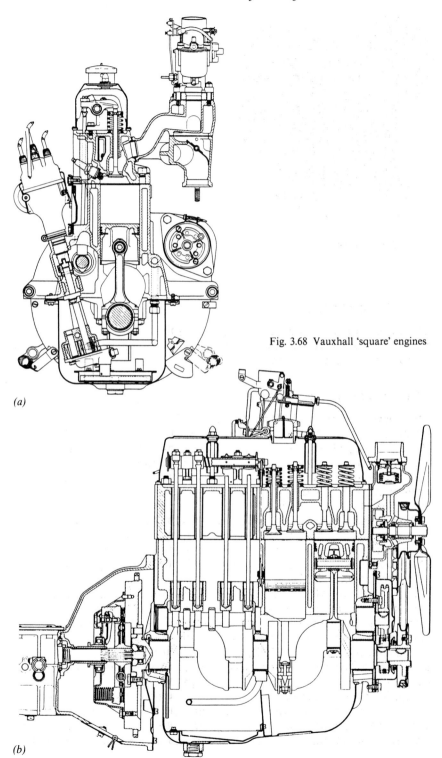

Fig. 3.68 Vauxhall 'square' engines

(a)

(b)

quench areas, incidentally, are avoided in modern engines because of the problem of exhaust gas emissions, discussed in Chapter 12.

A pipe of quadrilateral section, with slots in it, is inserted longitudinally in the cylinder head to direct coolant over the hot areas of the combustion chamber – the exhaust valve seats and sparking plug bosses. The thermostatic control of the hot-spot manifold is shown to a larger scale in Fig. 3.69.

A later version of this engine incorporated several changes, including cylinder head holding-down studs extending the full depth of the casting on both sides – increasing the cross-section, and so the stiffness, of the casting. Also, the air intake manifold branches slope down from the head, so that any liquid fuel drawn from the carburettor under cold conditions tends to collect on the exhaust heated hot-spot, where it is evaporated instead of being carried straight into the cylinder. This arrangement also enables the height of the bonnet to be reduced.

In general, with a square or over-square type of design, the rotational speed for a given piston speed can of course be higher, or *vice versa*. Although, by adopting a low rotational speed, it is possible to reduce inertia stresses and friction losses, the modern trend is towards high speeds, to obtain corresponding gains in power per litre.

Whereas, in 1935, stroke : bore ratios were commonly in the range of 1.36 : 1 to 1.75 : 1, they are now mainly less than 1 : 1, an extreme example being the 997 cm³ Ford *Anglia* engine, at 0.61 : 1. Exceptions are some of the transversely installed engines, where short length, and therefore a smaller bore, is the prime requirement.

Not only can the short stroke engine, for the same stresses, run about 14 to 15% faster than the long stroke unit, but also the breathing can be improved,

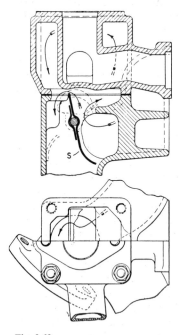

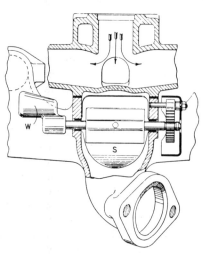

VAUXHALL MANIFOLD JACKET
WITH THERMOSTATIC CONTROL

Fig. 3.69

because of the larger area available for valve ports. However, there are certain disadvantages: at these higher speeds, the volume of the air drawn in per unit of time is greater and therefore a carburettor with a larger choke has to be used which, in turn, means that the velocity of air flow through the choke at low speeds of operation is correspondingly slower. This makes it more difficult to obtain satisfactory torque at low speeds, and some problems may be experienced in starting. In any case, the longer the stroke, and therefore crank throw, the greater is the mean torque applied to the crankshaft per unit of pressure on the pistons.

3.65   Austin-Rover O-series engine

The 1700 and 1994 cm^2 O-series engine (Fig. 3.70), at the time of its introduction, was an outstanding example of good modern design. Its bore is 84.45 mm, and the stroke are 75.87 and 89 mm respectively for the smaller and larger capacity versions. With a 9 : 1 compression ratio – an option of 8 : 1 is available for countries having low grade petrol – the 1.7-litre version develops 87 bhp at 5200 rev/min and the 2-litre unit 93 bhp at 4900 rev/min. Their maximum torques are 131.5 Nm at 3800 rev/min and 152.67 Nm at 3400 rev/min. (See Fig. 3.71.)

With vertical in-line valves, an inexpensive single overhead cast iron camshaft installation has been possible, the cams acting directly on inverted bucket-type cast iron tappets. The latter relieve the valves of all side and offset loading due to the motion of the cams. Three bearings carry the shaft, and their lower and upper halves are machined in the aluminium head and valve gear cover castings respectively. Because of the low inertia and high rigidity of this valve gear, the valve crash speed is 7000 rev/min. Since the camshaft is driven by a toothed belt, the overall length of the engine is a little greater than would otherwise have been possible, had a chain been used.

The cylinders are siamesed to keep the block as short as possible. This joining of all cylinders is sometimes frowned upon because, inevitably, it leads to uneven temperature distribution and variations in stiffness around them. However, the Rover Group has adopted this feature in a number of engines over many years and say that they have experienced no difficulties attributable to it.

The cylinder head is an LM8 WP aluminium alloy gravity die casting. It has a totally flat machined lower face, broken only by the valve ports, with their seat inserts, and the apertures for the sparking plugs, oil ducts and coolant transfer passages. This layout, again, is possible only by virtue of the use of vertical valves and the incorporation of combustion chambers in the form of shallow depressions in the crowns of the pistons. Its four inlet and four exhaust ports are arranged alternately along one side, to keep the temperature distribution as uniform as possible. The sparking plugs are along the opposite side, so access to them is easy.

Noise tests have shown this engine to be 6 dB quieter than the old B-Series which it replaces. This is attributed to the use of a cast aluminium valve gear cover – instead of pressed steel – the absence of push rods and rockers, the use of a toothed belt drive for the camshaft, the formation of the combustion chambers in the pistons, the resultant minimum variations of compression ratio from cylinder to cylinder, the stiffness of the cast iron cylinder block, and

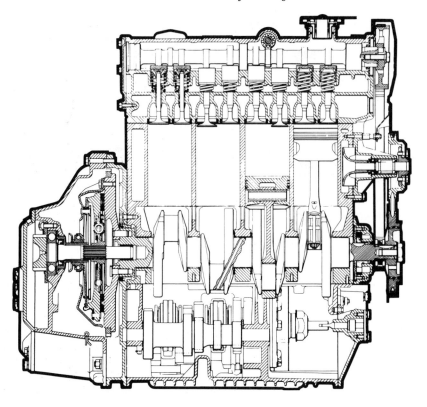

Fig. 3.70 Austin-Rover O-series engine and transmission for transverse installation

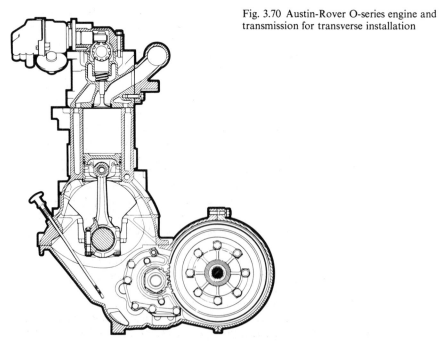

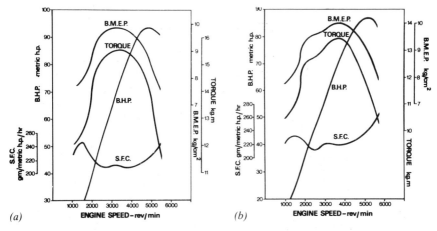

Fig. 3.71 Performance curves for Austin-Rover O-Series engine (*a*) 2-litre, (*b*) 1.7 litre versions

the use of aluminium alloy pistons having steel inserts to control their thermal expansion.

Because of the absence of a tappet chest and camshaft bearings in the crankcase, this component is of commendably simple and symmetrical design. This, together with a deep skirt, vertical external ribbing of the water jacket between the cylinders, the siamesing of the cylinders, and the profiling of the water jacket to wrap around each of them, has made the crankcase and cylinder block casting uncommonly stiff, again contributing to noise reduction.

The five-bearing crankshaft is of nodular cast iron. An increase in fatigue life of about 60% is claimed by virtue of fillet rolling the junctions between the ends of the pins and journals and the sides of the webs. Because the shaft is cast, it has been possible to make the counter-weights integral with it. They provide total primary balance on the 1.7-litre and 80% balance on the 2-litre unit.

Timing is checked electronically, by means of a light-emitting diode (l.e.d.) on one side, beamed towards a photoelectric cell on the other of a disc with a radial slot in its periphery. This disc is bolted to the crankshaft, but the l.e.d. and photoelectric cell are separate pieces of servicing equipment. The advantage of this system, using as a datum the impulses generated when the light shines through the slot on to the photoelectric cell, is that the ignition can be checked at all engine speeds. Such a facility is necessary where, to satisfy exhaust emission requirements, the spark advance does not bear a simple relationship to speed. The valve timing is: inlet opens 19° before top dead centre (BTDC), closes 41° after bottom dead centre (ABDC); exhaust opens 61° before bottom dead centre (BBDC), closes 15° after top dead centre (ATDC).

3.66 Austin-Rover M16 engine

For the Austin-Rover 800 Series cars, introduced in 1986, two new 16-valve versions of the O-Series engine were produced. One had single-point and the other multi-point injection, producing, respectively 88 kW at 5600 rev/min and 103 kW at 6000 rev/min. Their maximum torques were 162 and 178 Nm at 3500 and 4500 rev/min, respectively.

Few changes were made to the crankcase. An oil drainway down from the head, through the flywheel end of the crankcase, to the sump, was moved to shorten the block and crankshaft by 7 mm and to make room for four cylinder-head holding-down bolts equally pitched around all the cylinders, including the rearmost one. At the other end of the block, the water pump was moved to one side, instead of in front of the crankcase, so that the drive could be moved nearer to the block and thus reduce by a further 20 mm the overall length. Lengthening the connecting rods by 6.4 mm enabled the flat crowns of the pistons to be lowered, reducing their weight by 6 g relative to those of the turbocharged version of the O-Series leaving them only 34 g more than those of the EFI version.

Obviously, the one-piece, gravity sand-cast LM25 aluminium silicon alloy cylinder head, accommodating the four valves per cylinder and twin overhead camshafts, is totally different from that of the original engine. An interesting detail is the sealing of the two camshaft covers, by means of a steel plate with a rubber bead around its outer edge. The function of the steel plate is to control precisely the degree of compression of the rubber bead. The gap between the two covers is bridged by a shield, seen in Fig. 3.72, that protects the 14 mm sparking plugs and their leads from contamination and capacitance that might reduce the spark voltages.

With the engine installed transversely, the inlet valves, all lifting 8.83 mm, are set at the rear and the exhausts at the front. They are directly actuated through the medium of inverted bucket type hydraulic tappets designed for effective operation up to 7200 rev/min. With the multi-port injection system, the timing is inlet opens 18° before TDC, closes 46° after BDC; exhaust opens 52° before BDC and closes 12° after TDC. For the single-point injection version, however, these timing events are 14, 50 and 50, 14, respectively. The two cylinder heads, combustion chambers and porting arrangements are illustrated in Fig. 3.72, and the complete engines in Fig. 3.73.

This engine is of the high-compression lean-burn type, having a compression ratio of 10:1 and being capable of burning mixtures as lean as 18:1 air:fuel ratio, in the economical cruising speed range. To burn such lean mixtures the inlet ports are almost vertical and the incoming mixture is swirled about an axis diametrically orientated in each cylinder, instead of coaxially with it.

To obtain this effect, the pairs of valves are inclined symmetrically about the vertical, at an included angle of 55° and, as the ports lead towards the valve seats, their sections progressively narrow to increase the velocity of flow. The heads of the martensitic stainless steel inlet valves are 29 mm and those of the austenitic stainless steel exhausts 26.5 mm diameter. All have hard chromium-plated stems for good wear resistance. The port shapes are smoothed by computer-aided design to guide the flow without abrupt changes in velocity. They are left as-cast, because machining marks were found to leave ridges that upset the boundary layer flow. The outcome is consistency in manufacture and minimal cyclic variation in the power developed by each cylinder. As can be seen from Fig. 3.72, these valves are in double penthouse combustion chambers, in almost exactly the centre of each of which is the sparking plug.

(a)

(b)

Fig. 3.72 Cylinder head and manifolding arrangements for the Rover 800. (a) With throttle body injection and (b) with multi-point injection

Brico LAR 44-M-T inlet and X/W35 D-HT sintered powdered metal seats are shrunk into the cylinder head casting. These are complex alloys containing respectively 80% and 62% iron and 1% and 1.8% carbon, as well as chromium, molybdenum and 6% and 20% copper, the exhaust valve seats only having additions of 5% tungsten and 2.5% vanadium.

The 4 mm thick, gravity diecast, aluminium alloy manifolding is tuned to minimise the variation in induction pressure from cylinder to cylinder, particularly important in four-cylinder engines because they are inherently less well balanced than six-cylinder units. For the single-point injection version, the throttle-body injection unit is designed and developed by Austin-Rover Fuel (formerly SU Carburettors) department of the Austin-Rover Group. It is mounted on top of a plenum chamber, from which the four branch pipes, 270 mm long, pass down to the ports, Fig. 3.72, to give a mixture distribution to the cylinders that is consistent at part throttle to within 1 air : fuel ratio. The plenum chamber is water-jacketed to prevent ice formation within it in cold, humid conditions. In addition, a thermostatically controlled electric heater is mounted on its floor below the injector unit to facilitate vaporisation during cold starting.

To reduce exhaust emissions, this version of the engine has a thermostatically controlled valve in its induction pipe, for taking the air either directly into the induction system or from a shroud over the exhaust manifold. On its way to the cleaner, the air intake passes over a Helmholz resonator to reduce induction roar.

For the multi-point injected engine, the control unit, developed jointly by Austin-Rover and Lucas, is mounted on the flywheel end of the plenum chamber, and the branch-pipes are 292 mm long, Fig. 3.73. In this engine, the variations of mixture strength from cylinder to cylinder are: at full throttle 1 and, part throttle, only 0.5 of an air : fuel ratio.

Both versions have an electronic control system that combines a 2 × 64 point fuelling map with a 144-site ignition map and a knock-sensing system. It automatically compensates for varying altitude and for any leaks that might develop in service in the induction system. Additionally, it combines tight control over exhaust emissions with reliable starting and drive-away, regardless of engine temperature and ambient conditions, and an automatically operating, jerk-free, fuel cut-off during overrun.

Also common to both engines is the 3.5 mm thick nodular cast iron exhaust manifold comprising four primary branches, 340 mm long, joining two secondary branches, 860 mm long, into which Nos 1 and 4, and 2 and 3 merge, to give optimum extraction effect with the 1–3–4–2 firing order. A recess in the twin outlet casting takes, if required, a lambda sensor, Section 10.11.

3.67   ADO 15 power unit

The great success of the BMC – now Austin-Rover – *Mini* range heralded the adoption by many other manufacturers of the transverse engine installation, coupled with front-wheel-drive. This layout is extremely compact, especially as regards overall length of the car, so that exacting requirements of adequate space for both comfortable seating and luggage are successfully met.

The 848 cm^3 four-cylinder engine may be regarded as a derivative of the Series A 948 cm^3 unit (see *Automobile Engineer*, Vol. 50, No. 7); the following notes on the special features of the smaller engine are abstracted from a

(a)

(b)

Fig. 3.73 Rover 800 Series engine, (a) with throttle body injection and (b) with multi-point injection with its control unit on the end of the plenum chamber from which the four branch pipes sweep down to the ports

comprehensive description of the complete vehicle in *Automobile Engineer*, Vol. 51, Nos 4 and 5.

The Series A block and head are employed without alteration, a new crankshaft giving the reduced stroke of 68 mm. Thus with a bore of approximately 63 mm, the stroke:bore ratio is 1.08:1 which is nearly square.

To accommodate the larger bore in the limited length of block, siamesing of the cylinders in pairs and offsetting of the connecting rod big ends by 3 mm is resorted to, with no detriment to the running. The big ends of the connecting rods are diagonally split, and the gudgeon pins are clamped in the small-end eye, a construction which, it is considered, contributes to silence as there is only one clearance to develop possible knock.

The sturdy forged crankshaft carries integral balance weights, and the short stroke results in considerable crankpin/journal overlap. All these features make a valuable contribution to smooth running and may be seen in Fig. 3.74.

The compression ratio is 8.3:1, achieved by using taller pistons to compensate for the shorter stroke.

The joint between cylinder block and sump is on the crankshaft centre line, as this affords free access to the big ends and provides a deep and rigid housing for the transmission.

The camshaft is of normal design, driven by a single roller chain of 10-mm pitch. To take the driving load and provide for wear, the front end bearing is a steel-backed white-metal-lined shell, while the other two are of the plain type running in bores machined directly in the block. The Hobourn-Eaton oil pump is driven coaxially by a slot and peg in a counter-bore.

The bucket type cam followers and cup-ended push rods are now common practice, but the rockers are of novel construction, being formed from a pair of light steel pressings by projection and spot welding. This results in a considerable saving in weight as compared with the usual stamping, and may be compared with the pressed form shown in Fig. 3.63. The rocker bearings are Clevite 8 wrapped copper-lead bushes.

Curves of engine performance, measured at the flywheel, are reproduced in Fig. 3.75.

Those shown by full lines are for the engine to BL's standard specification. The gross performance, shown by dotted lines, is measured under conditions similar to standard, except that the air filter is removed and manual controls are employed for the carburettor and ignition.

Chain-dotted lines are used to indicate the performance as installed; the engine specification then differs from that for the standard curves in that the cooling fan is fitted, and the car's exhaust system replaces that of the test-bed installation. The results are obtained with an SU semi-down-draught carburettor.

It is of interest that the gross power output of 27.55 kW at 5500 rev/min is slightly greater than that of the Austin A40, which has a Zenith carburettor and produces 26.85 kW at 4800 rev/min, and less than that of the Morris *Minor 1000*, which is 29.8 kW at 5000 rev/min. The higher specific output of the ADO 15 unit, 32.6 kW/litre as against 28.4 for the A40 and 31.5 for the *Minor*, is mainly attributable to the proportionately larger valves, which have made possible an upward extension of the speed range. While the respective maximum piston speeds of the two 948 cm^3 engines are 1220 cm/s and 1270 cm/s, that of the ADO 15 unit is 1250 cm/s. The ADO 15 has a maximum gross bmep of 910 kN/m^2, about average for an engine of this type.

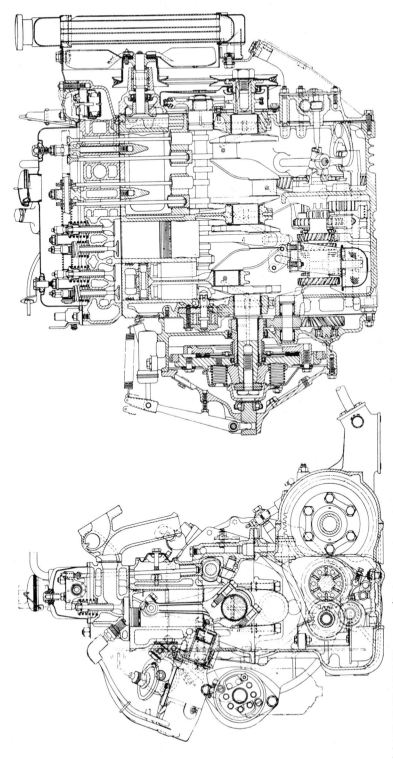

Fig. 3.74 BMC ADO 15 power unit

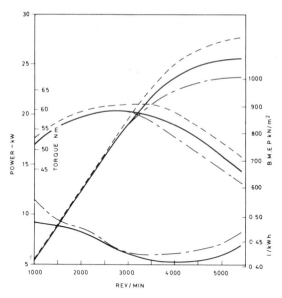

Fig. 3.75 BMC ADO 15 performance curves

3.68 Engine position

For small cars, in which space is at a premium, many manufacturers have abandoned the traditional front engine rear-wheel-drive layout. Before the Second World War, rear wheel drive began to attract attention, early examples including Mercedes experimental models and, of course, the VW *Beetle*. Several rear wheel drive arrangements are illustrated and described in Section 20.2. For cars, the principal advantage is the absence of intrusion of the engine and gearbox into the space available for the passengers. An incidental advantage is good traction, especially uphill, because the engine and transmission mass is approximately over the rear wheels or, in some instances, even behind them.

Disadvantages, however, include a severe tendency to oversteer, difficulty in preventing engine noise from being transmitted into the passenger space, and severely restricted luggage space. The latter arises not only because the engine occupies most of the space in the boot but also because of the limited space available between the front wheels when steered on full lock, and the need for a low bonnet profile for a good forward range of vision.

Front wheel drive, on the other hand, does not suffer any of these disadvantages, though if the engine is installed transversely, to provide maximum space for the occupants behind it, the extra width it occupies may, of necessity, dictate some restriction on the angle of the steering lock. An advantage, however, is that, with two passengers in the back seats and perhaps some luggage in the boot, the weight distribution tends to be equalised between front and rear. Front wheel drive also offers good traction except when pulling away from rest, especially uphill, when the weight transfer to the rear owing to both inertia and the angle of inclination of the vehicle may be a positive disadvantage. A major benefit derived from front wheel drive, however, is the fact that the traction is directed along the line on which the vehicle is steered, an aspect that is fully discussed towards the end of Section 20.3.

The so-called mid-engine installations are basically the front engine layouts, usually of the type in Fig. 20.9 though occasionally of Fig. 20.10, transferred to the back of the car. Where their engines are forward of the back axle, they offer the benefit, especially for two-seat sports cars, of significantly better weight distribution than either the front or rear engine layouts. Noise, heat and access, however, are major problems. All these alternatives to the conventional front engine with rear wheel drive have no propeller shaft, and therefore no obstruction of the floor, which can be an advantage not only for cars but also for some types of commercial vehicle.

3.69   Under-floor engines

The under-floor layout is of course widely employed for public service vehicles. Horizontally-opposed, and even V, layouts have been used for this type of installation, but more commonly a vertical engine is adapted by making special arrangements for all parts and accessories affected by the changed gravitational conditions. Principal among these, of course, are the sump and lubrication system.

Most of the well-known oil-engine builders have developed designs on these lines, and the arrangement would seem to have the merit that maintenance operations of any particular type can be confined to one or other side of the vehicle, with the additional facilities provided by access doors. An example is given in Fig. 3.76, which shows in cross-section the 8-litre, six-cylinder, direct-injection diesel built and operated by the Birmingham & Midland Motor Omnibus Company Ltd.

On the other hand, there is the Bedford YQR chassis, which has a vertically installed engine, under the floor midway between the front and rear axles. The advantages are: quietness, good weight distribution, the floor is flat and clear for placing the entrances wherever desired, and an engine common to a wide range of vehicles can be used. Access for servicing is gained through a hatch in the floor. This layout is suitable only for buses, coaches and, perhaps, pan-technicons and vehicles such as mobile shops.

3.70   Panhard Dyna 55 engine

Many years ago the Panhard company developed a 12-cylinder horizontally-opposed air-cooled engine for aircraft, and it was illustrated in *Automobile Engineer*, Vol. 43, No. 8. Because of its flat layout, it would have been eminently suitable for under-floor installation, possibly in an eight-cylinder form, for commercial vehicles. Although the advent of the gas turbine ruled it out for aircraft, the basic concept survived, but in twin-cylinder form, for a front-wheel-drive car, a racing version of which was successful in its class in the 1953 le Mans race.

A cross-section of the car engine is shown in Fig. 3.77. An interesting feature is the valve gear layout, derived from the 12-cylinder version. Torsion-bar springs and roller cam-followers are used. The rockers are similar to that in Fig. 3.63, but are of cast iron, instead of simple pressings, and the tappets are

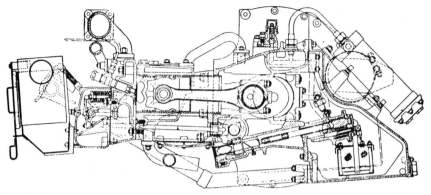

Fig. 3.76

more positively locked. Full details are in *Automobile Engineer*, Vol. 48, No. 4.

With a bore of 85 mm and a stroke of 75 mm, the swept volume is 851 cm^3. The maximum power output is 31.3 kW at 5000 rev/min, while the maximum bmep and torque are respectively 94 kN/m^2 and 63.7 Nm, at 3500 rev/min.

A feature of the aluminium alloy cylinders is that their heads are integral. Dry liners are fitted, and the fins are of generous area for good heat-transfer.

A three-piece crankshaft is employed. One piece is the central web, while the other two each comprise a journal, web and crankpin. After the roller bearing big ends have been threaded on to the pins, the whole crankshaft assembly is completed by pressing the ends of the crankpins into the holes in the central webs. Plugs pressed into the ends of the pins prevent them from deflecting radially inwards under dynamic loading and thus allowing the joints to loosen. An axial hole in each plug is tapped to receive the end of a taper-seating bolt used to tighten a dished pressed steel plug into the other end of the hollow pin, to seal it.

To balance the rotating couple due to the offset of the cylinder axes, masses are secured by cheese-head screws, and located by dowel pins, to the outer faces of the crank webs. These masses also partly balance the reciprocating inertia of the pistons. The whole crankshaft and connecting rod assembly is inserted through a large oval aperture in the front end of the crankcase, which therefore has no horizontal or vertical joints to give trouble by leaking.

3.71   Citroën *Visa* engine

A more modern horizontally-opposed twin-cylinder air-cooled engine is that of the Citroën *Visa* and *LNA* models. This is a 652 cm^3 unit, with a bore and stroke of 77 and 70 mm respectively. Its compression ratio is 9:1 and its maximum power and torque are 26.8 kW at 5500 rev/min and 51.5 Nm at 3500 rev/min.

An unusual feature is the use of three bearings to carry the crankshaft. As can be seen from Fig. 3.78, the third bearing is in fact a support for the rather long nose of the crankshaft, which is necessary for carrying the cooling fan. To have shortened the shaft and brought the fan closer to the engine presumably

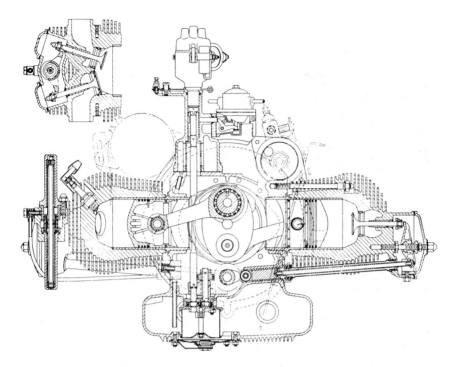

Fig. 3.77 Panhard Dyna engine

would have made it impossible to collect the air efficiently from the whole of the 360° sweep of its blades for directing it over the two cylinders. Unsupported, such a long nose would be liable to vibrate at certain crankshaft speeds, and this would have made the gear drive to the camshaft noisy.

The cylinders, which have separate heads, are of aluminium alloy. Their bores are spray-coated with Nicasil – a ceramet obviously containing nickel and silicon – which was originally developed for the Citroën version of the Wankel engine. There are, of course, no cylinder liners.

The ignition system is entirely electronic. It comprises two proximity detectors, a vacuum sensor, a computer and a coil. As can be seen from Fig. 3.79, the proximity detectors sense the passing of a metallic pin attached to the flywheel, to obtain an accurate indication of both the rotational speed of the engine over the critical period within which the spark must occur, and the datum point relative to which the spark must be timed. The vacuum sensor, connected to the carburettor choke, gives an indication of induction air flow and therefore engine load. All three sensors pass signals to the computer, which calculates the spark timing required. The computer also regulates the primary current to the coil, to ensure that the secondary voltage is appropriate under all conditions of operation.

With this system, there are no adjustments to be made. Nor are there any moving parts subject to wear. Consequently, the shape of the advance curve remains constant throughout the life of the engine. Moreover, this curve can be

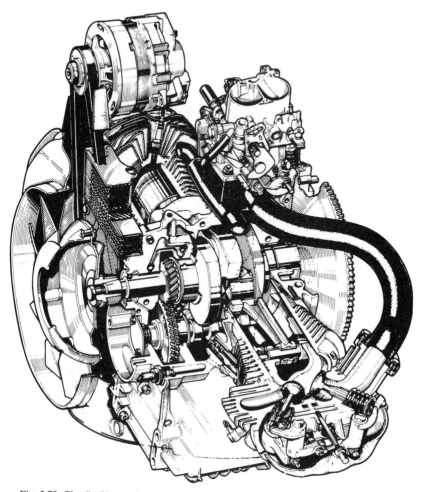

Fig. 3.78 Citroën *Visa* engine

exactly matched to the continually changing requirements of the engine. The outcome is low fuel consumption, minimum exhaust gas pollution, and good low-speed torque and acceleration.

3.72   Ford V-four engines

With the adoption of the over-square cylinder proportions, together with the need for longer main bearings to cope with higher ratings, there has been a tendency for in-line engines to become long. This, and a demand for low bonnet profiles – to improve forward range of vision and hence safety – has led to the adoption of the V-four layout by some manufacturers. Besides compactness, there are also advantages to rigidity of the cylinder block, crankcase and of the short crankshaft. Moreover, because of the wide spacing of the cylinders, there is plenty of room for coolant passages around them and for ample separation of the hot regions around the exhaust valves.

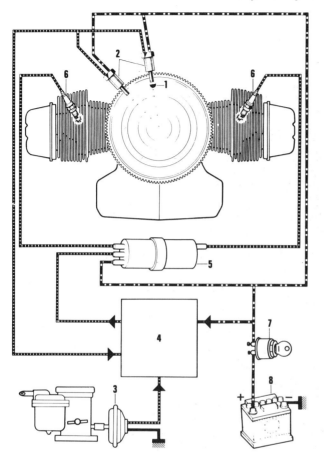

1 Peg on flywheel
2 Proximity sensors
3 Vacuum sensor
4 Micro-processor
5 Ignition coil
6 Spark plug
7 Ignition switch
8 Battery

Fig. 3.79 Citroën Visa electronic system

A good example is the Ford range, Fig. 3.80, comprising the 1662 cm³ and 1995 cm³ *Corsair* engines, with common bore dimensions of 93.66 mm and strokes of 60.3 and 72.4 mm, respectively. To enable connecting rods common to the whole range to be fitted – centre-to-centre length 143.3 mm – the gudgeon pin of the engine with the longer stroke is 60.3 mm closer to the crown and the piston has a shorter skirt than that of the other. Bowl-in-piston combustion chambers have been adopted, and the lower face of the cylinder head is entirely flat, except for shallow machined recesses around the valve heads. The combustion chambers are completely machined and therefore of accurately controlled form, finish and volume. Local recesses are machined in the cylinder crowns, to clear the valves at and near top dead centre.

The spheroidal graphite cast-iron crankshaft is carried in three 63.5 mm diameter aluminium-tin or copper-lead bearings. Crankpins 60.2 mm diameter are employed.

Between the cylinder banks, the angle of the V is 60°. With a firing order of 1R 3L 4L 2R, this gives equally spaced firing intervals and the primary forces are in balance, but a secondary couple remains. This couple is balanced by

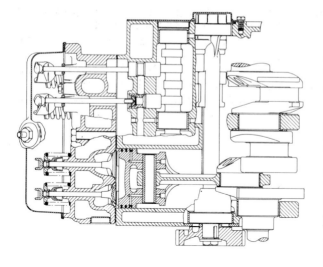

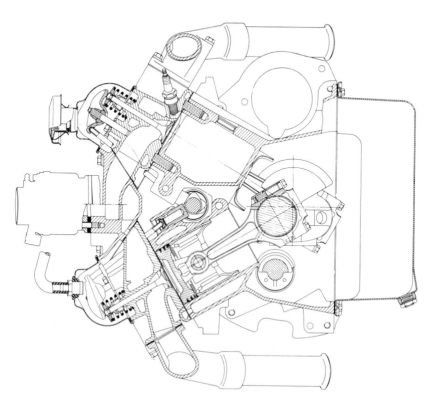

Fig. 3.80 Features of the Ford *Corsair* V-four engine include the balancer shaft, a shallow bowl-in-piston combustion chamber, and push-rod actuation of the valves

weights on a separate shaft, parallel to and driven at the same speed as the crankshaft, but in the opposite direction, Fig. 3.80. A 60° angle, of course, is ideal with the V-six engines, the blocks of which are made on the same production line. The balance of this engine is dealt with in Section 4.11.

In standard form, both engines have the Autolite 36 IV down-draught carburettor. There is, however, a GT version of the larger engine, which is equipped with a Weber 32/36 DF AV progressive choke twin down-draught carburettor, the two chokes feeding into a common riser bore.

Chapter 4

Six-, eight- and twelve-cylinder engines

The desire for a more uniform turning moment and better dynamic balance has led to the production of six-, eight- and even twelve-cylinder engines where questions of first cost are not of primary importance. The effective driving effort of an ordinary four-stroke single-cylinder engine extends, when friction is taken into account, over only about 150° of crank movement, and, as in a four-cylinder engine the power strokes are at intervals of 180°, or half a revolution, it follows that there are appreciable periods when the shaft is receiving no useful torque whatever, and the stored energy of the flywheel is called on to supply the deficiency. At low speeds the flywheel does not contain sufficient energy to tide over the idle periods without undue variation of speed and hence jerky running and high transmission stresses are set up.

Remembering that the number of power strokes per revolution is equal to half that of the cylinders, it will be seen that in a six-cylinder engine, the intervals between power strokes, if the cylinders and cranks are so disposed as to render them equal, have a duration of 120° of crank angle, while in an eight-cylinder engine the interval is 90°. There is thus an appreciable overlap in the former and a very considerable one in the latter. The result is great improvement in the smoothness of running and a lighter flywheel may be employed.

4.1 Six cylinders

Fig. 4.1 illustrates the normal arrangement of cranks in a six-throw crankshaft, the full number of bearings, namely seven, being here shown. The cranks are 120° apart in pairs, the halves being arranged to give 'looking-glass' symmetry, in order to give opposition of the inertia couples as explained in Chapter 2. The circle diagram shows the two alternative crank arrangements consistent with the above symmetry.

In what may be called a left-hand shaft, No. 1 crank is on the left when looking from the front of the engine, with 3 and 4 vertical. Cranks 1 and 6 may clearly be interchanged with 2 and 5 giving a shaft of the opposite hand. Both these arrangements are met with in practice.

4.2   Dynamic balance

Just as three equal and symmetrically distributed revolving masses in one plane will be in dynamic balance, since their combined centre of gravity is clearly at the centre of rotation, so six equal and symmetrically distributed revolving masses which do not all lie in one plane will be in dynamic balance provided that the couples due to the distances between the planes of revolution cancel out. This cancelling out is attained with the arrangement illustrated, provided that the corresponding throw spacing are the same and the throws uniform. Dynamic balancing of crankshafts is carried out by the majority of makers to correct for slight errors in weights and sizes of the various throws.

In like manner the primary reciprocating effects balance, since these represent the vertical components of corresponding revolving effects. This will perhaps be realised better if *standing* balance is considered, where the only forces are vertical ones due to gravity. A six-cylinder engine, even if quite frictionless, will remain at rest in any position of the crank provided the moving parts of all cylinders are uniform in weight. Further, the secondary effects cancel out, as may be ascertained from Fig. 2.9, by a careful combination of the forces in correct phase. The six-cylinder engine is, in fact, capable of complete balance for both primary and secondary forces and couples, thus eliminating vibration from this cause.

4.3   Firing order

To obtain good *distribution* of the fuel to all cylinders it is desirable and usual to arrange the inductions alternately in the front and rear halves of the engine. An examination of the circular diagram of Fig. 4.1 will make clear that the possible firing orders, to satisfy this condition, are 1 5 3 6 2 4 and 1 4 2 6 3 5 for the two arrangements of cranks. These are the most usual in practice, but others are sometimes met. Thus, from the point of view of torsional oscillation, it may be desirable in some cases to have two cylinders in one half of the engine firing in succession in order to prevent the 'unwinding' of the shaft between power impulses. This problem is referred to later.

4.4   Eight cylinders

Here the proper interval between power strokes is 90°. A 'straight eight' may

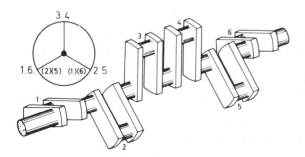

Fig. 4.1

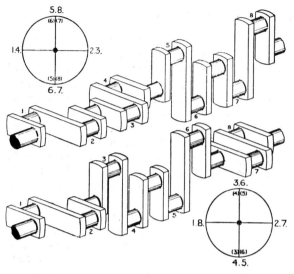

Fig. 4.2

clearly be formed by placing two fours end to end with their flat crankshafts at right angles, as in the upper diagram of Fig. 4.2. Each set of four cylinders is then in balance for primary forces and couples as explained in Chapter 2, but there remains the secondary effect. This secondary effect acts downwards when the cranks are horizontal and upwards when they are vertical; thus, while the downward secondary effect of one half of the engine opposes the upward effect of the other half there remains, with the 'double four' arrangement, an unbalanced couple, tending to pitch the engine in the longitudinal plane.

This may be eliminated by adopting the 'split four' arrangement shown in the lower diagram of Fig. 4.2. Here the downward secondary effect of two pistons at each end of the engine 1, 2 and 7, 8 opposes the upward effect of 3, 4, 5 and 6 without introducing a couple. This arrangement is in complete balance for primary and secondary forces and couples.

4.5　Firing order

There are clearly many possible orders of firing with eight cylinders arranged as shown in Fig. 4.2, but the more usual are 1 7 3 8 4 6 2 5, 1 5 2 6 4 8 3 7 and 1 6 2 5 8 3 7 4. Other crank arrangements are met with in special cases, particularly for racing purposes, where considerations of bearing loads may be of greater importance than exact dynamic balance. With adjacent cranks in phase, as in the case of 4 and 5 in the lower diagram (Fig. 4.2), the centrifugal and inertia forces of the moving parts of both cylinders act always together, producing very heavy loads on the intermediate bearing. If adjacent cranks are placed 180° apart as far as uniformity of turning moment permits, the mutual balance of opposing cranks is more direct, and bearing loads and crank case stresses are much reduced. Such arrangements result, however, in there being unbalanced couples.

4.6 Balanced webs and torsional oscillation

Another method of reducing bearing loads and crankcase stresses is to balance the individual throws for rotational effects by adding extended webs and balance weights, it being impossible, except in the 90° V engine, to deal with reciprocating effects in this way. But, as previously mentioned, while this expedient may cure shaft and crankcase *whip*, it is very liable to increase the risk of torsional oscillation of the shaft, because heavy masses revolving with a slender shaft make the natural torsional or twisting vibration a relatively slow one – slow enough in some cases to give rise to dangerous and uncomfortable resonance with the torque variations. An aircraft shaft of relatively slender proportions, with such attached balanced weights, is shown in Fig. 4.3.

To reduce torsional vibration, a vibration damper may be mounted at the front end of the crankshaft. A damper of the Lanchester type is illustrated in Fig. 4.4(a). It consists essentially of a flywheel member A, of 20 to 30 mm diameter, according to the size of engine, driven through a friction disc clutch, the inner plate member B of which is keyed to the front end of the crankshaft. The friction is adjusted by the springs acting on the presser plate C so as to be insufficient to transmit to the heavy rim A the high torsional acceleration during the winding up and unwinding of the crankshaft as the torsional impulses are applied to the cranks. Thus the torsional wind and rebound are absorbed in friction in much the same way as in the case of a spring damper,

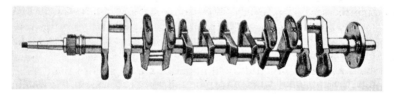

Fig. 4.3 Crankshaft with balanced throws

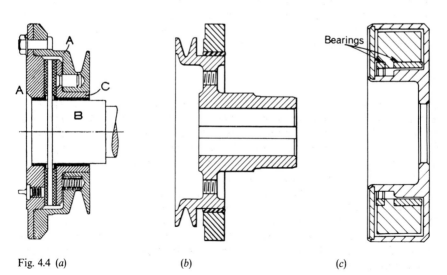

Fig. 4.4 (a) (b) (c)

and serious resonant vibrations are suppressed, reducing noise, premature wear and failure of chain drives, tensioners and even, possibly, the crankshaft.

Fig. 4.4(*b*) shows a simple bonded rubber insert type of damper. A more recent development in the UK is the introduction of dampers in which the rubber is held in place by compression between the inertia ring and the hub, instead of by bonding. Dampers of this type have long lives, and timing marks and V-belt grooves can be incorporated on the inertia rings. Rubber dampers are, in general, limited to engines of about 13 litres swept volume or less.

An alternative damper is the viscous type which, though more expensive than the rubber type, is extremely reliable. It has been described in *Practical Solution of Torsional Vibration Problems*, Vol. 4, by Ker Wilson (Chapman & Hall). Basically the unit comprises an annular flywheel, sealed in a metal casing. Fig. 4.4(*c*). The space between the two is filled with a silicone fluid of high viscosity. Since the casing rotates and oscillates with the crankshaft, while the flywheel tends to maintain a steady-state motion, there is a considerable viscous drag in the fluid between the two: thus, energy of vibration is dissipated as heat.

4.7   Difficulties met in design

Sixes and straight eights are liable to three important troubles in running and performance. These may be summarised as bad gas distribution, high fuel consumption and torsional oscillation. The second is partly a consequence of the first and partly due to the rather lower thermal efficiency obtainable in cylinders of small size, owing to the relatively greater losses to the jackets and other causes. Bad distribution is one of the most elusive difficulties with which designers have to contend. Although a correct and homogeneous mixture may leave the carburettor, it is highly improbable that it will reach all the cylinders in exactly equal quantity or of similar composition. Thus it may be necessary to employ an unduly rich carburettor setting, with correspondingly high consumption, in order to prevent starvation of certain of the cylinders. This trouble is present even in four-cylinder engines, and is accentuated in sixes and eights, where the induction manifolds present far greater difficulties. It is almost impossible for a designer to form a reliable forecast of the probable behaviour of any given arrangement of single or twin carburettor, and induction pipe arrangement, taken in conjunction with a particular firing order.

The third difficulty of torsional oscillation has already been referred to. These difficulties and the extra cost of manufacture of the six have enabled the reliable and economical, though less smooth-running, four-cylinder engine to maintain its popularity in the smaller powers, and for medium-sized commercial vehicles.

4.8   Humber *Super Snipe* engine

The layout of this engine (Figs 4.5 and 4.6) – like that of the Armstrong Siddeley *Sapphire, Automobile Engineer*, December 1953 – represented an intermediate stage in the trend of development from low to overhead camshaft. An interesting, but expensive, alternative is that of the Lea Francis 2.5-litre engine, *Automobile Engineer*, June 1952.

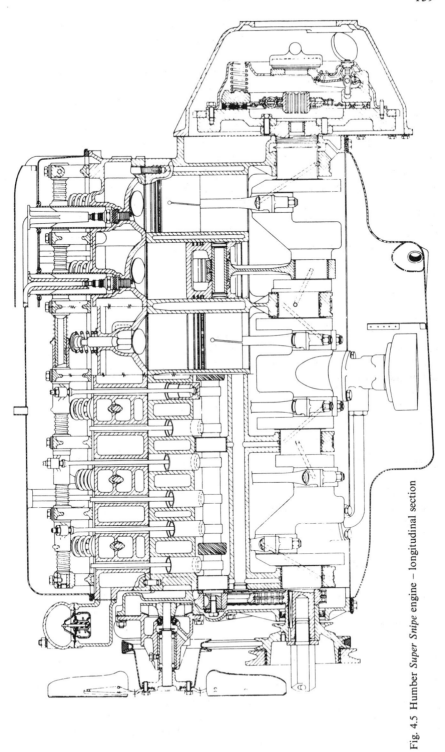

Fig. 4.5 Humber *Super Snipe* engine – longitudinal section

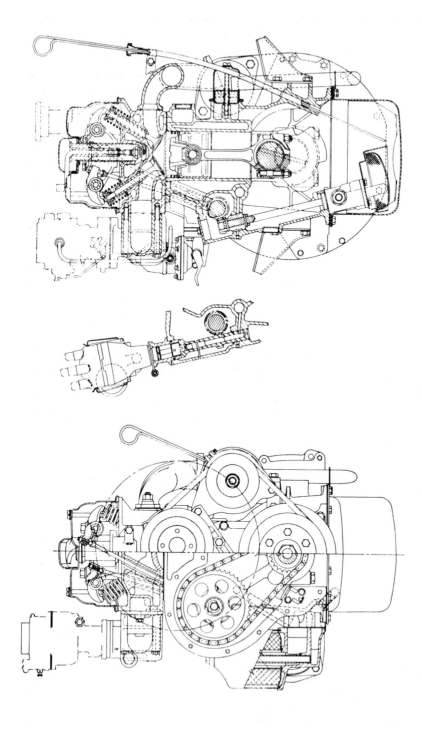

Fig. 4.6 Humber *Super Snipe* engine – transverse section

For the *Super Snipe* 2965 cm³ unit, the over-square bore/stroke arrangement was chosen to give a lower piston speed of 13.2 m/s at maximum power. It also permits a crankshaft with overlapping pins and journals, thus ensuring a very rigid crank free from torsional weakness. Both these points contribute to the smooth running of the engine. The combustion chamber is of the hemispherical type, shallowed down to give a flat topped piston for the chosen compression ratio of 8 : 1. With the shallow type of head, smooth combustion is assured as the flame front build-up is quenched by the piston top before roughness is initiated.

Large valves (3.556 mm inlet throat diameter, 3.048 mm exhaust throat diameter) are possible, and with the uninterrupted gas flow characteristic of this hemispherical type of head, good filling is achieved. The low expansion aluminium alloy piston has two compression rings and a scraper, the top ring being chromium plated, the second ring stepped.

Both rings are 1.98 mm thick and 4.06 mm wide, giving the fairly high wall pressure necessary with over-square engines. The 23.8-mm diameter gudgeon pin is fully floating and located with circlips.

A high camshaft position is a fundamental requirement of this type of engine. Short pushrods are necessary for high speed of operation to obtain high power output. Moreover, they have to be inclined to actuate rockers on two shafts – the alternative to overhead camshafts for actuating inclined valves in hemispherical heads. The camshaft layout adopted has the advantages of a relatively short drive and ease of tappet adjustment. Details of the valve timing – for good torque at low speeds and high power output at high speeds – are: inlet opens 20° BTDC, closes 46° ATDC; exhaust opens 52° BBDC, closes 14° ATDC.

A water-heated aluminium alloy inlet manifold is used to feed two separate galleries formed in the cast iron cylinder head. Each gallery feeds three cylinders.

The performance figures shown in Fig. 4.7 are obtained with a single Zenith 42 WIA down-draught carburettor.

Gross power output is 96.7 kW at 4800 rev/min giving 32 kW per litre and maximum torque of 219 Nm developed at 1800 rev/min.

The general arrangement drawings of Figs 4.7 and 4.6, and the performance curves, Fig. 4.7, give a very clear picture of the power unit, which may be amplified by the following notes.

The cylinder block is cast iron and the engine mounting point is rearward of the front face and therefore nearer to the centre of gravity of the unit. This, coupled with the deep section of the block, provides a rigid and vibration-free unit when installed.

To reduce noise and wear, the camshaft and bucket type tappets run in an oil bath. Particular attention has been paid to the valve gear to avoid unnecessary deflections. Short solid push rods operate cast iron rockers on induction-hardened shafts.

The four-bearing forged crankshaft has integral balance weights and bearings of 63.5 mm diameter, the crankpins being 50.8 mm diameter, giving a valuable crankpin/journal overlap of 15.875 mm. A Metalastik torsional damper is fitted at the front end.

The Hobourn–Eaton oil pump (see Section 16.18) has a delivery of $2\frac{1}{2}$ gallons per 1000 revolutions. A separate positive feed is taken to both the oil

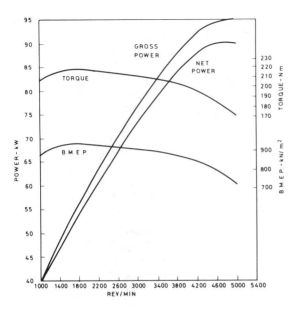

Fig. 4.7 Humber *Super Snipe* performance curves

pump drive gear and the distributor drive gear. This drains back to the camshaft gallery together with the rocker surplus. Standpipes and cast-in weirs maintain the level in the gallery and the overflow returns to the sump.

4.9   Jaguar AJ6 engine

When the oil crises of the early nineteen-seventies posed a threat to their V12, Jaguar originally set out to, in effect, remove two cylinders from each block to produce a V8 on the same machining lines. However, the characteristic secondary out-of-balance forces of the V8 could be counteracted only by the incorporation of a balance shaft, which so changed the shape of the crankcase that the enormous cost of introducing an entirely new one would have been entailed. Next, they produced, in 1971, prototypes of a slant six based on the V12. However, the power output of the resulting 2.6-litre unit was inadequate, and measures to remedy this by increasing the stroke from 70 to 90 mm would have made the blocks so tall that it could no longer be machined on the V12 line.

Following experience, in 1972, of a head with four valves per cylinder for the high performance version of the V12, a similar conversion was made, by 1975, of the XK engine to produce a 3.8-litre six-cylinder unit. However, although the head was new, the block remained an old design and would have restricted further development so, by 1976, Jaguar decided to start again and invest in a new machining line so that their design team could, beginning with a clean sheet of paper, produce an engine of modern design with potential for considerable further development, Figs 4.8 and 4.9. The original aim was two versions of a light alloy in-line six, one with the May head, Section 4.18, for emphasis on fuel economy and the other with a four-valve head and emphasis on high performance.

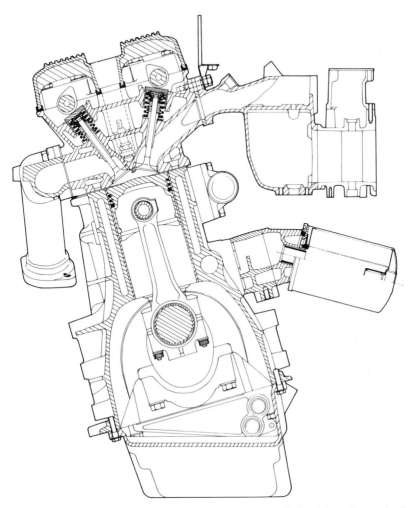

Fig. 4.8 The Jaguar AJ6 was the first six-cylinder, in-line, all-aluminium alloy engine in the world to go into quantity production

Although the resultant engine, the AJ6, was the first aluminium alloy in-line six-cylinder engine in the world to go into quantity production, the designers were able to make good use of their earlier experience with this material in their V12 unit, Section 4.17. The outcome was that the 2.9- and 3.6-litre versions of the AJ6 are, respectively, 25% and 30% lighter than Jaguar's earlier 3.4- and 4.2-litre XK engines.

Adequate under-bonnet clearance for the wide four-valve head and induction system of the AJ6 could be obtained only by installing the engine 15° from the vertical. Originally a toothed belt drive was employed for the camshaft, but, with its larger diameter pulleys, it not only added to the height of the installation but also made the engine unnecessarily long. Consequently, a sturdy twin-strand chain drive similar to that of the V12 was adopted. Although the seven bearings of the crankshaft endow it with great potential for

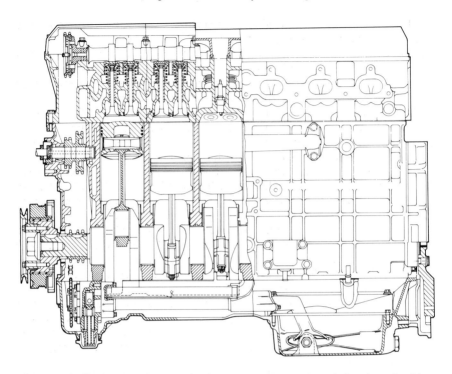

Fig. 4.9 Transverse section of the Jaguar AJ6 engine, which has been designed to take either the four-valve crossflow or May lean-burn combustion system head

future development, they also add to the length of the new engine, so to compensate for this, the oil pump was removed from the front and housed in the sump, where it is driven by a single-strand chain from the nose of the crankshaft.

The pump draws oil from a deep well at the rear of the sump, a baffle being interposed between the path swept by the big ends and the surface of the oil to prevent the lubricant from being churned up by them. An oil cooler and full flow filtration are features of the lubrication system. Sealant, applied in liquid form, is used for the metal-to-metal joint faces except the cylinder head and exhaust manifold, where conventional gaskets are employed.

The crankcase and cylinder block casting for the 2919 cm^3 engine is common to that of the 3590 cm^3 unit, the latter taking the 24-valve head and the former the 12-valve head with the May combustion system. Their manifolds of course differ, but in most other details the 2.9 is virtually identical to the 3.6-litre unit. Although the crankshafts are similar and have bearings of the same sizes, that of the 2.9-litre unit has a crank throw of 37.4 mm, giving a stroke of 74.8 instead of 92 mm: the 91 mm bore is common to both, so the 3.6 has virtually square and the 2.9-litre engine over-square bore : stroke dimensions. The power outputs and maximum torques of the 2.9- and 3.6-litre units are respectively 123 kW and 239 Nm, at 5600 and 4000 rev/min, and 165 kW and 337 Nm at 5000 and 4000 rev/min.

For the V12, an open-top deck design, with wet cylinder liners, was adopted for the cylinder block and crankcase, because it was originally intended to be diecast, though this idea was later abandoned. However, for the AJ6 gravity diecast unit a closed-top deck was adopted for greater inherent rigidity, and the use of thin-walled shrink-fit cylinder liners also saved weight. By virtue of the good thermal conductivity of the aluminium, the siamesed housings for the dry cylinder liners gave no trouble as regards thermal distortion.

On the production line there are two infra-red ovens through which the cylinder block castings pass for heating in two three-minute stages. As they emerge from the second oven the cast iron liners are slid in manually instead of with the more usual hydraulic press, after which they pass through a cooling tower so that the temperature can be reduced in a strictly controlled manner. Later, they are plateau-honed but left with a cross-hatch finish, Section 4.12, to give them the necessary oil retention priorities. The bore sizes are then checked and graded so that pistons can be allocated to them selectively.

During development a change was made from a forged steel to a nodular iron crankshaft, to reduce production costs and facilitate balancing. The accurate shell-moulding casting process used produced components requiring considerably less machining than is necessary on a forged shaft. Increasing the number of balance weights from four to eight reduced the internal couples and associated loading on the main bearings. On the production line, adjustment to the balance is effected by drilling up to four holes radially into each weight. Low temperature nitro-carburising hardens the journals. The bearings are 76.23 mm diameter, the same as those of the V12, and their lengths range from 24.33 to 30.48 mm.

Forged steel connecting rods, 166.37 mm from centre to centre, carry fully floating gudgeon pins on which are diecast aluminium pistons of the thermally compensating strut type. Each carries two compression rings and a Microland oil control ring. For Europe, the pistons in the 3.6-litre engine have slightly domed crowns, giving a compression ratio of 9.6:1, while engines to be exported to the USA have slightly dished crowns. The pistons in the 2.9-litre high-compression lean-burn engine give a compression ratio of 12.6:1.

A gravity diecast head is used for both versions of the engine: the 3.6 unit is of the crossflow type, and the 2.9-litre head is similar to that of the V12 with the May system. For the 3.6 unit the valves, each closed by a single spring, are actuated by twin, seven-bearing, cast iron camshafts with chilled cams. These shafts bear directly in their bores in the aluminium tappet blocks, and the tappets are of the squeeze-cast steel, inverted bucket type. On the assembly line, an electronic indicator presents a reading on a screen to inform the operator what thickness of shims he should use for setting the clearances for each of the valves.

The 35.3 mm diameter inlet valves, set 24° from the vertical, are of En52, while the 31.6 mm diameter exhaust valves, at 22.5°, making the included angle 46.5°, are of 21-4-N alloy. By virtue of the four-valve layout it was easy to site the sparking plug close to the centre of the double-penthouse combustion chamber.

A two-stage duplex chain drives the camshafts. The hydraulically tensioned primary drive is taken from a 21-tooth sprocket on the crankshaft to an intermediate one having 28 teeth. This, in fact, is a twin sprocket, the front portion of which has 20 teeth from which a secondary drive, again hydraulically tensioned, is taken round the two 28-tooth half-speed wheels, one on the front end of each camshaft. An auxiliary driveshaft is also driven by the primary chain. Near its front end is a spiral gear drive for the ignition distributor, while its rear end drives the pump for the power-assisted steering.

Since an electric drive for the eleven-blade fan would have made heavy demands on the battery it is belt-driven through the medium of a viscous coupling from the nose of the crankshaft. This coupling freewheels at medium speeds and over, so that neither power loss nor noise is generated when the car is cruising, when the natural flow of air through the radiator is adequate without fan assistance. A thermostatic control progressively regulates the characteristics of the coupling according to the temperature of the air flowing through the radiator.

For the 3.6-litre unit a Lucas electronic engine-management system is used, in which the fuel injection is regulated by a by-pass air flow, hot wire type meter. On the 2.9-litre engine, however, a Bosch system having a full-flow hot wire air flow meter is employed. In both instances, the electric current required to maintain the wire at a constant temperature is measured, as an indication of the quantity of air flowing past it. This type of meter, in contrast to most other meters, including venturis, which give indications of only the velocity, and therefore volume, of flow is sensitive to both the density and velocity, and therefore mass, of the air flowing through.

In the electronic control unit there is a 4-kilobyte memory. This control instantly compensates the idling power output to match changes of load, such as might be occasioned by the switching in of the air conditioning system or an electric heater for the rear screen, so that the idling speed can be set lower than would otherwise be practicable. The system also ensures that the mixture is exactly right for cold starting and driving away, as well as for transient conditions during warm-up. Overrun fuel cut-off is taken out of operation progressively to reinstate the drive without jerkiness. The minimum speed at which this cut-off comes into operation is 1500 rev/min, and reinstatement is effected either when the throttle is reopened or when the speed falls to 1050 rev/min.

4.10   Rover 2.3/2.6-litre E series engines

In this design, the aim has been to reduce the number of components to a minimum consistent with the achievement of a high level of performance and reliability. The alternative swept volumes have been obtained by the use of crankshafts having different throws. Consequently, the bore is 81 mm for both, while the strokes are 76 mm for the 2350 cm^3 version and 84 mm for the 2597 cm^3 unit. Their compression ratios are common at 9.25:1, the pistons having dimensions different from their crowns-to-gudgeon-pin axes. These engines are illustrated in Fig. 4.10, and their performance curves reproduced in Fig. 4.11.

The power outputs are 91.7 kW and 107 kW at 5000 rev/min, while the torques are respectively 182 Nm at 4000 rev/min and 203 Nm at 3750 rev/min. In both instances twin SU HS6 carburettors are employed.

Commendable features of the design are the head and the block. The layout of the low pressure diecast aluminium head is in many respects similar to that of the Triumph *Dolomite Sprint* but since high, rather than sporting, performance was required, only two valves per cylinder are used. Each pair is actuated

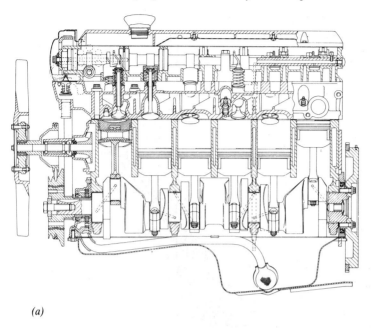

(a)

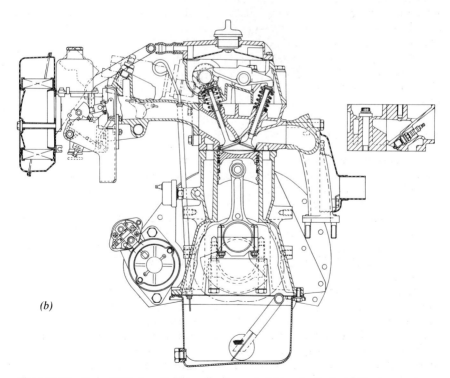

(b)

Fig. 4.10 The Rover 2600 six-cylinder engine shown in (a) longitudinal and (b) cross-sections

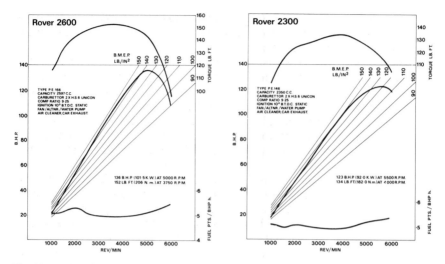

Fig. 4.11 Performance graphs for the Rover 2600 and 2300 six-cylinder engines. Net (DIN) installed condition

by a common cam, which bears directly on an inverted bucket type tappet over the inlet valve and on a rocker for actuation of the exhaust valve.

The rockers in this engine are of cast iron, so their bearing properties are such that no special provision has to be made for the lubrication of their ends, splash and mist being adequate. Valve clearances are adjusted – as on the *Sprint* and O-Series engines – by pallets, or thick shims, seating in recesses machined on top of the spring retainers.

The included angle of the valves is 40°, which gives a compact penthouse combustion chamber with good crossflow characteristics – the incoming wave of fresh mixture helps to sweep the burnt gases before it, out through the exhaust. Shallow dished piston crowns form the bottom portions of the combustion chambers. With such a large included angle between the valves, there is also plenty of space between the ports for the provision of adequate water cooling. Both the inlet and exhaust valves have identical springs except in that, to cope with the inertia of the additional mass of the rockers, a smaller supplementary spring is fitted inside the main one on each exhaust valve. The valve-head diameters are: inlet 42 mm, and exhaust 35.6 mm.

For simplicity of assembly and servicing, the camshaft bearings and tappet slideways are machined in a single high-pressure aluminium diecasting. Because of the precision with which this type of casting can be produced, oilways can be accurately cored in it and machining is reduced to a minimum. Unlike that in the *Sprint*, this casting extends the full width of the head, so the structural stiffness of the assembly is greater. It is bolted down on to a liquid sealant to prevent oil leaks. Having no flexible gasket in this joint face obviates the risk of tappet clearances closing due to settlement in service.

A steel-cored composite gasket is interposed between the head and cylinder block. The fourteen 12-mm set bolts that secure the head are tightened by the SPS Technologies system. With this system a machine with an electronic control and air motors tightens the bolts, sensing both the angle through which each is turned and the torque applied, stopping the tightening the instant the bolt begins to yield. A full explanation can be found in *Automotive Engineer*, Vol. 3, No. 3, 1978. Because the bolts are tightened so accurately, no retightening is required in service.

A toothed belt drives the cast iron camshaft, which runs in seven 51-mm diameter bearings line-bored directly in the diecast carrier. Since the contact breaker and distributor, mounted horizontally, is driven by a spiral gear machined on this shaft, it is high up and therefore extremely accessible.

The cast iron exhaust manifold is secured by two bolts per port to the head. No gasket is used, so a common source of failure has been eliminated.

A cast iron crankcase and cylinder block has been employed to give the engine a good degree of rigidity and durability. Since the contact breaker and distributor unit and camshaft are on the head, the Hobourn-Eaton multi-lobe eccentric rotor oil pump is on the front end of the crankshaft, and the water pump centrally positioned above it, the crankcase layout is symmetrical. This obviates thermal distortion problems, facilitates both casting and machining and contributes to rigidity and noise reduction. Uncommonly large vertical ribs on the outer faces of the water jackets, and extension of the skirt well below the level of the crankshaft further stiffen the unit. A plate extending rearwards from the 6-litre sump is bolted up to the clutch bellhouses to prevent sag of the engine and transmission unit.

An En16T forged steel crankshaft with four main journals is employed. Had seven journals been incorporated, it would have been possible to use a lighter shaft but, now that low fuel consumption is an important consideration, the additional friction and oil drag would have been unacceptable. The 50.5-mm diameter pins and 70.38-mm journals are induction hardened. They run in copper-lead bearings the surfaces of which are lead-indium flashed to assist running in.

A damper comprising a rubber spring element and a tuned annular mass is fitted to the front end of the crankshaft. Grooves are machined in the periphery of the annulus to take the V-belt drive for the fan and water pump, and to the alternator, which therefore form part of the damping system. To limit the power absorbed by the fan, it has a viscous coupling.

4.11   Ford V-six range

The transverse section of this engine is virtually identical to that of the V-four range, Fig. 3.80, except that there is no counterbalance shaft and the inlet manifold is, of course, different. Obviously, too, the considerations that led to the adoption of this layout are similar to those set out in Section 3.72, the need for compactness as regards overall length being even greater.

As in the V-four range, there are two sizes of V-six. They are the 2.5- and 3-litre units, and their bore and stroke dimensions are respectively the same as those of the 1662 and 1995 cm^3 V-four engines – 93.66 mm bore by 60.3 and 72.4 mm stroke. For the 3-litre unit, the mean piston speed at 4750 rev/min, maximum power, is 1145 cm/s.

How complete primary balance is obtained is illustrated in Fig. 5.12, showing the disposition of the counter-weights, the arrangement of the six crankpins, the rotating couples, and a vector diagram. The cylinders are numbered 1 to 3 from the front in the right-hand bank, and 4 to 6 in the left-hand bank, so the firing order is 1 4 2 5 3 6. Since this engine has a six-throw

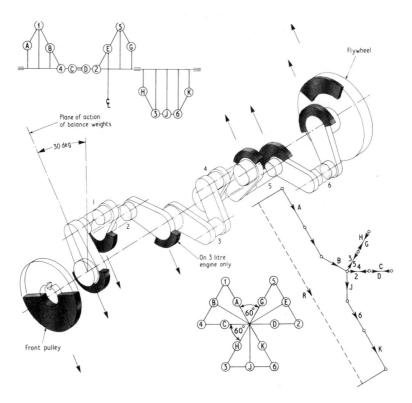

Fig. 4.12 Disposition of the balance weights on the crankshaft of the Ford V-six engine, with a diagrammatic layout of the arrangement of the crankpins (numbered 1 to 6) and the counterweights (lettered A to K). Top left is a diagram of the rotational couples; bottom right is a vector diagram of the couples

crankshaft, the cylinders in the right-hand bank are offset forward relative to those in the other.

The 2.5-litre engine has an Autolite 38 IV down-draught carburettor, with an automatic choke unit sensitive to coolant temperature. On the other hand, the 3-litre engine has a Weber 40 DF AI twin venturi down-draught carburettor: each venturi serves a three-branch section of the induction manifold. At the top of the riser bores is an air balance slot. A W-shape diffuser in the riser bores helps to promote uniform distribution of the charge. Again, an automatic choke unit sensitive to coolant temperature is employed. The use of down-draught carburettors offsets to some extent the advantage of low overall height of the engine – most in-line engines no have side-draught carburettors – but very thin pancake-shaped air filters and silencer units have been developed for this type of layout. A detailed description of this engine was published in the May 1966 issue of *Automobile Engineer*.

4.12   The V-eight

An objection brought against the straight eight engine, in addition to the liability to torsional oscillation of the crankshaft, is its great length, and an alternative arrangement of eight cylinders is the V-eight, employing two banks of four cylinders each, at right angles. This arrangement, long known and used in various applications, has established an overwhelming preponderance over alternative types in the USA for automobile purposes.

Following earlier pioneer designs, large-scale development was initiated by Cadillac in 1914, and aircraft and tank development produced air-cooled types. The flat or single plane crankshaft was used in these early constructions, but in 1926 Cadillac, and in 1932 Ford, introduced the 90° arrangement, the improved balance of which is described below. Side valves gave place to overhead valves from 1949 onwards.

The early Ford 30 hp and 22 hp engines are well known in this country, but their power output proved rather higher than the general demand required, with the result that the popular Ford models, as indeed of all makes, in Britain subsequently had the in-line engines. The situation in the USA has been different. The insatiable demand for increased power, and the earlier availability of 100 octane fuels, has resulted in the production by all the big groups of overhead-valve V-eight engines of rated maximum outputs from 160 to 300 hp, with compression ratios from 8 : 1 upwards, that of the Packard 300 hp being 10 : 1.

The cylinder dimensions are under square, with few exceptions, averaging about 95.25 mm bore and 82.55 mm stroke, giving a piston displacement of about 4.75 litres.

4.13 Balance and firing intervals of V-eight

With the single plane crankshaft, the 90° firing intervals are obtained by the disposition of the cylinders in two banks at right angles, whereas with the two-plane shaft, four of the intervals are due to cylinder disposition and four to crank arrangement.

The flat crankshaft is a simpler and, therefore, with comparable production methods, a less expensive form to make, but the dynamic balance of the engine is inferior to that obtained with the right-angle disposition of cranks. The former arrangement is, for balancing purposes, treated as two ordinary four-cylinder engines sharing the same crankshaft, each set of four pistons being self-balanced for primary forces and couples, while the secondaries remain unbalanced in each bank. This gives a combined resultant secondary force for the whole engine which is zero in the 'vertical' direction, but has, in the 'horizontal' direction a value 40% greater than that corresponding to one set of four pistons since the horizontal components combine in the ratio $\sqrt{2}:1$, while the vertical components neutralise each other.

When the right-angle disposition of adjacent cranks is adopted, the engine is treated for balancing purposes as four 90° V twins, and the primary forces are counteracted by means of revolving masses in the manner described in Section 2.2. The combined primary reciprocating effect of the two pistons, operating on the same crankpin and with their lines of stroke at right angles, is equivalent to the mass of one piston revolving at the crankpin, and the balancing problem is reduced to that of a revolving system.

In the V-eight crankshaft illustrated in Fig. 4.13 the thinner webs adjacent to the journals may be regarded as circular disc webs each corrected to neutralise half of the actual revolving mass at each crankpin, that is, half the pin and one of the big ends.

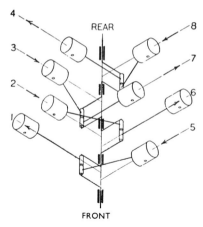

Fig. 4.13 Diagram of V-eight

The heavy masses B_1 and B_2 each incorporate in effect two of these corrected disc webs, together with further masses to balance both the adjacent *equivalent* revolving masses representing the effect of the two pistons on each pin.

If component couples in the plane of the paper for the lower view are considered, it will be realised that the arm of the couple due to cranks 1 and 4, which form a clockwise component couple, is greater than the arm of the component couple due to cranks 2 and 3, acting in the contrary sense. This is corrected by giving the masses B_1 and B_2, which act in intermediate planes, a bias to assist their opposition to cranks 1 and 4. This bias accounts for the unsymmetrical form of the masses.

Since the balance of the pistons involves masses incorporated in the crankshaft, it will be realised that not only should the piston masses be held to close tolerances among themselves, as in the four-cylinder engine, but also that their relation to the crankshaft must be carefully checked.

4.14 Secondary balance with two-plane shaft

The secondary balance with the right-angle shaft is superior to that of the flat shaft.

It will be found that adjacent pairs of pistons in each bank, moving in the same longitudinal plane, operate on cranks at right angles. Thus when one piston is at the position corresponding to $\theta = 0$ (see Fig. 4.9) its neighbour in the same bank has $\theta = 90°$, and the corresponding secondary forces will be opposed.

Fig. 4.13 shows the disposition of the cylinders and cranks, the shaft being indicated with five main bearing journals in order to make clear the relative disposition of the throws. The arrows represent the secondary disturbing forces in the configuration shown, and it will be seen that these are self-balanced in each bank, for both forces and couples.

4.15 Construction of V-eight

A cross-section of the early Ford side-valve engine is given in Fig. 4.14, which

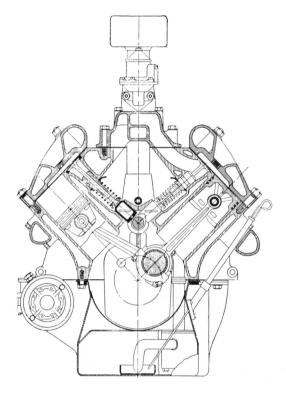

Fig. 4.14 Cross-section of early V-eight

shows the salient special features. The two banks of cylinders and the crankcase are formed in a single monobloc casting, the sump which forms the lower half of the crankcase being a light steel pressing.

Detachable heads with side valves operated from a single camshaft are conventional features, while the somewhat inaccessible position of the tappets and valve springs is mitigated by the special construction adopted. The tappets are non-adjustable, and the valve stems have a wide splayed foot which minimises wear at this point. The valve stem guide is split along its centre line for assembly around the valve stem, and the whole assembly may be withdrawn upwards through the cylinder block after removal of a retainer of flat horseshoe form. Precision gauging during assembly is claimed to render adjustment between periodical regrindings unnecessary, and so enables a simpler construction to be adopted.

A twin down-draught carburettor is fitted to a unit induction manifold and cover, rendering the whole assembly compact and of clean exterior form. Numbering the off-side cylinders 1, 2, 3, 4 and the near-side 5, 6, 7, 8, as in Fig. 4.13, the near-side choke feeds numbers 1, 6, 7 and 4 while the off-side choke feeds 5, 2, 3 and 8. The firing order is 1 5 4 8 6 3 7 2, resulting in a regular interval of half a revolution between cylinders fed from the same choke. The induction tracts are symmetrically arranged, but are not equal in length for all cylinders.

Mounted at the rear of the induction manifold may be observed the crankcase breather and oil-filler, up which passes the push rod for operating the AC fuel pump.

4.16   A British V-eight engine

A most interesting V-eight unit, because it is designed for production in large numbers and in conjunction with an in-line four-cylinder engine, is that used in the Triumph *Stag, Automobile Engineer,* July 1970. An in-line version, comprising in effect one bank of the V-eight unit, but incorporated in a different crankcase, is that of the 1708 cm³ Saab 99, *Automobile Engineer,* September 1968. This engine was subsequently used in the Triumph *Dolomite* range. The bore and stroke dimensions of the Saab version were originally 83.5 × 78 mm, but the bore was subsequently increased to 87 mm, giving a swept volume of 1850 cm³.

For the 2997 cm³ V-eight unit (Fig. 4.15), the bore is 86 mm and the stroke 64.5 mm. This gives a mean piston speed of 11.83 m/s at 5500 rev/min, at which the maximum power, 108 kW, is developed. The valve overlap and lift of the V-eight are larger than those of the in-line engine, helping to give a much higher maximum torque – 235 Nm at a speed of 3500, instead of 137.3 Nm at 3000 rev/min – but steeper flanks to the torque curve.

With a V-angle of 90° – the four cylinder version is canted over at 45° – and a two-plane crankshaft, complete balance can be achieved. The first and fourth crankpins are displaced 90° from the second and third. Cylinders 1, 3, 5 and 7 are in the right-hand bank and numbers 2, 4, 6 and 8 in the left-hand one, so the firing order is 1 2 7 8 4 5 6 3. In the right-hand bank, the cylinders are set 19.8 mm forward of those in the left. At the front end of the crankshaft, a Holset viscous coupling limits the torque transmitted to the fan to 6.56 Nm and its mean speed to 2400 rev/min, thus reducing waste of power when the engine is operating at high speed.

There are several other features of special interest. First, the auxiliary drive layout is common to both the V-eight and the four-cylinder units: a jackshaft rotating at two-thirds crankshaft speed and carried in the base of the V – an arrangement possible because of the use of the overhead camshaft layout – is driven by the timing chain for the camshaft of the left-hand bank, machined on the jackshaft are two spiral gears, one to drive the spindle for the water pump installed vertically in the V, and the other for the spindle driving the ignition distributor – inclined towards the left in the V – and the oil pump, which is on the left, near the base of the crankcase, Fig. 4.15. With this layout, the water pump does not add to the length of the engine, as it would if mounted horizontally in front. Apart from this, the drive to each single overhead camshaft is a simple run of chain from the crankshaft to camshaft pulleys with, in each case, a Renold hydraulic tensioner and a nitrile rubber-faced arcuate guide bearing against the slack run and a nitrile rubber-faced flat damping strip on the taut driving run.

So that the valve gear can be completely assembled on the head before it is mounted on the block, but without impairing accessibility for tightening the cylinder head on the block, five bolts and five studs are used to secure the two: the bolts are perpendicular to the joint face, but the studs are inclined at an angle of $16\frac{1}{2}°$ relative to the axes of the cylinders and, to sustain the component

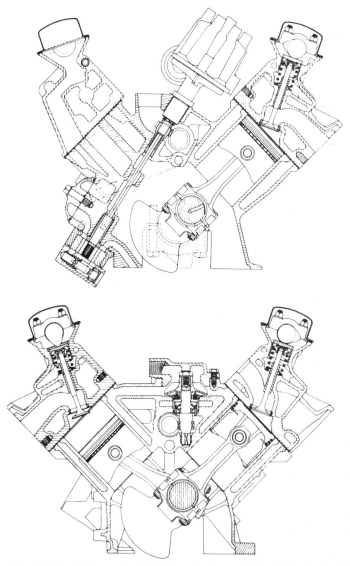

Fig. 4.15 The Triumph *Stag* V-eight engine has an inclined drive for the oil pump and ignition distributor, while the water pump, also driven from a spiral gear on the camshaft, is vertically incorporated between the banks of cylinders, thus economising on overall length and simplifying the drive arrangement

of the tightening force parallel to the joint face, they are a close fit in reamed holes in the head.

The in-line valves are inclined at an angle of 26° inwards, and wedge-shaped combustion chambers are employed. Each exhaust valve comprises a 21-4 N steel head welded to an En18 stem. These seat on the Brico 307 sintered iron inserts in the aluminium head – the use of sintered powdered components

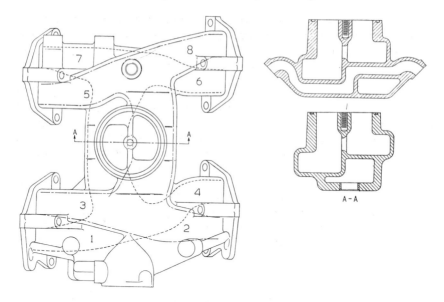

Fig. 4.16 Triumph *Stag* induction system. Section AA shows the water heating ducts and the upper section shows an alternative exhaust gas heating passage to meet US emission control requirements

saves a lot of machining. To clear the heads of the valves, and to form part of the combustion chambers, the crowns of the pistons are slightly dished.

Two Stromberg 175-CDS carburettors are mounted on top of the manifold. Each discharges into an H-shape tract, one serving number 2, 3, 5 and 8 cylinders and the other numbers 1, 4, 6 and 7, so that the induction impulses occur alternately in each, Fig. 4.16. A balance duct is cored in the wall between the two risers.

All the coolant leaving the cylinder heads passes through passages cored beneath the inlet tracts, leaving through the thermostat housing, which is integral with the manifold, Fig. 4.16, section AA. To satisfy emission regulations in the USA, an alternative exhaust heated manifold can be supplied. It has an 82.55-mm wide transverse passage, which communicates with the exhaust ports of numbers 3 and 5, and 4 and 6 cylinders, the gas leaving again through a port at the front. The alternative arrangement is shown in the scrap view above section AA, Fig. 4.16. In addition, a thermostatically-controlled warm air intake system is incorporated. This type of device is described in Section 12.11.

4.17   Jaguar 5.3-litre V-twelve

Obviously, success with any design depends on meeting the requirements of a distinctly identifiable sector of the market. A high proportion of Jaguar cars is sold in the USA and for this reason the 60° V-twelve layout was decided upon. First, it offers something different from the common run of V-eights in that country. Secondly, with a swept volume of $5343\,cm^3$, its potential is ample both to provide enough power for use with automatic transmissions and to offset limitations imposed by current or foreseeable future measures for avoiding atmospheric pollution by exhaust gas constituents. Thirdly, it is inherently in balance and, with six equally placed firing impulses per revolution, it is not only free from torsional resonances but also smooth running.

In this chapter, it will be possible to outline only a few interesting features of the design, Figs 4.17 and 4.18, but a full description was published in the April 1971 issue of *Automobile Engineer*. To save weight, aluminium castings are used for the cylinder block and heads, the sump, oil cooler, timing cover, coolant pump casing, tappet carriers, camshaft cover, induction manifolds, coolant outlet pipes, thermostat housing and the top cover of the crankcase. Although the crankcase was designed so that it could be diecast, sand casting is currently employed. With an open-top deck and wet liners, simple cores can be used and the sealing of the liners, by compressing them between the head and the block, is relatively easy; on the lower seating flanges, Hylomar sealing compound is used to prevent any possibility of leaking of water into the

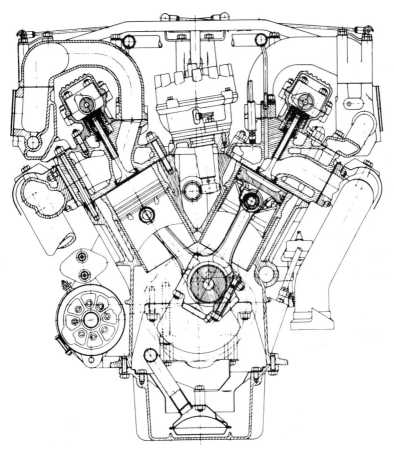

Fig. 4.17 Transverse section of the Jaguar V-twelve engine showing how the problem of differential expansion between the aluminium crankcase and the iron liners is minimised by incorporating the flange high up around the periphery of the liner

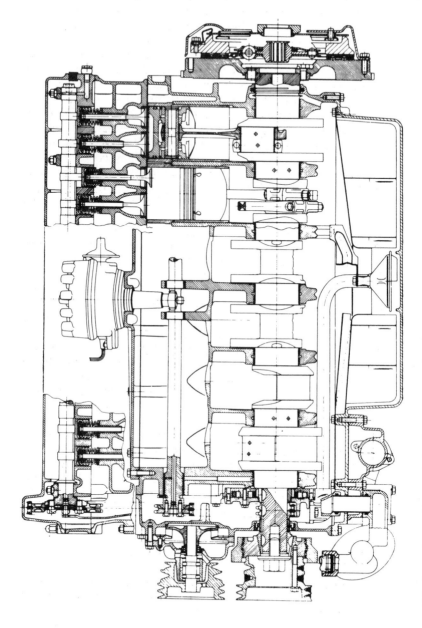

Fig. 4.18 On the Jaguar V-twelve engine, cast-iron main bearing caps are used, each being held down by four studs screwed into the aluminium crankcase

crankcase. The length of the liners between these flanges and the upper ends is only about 44.4 mm, so problems due to differential expansion of the iron liners and aluminium block are reduced to a minimum, while the hottest portions of the liners are in direct contact with the water.

Cast iron main bearing caps are used, and they are each held down by four studs. This ensures adequate rigidity, for avoidance of crankshaft rumble. It also reduces to a minimum variations in clearance due to thermal expansion.

A shallow combustion chamber depression in the crown of the piston, beneath the completely flat face of the cylinder head, was found to give a clean exhaust gas – originally, a deep chamber with clearances machined beneath the valves was tried. To reduce emissions it was also found necessary to lower the compression ratio to 9 : 1 from the 10.6 : 1 originally conceived. The stroke : bore ratio is 0.779 to 1 (70 to 90 mm) and the maximum torque is 412 Nm at 3600 rev/min – even between 1100 and 5800 rev/min the torque does not fall below 325 Nm. The gross bmep of 1110 kN/m^2 is quoted, and the maximum bhp is 272, or 202.5 kW at 5850 rev/min.

A single overhead camshaft is used for each bank. This is more compact, simpler, lighter and less costly than the twin overhead camshaft layout. Moreover, the timing drive is simpler: a single two-row roller chain passes round the drive sprocket on the crankshaft, the two camshaft sprockets and, in the base of the V, the jackshaft sprocket for driving the Lucas Opus contact breaker and distributor unit. A Morse tensioner bears against the slack run of the chain and a damper strip against each of the other runs between the sprockets. This tensioner is described in Section 3.51.

For the lubrication system, a crescent type oil pump is interposed between the front main journal and the front wall of the crankcase, its pinion being splined on a sleeve which, in turn, is keyed on to the crankshaft. The advantages of this type of pump are its short length and that the fact that axial clearance within the housing – in this installation, 0.127–0.203 mm – is much less critical than that of the more common gear type pump.

At 6000 rev/min, this pump delivers 72.7 litres/min. Normally, half of this output goes through a filter to the engine, while the other half passes through the relief valve, which lifts at 483 kN/m^2, to the oil cooler integral with the filter housing at the front end of the sump. The return flow from the base of the radiator to the water pump inlet passes through this cooler, lowering the temperature of the oil by about 22 °C and increasing that of the water by only just over 1 °C.

4.18   Jaguar with May Fireball combustion chamber

Early in 1976, the May Fireball combustion chamber was announced. Then, however, except for a few papers presented before learned societies in various parts of the world, little more was heard of it until mid-1981, when Jaguar introduced their V-twelve HE engine, the letters 'HE' standing for 'high efficiency'. This engine is exactly the same as that described in the previous section except in that it has the May Fireball combustion chamber, together with some recalibration of the petrol injection and modifications to the ignition system. Thus, Jaguar is the first manufacturer to develop the May system to the point of actual series production.

Basically, the May system is designed to burn very weak mixtures and thus, by ensuring that there is plenty of excess air, converting all the fuel to CO_2 and H_2O. This not only ensures that thermal efficiency is high, it also reduces exhaust emissions. It does this in three ways: first, as just mentioned, the formation of CO is prevented; secondly, piston crown temperatures are low – in fact down by about 100°C – partly because of the excess air, so the top land clearance can be kept small and this helps further to reduce hydrocarbon emissions which tend to be generated by quenching of the flame in clearances such as this; thirdly, because of the relatively low peak combustion temperatures, the generation of oxides of nitrogen is minimal.

To burn these weak mixtures, a high compression ratio is necessary and this, by increasing the thermodynamic cycle efficiency, also contributes considerably towards reduced fuel consumption. However, the high compression ratio alone is not enough: the other requirements are a controlled degree of turbulence of the charge, to distribute the flame, and a high-energy spark to start the combustion off vigorously.

With the May Fireball combustion chamber there are two zones in the cylinder-head casting. One is a circular dished recess in which is the inlet valve, and the other, extending further up into the cylinder head, accommodates both the exhaust valve and the sparking plug. Because the compression ratio is required to be high, the combustion chamber has to be fully machined, otherwise both the tolerances and clearance spaces would be too large. Below is the flat crown of the piston – instead of the previously slightly dished crown of the earlier version of this engine.

As the piston comes up to TDC in the final stages of compression, it forces the mixture out of the inlet valve recess through a channel guiding it tangentially into the deeper recess beneath the exhaust valve, Fig. 4.19. This generates a rapid swirling motion in that recess which, once the flame has been initiated by the spark, helps to spread it throughout the mixture. The spark plug, however, is screwed into a small pocket adjacent to the passage through which the mixture flows from the inlet to the exhaust valve regions. In this pocket, it is not only sheltered from the blast of swirling gas but also is in a position such that the fresh mixture that has just come in through the inlet valve is directed on to it. Consequently, it is supplied with an ignitable mixture; also, the nucleus of flame around the points will have time to develop and expand without being blown out, or quenched, before it can generate enough heat to be self-sustaining.

Most of the detail development work has been aimed at getting the guide channel between the inlet and exhaust valve pockets just right for inducing the optimum swirl. Incorporation of a ramp in this channel was found to give the best results.

At steady speeds at part throttle, with the compression ratio of 12.5:1, air:fuel ratios of 23:1 were burned consistently well, but richer mixtures were found to be necessary for transient conditions experienced on the road. Obviously, the Lucas digital electronic fuel-injection system had to be recalibrated to suit the lean burn requirements, but otherwise it is the same as used by Jaguar in the earlier version of the engine.

To ignite such weak mixtures, a high-energy spark is required. The reliable, constant energy ignition module introduced for the XK 4.2 engine was utilised, but given a more powerful amplifier to raise its output from 5 to 8 amp. Within the distributor, a magnetic pick-up has been incorporated. Then a twin coil

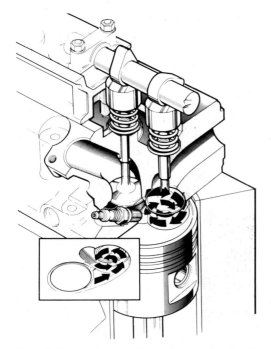

Fig. 4.19 Jaguar V-twelve HE combustion chamber. Inset: underside view of the swirl pattern

system – both coils being Lucas 35 C6 units – was developed, the secondary coil being used solely as a large inductor and mounted ahead of the radiator to keep it cool. The outcome of all this work is a system that will provide accurately timed ignition for 12 cylinders at 7000 rev/min – no mean achievement. On 97-octane fuel, the engine produces 223 kW (299 bhp) at 5500 rev/min. Its maximum torque is 44.05 kg at 3000 rev/min.

Chapter 5

Sleeve-valve and special engines

Interest in the reciprocating sleeve valve has persisted throughout the history of the automobile engine, and though not now so widely used as in the early days of the Burt-McCullum and Knight patents, the single sleeve still has many strong adherents, while later metallurgical advances made possible such exacting and successful applications of the single sleeve as the Bristol Perseus and Napier Sabre aircraft engines.

The double sleeve has become obsolete owing to its greater cost of manufacture and greater viscous drag as compared with the single sleeve and though the qualities of the Daimler-Knight, Minerva, and Panhard engines are proverbial, it does not appear likely that the double-sleeve arrangement will experience revival.

The sleeve valve, as the name implies, is a tube or sleeve interposed between the cylinder wall and the piston; the inner surface of the sleeve actually forms the inner cylinder barrel in which the piston slides. The sleeve is in continuous motion and admits and exhausts the gases by virtue of the periodic coincidence of ports cut in the sleeve with ports formed through the main cylinder casting and communicating with the induction and exhaust systems.

5.1 Burt single-sleeve valve

The Burt-McCullum single-sleeve valve is given both rotational and axial movement, because with a single sleeve having only axial reciprocation, it is impossible to obtain the necessary port opening for about one-quarter of the cycle and closure for the remaining three-quarters, if both inlet and exhaust are to be operated by the same sleeve. It will be found that a second opening occurs when the ports should be shut. The sleeve may be given its combined axial and rotational motion in a variety of ways, one of which is illustrated in Fig. 5.1. This shows an arrangement of ball-and-socket joint operated by short transverse shafts in a design due to Ricardo, and the same mechanism was used in the Napier Sabre engine. The ball B is mounted on a small crankpin integral with the cross-shaft A which is driven at half-engine speed through skew gears from the longitudinal shaft C. Clearly the sleeve receives a vertical movement corresponding to the full vertical throw of the ball B while the extent of the rotational movement produced by the horizontal throw of the ball depends upon the distance between the centre of that ball and the axis of the sleeve.

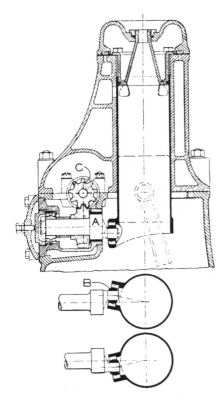

Fig. 5.1 Ricardo actuation mechanism for Burt-McCullum sleeve valve

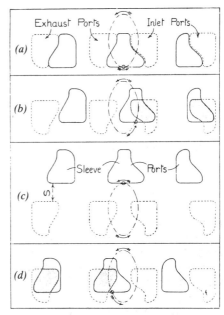

(a) Exhaust just closing and inlet about to open
(b) Maximum port opening to induction
(c) Top of the compression stroke, S being the maximum 'seal' or overlap of the ports and cylinder head
(d) Maximum opening to port

Fig. 5.2 Single-sleeve porting

5.2 Arrangement of ports

The form and arrangement of the ports are arrived at as the result of considerable theoretical and experimental investigation in order that the maximum port openings may be obtained with the minimum sleeve travel. This is important, as the inertia forces due to the motion of the sleeve, and the work done against friction, are both directly proportional to the amount of travel. Fig. 5.2 shows an arrangement of ports wherein three sleeve ports move relative to two inlet and two exhaust ports, the middle sleeve port registering in turn with an inlet and an exhaust port. The motion of the sleeve ports relative to the fixed cylinder ports is the elliptical path shown. In the figure, the sleeve ports are shown in full lines in their positions relative to the cylinder ports (shown in broken lines) at various periods of the cycle. In practice, it is usual to provide five sleeve ports and three inlet and exhaust ports.

5.3 Advantages and disadvantages of sleeve valves

The great advantages of the sleeve valve are silence of operation and freedom from the necessity for the periodical attention which poppet valves require, if the engine is to be kept in tune. Hence sleeve valves have been used in cars of the luxury class where silence is of primary importance. The average sleeve-valve engine has not shown quite such a good performance as its poppet valve rival in maintenance of torque at high speeds, owing chiefly to the somewhat restricted port openings obtainable with reasonable sleeve travel. Hence sleeve-valve engines have not figured prominently in racing, although some very good performances have been made from time to time. Sleeve-valve engines share with rotary valve types reduced tendency to detonation owing to the simple symmetrical form of the combustion chamber with its freedom from hot spots, and shortness of flame travel. The construction also lends itself well to high compression ratios without interference between piston and valves. The disadvantages of gumming and high oil consumption experienced with early sleeve-valve designs have been successfully overcome, but there is some element of risk of serious mechanical trouble in the event of piston seizure, which dangerously overloads the sleeve driving gear.

5.4 Rotary valve

Many types of rotary valve have been invented, either to act as distribution valves only, or to perform the double function of distribution and sealing.

Their great mechanical merit is that their motion is rotative and uniform, and the stresses and vibration of the reciprocating poppet or sleeve valve are eliminated. They are suitable for the highest speeds, and the limitation in this direction is determined by the inertia stresses in the main piston, connecting rod and bearings. A high degree of mechanical silence is obtained.

The performance shown by the two proprietary makes described in the following paragraphs has proved conclusively that from thermodynamic and combustion points of view they have outstanding qualities as compared with the conventional poppet valve construction. In both the Cross and Aspin engines in the single-cylinder motor-cycle form, exceptionally high compression ratios with freedom from detonation with fuels of quite low octane value have been obtained. The corresponding bmep reaches figures of the order 1100 to 1240 kN/m². The quite exceptional outputs per litre arising from the combination of extreme rotational speeds with these pressures should be considered in the light of the remarks in Section 3.24.

The remarkable freedom from detonation under a combination of high compression ratio and low octane fuel – a combination usually disastrous in a normal engine – is no doubt largely due to the general coolness of the combustion chamber, its smooth form and freedom from hot spots. Further, the high compression ratio probably results in complete evaporation of the fuel before ignition, and it is believed that this inhibits the formation of the peroxides referred to in Section 14.10. This may be consistent with observation that tetraethyl lead, which is also regarded as an inhibitor of such formation, appears to have little effect in these cool engines.

To offset these remarkable characteristics there are, unfortunately, considerable mechanical difficulties in pressure sealing and providing adequate lubrication of the valves without excessive waste to the cylinder and the exhaust ports.

5.5 Cross rotary-valve engine

The valve of the Cross engine normally runs at half engine speed, but by duplicating the inlet and exhaust ports through the valve it may be readily designed to operate at one-quarter engine speed.

The valve housing is split about the centre line of the valve, the halves being held in resilient contact with the valve. The bottom half of the valve housing is usually an integral part of the cylinder, which is not bolted to the crankcase, but allowed to press upwards against the valve, such pressure being proportional to the gas pressure in the cylinder.

In cylinder sizes above $200 \, cm^3$ a controlled valve loading scheme is adopted so that the pressure on the valve by the housing is only just sufficient for adequate sealing.

The lubrication of the valve is brought about by pumping oil on to one side of the valve and removing it with a scraper blade on the other side, an essential part of the mechanism being a non-return valve which prevents oil from being sucked into the induction side of the valve.

In cases where it is not possible to have a completely floating cylinder, the lower part of the valve housing is spigotted into the top part of the cylinder, suitable sealing rings being provided.

Cross engines usually use an aluminium cylinder without liner. The piston rings, being made from very hard steel, act not only as the means of pressure sealing, but as bearers to prevent the piston touching the bore. The cylinder bore wear with this construction is negligible.

Fig. 5.3 is a sectional view of an early type of valve for a $500 \, cm^3$ motor-cycle engine, having vertical shaft and bevel drive for the valve. V is the cylindrical valve operated by the dogs G on the half-speed shaft. The induction port is indicated at I, and S is the tunnel liner in which the valve rotates. In this design sealing is obtained by the resilient port edges E of the liner, which is so machined as to maintain an elastic pressure on the rotating valve.

This sealing pressure is transmitted through the valve body to the upper casting and back to the crankcase through two long hold-down bolts.

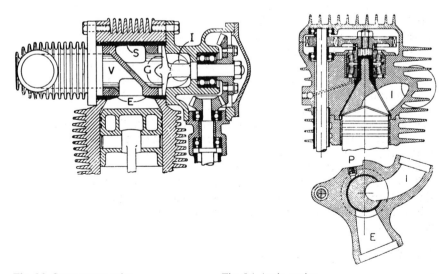

Fig. 5.3 Cross rotary valve   Fig. 5.4 Aspin engine

Various materials have been tried for the valve and liner, a combination of nitri-cast-iron valve running in a liner of bronze or nitralloy steel having given good results.

The makers report that this engine has developed a bmep of 1344 kN/m² at 4000 rev/min with a fuel consumption of 0.228 kg/kWh.

For further details of these interesting engines the reader should refer to a paper by R. C. Cross in Vol. XXX of the *Proceedings of the Institution of Automobile Engineers*.

5.6   Aspin engine

Like the Cross engine, the Aspin engine has shown the most striking performance in the single-cylinder, air-cooled motor-cycle form, an engine of 67 mm bore and 70.5 mm stroke with a compression ratio of 14 : 1 having developed 23.5 kW at the remarkable speed of 11 000 rev/min. When developing 15.64 kW at 6000 rev/min the fuel consumption was recorded as 0.1946 kg/kWh.

The general construction of this single-cylinder engine should be clear from Fig. 5.4. The valve consists of a nitrided alloy-steel shell partly filled with light alloy, within which a cell is formed to constitute the combustion chamber. As the valve rotates the cell is presented in turn to the inlet port I, the sparking plug P and the exhaust port E. During compression, ignition and combustion the cell is on the cool side of the cylinder, and after ignition the plug is shielded from the hot gases. These conditions play a great part in the thermodynamic properties of the engine. In this early engine a double-thrust Timken roller bearing was provided to take the bulk of the upward thrust due to the gas load, only a carefully regulated amount being carried direct on the conical surface.

A four-cylinder Aspin engine, of 4.6 litres, was developed for heavy duty, and in Fig. 5.5 are shown sectional views of its head construction. Water cooling

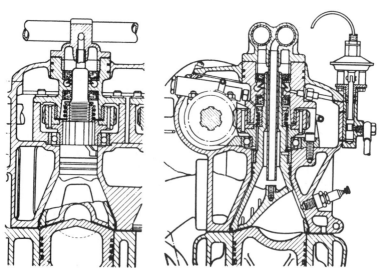

Fig. 5.5 Four-cylinder Aspin engine

was incorporated for the rotor, which was of fabricated steel construction, faced with lead-bronze alloy, and running in a cast-iron cylinder head.

For full descriptions and analysis of performance of these engines, the reader should refer to articles by Louis Mantell and J. C. Costello in Vols 34 and 35 of *Automobile Engineer*.

5.7   NSU Wankel rotary engine

The information that follows is an abstracted summary of the comprehensive technical articles by R. F. Ansdale, AMIMechE, in *Automobile Engineer*, Vol. 50, No. 5 and Dr-Ing. Walter Froede in Vol. 53, No. 8. An article by Felix Wankel on the performance criteria of this type of engine is in Vol. 54, No. 10 of the same publication. A survey covering the developments from the inception of this engine to the mid-nineteen-eighties, with special reference to work on it by Norton Motors Ltd, is given in an article by T. K. Garrett, in *Design Engineering*, December 1985.

Although the Wankel engine represented a major advance in the search for a rotary engine mechanism, it was not based on any new principle or thermodynamic cycle. The four events of the four-stroke cycle take place in one rotation of the driving member.

The general profile of the straight working chamber is of epitrochoid form, a group of curves of the cycloid family, the geometry of which is fully discussed in the articles under notice.

The general construction of a single-rotor type, known as the KKM version, with three-lobed rotor, is shown in Fig. 5.6.

The rotor provides three equal working spaces, and clearly an exhaust release will occur each time an apex seal overruns the leading edge of the exhaust port E, that is, three times per revolution of the rotor, and this exhaust will continue until the following seal reaches the trailing edge of the port.

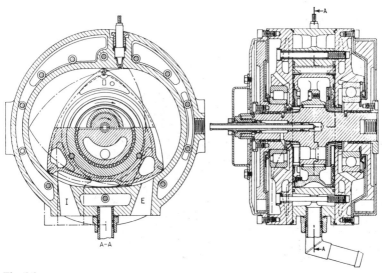

Fig. 5.6

Induction will have commenced in the same space about 60° of rotor movement earlier.

There are thus three complete four-stroke cycles per revolution of the rotor in different working spaces, but all fired by the same sparking plug as maximum compression is reached.

The stationary shell pinion is fixed in the casing, and the annulus mounted at the centre of the rotor, and carried on needle rollers on the periphery of the shaft eccentric or crank, engages and rolls around the fixed pinion. With 24 and 36 teeth on the pinion and annulus respectively, the main shaft will make three turns for one turn of the rotor, this giving a complete cycle for each revolution of the main shaft. The driving impulses transmitted to the shaft thus correspond to those of a normal two-cylinder four-stroke engine.

There is a rotating primary imbalance due to the eccentric path of the rotor, but this is readily dealt with by means of two symmetrically mounted flywheels, suitably drilled to provide the countervailing imbalance. Suitable balancing provision for the cooling water and oil is incorporated.

The most interesting and informed articles on which these notes are based give an extensive description of the construction and related experience with the various sealing devices, and also performance curves and figures of the pioneer KKM unit in comparison with a normal piston engine.

Fig. 5.6 shows longitudinal and cross-sections of a typical unit, and Fig. 5.7 gives views of three different forms of rotor. Recesses in the curved faces are provided to obviate strangling of the charge during passage from one zone to the next.

The form of these faces, subject to the necessary compression ratio being provided, is not limited to any particular profile.

Cooling has not presented many problems, because the complex movements of the rotor, and the resultant changing accelerations, tend naturally to circulate the oil and so to cool the interior. The circulation thus set up contributes greatly to the cooling.

Most of the development problems have been associated with reducing the rates of wear of the apex seals and bore, and improving the efficiency of combustion.

Fig. 5.7

As regards wear the problems have been solved. Tojo Kogyo has developed seals made of glass-hard carbon, which run in hard chromium-plated aluminium bores. NSU have used a proprietary metallic seal called IKA, and they have also used a cermet, called Ferrotic, which is mainly iron and titanium carbide sintered.

In the UK, Norton Villiers have used, for their *Interpol* motor-cycle and their aviation and industrial engines, a resilient gas-nitro-carburised steel, because it is more conformable than either the cast irons used for piston rings or the previously mentioned carbon, special alloy or cermet seals. In addition, however, these iron seals are of a self-tracking design, their faces remaining mostly parallel to the walls of the trochoidal chamber as they sweep over them. All the manufacturers mentioned have used a high silicon aluminium alloy for the chamber and plated it with Elnisil, which is nickel containing 4% by volume of fine silicon carbide particles.

Disadvantages of the Wankel engine include the fact that, at low speeds, the rate of leakage past its seals is five times that past the piston rings in an equivalent piston engine. For this reason the torque falls off steeply at low speeds. Norton Motors Ltd claim that their *Interpol* motor-cycle has been shown to give 1 mpg better fuel consumption than a competitive machine powered by a four-stroke reciprocating engine. Their twin rotor engine for the Cessna aircraft develops 90% of the power of the conventional engine that it was designed to replace, but at less than half the weight, even though it has to carry a 3 : 1 reduction gear. Moreover, it can be comfortably accommodated inside a 406 mm diameter tube. In the meantime, Toyo Kogyo continues to produce its Mazda cars powered by twin rotor Wankel engines.

A diesel version of the Wankel engine was developed by Rolls-Royce Ltd. It has been described in a paper by F. Feller, *Proc. I. Mech. E.* 1970–71, Vol. 185, 13/71. Basically, it comprises two units, a small one incorporated integrally in the casing above a larger one. The larger one acts as a compressor, supercharging the other, which is the power unit. With this arrangement, compression and expansion ratios as high as 18 : 1 can be obtained, and the surface : volume ratio in the combustion chamber is about the same as that of an equivalent reciprocating piston engine.

In this unit, the restriction, or throat, formed between the two portions of the combustion chamber as the rotor sweeps past top dead centre, is used to generate the turbulence required for burning weak mixtures in the pocket, or depression, in the periphery of the rotor. To obtain this effect, the pocket is shaped like a cricket bat, foreshortened with its lower end bifurcated like the section of the base of the Saurer combustion chamber, Fig. 6.7. As the channel represented by the handle of the bat passes the restriction, the compressed gas is forced along it, directing a jet at the base, which divides the flow into two turbulent eddies rotating in opposite senses. These eddies, of course, swirl one each side of the chamber represented by the foreshortened blade of the bat.

Another special feature of the Rolls-Royce version is its tip seals. These are shaped so that gas pressure forces the trailing seal into contact with the wall of the chamber. Specific fuel consumptions of the order of 0.232 to 0.2433 kg/kWh are expected. Although the useful speed range of this engine is narrow, this may be overcome by the use of automatic transmission.

A survey of various types of rotary combustion engine, including brief comments on development work by Renault on two-stroke and four-stroke versions, is given in a serial article by R. F. Ansdale, AMIMechE, in *Automobile Engineer*, Vol. 53, No. 11 and Vol. 54, Nos 1 and 2. The thermodynamics have been dealt with by D. Hodgetts, BSc, AMIMechE, in Vol. 55, No. 1 of the same journal.

Chapter 6

The compression-ignition engine

By virtue of its inherent durability, and high thermal efficiency and therefore low specific fuel consumption, the compression-ignition (ci) engine is by far the most favoured power unit for commercial vehicles and is beginning to encroach significantly into the private car field too. The thermal efficiency of an indirect (idi) diesel engine, Section 6.11, is about 25% higher than that of the gasoline engine, while that of a direct injection (di) unit, Section 6.10, is of the order of 15% higher still. A considerable disadvantage of both idi and di types is their low power output relative to both weight and cylinder capacity, compared with the spark ignition engine. However, to a large extent, this can be offset by turbocharging the ci unit and even more so if charge cooling is employed too.

The compression ignition type of power unit is sometimes called the oil engine but is more widely known as the diesel engine, after the German engineer, Dr Rudolph Diesel who, in 1893, published a book on the subject and, in 1897, produced the first practical engine running on fuels of higher specific gravity than gasoline. Other workers early in the field were W. D. Priestman and H. Ackroyd Stuart, both from Yorkshire. The essential features of compression ignition engines are the injection of the fuel into the cylinders as their pistons approach inner, or top, dead centre, and a compression ratio of not less than 12–13:1 and in some applications as high as 22:1 or more.

6.1 Ignition by the temperature of compression

If the compression ratio is 14:1, the initial temperature of the air in the cylinder is 60°C, and if compression is truly adiabatic (no heat loss to the surroundings), the temperature at TDC would be 675°C. At this temperature, the injected fuel ignites easily, because its self-ignition temperature, in air at atmospheric pressure, is between 340 and 350°C, and even lower at the high pressure, and therefore density, at the end of compression.

When starting from cold, however, problems can arise, since the ambient temperature is likely to be about 15°C and, according to geographical location and season, could be a great deal lower. Moreover, even at 15°C ambient, the heat loss at cranking speed may be great enough to prevent the temperature from rising above about 400°C.

Incidentally, temperature rise is affected by the initial temperature only in so far as the rate of loss of heat is influenced by the density of the charge. Therefore, for a given initial temperature and volumetric compression ratio, throttling of the ingoing air has an insignificant effect on the compression temperature. It was because of this that pneumatic governing was practicable. With this system of governing, the supply of air was throttled, as in a petrol engine, and the resultant depression in the induction system employed to actuate a diaphragm type control connected to the fuel supply rack in the injection pump. This type of governing, however, is no longer used, because it is not accurate enough for modern requirements, and the pumping losses due to throttling reduce thermal efficiency.

If induction is unrestricted except for the normal throttling effect of the inlet valve, the pressure at beginning of compression is between about 90 and $103 \, \text{kN/mm}^2$ absolute, depending on the speed and breathing characteristics of the engine. At the end of compression, the pressure will be between about 3100 and $3800 \, \text{kN/m}^2$ or more, again depending on the running conditions and design of the engine. Leakage past the piston rings tends to reduce the ultimate pressure and temperature. In a turbocharged diesel engine, the peak combustion pressure may be about $12\,000 \, \text{kN/m}^2$, as compared with $5500\text{–}6900 \, \text{kN/m}^2$ in a petrol engine.

From the foregoing it can be seen that the cycle of operations differs from that in a spark ignition engine in that the compression ratio is higher and only air is compressed, the fuel being injected late in the compression stroke. Methods of injection and forms of combustion chamber differ widely, while the basic combustion process, unlike the progressive burning of the homogeneous mixture of gasoline and air in a spark ignition engine, is complex, as described in detail in Sections 6.4–6.8.

6.2   Air blast injection

This method constituted the true diesel method as originally used in large stationary and marine engines, and involved the following features. The compression pressure was about $3450 \, \text{kN/m}^2$, and the fuel was measured and delivered by a mechanical pump to the annular space behind a small conical injection valve placed in the centre of the cylinder head and arranged to open outwards. To this space was applied an air pressure of 5516 to $6895 \, \text{kN/m}^2$ from air storage bottles charged by a compressor which was usually incorporated in the engine itself.

At the correct moment the injection valve was lifted off its seat and the high-pressure blast air drove the fuel in at a very great velocity, when it mingled with the combustion air in the cylinder and was ignited by the high temperature of this air caused by the high compression. It must be realised that the volume of liquid fuel delivered each cycle was extremely small, but the accompanying bulk of blast air, which was from 2 to 3% of the total air, lengthened the injection period with the result that the pressure did not rise during combustion, but was merely maintained at approximately the compression pressure as the piston moved outwards until combustion was completed. This led to the use of the expression *constant pressure cycle* to describe the diesel cycle.

The compression indicator diagram is shown at *a* in Fig. 6.1, the black dot

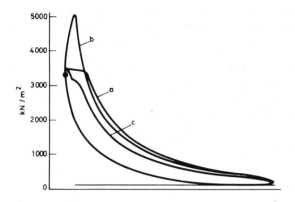

Fig. 6.1 Indicator diagrams of compression-ignition engine

indicating the approximate point of commencement of injection. If the weight of fuel injected is such that there is 30–40% air in excess of that required for complete combustion, the maximum and mean pressures will be respectively about 3450 and 690 kN/m². With reduced fuel supply, for part load, diagram *c*, the maximum pressure is the same but the rate of combustion falls behind that of the descent of the piston, which is why the pressure falls irregularly until combustion is complete.

The high pressure compressor needed for air blast injection was costly, troublesome in service, and absorbed considerable power, lowering overall efficiency. Moreover, the storage bottle installation was heavy and bulky, rendering it unsuitable for road vehicles. Indeed, it became obsolete, even for large industrial and marine power units, and was replaced by the jerk pump.

6.3   Mechanical injection

With mechanical injection, the oil is forced in from a pump through a sprayer or pulveriser, comprising one or more fine holes in a suitable nozzle. Very difficult conditions have to be met. The volume of liquid to be injected is very small and must be injected at a very high velocity in order that it may be thoroughly atomised and yet be capable of penetrating through the whole volume of air present. The jet must also be so disposed and directed that a stream of liquids is not likely to impinge on the cylinder wall or piston, where rapid carbonisation could occur – an exception is the system adopted for the MAN engines, in which the fuel jet is deliberately impinged on the hot wall of the bowl-in-piston combustion chamber to facilitate evaporation.

The injection of a small volume at high velocity implies a very short *period* of injection, and this results in the action approximating closely to an explosion with a more rapid rise in pressure and a much higher maximum pressure as shown at *b* in Fig. 6.1, which represents a typical full load diagram with mechanical injection. This higher maximum pressure makes the ratio of mean pressure to maximum pressure even less favourable than in diagram *a*, which already compares unfavourably in this respect with the petrol engine.

6.4   Power: weight ratio

The above ratio of mean to maximum pressure is the determining factor in the value of the power: weight ratio, since the power depends on the mean pressure while the sturdiness and therefore weight of the parts will depend on the maximum pressure to be provided for. Thus, the compression-ignition engine is inherently heavier than its rival the petrol engine, owing to the very factor which results in its superior economy, namely, the higher compression and expansion ratio. As the fuel consumption of the petrol engine is improved by increasing its compression ratio, exerting closer control over fuel supply, and other measures, prospects for its being, to a major extent, replaced by the diesel engine recede. However, the relative flammability, and therefore safety, of the fuels, and the specific performances of the engines are now becoming the critical factors. Improvements in materials and manufacturing techniques, of course, apply equally to both types of engine.

6.5   Injection and combustion processes

Extensive research has been and is being carried out to determine the best methods of injection and form of combustion chamber to give smooth and complete combustion of the injected fuel and suppression of the characteristic 'diesel knock' which gives rough and noisy running or the high speed mechanical injection engine.

The problem is to inject into the cylinder an extremely small volume of liquid fuel in such a manner and into such an environment that every minute particle of oil shall be brought into immediate contact with its full complement of heated air, in order that combustion shall be rapid and complete without being so sudden as to give rise to rough running.

There are two general alternatives – either to cause the fuel to penetrate by its own velocity to all parts of the combustion chamber and find the air required, or to give the air itself such a degree of swirl or turbulence that it will seek out the fuel as it enters. The former represents the process of direct, or open chamber, injection in which some movement of the air is generally induced deliberately to facilitate mixing, while the second covers the many designs of pre-combustion chamber arranged to give much more vigorous swirl or turbulence, or a combination of both (Sections 6.10–6.15).

6.6   Three phases of combustion

Ricardo has recognised and described three phases of combustion. These are illustrated in Fig. 6.2 which shows a form of indicator diagram differing from the form in Figs 1.4 and 6.1, in showing the pressures plotted on a continuous crank angle base instead of on a stroke base. With the crankshaft turning steadily at some measured speed, the crank angle base becomes also a *time* base and, as time is an important factor in combustion processes, very useful information can be obtained from such diagrams, though they cannot be used for the calculation of power output unless they are first reduced to a stroke base by a suitable graphical process. These crank angle diagrams can be obtained by means of indicators of the 'Farnborough' and 'cathode-ray' types.

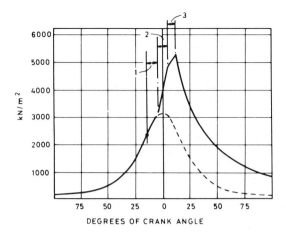

Fig. 6.2 Fundamental phases of combustion in mechanical-injection ci engine

6.7 Delay period

Referring to Fig. 6.2, the commencement of injection is indicated by the dot on the compression line about 15° before dead centre. The period 1 is the 'delay' period during which ignition is being initiated, but without any measurable departure of the pressure from the air compression curve which is continued as a broken line in the diagram as it would be recorded if there were no injection and combustion.

6.8 Second phase

This delay period occupies about 10° of crank angle and is followed by a second period of slightly less duration during which there is a sharp rise of pressure from 3100 to over 4830 kN/m^2. This period represents a phase of rapid flame-development and combustion of the whole of the fuel present in the cylinder and approximates to the 'constant volume' combustion of the vaporised fuel in the spark ignition engine. The steepness of this rise – here about 200 kN/m^2 per degree of crank movement – is a determining factor in causing diesel knock, or rough running. In general, though the form of the combustion chamber has an important influence, the longer the delay period the steeper will be the second phase and the rougher the running, as a greater proportion of fuel is present. The nature of the fuel, the temperature and pressure of compression, and initial rate of fuel injection are all factors in deciding the length of the delay period apart from the form of the combustion chamber.

With the majority of engines the running is rougher at light loads and idling, as the lower compression temperature and smaller quantities of fuel injected increase the delay period, but this is not universal. Ricardo has shown that this period tends to be constant in time, thus occupying a greater crank angle at higher speeds and calling for injection advance, while the second phase covers more nearly a constant crank angle, as in a given engine it is dependent on turbulent displacement, the speed of which will increase with engine speed.

6.9   Final phase of combustion

The third phase involves combustion of the fuel as it issues from the sprayer holes. With air blast injection this period could be largely determined by the blast pressure, and the characteristic flat top to the diagram could be obtained; but with mechanical injection most of the fuel is already in the cylinder before this stage is reached, so less control can be exercised over it.

It should be appreciated that the diagram in Fig. 6.2 is hypothetical, and that the three phases are not in general so sharply distinguishable one from the other.

6.10   Types of combustion chamber

Typical forms of combustion chamber are shown in Figs 6.3 to 6.8. These illustrations are diagrammatic and are a selection from the many forms that have been invented and introduced with varying degrees of success. In general, the tiny quantities of fuel injected in small high speed diesel engines have an exceptionally short time in which to burn, so such engines need small swirl or pre-combustion chambers incorporated in their cylinder heads. This is why they are generally termed the indirect injection type. Virtually all others are of the open chamber type, in which the combustion chamber is usually a hemispherical or toroidal bowl in the piston crown. These are termed direct injection engines.

6.11   Direct injection

Fig. 6.3 illustrates the direct injection or open chamber type, in which the fuel is sprayed through two or three fine holes at a high velocity, and requiring injection pressures of $27\,580\,kN/m^2$ or more. The resulting 'hard' jet enables the fuel to penetrate the dense air and find the necessary oxygen for combustion, aided in most cases by some residual swirl or turbulence set up during the induction stroke by the masking of the inlet valve, as is indicated in Fig. 6.7. The merits of this type are that a moderate compression ratio of 13 or 14:1 may be used, no auxiliary starting devices are necessary and, for certain industrial and marine applications, the smaller sizes of engine can be started by hand-cranking from cold, provision being made for releasing the compression until the engine is rotating at sufficient speed.

The form of the combustion chamber with a moderate ratio of surface to volume is favourable to the reduction of heat loss, and engines of this type show good fuel economy and high mean pressure. Maximum pressures tend to be high, however, and there is a somewhat greater tendency to rough running. The indicator diagram approaches in form to 'constant volume' rather than to 'constant pressure' combustion. The injectors with their minute spraying holes – about 0.2032 mm – and high pressures, require highly skilled technique in production and most careful provision for filtering the fuel.

Direct injection engines have not generally been considered capable of quite such high speed or so tolerant of poor fuel as the pre-combustion or ante-chamber types, but their easy starting and high thermal efficiency have produced many converts as the more difficult injection technique has been mastered.

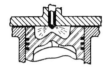 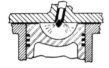 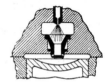

Fig. 6.3 Direct injection   Fig. 6.4 Benz

6.12   Pre-combustion chamber

Illustrated in Fig. 6.4 is a pre-chamber developed by Benz and Co. and used by several other manufacturers. The action gives an approximation to the characteristics of air blast injection and limits the maximum pressure. The pre-chamber represents about 40% of the total clearance space, and the fuel is injected into this air and partly burned, the spread of ignition being helped by the turbulence arising from the passage of the air through the communicating pepper-castor holes during compression.

The products of this partial combustion and the remaining fuel are then forced by the excess pressure in the pre-chamber back through the communicating holes at high velocity as the piston commences to descend. The high turbulence thus created aids the final flame-spread and combustion in the main cylinder, while the piston is protected from the high initial pressure which would arise from too rapid combustion of the whole fuel charge. This system has the disadvantage that there is considerable cooling loss during passage of the air through the communicating holes, and due to the high surface:volume ratio of the combustion chamber.

The cooling effect is mitigated by partially isolating the inner lining of the chamber from the main body of metal as shown in the figure, but even so a higher compression pressure is necessary to obtain satisfactory starting with (in the majority of designs) the further help of heater plugs – though engines employing this arrangement are in general free from rough running and high maximum pressures. To obtain a relatively smoke-free exhaust, however, the power output has to be limited. One reason for this is that in such a tiny combustion chamber some of the droplets of fuel injected at high pressure through small diameter jets are inevitably deposited on the walls, and this reduces their surface exposed to the air by a factor of 16. Another is that the mixture in the pre-chamber when ignition is initiated is inevitably very rich. An outcome of this relatively inefficient combustion is a higher rate of contamination of the oil by blow-by past the piston rings, and therefore shorter intervals between the need for oil changes than with a direct injection engine.

Comparable direct and indirect combustion chamber and induction systems are illustrated in Figs 6.6 and 6.7.

6.13   Controlled air swirl

Sir Harry Ricardo was the first to discover that, if the inlet ports are arranged tangentially to the cylinders, the incoming air forms a vortex about the axis of the cylinder, which persists throughout the compression stroke. Moreover, when the vortex is forced by the piston into the smaller diameter combustion chamber, its rotational speed increases. By injecting a finely atomised jet of fuel along an axis offset from but parallel to that of the swirl, as in Fig. 6.5, it can be rapidly distributed uniformly throughout the mass of air without resorting to very high injection pressures. Sir Harry used a single-cylinder engine with a single-sleeve valve, Fig. 6.5, so that he could most easily fit different forms of combustion chamber into the cylinder head.

6.14 Comet swirl chamber

A fairly genuine prejudice against the sleeve valve led to the development by Ricardo, in conjunction with AEC and other firms, of the Comet chamber, which is illustrated in Fig. 6.6.

Different designs vary in detail, and the twin turbulence recesses in the piston crown are a later development.

The usual shape of chamber is spherical with a single tangential entry passage. With this design heater plugs are unnecessary. An important feature of the Ricardo patents is the semi-isolated hot plug forming the lower part of the chamber, the purpose being the mitigation of the heat loss to the jackets, which loss gives rise to higher specific fuel consumption, as in the pre-chamber arrangement.

Comparative tests of AEC transport engines with open and Comet chambers showed lower fuel consumption and better low-speed torque with the former but better high-speed torque with the latter.

Another application to the Ricardo Comet chamber, enabling easy starting to be obtained without a heater plug, is the Pintaux injector nozzle developed jointly by CAV and Ricardo. This is illustrated and described in Section 6.20.

6.15 Suarer dual-turbulence system

This system combines the effect of what has been aptly termed the *squish* of the air trapped between the piston crown and cylinder head, with rotational swirl produced by masking the inlet valve, as indicated in Fig. 6.7. In the actual construction there are two inlet and two exhaust valves, and the two directions of swirl are superimposed in the doughnut-shaped recess in the piston head. The injector has four radially directed holes and only a moderate injection pressure is called for owing to the relatively large diameter of these. With a compression ratio of 15, no heater plug is required to obtain easy starting, and the delay period, owing to the good turbulence, is short. Similar systems are now widely used by most manufacturers.

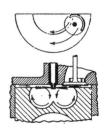

Fig. 6.5 Vortex   Fig. 6.6 Comet   Fig. 6.7 Saurer

6.16 Evolution of the Perkins range of diesel engines

Perkins Engines Ltd and the diesel engine division of Rolls-Royce merged in 1984, to form what was probably the most advanced diesel engine manufacturing group in the world, and then L. Gardner & Sons were acquired in 1986. Perkins had introduced their Aeroflow indirect injection system prior to the Second World War. This had a pre-chamber in which was a nozzle of the Pintaux type. One of the two jets was directed towards its centre and the other down the throat towards the cylinder. The latter pumped fuel into the incoming stream of air with which it mixed thoroughly, a small proportion of the fuel droplets briefly emerging into the main chamber, to ignite as in a direct injection engine: that is, before the air had been cooled by passing through the throat. This fraction of the mixture, already passing through the initial stages of combustion, was then immediately driven back up through the throat to help to ignite the remainder of that in the pre-chamber. With this arrangement, cold starting was better than with the simpler swirl chamber systems.

By 1980 Perkins, having already had many years' experience also with direct injection, had finally abandoned the Aeroflow system. The direct injection engine illustrated in Fig. 6.8 is the 6.354, introduced in 1957. One of its interesting details is the mounting of the CAV distributor type injection pump

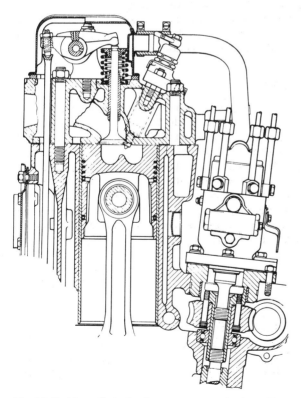

Fig. 6.8 Perkins cylinder head

on a flange that houses the bearing at the top of an almost vertical shaft driven by a spiral gear on an auxiliary shaft.

Then, in 1986, Perkins introduced two ranges of engines of considerable merit, both of the direct injection type. These were the Phaser and the Prima, the former were four- and six-cylinder units having swept volumes of 1 litre/cylinder, for commercial vehicle and industrial applications, while the latter was a four-cylinder unit, 0.5 litre/cylinder, developed in conjunction with the Austin-Rover Group. With a governed maximum speed of 4500 rev/min, this was the first 2-litre direct injection engine in the world designed to run at such a high speed for installation in cars. It is also offered for industrial applications.

Both the Phaser and Prima are available in naturally aspirated and turbocharged forms. Phaser was produced from the outset with the alternatives of either simple or charge-cooled turbocharging. Its power outputs ranged from 65 to 134 kW, at 2800 rev/min in naturally aspirated and 2600 rev/min in turbocharged forms, Fig. 6.9. The power outputs of the two versions of Prima first announced were 46 and 59.5 kW, in naturally aspirated and turbocharged forms respectively, both at 4500 rev/min. This engine is described in detail in Section 6.45.

6.17   The Phaser combustion chamber

Of major interest in the Phaser was the combustion chamber in its piston, which Perkins termed the Quadram design. This chamber, as can be seen from Figs 6.10 and 6.11, is of approximately toroidal form but with four lobe-like cavities equally spaced around it. Basically, the incoming air is directed tangentially into the cylinder to form a vortex swirling round its axis. Then, as the piston comes up to inner, or top, dead centre, this vortex is forced into the combustion chamber. Because the total energy content of the swirling bases must remain constant despite the fact that its diameter is reduced, its speed of rotation is increased as it is forced into the smaller diameter chamber. The four cavities around the periphery of this chamber generate a comprehensive turbulence system within the column of gas, into which the fuel spray is injected. This so improves the quality of combustion that it reduces the ignition delay period and smooths out the rate of pressure rise, Fig. 6.12, reducing the peak pressure by 10%. Consequently, there is less contamination of exhaust gas, quieter combustion, reduced fatigue loading on components such as pistons, connecting rods and bearings, increases of 13% in power output and 17% in torque, and an 8% improvement in fuel consumption.

An unexpected benefit was a marked insensitivity to manufacturing tolerances. For instance, the size of the gap between the head and the piston crown at top dead centre is usually critical because it has a major impact on squish. With the Quadram system, on the other hand, varying the gap between 0.18 and 0.36 mm was found to have no measurable effect on engine output. In consequence, the combustion chambers did not need to be machined after diecasting while, for selective assembly, the pistons had to be divided into only two grades, as regards height. When machining, tolerances tend to drift out of limits, whereas in diecasting, they do not and, moreover, diecasting in such large quantity production is less costly.

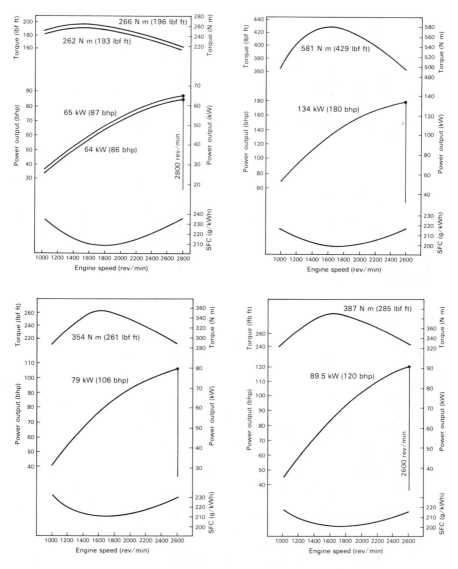

Fig. 6.9 Perkins Phaser engine performance curves BS Au 141a: 1971 and SAE J1349 4.2.4: (*top left*) the 90 unit; (*top right*) the 180Ti; (*bottom left*) the 110T; (*bottom right*) the 120 Ti

6.18 Injection equipment

The nature of the problem in providing fuel injection equipment for compression-ignition engines can be illustrated by quoting some of the parameters that have to be met with the small cylinder sizes now prevalent.

Engines with indirect injection into swirl-type combustion chambers are now as small as 0.35 litre/cylinder. The full load fuel delivery requirement is of the order of 20 mm³ per stroke and that for idling about 15% of this, or about 3

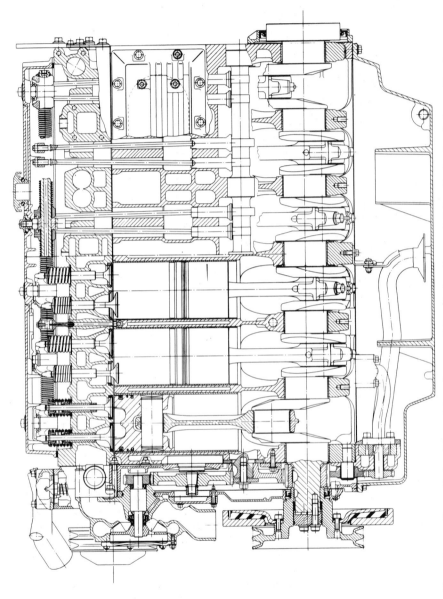

Fig. 6.10 The Perkins Phaser has an all-gear timing drive and, to reduce overall length, the oil pump is below the front main journal bearing

Fig. 6.11 In the Phaser piston is a slightly off centre swirl chamber with, spaced equally around its periphery, four turbulence-generating pockets into which the fuel is injected from a four-hole nozzle

to 4 mm^3 per stroke. For a four-cycle engine of this size running up to 5000 rev/min, these minute quantities have to be injected with precision and regularity at a frequency of up to 2500 per minute. Moreover, the duration of each injection, even at full load, must not exceed some 30 to 50° of crankshaft, implying a time period at maximum speed and load of only one-thousandth of a second.

Direct injection engines, although currently not quite so small, present even more difficult problems owing to the need for shorter injection periods and higher injection pressures. A direct injection engine of 0.5 litre/cylinder running up to 3000 rev/min would need a maximum injection pressure at full load speed of at least 500 bar to force the fuel through injection orifices of no more than 0.20 mm diameter – assuming that four holes are required to give adequate distribution. Moreover these holes need to be drilled to an extremely close tolerance to ensure equality of flow between individual nozzles.

Therefore, highly specialised manufacturing techniques not met with in any other industry are required, with the result that the majority of injection equipment is supplied by a handful of manufacturers scattered over the world, most coming from two major manufacturers – Lucas CAV and Robert Bosch – and their licensees. The only significant exceptions are two US engine makers – General Motors and Cummins – who produce their own equipment peculiar to their engines.

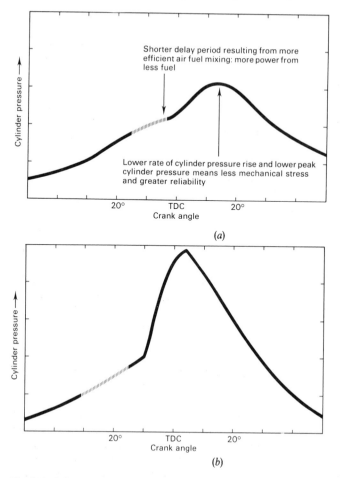

Fig. 6.12 Cylinder pressure diagrams for (a) the Phaser and (b) a conventional combustion chamber. Efficient air–fuel mixing in the Phaser gives a shorter combustion delay period followed by gentler rate of pressure rise to a lower peak pressure, and therefore less noise and mechanical loading

Broadly speaking, the fuel injection system used for both direct and indirect injection diesel engines comprises individual injectors – one for each cylinder – a high-pressure injection pump of rotary or in-line configuration and having an integral governor, and a filtration system to remove contaminants from the fuel. In the following sections we look in greater detail at these components.

6.19 Pintle type nozzle

This type of nozzle (Fig. 6.13) is used with pre-chamber and swirl types of combustion system, because these do not need the high energy jet characteristic of the hole type nozzle. The orifice through which the fuel is discharged into the combustion chamber is an annulus formed by an extension of the valve needle – the pintle – projecting into the single axial hole. The pintle has a biconical profile, shown clearly in the enlarged section, Fig. 6.14. This profile

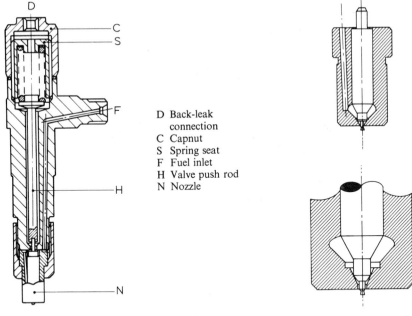

D Back-leak
   connection
C Capnut
S Spring seat
F Fuel inlet
H Valve push rod
N Nozzle

Fig. 6.13 Pintle type   Fig. 6.14

can be varied to produce the angular dispersion characteristics required of the emergent spray, to suit the particular combustion chamber.

When the valve is closed, a parallel portion of the pintle, just beneath the conical seat, protrudes into the hole, and there is a small clearance between it and the hole. During the first part of the valve opening motion – known as the *overlap* – while this parallel portion is lifting out of the hole, it restricts the area of the orifice. This limits the quantity of fuel injected during the combustion delay period, reducing the initial rate of combustion and therefore noise – especially beneficial under idling conditions. All pintle nozzles have some overlap, but those for the modern engines have an extended overlap, and are termed *delay type* pintle nozzles.

A particular advantage of the pintle nozzle for small high-speed engines is the orifice area increases with the valve lift so, in effect, the nozzle is self-adjusting over the wide range of flow rates required as load and speed change. This of course cannot be achieved with the fixed-area hole type nozzle.

6.20   Pintaux nozzle

The Pintaux nozzle was developed jointly by CAV Ltd and Ricardo & Co. Engineers (1927) Ltd, and is applicable only to the Ricardo Comet type cylinder head. This nozzle was designed primarily to enable cold starting to be obtained without the use of heater plugs.

With the normal type of pintle nozzle the spray is coaxial with the injector, and the direction of injection found most advantageous for all-round engine performance is that indicated in the diagrams of Fig. 6.15, which is downstream on the opposite side of the swirl chamber from the throat entry.

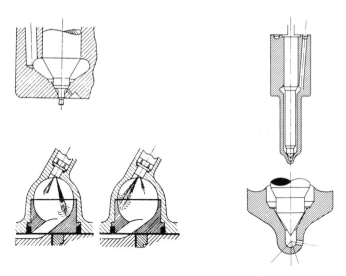

Fig. 6.15 Pintaux type   Fig. 6.16 Hole type

Investigation has shown, however, that under starting conditions the hottest zone of the chamber is outside this spray path and nearer the centre of the chamber on the opposite side. By directing the spray into this zone considerably improved starting can be obtained without the use of heater plugs.

This direction does not, however, give the best all-round engine performance, and the construction of the nozzle has been so contrived that at the low speeds of starting, when the injector needle is only partly lifted, the main orifice is largely blanked by a cylindrical portion of the needle point, and the bulk of the delivery is discharged through the oblique hole shown in the enlarged view of the nozzle. At normal running speeds the cylindrical portion is lifted clear, and about 90% of the delivery issues in the normal manner, but preceded by a small pilot charge which promotes quieter running.

6.21   Hole type nozzles

The hole type of nozzle, Fig. 6.16, is invariably used with direct injection combustion chambers, in which the fuel has to be distributed radially through a number of holes. Generally, too, the spray energy level is higher and the droplet density lower than in indirect injection engines: this is to compensate for the lower degree of swirl.

Most widely used with such combustion chambers is the multi-hole type. This has between two and six holes more or less symmetrically arranged on a wide angle cone, as shown in the inset to Fig. 6.17. These holes are drilled normally to the spherical surface of the tip, breaking through the inner surface, which too is of spherical form and is concentric with the outer one so that the lengths of the holes are equal. To avoid leakage of the fuel into the combustion chamber between injections, the volume enclosed between the valve seat and the tip has to be as small as practicable. This volume is known as the *sac*.

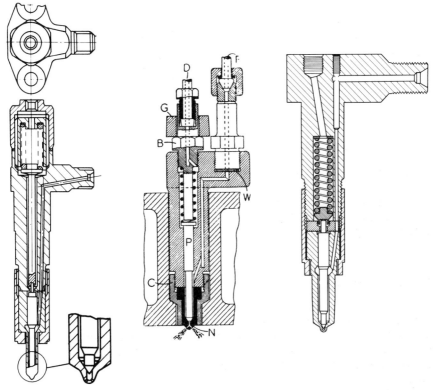

Fig. 6.19 CAV low-spring injector

Fig. 6.17 Multi-hole injector   Fig. 6.18 Gardner injector

In the 'closed' type of nozzle which, for vehicle engines, has now entirely superseded the 'open' type, the valve is a conically-ended needle, the cylindrical upper portion of which is close fitting in the nozzle body, to guide it concentrically with the valve seat. Alternatively, as with the Gardner design shown in Fig. 6.18, it is guided in a separate nozzle holder. In each instance the valve is closed by a spring mounted in the nozzle holder, Figs 6.14, 6.17, 6.18 and 6.19.

A valuable feature of this design is the differential lifting force obtained because the diameter of the guide is larger than the effective diameter of the seat. The valve will begin to lift only when the fuel pressure, acting on the area comprising the difference between that of the guide and the seat, overcomes the force of the spring. As soon as the lift begins, the pressure acts instantly upon the whole of the guide area, so the valve rises rapidly to its limit stop and the flow to the nozzle orifice is unrestricted. A less favourable effect of this differential feature is that the valve will not return to its seat until the pressure has fallen to a value such that the spring force can overcome the pressure acting upon the whole guide area. This means that the closing pressure will always be lower than that at which the valve is opened. The closing pressure in fact often becomes the controlling parameter, since too low a setting may allow combustion gases to enter the nozzle at the end of injection and this can lead to carbon fouling of the interior. This effect can be minimised by reducing the mass of the moving parts, as has been done with the low spring injector, Fig. 6.19.

With the now obsolete 'open' type nozzle, there was no valve to control injection, which began as soon as the fuel pressure at the open end of the hole exceeded the gas pressure in the combustion chamber and ceased when it fell again below that of the gas pressure. One or more non-return valves prevented gas pressure from forcing the fuel to flow in the reverse direction.

6.22   Injector assemblies

Some typical injector assemblies are illustrated in Figs 6.13, 6.17, 6.18 and 6.19. In each, provision is made for collecting and returning to the supply pipe any leakage of fuel between the differential valve and its guide.

This leakage is reduced to a minimum by the extreme precision and lapped finish of the stem and guide assembly, tolerances of the order of 0.0025 mm being common. Valves and bodies must be kept in pairs, never being interchanged.

In Fig. 6.13, the nozzle N is of the pintle type. H is the spindle transmitting the spring load to the differential valve and, on removal of the protecting screw-cap C, the spring load can be adjusted by the screw S which is locked simply by tightening the cap C against the body.

In the Gardner injector (Fig. 6.18) the spring acts directly on the differential valve or plunger P. The upper end of the spring seats on the breech plug B, on top of which bears the holding down yoke G. N is the nozzle and C the nozzle cap, while W is a very fine gauze filter washer provided as a final precaution against choking of the tiny spray holes.

In both, the fuel enters at F, and leakage past the valve is led away at D, to be returned to the main fuel intake.

Turbocharging, which commonly implies cylinder pressures higher than with natural aspiration, can result in gas entering the nozzle, unless opening, and hence closing, pressure is set undesirably high. This problem can be largely overcome using the 'low spring' injector, Fig. 6.19, the moving parts of which have a smaller mass. The effect is a more rapid closure of the valve, as can be seen in the comparative needle lift diagram, Fig. 6.20.

6.23   CAV Microjector

The CAV Microjector first went into large-scale production in 1981. It was developed because of the rapidly increasing demand for small diesel engines, especially for cars, and because the existing types of injector were too bulky for this type of installation and were unsuitable for metering and spraying satisfactorily the small quantities of fuel required per stroke. Such engines, ranging from about 1.6 to 3 litres swept volume, can have cylinders as small as 0.4-litre capacity, and they are virtually all of the pre-chamber type.

With both the pintle and Pintaux type injectors, a major disadvantage is the need for a back-leak pipe to allow fuel to be displaced back as the pintle lifts inwards, in a direction opposite to that of the flow of injected fuel. In the Microjector, this requirement is obviated by the use of a poppet valve opening in the direction of flow. Also, the whole injector has been made smaller, since

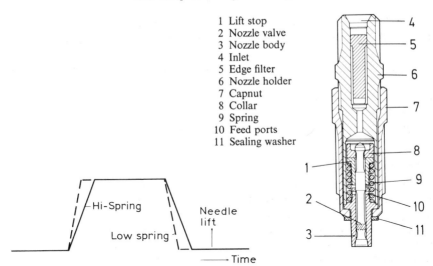

1 Lift stop
2 Nozzle valve
3 Nozzle body
4 Inlet
5 Edge filter
6 Nozzle holder
7 Capnut
8 Collar
9 Spring
10 Feed ports
11 Sealing washer

Fig. 6.20 Needle lift diagram

Fig. 6.21 General arrangement of the CAV Microjector

gas pressure in the cylinder assists the spring in holding the poppet valve on its seat.

As explained in Section 6.19, the Pintaux type injector was developed to facilitate cold starting with the Ricardo Comet combustion chamber. Consequently, an experimental predecessor to the Microjector had the auxiliary hole for injection into the hottest part of the combustion chamber during cranking. However, with the development of greatly improved glow-plugs for cold starting assistance, the auxiliary hole became unnecessary and, with a single hole closed by a poppet valve, all risk of entry of foreign matter into the injector was obviated and production was simplified.

The construction of the Microjector can be seen from Fig. 6.21. Fuel from the injector pump flows through the edge-filter into the nozzle holder, and on into the spring chamber. When the pressure rises sufficiently to overcome the spring and gas forces on the poppet valve, this valve is lifted, and the fuel flows through the spring and the diametral holes, or feed-ports, in the nozzle body, past the helical grooves in the lower guide, and on past the valve seat. If the pressure is high enough, the valve will continue opening until the lift stop has descended on to the upper end of the nozzle body; otherwise it will float in an intermediate position. When injection ceases, the valve closes under the influence of both the spring and the gas pressure, the latter of course being the combustion pressure and therefore much higher than the compression pressure against which it was initially opened.

With this design, the reciprocating masses are small and gas blow-by into the nozzle is not a problem. Also, the jet pattern is such that the droplet size is smaller and the fuel more finely and widely distributed within the chamber than with the Pintaux type. Consequently, the vaporising time, and therefore ignition delay, are shorter, so both the mechanical and combustion noises are less too.

Some features not immediately apparent in Fig. 6.21 are as follows. The collar by means of which the spring assembly is retained, is fitted by passing the top of the valve stem through a hole not shown in the illustration, drilled eccentrically in the collar, and then sliding it radially to centre it on to its smaller-diameter seating, which can be seen in the illustration. The poppet valve is guided in the nozzle body by both the central land between its two waisted portions and the lower land in which spiral grooves are machined to allow the fuel to pass through. Had these grooves been parallel to the axis of the valve, they would have tended to have worn grooves in the nozzle body. The lower end of the valve is a close fit in the body, to prevent combustion products from getting near or entering the conical seat a few millimetres above. Between the seat and the lower end of the valve, the profile is such as to maintain the required spray characteristics as the valve lift is progressively increased. Finally, the thread around the lower portion of the capnut is similar to that of a 14-mm sparking plug so the unit is very compact and easy to install on the cylinder head.

6.24 Injection control

For the high-speed automobile engine, the 'jerk' type of constant stroke plunger pump is widely employed. The term *jerk* indicates the sudden impulse given to the fuel at the instant of injection, followed at the appropriate moment by a sudden collapse of pressure, causing injection to cease in an abrupt manner to prevent dribble.

The volume of the high-pressure system and the inherent compressibility of the fuel, however, leads to a complicated sequence of events in which the pressure developed by the injection pump is propagated along the high-pressure pipe. This pressure wave has a travel time that is significant in terms of the whole injection period. Moreover, the pressure waves are reflected backwards and forwards along the system from its closed, or partially closed, ends both during and after the injection. Therefore it is not possible to predict the behaviour of an injection system by simple calculations. In fact, complicated mathematical solutions, rendered feasible by the use of digital or analogue computers, are now used to supplement and assist in the protracted experimentation necessary to develop an injection system for each particular application.

6.25 The in-line injection pump

Since the merger of the CAV and Simms, the range of classic in-line pumps manufactured in the UK has been rationalised into three basic types, covering the requirements of the smallest to the largest engines for vehicles. These are the Minimec, the Majormec and the Maximec, Fig. 6.22, in increasing order of size and output capability. The maximum horsepowers per engine cylinder obtainable with these pumps are respectively 40, 60 and 100. Throughout this rationalised range the basic constructional features are similar, so it will be necessary to describe only one – the Minimec pump.

6.26 Construction of the Minimec pump

The six-cylinder version, with an integral governor, is shown in Fig. 6.23. It is also available in three-, four- and eight-cylinder forms. For some applications, however, the pump and governor are in separable housings.

Fig. 6.22 Minimec, Majormec and Maximec in-line injection pumps

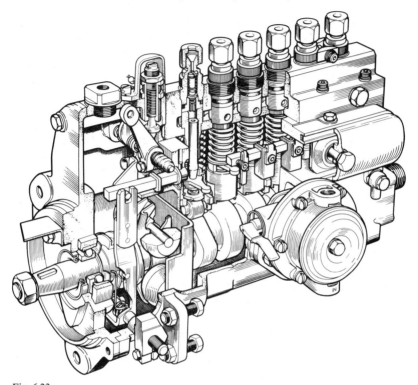

Fig. 6.23

Although the main housing is a light alloy casting in which the camshaft, tappet assemblies and control gears are contained, the pumping element assembly is carried in a one-piece steel pumping head. This design feature provides good sealing capability and stiffness, as compared with the traditional all-aluminium arrangement. Moreover it obviates relaxation problems due to differential thermal expansion.

The alloy steel camshaft has a large base diameter, for longitudinal stiffness. It is carried in ball or roller races mounted in the cast iron end-covers. In the example illustrated, the cover at the governor end is in the form of a flange for direct mounting on the engine timing case, so that the drive gear can be fitted directly to the camshaft. Alternatively, the pump may be mounted on its base, in which case the drive will be transmitted through a jackshaft and couplings. Pumps with separable governors do not have through shafts, and therefore may be driven or flange-mounted only at the end remote from the governor.

To reduce the length of the pump to a minimum, the cams are as close together as possible, and they actuate roller tappets reciprocating in bores machined in the main housing. These tappets are prevented from turning by flats on their peripheries, against which register steel keep-plates, one between each pair. Their upper ends carry shims for adjustment of angular phasing between the delivery lines, to make good any differences arising out of manufacturing tolerances.

In the upper steel body, each pumping element assembly comprises the plunger and barrel, the latter being held down against a shoulder by the screwed-in delivery valve holder. The feed gallery is drilled longitudinally to interconnect the outer ends of the radial inlet ports in the barrels, and the fuel passes into the gallery from the free end. In the body, the plunger and tappet return-springs are retained during assembly by spring plates, which then rest upon the upper faces of the tappet shims.

For controlling fuel delivery, arms pressed on to the lower ends of the plunger register in vertical slots, or forks, in which they slide during the plunger stroke, Fig. 6.24. These forks are mounted on the square section control rod, which slides axially in bearings at each end of the housing. When the control rod is moved by the governor output lever, it rotates each plunger to control the delivery volume, as described in detail in Section 6.26. On assembly, each fork can be adjusted relative to the others, to equalise the delivery from all of them, and they are finally clamped to the control rod with set-screws. Access for this adjustment is gained through the front cover.

The delivery valve holder contains the valve and its spring, the function of which is described in Section 6.28, and at its upper end is the conical seat and screw connection for the high-pressure pipe line. A clamp between each pair of delivery valve holders presents them from rotating during attachment of their pipes.

In Fig. 6.23, the assembly is shown with a feed pump driven by an eccentric on its camshaft. Alternatively, the feed pump may be driven from the engine camshaft. In either case, its outlet is connected through a filter, described in Section 6.32, to the pump inlet.

The camshaft and tappets are lubricated by engine oil. All the larger pumps, and increasing numbers of smaller pumps, are connected to the engine lubricating oil supply. The remaining small pumps have an initial fill of engine oil, which needs to be changed periodically owing to gradual dilution with fuel oil. Measures taken to reduce this leakage, past the elements, are described in Section 6.27.

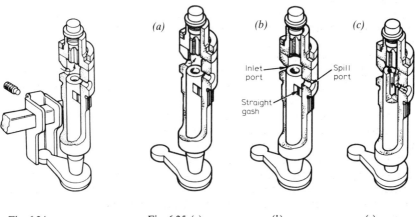

Fig. 6.24   Fig. 6.25 *(a)*   *(b)*   *(c)*

6.27   Operation of the pumping elements

Towards the upper end of the plunger, a straight gash, inclined relative to the axis of the plunger, Fig. 6.25, communicates through radial and axial drillings with its upper end. In the barrel are two radial ports, one termed the *inlet port* and the other the *spill port*. They are not necessarily diametrically opposite one another, and are positioned so that the top face of the plunger, as it moves upwards, covers the uppermost port (or both if they are diametrically opposite) at a point when it approaches its maximum velocity, as determined by the cam profile. Consequently, the fuel that has entered the pumping space during the downward stroke is then cut off from the feed system so, during the remainder of the stroke, it is pressurised and forced through the delivery valve to the high pressure pipe. Displacement continues until the previously-mentioned straight gash begins to uncover the spill port whereupon, because of the interconnecting holes between the gash and top end, pressure will start to collapse and displacement of fuel therefore cease. Fuel displaced during the remainder of the stroke is returned to the fuel system.

The instant when the port, or ports, are just closed by the top face of the plunger is known as the *cut-off point* – in terms of plunger lift or cam angle – and the instant of just uncovering of the port by the plunger gash is known as the *spill point*. Obviously, the volume displacement of fuel, between cut-off and spill, can be varied from zero to a maximum, by rotating the plunger relative to the barrel, since this changes the spill point by moving it up or down the inclined edge of the gash. The rotation is effected, independently of the reciprocation, by the mechanism described in Section 6.25. With the arrangement described, the cut-off point of course is constant, regardless of fuel delivery. When the control rod, actuated by the governor, rotates the element, the delivery is increased or decreased by retarding or advancing the spill point in terms of lift or cam angle.

What has just been described is the normal arrangement. However, it is also possible to obtain constant termination of displacement of fuel by inclining the top edge of the plunger and having a circumferential spill groove. Intermediate effects can be obtained by inclining both edges. A slot can be milled in the upper face, in a position beyond that for the normal maximum displacement so that, by retarding the spill point, the injection can be retarded for cold starting. With such an arrangement, provision must also be made for supplying excess fuel, as described in Section 6.40.

6.28   Control of element leakage

At this point it is appropriate to describe the method of restricting leakage past the pumping elements. A circumferential groove is machined below the gash, in the plunger, Fig. 6.26. This groove connects with a spiral groove extending up towards but not breaking through the top face of the plunger. Fuel leaking downwards from the upper part of the plunger during the pumping reaches this groove, which is so placed that, during the high-pressure injection part of the stroke, it is in communication with the feed system through the inlet port. The leakage is therefore prevented from progressing to the lower part of the barrel where it would pass into the cambox and dilute the lubricating oil.

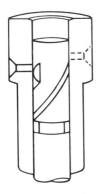

Fig. 6.26 CAV pumping element, spiral groove design

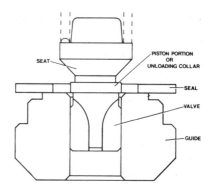

Fig. 6.27 Delivery valve and guide, with valve open

6.29 Delivery valve

The delivery valve, shown in Fig. 6.27, serves an important purpose. It is a spring-loaded mitred-seating non-return valve, guided in its body by a close-fitting grooved shank. Its function is to prevent the pressure waves reflected back from the injector, during or after injection, from interfering with the subsequent injection cycle. The large fluctuations in pressure due to these waves would cause the pipe line to be partially emptied between injections, or would prevent complete filling of the pumping space in the element in readiness for the next injection.

If, however, a simple valve were to be used it would, after closure, reflect the same pressure waves forward to the injector, probably causing secondary injections. Therefore, a plain cylindrical portion known as the *unloading collar* is machined on the valve, immediately below the seat. This, on closure of the valve, withdraws a predetermined volume of fuel from the line, thus moderating the forward reflections sufficiently to avoid secondary injection. The magnitude of this unloading volume has to be carefully chosen to match the parameters of the system.

Unloading also significantly affects the quantity of fuel that is delivered over a speed range with the pump control rod in a fixed position. So, to modify the shape of the torque-speed curve of the engine, a flat or other restricting passage is sometimes machined across the unloading collar.

6.30 Advance devices

An optional feature is a *centrifugal advance device*. This may be required for engines with a wide speed range, to improve performance by optimising injection advance. An example, mounted on the drive end of the Minimec pump, is illustrated in Fig. 6.23. It angularly displaces the camshaft relative to the drive shaft, or gear, as a function of speed. This displacement is determined by the centrifugal force produced by masses acting through cams or wedges to vary the relative angle between driving and driven members.

6.31   Majormec pump

As stated before, this pump is basically of the same construction as the Minimec. The pump and governor housings, however, are always separable to accommodate a variety of governors, and the pump must be driven from the end remote from the governor. Six- and eight-cylinder models are currently available.

6.32   Maximec pump

This is an up-rated and strengthened, long-life version developed from the Majormec unit. It was introduced to meet the requirements for engines of the largest cylinder sizes and highest ratings. To extend the maximum pressure capability and to reduce radiated noise, the cambox has been made extremely rigid. The length and diameter of the tappets have been increased, and the camshaft is carried in large taper roller bearings. This pump is available in four-, six-, eight- and twelve-cylinder forms. The six-cylinder unit is shown in Fig. 6.28.

6.33   CAV paper element filter

Filter elements of cloth, felt and other materials have been used, but an extremely efficient unit is the paper element filter produced by CAV. Its element comprises a double spiral of creped paper, V-convoluted and wound on to a core. This is contained in a cartridge, or metal canister.

The special filter paper is treated with resin, and is of high strength when

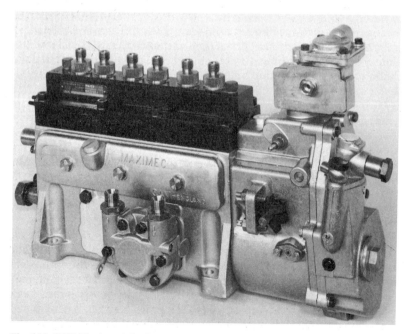

Fig. 6.28 CAV Maximec injection pump

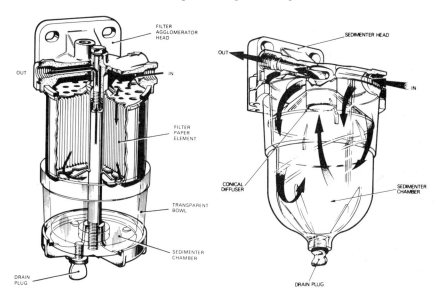

Fig. 6.29   Fig. 6.30

wet. It has been shown that by reducing abrasive wear, the use of such a filter can increase the life of a fuel injection pump element by a factor of five. The spiral convoluted construction gives a large filtering area, and the life – limited by choking – is comparable to that of the older cloth or felt types.

This CAV filter is designed also to trap water droplets, which agglomerate on the clean side of the element. It does so by virtue of the fact that the flow is downwards, so that the water falls into the base which, if required, can be fitted with a transparent bowl, or sleeve, of large volume, as shown in Fig. 6.29.

An additional water separator, for interposing upstream in the feed system, is also available for applications where severe water contamination is likely. This is shown in Fig. 6.30.

6.34   Control of engine speed

Generally some form of governor is used to control both the maximum and idling speeds of compression-ignition engines. This is necessary to limit the inertia stresses that arise from the acceleration of the massive moving parts and to maintain idling stability – since the system is inherently unstable at a fixed pump control setting.

With the petrol engine, a single 'authority' – the carburettor – is of course responsible for delivering a correct mixture, while the air flow, in effect, controls the petrol flow. In the compression-ignition engine, on the other hand, the air drawn in is normally dependent on the breathing characteristics of the cylinders, and varies with speed and operating temperature in a manner characteristic of the engine – the actual quantity of air drawn in per induction stroke decreases with increase of speed, as in a petrol engine.

The fuel delivered per injection is determined, quite independently, by the characteristics of the pump. Therefore, for a fixed control position, it will not fall sufficiently, as speed increases, to match the reduction in quantity of air drawn in per stroke. Consequently the 'matching' of the characteristics of engine and pump is difficult where it is necessary to provide for running at varying speeds at any fixed accelerator position or where, without limiting maximum torque at any speed, excessive fuel must not be injected – with its attendant waste, smoke and smell – at any combination of speed and accelerator position.

6.35   Mechanical or centrifugal governor

If the governor is designed merely to limit maximum speed, movement of its sleeve cannot take place until the centrifugal force developed by the revolving masses is sufficient to overcome the load of the control spring. Therefore it cuts down the fuel supply only at the predetermined maximum speed.

Below this maximum, the accelerator pedal directly actuates the pump control rod. Lost motion arrangements allow the governor to override the foot control at maximum speed. Another spring comes into action when the accelerator pedal is raised and the governor idling stop is reached. This type is known as an *idling-maximum* or *two-speed governor*.

Alternatively, direct connection between the accelerator pedal and pump control rod is dispensed with, so that all movements of the rod are performed by the governor under the control of a governor spring, the compression in which is determined by the position of the accelerator pedal. With such a system, the speed of the engine and vehicle is fixed by the position of the pedal, and the governor actuation spindle then moves the control rod to the position required to maintain, under the prevailing road conditions, the speed corresponding to the compression of the governor spring. This constitutes an *all-speed governor*.

In the Gardner execution of this type of governor, shown in Fig. 6.31, the

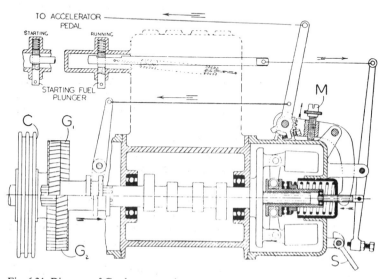

Fig. 6.31 Diagram of Gardner control

mechanism is linked with an injection advance control which gives to one of a pair of helical driving gears a slight axial movement. This correspondingly changes the phase between the pump shaft and the driving shaft. The result is injection advance dependent on pedal position – which is not necessarily the same as on engine speed.

In Fig. 6.31, the arrows indicate the movements that would increase speed and injection advance. C is the chain sprocket on the engine camshaft, G_1 the wide camshaft gear and G_2 the sliding gear on the pump-shaft. S is a stopping lever acting directly on the pump control rod and M is a speed screw operated by hand, or remote control, to maintain any steady speed desired. The last-mentioned feature is convenient for stationary or marine applications, or for testing.

6.36   Vacuum governor

Although formerly popular on the smaller engines, the vacuum governor has now become virtually obsolete. It fell out of fashion owing to inconsistency and ease of unauthorised adjustment. Moreover, it was associated with a risk of uncontrolled runaway should the engine inadvertently be caused to run backwards, for instance when stalled on a hill. Now, such engines have simplified forms of mechanical governor or, with distributor-type pumps, in-built hydraulic or mechanical governors.

6.37   Hydraulic governor

An hydraulic governor for in-line pumps was, until recently, manufactured by CAV but it has now been replaced by cheaper mechanical governors in association with the rationalised pumps described in Sections 6.24 to 6.31. Its unique feature was an ability to give very low idling speeds which, owing to the large amplitudes of rocking motion at low resonant speeds following the universal adoption of flexible mountings for road transport engines, became impracticable without it.

6.38   CAV governors for in-line pumps

The range of CAV governors covers the requirements for all the types of pump already described, and some are suitable for use with more than one of them. Each is basically an all-speed type, but those for road vehicles have the special characteristics required for this duty.

Fitted only to the Minimec pump and shown with it in the cut-away view, Fig. 6.23, is the GE. This is the simplest of the range. The centrifugal weight unit consists essentially of two dumb-bell like masses, driven by a U-shaped member attached to the camshaft, carrying slippers sliding on inclined faces of a member that moves axially on an extension of the camshaft.

More recently, CAV has introduced an alternative governor, known as the C type. Its lever and spring-loading system, common with that of the GE unit, can be seen in Fig. 6.32, but the components peculiar to the C unit are shown more clearly in Fig. 6.33.

A cage having six pockets drives the weight unit. It is either riveted directly to the camshaft flange or, as in Fig. 6.32, has a cush drive to isolate it from

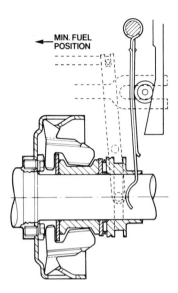

Fig. 6.32 CAV C type governor

Fig. 6.33 Components of the CAV C type governor

torsional oscillations of the camshaft. Two, three, four or six weights can be fitted, pivoting on the outer inside corner of the pocket. A finger on each weight engages the sliding sleeve, which transmits the centrifugal force through a thrust race to the spring loading mechanism.

A compound leaf spring is employed. It is loaded against the thrust sleeve by a roller assembly which, moved up and down by a forked lever, runs between the spring and a fixed ramp attached to the governor cover. The forked lever is actuated by a linkage system connecting it to the driver's pedal.

Movement of the roller not only regulates the reaction to the centrifugal force, and hence the position of the thrust sleeve at a given speed, but also varies the rate of the spring by shortening its effective length as the load is increased. By appropriate selection of ramp angle and position, the governor characteristics can be made to suit individual requirements at idling, inter-mediate and maximum speeds, without the need for auxiliary damping devices.

The movement of the thrust sleeve is transmitted to the pump control rod through the pivoted lever shown dotted in Fig. 6.32. Thus, this governor is of the all-speed variety, the driver setting the equilibrium speed by directly selecting the load on the spring. The variable-rate feature compensates for the normally unfavourable square law characteristic of a centrifugal governor, otherwise the governing would become progressively sharper as speed is increased, which is undesirable for road vehicles.

A former variant of the GE, known as the GX unit, had a conventional swinging weight unit to give additional governor force for some duties. The GM unit is a somewhat larger governor with two or four swinging weights, for the Majormec and Maximec pumps. It can have the leaf-spring loading system, as used with the GE and C, or an alternative tension spring, both having the variable spring-rate feature.

More recently, the GCS governor has been introduced for the more exacting duties of the Majormec and Maximec pumps. It is shown, with a boost control above, mounted on the Maximec pump in Fig. 6.28. The weight unit is that of the C type already described; however the thrust sleeve is not directly connected to the pump control rod but actuates it through an hydraulic servo, using engine oil as the pressure medium. In this way, high control rod forces are obtained without the need for large centrifugal forces. As a result, the compact and high quantity production C unit can be utilised.

6.39 Boost control

The incidence of turbocharging of vehicle engines has led to the need for a boost control. Such devices are available as an option for most of the previously described governors.

Turbocharger characteristics are such that the maximum fuel the engine can take at low speeds is less than that at higher speeds. It is difficult to match this requirement with the natural delivery against speed characteristic of the injection system, except with very moderate degree of turbocharging. More-over, the slight lag in the response of the turbocharger, when accelerating the engine rapidly from idling, causes an unacceptable puff of smoke owing to excessive fuel delivery.

Both these difficulties may be overcome by replacing the fixed maximum fuel stop in the normal governor by a moving stop whose position is a function of boost pressure. This can be arranged to match the maximum delivery of the pump to the ability of the engine to consume fuel without smoke, both at steady speeds and when accelerating.

Such a device is attached to the top of the CAV GCS governor on the

Maximec pump illustrated in Fig. 6.28, and sections are shown in Fig. 6.34. It is known as the BC50. Manifold pressure, applied to the upper surface of a rolling diaphragm attached to a guided spindle, acts against a spring. Movements establishing equilibrium are transmitted by a rocking lever to another spindle, which replaces the normal fixed fuel stop. The limits of the movement and the ultimate maximum position can be adjusted. Also shown in the section is an optional spring-loaded stop for the maximum fuel rod, which varies the maximum fuel as a function of governor force and therefore of speed, thus providing a facility for shaping the delivery curve over the upper part of the speed range – after the boost control has reached its maximum position.

6.40   Torque control

Shaping the curve of maximum fuel against speed, so that the torque curve falls as speed increases, can also be effected with any of the governors by replacing their usual fixed maximum fuel stop by a spring-loaded stop similar to that of the boost control. This is a common requirement for tractor engines.

6.41   Excess fuel

An excess fuel feature is commonly provided to improve the cold starting of the engine. Legislation requires that, to prevent excessive smoke emission, this facility cannot be used by the driver whilst the vehicle is moving.

The additional movement of the control rod for obtaining extra fuel is effected by a latch mechanism in the governor, which allows the normal maximum fuel stop to be overridden. However, it can be brought into operation only if the driver leaves his seat to operate it. This device, usually on the pump, is designed to latch out automatically as soon as speed rises after starting. Additionally, the design is usually such that the driver cannot set the device permanently into the excess fuel position. To meet all these requirements, several ingenious mechanisms, too detailed to describe in a few words, have been produced.

To satisfy requirements for excess fuel conditions, it is possible to retard the beginning of fuel injection, by modification to the pumping elements. This is done by milling a slot in the appropriate position in their upper faces.

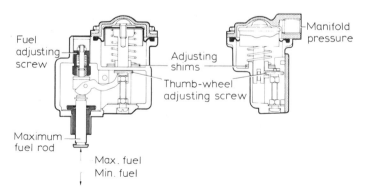

Fig. 6.34 The BC50 boost control

6.42 CAV DPA pump

With the development of the smaller diesel engines, manufacturers soon realised that the cost of a multi-line fuel injection pump of the classic type was becoming a prohibitive proportion of the total engine cost, and that progress depended on the development of smaller, simpler and less costly equipment. Many other advantages are offered by the distributor type of pump, on which much development work has been done since its introduction. A schematic diagram of the complete system is shown in Fig. 6.35.

The DPA pump, manufactured by CAV and its licensees for some years, has more recently been developed further to increase its capability and to meet the special demands for compression-ignition engines for light vehicles and cars. It has a single pumping element, which serves all cylinders. Distribution to the individual injectors is effected by a rotor having a single inlet, and delivering in turn to the appropriate number of outlets. This ensures both uniformity of delivery to all injectors and in-built and exact uniformity of interval between successive injections. There are no adjustments to be made or maintained.

The pumping element consists of two plain opposed cylindrical plungers in a diametrical hole in the rotor head, an extension of which forms the distributor. An axial hole drilled in this extension connects the pumping chamber with a radial hole which registers in turn with radial delivery ports, one for each cylinder of the engine, in a steel sleeve in the 'hydraulic head', which is an aluminium housing containing also the governor and control mechanism.

Fuel is delivered by a transfer pump through the pressure regulating and metering valves and an inlet port to force the pumping plungers apart with a

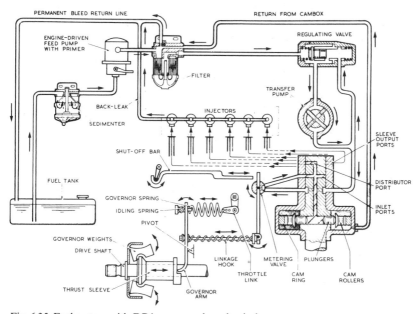

Fig. 6.35 Fuel system with DPA pump and mechanical governor

metered volume of fuel. They are returned, to effect the injection, by lobes in an external cam ring, which is stationary except for a small angular movement within the pump housing to give automatic advance to the injection timing when required. The arrangement of the ducting for the incoming fuel is the inverse of that for delivery: that is, there is one radial hole in the hydraulic head and there are as many radial holes in the distributor rotor as there are engine cylinders, each passing the fuel into the axial hole, as shown in Fig. 6.36.

Hardened rollers carried in sliding shoes are interposed between the pumping plungers and cam ring. Cam-shaped lugs on the sides of the shoes register in corresponding slots in two rotatable side plates, to provide the adjustable 'maximum fuel stop' by limiting the outward movement of the pumping plungers, as shown in Figs 6.37 and 6.38.

No ball or roller bearings, or return springs are required for the location or

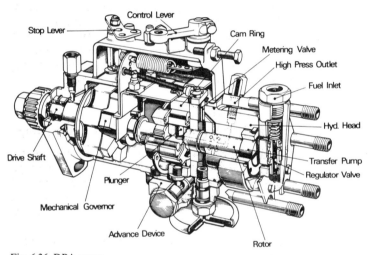

Fig. 6.36 DPA pump

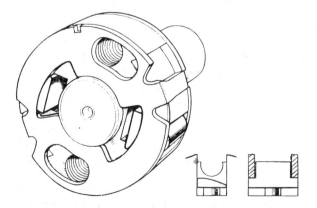

Fig. 6.37 The adjusting unit assembled for fitting

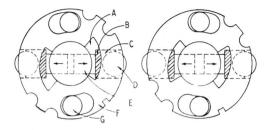

A	Cam-shaped slot
B	Lug on roller shoe
C	Roller shoe
D	Roller
E	Plunger
F	Adjusting plate
G	Locking screw hole
	in shoe carrier

Fig. 6.38 The method used to adjust the maximum travel of the plungers

control of the moving parts, which are lubricated by the fuel oil delivered under controlled pressure by the transfer pump.

By virtue of these features of construction, significant reductions in cost and gains in performance have been obtained. The small mass of the pumping elements and absence of return springs for the pumping plungers enable higher speeds to be attained. Currently, the limit is about 12 000 deliveries per minute, the controlling factor being the ability to recharge the pump between successive injections. The only adjustment to be made by the engine builder is that of the maximum delivery stop, but even this is normally preset during testing of the pump by its manufacturer.

6.43   Governing metering and timing adjustment of the DPA pump

Variation of delivery with load is effected by the metering valve under the control of an 'all-speed' governor of either the mechanical or hydraulic type, the former having important merits of precision and flexibility of control, while the latter, Fig. 6.39, is more compact, simpler and less costly. A variant of the mechanical governor is available, which governs at idling and maximum speeds only.

The standard all-speed governors operate on the principles referred to in Sections 6.36 and 6.37. The accelerator pedal movement does not control directly the quantity injected, but only the *position* at which spring-load balances centrifugal force or hydraulic pressure. The required speed of the engine is then maintained automatically, no matter what the load, gradient or gear ratio of the transmission.

A mechanical governor is most commonly used. It is shown diagrammatically in Fig. 6.35 and, annotated appropriately, in Fig. 6.36. The hydraulic version differs only in that the mechanical governor and its attachment to a rotating metering valve is replaced by a reciprocating metering valve – spring-loaded by the control lever – which responds to a speed-sensitive pressure generated by the transfer pump.

With both types of governor, metering is effected by opening communication through a variable orifice between the transfer pump – under regulated pressure – and the injection plungers for an interval of time determined by the period of registration of each rotor inlet port, in turn, with the single port in the hydraulic head. This period of course varies with rotor speed, but the governor – mechanical or hydraulic – automatically adjusts the variable orifice so that, within that period, the required volume of fuel is delivered.

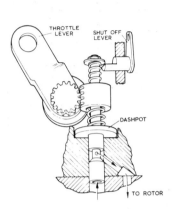

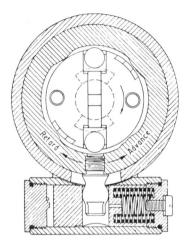

Fig. 6.39 Hydraulic governor

Fig. 6.40 The injection timing is controlled
automatically by moving the cam ring

With either mechanical or hydraulic governoring, the pump is basically
stable in the fixed metering valve condition – equivalent to fixed rack with the
in-line pump. This is because, during the fixed angle of registration of the filling
ports, the volume of fuel metered is inversely proportional to speed: for
example, as speed is reduced by an increase in load, the time period for filling
or metering is increased, and more fuel is metered. By virtue of this
characteristic, good governing is obtained with relatively light forces.

The speed-dependent automatic injection timing advance device consists of
a piston, subjected to transfer pressure, moving in a housing mounted under
the pump body. Its motion, transmitted through a ball-ended screw, rotates
the cam ring. Equilibrium is maintained by coaxial springs opposing the force
of transfer pressure on the piston, Fig. 6.40. This timing device can be fitted to
both the hydraulically- and mechanically-governed versions.

Figs 6.41(*a*) and (*b*) show diagrammatically the operation of the regulating
valve during priming, both when the transfer pump is at rest and during
normal running. For priming, the separately-mounted fuel-feed pump is
operated by hand and the free plunger piston is in equilibrium between the low
pressure feed and the light priming spring. When the transfer pump becomes
operative, the stiffer regulating spring is in equilibrium between it and the
high-pressure feed.

The spring characteristics and orifice arrangement determine the final
torque/speed and advance/speed characteristics of the engine, and can be
varied to suit the application. For diagrammatic convenience the valve is
shown in the horizontal position. The actual setting in current models may be
seen in Fig. 6.36.

6.44   Additional DPA features

A more recent development is the four-plunger pump, in which the two
plungers of the standard pump are augmented by a further two for obtaining a
higher rate and more rapid termination of injection. This requirement has

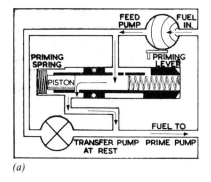 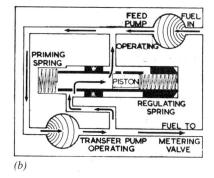

Fig. 6.41 Regulating valve

arisen, in particular, to enable engines to meet stringent smoke, noise and emission legislation without sacrifice of power output.

The pump offers greater flexibility in choice of the injection-pressure pattern, by increasing the displacement rate without increasing mechanical stresses. For example, it has been found that engine timing can be retarded up to 6° without increase in smoke and with advantages in respect of both noise and emission levels. Alternatively, where stringent legislation does not apply, this pump can be used to give an increase in performance by providing higher injection pressure and increased fuel delivery within established reliability limits.

It is available for either four- or six-cylinder engines. The major difference by comparison with the standard pump is in the rotor head assembly. From Fig. 6.42 it can be seen that an additional pair of plungers is provided, and the spacing is 90° for four-cylinder engines, or 60° and 120° for six cylinders. This of course is to suit the spacing of the cam lobes. The four-plunger pump is suitable for engine applications up to about 20 kW/cylinder.

The standard device previously described advances injection as a function of speed, as indicated by the transfer pressure. It is commonly arranged to give a sharp retard when the engine, and therefore the pump, is rotated at cranking speeds. This is to optimise timing for cold starting – an alternative to excess fuel.

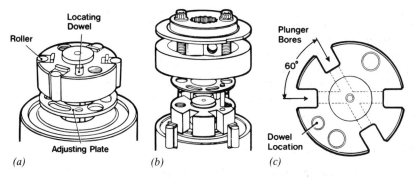

Fig. 6.42 DPA pump with four-plunger head and rotor. (*a*) Roller and shoe assemblies, (*b*) head and rotor assembly, (*c*) end view of rotor

This same device can be arranged to advance injection with reduction of load, instead of speed, when necessary to counter the natural characteristic of the pump. This is done by utilising a pressure signal dependent on metering valve position.

To meet the requirements laid down by legislation, particularly regarding gaseous emissions, an injector-advance device sensitive to both speed and load has become necessary. A more recent development, therefore, has been the introduction of advance devices sensitive to both speed and load signals independently. In some instances these devices are designed to have non-linear characteristics.

With the DPA pump, provision can be made for delivering excess fuel. Although the optimisation of timing for cold starting is usually sufficient for direct injection engines and all but the smallest indirect injection units, generally the latter have either combustion chamber heater plugs or external devices for heating induction air during cold starting. With the introduction of very small indirect injection engines for cars and light vehicles came a need for both optimum timing and excess fuel, since the high surface : volume ratio of the small cylinders leads to particular difficulty in cold starting. Features are therefore incorporated for excess fuel to be both supplied and cut off automatically.

We conclude this chapter with some examples of actual engine designs that demonstrate the application of the fundamental principles in practice. They also show how the components are assembled to form compact and efficient power units, and the way in which the problems that have been outlined previously are overcome.

6.45　Perkins P3 diesel engine

The application of the compression-ignition engine to lighter vehicles was originally delayed by the fact that injection and combustion difficulties increase greatly with diminution in size of cylinder. This has stimulated interest in the use of a reduced number of cylinders for the lower powers.

The Perkins P3 engine has been developed as a substitute or conversion unit for certain tractors and light commercial vehicles where the flat torque characteristics and operational economy of the diesel engine are decisive factors in the choice of the power unit. Suitably limited power is obtained without the difficulties that are experienced with small cylinder and injection equipment, and a simplified servicing system and rationalised manufacture are made possible by the use of standardised general engine parts in the P3, P4 and P6 engines.

The P3 is half the P6 engine, the special parts required being confined to those depending on the length of the engine, such as crankshaft, cylinder block and sump.

The overall dimensions have proved very convenient for installation in the space occupied by the alternative petrol engine, and the high reputation of the engine has led to a considerable demand for the P3 as a conversion unit both in the UK and abroad.

Longitudinal and cross-sectional views of the engine are shown in Fig. 6.43, from which the sturdy and yet compact design will be noted. Dry cylinder

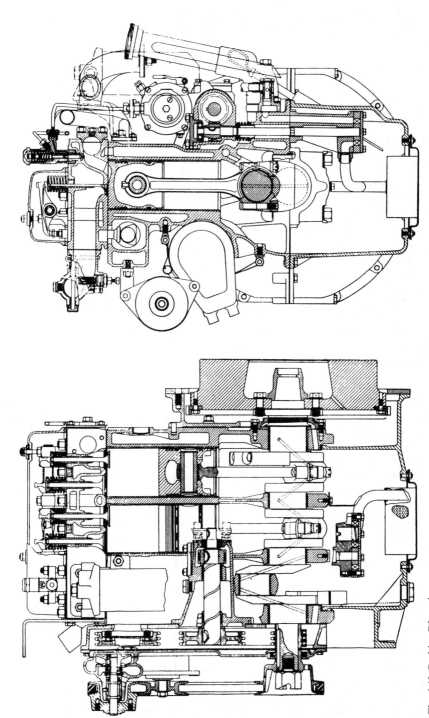

Fig. 6.43 Perkins P3 engine

liners are fitted into a nickel or chromium cast iron block which extends from the head face to the crankshaft centre line in the conventional manner.

The bore and stroke are 89.9 and 127 mm respectively and the connecting rods are 228.6 mm long between centres.

The sturdy four-bearing crankshaft is of forged steel, the four main journals being Tocco hardened.

With three cranks at 120° pitch there are no primary or secondary unbalanced reciprocating forces, and the rotating couple is balanced by the two attached balance weights on the end crank webs. Primary and secondary couples remain to be absorbed by the engine mounting, and this disadvantage and the massive flywheel required to absorb the variations in turning moment are the penalties to be paid for the convenience of the three-cylinder lay-out.

6.46 Perkins Prima DI engine

The alternative to reducing the number is, of course, to reduce the size of the cylinders. This had, until 1986, been impracticable for the reasons given in Section 6.12. Though, as mentioned previously, Perkins was first with its Prima engine, Ford had about a year earlier introduced a 2.5-litre DI diesel engine rated at 4000 rev/min for their *Transit* van. Some of the disadvantages of indirect injection are outlined in the last paragraph of Section 6.12 and the problems to which it is a solution in Section 6.18.

As can be seen from Fig. 6.44, the Prima is a four-cylinder unit with an enclosed, 30 mm wide, HTD toothed belt drive from the crankshaft to the injection pump and overhead camshaft. The alternator on the side of the crankcase, and the spindle of the fan and water pump above the crankshaft, are both driven by a V-belt from a pulley on the front end of the crankshaft. On the other hand, the oil pump is interposed between the toothed wheel for the HTD drive and the crankcase wall and driven by a gear on the crankshaft.

It is of interest that, unlike the Prima, the larger and more powerful Phaser engine has an all-gear drive for its auxiliaries, Sections 6.16 and 6.17. On the Prima, a toothed belt is more suitable because it is lighter yet adequate for the loading and, particularly important for a car engine, quieter. Another factor that may have influenced the choice is the fact that the crankcase was designed to be machined by Austin-Rover on the same production line as the O-Series petrol engine described in Section 3.65, which also has a toothed belt drive.

Structurally, the engine is similar to the O-Series, with siamesed cylinders and a fully balanced SG iron crankshaft carried in five main bearings. However, whilst the crankcase, because of the need for good bearing properties in the cylinder bores, is of the same high quality, flake graphite cast iron as that of the O-Series, the main bearing caps are of the stronger SG iron to react the higher peak gas pressures of the diesel engine. In the turbocharged version, peak combustion pressure is of the order of 12 000 kN/m².

One of the advantages of DI head, as compared with one for an IDI engine, is the freedom to position generous coolant passages all round the valve seats, owing to the absence of a pre-chamber, and the consequent reduction in thermal fatigue loading. The eight valves are in line, with their axes in a vertical plane slightly offset to one side of that containing the axis of the crankshaft, so that the injector nozzles, on the other side, can be sited appropriately relative

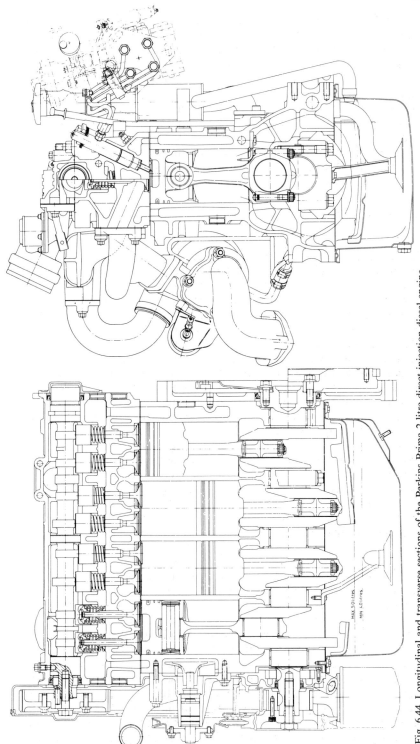

MAX 5.0 litres
MIN 4.0 litres

Fig. 6.44 Longitudinal and transverse sections of the Perkins Prima 2-litre direct injection diesel engine

to the bowl type combustion chambers in the pistons. Sintered iron valve seats are shrunk into the head.

The pistons have steel inserts for expansion control. They are of the three-ring type, the top ring being armoured, which means that its groove is machined in a steel ring bonded in the periphery of the piston, the appropriate distance below its crown. Armoured ring grooves, which are a useful aid to reducing the rate of wear of the very hot top grooves, especially in the more severely loaded turbocharged engines, are uncommon in light diesel engines for cars. The turbocharged engine also has jets in the main gallery to squirt oil up into the pistons to help cool them. Centrally positioned in the piston crowns are shallow cylindrical combustion chambers. These have flat bases around which are fillets of large radii to merge them smoothly with their vertical walls. Small lips inside the upper edges of these walls trigger micro-turbulence in the air as the squish spills over them into the chambers below.

The chilled cast iron camshaft is carried in three bearings, the upper halves of which are formed in the cast aluminium alloy cover and the lower halves in the head casting which, by virtue of the absence of the need to core in it a pre-chamber capable of resisting very high temperatures, is also of aluminium alloy. Single springs for each valve are enshrouded by inverted bucket type steel tappets, discs of appropriate thicknesses being used for setting the valve clearance. Two concentric springs would have been unnecessary for this engine because its valve train is of relatively light weight compared with those of the heavier engines for commercial vehicles.

High volumetric efficiency was obtained by curving the inlet ports to give the air an initial swirl before directing it tangentially into the cylinders, instead of masking the valves, or restricting their throats to increase the velocity of flow. Then, as the piston rises to TDC, micro-turbulence is introduced, as described above. This system was found to be much more appropriate to such a small engine than any of those of the larger, slower speed diesel engines, many of which have combustion chambers of part spherical, or even toroidal, form to introduce a secondary swirl, as distinct from micro-turbulence, into the primary one.

As regards fuel supply, a prime requirement was to shorten the ignition delay period. This was necessary in order that injection could be retarded, to limit the length of time during which the combustion products dwell in the very high temperature range, thus reducing emissions of NO_x yet still enabling the engine to operate efficiently over a speed range adequate for a private car installation. In the event, it was found that injection at 10° before top dead centre was the optimum at the maximum speed of 4500 rev/min. As the speed of the engine falls, it is automatically retarded further, because there is then more time in which to complete combustion before the exhaust valve opens. The relationship between engine speed and injection retard, however, is not linear.

The next problem was intimate mixing with the air to ensure that all the fuel would be completely burned without leaving black smoke or other undesirable emissions in the exhaust. In addition to atomising the fuel as finely as possible as it was injected, this was effected by inducing a degree of turbulence in the air, in the way previously described, such that it would evaporate the fuel from the droplets without being so fierce as to quench the combustion locally during the early stages of combustion. Most difficult of all was to obtain consistent results over the whole range of load and speeds. This was achieved only by patient development, using high technology equipment which, in recent years, has become available for meeting demands for clean exhaust gases and low fuel consumption. With such equipment it is now possible to study velocities and patterns of movement of the gases and their combustion in the cylinders of an engine whilst it is running under load.

By injecting at a very high rate the fuel is introduced into the air in the shortest possible time and in the most finely atomised spray practicable and, at the same time, its kinetic energy is available for conversion into heat energy for evaporation. Rapid injection in a fine spray implies small spray holes in the injector nozzles and high fuel pressures. In fact the naturally aspirated Prima has four holes and the turbocharged version five holes, all 0.24 mm diameter, and the maximum injection pressure is $65\,000\,kN/m^2$.

Features of the Bosch distributor type pump include governing at idling and maximum speeds and control over torque by regulating the output of the pump in relation to the hydraulic pressure in its internal fuel supply, which is proportional to engine speed. In turbocharged engines an additional control senses boost pressure and regulates the pump output accordingly up to maximum boost, after which the output is determined, by the hydraulic control, on the basis of engine speed only. The function of this control is to prevent the emission of black smoke momentarily, between the instant that the rate of fuelling is increased for acceleration and the time the turbocharger takes to accelerate to the required speed. If the throttle is opened suddenly at no load, the engine speed will rise to about 5100 rev/min, where it is held by the governor. The design safe speed for the engine is 6000 rev/min.

For cold starting, there is a glow plug in each cylinder, and these plugs are brought into operation automatically by a thermostatic device when the ignition key is turned to start the engine. They then remain on for a predetermined period, the length of which is dependent upon the temperature of the engine when the starter was actuated. At these low temperatures a wax element thermostat actuates a valve to modify the hydraulical pressure in the timing system, which advances the injection timing. At the same time, a solenoid is activated to obtain fast idle. Performance curves are illustrated in Fig. 6.45.

6.47 Gardner LW

An outstandingly successful example of the direct injection type is the Gardner LW engine, built in three-, four, five- and six-cylinder forms, all of 107.95 mm bore and 152.4 mm stroke to develop about 13.67 kW per cylinder at maximum governed speed of 1700 rev/min with a best bmep of about $703.27\,kN/m^2$.

The meticulous care and skill devoted to the manufacture of the engines and injectors have resulted in an engine of proved reliability which has been adopted as the power unit by a large number of commercial vehicle manufacturers. To meet the challenge of the demand for a still lighter and higher speed engine, the makers, Norris Henty & Gardners Ltd, later produced the 4 LK engine of 95.25 mm bore and 133.35 mm stroke, which develops 39.5 kW on a rising curve at 2000 rev/min governed speed.

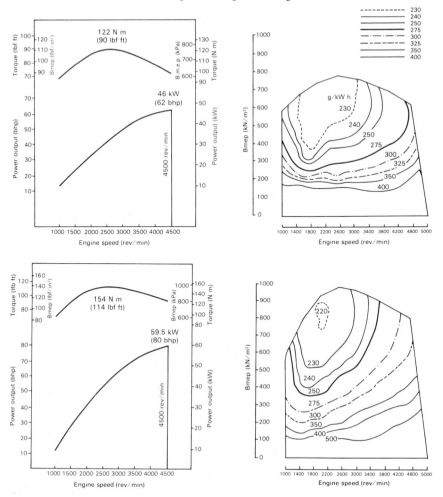

Fig. 6.45 Performance curves (BS Au 141a: 1971) and fuel consumption maps for (*top*) the Perkins Prima 65 and (*bottom*) 80T direct injection diesel engines

At the bare engine weight (without electrical equipment) of 261 kg, the specific weight is 6.61 kg/kW. This engine is in successful use in the lighter types of commercial vehicles, replacing the 3 LW unit which develops about the same power but at a lower speed. The fuel pump used on all Gardner engines was originally of a type modified to incorporate special priming levers. The arrangements for altering the compression at starting are as follows.

A compression control lever is provided for each pair of cylinders as shown at L in Fig. 6.46. This operates a gear quadrant meshing with a gear pinion mounted on the end of a control shaft C lying under the push rod end of the valve rockers.

The shaft C carries radial cams for lifting each inlet valve rocker through the adjusting studs S and a face cam F, which in one position of the lever L moves the rocker of No. 1 inlet valve to the left against the coil spring R, thus bringing the offset portion O of the rocker end over the valve stem. This offset end is

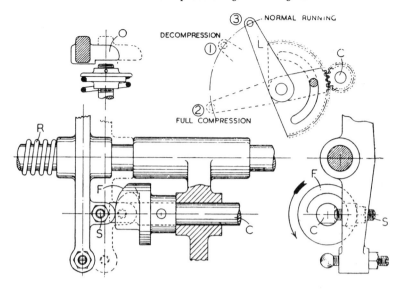

Fig. 6.46 Gardner compression control

stepped so as to increase the tappet clearance to about 1.5 mm. There is a further slight increase due to the tilting of the push rod as its cupped upper end moves over with the rocker.

The effect of this increase of tappet clearance is to make the closing of the inlet valve earlier, and thus give an effective compression ratio corresponding to the full swept volume of the cylinder.

The accompanying later opening and reduced lift of the valve do not, at the low speed of cranking, have any appreciable strangling effect on the induction.

The radial cams are provided for all cylinders, and consist simply of clearance flats on the control shaft C, rotation of which causes the adjusting studs S to ride on to the cylindrical portion and so raise the rockers. The corresponding positions of the control levers are shown in the diagram in Fig. 6.47.

The first position (1) gives complete decompression by holding the inlet valves off their seats during initial cranking. Position (3) is the normal running position with the starting control cut out.

The face cam is ordinarily provided for No. 1 cylinder only, and is brought into action in position (2) of the lever to give higher compression in that cylinder in order to obtain the first impulse. The control levers may be operated by a common grouped control or independently, as may be most suitable for the particular installation. It may be found convenient to keep a pair of cylinders decompressed until the starting cylinders have got well away.

Fig. 6.47 shows an exterior view of the near side of the 5LW engine. A feature which should be noted is the deep and rigid crankcase structure extending well below the crankshaft centre line. The sump is an electron casting. The cylinder block is in two separate portions of three and two cylinders, and the special CAV injection pump is assembled from corresponding units. On the injection pump are five priming levers, one for each cylinder, to actuate the plungers in

Fig. 6.47 Gardner 5LW engine

the pump. The delivery pipes to the injectors are of approximately equal length.

6.48 Unit injectors

So far, we have been describing systems in which the pump and injectors are separate, on the basis that the cost is lower if we have only one engine-driven pump serving all cylinders. However, there are advantages in having one combined pump and injector, a *unit injector*, per cylinder, without a separate pump. These individual injectors each incorporate an injection nozzle and a plunger type pump, usually actuated through the medium of a rocker and pushrod by an extra cam on the camshaft for the valves. However, because of the very heavy loading, a roller follower is usually used with the cam.

The advantages are: a single unit, with only minor variations in detail design, can be used for many different engines over a wide range of power outputs; since removal and replacement does not entail retiming, servicing is simple; only one pipeline is required for supplying all the units on one bank of cylinders, since it can take the fuel to a gallery, preferably beneath the rocker cover on the head, whence it can be distributed through short branch pipes to each unit; the fuel in the supply line to the injectors is at a low pressure, so lightweight pipes can be employed throughout and risk of leakage is reduced; indeed, the gallery, or even a manifold comprising gallery and branches, can be drilled in the cylinder head, thus further reducing both weight and vulnerability to leaks; not only are all the high pressure ducts to the injector nozzles actually within the injectors and therefore both protected and short, but also they are of equal length. The significance of short passages of equal length is that injection pressures can be much higher, and injection timing more accurate and consistent from cylinder to cylinder. Moreover, hydraulic pulsations caused by vibrations due to the elasticity of long high pressure pipelines are avoided, as also are those due to reflection of pressure waves from end to end of the pipe runs.

The penalties are: the higher cost of as many injection pumps as there are cylinders, the fact that they occupy more space than simple injectors on the cylinder head, and the need for a camshaft that is much sturdier than if it had to actuate only the valves. By comparison with the short shaft in a separate injection pump, such a long, large diameter camshaft is very heavy and can more than offset the advantage of lighter fuel lines. On the other hand, because it is of such large diameter it can more easily be cored during casting, to make it hollow and thus partly reduce the increase in weight.

6.49   The Cummins PT system

Cummins has been using this system since 1924. The initials PT, standing for pressure–time, imply that the quantity of fuel flowing through the orifice into the injector cup is determined by its pressure and the time the orifice remains open. The layout of the system is illustrated diagrammatically in Figs 6.48 and 6.49. Fuel is drawn from the tank, through a filter to a gear type pump and thence into the governor, whence it passes through a throttle valve and a shut-down valve, to the pipeline that delivers it to the injectors. Of these components, all between the pipelines from the tank and to the injectors are actually grouped in a single unit, Fig. 6.50, into which both the spin-on filter may be screwed and the drive taken, either directly or in tandem with another auxiliary such as the compressor, from the engine to the gear type pump. Delivery pressure from the fuel pump will be subsequently boosted to the injection pressure by the cam and rocker mechanism, so it does not have to be more than $1750 \, kN/m^2$ as compared with well over $70\,000 \, kN/m^2$ for injectors in which the valves have to be opened by hydraulic pressure supplied from an external pump.

The governor, which is of the rotating twin bob-weight type, regulates only maximum and idling speeds. It does this by moving a spool valve axially between stops to limit the rate of supply of fuel at its two extreme positions. From zero load up to maximum speed at any load, the driver effects control through the accelerator pedal, which actuates the throttle in the fuel delivery line. When maximum speed is attained at full load (maximum power output), the throttle lever is in the maximum fuel position, so the pressure, and therefore quantity of fuel delivered, is at its maximum. If the load is then increased, the engine speed and, with it, the fuel pressure from the gear type pump will fall. This fall in speed causes the mechanical governor to relax its axial pressure on its return spring, called the *torque spring*, thus allowing the spool valve to move to the left, in Fig. 6.48, to reduce the quantity of fuel recirculating back to the induction side of the pump. Consequently, more fuel is delivered through the driver-controlled throttle in the delivery line to the injectors. Another, but natural, consequence of a fall in engine speed is that the duration of opening of the injector orifice increases, so more fuel can enter the injector cup. Both effects increase the engine torque as the speed and power fall off.

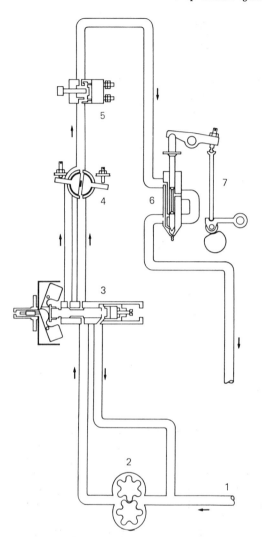

1 Fuel from tank
2 Gear type pump
3 Governor/pressure regulator
4 Hydraulic throttle
5 Shut-down valve
6 Injector
7 Cam, roller follower and pushrod for actuating the injector

Fig. 6.48 Diagram of the Cummins PT injection system hydraulics

The shut-down valve simply cuts off the fuel supply. It is actuated either electrically, pneumatically or manually.

For turbocharged engines, an air–fuel control (AFC) valve is introduced into the main control unit, Fig. 6.50. This is a spool valve actuated by a diaphragm exposed to the boost pressure, and it is interposed between the throttle and shut-down valves. If the accelerator pedal is suddenly depressed, and throttle valve in the fuel supply system thus opened, the passage on to the injectors is restricted by the AFC valve which, progressively opening, limits the rate of increase of flow to match that of the boost pressure. This avoids the emission of black smoke while the turbocharger is accelerating to catch up to supply enough air for combustion for coping with the extra load.

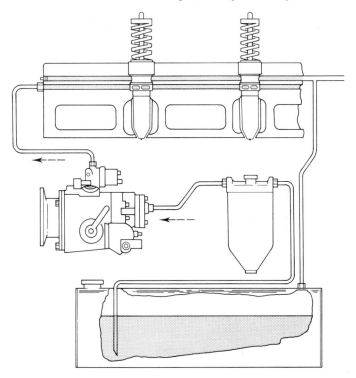

Fig. 6.49 Diagram showing layout of Cummins PT system

Other components in the main control unit include a magnetic screen between the gear type pump and the governor, to take out any particles of metal that might damage or impair the operation of the unit injectors; a pulsation damper to smooth out the delivery from the pump; and a spiral gear for driving a tachometer. A screw on the end remote from the bob-weights on the governor shaft limits the axial movement of the governor sleeve away from it, for setting the idling speed.

The injectors are illustrated in Fig. 6.51. At the beginning of the upstroke, in preparation for the next injection, fuel from the low pressure manifold enters at A, passes through the inlet orifice B, and on down through a series of drilled holes, turns up to pass through a check valve F, and then down again to an annular groove in the top end of the injector cup, whence it flows up yet again through passage D into the waisted portion of the stem of the injector. From there it flows out and up through passage E on its way back to the tank. This fuel flow cools the injector and tends to warm the fuel in the tank, thus helping to prevent wax formation in very cold weather. The quantity of fuel flowing is a function of its pressure which, in turn, is primarily a function of engine speed but modified by the restrictions imposed by the governor, throttle valve and, in the case of a turbocharged engine, the AFC valve.

As the upstroke is completed, the metering orifice C is uncovered, and the circulation back to the tank is interrupted by the closure of the passage D. Pulsations in the supply from the fuel pump are absorbed by the pulsation

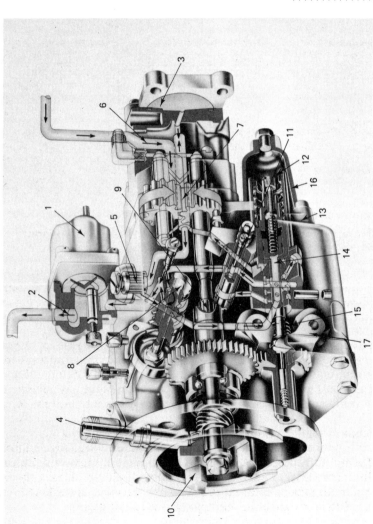

1 Shut-down valve
2 Fuel to injectors
3 Pulsation damper
4 Tachometer shaft
5 Filter screen
6 Fuel inlet
7 Gear pump
8 Air–fuel control barrel
9 Main shaft
10 Drive coupling
11 Throttle shaft
12 Idle speed adjusting screw
13 By-pass 'button'
14 Governor plunger
15 Torque spring
16 Idle spring pack
17 Governor weights

Fig. 6.50 The combined control, governor and pump unit of the Cummins PT system

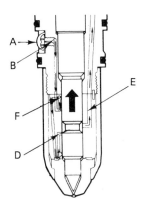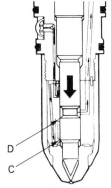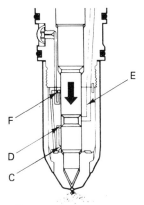

**Start upstroke
(fuel circulates)**

Fuel at low pressure enters injector at (A) and flows through inlet orifice (B), internal drillings, around annular groove in injector cup and up passage (D) to return to fuel tank. Amount of fuel flowing through injector is determined by fuel pressure before inlet orifice (B). Fuel pressure in turn is determined by engine speed, governor and throttle.

**Upstroke complete
(fuel enters injector cup)**

As injector plunger moves upward, metering orifice (C) is uncovered and fuel enters injector cup. Amount is determined by fuel pressure. Passage (D) is blocked, momentarily stopping circulation of fuel and isolating metering orifice from pressure pulsations.

**Downstroke
(fuel injection)**

As plunger moves down and closes metering orifice, fuel entry into cup is cut off. As plunger continues down, it forces fuel out of cup through tiny holes at high pressure as fine spray. This assures complete combustion of fuel in cylinder. When fuel passage (D) is uncovered by plunger undercut, fuel again begins to flow through return passage (E) to fuel tank.

Fig. 6.51 Sequence of operations of Cummins unit injector: (*left*) start; (*centre*) upstroke; (*right*) downstroke

damper in the control unit so, with the closure of passage D, the flow through orifice C is steady. Therefore the quantity of fuel passing through this orifice into the injector cup is a function of its pressure. Any back-flow will close the check valve F.

On the next injection stroke the downwardly moving plunger first shuts off the fuel supply coming through the metering orifice C and thus traps the metered quantity of fuel in the injector cup. Since no more fuel can subsequently pass in from the metering orifice, there is no possibility of dribbling through the injector holes after the injection stroke has been completed.

Continuing down, the plunger pressurises the fuel in the cup and forces it out through tiny holes in the nozzle, spraying it into the combustion chamber. Toward the end of the stroke, the passage D is once more uncovered, and the cooling flow of fuel back to the tank resumed. On completion of injection, the tapered end of the plunger momentarily remains on its seat, in the bottom of the cup, until the next metering and injection sequence begins.

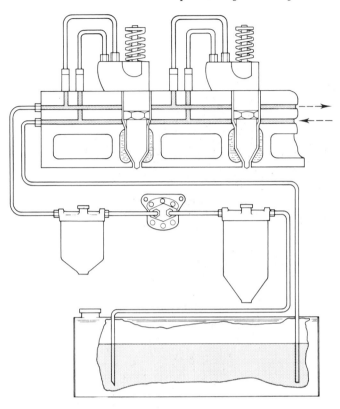

Fig. 6.52 Diagram showing layout of the General Motors unit injection system

6.50   The GM unit injection system

In basic concept, the GM unit injection, Fig. 6.52, bears some similarity to the Cummins PT system just described, but it differs in many respects. First, there is no separate unit housing all the control functions: instead, each injector, Fig. 6.53, houses what is virtually a single element of a jerk pump, such as that illustrated in Fig. 6.24, and injection is controlled by a multi-segment toothed rack that extends the full length of the head from the foremost to the rearmost injectors.

From the tank, the fuel is lifted by a transfer pump, through first a strainer and then a fine filter, up to the gallery and on into branch pipes connecting it to the unit injectors. As the fuel enters each injector, at A, Fig. 6.53, it passes through an additional, small, filter from which ducts take it down through B into a sleeve in the casting around the injector barrel and plunger. Thence it flows through the radial port F in the barrel, into the chamber below the end of the plunger. As the plunger descends, the fuel beneath it is forced up the axial hole in it and out through a radial hole into the spill groove. From the spill groove, it flows through the radial port E, on the left of the barrel, out into the sleeve in the housing. The return passage from the housing, delivering to the outlet H, is behind that for the inlet. It is of smaller diameter than the inlet, so

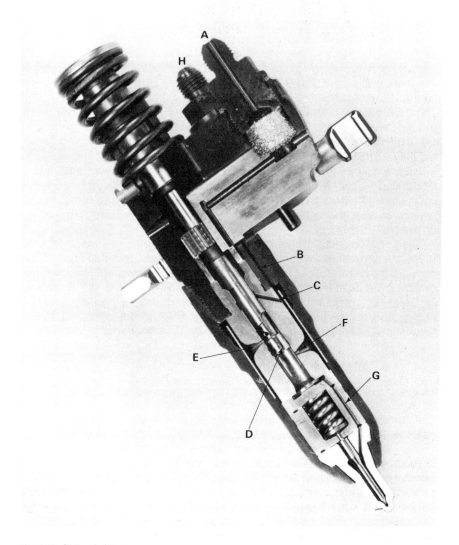

Fig. 6.53 GM unit injector

that the fuel in the housing remains always under pressure. The function of the surplus fuel flow is to cool the unit during its passage through the barrel.

As the plunger is lifted by the return spring at its upper end, it shuts off the spill port on the left in Fig. 6.53, and then draws fuel through the radial hole on the right, in the barrel, into the chamber beneath it. Incidentally, higher up on the right, there is another hole C sloping upwards, to allow fuel to run into an annular groove in the bore of the barrel, for its lubrication.

When the cam actuates the rocker mechanism, it pushes the plunger down again, so that its lower end D first shuts off the inlet hole, after which the upper edge of its spill groove shuts off the spill port E. The closure of the latter traps a metered quantity of fuel beneath the plunger which, continuing down, forces this fuel, at increasing pressure, through hole G in the wall of the cylindrical housing for the needle return spring, whence it passes into the nozzle. On the pressure of this fuel reaching a predetermined value, it lifts the piston on which the needle return spring seats and, with it, the needle from its conical seat, whereupon the fuel sprays out through the holes in the nozzle into the combustion chamber.

As the plunger returns, the spiral upper edge of the spill groove in the plunger uncovers the spill port in the barrel, suddenly releasing any pressure in the fuel remaining in the nozzle so that, subsequently, there can be no dribble through its spray holes. The surplus fuel flows back through the axial and radial holes in the plunger into the spill groove, whence it passes out through the radial hole, on the left in the illustration, back into the main housing. On completion of the injection cycle, the plunger comes back up to its original position, with both the inlet and spill ports open, for resumption of the cooling flow.

The upper edge of the spill groove around the plunger is of spiral form, so that the spill timing, and thus the metering of the quantity of fuel injected, can be regulated by rotation of the plunger. This is done by means of the previously mentioned rack. To stop the engine, the rack is moved to the right-hand extreme of its travel, rotating the plunger clockwise to the position where, as can be seen in the illustration, the spill port is at no point shut off by any vertical displacement of the plunger between the limits of its operation.

6.51   Cummins 10-litre diesel

This is a power unit designed for eight-wheel rigid vehicles and lightweight tractors. Because it is only 280 mm high it can be installed vertically in cabs that are too low to accept the earlier Cummins in-line engines. Additionally, its compactness and light weight, 850 kg, render it particularly suitable for rear engine installation in buses or coaches for which, because they are not so heavy as the trucks, it is normally derated from its 186 kW to either 164 or 134 kW. Derating, of course, has the incidental advantage of further increasing both reliability and life: in truck operation, a life of up to 800 000 km is claimed before overhaul is necessary. A charge-cooled version developing 216 kW is also produced, mainly for sale in the USA.

In all its forms, this 10-litre engine was introduced as a turbocharged unit. The objectives were to make use of energy that would otherwise go to waste, to gain the slight advantage in the fuel economy generally associated with turbocharging. At 186 kW rating, the brake specific fuel consumption is 0.207 kg/kWh at 2100 rev/min and, at maximum torque, 0.199 kg/kWh. The latter is equivalent to 43% thermal efficiency.

Light weight has been obtained partly by computer-aided design, using the finite element technique of dividing the structure up into small interacting elements. This technique is based on the fact that the strains and loads carried over from element to element must balance. Other factors helping to reduce weight include restriction of the water jacket to the upper ends of the cylinders and the low height of the engine, associated partly with the use of short connecting rods. Also, by incorporating the induction manifold partly in the head casting, Fig. 6.54, and partly in the rocker cover, Fig. 6.55, not only is weight again saved but compactness is achieved too.

Fig. 6.54, in which half of a sectioned cylinder head is illustrated, merits careful examination. If we imagine four lines joining the centres of the valves, as viewed from above, we see that the resultant square is arranged not with its sides parallel to the sides and ends of the cylinder block, but rotated 45° to form a diamond shape on the head. On the top face of this half of the head, in the illustration, we can see one-and-a-half very large, pentagon-shaped inlet ports and, on the side, three circular exhaust ports. The inlet ports each serve the four adjacent valves, two each side of them, while each of the exhaust ports serves the two valves beyond the adjacent pair of inlets, the layout of the porting for the latter being visible through the sectioned end of the head.

As can be seen from Fig. 6.55, the turbocharger is mounted just below the exhaust manifold, and delivers air up to the inlet trunk in the rocker cover above. All the exhaust ports are of equal length, so that maximum benefit can be derived from the pulse effect for driving the turbocharger. Immediately below the turbocharger is the oil cooler, below which again are an oil filter and, an unusual addition, a water treatment filter to keep the radiator and oil cooler clean. With this layout, all the passages for oil, water to the cooler, and the air and exhaust gas are very short and mostly relatively straight. Thus, energy losses, and in particular those from the gases entering and leaving the turbocharger, are minimal and the need to take a bulky induction trunk over the top, as would have been necessary with a crossflow head, has been obviated. The fuel filter is immediately behind the oil cooler.

Fig. 6.54 Arrangement of the valves and ports on the Cummins 10-litre engine

Fig. 6.55 The exhaust manifold of the Cummins 10-litre unit is designed to take advantage of the pulse effect in serving the turbocharger

A unit injection system, described in principle in Section 6.49, is used by Cummins. The cam-actuated combined injector-and-nozzle units are accommodated in the centre of the previously mentioned diamond pattern valve layout, to inject into the centre of the combustion chamber. Each injector is held down by a saddle piece, as can be seen from Fig. 6.54. In essence, the fuel is taken to the air, rather than vice versa: there is virtually no induction swirl, the fuel being injected at a pressure of $138\,000\,kN/m^2$, through ten 0.12 mm diameter holes per injector.

This very high pressure is practicable mainly by virtue of the use of a very stiff actuation mechanism for the valves and injectors. A single, high mounted, 72 mm diameter camshaft, with very short pushrods, serves all the valves and injectors, so there are three rockers per cylinder, Fig. 6.56. The shaft that can be seen above the camshaft carries rocker type cam followers for actuating the pushrods. On top of the cylinder head three very sturdy rockers for each cylinder pivot on a shaft carried on pedestals: the central rocker actuates the unit injector, while the outer ones bear down on a saddle bridging the ends of the pairs of valves. These outer saddles slide vertically on the guide-posts which, in Fig. 6.54, can be seen between the pairs of valves in the central group, where the saddle over the injector has been removed.

The crankshaft, connecting rod and piston assembly is shown in Fig. 6.57. Copper-lead bearings with a lead-tin flash carry the journals, which are induction hardened. The fillet radii are not hardened, though for engines developed to produce over 224 kW, they may have to be.

Fig. 6.56 On the Cummins engine, the two outer rockers actuate the valves and the central one the injector. Rocker type cam followers actuate the very short pushrods

Fig. 6.57 As can be seen on the left, the piston skirts are shortened locally to clear the crankwebs

Fig. 6.58 The intermediate flange forms the seating for the liner, while the upper flange is a press fit in its aperture in the block, which therefore does not have to be counterbored

Each piston has one oil control and two compression rings, the top ring groove being machined in a Ni-resist insert. The top land is only 4.76 mm deep, to help to avoid exhaust gas pollution. Such a narrow land would have been impracticable without oil cooling of the underside of the piston crown. This is done by means of a metal jet in a plastic moulding in the form of two bosses joined by an integral bridge-piece. Both bosses are spigoted into holes in the crankcase, one of which is blind and is only for location of the moulding and thus for aiming the jet, while the other communicates with the oil gallery. Oil passes from the gallery into a blind hole cored axially into its spigot, and thence upwards through a radial hole into the outer end of which is snapped the metal jet.

To keep friction to a minimum, only the thrust faces of the piston skirts bear against the cylinder walls. Because the connecting rods are short, the lower ends of the skirts have to be cut away to clear the balance weights on the crankshaft as the piston passes bottom dead centre.

To keep vibration, and thus cavitation erosion, to a minimum and to confine the water jacket to their top ends, the seating flanges of the wet liners are only about 75 mm from the top, Fig. 6.58. There is no joint ring beneath these flanges, but an O-ring is carried in a peripheral groove around them, just in case a piece of swarf or dirt happens to be trapped between the metal-to-metal joint faces when the liners are replaced in service. The upper end of the liner stands proud of the block to seal tightly against the cylinder head gasket.

All the auxiliaries are gear-driven, as can be seen in Fig. 6.59. The fan drive incorporates an oil-actuated multi-plate clutch which, controlled by a thermostat, disengages for as much as 95% of the running time, engaging only when the coolant temperature approaches its upper limit.

A Holset H2C turbocharger is fitted. It has a twin entry for the exhaust gas and a divided nozzle, Fig. 6.60. The mechanism that can be seen on top in the illustration is an exhaust brake, which is a twin valve of the barrel type, with a pneumatic actuator on the right. When the actuator rotates the valve through 90°, to close the twin passages through it to the turbocharger, the slot that can be just seen on top of the barrel interconnects the exhaust manifold branches from the front and rear sets of three cylinders, so that the engine pumps exhaust into a larger volume than it would otherwise be able to. This helps to make the exhaust brake smooth in operation.

6.52   Relative merits of spark ignition and ci engines

In spite of the inherent disadvantages of greater weight and bulk per horsepower and rougher running, the ci engine has fully consolidated its position. Greater economy, greater security from risk of fire, and with modern bearing materials and methods of manufacture a degree of general reliability at least as good, if not better, than that of the petrol engine, are definitely attained.

The injection equipment, provided proper care is taken with filtration of the fuel, is proving itself more reliable than the electrical equipment of the spark-ignition engine. Overheating troubles are less for, owing to the higher thermal efficiency, the heat losses, both to the jackets and to the exhaust, are smaller than with the petrol engine. Flexibility, silence and smooth running

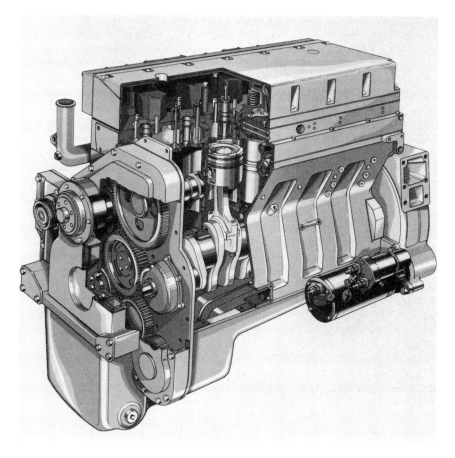

Fig. 6.59 Clean rectangular lines characterise the Cummins 10-litre diesel engine, and the panelling of the crankcase reduces noise output

still leave something to be desired, but as knowledge of the injection and combustion processes increases, so will the smoothness of performance improve.

Because of the greater gas-loadings, diesel engine components are heavier, and therefore the engines themselves bigger, than comparable petrol engines. Moreover, in a diesel engine cylinder, since there is only an extremely brief interval after the start of injection for the fuel to mix with the air, only about 75% of the total throughput of air can be burnt. Otherwise, because of local concentration of rich mixture, black smoke will be emitted from the exhaust. Consequently, for a diesel engine to produce the same power as a petrol unit, either it must have a larger swept volume or, if it is required to be of the same size, a larger charge must be forced into its cylinders. This is usually done by turbocharging.

The injection pump is more difficult both to accommodate and to drive than the ignition contact breaker-distributor unit. Another factor that makes the diesel engine larger is that, in commercial vehicle operation, it has to operate

Fig. 6.60 A pneumatically controlled exhaust brake, top, integral with the turbocharger is an option

with even greater reliability for long distances, and therefore must be sturdier. The problem of compactness has been solved by some manufacturers by adopting the V-six or V-eight layout. Since the arguments for the use of the V layout have been outlined already, in connection with the petrol engines, Section 3.72, there is no need to repeat them here. For commercial vehicles, the main difference is that the compact form is required in order to leave the maximum possible space free for the load-carrying platform. There are also some types of installation, for example, transverse rear engine, where a power unit of short length reduces the angularity required in the drive line to the rear axle. Among the engines of this type are the Gardener five-cylinder, the Cummins V-six and V-eight engines, and the Perkins V-8.510 unit, described in *Automobile Engineer*, January 1967.

First cost of both engines and injection equipment remains high, owing to the meticulous care required in manufacture, but for the commercial user whose vehicles cover a high annual mileage, particularly on long runs, saving in fuel costs results in a rapid recovery of initial expenditure.

In special cases it may be possible to approach, by raising the compression ratio of the petrol engine, the efficiency of the ci engine, but this is possible only by the use of very expensive and special fuels, as the requirements of the vapour compression (spark-ignition) engine become increasingly exacting with increase of compression ratio, though rotary valve engines appear to be exceptions to this generalisation.

For equal thermal efficiency the spark-ignition engine does not require quite such a high compression ratio as the ci engine, the maximum pressures would be about the same. Thus weight per unit of cylinder volume would tend to be the same.

The question of fuels for compression-ignition engines is dealt with in Chapter 15.

Chapter 7

The two-stroke engine

Both the specific power output and the potential for smoothness of torque at any given speed are restricted with the four-stroke engine, because it has only one power stroke every two revolutions. This has led to a quest for a cycle giving one power stroke per revolution. The answer was to exhaust the cylinder as the piston approached and passed the bottom, or outer, dead centre, and to use the depression caused by the inertia of the high speed flow of the outgoing gases to assist the induction process. Induction, therefore, had to be timed to begin shortly before the exhaust ports closed and to continue for a brief period during the subsequent upstroke of the piston.

The objective was to complete both induction and exhaust within the period that the piston was swinging over BDC, and thus detract very little from either the exhaust or compression strokes. In fact, this is not too difficult because the piston dwells momentarily at BDC, and the quarter revolution from 45° before to 45° after represents less than one-eighth of its displacement from the bottom end of its stroke. It follows that, in a two-stroke engine, the burnt gas is exhausted from the cylinder primarily by the pressure difference between it and atmosphere, rather than by the motion of the piston.

Thus, the two-stroke cycle, starting at top, or inner, dead centre firing stroke, can be said to compromise first a combined power and exhaust stroke as the piston moves down, and then induction and compression as it moves up again. However, because of the overlap of these functions at BDC, this is perhaps a slight over-simplification.

Doubling the number of power strokes per revolution might be thought to offer potential for a power output double that of the four-stroke engine, but it does not. Indeed, the outputs of two-stroke engines range from only about 10 to 40% above those of equivalent four-stroke units. This situation arises partly because the pumping losses in the two-stroke engine are generally higher, but mainly because, for reasons to be explained later, it is not possible to develop such high mean effective pressure as with the two-stroke cycle.

Because both induction and exhaust occur around BDC, the inlet and exhaust ports can be situated near the bottom end of the cylinder and can be covered and uncovered by the piston. This obviates the need for valves and their actuating gear, so one of the major attractions of a two-stroke engine of this layout is its extreme simplicity, and therefore low cost.

It also, however, leads to one of its principal disadvantages, which is that its fuel consumption is high because over most, if not all, of its speed range some of the incoming charge inevitably is lost through the exhaust ports during the overlap period. Although both efficiency and specific power output can be improved by measures such as injection of the fuel after the exhaust ports are closed, incorporating poppet type exhaust valves into the head, scavenging the exhaust gases more effectively by supercharging, or even incorporating extra cylinders for scavenging by providing extra air, all involve increasing the complexity of the engine, which reduces its attractiveness relative to a four-stroke unit.

Even if all the advantages (mechanical simplicity, low cost, greater mechanical silence, smooth torque owing to the shortness of the intervals between combustion impulses, and consequently the small flywheel and therefore light weight) were valid, they would still have to be set against the apparently inescapable disadvantages. These are: greater noise due to the sudden uncovering of the ports by the pistons, high specific fuel consumption, excessive hydrocarbon content of the exhaust gas, and some more, including difficulty of starting and irregular firing at idling and light load with some types of two-stroke engine.

Together, these disadvantages have, in fact, led to the abandonment of this type of engine for cars. Moreover, although in diesel engines, injection after the inlet ports have closed obviates the fuel consumption problem, two-stroke diesels are still widely regarded as too noisy for commercial vehicles. Noise is of course a major disadvantage for an engine that may have to be offered for use in buses as well as trucks.

It follows that, at the time of writing, the contents of this chapter are mainly of historical interest. However, as the use of superchargers and turbochargers on diesel engines, and injection on petrol engines, is now becoming the norm, the two-stroke unit with fuel injection and blown scavenging no longer compares so unfavourably, so far as complexity is concerned, with its four-stroke equivalent. So, if the silencing problem can be overcome, we could see its revival.

7.1 Three-port two-stroke engine

Fig. 7.1 shows in simple diagram-form the Day three-port engine. The exhaust port is shown at E, this being uncovered by the piston after completion of about 80% of its stroke. The transfer port T, through which the charge is pumped from the crankcase, opens slightly later than the exhaust port, as shown in 1, to reduce the risk of hot exhaust gas passing into the crankcase and igniting the new charge. It follows that the transfer port is closed by the rising piston slightly before the exhaust port, so that the final pressure in the cylinder, and therefore the total quantity of charge (consisting of a mixture of burnt gases, air and fuel vapour) is determined not by the pump delivery pressure but only by the extent to which the throttling and pulse effects of the exhaust pipe, silencer, etc., raise the cylinder pressure above that of the atmosphere. The piston head is specially shaped to deflect the entering gases to the top of the cylinder. This is known as *cross-flow scavenge*.

The piston rises and compresses the charge, after which it is ignited and expands in the usual way. The indicator diagram takes the form shown at (a) in Fig. 7.2, which differs from that of the four-stroke cycle only in the rather more

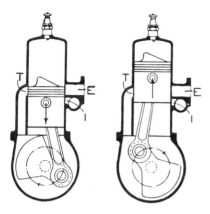

Fig. 7.1 Three-port two-stroke engine

sudden drop of pressure as the exhaust ports are uncovered and the elimination of the 'bottom loop' showing the exhaust and suction strokes. This bottom loop is replaced, of course, by the indicator diagram, shown at (*b*), obtained from the crank case or scavenge pump cylinder. There is no possibility of eliminating this pump work from either the four-stroke or the two-stroke cycle – in one case it is done in alternate revolutions in the main working cylinder, and in the other in every revolution in the scavenge pump cylinder. Indeed, the 'phased pump' type of two-stroke engine, of which a later example is the Trojan design as shown in Fig. 7.5, may be regarded as a V-twin four-stroke engine in which the positive work is concentrated in one cylinder and the negative pumping work is done in the other, instead of each cylinder doing half of both.

To return to the Day type engine (Fig. 7.1), it is necessary now to describe how the charge is drawn into the crankcase from the carburettor.

As the piston rises, a partial vacuum is formed in the crankcase, the pressure becoming steadily lower until, near the top of its stroke, the rising piston uncovers the induction port 1, which communicates with the carburettor, as shown in 2. Air rushes in to fill the vacuum and carries with it the petrol from the jet necessary to form a combustible mixture. It will be realised that the suction impulse on the jet is a violent one of short duration – the very worst from the point of view of obtaining a correct and homogeneous mixture – while the time interval during which the induction port is open is also unduly short from the point of view of the inspiration of a full charge of air. Fig. 7.2(*a*) shows a typical timing diagram of the various port openings and closings expressed in degrees of crank angle.

An alternative to the piston-controlled induction port which is used in many two-stroke heavy-oil engines is the light spring-controlled automatic air valve. This valve remains open for practically the whole of the upward or suction stroke of the piston, ensuring higher volumetric efficiency though at the cost of an additional moving part. The American Gray motor, extensively used for marine work in the water-cooled form, employs both port and automatic valve.

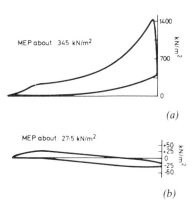

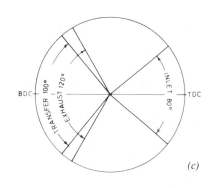

Fig. 7.2 Two-stroke indicator diagram

7.2 Reverse-flow scavenge DKW engine

An interesting and successful example of the three-port engine using 'reverse-flow' or 'inverted' scavenge is the German DKW small car engine. This is a twin cylinder unit of 76 mm bore and 76 mm stroke, using crankcase displacement, but having twin transfer ports, one each side of the exhaust port, and imparting a tangential and upward flow to the entering charge.

The piston deflector is dispensed with, thus eliminating pressure and inertia tilt due to lack of symmetry – a source of two-stroke rattle – and the direction of scavenge flow becomes somewhat as indicated in Fig. 7.3. By the provision of ports through the piston skirt the transfer passages are kept short, but as an additional change of direction is involved it is doubtful if the transfer resistance is appreciably reduced as compared with the arrangement of Figs 7.1 and 7.4.

The makers carried out comparative tests of the two methods of scavenge, and at 3000 rev/min they improved the bmep by 14% by converting to the reverse-flow system.

The Villiers Company successfully applied this principle, but in a modified form, using two pairs of transfer ports and two exhaust ports in their well-known air-cooled motor-cycle engines. From their 249 cm³ engine 8.95 kW has been obtained at 5000 rev/min, which represents a bmep of 427.5 kN/m².

7.3 Special constructions of two-stroke engine

The loss of efficiency arising from the loss of new charge through the exhaust ports, which occurs in these simple port constructions, has led to the use of a double-piston form of engine in which two pistons working in twin barrels side by side share a common combustion chamber and sparking plug. There is, of course, only one impulse or firing stroke per revolution, no matter what particular arrangement of connecting rods and cranks is employed. The Lucas engine (Fig. 7.4) is now no longer made but it had interesting and ingenious features.

The two pistons drive independent connecting rods operating on separate

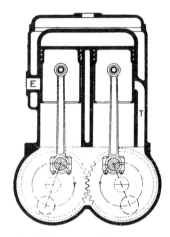

Fig. 7.3 DKW engine   Fig. 7.4 Lucas engine

cranks, which are arranged to run together by means of gear teeth cut in the periphery of the circular crank webs. The flywheel and clutch are connected to one of these cranks only, so that half the power necessarily passes through the gear teeth. The cranks clearly revolve in opposite directions, and use is made of this fact to obtain complete primary balance. Each crank assembly is provided with revolving balance weights corresponding to the whole of the revolving and reciprocating parts for one line. This ensures primary balance in the vertical plane as explained in Chapter 3, while the unwanted horizontal effects of the two revolving balance weights neutralise each other, exactly as in the Lanchester harmonic balancer, which, however, deals with secondary unbalanced forces.

With the twin cylinder-barrel in this arrangement one piston may be used to control the transfer port T while the other uncovers the exhaust E, and in this way the new charge has to traverse the full length of both barrels before there is any possibility of its escaping out of the exhaust port. Further, it is possible, by suitably meshing the gear wheels, to arrange that the exhaust port is given a 'lead' which results in its opening *and* closing somewhat ahead of the corresponding events of the transfer port, thus tending to give a more complete cylinder charge. If this is done, however, the advantage of exact primary balance will be affected.

7.4   Separate phased pump

The reader will have realised, in considering the characteristics of simple crankcase displacement, that the swept volume available for induction is the same as that of the main cylinder, namely, the piston area multiplied by the stroke, and, moreover, that the charge is not transferred to the cylinder simultaneously with the pump displacement, but in a rush after pre-compression. This causes the indicated pumping work to be high, and though some of this energy is no doubt usefully applied to producing a warmed and homogeneous mixture as a result of the turbulence, there is risk of admixture with the old charge and loss of fuel through the exhaust port.

Both the quantity and, especially at light loads, the quality of the mixture suffer, leading to limited power output and high specific fuel consumption.

These considerations have led many designers to have recourse to the separate charging pump, which is driven usually by a separate crank placed in appropriate phase relationship with the main crank.

7.5   Trojan engine

The redesigned Trojan engine retained a twin-barrel construction successfully used in the original model, but, instead of crankcase compression scavenge, the phased charging pump was adopted, in combination with piston-controlled transfer and exhaust ports.

As result of this change, a normal lubrication system can be used, and the necessity for pressure sealing of the crankcase is obviated.

In early development of the new engine, a mechanically-driven, cylindrical distribution valve was used, but this was subsequently replaced by light, spring-blade automatic valves.

The general construction should be clear from the two sectional views given in Fig. 7.5, while Fig. 7.6 shows to a larger scale the details of the interesting automatic valves.

The bore of the working barrels is 65.5 mm and of the charger cylinders 96.8 mm, the stroke of all being nominally 88 mm, but owing to the offset cylinder axes, the actual stroke is fractionally greater than twice the crank throw in each case, different offsets being used for the three bores.

The nominal swept volume of the power cylinders is therefore 1186 cm^3, or the equivalent of two cylinders of 92.6 mm bore, while that of the pump cylinders is 1293 cm^3, thus providing a 9% displacement margin.

The effect of the cylinder offsets, combined with the relative position of the leading edges of the ports, is to give the exhaust period, with a duration of 106°, a lead over the 104° inlet period, of about 22°, the overlap being about 83°.

The offset of the pump axis not only reduces the obliquity of thrust on the delivery stroke, but advances the stroke timing by about 8° relatively to the 90° at which the cylinders are set. By the use of a single transfer pipe and twinned inlet ports for the two sets, No. 1 charging cylinder delivers to No. 2 pair of working bores, and *vice versa*.

All the six connecting rods are identical, and each pair of aluminium alloy main pistons with their small ends are made identical in weight with the corresponding cast iron piston of the charging cylinder. Dynamic balance is thus secured on the principle described in the latter part of Section 2.2, the necessary balance masses being incorporated in the cast crankshaft of Mehanite iron. There is a slight discrepancy due to the offset bores, and the secondaries remain unbalanced, but neither of these gives rise to serious vibration.

One cage of the automatic valve assembly is shown partly exploded in Fig. 7.6. The upper inlet cage is identical with the lower delivery cage, the 'air-flow' being radially outwards in both cases.

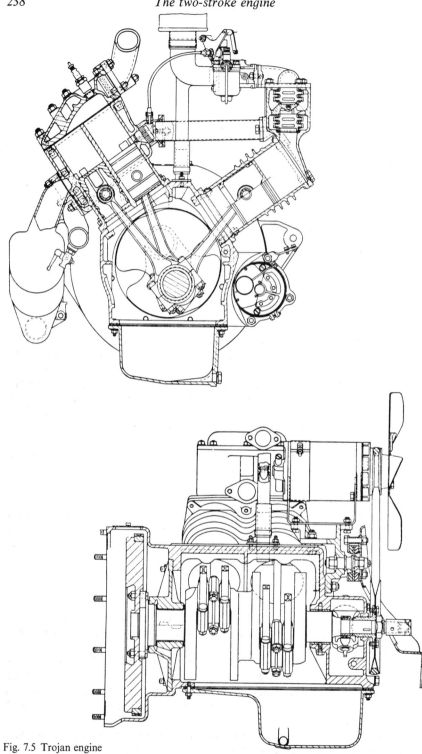

Fig. 7.5 Trojan engine

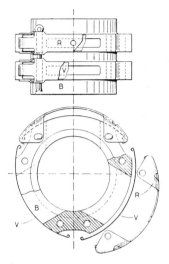

Fig. 7.6 Trojan automatic valves and cage

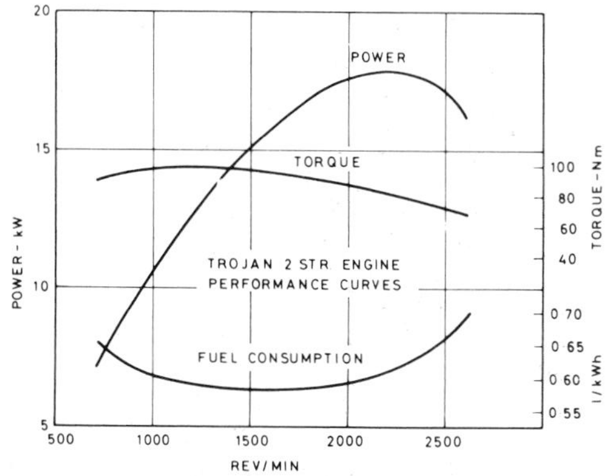

Fig. 7.7 Trojan performance curves

Two tiers each of three spring blades V form the valves, which bear on faces machined to an eccentric radius on the outside of the aluminium alloy body B. The pressed steel retainers R exercise a minimum constraint for the location of the spring blades, which have less free curvature than have the retainers, with the result that when blown open by the air flow the blades are strained against the retainers, and spring back sharply to the closed position on reversal of the flow. The rolled ends of the blades minimise friction under the relative motion. The total area of the six ports is about 6.45 cm^2.

The flame trap, consisting of four thicknesses of gauze, may be noticed at the junction of the transfer pipe and transfer port, with a petrol priming pipe to aid starting.

A representative set of performance curves is shown in Fig. 7.7, a most valuable feature being the excellent maintenance of torque at low speeds. This results in very good top gear performance, reflected in satisfactory mpg consumption figures, though the specific fuel consumption on the bench is higher than normal four-stroke figures.

7.6 Kadenacy system

The scavenging system of large two-stroke diesel engines, essentially of the stationary constant speed type, have been the subject of intensive theoretical investigation and practical experiment.

An authoritative work by P. H. Schweitzer, *Scavenging of Two-stroke Cycle Diesel Engines* (published by Macmillan, New York) deals exhaustively with the mathematical and practical aspects of design involved in the efficient utilisation of exhaust gas energy in scavenging and charging engines with and without blowers.

The phrase 'Kadenacy effect' has been extensively and somewhat loosely used to describe pulse phenomena that have been utilised with varying degrees

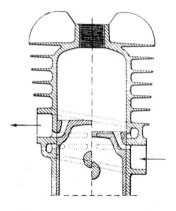 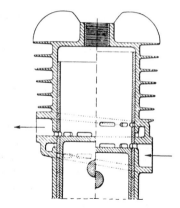

Fig. 7.8 Kadenacy patents

of success in different designs, often showing little similarity with the patent specifications.

In Schweitzer's work the Kadenacy system is described as basically the development of a high degree of vacuum in the cylinder by the sudden and very rapid release of the charge through sharp-edged exhaust ports of large area, this depression being sufficient to aspire the new charge without the necessity of a positive blower or pump.

Kadenacy British Patents Nos 308,593 and 308,594, which relate to small engines of the automobile type, cover a variety of claims relating to cylinder porting, a somewhat elaborately ported piston working in a cylinder provided with inlet and exhaust belts each ported round the full circumference of the cylinder as shown in Fig. 7.8(*a*) and several claims of a different nature involving the use of light automatic non-return valves in both inlet and exhaust ports, a mechanically-operated sleeve valve as shown in Fig. 7.8(*b*) and a mechanically-operated inlet valve.

An opposed-piston design is also covered.

The early type of deflector piston, liable to distortion and to cause two-stroke rattle, is replaced by flat-topped pistons in all cases.

The possibility of dispensing with a scavenge blower is suggested in the valve operated designs, but not with the simple ported constructions, except under favourable conditions of steady speed and load.

Poppet exhaust-valve engines such as the General Motors and Foden designs can hardly be said to use the above fundamental Kadenacy principle, the Kadenacy effect in these instances being rather a sustained extractor effect due to the moving column in the exhaust, as in the four-stroke engine with valve timing overlap.

7.7 Loop scavenge, Schnuerle system

The term *loop scavenge* is associated with Schnuerle and other patents, and is usually indicated diagrammatically as shown in Fig. 7.3 applied to the DKW engine, in which case the less common designation *reverse flow* is used.

Schnuerle patents cover also a variety of exhaust pipe systems in which the extractor and non-return effects of a volute form initiating vortex flow are described.

These ideas were developed and exploited in the small two-stroke engines used in Continental Europe for the popular family car of small cost.

7.8 Exhaust pulse charging

Though not fully applicable to the variable speeds and loads of automobile engines, this important development by Crossley Bros and H. D. Carter, described in *Proc. Inst. Mech. E.*, Vol. 154, should be mentioned.

In multi-cylinder stationary diesel engines, by providing a 'tuned' exhaust system, the high-pressure pulse from the initial exhaust of each cylinder is utilised to 'pack up' the fresh charge in an adjacent cylinder which has reached the last portion of the exhaust port opening period.

Thus the first portion of a new, relatively low-pressure charge may be caught as it is entering the exhaust pipe, where the mean pressure again may be low, and be forced back into the cylinder to raise the density of the charge finally trapped. In a single-cylinder engine it may be possible to produce a suitable wave pulse by reflection from an appropriate face of the exhaust manifold.

7.9 Uniflow scavenging: opposed-piston engines

Uniflow scavenging occurs in any design in which the inlet and exhaust ports or valves are situated at opposite ends of the cylinder bore, as in the poppet exhaust and sleeve-valve constructions, but more particularly, and with simpler cylinder construction, in the opposed-piston engine.

7.10 Compression-ignition two-stroke engine

Since in the compression-ignition engine no fuel enters the cylinder until all ports and valves are closed, and cannot therefore be lost through the exhaust ports, one of the greatest objections to the two-stroke cycle is eliminated, and it is with this type of engine that the cycle tends to make greatest progress.

There is another important merit of the high-compression ci engine in its application to the two-stroke form. The high expansion ratio results in lower exhaust temperature and reduced waste heat, and thus the inherent cooling difficulties of the cycle arising from doubled heat flow are less than with the petrol engine.

This factor, combined with the unidirectional scavenge with cool air over piston head and valves, appears to have justified the development work that has been devoted over the years to the port scavenge, poppet valve exhaust two-stroke engine, which attained a considerable degree of success.

At first sight the type appears to perpetuate one of the chief weaknesses of the four-stroke engine, the highly stressed and heated exhaust valve, in addition to the mechanical complication of camshaft and valve operating gear. Experience seems to show that, owing to the favourable conditions described above, the valves give remarkably little trouble

7.11 GM diesel with rotary blower and poppet exhaust valves

An interesting example of a two-stroke ci engine is the General Motors three-cylinder engine shown in Fig. 7.9.

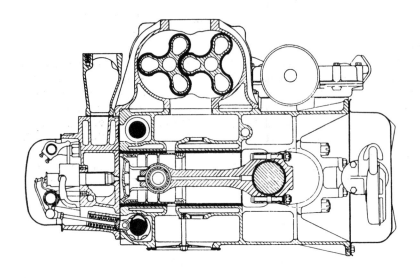

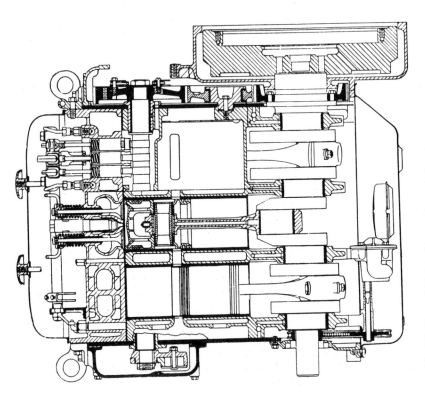

Fig. 7.9 General Motors blower-charged ci engine

The engine illustrated has a bore and stroke of 107.95 and 127 mm, and a compression ratio of 16 : 1. Its continuous rated output is 47.7 kW at 1600 rev/min with a fuel consumption of 0.2735 kg/kWh while the maximum output is given as 59.6 kW at 1800 rev/min.

The corresponding values of the bmep are 482.6 and 572.3 kN/m² respectively. These ratings would be applicable to marine propulsion or similar duties.

The high camshaft and the three-lobed Roots blower are driven by gearing at the flywheel end of the engine, damping and cushioning devices being incorporated. The camshaft is duplicated by a balancing shaft, both carrying bob-weights which produce a reciprocating effect in a similar manner to that of the Lanchester harmonic balancer, but in this case arranged to produce a couple to balance the primary reciprocating couple due to the three pistons. In the figure the bob-weights are shown in false phase for convenience.

The direct injection system is employed, a combined pump and sprayer unit (see Section 6.50) being mounted between the two exhaust valves and operated by a rocker from the camshaft.

Very complete information on this engine is given in the work by Schweitzer, referred to in Section 7.6.

7.12 Foden six-cylinder two-stroke ci engine

Similar in principle is the Foden FD 6 engine, described in detail in *Automobile Engineer*, Vol. 39, No. 4. Its injection pump is a high speed version of a CAV unit with an H-type hydraulic governor giving remarkably sensitive and accurate control. In the later versions of this engine the bore was increased from 85 to 92 mm, thin shell aluminium-tin bearings and the cylinder head and liner design was altered so that each cylinder had a separate head. Also, the Mark VII six-cylinder engine was exhaust turbocharged and intercooled.

As a measure of the advance in design of this particular engine, a comparison of the later performance figures with those obtained from the Mark I engine is interesting. Although the physical dimensions of the engine remained practically unchanged, the power increased by 79%, the torque by 82%, the specific weight decreased by 40% and the specific fuel consumption improved by 10.7%.

An output of 35 kW per litre at 2200 rev/min, a specific weight of 3.837 kg/kW, an overall length of less than 1.2192 m and weight, without electrical equipment, of 635 kg, compare most favourably with most advanced designs of four-stroke diesel engines.

The maximum value of bmep is 1069 kN/m² at 1300 rev/min and the best specific fuel consumption is 0.218 kg/kWh.

At maximum rev/min the piston speed is about 6% less than in corresponding four-stroke types, and this, combined with the unidirectional loading of the connecting rod, makes the rod and bearing load-factors more favourable. Since every upward stroke is a compression stroke, and the gas load exceeds the inertia load up to about 3500 rev/min, there is no reversal of stress in the connecting rod, or big-end cap loading. The fluctuations of torque throughout each revolution are consequently much less, and a lighter flywheel can be used, with direct gain in liveliness and ease in gear changing.

The general construction of the FD6 Mk I engine may be studied from the longitudinal and cross-sections of Fig. 7.10.

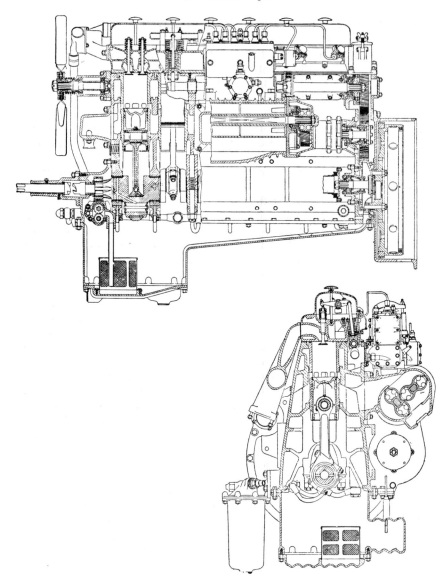

Fig. 7.10 Foden two-stroke ci engine

Centrifugally-cast wet liners are used in an aluminium alloy monobloc casting, with two cast iron heads, each covering three bores.

The various pressure and water joints are made with copper gasket rings and synthetic rubber sealing rings respectively, the joint faces of the liners standing proud of the block to ensure the proper degree of nip on the various washers.

The air delivery gallery supplies air to each bore through a series of ports formed to impart tangential swirl in a clockwise direction, seen from above, the injector delivering fuel from a single-hole nozzle downstream into the air. The toroidal recess in the piston crown has the effect of increasing the rotational swirl when the piston reaches the top of the stroke, on the principle of the conservation of angular momentum in a space of reduced diameter, as utilised by Ricardo in the cylindrical head of sleeve-valve engines.

Tin-plated cast iron pistons are fitted, and the great length of these above the gudgeon pin will be noted. This makes them admirably fitted to perform the crosshead function of the piston without risk of tilt, and provides generous bearing area with an excellent wearing material. It also helps to minimise small end bearing temperatures.

The arrangement of piston rings is of interest as illustrating two-stroke requirements.

The top ring is a composite 'flame' ring designed to give gas sealing for the full depth of the top land and incorporating the ring groove for the first normal pressure ring, which is provided with a snug to prevent relative rotation. The flame ring serves to prevent the top edges of the scavenge ports being exposed to the cylinder gases earlier than the nominal instant when they are overrun by the top of the piston and the gas pressure is sufficiently reduced. Two further pressure rings – they are all of taper section – are provided above the gudgeon pin, and below are the air-chest seal ring to prevent blow-by of scavenge air to the crank chamber as the piston rises, and a normal oil scraper ring. All the rings have integral cast snugs or dowel pegs to prevent circumferential movement and ensure that the joints ride along one or other of the port bars.

The continuous compression loading of the connecting rod is reflected in the arrangement of the gudgeon pin and piston bosses, the latter being stepped in order to give the maximum bearing area on the underside combined with the fullest possible support of the pin against bending. Provision is made for pressure lubrication of the pin through central drilling of the rod, to ensure continuous maintenance of an oil film under the somewhat more difficult conditions when no reversal of load occurs. An oil-jet for piston cooling is incorporated, and in view of this provision, the thickness of the piston crown is limited in order to reduce the heat flow to the piston rings and lands.

The diagonally split big ends may be noticed. This is now general practice in ci engines, and reflects the continuous increase in crankpin and main journal diameter in the search for improved crankshaft stiffness and reduced intensity of bearing loads. The diagonal split enables the lateral width of the big end to be made small enough for upward withdrawal through the bore.

7.13   Blower and scavenging

The blower is of the Roots type with two-lobe rotors and runs at twice engine speed, average boost pressure is about $34.474\,\mathrm{kN/m^2}$, sufficient with the Kadenacy extraction effect to give thorough scavenge and exhaust valve cooling, followed by a degree of supercharge which enables four-stroke values of the bmep to be maintained.

The twin exhaust valves, of moderate diameter, are driven by pushrod and rocker gear of the normal type, but roller cam followers, running on needle-roller bearers, are employed.

Fig. 7.11 shows the interesting timing of the valves and cylinder ports, the early opening of the exhaust valves aiding the thorough scavenge without

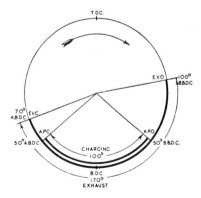

Fig. 7.11 Foden ci engine valve and port timing

unduly delaying the commencement of compression. The effective expansion ratio is appreciably less than in the average four-stroke compression-ignition engine, but some of the remaining energy is usefully employed in the wave extraction due to the carefully designed exhaust ducts, which extend independently for about 15.46 mm from the ports.

The camshaft and auxiliary driving gear is at the rear of the engine and, having helical gears, is of similar construction to that described in Section 3.62, though with a 1 : 1 ratio for the camshaft and injection pump shaft. To conform with the best practice, ball-and-roller races are housed in spigotted bronze castings fitted on studs, instead of being carried directly in the aluminium alloy casting. The whole timing gear is robust and well designed.

Careful arrangements are made for vigorous circulation of the cooling water, with jets directed to the potential hot spots around exhaust valves and injectors. The direct cooling of the valve guides will be noticed.

An indication of the remarkable coolness of the exhaust is the successful use of an aluminium alloy exhaust manifold.

7.14 Crankshaft balance and firing order

The crank throws are phased 120° apart in two sets of three, the two sets being out of phase by 60° to give 60° firing intervals, six in each revolution.

This sets the opposing primary and secondary couples slightly out of phase, as the shaft is not in exact mirror symmetry. However, the resulting slight tendency to pitch and yaw is of negligible effect, though of some technical interest. The engine is quite free from critical vibration within the normal speed range.

The firing order is 1 5 3 4 2 6, repeated each revolution. Performance curves for the FD.6 Mk I, Mk VI and Mk VII engines are given in Fig. 7.12. They show exceptionally good maintenance of torque at high speed, and the improvement in bmep and power of the late marks is an outstanding achievement.

As stated earlier, all marks of engines after the Mk I were fitted with individual heads. Mark III and all later engines have thin shell bearings and a larger diameter crankshaft. At this point the fuel injection pump was also modified for engine oil lubrication in place of the fuel oil lubrication which had

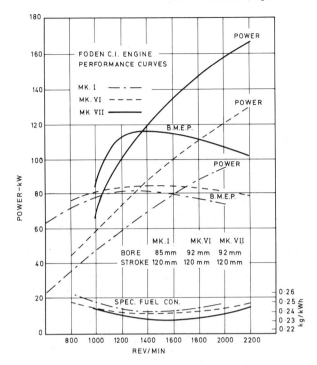

Fig. 7.12 Foden ci engine performance curves

been used hitherto. The lubricating oil pump and water circulating pump speeds were also increased at this stage. Mark VI and VII engines had the cylinder bore increased to 92 mm, these two engines being similar in all mechanical details except that the Mk VII was fitted with a CAV turbocharger and intercooler.

An external view of the FD.6 Mk VII engine is given in Fig. 7.13.

The intercooler and the connecting trunk between the supercharger and the Roots blower have been removed so as not to mask the engine details.

The authors are much indebted to the makers of the engine for information and for drawings from which the illustrations were prepared.

7.15 GM two-stroke diesel

For automotive applications, General Motors offer their 53 and some of their 71 and 92 Series two-stroke diesels, these figures representing, in each instance, the capacity (in^3) of one cylinder. The 53 Series are available as three- and four-cylinder in-line, and V6 units, and the 71 Series as three-, four- and six-cylinder in-line, and as six-, eight-, twelve- and sixteen-cylinder V units, while all those with larger cylinders are available in the V layout only. The power outputs of the 53 and 71 Series range from 48 to 597 kW, the larger engines, of course, being for industrial applications. GM claims that the specific power output typical of their heavy duty turbocharged diesels is 5.57 lb/bhp, as compared with 7.78 lb/bhp for an equivalent four-stroke diesel.

Fig. 7.13 External view of Foden two-stroke ci engine

Fig. 7.14 illustrates the three-cylinder version of the 53 Series, while Fig. 7.15 is a cross-section of the 92 Series with a charge cooler between the banks of cylinders. Although the V-type engines do not necessarily have charge cooling, the latter is nevertheless in most other respects typical of the range.

Mechanical features of special interest include the fact that the crankshaft is a drop forging with journals hardened to a depth such that they can be reground in service, and with fillets that are either hardened or rolled, according to the duty. The camshaft, too, is a drop forging since, under the heavy loading imposed by the unit injection, cast iron would not be adequate. Its cams and journals are, of course, hardened.

To obtain maximum areas of porting for both induction and exhaust, the whole of the periphery of the lower end of the cylinder is taken up by the inlet ports, while the exhaust gases leave through large diameter poppet valves in the cylinder head. These valves are heat-treated nickel steel forgings with hardened stems. With uniflow, instead of loop type, scavenging assisted by a Rootes blower and, in some instances a turbocharger too, the incoming air is driven up the cylinder and out past the exhaust valves in the head, sweeping the exhaust before it. This has the advantage of providing cooling additional to that afforded by the water jackets. The wet liners are of heat-treated cast iron and are easy to replace.

A recent development is the use of a turbocharger which, for cruising, by-passes its output around the casing of a Roots type blower, the latter being used for scavenging at light load in the low speed range. With this arrangement, a smaller blower can be used, saving both weight and parasitic power losses.

7.16   Opposed-piston engine

The Lucas and Trojan engines, with parallel cylinder barrels, may be described as *uniflow opposed-piston engines* since the scavenge air flows continuously

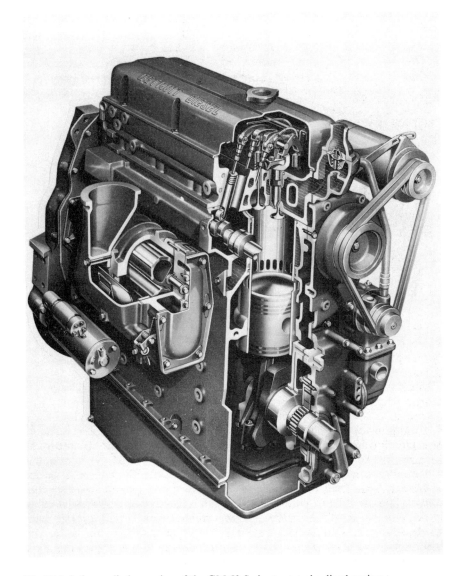

Fig. 7.14 A three-cylinder version of the GM 53 Series two-stroke diesel engines

from the inlet ports uncovered by one piston to the exhaust ports uncovered by the other, the two pistons moving towards each other for compression and away from each other for expansion.

More usually, however, the description is confined to the construction incorporating a single straight cylinder barrel in which the two pistons move in opposite phase towards and away from each other.

This involves either two crankshafts phased by gearing or equivalent means, or a single crankshaft to which the power of one piston is transmitted directly, and that of the other by means of a pair of side connecting rods, or alternatively by means of a symmetrical reversing rocker system for each piston.

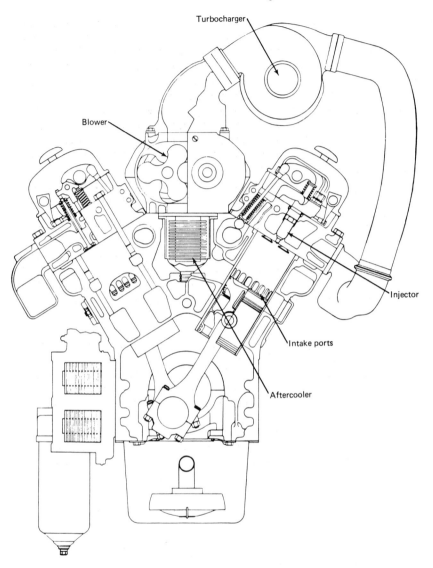

Fig. 7.15 This GM 92 Series V6 two-stroke diesel unit has a charge cooler between its banks of cylinders

Side return rods involve either a three-throw crankshaft, a crank and two eccentrics with smaller throws (Harland & Wolff), or the special arrangement used in the successful Fullagar marine diesel, in which, in each unit pair of cylinders, oblique tie rods connect the pairs of pistons which move in phase, suitable crossheads and slides being provided to take the side thrusts. Thus two cranks and two connecting rods serve four pistons. In the return rocker arrangement a two-throw shaft with offset rockers is sufficient for each cylinder.

These have been exemplified in (1) the Junkers Jumo aircraft engine, (2) the

Oechelhauser horizontal gas unit even earlier and the more recent Peugeot-Lilloise single-line stationary type made under Junkers licence and (3) a number of recent designs and proposals of which the Commer TS.3 is the best known. The return connecting-rod method, if applied to a multi-line design, would clearly involve a crankshaft of prohibitive complexity and flexibility, with a lack of symmetry leading to balancing difficulties.

When first installed for electric generating purposes in Johannesburg, the Oechelhauser gas engines gave considerable trouble through the distortion of crankshafts and side rods arising from pre-ignition due to dirty gas.

The large gas engine, though important in its day in the utilisation of blast-furnace gas, is obsolete as a prime mover.

The third of the constructions listed above, namely the return rocker arrangement with a single crankshaft, has been the subject of many patents and designs.

Two of these, which are of special importance and interest, are illustrated in Figs 7.16 and 7.17.

Fig. 7.16 shows the Sultzer engine produced in about 1936 which exhibited features of historic interest as representing a design by an experienced firm of repute, in which the opposed-piston rocker arrangement is combined with a phased reciprocating scavenge pump, in a manner that will be clear from the illustration.

The admission and discharge of air to the scavenge pump, the capacity of which is considerably in excess of the displaced volume of the main cylinders, is by automatic valves of the reed type.

The bore and combined stroke are 90 and 240 mm respectively, giving a volume of 1527 cm^3 per cylinder, the maximum rating under intermittent full load conditions being 22.4 kW per cylinder at 1500 rev/min. This corresponds to a bmep of 586 kN/m^2.

Phasing of the pistons, to give a lead to the exhaust, is accomplished by the disposition of the rocker pins and their arc of movement. With clockwise rotation of the crankshaft, it will be noticed that the right-hand rocker has

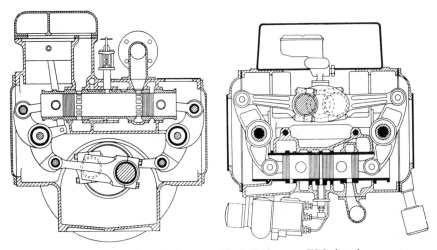

Fig. 7.16 Sultzer opposed-piston engine   Fig. 7.17 Commer TS.3 ci engine

passed its dead point while the left-hand one is just on it, the 180° disposition of the cranks being maintained for balancing purposes.

The Commer TS.3 compression-ignition engine was a post-Second World War example of the opposed-piston and rocker construction which was in production and successful use for about 20 years.

Instead of the phased pump, a Roots blower is used for scavenging, this being more in line with later two-stroke practice. The engine has three cylinders of 82.55 mm bore and a combined stroke of 203.2 mm giving a swept volume of 3262 cm^3.

At 2400 rev/min the maximum output of 78.3 kW is obtained, corresponding to a bmep of 414 kN/m^2.

The maximum value of the bmep is 724 kN/m^2 attained at 1350 rev/min. The minimum specific fuel consumption claimed is 0.232 g/kWh, an excellent result.

The forged crankshaft has four main bearings and six crankpins in pairs opposed at 180°, and for each crankpin there is a normal connecting rod, a rocker, and a short swinging piston link, and piston. Except for the couple arising from the distance between the planes of the two connecting rods of each pair, and the corresponding offset of the two arms of the rocker, the two sets of reciprocating masses are in dynamic balance (see Section 2.4), but heavy pin loading in each set of reciprocating parts is involved in all examples of this construction because of the large masses. Reversal of loading due to inertia forces is less likely to occur than in a four-stroke engine, as compression is met on each inward stroke. Thus running should be quiet at some sacrifice of the benefit to lubrication of thrust reversal. It will be noticed that the rockers in the Commer design are longer than the centre distance between cylinder and crankshaft, but owing to the reversal of the porting relative to the direction of rotation, lead of the exhaust piston is again provided. Figs 7.17 and 7.18 show the general layout of this engine arranged in such a manner as to make it very suitable for under-floor installation.

Full details will be found in *Automobile Engineer*, Vol. 44, No. 8.

There is also an interesting article by R. Waring-Brown in Vol. 47 of the same journal.

7.17   Comparison of advantages

From the foregoing descriptions it will be realised that the high performance two-stroke engine cannot claim greater mechanical simplicity than its four-stroke competitor, and it is still doubtful whether it will ever achieve equally favourable specific fuel consumption.

Approaching 100% higher output for a given space and weight appears to be achieved, though progress in supercharging of the four-stroke engine may reduce the advantage once more. Waste heat difficulties may well be the ultimate determining factor with both types.

Smoother torque and more favourable bearing load factors are inherent advantages which the two-stroke cycle will always be able to claim, and these should react favourably on maintenance and operational acceptability. Higher rotational speeds and smoother torque will react favourably on transmission design and endurance, and the advantages of a lighter flywheel in facilitating gear changing and aiding rapid acceleration have already been mentioned.

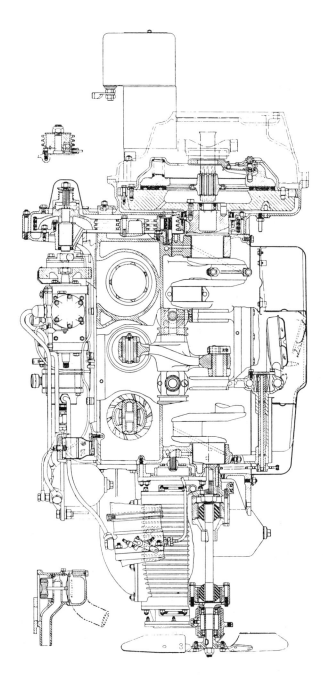

Fig. 7.18 Commer TS.3 opposed-piston ci engine

Chapter 8

Fundamentals of carburation

Most internal combustion engines introduced during the 1880s were for industrial applications running mainly at constant speed, so the earliest carburettors tended to be very simple. Three types – wick, diffusion and surface – predominated.

In the first mentioned, the lower end of the wick absorbed fuel from a small reservoir immediately below the air intake, so the air flow past its upper end evaporated the fuel and took the mixture thus formed on into the cylinders. In some later versions, a manual control was added to raise and lower the wick into and out of the air stream. This was primarily for increasing the rate of fuelling for starting from cold.

The diffusion-type carburettor comprised a small reservoir of fuel, which was warmed by exhaust gas flowing along a tube passing through it. Air entered through a second tube, parallel to and above that carrying the exhaust gas, but with its far end closed by a blanking plate. The air issued through perforations in the walls of this tube and bubbled up to the surface of the petrol, thus carrying with it fuel vapour through the engine induction system to the cylinders.

In 1885, Gottlieb Daimler and Karl Benz introduced the surface-type carburettor. Again an exhaust-heated pipe ran through the base of the reservoir, but the air passed vertically downwards through a tube the lower end of which opened into a very shallow, large diameter inverted dished plate. The edges of this plate dipped just below the surface of the fuel, which was maintained at a constant level by a float. Air flowing through the inlet pipe was therefore distributed radially outwards beneath the plate, up around its edges and through the fuel. After breaking surface it carried the fuel vapour that it had collected out through a pipe to the induction system. The flow of mixture into this pipe was controlled by a manually actuated rotary valve.

Although this invention was a step in the right direction, much more accurate matching of the fuel supply to the requirements dictated by continuously-varying loads and speeds were necessary for automotive applications. Moreover, none of these three early systems was really satisfactory for starting from cold.

The carburettor that eventually evolved comprised essentially an *air intake*, sometimes called the *air horn*, through which the air passed into a *venturi*. A venturi is a tube in which is a *throat* of a streamline section such that the velocity of the air passing through the tube increases as it flows towards its narrowest section, and then decreases as it passes to the other end. Since the energy content of the air flow must remain constant, except for the virtually negligible losses due to drag in the layers nearest to the walls of the tube, increasing velocity of flow is accompanied by a decreasing pressure, and *vice versa*. This is sometimes called the *Bernouilli effect*, after the man who first identified it.

Fuel jets, subjected to the depression in the throat of the venturi, supply fuel to the engine at a rate proportional to the value of that depression which, in turn, is a function of that of the air flow. Control over the flow of the air and fuel mixture into the engine is effected by a *throttle valve* downstream of the venturi. Since this simple system was introduced, it has been greatly enhanced and consequently become increasingly complex.

8.1 The basic requirements

In a spark ignition engine, torque and power output are regulated by varying the quantity of combustible mixture supplied to the cylinders. This is, of course, done by means of a throttle valve, now mostly the butterfly type, though slide, sleeve and rotary valves have been used in the past.

Throttle valves must be installed downstream of the main jets of the carburettor, and therefore of the venturi. This is because, if it were installed upstream, it would obstruct the flow and thus cause wide variations in the depression over the fuel jets. Moreover, as it opened and closed, a throttle valve would change the direction of the air flow relative to the point at which the fuel from the jets issues into the venturi, and therefore adversely affect mixing.

For complete combustion, a stoichiometric mixture is needed, such a mixture being defined as one which contains the precise proportions of fuel and air required for complete combustion of both the fuel and, equally importantly, the air. With petrol, the stoichiometric air : fuel ratio is approximately 14.7 : 1 by weight, though such a mixture will not burn as completely as one might expect. This is because, in the extremely short time available, the vaporous and gaseous constituents are unable to form a perfectly homogeneous, or intimate, mixture in the combustion chamber. As a result, the exhaust gas tends to contain traces of oxygen, carbon monoxide and some unburned and partially burned hydrocarbons.

The weakest mixture ratio that is ignitable by a spark in a conventional petrol engine is about 18 : 1, though in engines specifically designed to burn weak mixtures (Section 18.15) ratios as high as about 23 : 1 are practicable. At the other end of the scale, mixture ratios of about 7 : 1 are too rich to ignite.

8.2 Requirements for metering and mixing

Considerations such as exhaust emission control, the need for fuel economy and for a reserve of extra power to meet needs arising in special circumstances ultimately dictate what is required of a carburettor. Ignoring these considerations for the time being, however, it can be said that the carburettor must supply continuously to the engine a mixture of fuel and air in proportions such that—

(1) In all circumstances, it can be easily ignited by the spark.
(2) The maximum possible amount of chemical energy can be extracted from it and converted by the engine into mechanical energy.
(3) All the fuel will be oxidised completely: that is, without producing carbon monoxide in the exhaust.

As regards emissions, oxides of nitrogen are generated by high temperatures in the combustion chamber, so their output can be influenced by some aspects of carburation. For instance, if heat is supplied to assist vaporisation of the mixture, this will be reflected in higher combustion temperatures, and weak mixtures burn hotter than rich ones. However, control of these nitrides is effected predominately by other measures such as adjustment of the spark timing, exhaust gas recirculation and, in turbocharged and supercharged engines, cooling the charge after it has been compressed.

The actual process of carburation comprises three stages—

(1) Metering the fuel through the jets in proportion to the air flowing into the engine.
(2) Breaking up the liquid into fine droplets, or *atomising* it, to assist evaporation.
(3) Distributing the evaporating fuel uniformly into the air flow to form a homogeneous mixture, preferably before it enters the cylinders, though in some circumstances, this process may be completed within them.

As previously indicated, the mixture is generally heated as it leaves the carburettor, to improve evaporation and uniformity of distribution. This is especially desirable in very cold ambient conditions, and for starting the engine. A supply of heat can also prevent the formation of ice in the carburettor due to the cooling effect of the latent heat of evaporation of the fuel.

8.3   Mixture quality

The quality of mixture required depends on the circumstances. In general—

(1) A rich mixture is needed for starting, especially in cold conditions when a high proportion of the fuel condenses out on to the cold walls of the induction manifold.
(2) Enrichment of the mixture is necessary for idling because the quantities of fuel being consumed are so small that that which condenses out on the walls of the induction manifold represents a high proportion of the total supplied, so the quantity of fuel actually reaching the cylinders is correspondingly reduced. Under idling conditions, mixture formation tends in any case to be poor, so the risk that the mixture in and around the spark plug gap will be too weak to burn is increased. All these factors mitigate against consistent firing of all cylinders unless extra fuel is supplied. The generation of unburnt hydrocarbon emissions can be worse with intermittent firing than with a slightly rich mixture.
(3) A slightly weak mixture for cruising, at part throttle, ensures that there is enough air to burn completely all the fuel, and therefore that all of its chemical energy is converted by the engine into mechanical energy.
(4) An extra supply of fuel for acceleration is essential because, when the

throttle is suddenly opened, the flow of air increases more rapidly than that of the fuel. This is partly because the fuel is more viscous and partly because it is heavier and therefore its inertia is greater. Moreover, the sudden opening of the throttle raises the pressure in the manifold, and therefore temporarily increases condensation of fuel out of the mixture.

(5) To obtain the maximum possible power output from an engine of any given size, the maximum possible quantity of chemical energy must be supplied to it, so the mixture must be enriched. However, this is achieved at the expense of higher brake specific fuel consumption and the presence in the exhaust of carbon monoxide, oxygen and incompletely burned hydrocarbons.

In a nutshell, for economical cruising all the fuel must be burned; on the other hand, to produce maximum power, all the air has to be burned, and economy has to go by the board. However, emissions regulations, the consequent use of catalytic conversion, and the need to conserve rapidly dwindling reserves of energy, have caused the emphasis to move towards the maintenance, so far as is practicable, of a mixture strength slightly lower than stoichiometric. The reason for this is explained in Sections 10.7 and 12.17, and mixture requirements in general are discussed in greater detail in Section 8.17.

Mixture formation and preparation present severe problems, which arise not least because the boiling points of the constituents of the liquid hydrocarbon fuel are spread over a very wide range and they all have to be metered into and blended homogeneously with the gaseous mixture of oxygen and nitrogen comprising the air. Distribution from the carburettor in equal quantities and identical qualities to each of the cylinders is another difficulty. Moreover, not only do ambient temperatures, pressures and humidities vary widely, but so also does the temperature of the engine, from cold starting to continuous operation under heavy load. The situation is further complicated by pulsations in the flow due to the reciprocation of the pistons and the opening and closing of the valves.

8.4   Induction of the mixture

Whether the engine is being cranked or driven under its own power, its theoretical rate of consumption of mixture is the volume swept by one of its pistons, between top and bottom dead centre, multiplied by the number of its cylinders and the speed of rotation of the crankshaft in rev/min. As the speed of rotation increases, so does the aerodynamic drag tending to reduce the rate of flow into the cylinders. The force pushing the mixture into the cylinders at any given instant is the pressure difference between ambient, or atmospheric, pressure and that caused by the downward motion of the pistons in their cylinders while the inlet valves are open. Note that the term *depression*, not *vacuum*, is for obvious reasons the correct term to use for describing the induction pressures in the manifold and cylinders.

The value of the depression in the cylinders depends primarily upon the speed of rotation of the crankshaft and the size and discharge coefficients of the series of apertures through which the mixture has to pass on its way into them. Some of these apertures, such as the gap between the valve head and seat and that between the throttle valve and barrel, are of variable size, but most, such as the valve throats, venturi or choke, air intake and various constrictions in the induction pipes, are fixed. Some carburettors, however, have chokes the sizes of which are varied automatically to maintain a constant depressions over the jets of their carburettors, as described in Sections 9.12 to 9.17. The valve throat must be large enough to ensure that, except when the throttle is fully open for maximum power output, it is not the determining factor as regards the depth of the depression generated in the cylinder, and therefore the rate of inflow of mixture. In all other circumstances, the degree of throttle opening is the controlling factor.

A secondary factor that influences the value of the depression is the ratio of the clearance volume above the piston at top dead centre to its displacement volume. Others include any leakages that might occur between the inlet valves and their guides, and through induction manifold joints or between the throttle valve spindle and its bearings.

8.5   Volumetric efficiency

Because induction occurs in each cylinder only once in every two revolutions of the crankshaft, a cylinder of 1 litre piston swept volume of an engine running at 2000 rev/min would ideally consume 1000 litres of mixture per minute. This can happen, however, only if the speed of rotation of the crankshaft is so low that the rate of flow of mixture through the valves and past the throttle is so slow as to be virtually frictionless and unrestricted, the clearance volume at top dead centre zero and, on completion of the induction stroke, the mixture is at atmospheric temperature and pressure.

It is important to remember that the fuel–air mixture is based on the masses, not volumes, of the two constituents: obviously volume, and therefore density, vary widely with both temperature and pressure. To obtain maximum power output, the efficiency of filling the cylinders would have to be 100% and the air : fuel ratio the optimum. In these circumstances, the average force exerted on the piston during the combustion process, that is the *mean effective pressure* multiplied by the projected area of the bore, would be the maximum attainable.

As previously indicated, the actual efficiency of filling of the cylinders is affected by several factors. The influence of the clearance volume between the crown of the piston and roof of the combustion chamber arises because, under the influence of the depression, the gas remaining in it after the exhaust valve has closed expands into the piston swept volume. Additionally, heat from the residual gas in that volume is transferred to the incoming gas, expanding it and thus reducing its density.

Fixed valve timing also has an adverse effect on filling of the cylinder. Normally, the valve overlap is set for maximum efficiency at the speed at which maximum torque is developed, so it will be too large for operation at idling speed. Consequently, under these conditions, exhaust gas will puff back into the inlet manifold and further dilute the charge when the flow into the cylinder begins.

Flow of mixture into the cylinder is also impeded, and therefore the density and mass of the charge reduced, by not only the throttle valve, but also by the previously mentioned fixed restrictions and those that vary, such as the valves. The effects of all these increase with engine speed.

The consequent loss of charge is expressed in terms of *volumetric efficiency*, defined as the ratio of the charge actually drawn in to that which would be inducted if its temperature and pressure could be maintained at ambient atmospheric and the clearance volume were zero. Volumetric efficiency generally rises to a peak at or around the speed at which maximum torque is obtained, when it is rarely much higher than 80%. The speed range over which maximum volumetric efficiency is obtained is influenced to a major degree by valve timing. This accounts for the widespread interest in variable valve timing, despite the extra cost entailed, Section 3.36.

8.6 Throttling

As previously indicated, the power output of an engine is controlled by throttling the flow of mixture into the cylinders to reduce the weight of charge according to the power output required. The smaller the degree of throttle opening, the greater is the depression induced in the cylinders, by the descending pistons. This depression is transmitted all the way back to the throttle valve. Because of the fall in volumetric efficiency with speed, maximum torque is obtained at a speed lower than that at which maximum power is developed. Although the consumption of mixture continues to increase with speed above that of maximum torque, the rate of increase progressively falls to zero as the flow becomes increasingly choked by the resistance at the valve throat or venturi.

The variations in pressure along the induction system under differing conditions of speed and throttle opening are illustrated in Fig. 8.1, but, for clarity, the degree of depression below atmospheric, in relation to the absolute-zero line, has been exaggerated. A venturi, sometimes called a *choke*, of the ideal form is assumed to be incorporated, allowing complete recovery of pressure from the relatively low value in its throat by the time the flow reaches its downstream end. Another assumption is that the flow is steady and, at any given speed, the depression remains constant. In practice, of course, the flow pulsates, the worst condition arising in a single cylinder four-stroke engine when the duration of the suction impulse is only approximately 25% of the total cycle time. Even in a six cylinder engine, pulsations still exist, though at a higher frequency and lower amplitude. The adverse effect of pulsating flow is due mainly to the relative differences between the masses, viscosities and inertias of fuel and air, which lead to fluctuations in the flows of the two fluids relative to each other.

The conditions illustrated in Fig. 8.1 are as follows—

(1) Low speed, small throttle opening (engine idling).
(2) Low speed, large throttle opening (at the beginning of acceleration, or when the engine is labouring during the ascent of a hill).
(3) High speed, small throttle opening (descending a hill).
(4) High speed, large throttle opening (fast motoring on level ground).

The volume of air in a stoichiometric mixture is about 97.5% of the total, so the calculations made are reasonably close approximations, even though the assumption has been that all the fluid passing through the system is air. Indicated on the left of the diagrams is the flow of air in litre/min. The example that has been taken is a 2 litre four-stroke engine, and the displacement of its

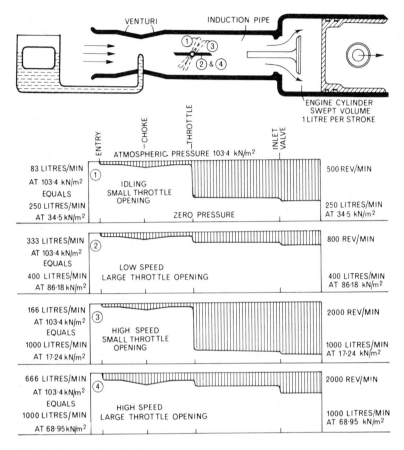

Fig. 8.1 Pressures in induction system

pistons at 2000 rev/min is assumed to be 1000 litre/min.

Varying solely with rate of flow through it, the relative depression in the venturi would normally range between about 1.7 to 3.5 kN/m² , but for the sake of clarity it has been exaggerated in the diagram. On the other hand, that in the manifold (or downstream of the throttle) is mostly considerably higher and, even for any given throttle opening, varies much more widely because it is directly related to both engine speed and load.

For idling, an extremely small weight of charge is needed. This is obtained by closing the throttle, and thus creating the very large pressure differential needed for metering a small volume of air through the tiny gap between the throttle and its barrel. Both the mass and density of charge inducted are therefore dramatically reduced. As the throttle is opened, the initial result is an increase in mass flow and, therefore, power output. This causes either an increase in speed or, if the vehicle is, for example, ascending a steepening incline, prevents the speed from falling off.

From the illustration, it can be seen that the depression in the venturi, which is the region into which the fuel from the main metering jets is delivered, increases steadily with the velocity of the air flow, which is related only indirectly to the throttle opening. This is why, although the throttle openings in diagrams (2) and (4) are the same, the depressions in the venturi are different while, in diagrams (2) and (3), the depressions in the venturi do not differ but the throttle openings do.

8.7 Fuel and air metering

Atmospheric pressure forces not only the air, but also the petrol, to flow through their respective orifices. The depression in the induction system simply reduces the pressure opposing these flows: it does not suck the fuel through the jets and the air through its intake or horn. Therefore, if the vent to atmosphere at the the top of the float chamber were to become blocked, the pressures above the fuel in that chamber and over the jet would rapidly equalise and flow would cease.

Where a supercharger is mounted upstream of the carburettor, the pressure generated by the supercharger, instead of atmospheric pressure, is vented to the float chamber. However, this layout presents problems, Section 14.2, so the supercharger is generally mounted downstream of the carburettor, and the fuel therefore does not have to pass through the jets into the pressure-boosted region of the incoming air flow. In a simple carburettor, a float-actuated needle valve maintains the fuel at a constant level, usually just below that of the jet, so it cannot run out and flood the manifold when the engine is stopped.

In more complex carburettors, however, either a hydrostatic pressure head of fuel or a combination of pressure head and atmospheric or boost pressure may be utilised to force the fuel through the jets. Such arrangements are in many instances adopted to compensate for the differences between the laws governing fuel and air flow, explained in Section 8.10.

The term 'choke', although widely used instead of 'venturi', is unfortunate because it is commonly employed also to describe the valve installed upstream of the venturi to increase the depression over the jets for enriching the mixture for cold starting. It is therefore better to use the term *strangler* to describe the cold start enrichment valve.

Regardless of how the pressure difference is generated, it is expressed in terms of a pressure head, so that the formula $v = \sqrt{(2gh)}$ for a freely falling body in a vacuum, can be applied to calculate the flows of both fuel and air. Owing to air resistance, however, the velocity of a body falling freely in the atmosphere differs from that indicated by this formula. Even so, it is a reasonably close approximation and, by using it, flow testing is simplified, because velocity of flow v through an orifice is a function of the pressure difference corresponding to a head h, while g is the constant gravitational force. Although SI units are used nowadays, any other system of units can be employed provided consistency is maintained throughout all the calculations.

Pressures and corresponding heads and velocities for water, petrol and air are given in Table 8.1. The density ρ for air is that at 101.3 N/m^2 and $0\,^\circ\text{C}$, while the velocities, which are those for the conversion of the potential energy of the pressure head to the kinetic energy of the velocity are calculated, ignoring losses, from $v = \sqrt{(2gh)}$. In practice, because of the viscosity, surface tension and friction effects opposing the flow of the liquids, the velocity of flow of fuel through a jet follows a slightly different law. Air, however, more closely approaches the ideal of a perfect fluid, so the formula serves with reasonable

Table 8.1—PRESSURE, HEAD AND VELOCITY

Pressure		Head and corresponding velocity					
		Water $D = 1000\,kg/m^3$		Petrol $D = 753\,kg/m^3$		Air $D = 1.298\,kg/m^3$	
kN/m^2	mm(Hg)	Head m	$v = \sqrt{(2gh)}$ m/s	Head m	$v = \sqrt{(2gh)}$ m/s	Head m	$v = \sqrt{(2gh)}$ m/s
101.3	760	10.33	14.23	13.72	16.4	7955.0	395.0

accuracy for calculating its velocity of flow, where the head h is determined from the following formula, where P is the pressure—

$$h = P/\rho g \qquad (8.1)$$

During normal operation, the depression in a fixed carburettor ranges from about 100 to 500 mm of water column, corresponding to air speeds of 39.6 to 85.3 m/s. For petrol, the corresponding velocities are, of course, much lower.

8.8 Volume and mass flow

The volume flow through an orifice is obtained by multiplying the velocity by its cross-sectional area—

$$\text{Volume flow} = A \times \sqrt{(2gh)} \qquad (8.2)$$

All carburettor jets are flow-tested under a standard head of 500 mm of fuel and the rates of flow through them stamped on one of their ends. This method of identifying jets is useful for engine tuning as well as carburettor assembly. Following the imposition of emissions regulations, most manufacturers flow-tested all their carburettors under normal operating conditions after assembly.

For combustion calculations, mass flows are required, for which it is necessary to multiply the volume flow, as given by Eq. (8.2), by the density D of the fluid—

$$\text{Mass flow} = A \times \sqrt{(2gh)} \times D \qquad (8.3)$$

From Eq. (8.1) it follows that the head h can be expressed in terms of the pressure difference—

$$gh = \frac{p_1 - p_2}{D} \qquad (8.4)$$

where p_1 and p_2 are the absolute pressures at the ends of the orifice and D is the density of the air or petrol. So the formula for mass flow becomes—

$$\text{Mass flow} = A\sqrt{[2g \times (p_1 - p_2) \times D]}$$

$$= K \times A\sqrt{[D(p_1 - p_2)]} \qquad (8.5)$$

where K is a constant which is needed to take into consideration the resistance to flow. This constant covers the effects of the viscosity of the fluid, friction, surface tension and the form and proportions of the orifice, and is termed the *coefficient of discharge*. For most carburettor jets, it ranges from about 0.6 to 0.8, varying little with jet size.

Although the density of petrol does not change appreciably with temperature, that of air does, resulting in enrichment of the mixture on warm days. Enrichment also occurs at high altitudes, owing to the reduced density of the air. Disregarding, these factors, the only two variables in (8.5) are the pressure difference and orifice area.

8.9 Fixed- and variable-choke carburettors

There are two main categories of carburettor: fixed and variable choke. The latter, in which the size of the choke (venturi) is controlled on the closed-loop principle, is generally termed the *constant-depression* type. In the former, the venturi is of fixed cross-section so the depression in it varies with the velocity of flow through it. This depression draws the fuel through the jets, metering it in proportion to the rate of air flow.

In the variable choke type, the area of cross-section of the venturi is regulated (on the closed-loop principle) to maintain that depression constant, usually by either a slide valve actuated by means of a piston, or diaphragm device subjected to the depression in the venturi. In this type, as the piston or diaphragm opens and closes the slide valve, it simultaneously moves a tapered needle in a single jet, to meter the flow of fuel accurately in relation to that of the air. The needle can be tapered, or profiled, in such a way as to compensate for the previously explained divergence in flow characteristics between the fuel and the air.

8.10 The fixed-choke type

There are three broad categories of fixed-choke carburettor: updraught, downdraught and sidedraught, though one occasionally comes across the terms semi-sidedraught and semi-downdraught, describing carburettors in which the venturis are not absolutely horizontal or vertical respectively. Sidedraught carburettors and their air cleaners are difficult to accommodate in most modern engine installations. In the updraught type, gravity hinders the intake of fuel, sprayed at relatively low velocities, into the induction system, so a smaller venturi must be employed than in any of the others. With the downdraught carburettor, the larger venturi increases the volumetric efficiency of the engine, so this type has largely displaced the others. The air cleaner can be mounted on top of or to one side of the carburettor and can be easily removed, for access to the carburettor.

Fixed-choke carburettors having only one jet, Fig. 8.2, have been obsolete for very many decades but, by virtue of their simplicity, clearly demonstrate the basic principles. The fuel level in the float chamber is maintained, by the float-and-needle valve mechanism, constantly just below the jet orifice.

When the engine is running, both orifices (fuel and air) are subject to the depression in the throat of the venturi. Consequently, given perfect fluids, it

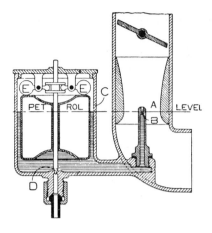

A   The throat of the venturi, or choke
    (air orifice)
B   Fuel jet (fuel orifice)
C   Float chamber
D   Needle valve
E   Levers translating upward motion of
    the float to downward motion of the
    needle valve, and *vice versa*

Fig. 8.2 Elementary carburettor with only a single jet, together with the bottom feed needle valve and its actuation mechanism. This became obsolete long ago

would be possible to provide the air:fuel ratio appropriate for complete combustion simply by arranging for the ratio of the area of the venturi to that of the fuel jet to be 14.7:1. Since the density of petrol is about 753 kg/m³ while that of air at normal temperature and pressure is 1.28 kg/m³, the ratio of their densities is 587:1. It follows that, for all values of depression in the throat of the venturi, the ratio of the areas of the air and fuel orifices would have to be—

Area of choke/area of jet = 14.7/1 × √587/1 = 363.4
Diameter of choke/diameter of jet = √363.4 = 19.06

As previously indicated, while air approximates to a perfect fluid, petrol does not. In Fig. 8.3, the full-line curve represents the mass flow through a No. 110 jet (110 ml/min flow beneath a head of 500 mm), as determined experimentally at a temperature of 0°C, of a petrol the specific gravity of which was 0.75. The depression was measured in centimetres of water column, but scales of air and petrol columns are also shown.

Three other parabolic (square-root) curves, dotted and chain dotted, have been added to represent the air flow through alternative venturis that would be suitable for use with the same petrol jet. Scales of velocity of flow for air and petrol have been added to the diagram. That of the air was taken to approximate to the theoretical relationship, $v = √(2gh)$, while that of the petrol was calculated from the experimentally determined mass flow, which was between 65 and 70% of the velocity given by the formula. From the illustration, it can be seen that, because of the slight difference between the flow characteristics of air and fuel, a stoichiometric fuel:air ratio is obtainable at only one value of depression for any given size of venturi. So, when tuning an engine, a fuel jet the size of which is appropriate relative to that of the venturi is selected.

The mixture tends to become richer as the depression increases. From the illustration, it can be seen that, with the large choke, the mixture is weak throughout the whole range up to 50 cm depression and, with the smallest choke, it is weak below and rich above about 20 cm depression.

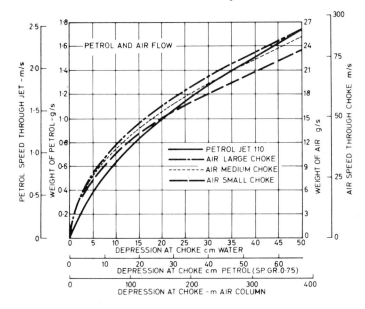

Fig. 8.3 The full line represents the flow of fuel through a No. 110 jet in three different sizes of venturi. The corresponding air flows through each of these venturis of course differ for each value of fuel flow, and are represented by the three dotted lines

On a level road, the rate of air flow through the venturi depends on both the engine speed and degree of throttle opening. Consequently, on a varying incline the velocity of flow can be held constant either by varying the opening of the throttle to keep the speed constant, or allowing the speed to vary but keeping the throttle opening constant. If the fuel supply is properly compensated, as described in Section 8.11 *et seq.*, neither action will alter the mixture strength. On the other hand, varying the velocity of flow will alter the quality of the mixture: as it is reduced, atomisation will be adversely affected and droplets of fuel will tend to fall out of suspension in the induction manifold. At the same time, the associated rise of pressure in the venturi will reduce the rate of vaporisation.

8.11 Fuel: air ratio compensation for fixed-choke carburettors

From the foregoing, it can be seen that compensation is necessary: the mixture strength must be adjusted at all depressions above and below that which gives the correct air : fuel ratio. This can be done either by weakening the mixture as the depression increases above, or enriching it as the depression falls below that selected for giving the chemically correct value. The most commonly used devices for doing so are as follows—

(1) A valve for introducing extra air into the induction system downstream of the venturi. Although manually actuated valves have been used, most are automatic.
(2) The use of compensating jets for hydraulic rectification of the air : fuel ratio.

(3) Air bleed into the fuel jet system to relieve it partially of the depression in the venturi towards the upper end of the range.

Of these, the third has been virtually universally preferred, because not only can it correct over a wide range of depressions, but also the bleeding of air into the fuel contributes to atomisation, vaporisation and uniformity of distribution. Manually actuated extra air valves, which have been mostly of the slide-valve type, are no longer in use except on a few motor-cycles and small industrial engines. The additional complication of the control system entailed is one disadvantage, and it is doubtful whether many motor-cycle drivers have the skill and experience to use them properly. In recent years, pressures aimed at reducing emissions have, in any case, rendered such devices unacceptable.

8.12   Compensation by compound and submerged jets

One of the earliest successful methods of compensation was the Baverey compound jet illustrated in Fig. 8.4. The fuel is drawn through two jets, that at A being the main supply and the other, at that at B, the compensating jet. Originally only the compensating jet orifice was situated in the base of the float chamber, but it soon became common practice to submerge both jets. The advantage of submerged jets is that, because they are lower and therefore subjected to a weaker depression than if they were in the venturi, they have to be of larger bore. Consequently, they are easier to manufacture to close tolerances and less likely to become clogged by debris or dirt in the fuel, and the flow characteristics of two jets totally submerged at approximately the same level are more favourable for mixture correction than if they were discharging into the air. In the illustration, the main jet is shown discharging fuel from the float chamber directly into the air, while the submerged compensation jet discharges into the well D, whence the fuel is drawn on up to the discharge nozzle C.

When the engine is started, the level of the fuel in the well is initially the same as that in the float chamber, so there is no flow through the compensating jet and discharge nozzle C: since atmospheric pressure, h_2, acting on the surfaces of the fuel, is transmitted to both ends of the jet, there is no pressure differential across it. As the throttle is opened and the rate of flow of fuel through both jets

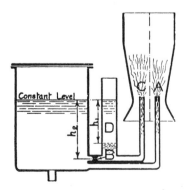

Fig. 8.4 A diagrammatic representation of a main and compensating jet system

increases, the level of the fuel in the well falls and the pressure differential, $h_2 - h_1$, across the compensating jet progressively increases, h_2 being the maximum head above the jet. Consequently, the flow through the compensating jet similarly increases until, at the point at which the well is totally empty, it remains constant despite increasing air flow and therefore a weaking mixture. In the illustration, h_1 is the head when the level of the fuel in the compensating jet well has fallen to an intermediate value.

Over the steeply rising portion of the air flow curve in Fig. 9.3, the increasing depression progressively increases flow from both the compensating well through the tube C, and that replenishing the well through the compensating jet. This initially maintains the mixture at a constant value. Then, when the well is empty and the flow from the compensating jet therefore remains constant while the air flow continues to rise, the tendency towards increasing richness of the mixture is countered, so it again remains constant. Over the upper part of the range, when the basic enrichment curve is straighter and less steep, some further compensation may be effected by the passage of air from the well up through the discharge nozzle. This method of compensation has been widely used in carburettors, though usually incorporating an additional air bleed principle, to be described in Section 8.13.

In the meantime, Fig. 8.5 demonstrates the principle graphically and in more detail. In this illustration, mass fuel flow curves for main jets No. 90 and 100, and compensating jets Nos 110 and 135 are plotted together with the curve for the air flow through the largest choke from Fig. 8.3. It can be seen that, as the head above the two compensating jets increases from zero to $h_2 - h_1$, the latter being in this instance 3 cm of water (or 4 cm of petrol), the rate of flow from them lies above that for the main jets. Beyond this point, however, the fuel flow from these jets remains constant.

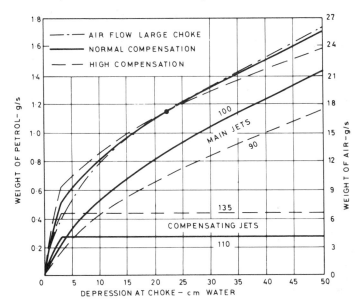

Fig. 8.5 Fuel flows through different sizes of main and compensating jet plotted together with the air flow through the largest venturi plotted in Fig. 8.3

It will be found that, if the flow from the main jet 100 is added to that from the compensating jet 110 and that from the main jet 90 to that from the compensating jet 135, the curves produced by plotting the totals from both will, at 20 cm depression, coincide with the air curve, indicating that, at this depression, the mixture is chemically correct. Indeed, the jets were selected to give this result. On the other hand, the larger main jet combined with the smaller compensating jet gives reasonably good compensation throughout, with only slight over-compensation.

Although the other combination (100 and 135) gives considerable over-compensation, this may give the result desired, since we need rich mixture for starting and idling and then, for good fuel economy, a weak mixture which can be enriched as and when necessary, for example for acceleration and obtaining maximum power, by means of other devices to be described later. It can be revealing to plot the curve for over-compensated mass fuel flow from Fig. 8.5 together with the three mass air flow curves from Fig. 8.3, all against mass air flow, to show quality of mixture (both as a percentage and a ratio). This has been done in Fig. 8.6.

8.13 Air bleed compensation

As has been previously stated, air bleed compensation can be incorporated in the compensating jet well. However, to simplify the explanation, we shall describe how it functions as the sole method of compensation in a main jet. A good example, which was incorporated in some of the early Solex carburettors, is illustrated in Fig. 8.7.

An advantage of using this device is that only one jet is needed. As can be seen from the illustration, the jet D is drilled in the lower end of a thimble-shaped tube, where it is fully submerged. The open upper end of this

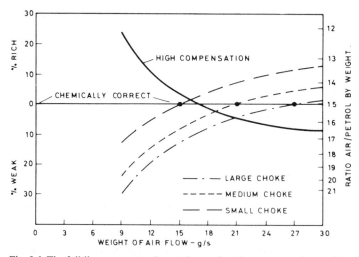

Fig. 8.6 The full line represents the total mass fuel flow through the small main and large compensating jets in Fig. 8.5, while the other three lines are the corresponding mass air flows depicted in Fig. 8.3, but all have been re-plotted against mass air flow instead of depression in the venturi

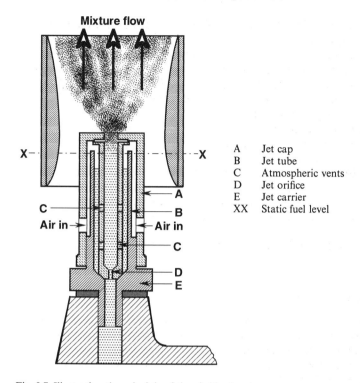

Fig. 8.7 Illustrating the principle of the air bleed system

A	Jet cap
B	Jet tube
C	Atmospheric vents
D	Jet orifice
E	Jet carrier
XX	Static fuel level

tube is subject to the depression in the venturi. Solex called this the *jet tube*, though terms such as *emulsion tube* or *diffuser tube* are used by other manufacturers to describe similar components.

Drilled radially into the jet tube are as many air bleed holes as may be necessary to perform the compensation function: in the illustration, two diametrically opposed pairs of holes are shown. The open lower end of a deep thimble-shaped cap A is screwed down over the jet carrier E, the lower end of which is screwed into the carburettor body. In the otherwise closed upper end of the cap, which seats firmly on the upper end of the jet tube to retain it, there is a small hole.

Interposed concentrically between the jet tube and cap is a tubular upward extension of the jet carrier, forming a well between it and the jet tube. Between the upper end of this extension tube and the inner face of the upper end of the cap is a clearance, so that air at atmospheric pressure passing in through the two radial holes near the base of the cap can flow up and over into the top of the well.

When the engine is started, the jet tube and the well are full of petrol up to the level XX. As the throttle is opened and the well empties, the level of the fuel in the well falls. Because this progressively uncovers the air bleed holes C, the depression over the upper end of the jet tube is correspondingly weakened, thus offsetting the tendency towards enrichment of the mixture. At the same time, the air begins to bubble through the fuel in the jet tube and emulsifies it, thus assisting evaporation. At high levels of depression, jets of air squirt

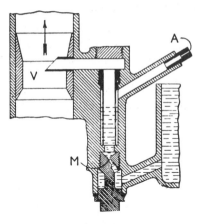

Fig. 8.8 Claudel–Hobson air bleed

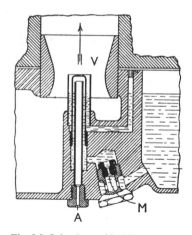

Fig. 8.9 Solex 'assembly 20'

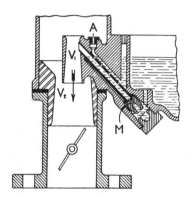

Fig. 8.10 Stromberg air bleed

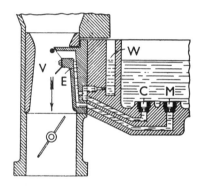

Fig. 8.11 Zenith air bleed

through the holes in the bleed tube into the rising column of petrol, emulsifying it even more effectively.

In some of the Weber downdraught carburettors air bleed systems are used in conjunction with rotary valves. The latter are actuated by the throttle controls, to reduce the air supply as full throttle is approached, thus enriching the mixture for the development of maximum power.

Figs 8.8 to 8.11 are diagrammatic sections through an early Claudel–Hobson and three later carburettors, showing variations on the air bleed theme. In these, A is the air metering bleed plug, M the main jet, and V the venturi, or choke. Also, a single air bleed orifice has supplanted the bleed holes in the jet tube that was shown in Fig. 8.7, so the degree of compensation can be adjusted simply by fitting an air bleed plug having an orifice of a different size. As indicated in Section 8.12, an additional air bleed effect can be obtained by virtue of the flow of air through the idling jet when the well or feed passage becomes empty.

8.14 Multiple venturis intensify air bleed compensation

In the Stromberg downdraught carburettor, Fig. 8.10, the exit from the venturi V_1 is in the throat of V_2. Consequently, because the pressure difference across V_1 is higher than it would be if only a single venturi were used, the velocity of flow through, and depression in, it are also higher. Double venturis not only increase the depression over the diffuser but also introduce a two-stage mixing process: the primary stage occurs in V_1 and the secondary one in V_2, where the rich mixture issuing from V_1 is blended with the air emerging from V_2. Incidentally, a triple venturi can be seen in Fig. 9.22.

8.15 The Zenith V-type emulsion block

In Fig. 8.11, C is the compensating jet, which is installed in the bottom of the float chamber, next to the main jet. The passage leading from it takes the compensating fuel supply into an intermediate chamber in the emulsion block E, instead of directly into the well W. In this chamber, the level of depression is intermediate between those in the well and venturi, its actual value depending on the sizes and characteristics of the communicating ducts and the resistance to flow through the main passage in the emulsion block. The general principle is discussed in detail in Section 8.18.

8.16 Secondary suction effects

In most carburettors, secondary suction effects are introduced to enhance atomisation and mixing. These may arise from chamfering the ends of the tubes, usually termed *spray tubes*, that deliver the fuel into the venturi, as in Figs 8.8 and 8.11. Another device, which is shown as a black dot in Fig. 8.11, is the introduction of a *spray bar*, usually mounted diametrically across the throat of the venturi. This has two effects: first, it increases the depression over the end of the spray tube; and, secondly, the turbulence that it generates downstream enhances mixing. The effects of these devices are generally evaluated experimentally or by modelling.

8.17 Mixture requirements in more detail

So far, we have expressed mixture strength in terms of air : fuel ratio. However, over the past 40 years, the symbol λ has been increasingly widely used, mainly because λ values represent percentage mixture strength. Since $\lambda = 1$ is the stoichiometric, or chemically correct, mixture, the 10% weak and 10% rich values are respectively $\lambda = 1.1$ and 0.9 which, expressed as percentages, become respectively 110 and 90%.

Although a 10% weak mixture is the normal average for maximum economy and 10% rich for maximum power, several factors complicate the situation. At wide throttle openings, the cooling effect due to the evaporation of the extra fuel increases the density of the charge and, at small throttle openings, a slight enrichment is necessary to compensate for the effects of the residual exhaust gas in relation to the small quantities of air getting past the throttle and passing into the cylinders. Furthermore, to satisfy modern emissions regulations, extremely accurate control over mixture strengths is

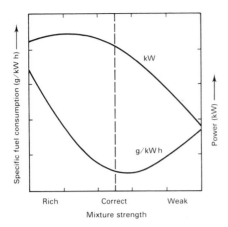

Fig. 8.12 The influence of mixture strength on engine performance

needed. The influence of mixture strength on engine performance is illustrated in Fig. 8.12.

Electronic control of carburettors would appear to be necessary to meet modern requirements. So far, however, it has unacceptably increased their overall complexity and, therefore, cost. Since fuel injection is easier to control electronically and the product is ultimately less costly, the carburettor is rapidly moving towards obsolescence.

Referring back to Fig. 8.1, it can be seen that the mass air flows through the venturi under conditions (2) and (3) are identical, despite their widely differing throttle openings. The mixture requirements also differ widely: for (2) a rich mixture is required to obtain high mean effective pressure and torque, to cater for acceleration; on the other hand, (3) must have a weak mixture persisting up to 80 to 90% full throttle, for economy at moderate loads. Condition (4) again calls for a rich mixture, for maximum power as full throttle is approached. Mixture requirements in relation to load and to engine speed at constant load respectively are illustrated in Figs 8.13 and 8.14.

If we plot the requirements against mass of air flow, as was done in Fig. 8.6, we have the curves illustrated in Fig. 8.15. These are an elaboration on the curve in Fig. 8.6 in that the ideal mixture requirements are plotted for part and full load (for simplicity, emissions regulations have been ignored). From this and the previous paragraph, it is clear that a carburettor in which the mixture strength is determined only in relation to the depression in the venturi could not meet these requirements.

8.18   Principle of the intermediate chamber

At this point it is appropriate to explain the principle of the intermediate chamber, a device that is applied in various ways to carburettors. The fuel discharge orifice is situated in the intermediate chamber instead of the venturi, and the areas of both the inlet and outlet through which the air passes are

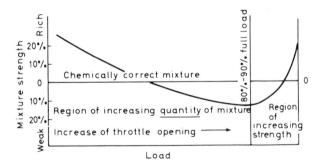

Fig. 8.13 Ideal carburettor characteristic

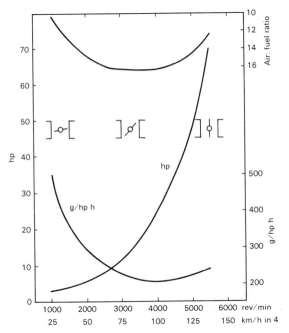

Fig. 8.14 Engine performance and mixture strength requirements at increasing throttle opening increments in top gear

controllable, for example by the air and throttle controls on a motor-cycle or by a strangler valve, as described in Section 8.19.

As can be seen from Fig. 8.16, the value of the depression p_2 in this chamber, depending on the mass flow and relative areas of the inlet and outlet, is intermediate between p_1 and p_3. At (*a*) the inlet is larger than the outlet, the sizes being such as to cause the depression over the fuel discharge orifice to be less than half that in the induction pipe. At (*b*) the relative sizes of the openings are reversed but adjusted so as to maintain the same rate of air flow. In the latter instance, the depression over the fuel discharge orifice is more than half that in the induction pipe, so the mixture is richer. With such a device,

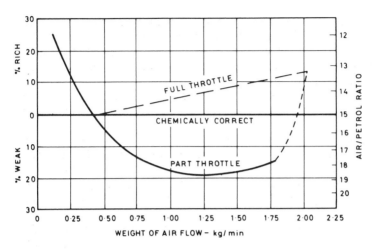

Fig. 8.15 Dual carburettor characteristic

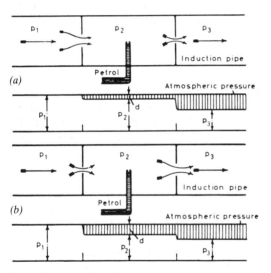

Fig. 8.16 Intermediate chamber

therefore, provided control over the adjustments is fine enough, any change in mixture strength can be put into effect with constant air flow, or the mixture strength can be held constant while the air flow is varied.

8.19 Starting and idling enrichment devices

When the vehicle is stationary and the engine *ticking over, idling,* or *slow running,* the velocity of flow though the carburettor is such that the fuel drawn from the jets is sufficient to develop only enough power to overcome the resistance of its moving components. In this condition, the engine should be running as slowly and quietly as practicable, but be capable of being accelerated instantly, for instance for moving away from traffic lights.

However, in a fixed-choke carburettor, when the throttle is closed, for either starting or idling, the flow through the venturi is too slow to create the depression needed to lift the fuel from the jet, let alone to atomise and mix it. To cater for these conditions, therefore, it is necessary to provide an alternative, which takes the form of an additional jet and air orifice. Since, in these circumstances, the highest velocity flow is that through the gap between the edge of the butterfly valve and the throttle barrel, this is obviously the best place for siting the idling system.

Because the depression at this point is strong and the velocities of flow of both air and fuel high, atomisation and mixing will be good. On the other hand, air leaks, for example past gaskets downstream of the idling system, will tend to represent a high proportion of the total flow. Therefore, because they would dramatically reduce the mixture strength, they must be carefully avoided.

Even when the engine is warm, the idling mixture must remain rich. If it is not, it is impossible to be sure that an ignitable mixture will exist adjacent to the points of the plug at the instant the spark is passed. This is because of the low density of the charge and difficulty of obtaining homogeneity when the jet is discharging down one side of the induction pipe. When the engine is cold, only a small proportion of the fuel issuing from the idling jet will evaporate, so the situation is even more difficult and therefore the mixture has to be richer still.

In the early days, a spring-loaded plunger was provided on top of the float chamber, so that the driver could, by depressing it and thus pushing the float down momentarily, raise the fuel level in that chamber to enrich the mixture for cold starting. Subsequently, manually actuated stranglers were installed to perform the enrichment function. They comprised a cable-controlled valve, usually of the butterfly or flap valve type, upstream of the venturi. Partially closed, they increased the depression over the jets, and therefore enriched the mixture. However, if the drivers forgot to open them again, the engine could run for long periods with rich mixtures, causing sparking plug points to become fouled with soot. Even worse, in very cold weather, neat fuel could be drawn into the cylinders, wetting the sparking plug points and draining down into the sump, thus diluting the lubrication oil.

Subsequently, a demand arose for automatic stranglers, which in the first instance were actuated by bimetal strips or other thermostatic devices. Some drivers, however, because they were unlikely to be aware whether the device was malfunctioning, did not entirely approve of automatic chokes. Another method of overcoming the problem was to use combined cold-start and warm-up devices which, although independent of the carburettor, were in most instances mounted on it. Some of these will be described in the following sections. More recently, to satisfy the exhaust emissions regulations, electronically controlled cold- and warm-start enrichment devices have become virtually mandatory. These are mostly used with fuel injection systems and therefore will be described in the Chapter 10.

8.20 Separate starting and warm-up enrichment devices

The separate starter devices in use after the Second World War were brought into operation manually, but semi- or fully-automatic adjustment followed progressively as the engine warmed up. In many instances, however, when the engine was warm, the device had to be disengaged manually. The enrichment fuel was delivered either through the idling jet system or independently through ducts that were opened and closed by the manual control.

Progressive weakening of the mixture as the engine becomes warm can be effected by regulating the temperature, depression or velocity of flow of the ingoing air. It should be borne in mind, however, that the velocity can be increased only by increasing the depression. Since this can be simply a result of increasing engine speed, it is not strictly an independent means of control. Lowering the pressure or raising the temperature increases the rate of evaporation, so both tend automatically to enrich the mixture.

Interconnection of the strangler and throttle controls is widely used for increasing the idling speed to prevent the engine from stalling after it has been started from cold. Such devices, actuated automatically by movement of the strangler control, usually comprise a mechanism for moving the throttle stop also. To this end, links are employed to actuate either a swinging arm on which the throttle stop is mounted or to rotate a cam, in both instances to open the throttle.

The closing of the strangler and opening of the throttle increases the depression over the main jet, on the principle illustrated in Fig. 8.16. This causes fuel to be drawn from the main in addition to the idling jet, to enrich the mixture. On the Zenith VIG and VIM models, mounted on the strangler valve was a spring-loaded flap valve which would be opened by increasing depression. From the comments in Section 8.18, it can be deduced that such a valve could be spring loaded to admit an increasing volume of air to maintain either a constant or an increasing depression in the venturi.

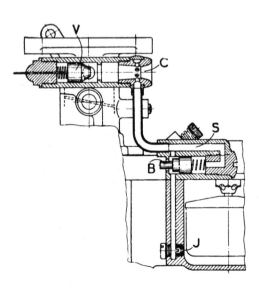

Fig. 8.17 Zenith starting device

8.21 Zenith VE starter carburettor

A good example of a separate semi-automatic starting device is the Zenith VE unit, Fig. 8.17, production of which ceased several decades ago. It is simply a fixed choke, single-jet, miniature carburettor drawing its fuel from the float chamber of the main carburettor, the output of which is delivered directly into the induction pipe. It is calibrated to provide the appropriate rich mixture for cold starting and has an automatic air bleed valve actuated by the depression in its own venturi.

In the illustration, the cone valve V is opened by a cable connected to the pull-out control on the dash to bring the device into operation. The depression in the venturi C causes fuel to flow from the enrichment jet J in the base of the main float chamber, to mix with the air passing through the venturi and the open valve C, out through a discharge orifice in the wall of the induction pipe just downstream of the throttle valve, shown dotted in the illustration.

As the engine speed increases, so also does the depression in the space S, until it is high enough to unseat the air bleed valve B. This allows air to bleed through the annular space around the stem of the valve into the fuel duct. The stem has parallel sides, but its length is limited so that a fast idling speed is maintained until the engine is warm, after which there is no risk of neat petrol entering the cylinders, even if the driver forgets to return the control to the position for normal running. Valves having stems of different dimensions were available for matching the device to different engines.

8.22 Thermostatic control for starting and warm-up

A good example of fully automatic control of the mixture for starting and warm-up is the Zenith Autostarter, Fig. 8.18, fitted to the CD4T and CD5T constant depression carburettors. Its functions are to open the throttle slightly

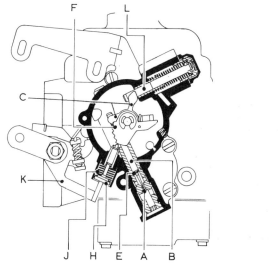
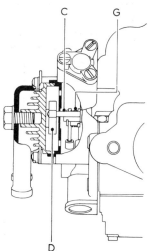

Fig. 8.18 Zenith Autostarter

for cranking and the subsequent idling and warm-up, and to enrich the mixture not only during both these operations but also for acceleration while the engine is still cold.

Fuel for enrichment, drawn from the float chamber, is delivered though orifice A to the mixing chamber of the carburettor, described in Section 9.15. Its rate of flow is regulated by the tapered metering needle B, the position of which is regulated by the bell crank lever C on a spindle to which is secured the inner end of a coiled bimetal strip D. At its outer end, this strip is secured to its cast aluminium housing, which is a dished cover over the assembly comprising the cam F and lever C. The outer face of the housing is finned and, in turn, is covered by another dished casting which is in fact a water jacket through which flows engine coolant.

Also part of the actuation mechanism is the spring-loaded piston and cylinder assembly sloping up towards the right, at about 30° from the horizontal, as viewed in the illustration. The spring tends to push the piston to the upper end of the cylinder. This motion is opposed, however, by manifold depression transmitted through a duct from induction pipe to the space below the piston.

The follower for cam F limits the motion of the bell crank lever K, which floats freely on the throttle spindle. Carried on one of the ends of the bell crank is the pin that rests on the follower assembly. This assembly includes a coil spring, which tends to retract the follower and thus swing the bell crank clockwise about its pivot as the throttle is opened. On the other end of the bell crank is a movable throttle stop which, as the bell crank is allowed to rotate progressively clockwise by the cam and follower, moves away from the adjustable screw on the throttle control lever until the butterfly valve can be closed down to its hot idle position. If rotated anti-clockwise, it tends to open the throttle towards the cold idle position.

The mechanism functions as follows. When the engine is fully warmed up, a sealing ring E on the shoulder of the needle closes the enrichment valve orifice completely. Then, as the warm engine is throttled back to idling speed, prior to being stopped, the increasing manifold depression acting on the piston rotates the lever C, anti-clockwise. Since the lever C and cam F are interconnected by the torsion spring G, the cam is also rotated anti-clockwise. Therefore its follower, which is part of the pin and spring assembly H, ultimately rests on one of the lower steps of the cam F. When the engine is finally switched off, and the depression collapses, the spring pushes the piston L to the end of its cylinder, pulling lever C with it to move the needle to the rich setting in the valve orifice, ready for cold starting. However, because the follower is on one of the lower steps, the cam cannot rotate with C, which therefore winds up the torsion spring.

Because the cam follower was locked on its lowest step when the engine was switched off, before the driver can start the engine, he must depress the throttle momentarily to lift the follower. This releases the cam so that the torsion spring can rotate it to the position at which, as soon as the throttle pedal is released again, the follower will drop on to its highest step. In this condition, which is that shown in the illustration, the engine is ready for cranking.

As soon as it fires, manifold depression pulls the piston L down to the left, leaving the end of the lever C free to float over the full range of its movement in the lost motion slot in the piston rod, except in that its position, and therefore that of the needle in the enrichment valve orifice, is in fact regulated by the bimetal strip. Consequently, each time the throttle is closed, the follower descends on to the cam. Which step it lands on is determined by the temperature of the engine coolant in the jacket. As the engine becomes warmer, the bimetal strip progressively rotates the cam lever, until it is aligned with the lowest step and the throttle therefore can close to its hot idling position.

In the meantime, if the throttle is opened wide while the engine is still cold, the depression in the manifold will collapse. This will allow the return spring to move the piston to the end of the cylinder, thus rotating the lever C clockwise and withdrawing the tapered needle from the valve orifice, to enrich the mixture for acceleration with a cold engine. Among the alternative stepped cam arrangements is the Solex system illustrated in Fig. 9.13.

8.23   Solex progressive starter

There are several variants of the progressive starter first introduced by Solex. All have been actuated either manually by a pull-out control on the dash or automatically by a thermostatic device such as a bimetal spring. The original Solex unit had simply an on–off control, while later versions were progressive, having positive-feel two- or three-position rotary disc valve controls. A slightly different application of the multi-hole disc valve principle, can be seen in Fig. 9.27.

In Fig. 8.19, the main view is of the early version, while the scrap view (left) shows the modification for the two-position version. In the original version, the orifice Ga metered the flow of air from immediately upstream of the venturi into the starter chamber. The disc valve opened and closed two ducts simultaneously. One of these was the small duct D, which drew fuel from the starter well integral with the float chamber shown in the sectioned view on the right. The other was the much larger duct leading from the bottom of the starter chamber to a point downstream of the throttle valve.

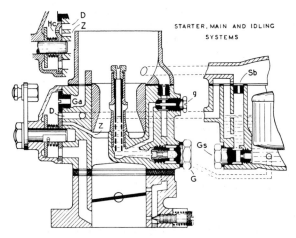

Fig. 8.19 Solex B32-PBI-5 carburettor

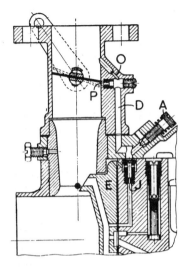

Fig. 8.20 Zenith idling system

When these two ducts were open, an extra supply of fuel was drawn from the starter well, mixed with the air in the starter chamber and delivered down through the large duct directly into the induction pipe. The strength of the mixture was determined by the sizes of the slow running jet Gs and the orifice D. This system, of course, can operate only when the throttle valve is closed. In the diagram, Sb is the idling mixture air bleed, Gs the slow running jet, G the main jet and g the air bleed orifice for the main jet emulsion system.

In the modified version, scrap view on left, part of the upper portion of the disc is dished to embrace both the fuel inlet D and two ducts. One is from the slow running well and the other an extra air bleed Z, from immediately downstream of the venturi where, when the throttle is closed, the air pressure is atmospheric. In the crown of the dished section is the metering hole Hc, through which is drawn an emulsified mixture of fuel and air from these two ducts. This emulsion is then mixed with the air in the starter chamber and passes on, as before, down through the duct into the induction pipe. The outcome is that a slightly larger quantity of fuel and air, though better mixed, is fed to the engine induction system. As the pull type control is actuated to bring the system into and out of operation, the edges of the dished section progressively open and close the ports D and Z, but the hole over the delivery port to the induction system is elongated so that it will continue to allow the mixture to pass into the induction system as long as the input ports are open.

In a later variant, the disc valve was again flat, but had a series of fuel inlet holes in it. Also, the air orifice Ga was moved to the opposite side of the starter chamber, where it delivered its air through a port and an elongated hole in the disc at the outer end of the spring that loads the inner disc valve. In other words, there were two disc valves: that on one side for air and the other for fuel.

8.24 Idling systems and progression jets

As was explained in Section 8.18, the idling mixture has to be discharged into the region of low pressure generated by the rapid air flow adjacent to the edge of the throttle valve. However, if only a single discharge orifice were to be placed there, it would become ineffective as soon as the throttle was opened, so the engine would hesitate, or even stall, before the depression over the main jet had risen sufficiently for it to take over.

This problem is usually overcome by having two discharge orifices, one adjacent to the edge of the throttle valve when it is closed and the other a short distance downstream. Such an arrangement, for example that of the Zenith VE updraught carburettor shown in Fig. 8.20, and in the Stromberg DBV carburettor, Fig. 8.26, is termed the *progression system*. In Fig. 8.26, A is the adjustment screw for regulating the flow of air from above the venturi to emulsify the fuel entering below, D is the delivery duct for fuel passing from the idling jet into the passage that takes the emulsified mixture down to the progression holes H.

When the engine is idling, fuel from the float chamber flows through the main and compensating jets into an idling well and on into the emulsion block E in Fig. 8.20. Under the influence of a depression, which is determined by the size of the hole O, this fuel is sprayed through the idling jet J into the large diameter duct D, which serves as an intermediate chamber.

As the fuel issues from the jet, it is mixed with air bleeding through three separate orifices. One is P, and the second comprises a series of holes from the venturi where, because the throttle is closed, the pressure is atmospheric. These radial holes feed air into the fuel jet, to emulsify the mixture. Further emulsification is effected by air entering from the third bleed orifice, which is equipped with an adjustment screw A. The size of this orifice, relative to those of the others, is such as to enable the overall rate of bleed to be accurately adjusted.

As the throttle is opened, and the depression over the hole O reduced, the resultant shortfall in fuel supply is made up by an additional flow of fuel through what was previously the air bleed orifice P. With further opening of the throttle, and a consequent significant increase in the air flow, the depression in D is reduced, and with it the quantity of fuel supplied through the idling system. This, together with the progressive draining of the well, which ultimately starves the idling system of fuel, provides effective compensation right up to the point at which the main jet takes over. From this point on, extra air continues to bleed through the idling system into the emulsion block through the main jet, to contribute to compensation.

There are various other ways of progressively increasing the supply of mixture and providing compensation during idling and warm-up. One is illustrated in Fig. 9.8 and another in Fig. 9.12.

8.25   Requirements for acceleration

If, after a period of operation at low speed and light throttle, the accelerator pedal is suddenly depressed, the mixture suddenly becomes very weak. This is partly because, although the depression that previously existed in the induction pipe is momentarily applied to the venturi, the sudden rush of air that this induces is too short lived to overcome the inertia and drag of the fuel in the jets. In any case, the inertia of the fuel in the delivery system from the jets will cause delivery to lag behind the increase in depression. Furthermore, the

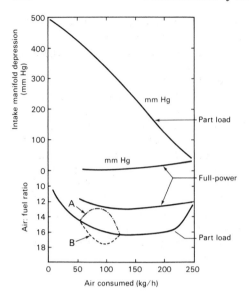

Fig. 8.21 Air:fuel ratio requirements and typical levels of manifold depression as the throttle is opened incrementally

opening of the throttle may have cut the idling system out of operation.

Since the pistons will have had neither the combustion pressures nor the time required for them to accelerate, the rate of flow of air through the venturi will rise relatively slowly so, temporarily, the depression over the main jet will not be high enough to atomise the fuel adequately. Moreover, the sudden collapse of the depression in the manifold will reduce the rate of vaporisation and, if the engine is cold, some that has already evaporated may condense out on the manifold walls.

In Fig. 8.21, the air:fuel ratio requirements and levels of manifold depression experienced as the throttle is opened progressively are plotted against air consumption. Also, the plot at A shows the air:fuel ratio required for producing the acceleration, and that at B shows what the air:fuel ratio is if the mixture is not enriched.

8.26 Provision for acceleration

The simplest method of enrichment is to insert a well between the discharge end of the spray tube and the main jet, so that the fuel in it is instantly available for acceleration. However, this measure is rarely, if ever, adequate to provide the enrichment needed during the initial snap acceleration period in automotive applications. On the other hand, as already explained, it is used almost universally in association with an emulsion tube, for general mixture compensation.

If, however, the fuel supply for an *acceleration pump* is taken from a point between the main jet and the well of the compensating system, the contents of that well are available for complementing the flow from that pump. This flow is dependent on the acceleration pump spring rate, as explained in the next paragraph but one, whereas that from the compensating well is at least partly dependent on the value of the depression over the main jet.

Most carburettors have an acceleration pump. This is a simple plunger- or diaphragm-type pump the control linkage of which is interconnected with that of the throttle. As the throttle is opened the pump plunger or diaphragm is depressed, spraying a small dose of fuel directly into the induction system, usually just above the venturi, the low pressure in which assists evaporation.

To prolong the spraying process during the acceleration, the piston rod generally incorporates a lost motion device, so that the control does not instantly move it but first compresses a spring around the rod. This spring pushes the piston down its cylinder to discharge the fuel progressively through the *acceleration jet*. To avoid over-enrichment and waste of fuel during slow movements of the accelerator pedal, there may be a controlled leak-back, usually through a small clearance between the piston and its cylinder walls, though sometimes through a restricted orifice or a by-pass duct. This leak-back may be adequate to avoid supplying fuel in excess of the requirement when the throttle is opened only very slowly.

8.27 Mechanically actuated acceleration pumps

Two examples of acceleration pump mechanisms are that in the Zenith IV, Fig. 8.22, and the Stromberg DBV carburettor, Fig. 8.23. In the Zenith unit, there are two concentric springs over the pump. Both are compressed by the lever connected to the throttle control. However, the inner one, by pushing the piston down after the piston rod has been slid through the hole in its crown by the actuation lever, performs the delaying function; the other returns the piston after the throttle has been closed again.

The throttle control is linked to the acceleration pump actuation lever, so when it is closed the piston B is lifted, drawing fuel up through the inlet valve in the base of its cylinder. As the throttle pedal is depressed, for acceleration, so also is the pump piston rod A which, while compressing the delay action spring, slides down through the hole in the piston. This pressurises the fuel below, closes the inlet valve, and opens the non-return valve D, through which

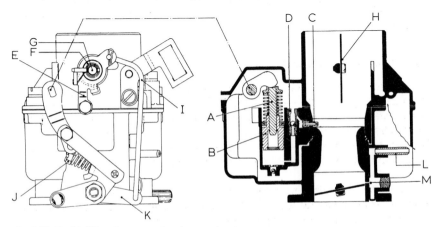

Fig. 8.22 Zenith IV carburettor, showing accelerator pump

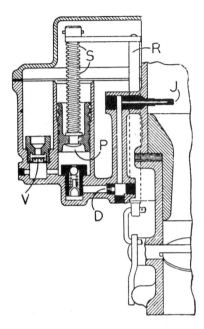

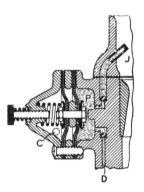

Fig. 8.23 Stromberg mechanical pump   Fig. 8.24 Solex membrane type acceleration pump

it delivers the initial charge of fuel for acceleration through the spray jet C. During subsequent closure of the throttle, valve D closes to prevent reverse flow as the piston rises. The vent above this valve prevents fuel in the pump from being siphoned out through the spray jet.

The Stromberg accelerator pump in Fig. 8.23 functions in a similar manner but, because there is a direct link connection R between the pump P and throttle control, there is only a single spring S: the return spring having been omitted. A plate valve V is used instead of a ball-type inlet valve and the delivery valve is in the base of the cylinder. Another difference is the interposition of a discharge reducer D between the delivery valve and the spray jet J. Again, a clearance between the piston and cylinder obviates wastage of fuel when the throttle is opened only slowly.

8.28 Depression actuated acceleration pumps

Acceleration pumps can also be actuated by manifold depression. The device illustrated in Fig. 8.24 was fitted to the Solex AIP carburettor. When the engine starts, the high depression in the manifold is communicated through the hole C, to pull the double diaphragm to the left and compress its return spring. This draws fuel through the inlet valve D into the pump chamber P. Subsequently, when the throttle is opened suddenly, causing the depression in the manifold to collapse, the diaphragm is pushed to the right by its return spring. This forces the fuel past the delivery valve and through the acceleration jet J, which sprays it into the air flow upstream of the venturi. If one of the two diaphragms leaks, there is still no possibility of fuel being drawn continuously

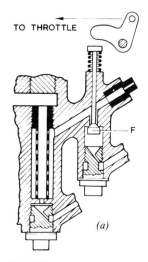

TO THROTTLE

F

(a)

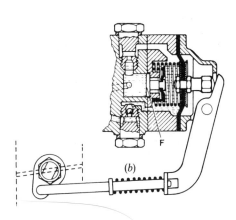

F

(b)

Fig. 8.25

through the device into the induction manifold. Hopefully, the consequent deterioration of the functioning of the pump would be noticed before the second diaphragm leaked.

For adjusting the stroke of the pump there is a screw on the left. The level of manifold depression at which the pump will begin to draw fuel into the chamber P is determined by the pre-load of the return spring which, in turn, sets the degree of throttle opening beyond which the pump ceases to become effective. At large throttle openings, the depression over the jet J is high enough to draw fuel continuously through the device and thus enrich the mixture for maximum power.

8.29 Enrichment for maximum power

In some instances, the air flow over the discharge orifice when the throttle is wide open generates a depression sufficient to draw fuel through it for producing at least some of the enrichment needed for developing maximum power. However, more is needed. The earliest devices for automatic power enrichment were mechanically actuated, by means of linkages connected to the throttle control. Two such mechanisms are illustrated in Fig. 8.25. That at (a) is from the early Claudel-Hobson carburettor of Fig. 8.8, in which a lever connected to the throttle valve mechanism opens the power enrichment valve F over the last few degrees of throttle opening. This allows fuel to pass from the float chamber, through the power jet into the emulsion system. An air bleed hole in the plug in the end of the passage delivering the fuel to the emulsion tube not only helps to emulsify the extra fuel supplied and to regulate the flow according to the degree of depression in the venturi, but also serves as the air bleed for economy when the power jet is not in operation. The section at (b) is a Solex device, in which F is again the power enrichment valve and is actuated through a lost motion device by the throttle control.

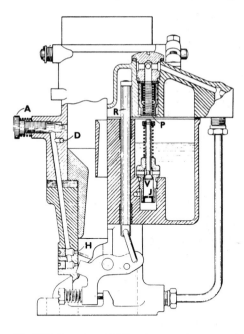

Fig. 8.26 Stromberg DBV carburettor by-pass valve and jet

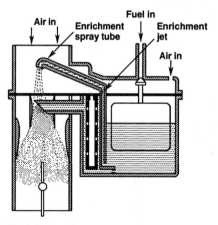

Fig. 8.27 In this static power enrichment system, the extra fuel is drawn from the float chamber and discharged at a point well above the twin venturi

The next development was of manifold depression actuated devices for bringing the power jet into operation. In Fig. 8.26, R is the rod that actuates the acceleration pump, which is not shown in this illustration. To the left of it is the idling system previously described, while to the right is the power-enrichment device. Manifold depression is taken through the external pipe to a connection above the piston P, within which is its return spring. When the depression is high, it lifts the piston against its return spring and the conical valve V closes.

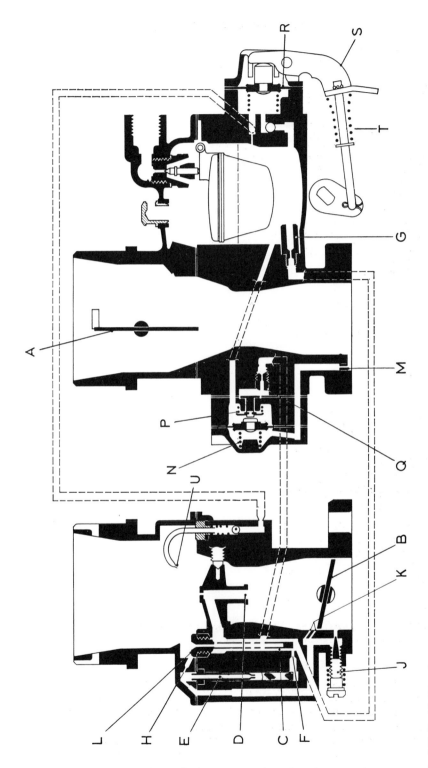

Fig. 8.28 Zenith IZ carburettor

288

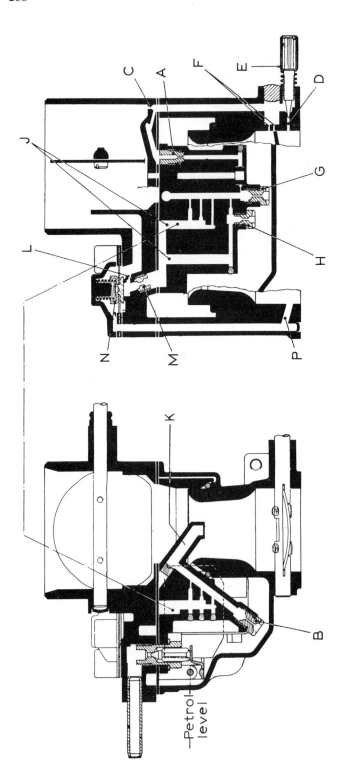

Fig. 8.29 Zenith IV carburettor

As the throttle approaches the fully-open position, the depression largely disappears, so the spring pushes the piston down. This opens the valve and thus allows fuel from the float chamber to pass through it and the power jet J, whence it flows up again to pass through a duct to the left of J, ultimately supplementing the flow through the spray tube into the venturi.

Other systems such as a mechanical connection with the throttle control, incorporating either a lost-motion device or a cam to actuate an enrichment valve have been used. Also, needle valves, tapered to provide the required fuel-flow characteristics, have been linked to the throttle control. Another method is simply to place the enrichment discharge orifice upstream of the venturi, where the depression to which it is subjected is calculated to be sufficient for drawing fuel from it only when the throttle is wide open.

8.30 Static power enrichment

Mechanisms are potential sources of unreliability and wear and therefore are undesirable. Static devices, such as that in Fig. 8.27, tend to be more attractive. Extra fuel is delivered through a duct in the top of the spray tube into the venturi. This fuel is drawn directly from the float chamber and discharged well upstream of the twin venturis so that it is evaporating in the air stream for as long as is practicable. Evaporation is further increased by both its passage through the low-pressure regions in the venturis and the turbulence generated around the end of the spray tube. To ensure that this device comes into operation only when the throttle is wide open and the depression upstream of the venturi therefore large, the position of the discharge orifice is such that the head of fuel against which the depression must lift the fuel is fairly large.

8.31 Economiser devices

Actually, there are two different approaches to providing for maximum power operation: one is the provision of extra fuel, as just described, over the last few degrees of throttle opening, while the other is to reduce the strength of the mixture throughout most of the range, leaving the fuel to flow more freely over the last few degrees of throttle opening. Economy devices that have been used in Zenith carburettors are illustrated in Figs 8.28 and 8.29. In Fig. 8.28, a depression-actuated diaphragm valve closes to cut off the fuel supply to what is termed an 'economiser jet' for part throttle cruising. On the other hand, in Fig. 8.29, a similarly actuated diaphragm opens a valve to supply extra air to weaken the mixture under these conditions. Yet another system is used in the Zenith IVEP carburettor: an economy valve similar to that of the IZ, Fig. 8.28, is used, but regulating the fuel supply to the power jet.

Chapter 9

Some representative carburettors

To demonstrate how the principles outlined in Chapter 8 are put into practice, and all the different devices previously described are brought together to produce a comprehensive control and metering system, a representative selection of actual carburettors will now be described in detail. From these, it will become clear that the majority of carburettors comprise six operating systems. They include the float chamber and one for each of the five functions listed at the beginning of Section 8.3, namely: starting, idling, part throttle, power and acceleration.

Explanations will be given, too, of some the measures that were taken to tighten the tolerances on metering to meet the requirements for exhaust emission control. In this context the constant depression carburettor fell from favour because, with only one jet and dependence on moving parts for accuracy of metering, it was virtually impossible to meet emission control requirements over long periods in service. However, it is described because of the inherent interest of the principle and because many cars equipped with this type remain.

9.1 Venturi diameter

A problem that engine designers face is selection of the most appropriate diameter venturi, or choke. Two charts issued by Weber for the selection of a venturi are illustrated in Fig. 9.1. That at (a) is for in-line engines having between one and six cylinders, while that at (b) is for sports car engines with one carburettor per cylinder.

It can be seen that, for a 1-litre, four- or six-cylinder engine for a saloon car, the diameter would have to be between 19 and 22 mm. However, both single- and twin-cylinder engines inhale more mixture per cylinder because the manifolding is shorter, less complex, and its walls smaller in area so, for either, we have to use a choke appropriate for a multi-cylinder engine of double the swept volume. For a 1-litre twin-cylinder engine, therefore, the venturi diameter would be between 27 and 32 mm. Where the precise choice will fall, between the upper and lower limits, is dictated by whether the designer wants to place most emphasis on torque at high or low speed, in other words on

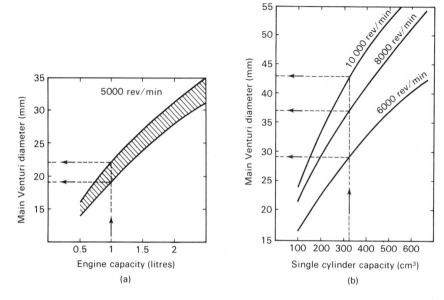

Fig. 9.1 Weber charts for the selection of venturi diameter (*a*) for engines having between one and six cylinders and (*b*) for sports car engines having one carburettor per cylinder

either a sporty performance or good flexibility for ease of driving at lower speeds in traffic in urban conditions.

In Fig. 9.1(*b*) we see that only one size of choke is indicated for any given size of cylinder. The reason, of course, is that for engines having such wide speed ranges it would not be possible to select any one size of choke that would be satisfactory for operation at both maximum power at high speeds and light load at low speeds. In a sports or racing car engine, however, the performance in the upper portion of the speed range is all that matters because its driver, almost invariably highly skilled, will maintain high rev/min by using his gears. Cold starting problems do not arise, because the engine is normally fully warmed up before a race begins.

9.2 Zenith W type carburettors

Many of the basic features of the Zenith WIA carburettor, Fig. 9.2, have been developed from Stromberg designs, including bottom feed to the float chamber, double venturi and air bleed emulsion tube as illustrated in Fig. 8.10, mechanically-operated accelerating pump and economiser or power enrichment valve, which is opened for full load conditions.

At (*a*) is a section through the float chamber, twin venturis, main jet M and emulsion tube T, with air bleed holes on its underside. Section (*b*) illustrates the idling system with, bottom right, progression holes adjacent to the edge of the throttle, the screw adjustment for quantity of mixture and, shown dotted, the duct taking the manifold depression up to the spring-loaded diaphragm that actuates the economiser valve, which can be seen in (*c*). This valve differs from that in Fig. 8.26 only in that it is diaphragm instead of piston actuated. The

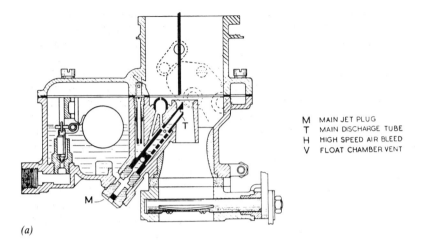

(a)

M   MAIN JET PLUG
T   MAIN DISCHARGE TUBE
H   HIGH SPEED AIR BLEED
V   FLOAT CHAMBER VENT

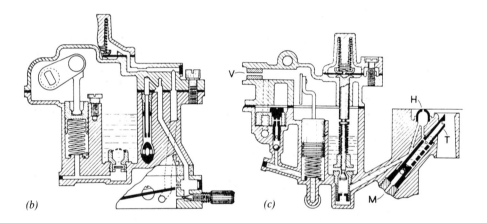

(b)

(c)

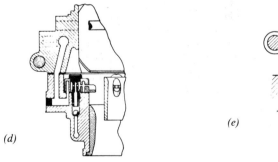

(d)

(e)

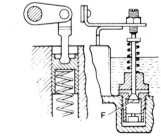

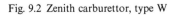

Fig. 9.2 Zenith carburettor, type W

section at (*b*) also shows the mechanism actuating the acceleration pump, and the pump inlet valve: the delivery valve can be seen in both (*c*) and (*d*).

For the WI carburettor, a mechanical control, shown at (*e*) was employed for enrichment, a lost-motion device in the linkage between it and the throttle control bringing it into operation over the last few degrees of throttle opening. In both the WI and the WIA, the acceleration pump is mechanically actuated but there is no progressive delivery spring, the piston having only a return spring. Seasonal adjustment of the pump stroke can be made by transferring the pin in the end of the interconnecting link into the appropriate hole of the three in the end of the pump actuation lever. Two of these holes can just be seen in Fig. 9.3. In a larger version, the 42W, an acceleration pump like that in Fig. 8.22 is installed, but its valve arrangement is different.

The strangler flap is closed by a torsion spring around the spindle, and opened by a cam rotated by a pull-out control on the dash. This cam and its follower pin on the strangler actuation lever can be seen clearly in the middle of Fig. 9.3. A lug on the strangler control lever bears down on the throttle lever to open it slightly, to its position for cold starting.

9.3 Zenith IZ carburettors

Each carburettor in the IZ series, Fig. 8.28, has an offset manual strangler, for cold starting, a prolonged-action-type accelerator pump, a depression-actuated economy device and volume control of idle mixture. Other refinements include a filter for the slow running tube, and jets and passages designed for the avoidance of fouling by foreign matter in the fuel.

For starting from cold, operation of the choke control closes the strangler

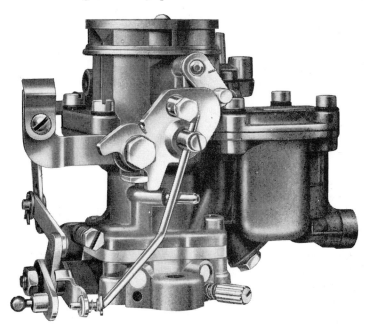

Fig. 9.3 Zenith carburettor, type W

flap A. Simultaneously, a cam interconnection opens the throttle a predetermined amount to allow the manifold depression to reach the choke tube and mixing chambers for drawing off the fast idling mixture from the main well C. This mixture is discharged through the orifice D. As soon as the engine fires, the increased depression partially opens the strangler against the closing force applied to it by the torsion spring connecting it to the choke control. The degree of opening of course depends upon the position of the throttle. However, the choke control must be fully released as soon as the engine temperature has risen sufficiently.

In normal idling conditions, without the strangler, the mixture is supplied by the slow running tube E, which is enclosed in a gauze filter. The fuel is drawn initially through the restriction F, from the metered side of the main jet G, and the air enters, to emulsify it, through the calibrated bleed orifice H, from the air intake. Ultimately, the emulsified mixture is drawn down a vertical channel to the idle discharge hole, into which projects the tapered end of the volume adjustment screw J. While the throttle stop screw is used to set the idling speed, the volume adjustment screw regulates the quantity of emulsified idling mixture supplied for mixing with the air passing the throttle. Smooth transfer from the idle to main circuits is obtained by the two progression holes K, which come in turn under the influence of the local venturi effect caused by the proximity of the edge of the throttle to them.

As the throttle is opened further, the increasing depression in the venturi brings the main system into operation. From the main jet G, the fuel passes into the well C. Air, metered through the orifice L, passes down the emulsion tube and, passing through radial holes in it, mixes with the fuel before it enters the main discharge orifice D in the narrowest part of the venturi. As the engine speed increases, the fuel level in the main well falls, uncovering more radial holes in the emulsion tube so that an increasing quantity of air can mix with the fuel to correct the mixture strength.

A depression-actuated economy device is attached by three brass screws to one side of the float chamber. At cruising speeds, the relatively high induction manifold depression is transmitted, through a calibrated restriction M, to the chamber between the diaphragm and its outer cover. This overcomes the load in the return spring N and moves the diaphragm to the left, as viewed in Fig. 8.28, allowing the chamber between the diaphragm and the main body of the device to fill with fuel, and the spring-loaded valve P to close. Since closure of this valve puts the jet Q out of action, fuel can now be drawn only from the main jet.

As the throttle is opened further, and the depression in the manifold becomes less intense, the diaphragm return-spring extends, moving the diaphragm to the right and opening the spring-loaded valve P. Fuel, metered by jet Q, then passes into the main well to enrich the mixture for increasing the power output.

For acceleration, especially from cruising speed at the weak mixture setting, a prolonged action diaphragm pump R is incorporated. This pump functions on principles similar to those described in Sections 8.26 and 8.28. In detail, however, it differs in several respects. The prolonged action is obtained by arranging for the link with the throttle control to slide in a hole in the pump actuation lever S, while transmitting the motion through a compression spring T interposed between them. From Fig. 8.28 it can be seen that there is a small back-bleed hole interconnecting the pump delivery chamber and the float chamber. This is to prevent discharge of fuel through the pump jet U, owing to thermal expansion of the fuel if the carburettor castings become very hot – for example, if the engine is stopped immediately after a period of operation under high load.

9.4   Zenith IV carburettors

The IV series of carburettors is a development of the V type. Among the improvements is the incorporation of twin floats, set one each side of the choke and with their centroids and that of the float chamber itself as close as practicable to the jets, Fig. 8.29, so that the fuel level above the jets is virtually unaffected by changes in inclination of the vehicle, or by acceleration, braking and cornering.

All the jets, and the accelerator pump are carried in an emulsion block, which can be readily removed with a screwdriver and a $\frac{7}{16}$ in spanner. The outlet from this emulsion block, or jet carrier, passes into a spray tube K, which is cast integrally with the block and which takes the mixture to the venturi. Because the venturi and float chamber are cored integrally in a single casting, there are below the fuel level neither screws nor plugs past which leakage could occur.

The principle of the accelerator pump has been described in Section 8.26. For other conditions of operation, the jets and systems that come into operation are as follows: on starting from cold, operation of the choke control pulls lever E, Fig. 8.22. This lever, through the medium of a torsion spring F, rotates the strangler G and closes the strangler H. Simultaneously, the rod I, interconnecting the strangler and throttle, opens the latter to set it for fast idle. After the engine has fired and is running, the increased depression opens the strangler against the torsion applied by spring F, to prevent over-choking. As the engine warms up, the choke control must of course be released to reduce the idling speed to normal.

For idling, the mixture is supplied through the slow running jet A, Fig. 8.29. Fuel reaches it from the main jet B – that is, from the base of the emulsion block – through a calibrated restriction. After discharging from the slow-running jet, the fuel is emulsified by air bleeding from orifice C into the vertical channel which takes it down to the idle hole D, through which it is discharged downstream of the throttle.

The tapered end of the volume control screw E projects into the idle hole D. Adjustments to idling speed are made as follows: turn the throttle stop screw, J in Fig. 8.22, until the required speed is obtained – clockwise to increase, anti-clockwise to decrease. Then turn the volume control screw, E in Fig. 8.29, to obtain the fastest possible idling speed at that setting of the throttle stop. Repeat both operations as necessary. Where stringent emission controls are in force, it may be necessary for the volume control to be set – clockwise rotation weakens the mixture – by the vehicle manufacturer, by reference to an exhaust gas analysis, and then sealed.

As the throttle is opened, the local venturi effect between its edge and each of the progression holes F, in turn, draws additional fuel through them until the main jet system can take over. The size and positioning of these two holes is of course critical and no adjustment is allowed. Incidentally, connection L in Fig.

8.22 is for the automatic ignition advance, and the small hole M, through which it communicates with the throttle bore, is carefully calibrated.

With further opening of the throttle, and the consequent increase in the depression in the waist of the choke tube, fuel is drawn from the outlet from the emulsion block. This fuel comes from the main jet G and compensating jet H. As the level of the fuel in the channels above these jets falls, air takes its place in the capacity wells J, above the main and compensating jets, and then bleeds through the emulsion holes into the outlet K. The rate of flow of this air is controlled by the full throttle air bleed hole L and, at times, by the larger orifice in the ventilation screw M, which depends for its extra air supply on operation of the economy diaphragm valve N. Fuel, already emulsified by the time it leaves the outlet, is atomised as it is swept away by the air flowing through the choke tube.

The arrangement of the economy device is as follows: it is housed in a small casting secured by three screws on top of the float chamber, adjacent to the air intake, and the diaphragm valve N is held on its seat by a spring. The chamber above the diaphragm is connected to an outlet P downstream of the throttle butterfly valve.

At part throttle, when the depression downstream of the butterfly valve is high, the diaphram valve is lifted off its seat, allowing extra air to flow from the air intake through the ventilation screw M, to increase the emulsification of the fuel and thus to weaken the mixture, for economical cruising. When the throttle is opened further, calling for high power output, the manifold depression falls, allowing the spring to return the diaphragm valve to its seat, and the mixture is therefore enriched.

9.5   Adaptation for emission control

Zenith fixed-choke carburettors adapted for emission control regulations, mostly up to the end of 1992, carry the suffix E on their designations. These include the IZE, IVE and WIAET. The letter T, incidentally, is used to indicate that an automatic strangler is incorporated.

Among the features incorporated is a solenoid-actuated slow-running cut-off – used on some of the IVE carburettors. Because of the weak setting of the idling and slow-running mixtures on emission-controlled engines, the resultant abnormally high temperatures of combustion can cause auto-ignition when the engine is switched off. To avoid this, the slow-running supply is automatically cut off with the ignition. The device used is simply a conical ended plunger, which is forced on to a seating by a spring. When the engine is switched on, the solenoid is energised, to lift the plunger off the seating, thus opening the slow running supply system.

On the IZE, instead of the drilled hole and dust cap on the float chamber, there is either a two-way venting system or a simpler internal vent. The simpler system is a vent channel running within the float chamber cover casting and breaking out into the upper part of the carburettor air intake. This satisfies requirements in respect of evaporative emission control and, by subjecting the float chamber to air intake pressure, obviates all possibility of enrichment of the mixture as a result of abnormally high depression over the jets due to a clogged air intake filter element.

A disadvantage, however, is that when the engine is idling, fumes from the float chamber vent can enrich the mixture and adversely affect emissions. Additionally, because of the accumulation of fumes in the air intake after a hot engine is switched off, restarting may be extremely difficult.

For these reasons, in some applications, the dual venting arrangement may be necessary. With this arrangement, the internal vent A, Fig. 9.4, is permanently open, but an external vent is brought into and out of operation by a plunger type valve actuated by the accelerating pump lever.

The valve assembly is a press fit in a boss on one side of the float chamber cover. Its plunger is spring-loaded towards its outer position, in which the float chamber is freely vented to atmosphere through hole B. A spring steel blade C, attached to the accelerator pump control, is used to close the valve when the the throttle is opened. This leaves the float chamber under the influence of air intake depression, through the internal vent. The proximity of the steel blade to the plunger is set, using screw D, during manufacture and should not be altered subsequently.

Another device used, in one form or another, in several carburettors including the Zenith IVE, is an over-run control valve. This is necessary because, on sudden closure of the throttle, the intense depression draws into the engine all the condensed fuel clinging to the walls of the manifold. This initially enriches the mixture and, subsequently, leaves it over-weak. In each condition, the hydrocarbon emissions in the exhaust become unsatisfactory.

The over-run-control device is a spring-loaded, poppet-type, non-return valve in the throttle butterfly, as shown in Fig. 9.5. For normal operation, including tick-over, this valve is kept closed by its spring but, if the throttle is closed for deceleration and the manifold depression is therefore high enough to suck it off its seat, two things happen. First, the depression is relieved sufficiently to avoid the over-enrichment phase and, secondly, extra mixture bleeds through holes beneath the head of the poppet valve, to maintain proper combustion in the cylinders and to relieve the depression slightly.

Because the idle and light throttle opening positions are critical as regards emission control, it is sometimes desirable to have a pre-set relationship between the edge of the throttle valve and the idling progression holes and, on some carburettors required for meeting the US emissions regulations, a suction retard port for the ignition. Consequently, on some IZE carburettors,

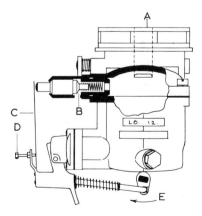

Fig. 9.4 Zenith dual venting system

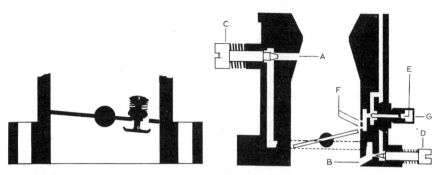

Fig. 9.5 Overrun air valve   Fig. 9.6 Throttle by-pass and air control

the throttle stop is adjusted during manufacture and thereafter sealed, so another method has had to be introduced for adjusting idling speed in service.

For this purpose a throttle by-pass system has been introduced. As can be seen from Fig. 9.6, a channel runs from A below the choke to an outlet B downstream from the edge of the throttle. Air flow through this channel is adjusted by means of a taper-ended screw, C, near the inlet – if turned clockwise, it reduces the idling speed, and *vice versa*. The idling mixture is controlled by the volume control screw D, in the outlet. When the throttle is opened, the depressions at the inlet and outlet of the by-pass channel become much the same, so it ceases to function, the progression holes taking over the function of supplying a suitable mixture.

It is of interest that the emission-controlled versions of the W type carburettors are adequate without any of the devices described in this section. Their emission control is effected by close tolerances in production, and subsequent testing.

9.6   Multi-barrel carburettors

The performance of an engine designed to run over a wide speed range with only one carburettor is considerably improved by installing two- or four-barrel carburettors. *Twin barrel* should not be confused with *twin carburettors*, the latter of course referring to an installation comprising two separate carburettors. The latter have been used on four-cylinder engines but, because they can obviate inter-cylinder robbery of charge, as explained in Sections 11.6 and 11.7, they are of greater benefit when supplying groups of three cylinders. A disadvantage of a multi-carburettor installation is that the throttle controls may fall out of synchronisation in service, leading to uneven running and loss of power and efficiency. Starting and idling may be adversely affected too. Furthermore, provided an appropriate twin- or four-barrel carburettor is produced in quantities large enough, it can be less costly than separate carburettors because single components such as housings and float mechanisms can serve all barrels.

As will be seen in Sections 9.7 and 9.8, the *primary* throttle or, throttles, in a four-barrel carburettor is kept closed for starting and operation at light load until the velocity of the air flow through the venturi is high enoug h for the *secondary* throttle to begin to open. An alternative arrangement is to synchronise the throttles and arrange for them to deliver separately into two or more channels in the manifold, each serving a different group of cylinders.

The throttles may be either interlinked mechanically, Fig. 9.7(a), or the secondary ones actuated by manifold depression, as in Fig. 9.7(b). With mechanical actuation, the simplest course is to link the primary throttle, or pair of throttles, directly to the throttle control, and to open the second indirectly by linking it to the former. For sequential operation a lost-motion slot is sometimes machined in the interconnection, so that the secondary throttle can be held closed by a spring until the primary throttle has opened far enough to take up all the lost motion, at which point the secondary throttle begins to open. Alternatively, the secondary lever can be held closed by a spring and opened by a lug on the primary throttle lever, as in Fig. 9.7(b).

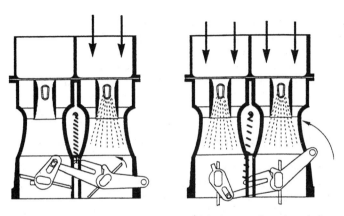

Fig. 9.7(a) A twin-barrel carburettor in which the secondary throttle is opened by a lost motion mechanisms interconnecting it with the primary throttle

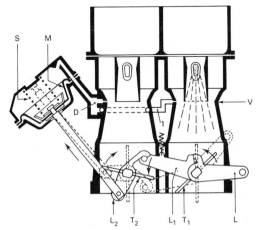

L	Intermediate lever	S	Diaphragm return spring
L₁	Primary throttle lever	T₁	Primary throttle
L₂	Secondary throttle lever	T₂	Secondary throttle
M	Diaphragm		

Fig. 9.7(b) Here, the secondary throttle is pneumatically actuated

In this illustration, both venturis are of equal size, an arrangement that is suitable for engines designed for high efficiency and power output over the middle to upper range of throttle opening. The depression in the primary venturi V is communicated through duct D to the upper face of the diaphragm M, which is held down by spring S. Opening the primary throttle lowers lever L_1, and thus frees lever L_2. Then, when the velocity of the flow through the primary venturi V creates a depression high enough to overcome the force exerted by spring S, atmospheric pressure acting on the lower face of diaphragm M will lift it, opening the secondary throttle T_2 by an amount dependent on the rate of flow through the primary venturi. When the primary throttle is closed, the stop on the left-hand end of intermediate lever L ensures that the secondary throttle is returned without delay to its closed position.

9.7   A three-stage throttle mechanism

On some of the 1990 GM Vauxhall/Opel *Senator* and *Carlton* models, an ingenious three-stage throttle mechanism, Fig. 9.8, was introduced. The benefit obtained with this arrangement is that the depression under all conditions in the manifold is high enough to assist evaporation of the fuel and thus to optimise the torque characteristics.

The carburettor has twin barrels, the primary barrel having a bore of 25 mm and the other 64 mm. Over the first 20 mm of pedal travel, the primary throttle valve opens 27°. Up to this point, because there is a lost-motion mechanism in the linkage interconnecting the two throttles, the secondary valve remains closed. From 20 to 30 mm pedal travel, the primary throttle valve opens up to 46°. Then, as the pedal is deflected further, to 55 mm, it opens up to 90°, as also does the secondary throttle.

Each butterfly valve has its own spiral torsion spring so, as the transition

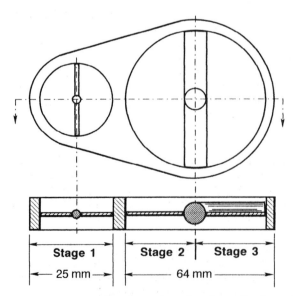

Fig. 9.8 Three-stage throttle opening: when the small primary throttle is almost fully open, the left-hand edge of the larger, secondary throttle cracks open and then, 9° later, because it is so thick, its right-hand edge cracks open

from one to two barrels occurs, the driver feels a slight increase in resistance to the movement of the pedal. This is to indicate to him that his throttle is opening beyond the maximum economy into the high performance range of operation.

In effect, the secondary throttle opens in two stages. This is because one half of the butterfly valve is of wedge section, the thinnest end of the wedge being along the line at which it joins the cylindrical section housing the pivot pin that carries the valve. As the butterfly valve rotates about its pivot, only one edge opens, because the thick end of the wedge section has to rotate 9° further until it begins to crack open. From that point on, the valve opens progressively until, at 90°, both halves are wide open. The peripheral section of the wedge is not thicker than the diameter of the pivot, so it does not significantly impede the flow through the valve.

9.8 Solex MIMAT carburettor

The MIMAT is a downdraught, twin-choke carburettor, with throttles compounded to open one after the other – the throttle in the secondary choke does not start to open until that in the primary is about two-thirds open. Both chokes of course are the same diameter. This twin-choke arrangement mitigates the disadvantages of a fixed-choke carburettor, which are a tendency towards poor atmomisation at low air speeds and strangulation at high speeds. Although careful design and setting is necessary to obtain smooth change-over from single- to twin-choke operation, the difficulties associated with the maintenance of synchronisation of two carburettors throughout the range in service are avoided.

From Fig. 9.9, the arrangement of the main jets can be seen, while the slow-running system is shown in Fig. 9.10. The method of operation of the main jets is obvious from the illustration, but one or two details need clarification.

In Fig. 9.9, the nozzles A through which the mixture is delivered to the choke are held in position by spring retaining devices, which can be seen one each side of the siamesed central portion of the choke tube. The main jets are at B, and the illustration shows only the primary choke in operation. There are two air-bleed passages to each diffuser tube assembly, which comprises a central tube drawing its air supply from C, and an outer tube the air supply to which comes from D. These air supplies pass respectively through calibrated restrictors E and F. Fuel enters the outer tube through its open lower end, into which air bleeds through radial holes in the inner tube to the annular space between the two.

The idling and slow-running progression system is more complex, because it comprises three different circuits: two are identical, one serving the primary and the other the secondary choke tube, while the third supplies only the primary system. The first two draw their fuel from the metered supply from the main jets B, in Fig. 9.10, while the third takes it also from a jet B but only the one serving the primary choke.

In Fig. 9.10, both throttles are shown closed. Consequently, the edge of that in the secondary choke tube is downstream of the idling mixture outlet A_2, which is circular, and therefore renders the idling system for that choke inoperative. Although the outlet A_1 in the primary choke tube is a slot, for obtaining the required progression as the throttle is opened, its lower end is closed by the edge of the throttle, so this outlet too is unaffected by any local depression.

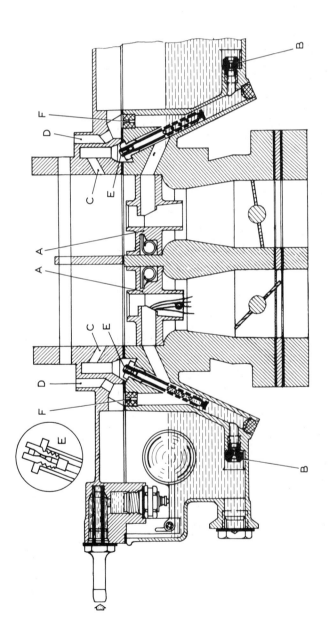

Fig. 9.9 Solex MIMAT carburettor, main jet system

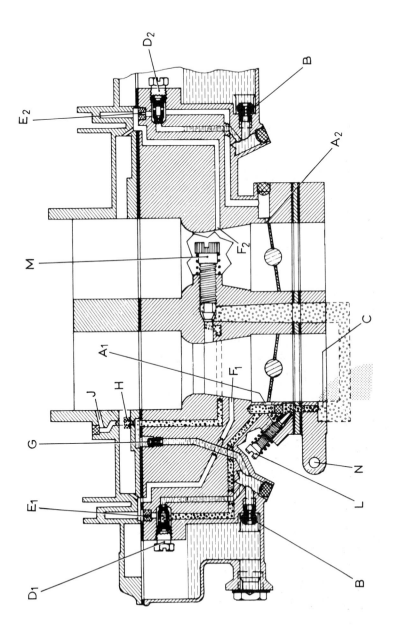

Fig. 9.10 Solex MIMAT, slow-running system

In these circumstances – hot idling – the mixture is supplied through a throttle by-pass system serving only the primary choke tube. Manifold depression is transmitted through the mixture outlet C, in the bottom flange of the carburettor, and draws fuel from two sources: one is the slow-running jet D_1, into which air is bled through orifice E_1 from the inlet F_1 below the waist of the primary choke, while the other is the jet G, into which air is bled through orifice H from both J, in the primary air intake, and K, just below the waist of the primary choke. The relative positions of all these inlets and outlets are such that, when the throttles in their chokes are open, the pressures in the mixture passages are not low enough to draw any fuel from the jets.

Jets D_1 and D_2 are not for adjustment in service, nor is screw L, which is adjusted in the factory for setting the idling air:fuel ratio. On the other hand, screw M is a volume adjuster for regulating the quantity of mixture passing, and thus the idling speed in service.

As the primary throttle is opened, the progression slot A_1 comes into operation, under the influence of the local depression, until the edge of the throttle valve swings clear of it. At about two-thirds primary throttle opening, the secondary throttle begins to move and, therefore, the progression hole A_2 comes in to play, drawing fuel from jet D_2 and air from orifice E_2.

For cold starting, an automatic strangler and fast idle system, Fig. 9.11, operates. The strangler is opened by a spiral bimetal strip A and closed by a diaphragm B, actuated by manifold depression. For opening the throttle for fast idling, there is a stepped cam rotated by a second spiral bimetal strip C. The first mentioned bimetal strip is subjected to engine coolant temperature by a water jacket D, while the second is under the influence of ambient air temperature.

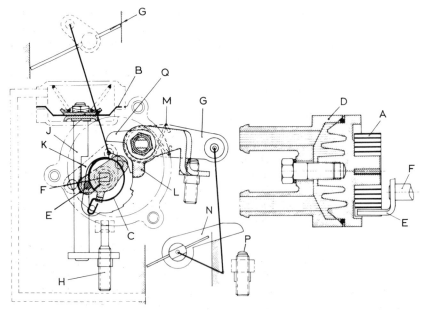

Fig. 9.11 Solex cold-starting device

One end of the first bimetal strip is anchored to the water-jacket casting, while the other is connected to a lever E on a spindle F, which is linked, by another lever K and a rod, to the strangler G. The setting of the bimetal strip A is such that it tends to open the strangler above a preset low temperature, generally about $-20°C$.

The bimetal strip C, subject to ambient air temperature, is anchored to the spindle D, its other end being connected to a stepped cam, which is free to rotate on that spindle except that it tends to be rotated anti-clockwise by the bimetal strip with decreasing temperature. There are two sets of steps on the cam: one is for limiting the closure of the primary throttle by lever G, and the other for limiting the upward motion of the stop-screw H at the bottom of tie rod J.

The tie-rod is actuated by the diaphragm B which, prior to starting, is pulled down by its return spring. This allows lever E, a pin which projects into the long notch in the tie-rod, to be pulled downwards by bimetal strip A. The strangler can be opened again by either manifold depression or increasing engine coolant temperature, or by override devices to be described later, but its upward motion is limited by the cam, the position of which is determined by ambient air temperature acting upon bimetal strip C.

This cam also limits the closure of the primary throttle in cold conditions, by acting as a stop for lever L, which is secured to lever G by a spring M. The positions of the levers G and L, relative to each other, are adjustable by means of the screw stop on the latter, and the closure of the throttle in warm conditions is limited by stop P acting on lever N.

Before starting from cold, the driver presses the throttle pedal down once and then releases it. This allows the cam, under the influence of bimetal strip C, to take up a position appropriate to the ambient air temperature. In Fig. 9.11, the ambient temperature is assumed to be $-20°C$, so the strangler is appropriately closed by bimetal strip A.

When the engine starts, the increasing manifold depression, acting on the diaphragm, opens the strangler an amount limited by the stop at the lower end of its tie-rod, which is pulled up against the cam. Should the ambient air temperature subsequently rise, the cam will be progressively rotated clockwise by the bimetal strip C, so that the strangler can be opened further step by step.

As the engine coolant temperature rises, rotation of the spindle F, by the bimetal strip A, of course rotates the strangler actuating lever – until it is fully open – its end moving upwards within the notch in the diaphram tie-rod.

If the engine is to be started from cold, but in a moderate ambient temperature, perhaps about $+10°C$, a larger throttle opening than that set by the position of the cam might be required. This opening is set by a peg projecting from the end of the lever K. As the strangler is closed, this peg comes into contact with the peg that connects the cam to its bimetal strip, pushing it round to set the throttle stop accordingly. When the engine starts, the first light pressure on the accelerator pedal releases the cam, which then takes up its normal position.

Should the engine fail to start, owing to an over-rich mixture, the accelerator pedal should be pushed right down. This causes the strangler to be opened by contact between projection Q, from the throttle actuation lever G, which rotates lever K against the loading applied to it by the bimetal strip A. The engine is then rotated to ventilate the cylinders, before attempting to start it again in the normal manner.

To avoid icing in cold and humid operating conditions, an engine coolant supply is taken straight through the bottom flange of the carburettor by connections to each end of the hole N in Fig. 9.10. This heats the region of the butterfly valves, the slow running outlets and the throttle by-pass circuit.

The accelerating pump, Fig. 9.12, is diaphragm actuated by a lever, the end A of which bears on a cam B mounted on the spindle carrying the secondary throttle. It therefore operates only when the secondary throttle is in use, which helps as regards fuel economy. Spring C is provided to prolong the action of the pump.

Another device that operates only in the secondary choke tube is the compensating jet A, Fig. 9.13, which is controlled by a diaphragm valve subject

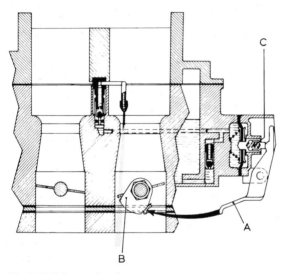

Fig. 9.12 Solex acceleration pump

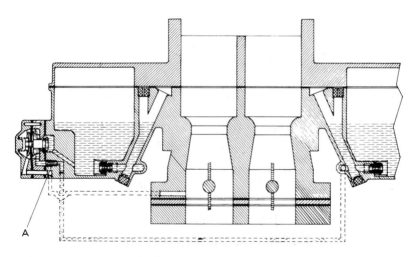

Fig. 9.13 Solex compensating jet system

to manifold depression. This valve is held open by a return spring but, when the manifold depression is high – light throttle operation – it is closed and sealed by an elastomeric ring on its seating face. As can be seen from the illustration, when the valve is open, the fuel flows through jet A into the diffuser well serving the secondary choke. Solex also can supply a similar valve, but operating in the opposite sense – that is, enriching the mixture at wider throttle openings, instead of weakening it at light openings.

These diaphragm-actuated devices should not be confused with the Econostat power-jet system, Fig. 9.14. In this, the fuel, coming directly from the float chamber, is metered through jets A and discharged through nozzles B into the air intakes upstream of the chokes. By virtue of their position and the fact that fuel can be drawn off only when the depression in the air intake drops to a certain value, they cannot discharge fuel except at high speed and load. The level of depression at which these jets come into operation is determined by the size of the air-bleed orifices C.

One more detail of interest on this carburettor is the two-way vent system for the float chamber, Fig. 9.15. Under slow-running conditions, the valve is spring-loaded on to its innermost seat so that the float chamber is vented externally at A. However, when the throttle is opened, an interconnecting linkage moves the valve to the right, as viewed in the illustration, on to its other seat. This opens the port to the internal ventilation system connected at B.

9.9   An electronically controlled four-barrel carburettor

The Rochester Quadrajet 4M carburettors are four-barrel, two-stage down-draught units. In the E4M series, the letter E indicates that the carburettor is electronically controlled. These carburettors have a choke (strangler) regu-lated by a bimetal coil which is exposed, in the E4MC model, to air ducted to it from over the exhaust manifold or, in the E4ME, to an electric element. The

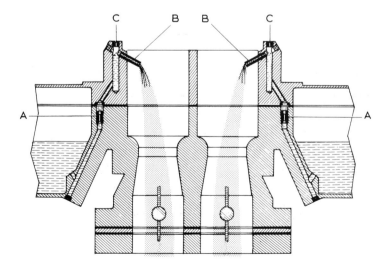

Fig. 9.14 Solex Econostat power-jet system

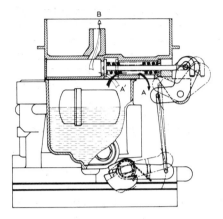

Fig. 9.15 Solex two-way vent

two primary barrels are situated each side of the float chamber, in front of the larger secondary barrels, Fig. 9.16.

A study of what follows should dispel any doubts readers might have as to the magnitude of the extra cost and complication inevitably incurred in adapting carburettors for strict control of exhaust emissions.

The electronic control system. Electronic control is effected by means of a small on-board computer termed the electronic control module (ECM). The main input to the ECM is from an oxygen sensor in the exhaust, in conjunction with which the carburettor can be operated on the closed-loop principle. However, it can also operate on the open-loop principle, for which purpose similar or alternative sensors, for instance a timer, are brought automatically into circuit to regulate it in relation also to signals from coolant temperature and throttle position sensors.

The output from the ECM is a stream of timed-pulse electric signals, issued at a rate of ten per second, which modulate the action of a single, mixture-

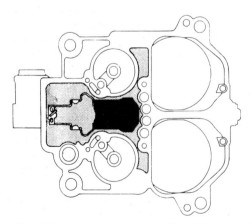

Fig. 9.16 Arrangement of the barrels and float chamber in the Quadrajet 4M carburettor. The smaller barrels are the primary ones

control solenoid. Movement of the core of this solenoid is transmitted by a yoke to the upper ends of the two metering needles, or rods, in the jets serving only the primary barrels. The lower ends of these rods are stepped so that, against the background of modification by the electronic control, the fuel supply is modulated from two distinct datum bases, the first of which serves throughout idle and off-idle, and the second the remainder of the movement up to full throttle. For the secondary barrels, on the other hand, the jet needles are cam actuated by interconnection with the primary throttles, and their lower ends are tapered to increase the fuel supply progressively as maximum power output is approached.

The yoke over the primary rods pushes them down, against the resistance from their coil return springs, into the main metering jets to weaken the mixture, while simultaneously withdrawing another rod with a tapered end from an orifice, above, increasing the idle air bleed to the two primary barrels. If the mixture becomes too lean, the excess of oxygen in the exhaust is immediately detected, and the computer signals to the solenoid to allow the yoke over the jet needles to lift again and this, by lifting also the overhead needle, simultaneously reduces the idle air bleed. Since the primary needles are stepped rather than tapered, the mixture strength at any specific throttle opening, especially in the idling condition, is modified solely by changes in the rate of air bleed.

Stored in the memory of the ECM is a pattern of operating conditions, with the ideal air : fuel ratio for each. This memory is continuously updated so that, if the ideal air : fuel ratio changes, for instance owing to temporary operation at high altitude or, over the longer term, due to wear of components, the most recently entered data is the basis on which the mixture strength is set. During wide open throttle operation the throttle position sensor signals the ECM to enrich the mixture for obtaining maximum power output. Although in this condition all four barrels are in operation, this component of the enrichment occurs only in the two primary barrels, since the tapers on the ends of the needles serving the secondary barrels are profiled for appropriate enrichment of their mixture supply.

Float chamber. Fuel entering the float chamber passes through a check valve, comprising a Viton plunger and steel coil spring, housed in the open end of a thimble-shaped, pleated paper filter. A second coil spring bears on the closed end of the filter to seat the complete assembly permanently against the end of the fuel inlet connection, Fig. 9.17. Neither of these springs performs a pressure-relief function. The check valve is for preventing leaks should the car roll over and the engine stop.

The closed-cell rigid foam plastics float is integral with the lever on which it pivots. Secured by means of a spring clip to the other end of this lever is the float valve with an elastomeric-faced conical lower end. It closes on to a brass seating the face of which is ground at two angles, the lower of which is the more obtuse, to enable fuel under pressure from the lift-pump to enter and facilitate opening, especially in the event of the valve's becoming fouled by sticky gum deposited out of the fuel.

A small diameter duct in the body passes from the top of the float chamber to the intake, widely called the *air horn*, to vent it to the clean side of the air intake filter, and thus to equalise the static, i.e. the datum, pressure over the fuel in the float chamber and jets. However, with the throttle closed, fumes issuing

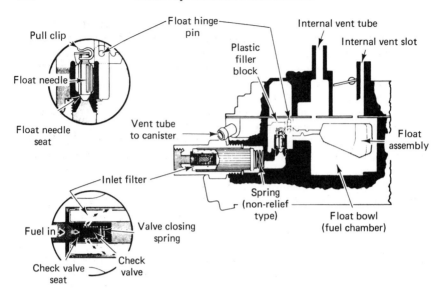

Fig. 9.17 Layout of float chamber, with scrap views of the needle valve and the check valve in the fuel intake filter

from this vent could contravene regulations regarding hydrocarbon emissions or, by over-enriching the mixture, lead to difficulties in restarting a hot engine. Therefore, a vent that is much larger and therefore operates preferentially except when it is closed by a valve actuated by depression in the induction manifold, is taken to a canister containing carbon granules. The depression-actuated valve is closed when the engine is running but opened by a spring when it is not. Provision may be made, also, by briefly opening the valve automatically when the engine is started, to purge the canister of vapour that has collected in it, by drawing fresh air through it into the manifold.

Throttle stop. During idling, both secondary barrels are locked in the closed position, leaving in operation only the two primary barrels. These are served by a pair of identical idling mixture supply systems, one for each barrel, Fig. 9.18. The throttle is automatically held open far enough to cope with whatever load may be applied to the engine by, for example, the air conditioning system or a rear screen heater.

Depending on the application, this is done by either an idle speed control (ISC) or an idle load compensator (ILC), or by an idle speed solenoid (ISS). The ISC is, in effect, a movable throttle-stop that is electronically controlled to maintain a constant idling speed. For the ILC the aim is the same but actuation is by a spring-loaded diaphragm sensitive to manifold depression. ISS, on the other hand, is a device based on a solenoid that is energised either automatically to set the idling speed to a level such that the engine will not run-on when switched off or, alternatively, it is wired through the air conditioning control circuit so that, when this is switched on, the engine will not stall.

Idling system. In each barrel, the hole through which the idle mixture discharges is downstream of the throttle valve when shut. With the throttle closed and the float chamber vented to atmosphere, manifold depression is transmitted through this idle discharge hole to the idle jet. As a result, atmospheric pressure above the fuel in the float chamber forces it through the main jet into the main fuel well, from which it passes into the idle tube. As can be seen from Fig. 9.18, an idle jet meters it into the bottom of that tube, whence it then passes up to the top end, at which point it is mixed with air entering from the solenoid-controlled idle air bleed valve.

The resultant air fuel emulsion passes over the upper end of the idle tube and down through a calibrated restriction into the idle mixture channel, in which more air is bled into it through both the lower idle bleed hole and the slotted off-idle port before it reaches the idle mixture discharge port through which it passes out into the manifold. An idle mixture adjustment screw, with a conical end projecting into the port, limits the flow through it. The adjustment, however, once set in the factory, is sealed by pressing a hardened steel plug into the end of the counterbore in which the head of the screw is housed. This is necessary to discourage unauthorised tampering, probably leading to contravention of the emissions regulations.

As the throttle in each primary barrel is opened its edge traverses the slotted off-idle port, progressively exposing a larger proportion of the slot to the depression in the manifold. Consequently, mixture is increasingly drawn out of its lower end and less air sucked in its upper, until ultimately the emulsified fuel supply is passing out through both ports. Thus, as the mass flow of air past the throttle valve into the engine increases, so also does the metered fuel supply from the idle system until the mounting depression generated by the rising flow

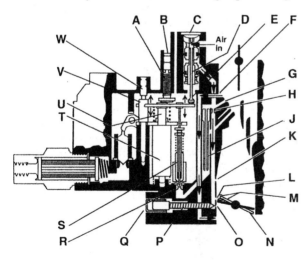

A	Rich mixture screw	H	Idle channel restrictor	P	Idle mixture control screw
B	Plug	J	Idle tube	Q	Plug
C	Riveted cover	K	Lower idle air bleed	R	Main metering jet
D	Idle air bleed	L	Off idle port	S	Main metering rod
E	Valve stem	M	Exhaust gas recirculation,	T	Mixture control solenoid
F	Fixed idle air by-pass,		timed depression ports	U	Solenoid plunger (upper
	only on some models	N	Throttle valve		position)
G	Idle air bleed	O	Idle discharge hole	V	Lean mixture screw
				W	Plug

Fig. 9.18 In the Quadrajet carburettor, the idle system is in the primary barrels only

through the venturi gradually takes over by progressively drawing more from the main jet.

For some applications a fixed idle air by-pass system is added, to take air from above the venturi to a point downstream of the throttle valve. This enables the throttle valve to be closed further without reducing the idling speed to too low a level. It is, in some instances, necessary because the triple venturis in the primary barrels are so sensitive to air flow that too much fuel would otherwise be drawn off through the main jets, even though the throttle was almost closed.

Rough running may be experienced owing to dilution of the very small idling charge by the exhaust gases recirculated to prevent excessive emission of oxides of nitrogen under open throttle conditions. To overcome this, manifold depression is used to close the exhaust gas recirculation (EGR) valve, so that it is inoperative when the engine is idling or in the overrun mode. According to the application, the depression for this purpose is taken off through either one or two additional ports punched in the throttle barrel, immediately upstream from the throttle valve. As the throttle is opened and level of the depression therefore reduced, this valve is re-opened by a return spring. A similarly timed port, not shown in the illustration, can be provided for actuating a clutch for unloading the torque converter of an automatic transmission.

Main metering system. The main metering system, Fig. 9.19, discharges into the two primary barrels. It takes over as the throttle is opened and the depression over the idle system correspondingly decreases, causing the flow through it to tail off. Fuel flows past the needles in the main metering jets to the main wells, into which air is bled through two holes upstream of the venturis. From these wells the air–fuel emulsion passes up two spray tubes, discharging one into each central element of the two triple venturis. The overall mixture strength continues to be regulated throughout, as previously described, by the electronic control.

On some versions of this carburettor, what is termed a *pull-over enrichment* (POE) device is automatically brought into operation during high speed operation under heavy load. In such conditions the flow of air through the four bores is so rapid as to generate a significant degree of depression above the venturis. This is used to draw a supplementary supply of fuel, directly from the float chamber, through a calibrated orifice in the bore of the horn, just above the strangler.

Secondary system. When the power output reaches the limit beyond which the primary barrels cannot support any further increase, the two barrels of the secondary system come into operation. Up to this point, the two secondary barrels, Fig. 9.20, have been blanked off by two centrally-pivoted plate valves. These are termed the *air valves*, and are situated upstream of the throttle valves. Each is held shut by a helical spring acting on its pivot, and opened by the differential between atmospheric pressure above and manifold depression below them.

The secondary throttle valves are opened by a link actuated by a lever on the primary throttle spindle. As they open, they release manifold depression to the underside of the air valve to open it against the action of its return spring. However, if the engine is cold, this opening of the air valve is baulked by a lock-out lever actuated by the automatic strangler mechanism. When the engine temperature rises to a predetermined level the lock-out lever is retracted by the bimetal coil of the automatic strangler.

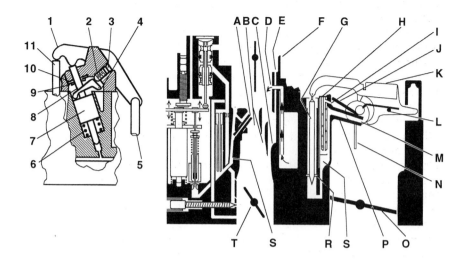

Fig. 9.19 Main metering system with, left, a scrap view showing the throttle position sensor

1	Accelerator pump lever	D	Pull-over enrichment feed
2	Factory adjusted screw	E	Internal vent slot
3	Plug	F	Baffle
4	Adjustment lever	G	Secondary metering rods (two)
5	Link between throttle control and pump	H	Accelerator wells and tubes
		J	Air valves (closed)
6	Sensor spring	K	Metering rod lever (lowered)
7	Throttle position sensor	L	Hinge pin
8	Sensor stem	M	Eccentric
9	Retainer	N	Baffle
10	Seal	O	Secondary throttle valves
11	Pump stem	P	Main discharge nozzle
A	Main discharge nozzle	R	Metering discs
B	Boost venturi	S	Main fuel wells
C	Main venturi	T	Primary throttle valves (partly open)

A dashpot regulates the rate of opening of the air valve, to ensure a smooth transition from operation on the primary system alone. As can be seen from Fig. 9.21, the dashpot is linked to a slotted hole at the end of a lever on the air valve spindle. It contains a diaphragm that is held down, against the action of a return spring, by manifold depression, in excess of about 127 to 152 mm (5 to 6 in) of mercury. If the depression drops below this value the spring takes over. However, the rate of return of the diaphragm is limited by a calibrated orifice in the tube conveying the depression from the manifold. In conditions of, for example, zero depression, the rod from the dashpot will be right forward in the slot in the lever, so the air valve will be free to open rapidly when the accelerator pedal is depressed. At 127 mm depression or more it will be withdrawn to the other end of the slot and therefore the opening of the air valve will be regulated by the dashpot, as described. The details quoted above may, of course, vary from application to application.

On most models, as the air valve opens, its upper edge passes over an acceleration well port in each of the bores. Since these ports are then exposed

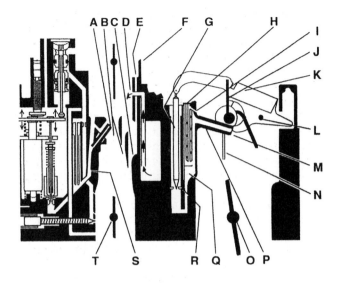

Fig. 9.20 The power system. All keyed details are the same as in Fig. 9.19, except in that the air valves J and secondary throttle valves O and T are open and the metering rod G and its lever K are in the up position

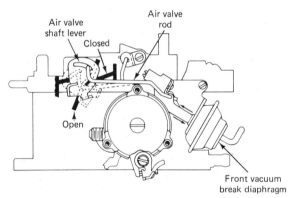

Fig. 9.21 Air valve dashpot operation

to manifold depression fuel is immediately drawn up from the acceleration wells and discharged, through calibrated orifices in them, into the barrels, to compensate for any lag, due to inertia, in the supply from the secondary discharge nozzles.

The secondary barrels do not contain venturis. Instead, manifold depression, maintained by the action of the air valve upstream of the throttle valves, draws the fuel from the secondary metering jets. The metering function is performed by the above-mentioned secondary rods, or needle valves, which are actuated not by the ECM-controlled solenoid but by a plastics eccentric cam on the spindle of the air valve. As the spindle rotates, the eccentric lifts a cam follower and thus a lever termed the *secondary rod hanger*, which in turn lifts the tapered ends of the secondary needles out of orifices in metering discs interposed in the supply line between the float chamber and the secondary fuel wells. This progressively increases the supply of fuel as the air valve opens. From the metering orifices the fuel flows into the secondary main wells, where air is introduced by air bleed tubes the lower ends of which extend well below the normal fuel level. The emulsified mixture then travels up the secondary nozzles, or spray tubes, which discharge it into the barrels. All automatic adjustment of mixture strength in response to signals from the oxygen sensor in the exhaust is therefore still done by the action of the ECM in controlling the supply of fuel to the primary barrels. Although the design of the secondary air valve actuation lever determines the maximum air flow, in some instances the capacity of the carburettor can be adjusted by means of a stop limiting the movement of these valves.

Some other features of the secondary system are as follows. When the air valve is fully closed the shoulder at the lower end of the largest diameter portion of the rod may almost seat on the edges of the orifice of the metering jet. Therefore, in some instances a slot is milled in the shoulder so that a small amount of fuel can still flow from the float chamber into the main wells, to keep them constantly full, ready for instant discharge into the secondary barrels, as well as for the replacement of any fuel that may boil out of the secondary wells owing to conduction of heat to them after the engine is switched off, or during hot idle.

In some versions a baffle plate with horns extending up and around the discharge ends of the nozzles, or spray tubes, helps to equalise the distribution of mixture to all cylinders at low rates of air flow. On all models a baffle plate is also fixed on to the underside of the air valve itself, to improve mixture distribution at high rates of flow. In other applications there is yet another baffle, extending from over the secondary main well air-bleed tubes up into the air horn between the primary and secondary bores. This prevents incoming air from forcing down the fuel level in the secondary wells otherwise, during rapid acceleration, delivery from the secondary nozzle might lag.

Acceleration pump. The principle of operation of acceleration pumps is covered in Sections 8.26 and 8.27. This pump, discharging into the primary barrels, is illustrated in Fig. 9.22. It has a coil type garter spring in the cup seal around the piston to maintain constant contact between cup and bore. The seal assembly floats on the plunger head against which it seats during the downward movement but, on the return stroke, unseats to allow fuel to flow through into the delivery well below it. When the plunger is not in operation it remains unseated, to allow vapour to vent from the well. As mentioned previously, the pump lever actuates the throttle position sensor.

In conditions of rapid air flow through the primary barrels a depression may develop over the pump delivery port, so to break it a vent is taken from a point just behind the pump jet up to the intake horn. This not only prevents fuel from being drawn out during steady throttle operation at high speed but also ensures that the pump is at all times in readiness for providing a full charge of fuel when the throttle is opened for sudden acceleration.

Strangler control mechanism. A strangler valve is situated upstream of the venturis in the primary barrels. It is eccentrically pivoted so that atmospheric pressure above will tend to push it open as the manifold depression below increases. For starting from cold the secondary air valve is, as previously

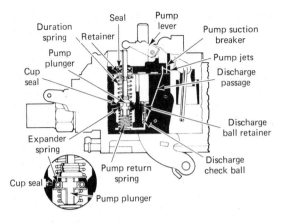

Fig. 9.22 Acceleration pump system with, in the scrap view, details of the head of the plunger and its seal

described, automatically locked in the closed position and, moreover, it cannot be opened until the engine has been thoroughly warmed up and the strangler valve is wide open.

The strangler control mechanism is conventional in that it incorporates a fast-idle cam actuated by a bimetal thermostatic coil heated by either air drawn from a jacket over the exhaust manifold or, on a timed basis, by an electric heating element. This bimetal coil rotates the cam, which is stepped, and thus allows the throttle to close stage by stage on to successively lower steps as the engine becomes warm.

However, this is a precision system designed specifically to meet requirements regarding both driveability and exhaust emissions regulations. As previously mentioned, the strangler valve is opened by the pressure differential between the air above and below it, but it is closed by the bimetal coil, the action of which is moderated by the tendency of a depression actuated diaphragm unit, or units, to open it.

According to application, the diaphragm assemblies are of either the positive action or depression- or air-restricted type, the latter being similar to that previously described for regulating the rate of opening of the secondary air valve, except that the restrictor is in the air inlet instead of the depression inlet to the diaphragm chamber. In some applications two are used, one positive action and the other of the restricted type. For other applications a thermal switch in the air cleaner housing simply prevents the manifold depression from reaching one or both of the diaphragm chambers until the carburettor inlet temperature, under the influence of a heated intake valve for emission control, rises to a suitable level.

Yet another variant of the restricted depression-controlled diaphragm system has what is termed a *bucking spring* to modify its operation. This is a cantilever leaf spring, the free end of which pushes down on the end of the plunger of the depression-actuated diaphragm in a manner such as to oppose the action of the bimetal coil. The principle of operation is as follows.

During warm weather the tension in the bimetal coil is only light, so the plunger of the diaphragm unit is held down mainly by the action of the bucking spring, against which the manifold depression is therefore able to open the strangler relatively easily. During extremely cold weather, on the other hand, the combination of a tightly tensioned bimetal coil and the bucking spring tend to oppose the opening of the strangler valve much more strongly, giving a richer mixture for cold starting. As the throttles are opened wider the strangler valve will tend to open further because of the combined influences of the depression on both the diaphragm unit and the differential pressure on the eccentrically pivoted valve.

Some diaphragm units have, in addition, a clean air purge system. During engine operation clean air is drawn through a filter before passing to a restricted orifice and on through the unit, to clear it of fuel vapour and dirt that might have accumulated in it and which could adversely affect its operation. This venting of the diaphragm chamber also allows the diaphragm to be retracted rapidly by its return spring when the depression in the manifold falls precipitously to a very low level or zero.

Strangler operation. Prior to starting from cold the accelerator pedal must be pressed down to the floor for two reasons. First, it pumps some fuel mist from the acceleration system into the venturi to facilitate starting; secondly, and more importantly, it also allows the fast-idle cam follower to lift clear of the stepped cam. This allows the cam to be rotated by the bimetal coil, so that when the throttle is closed again for starting the engine its stop comes down on to the highest step.

During cranking, the depression in the manifold draws fuel through both the idle and main discharge systems. Up to 200 rev/min, the ECM does not exercise control through the solenoid, so the spring lifts both the jet needles and the air bleed needle up to provide a rich mixture. Above 200 rev/min, the ECM takes over, but is programmed in a rich mode of operation and, since it is influenced by not just the oxygen sensor but also the engine temperature, it operates on the open-loop principle.

Manifold depression is applied to the positive action diaphragm unit to open the choke, against the action of the bimetal coil, to precisely the right degree for supplying a mixture rich enough for the engine to operate at that temperature without either stalling or over-enrichment. The function of the depression- or air-restricted diaphragm unit, if fitted, is to reduce the rate of the opening of the choke to prevent the engine stalling owing to sudden weakening of the mixture.

9.10 Constant-depression carburettors

In Sections 8.10 and 8.11 an explanation was given as to why compensation devices and multi-jet systems are necessary with a carburettor having a choke of fixed diameter. With this type, the choke diameter in any case is a compromise: if it is too small, engine performance is limited by the restriction on its breathing; on the other hand, if it is too large, the accuracy of metering of the fuel and atomisation and mixing are poor at the lower end of the range.

In principle, these disadvantages can be overcome by using an automatically variable choke carburettor, such as the SU or Zenith-Stromberg units. The object is to obtain a constant air velocity, and therefore constant depression – generally about 230 mm water head – over the jet, the size of which is varied to control the air–fuel ratio. Thus a single jet can be employed, and additional compensating devices are generally unnecessary.

Automatic variation of the size of the choke to maintain constant air velocity is effected by having a rectangular choke, in which a piston, or diaphragm actuator, is used to raise or lower its upper portion termed the *air valve* – away from or towards – its fixed bottom portion, termed the *bridge*, in which is mounted the jet. The lowering is effected by the weight of the piston plus a very light return spring, while the raising operation is done by the depression just downstream of the choke, which is communicated to the top of the piston or diaphragm.

A pendant-tapered needle fixed to the moving portion of the choke rises and falls with it and thus varies the size of the jet orifice – into which it dips – and therefore the rate of flow of fuel. The taper of the needle is not straight, but is profiled to vary the air : fuel ratio appropriately to the throttle opening – that is, for starting, idling, cruising and maximum power.

From Fig. 9.23, it can be seen that, as the throttle butterfly is opened, the manifold depression is communicated increasingly to the body of the carburettor and, through the duct V, to the chamber above the piston. The piston will rise, allowing increasing quantities of air and fuel to pass beneath it, until the depression is just sufficient to balance the weight of the piston plus the force exerted by its low-rate return spring. Consequently, since the height of the piston is a function of the mass of air passing beneath it, the depression over the jet in the bridge is maintained constant regardless of the demand. The designer of the carburettor arranges that the value of this depression is

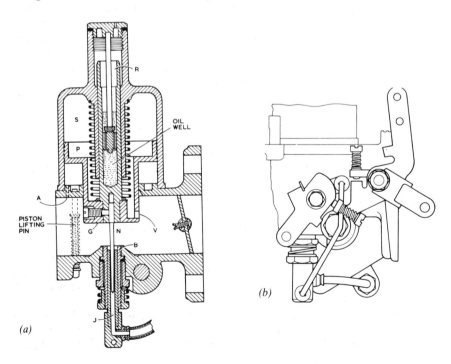

Fig. 9.23 SU carburettor, type HS

adequate for good atomisation of the fuel, but not so great as to restrict the breathing or filling of the cylinders to an unacceptable degree.

Enrichment, to compensate for inertia of the fuel – which causes its flow to accelerate too slowly to cater for sudden throttle opening – is effected simply by damping the upward motion of the piston, so that the suction over the jet is temporarily increased until the piston finally arrives at the equilibrium height. Downward motion of the piston is virtually undamped.

Mechanically-controlled variable-choke carburettors have been produced in the past, but of course are no longer used on cars. Their controls have been either manual or interlinked with the throttle. Such controls in any case would be unsuitable for meeting modern requirements.

The advantages of an automatic variable-choke, or constant-depression, carburettor can be summarised as follows: mixture compensation is unnecessary so only a single jet is needed. Consequently, there is no possibility of the flat spot that could otherwise occur during change-over from one jet to another. A separate idling system may also be unnecessary, as is an accelerator pump. Optimum filling of the cylinder at the top of the speed range is obtainable without detriment to the low speed performance. Moreover, metering of the fuel can be precise throughout the whole of the range.

9.11   SU constant-depression carburettor

The SU carburettor type HS, without its extra emission control features, is illustrated in Fig. 9.23. An air valve G is integral with the piston P and securely attached to the piston rod R, which slides in the close-fitting bore of the suction chamber S. The depression just downstream of the air valve is transmitted through a pair of holes, V, to the suction chamber, above the piston. Because of the very low rate of the return spring, the load to be supported by the vacuum is practically constant, being mainly the weight of the piston. The underside of the piston is vented through hole A to the entry to the carburettor.

Also rigidly attached to the piston assembly is the tapered needle N which therefore, as the piston moves, simultaneously varies the annular orifice between itself and the jet. The jet sleeve J can be raised or lowered for overriding the setting, for idling or starting. This is done by means of a cam and link connection to the throttle and cold start controls, 9.23(b). Petrol is delivered through a nylon tube to the lower end of the jet sleeve, from the float chamber – which can be bolted on to either side of the carburettor body, to suit the particular installation.

In Fig. 9.23, the jet tube J is a close fit in the flanged bush B, which is a clearance fit in the bridge. On assembly in the factory, the jet tube and bush are together centred relative to the bore of the suction chamber – and thus the piston and needle assembly – and then the upper of the two nuts is tightened to lock the setting by clamping the flange of the bush securely against the shoulder beneath which it seats. The lower nut is used subsequently for mixture adjustment.

For damping the motion of the piston, a plunger is secured beneath the screw-cap on top of the suction chamber. This plunger therefore remains stationary as its cylinder, which is the oil-filled hollow rod of the air piston, moves up and down.

With the advent of emission control regulations, additional features have had to be introduced. These, which are also incorporated in the HIF series of carburettors, are as follows: compensation for changes in viscosity of the fuel with temperature; spring-loading of the needle to one side of the jet to avoid variations in orifice coefficient, and therefore fuel flow, owing to concentricity errors; an overrun limiting valve, for maintaining proper combustion in the overrun condition; integral crankcase ventilation control; and a linear ball bearing, interposed between the stem of the piston and its housing, for minimising the sliding friction. These features are detailed in the next section.

9.12   SU carburettor type HIF

The HIF (Horizontal, Integral Float Chamber) carburettor was designed specifically to meet the requirements for exhaust emission control. At the same time, a noteworthy degree of compactness has been achieved by arranging the float and float chamber concentrically with the jet tube, as can be seen from Fig. 9.24. With this arrangement, too, the petrol level in the float chamber, immediately beneath the jet, is only minimally disturbed by changes in inclination of the vehicle, braking, acceleration and cornering. The principle of operation of this carburettor is the same as of that previously described and, in detail construction, it is much like the HS series. However, certain additional features have been incorporated.

Among these is the spring-loading of the needle against one side of the jet, so that it is positively located radially and therefore the orifice coefficient invariable – with a centrally-disposed needle, any movement off centre alters the orifice coefficient. A coil spring is used; it is axially in line with the needle and bears on a collar around its upper end. A peg, screwed radially into the piston to project beneath one side of this collar, tilts the needle by virtue of the resultant offset reaction to the spring force until it is stopped by its lower portion contacting the side of the jet orifice. The direction of bias of the needle can be either up- or downstream. Since the flanged bush in which the upper end of the jet tube is carried does not have to be centred, it is a close fit in its hole in the bridge. The jet is pressed into the top end of an aluminium tube, the lower end of which is pressed into a plastics moulding, known as the *jet head*. Fuel entering the open lower end of this moulding passes up the tube.

On the throttle butterfly, there is an overrun limiting valve, which is a small spring-loaded non-return plate valve. Under overrun conditions, the intense depression in the manifold opens this valve, as described in Section 9.4 and illustrated in Fig. 9.24, so that a small quantity of air–fuel mixture can pass through to maintain regular combustion in all the cylinders – irregular combustion of course would cause the discharge of unburnt hydrocarbons into the exhaust.

For provision of a good quality mixture at small throttle openings, there is a part throttle by-pass emulsion system, taking mixture directly from the jet. At its outer end, the by-pass passage breaks out into a slot (G in Fig. 9.25) – immediately downstream of the jet – in the bridge, while its other end breaks out in line with the edge of the slightly open throttle, so that the local high velocity of flow, and therefore depression, will atomise the fuel and mix it more thoroughly than it would otherwise.

Additional mixture for cold starting is supplied by an enrichment device, Fig. 9.25. This is a small rotary valve through which fuel is drawn from the float

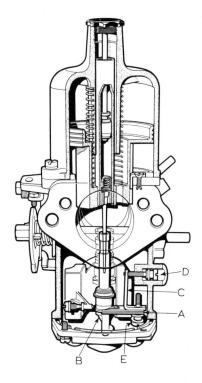

Fig. 9.24 A sectioned HIF4 carburettor

1   End seal cover
2   End seal
3   Starter valve body
4   'O' ring
5   Valve spindle

A   Fuel supply
B   Air bleed
C   Fuel delivery
D   Commencement of enrichment
E   Maximum enrichment
F   Enrichment outlet
G   Part throttle by-pass slot

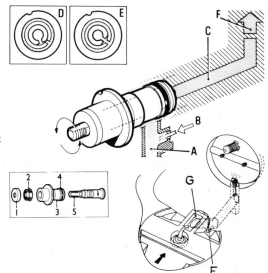

Fig. 9.25 Cold start enrichment device

chamber, mixed with air from a bleed hole, and discharged through an outlet – immediately downstream of the bridge. Housed in a hole in the side of the carburettor body, the device comprises a cylindrical valve body with a seal at each end and a spindle passing through it. The outer end of the spindle is threaded to receive the cam follower lever by means of which the valve is actuated.

Fuel is drawn by the depression over the outlet hole F, up a passage in the wall of the float chamber, past a bleed hole through which air is drawn in to emulsify it. This mixture is then taken on up into an annular space between the housing and the cylindrical body, and thence in through radial holes to a counterbore in the inner end of the valve spindle and to the outlet hole behind the bridge. The flow is of course cut off when the spindle is rotated so that its radial hole no longer aligns with that in the cylindrical body. A groove is machined tangentially across the end of the hole in the spindle so that, as the spindle is rotated, the fuel flow is started and stopped progressively, as shown at D and E in Fig. 9.25.

To compensate for variation of viscosity of petrol with temperature, a bimetal strip regulates the height of the jet tube assembly relative to the needle. The arrangement can be seen from Fig. 9.23. A bimetal strip A carries at one end the jet head B. It is pivot mounted at the other, where it is riveted to a bell-crank lever C. A spring beneath it, around the screw E, pushes it so that the upper end of the bell crank moves outwards until it is stopped by the adjustment screw D. The adjustment is made on assembly – clockwise rotation of the screw D lowers the jet, and therefore enriches the mixture. When adjustment is complete, a plug is fitted over the screw head, in its countersink, to seal it.

For crankcase emission control, the constant depression chamber, between the throttle valve and air valve, is connected by a hose to the engine breather outlet. This draws crankcase fumes into the carburettor and thence, with the air–fuel mixture, to the cylinders. The fumes are replaced in the crankcase by fresh air entering through the oil filler cap breather or, possibly on cars equipped with evaporative-loss emission control, through an absorption canister. Because the connection is made to the constant depression chamber, no valve is needed to safeguard against either excessive depression in the crankcase or interference with the slow-running mixture supply.

9.13   Zenith-Stromberg CD and CDS carburettors

The design of the Stromberg carburettors is basically similar to that of the SU types, but they differ structurally, as can be seen from Fig. 9.26. Among the major differences is the use in the Stromberg units of a diaphragm-, instead of piston-actuated, air valve. The concentric float chamber beneath the jet assembly is a feature of all the current Zenith-Stromberg designs.

On the CD carburettor, provision for cold starting is as follows: a semi-cylindrical bar seats in a groove across the bridge, to conform with its profile, as shown in Fig. 9.26. For cold starting, however, it is rotated by the choke control, to lift the air valve slightly and, at the same time, mask the choke opening and thus restrict the air flow. The consequent lifting of the needle of course enriches the mixture. Simultaneously, because of interconnection with the throttle control, by means of a cam, the throttle is opened an

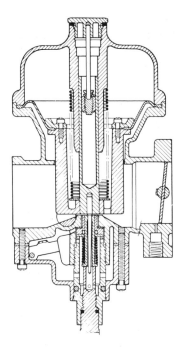

Fig. 9.26 Stromberg CD carburettor

amount determined by the setting of a fast-idle stop screw, which serves as a follower on the cam. This carburettor is now superseded by the CDS unit.

The CDS is the CD unit with, instead of the semi-circular bar just described, a rotary disc valve for supplying additional fuel for cold starting. This is described in more detail in Section 9.15.

9.14   Zenith-Stromberg CDSE emission carburettor

This is basically the CDS unit, but modified to tighten the tolerances on metering and fuel flow. Idle adjustment is effected by what is termed the idle trimmer screw, instead of by raising or lowering the jet – the latter is now fixed. The end of the idle trimmer screw is in the form of a needle valve, which regulates the amount of air passing through a duct from the intake into the mixing chamber: this relieves the depression in the suction chamber above the air-valve control diaphragm, and thus allows the needle to fall and weaken the mixture. It has a limited range of adjustment, because it is intended only to cater for variations in the stiffness of rotation of new engines. The adjustment is made to limit the output of carbon monoxide, as indicated by an exhaust gas analyser.

There is also a temperature compensator. This is a flat bimetal thermostat spring. It senses the temperature of the body of the carburettor and regulates the axial position of a tapered needle in an orifice, through which extra air passes to the mixing chamber. The effect of this air bleed is similar to that of the idle trimmer.

To prevent an increase in the flow of unburnt hydrocarbons through the engine during deceleration, there is a throttle by-pass duct. Through this duct, a small quantity of mixture is delivered downstream of the throttle, and this ensures continued combustion in the cylinders. The air is drawn from upstream of the throttle, and fuel is added through a jet orifice in the duct. This device is brought into operation by manifold depression, acting on a small spring-loaded diaphragm and disc valve.

The jet needle is biased, by spring loading, to one side of the orifice. This and the other features of the CDSE are similar in principle to, though differing in detail from, those of the CD4 and CD5 described in Section 9.15.

9.15 Zenith-Stromberg CD4 and CD5 carburettors

These carburettors were designed specifically to meet the then current emission requirements. They are similar to the earlier CD units but differ slightly in layout, as can be seen from Fig. 9.27.

Where the Auto Starter is fitted, Section 8.22, the letter T is appended to the type designation. It is in fact fitted on the CDSET, CD4T and CD5T. Apart from this, the principal new features of the CD4 and CD5 are as follows: for starting from cold with the manual starter, the choke control actuates a lever on the side of the carburettor, which rotates a disc drilled with a series of holes of different diameters, A in the scrap view Fig. 9.27. In the fully rich position, all the holes are in communication with the starter circuit and, as the choke control is released, they are progressively blanked off.

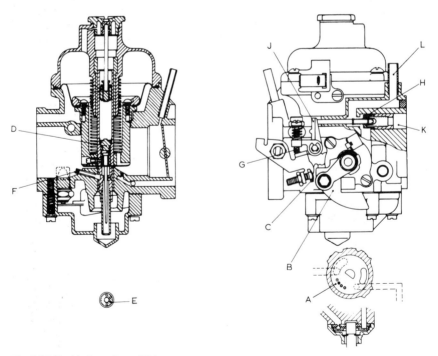

Fig. 9.27 Zenith-Stromberg CD4

Petrol is drawn from the float chamber, up a vertical channel adjacent to the main jet, through the disc device, and thence to the mixing chamber – between the air valve and throttle plate. In the meantime, the cam B on the starter lever, acting on the fast-idle screw C, has opened the throttle sufficiently to prevent stalling of the cold engine.

Once the engine has been started and is warm, the operational principle of the carburettor is similar to that of the earlier CD series. On the CD4, however, provision is made for adjustment of the mixture strength by raising or lowering the metering needle relative to the jet orifice. This is done from above, by lowering a special tool down the damper tube, to rotate the adjustment screw D, Fig. 9.27. The method of biasing the needle to one side of the jet orifice is similar to that in the SU carburettors, except that the needle is spring-loaded coaxially in a carrier in the piston assembly. The light bias spring is retained by the head of the needle which is tilted by either a swaged tag or a small pin. This arrangement is used also on many CDSE and CD3 carburettors, the latter being the CDSE without a temperature compensator.

A commendably simply device is applied to avoid the need for a mechanical device to compensate for variations in fuel viscosity with temperature. Four emulsion holes are drilled radially in the metering orifice, as shown at E in the scrap view. They are in the semi-circular segment opposite to that in which the biased needle seats. Air is metered to them through a calibrated jet F. Without this device, enrichment due to several factors, mainly reduction in viscosity with rising temperature, would adversely affect the metering performance and stability of running, especially under idling conditions.

An additional feature has been introduced to improve the mixture control for idling. It is termed the *downstream discharge circuit with idle air regulator*. A small proportion of the metered fuel from the main jet is sucked by manifold depression into a calibrated hole A, Fig. 9.28, and discharged downstream

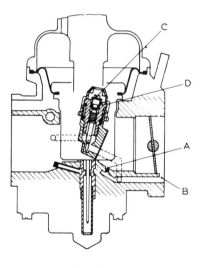
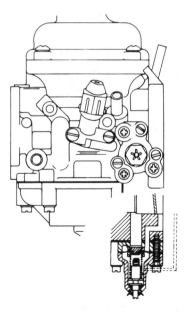

Fig. 9.28 Downstream discharge circuit

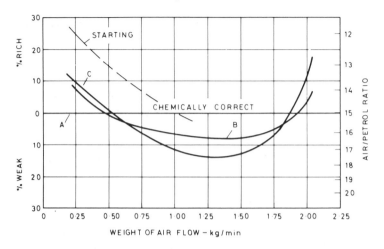

Fig. 9.29 Mixture ratio curves for constant-vacuum carburettors

from the throttle butterfly, through a port B. The position of this port is such that, once the throttle valve has been opened beyond a certain point, the system ceases to operate. Air is introduced into this system past an adjustable metering screw C, so that the air : fuel ratio of the mixture can be regulated over the idling range. This screw, which is for fine adjustment, is carried in a coarse adjustment screw D. The latter is set and locked in the factory, and the further fine adjustment can vary the mixture strength by no more than plus or minus half an air : fuel ratio, depending on the engine layout. Idling speed is adjusted by the throttle stop screw G in Fig. 9.27.

Float chambers are generally vented to the air cleaner. However, if the vent is taken to a point downstream of the filter element, fuel vapour can collect there when, in a recently stopped engine, heat has soaked through to the float chamber and evaporated fuel from it. This can tend to inhibit restarting the engine when hot. On the other hand, venting to the outer side of the filter element can lead to enrichment of the mixture if the element becomes clogged, since this will increase the depression over the jet.

In the CD4 and CD5, therefore, a ventilation valve H is actuated by the throttle linkage through lever J, Fig. 9.27. Under normal running conditions, the valve closes the external channel L and opens the float chamber to the internal ventilation channel K, but under idling conditions the valve moves to the right, as viewed in the illustration, to close the internal vent K and open the external one L. This system is also used on the CDSE V carburettors, the V indicating incorporation of the vent valve.

In these carburettors, the overrun valve is generally in the body casting instead of on the butterfly valve, though there are some instances of the use of a poppet valve on the butterfly. It is simply a depression-actuated non-return valve in the base of what is, in effect, a U-shaped channel the upper ends of the arms of which break out, one upstream and the other downstream of the butterfly valve, into the main induction passage. This allows enough mixture from the mixing chamber to by-pass the throttle valve and thus maintain regular combustion in all cylinders in overrun conditions. Additional information on the CD5 version is given in *Automotive Fuels and Fuel Systems*, Vol. 1, by T.K. Garrett, Wiley.

9.16 Mixture ratio curves

Mixture ratio curves may be drawn for these carburettors of the same form as those in Fig. 8.15, the weight of air flow being plotted horizontally, but with no dual characteristics.

This weight of air flow may be taken as directly proportional to the lift of the cylinder, since the rectangular choke area is proportional to this lift, and is subjected to a constant depression or head.

For each position, that is for each value of the flow, there will be a definite ratio of jet area to choke area and therefore a definite mixture ratio dependent on the diameter of the needle. The anular form of the jet area, it should be noted, renders it subject to viscosity effects, and the flow is not exactly proportional to this area since it is not of geometrically constant form. Fig. 9.29 illustrates the manner in which any desired curve can be obtained by modifying the profile of the needle.

At A the profile is so chosen as to make the petrol flow just proportional to the air flow and in the chemically correct ratio. If now the needle were changed for one having a slightly larger diameter than the first at all points near the middle of its length, but reduced at its two ends, curve B would be obtained. Curve C, giving a 12% weak mixture over the middle range and considerable enrichment at the two ends, would result from still further reduction at the ends of the needle and enlargement of its centre portion. It should be realised that, just as in the ordinary fixed choke carburettor, the mixture ratio is a function of flow only, and not of throttle opening, so that the double characteristic of Fig. 8.15 is obtainable only by use of the overriding jet control shown in Fig. 9.23(*b*). In certain special models thermostatic and throttle controls are provided.

9.17 Automatic governor

The dangers of excessive speed will have been appreciated by the reader in studying Chapter 2, in which the nature and magnitude of the inertia forces were discussed. These dangers are intensified in vehicles with specially low emergency gear ratios, and it is possible for modern high-performance engines to suffer serious damage from over-speeding without the driver appreciating the danger, owing to the relatively low speed of the vehicle.

On commercial vehicles powered by diesel engines, the fuel pump governor limits engine speed. For spark ignition engines, however, additional mechanisms have to be incorporated. In the past, these have not been favoured because they could, in critical circumstances, present a risk of accidents, owing to sudden loss of response to the accelerator pedal at the governed engine speed. Moreover, they were costly, the driver could tamper with them and the potential for failure was increased. With the introduction of electronic engine management systems, however, speed limitation has become more practicable.

One mechanical device that met the requirements of the 1960s was the Solex Velocity Governor, which was later adapted to fit some of the Zenith carburettors. The standard throttle valve was replaced by a butterfly valve of

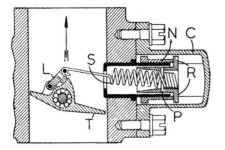

Fig. 9.30 Solex automatic governor

special form, Fig. 9.30, and a control mechanism added to close it automatically at a preset engine speed. To reduce frictional resistance to rotation, the butterfly was offset pivoted on a hardened and ground spindle, the ends of which were carried in needle rollers in an outer race in the throttle barrel. The butterfly was free to rotate towards the closed position, except that the coil spring S tended to hold it back against a dog on the spindle.

Three factors, therefore, determined the equilibrium position of the butterfly valve as follows—

(1) The spring tended to open the throttle.
(2) A limit to the degree of throttle opening was, however, determined by the position of the accelerator pedal.
(3) Owing to the offset of the spindle, the forces exerted by the air flow on the larger area of the trailing portion T of the throttle tended to close it.

With the accelerator pedal fully depressed at a speed below the governed level, the throttle is wide open and the forces (manifold depression and aerodynamic) tending to close it are therefore very light. As engine speed rises, so also do these forces. The tension in the spring is set so that it is overcome at the desired limit to the engine speed, after which the throttle valve closes progressively as, for example, the speed of the vehicle tends to increase during the descent of a gradient. The spring rate is such that the throttle can neither hunt nor, except possibly under exceptional conditions, snap shut.

The tension in the spring is set initially by adjusting the nut N. Subsequently, further adjustment can be made by means of the sleeve R through which is fitted diametrically the pin P, which forms the anchorage for the spring. Rotation of this sleeve varies the number of active coils of the spring. When adjustment is complete, the cover C is fitted to seal the mechanism. The geometry of the link L is such that the leverage exerted by the spring is relatively constant. Typical values for the torque exerted by the spring are 0.113 Nm at 6.33 mm throttle opening to 0.288 Nm with the throttle closed.

Chapter 10

Petrol injection systems

The first fuel injected engine was built in 1883, by Edward Butler, of Erith, Kent, and the next appears to have been built by Henri Tenting, of Paris, in 1891. Most of the other early examples were applied in aviation. These included Wilbur and Orville Wright on the first heavier than air machine to fly, in 1903, and on the Antoinette engine designed by Leon Levasseur for powering the Santos Dumont aircraft in 1906. The history of fuel injection, and its development from carburation, up to and including the Bosch Motronic system and Bosch/Pierburg electronic carburettor, are described in detail in an excellent book entitled *Automotive Fuel Injection Systems, a Technical Guide*, by Jan P. Norbye, published by Motorbooks International, Wisconsin.

The first application of petrol injection was on a car, the Gobron Brillé, on which a splined shaft drove a rotary injection pump. Although injection was not considered very seriously for production cars until 1946, when Mercedes was developing it for such applications, a Caproni Fascaldo electrically controlled system was installed experimentally on an Alfa Romeo car in 1940. Among the early racing applications were the Hilborn Travers continuous injection system installed on an Indianapolis engine in 1948, and an indirect injection system on the Connaught racing engine in 1953. Incidentally, electric control had already been used experimentally by the Atlas Diesel Co. in 1940.

The first quantity produced car to have direct injection was the Goliath introduced in 1951, with a two-stroke engine. By 1954, Mercedes had advanced far enough to introduce their 300SL and some racing cars with a Bosch direct injection system based on the use of a cam-actuated plunger, or jerk-type pump. In 1956, a pumpless system was invented by Bond, with the aim of avoiding the high costs of the other systems, but it never went into quantity production.

Of greatest significance in the history of petrol injection was the introduction in 1956 of the Bendix Electrojector continuous injection system, which by 1953 had been developed and installed on a Buick V8 engine. The Electrojector, described in *Automobile Engineer*, Vol. 47, No. 6, 1957, sparked off all the subsequent developments. In 1965, Bosch signed an agreement with Bendix, for access to its patents, and introduced the D-Jetronic system, which began to make a major impact in 1967. Some years later, Weber, Ford and GM developed electronic systems that were also based mainly on the principles first established by the Bendix Electrojector.

A Lucas shuttle valve system introduced in 1956 was on a Maserati V8 engine in 1958. This, however, was a hydro-mechanical system, and therefore could not compete with the rapidly evolving hydraulically, electrically and electronically actuated and controlled systems, and the Bosch system was adopted for a Ferrari Formula 1 racing car in 1962. The only other contender was Kugelfischer, who in 1961 introduced a jerk-pump system controlled by a profiled conical cam and follower described in the December 1961 issue of *Automobile Engineer*. By 1964, this was on some Peugeot production cars. Other contenders in the 1960s were Tecalemit Jackson, described in *Automobile Engineer*, July 1964, and AE Ltd, around 1966, described in *Automobile Engineer*, October 1966. Many of the other early systems mentioned are described in *Carburation*, Vol. 3, by Charles H. Fisher, published by Chapman & Hall, London.

Work resulting in another device of major significance in the history of petrol injection, the lambda sensor for detecting the presence of oxygen in the exhaust gases, Section 11.4, was begun by Bendix in 1970. This enabled control to be exercised on the closed-loop principle, without which satisfying the requirements of modern emissions regulations would be virtually impossible. It was patented in 1974 and the first car to be equipped with it was produced in 1976 by Volvo, for the American market, but with equipment patented and produced by Bosch.

10.1 Basic considerations

First the weight of the charge delivered to the engine is controlled by the throttle valve, and then the fuel required to make up the appropriate air : fuel ratio is metered, by the injection control system, through the injection nozzles into the air. To meet emissions requirements, the control system is, inevitably, electronic.

In the very early days, jerk pumps were envisaged for injecting directly into the cylinders. However, experience with the diesel engine had shown that such pumps were heavy, costly and noisy. Moreover, with petrol instead of diesel fuel, lubrication of the pumping elements was difficult, as also was control in relation to engine parameters.

Injection into the manifold was much easier, principally because of the much lower pressures needed. Indeed, fuel is delivered from a pump against back-pressures which, in naturally-aspirated engines, are lower than atmospheric.

Generally, the required air : fuel ratio is maintained under all the varying conditions of operation by regulation of the timing of the start and end of injection. An advantage of such systems is that they do not require a mechanical drive. Mixture strength requirements have been outlined in Section 8.3, so there is no need to comment further.

The pump must deliver the fuel reliably and continuously (without pulsation) and, in most systems, at a closely controlled constant pressure. Injectors must be capable of metering and delivering the fuel in the form of a fine spray into the manifold, without either pre- or post-injection dribble. Sensitivity of the electronic control unit is critical. So also are both the accuracy of the activators that it controls and the sensitivity of the sensors that signal to it the engine parameters on the basis of which it exercises control. These parameters include some or all of the following: mass air flow into the engine, ambient and engine temperatures, engine speed, throttle position and the oxygen content of the exhaust gases. The latter of course indicates whether the mixture is too rich or too lean. In some systems, sensors for other parameters such as crank position, induction manifold pressure and temperature, combustion knock and ignition timing are also employed.

10.2 Injection system types and layouts

Direct injection into the cylinders is not an impossibility but it suffers, in addition to the previously mentioned high back-pressure, other severe disadvantages. First, because the fuel has to be injected progressively to allow time for it to be atomised and mixed thoroughly with the air before the spark occurs, it has to enter the cylinder against a rising back-pressure. Secondly, in such a short time, complete evaporation and the formation of a homogeneous mixture for maximum power output are virtually impossible and smoke would be emitted from the exhaust. Thirdly, since the injector nozzles are exposed to the combustion process, carbon build-up is liable to occur. Similar problems arise, though perhaps to a lesser degree with indirect injection into a swirl chamber, but the cylinder head casting would be complicated and, as in the case of the diesel engine, thermal losses higher.

Low pressure injection into the induction system has therefore been adopted virtually universally. At such low pressures, about 3 to 6 bar, as compared with 900 to 1300 bar for diesel engines, dribble is not too difficult to avoid. Moreover, if it does occur, the consequences in terms of carbon deposit formation are not even remotely so severe.

Single-point injection (SPI) into the throttle body, sometimes termed *throttle body injection* (TBI) or *central fuel injection* (CFI), is attractive on account of its simplicity, and therefore low cost, as compared with *multi-point injection* (MPI) into the individual valve ports. A single jet of fuel from an injector just downstream of the throttle valve is less likely to be deflected to one side by that valve than if it were above. An alternative is to inject two twin jets from above, one passing each side of the butterfly valve. Single-point injection into the manifold is a possibility but is more complicated than if the injector is accommodated in the throttle body casting. All single-point systems, however, have three severe disadvantages: fuel tends to condense on the walls of the induction manifold, subsequently evaporating off in an uncontrolled manner; consequently, it is virtually impossible to obtain accurate distribution of mixture equally to each cylinder; and, thirdly, a hot spot must be provided in the throttle body to facilitate evaporation and prevent icing. For these reasons, and in view of the increasing stringency of emissions regulations, single-point injection is, in all probability, obsolescent.

The alternative, multi-point injection, is almost universally employed on all but the smallest, low-cost cars. The fuel is now injected into either each of the induction manifold branch pipes leading to the inlet valves or, more commonly, directly into the ports in the cylinder head. This avoids all the previously mentioned disadvantages of throttle body injection.

Accommodation of the injectors complicates manifold castings but, as regards cost, there is little to choose between port and manifold injection. Assembly on to the manifold is in some circumstances easier. Moreover, less heat will be conducted in it than from the cylinder head to the injectors. On the other hand, some heat may be desirable for assisting evaporation.

If the fuel spray is aimed directly at the hot inlet valve, not much will condense in the port. Moreover, there will be less risk of some of the charge being drawn off into an adjacent cylinder by the negative pressure pulse generated by the opening of its inlet valve. Aiming the spray at the inlet valve, Fig. 10.1, is relatively easy with port injection. Even with manifold injection it is not impossible, provided the injector nozzle can be accurately aligned in the manifold and the latter similarly aligned relative to the induction port. Incidentally, if fuel is to be injected on to the valves, care has to be taken to ensure that the valve material is corrosion resistant.

10.3   Injection strategies

The simplest, and therefore potentially the least costly, of all the strategies is *continuous injection*. This is applicable to both single- and multi-point systems. In either case, a cloud of mixture is formed in the manifold, ready to be drawn into each cylinder when its inlet valve opens. With port injection, swirl and turbulence in the cylinder play a major role in the production of a homogeneous ignitable mixture.

With continuous flow, the air:fuel ratio can be controlled by varying the pressure of the fuel delivered to the injectors, as in the Bosch K-Jetronic and KE-Jetronic systems, Sections 10.12 and 10.14. On the other hand, if overall electronic control is required, as in for example with the Bosch L-Jetronic

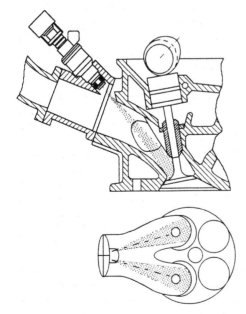

Fig. 10.1 Twin injectors directing fuel towards the inlet valves of the 1990 Vauxhall/Opel 3 litre engine

system, it is difficult to avoid an undue time lag between receipt of the signals from the control and effecting the required changes in air : fuel ratio. Therefore the fuel is maintained at constant pressure but the duration of delivery varied.

An alternative to continuous injection is *timed*, or *sequential*, injection, in which injection takes place over a limited period usually, though not always, once per revolution of the crankshaft. The timing of the opening of the solenoid-actuated valve in the injector is generally fixed, and control over the duration of delivery exercised by varying the timing of its closing. An example is the Bosch L-Jetronic system, Section 10.15.

Thus, the volume of fuel injected is controlled, but both the start and end of each injection occurs simultaneously in all the injectors, regardless of the position of the inlet valves. When injection occurs, a cloud of mixture forms in each induction manifold branch pipe, ready to enter the cylinder as soon as the valve opens. Timed injection is suitable for both single- and multi-point systems. Because the response to the signals from the electronic control is virtually instantaneous, extremely close control can be exercised over the air : fuel ratio, though at a higher cost than with continuous injection. A disadvantage is the risk of part of the charge being drawn into an adjacent cylinder by a negative pressure wave generated by the opening of its inlet valve.

A later development was termed *simultaneous double-fire* injection, or phased *injection*, which, with electronic control, enables the air : fuel ratio to be regulated extremely accurately. It entails the injection of fuel into the individual ports as the relevant inlet valve opens, and therefore once every two revolutions of the crankshaft, hence the term 'simultaneous double-fire'. This considerably reduces the risk of some of the mixture being drawn off into an adjacent cylinder, but is more costly than sequential injection.

An advantage of both timed and phased injection is the ease with which they can be used in conjunction with electronic ignition control to avoid detonation. Either a single detonation sensor can be mounted on the cylinder head or block or there can be one for each cylinder, or pair of cylinders. The sensors are generally mounted on the head. Signals from them and, of course, a crankshaft angle sensor are the basis on which the *central processing unit* (CPU) modifies either the ignition timing or the fuelling, or both, to stop the detonation. If more than one detonation sensor is installed, corrections can be made directly to the fuelling of individual cylinders or pairs of cylinders. Clearly, applying corrections to individual cylinders offers optimum performance as regards both fuel consumption and emissions.

To sum up, the benefits of petrol injection include the elimination of both the venturi and, with multi-point injection, the need for throttle body heating. Hence, volumetric efficiency is potentially higher than with carburation. Manifold design in general and, in particular, tuning to take advantage of pulse energy, Sections 11.12 to 11.19, are facilitated. Adverse effects due to the movement of fuel in the float chamber, especially during cornering, braking and acceleration, are obviated. By virtue of accurate control over fuelling, fuel consumption is lower and torque and power output higher, and the fuelling requirements for transient conditions can be satisfied much more easily than with carburation. Furthermore, with multi-point injection, deposition of fuel on the walls of the induction system is virtually eliminated, so the need for enrichment for cold starting and during warm-up is greatly reduced. Because of the substitution of electronic regulation of fuelling for manual or automatic strangler control, cold starting is easier. Finally, exhaust emissions can be reduced much more effectively under all conditions of operation.

10.4 Injector design

Because of the low pressures involved, pintle-type injectors lifted off their seats by the fuel pressure are impracticable for petrol injection. Most of those currently employed are illustrated in Figs 10.3 to 10.8.

All are electronically controlled and, since extreme accuracy of metering is essential, a prime requirement is a linear relationship between quantity of fuel delivered and the pulse width of the control signals. Moreover, this linearity has to be maintained throughout, from maximum to zero fuel delivery. This is clearly impossible, since it implies instantaneous starting and stopping of delivery, including from the full flow condition.

The plot of fuel delivery against pulse width in Fig. 10.2 illustrates the principle. Because of the delay, the delivery point is offset along the time axis or, in this case, from the start of the pulse. Following the delay, delivery can be linear up to but not including the repetition point, except if the pulse is very narrow. If injection is continuous, the pulse width extends over two revolutions of the crankshaft so, in effect, there is no offset at the start of each pulse.

Basically, the maximum duration of injection is primarily a function of the rotational speed of the crankshaft. In practice, therefore, it is determined by injector design and the consequent offset of delivery. The higher is the ratio of the actuation force to the mass of the delivery valve, and the closer the matching of the injector characteristics to the electrical and time constants of the solenoid circuit, the better is the linearity of delivery.

In the Lucas D Series injector, Fig. 10.3, the mass of the delivery valve is kept to a minimum by employing a disc valve, instead of a pintle, as the armature. The mass of the disc may be as little as one-eighth of that of some pintles, Fig. 10.4. Clearly, the lighter the armature the more rapidly will it accelerate between the open and closed positions.

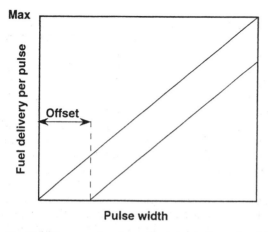

Fig. 10.2 The upper line shows the theoretically perfect delivery characteristic, and the lower one the same but offset due to inertia and consequent lag in delivery following the electronic triggering signal

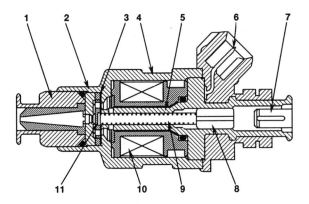

1 Nozzle holder	5 Core	9 Spring
2 Valve seat	6 Electrical connection	10 Winding
3 Shim and spacer	7 Filter	11 Armature
4 Body	8 Calibration slide	

Fig. 10.3 In the Lucas D-Series injector, alternative nozzles, for delivering single, twin or three sprays, can be fitted in the nozzle holder

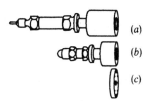

Fig. 10.4 Typically the masses of these three types of injector valve would be: (a) pintle 4.0 g, (b) ball 1.8 g and (c) Lucas armature 0.5 g

An electric pulse triggered by the CPU energises the solenoid. When the magnetic force has built up to a value exceeding that exerted by the return spring plus the hydraulic pressure, the armature is lifted from the seating ring around the valve port until it is stopped by the shim above, Fig. 10.5. When the solenoid circuit is broken, the current decays until the spring pushes the armature again on to the sealing ring. At this point, the additional force exerted by the fuel pressure ensures that it will not bounce.

The Lucas injector, has been designed for fuel pressures from 2 to 4 bar and delivery volumes of between 80 and 400 g/min, and can be used with simultaneous double-fire systems, Section 10.3. With static flow rates of less than 250 g/min, and a pulse repetition period of 10.0 ms, it has a dynamic range of 20:1. Reducing the repetition periods below 7.7 ms has been found to increase sensitivity to variations of supply voltage without giving significant increases in dynamic range. Typical values for the resistance of the solenoid coil are 2.3 ohm for the current mode and 16.2 ohm for the voltage mode systems.

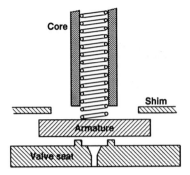

Fig. 10.5 Armature deflection is limited to the distance between the valve seat and the shim

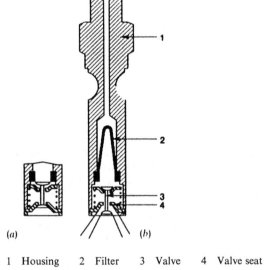

(a)   (b)

1 Housing   2 Filter   3 Valve   4 Valve seat

Fig. 10.6 Injector for the Bosch K-Jetronic system: (a) valve closed; (b) valve open

10.5   Some other injectors

The lightweight valves in the injectors for the Bosch K- and KE-Jetronic systems, Fig. 10.6, are opened by relatively low fuel pressures (3.3 to 3.6 bar). In fact, they vibrate axially at a rate of about 1500 Hz, to atomise the fuel as it is injected continuously into the manifold. When the engine, and therefore the electric pump, is switched off, the valve is seated by its light spring. As the valve opens, the area exposed to the fuel pressure suddenly increases, allowing the fuel to gush through. The consequent equally sudden release of pressure seats the valve again. This sequence is repeated cyclically.

The layouts of the injectors for the Bosch L-, LH-Jetronic and Motronic systems are all broadly similar, Fig. 10.7. Because their pintle-type valves and

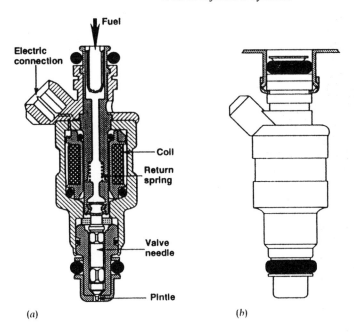

Fig. 10.7 L-Jetronic injector Mark I: (*a*) cross-section (*b*) showing how the upper end is inserted into a sleeve on the fuel rail

armature assemblies, with their return springs, are inevitably heavier than the valves in the K- and KE-Jetronic systems, their solenoids, which open them for timed periods, are energised by electric pulses triggered by the electronic control unit. The valve lift is about 0.1 mm and the release time is between 1 and 1.5 ms.

Various end fittings can be supplied, though usually the nose is a push fit in the boss in the manifold or port, sealing being effected by the elastomeric ring shown in the illustration. The ring at the upper end seals within a thimble-shaped connector integral with the fuel rail that supplies all the injectors for the cylinder head. These rings are large enough to help isolate the injector from vibrations and heat, either of which could cause bubbles to form locally in the fuel and thus adversely affect metering. Hexagons for receiving a spanner are machined near the upper ends of the bodies so that, if a screw-type end fitting is used, rotation of the injector can be prevented when it is being tightened.

In principle, the Weber injector, illustrated in Fig. 10.8, is similar to those just described. However, it has a tubular instead of solid armature. This would appear to offer greater stability and better guidance though possibly, makes it heavier and introduces more friction.

A shortcoming of all these injectors is that, when the quantities of fuel delivered are extremely small and the velocity of air flow low, especially during idling and at light load, atomisation tends to be very poor, Fig. 10.9. This can be overcome by using what is termed *air shrouded injection*, Sections 10.35 and 10.33. Under the conditions just mentioned, the pressure in the manifold is very low and therefore can be utilised effectively for drawing air through a duct, possibly cored in the manifold casting, and passing it out around the jet

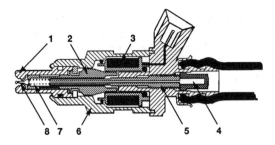

1 Nozzle   5 Threaded adjuster for spring compression
2 Armature   6 Body
3 Winding   7 Spring
4 Filter   8 Needle

Fig. 10.8 The Weber injector

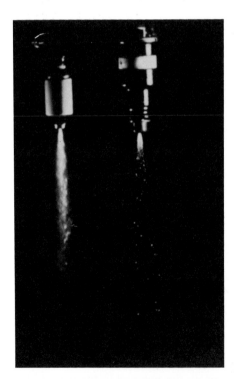

Fig. 10.9 Right, conventional injector; left, Pijet air-shrouded injector operating in the conditions obtaining at light load (see Sections 10.29 to 10.32). Flow rate $25\,\text{cm}^3/\text{min}$; induction manifold depression $550\,\text{mmHg}$

of fuel to help to vaporise and atomise it as it is delivered into the manifold or induction port. Among the earliest examples is that employed in the Piper Pijet system developed by Mechandyne Ltd for a military application in which low fuel consumption was a major consideration. Such systems, though not yet widely used, appear to be attractive as a means of reducing both emissions levels and fuel consumption under the relevant conditions.

10.6 Start valves

Many systems have start valves, through which the extra fuel needed for cold starting is delivered into the manifold. This is a form of injector, though of less sophisticated design. With electronically controlled systems, however, it is more satisfactory to signal the injectors to supply the extra fuel needed. A Bosch start valve, for the KE-Jetronic system, is illustrated in Fig. 10.10.

Fuel entering at the top passes through a filter and on through the hole in the end of the thimble-shaped valve, when it is lifted by the solenoid. Delivery into the manifold is effected through a swirl nozzle. When the solenoid is de-energised, the valve is closed down by its return spring on to its seat.

10.7 Air-flow metering

We now turn to the methods of measuring the air flow. Although the flow is controlled by the throttle, we need to be able to measure its mass, so that we can meter the appropriate mass of fuel into it. In some applications, the air flow is measured simply by using a sensor detecting throttle angle, which signals

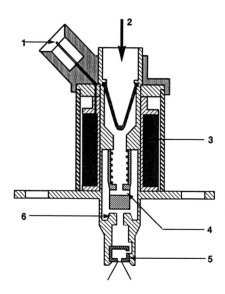

1	Electrical connection	
2	Fuel supply	
3	Solenoid winding	

4	Armature with valve seating face on its lower end	
5	Swirl nozzle	
6	Valve seat	

Fig. 10.10 Bosch KE-Jetronic cold start valve

volume flow to the electronic control unit (ECU). These signals, in association with others from manifold temperature and pressure sensors, are the basis on which the ECU calculates the mass air flow into the engine. These are termed *volume–density* computations. An alternative is to use *speed– density* computations, which are based on signals from engine speed and air temperature and pressure sensors.

A more accurate method, however, is to use an air-flow meter which, because its reading is a function of the degree of opening of the throttle, is in effect a load meter. Three types of air-flow sensor are employed: two meter the volume and the third mass flow of the air. Where volume is metered, the ECU has to convert it into mass, on the basis of its temperature and pressure, as indicated by signals from sensors in the induction manifold.

10.8   Suspended-plate-type flow sensor

For the Bosch K-Jetronic system the suspended-plate-type flow meter, Fig. 10.11, was developed. Essentially, it comprises a circular plate on one end of an arm swinging about a pivot at its opposite end. The plate is suspended in a horizontal plane in a circular throat of complex tapered form. This plate-and-lever assembly is balanced by a small weight, sited beyond the pivot, at the end of the lever opposite to the plate. Consequently, when the engine is switched off, the plate settles down in its equilibrium position, which is in the narrowest section of the throat.

Air entering from below pushes the plate up against the resistance offered by a hydraulically actuated *control plunger*, tending to push it down. This plunger bears down on a roller carried by a much smaller lever above and approximately parallel to the main lever. The roller rotates on a pin near the pivot common to both levers. At the other end of the smaller lever is a screw stop by means of which the idling setting of the plate can be adjusted. The plunger is part of the system metering fuel in proportion to the upwards deflection of the plate. From it the fuel is delivered to the continuous flow injectors. This system is described in detail in Sections 10.12 and 10.13.

To release the considerable pressure occurring in the event of a backfire, the lever is free to swing down, beyond the narrowest portion of the throat, onto a rubber stop, after which it is gently returned by a leaf spring. If the throat were of simple conical form, the air : fuel ratio would be constant over the whole of the range of deflection of the plate. The more obtuse the cone angle the leaner would be the mixture, and *vice versa*. In fact, the angle is changed in three stages, to vary the mixture from rich for idling, to a lean for cruising, and back to rich again for maximum power output. When the throttle is opened suddenly for acceleration, the plate over-swings momentarily, increasing the rate of supply of mixture, and then returns to its equilibrium position. Owing to inertia, the rate of flow of air into the engine does not increase rapidly enough to maintain a constant air : fuel ratio: in other words, enrichment for acceleration is automatic. This of course was adequate when exhaust emissions regulations were much less stringent than they are today: now systems that are electronically rather than hydraulically controlled are widely employed.

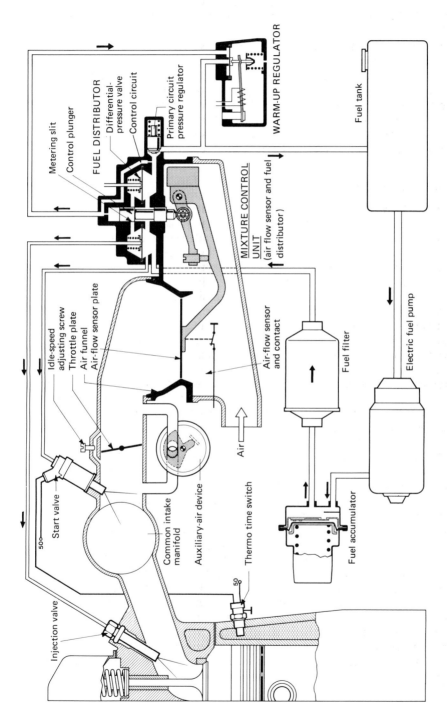

Fig. 10.11 Diagrammatic representation by Bosch of their K-Jetronic petrol injection system

10.9   Swinging-gate-type air flow sensor

This type of air sensor, Fig. 10.12, is especially convenient for use in conjunction with electronic control, which is why it forms part of the Bosch L-Jetronic and Motronic systems. It has two other advantages: first, the rotational inertia of the gate is less than the linear inertia of the suspended-plate-type sensor, so its response to changes in throttle opening is slightly faster; and, secondly, there is still a momentary lag between the opening of the gate and the inflow of air into the manifold so, as in the case of the suspended-plate type, a degree of automatic compensation for the inherent weakening of the mixture during acceleration is available.

Incoming air swings the gate open against the torque exerted by a coil spring around its spindle. As can be seen from the illustration, it has two leaves, or flaps, the second is set at slightly more than 90° to the first. Because the two opposed faces of the leaves are subject to the same pressure, pulsations in the air flow generate equal and opposite forces on the gate assembly, which is therefore unmoved by them. The profile of the passage in which the gate swings is such that there is a logarithmic relationship between its angle and the volume of air passing through. Consequently, at small throttle openings, when small errors of measurement would represent a large proportion of the total flow, a relatively high degree of accuracy is in fact obtained.

A potentiometer senses the angle of deflection of the gate and converts it into a voltage inversely proportional to the air throughput. The potentiometer brushes are of fine wire. By virtue of the multiplicity of wires, the rate of wear of the track is low and electrical contact is continuous, even though the surfaces of the tracks may not be absolutely flat.

The idling air:fuel ratio can be adjusted by a screw in the channel, by-passing a small quantity of air around the gate. Similarly, the idling speed is set by another screw, in the throttle by-pass. A large diameter non-return valve

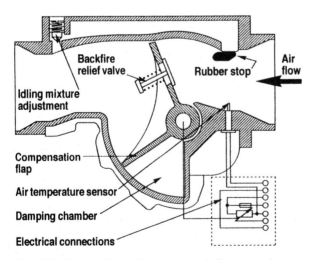

Fig. 10.12 The Bosch swinging-gate-type air flow sensor has a two flaps, one a sensor and the other a compensator, or damping, flap. The backfire pressure relief valve is optional

in the metering leaf of the gate relieves blow-back pressure in the event of a backfire.

10.10   Mass-flow sensors

Clearly, it is better to measure the mass flow directly than to measure volume flow and density separately and compute the mass flow from these two parameters. Mass air flow meters operate on the basis that the temperature loss from the heated element is a function of the velocity and density of the fluid flowing past it. The ECU calculates mass flow from the diameter of the passage, the velocity of flow and its density, the last two factors being indicated by a voltage in the electric circuit. There are two types of sensor: hot wire and hot film. In each, the electrically heated resistor (wire or film) exposed to the air flow is maintained at a constant temperature by the control circuit, which varies the current appropriately. The associated varying voltage, which is signalled to the ECU, is a function of the mass flow.

The hot-wire type is the simplest, but deposits accumulate on the wires and, to avoid inaccurate readings, need to be cleaned off. Cleaning is done by momentarily raising the temperature of the wire to a high level, each time the engine is switched off.

Hot-film elements, on the other hand, are on ceramic plates aligned parallel to the air flow. The edges of these plates are shaped to shed the deposits, which therefore do not interfere with the air flow past the sensing elements mounted downstream of them. Consequently, cleaning is unnecessary.

The control circuit for each is a Wheatstone bridge in which there are two sensors: one for the mass air flow and the other to compensate for variations in the temperature of the incoming air. That for the hot-film type is mounted together with the film on the ceramic plate. Good examples of both types are those produced by Bosch: the hot-wire type and its control circuit are illustrated diagrammatically in Fig. 10.13, and those for the hot-film type are shown in Fig. 10.14.

10.11   Lambda sensor

This is another sensor of major importance developed by Bosch for the internal combustion engine, and first used by Volvo in 1976. Its name is derived from the fact that the Greek letter lambda, λ, is used as the standard symbol for the air:fuel ratio, which is the ratio of the actual mass of fuel supplied to that required for complete combustion of all the air present in the mixture. The function of this sensor, which is screwed into the exhaust manifold, is to detect departures of the air ratio from $\lambda = 1$, by assessing the oxygen content of the exhaust gas.

From Fig. 10.15, it can be seen that the oxygen-sensitive component is a thimble-shaped piece of zirconium oxide, with its inner and outer surfaces coated with a thin layer of platinum in a manner such that it is permeable to gas. To protect the outer surface of the end of the thimble, which is exposed to the hot exhaust gas, a layer of porous ceramic is applied over it.

The thimble behaves like an electric cell in that, when the concentration of oxygen inside differs from that outside, an electric potential develops between the inner and outer platinum coatings. Thus, the voltage between the outer

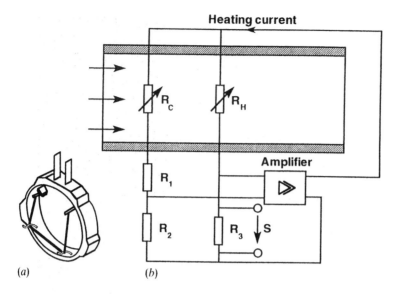

(a)   (b)

R_H	Hot wire	R_1, R_2	Impedance resistors
R_C	Temperature compensation sensor	R_3	Precision resistor
		S	Signal voltage representing air flow rate

Fig. 10.13 Diagrammatic representation of the hot-wire installation of the mass-flow sensor at (a) and its bridge circuit at (b)

coating, which is earthed, and the inner one, from which a connection is taken, is a measure of the difference between the two oxygen concentrations. The inside of the thimble is of course open to the atmosphere.

Even when excess fuel is supplied, there is still some oxygen in the exhaust. For example, at $\lambda = 0.95$ the unburnt oxygen is between 0.2 and 0.3%, by volume, of the exhaust gas. By careful development of the materials used, Bosch has been able to make a probe that is particularly sensitive to changes in oxygen content over a small range about $\lambda = 1$, Fig. 10.16. Because of this high sensitivity, the voltage changes can be used as signals and passed directly to the electronic control, so the probe becomes part of a closed-loop control system. This means that if the injectors are not supplying the amount of fuel required to form the correct air : fuel ratio, the sensor in the exhaust signals this back to the control unit, which instantly changes the duration of injection appropriately to correct the mixture strength. Such a device is especially valuable where catalytic systems are used to purify the exhaust gases, since these systems become overloaded and deteriorate rapidly if the air : fuel ratio is not correct.

A disadvantage of this type of sensor is that it does not function effectively until the exhaust gas temperature rises to about 350°C. This, however, has been largely overcome by the introduction, by Bosch, of a lambda sensor with an electrically heated ceramic rod extending coaxially into the platinum-coated thimble. Thus, reliable control is obtained from 200°C upwards. Consequently, the delay between starting the engine from cold and the establishment of closed-loop control is reduced, and reliable control is

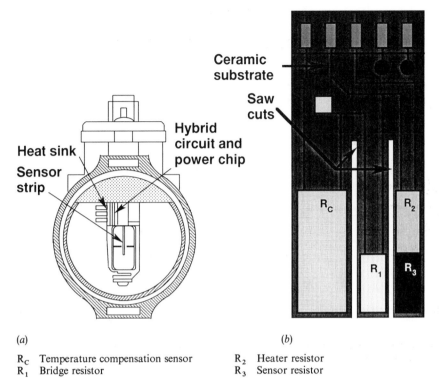

Ceramic
substrate

Saw
cuts

**Hybrid
circuit and
power chip**

Heat sink

**Sensor
strip**

R_C

R_2

R_1

R_3

(a)

(b)

R_C	Temperature compensation sensor	R_2	Heater resistor
R_1	Bridge resistor	R_3	Sensor resistor

Fig. 10.14 Diagrammatic representation of the Bosch hot-film mass air-flow sensor

exercised when the exhaust gas temperature is low, for example during idling. Moreover, the positioning of the sensor in the exhaust is less critical. An alternative approach is to install an electronically controlled and timed, fuel-fired heater in the exhaust manifold, as has been done by Ford.

10.12 Bosch K-Jetronic system

In detail, the operation of the K system, Fig. 10.11, is as follows: fuel is drawn from the tank by the electrically-driven pump and delivered into what is termed the *primary circuit*, at a pressure of 4.7 bar. It passes first into a hydraulic accumulator, the function of which is to maintain the pressure in the system for a long time after the engine has been switched off. This is to avoid vapour formation in the system while the engine is still warm, which would tend to inhibit hot starting.

From the accumulator, the fuel goes through a filter to the mixture control unit which houses, among other things, the primary circuit pressure regulator valve. Fuel in excess of requirements is dumped by this valve straight back to the tank. An extension of the primary circuit goes up to the cold-start valve, which sprays extra fuel into the manifold only during cold starting, Fig. 10.10. Operation of this valve is controlled electrically by a thermo-time switch screwed into the water jacket of the engine or, in an air-cooled engine, into the

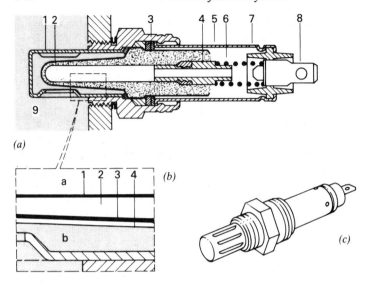

(a) *Cross-section of the probe*
1 Protective tube
2 Ceramic body
3 Housing
4 Contact bushing
5 Protective sleeve
6 Contact spring
7 Ventilating opening
8 Electrical connection
9 Exhaust gas

Fig. 10.15 The Lambda probe

(b) *Section from* (a)
a Air side
b Exhaust gas side
1 Electrically-conductive layer
2 Ceramic body
3 Electrically-conductive layer
4 Porous ceramic layer

(c) *The complete unit*

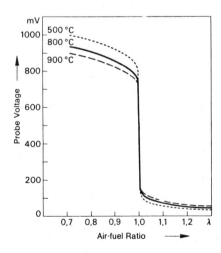

Fig. 10.16 Operating characteristic of the Lambda probe

cylinder head. The solenoid actuating the valve is energised to open it when the engine is cold. It is closed by a return spring as the engine becomes warm enough for the sensor to signal de-energisation of the solenoid. A swirl device in the nozzle helps to atomise the fuel passing through. Should the engine fail to start within a predetermined time, the thermo-time switch – actuated by either heat or, in this situation, by an electrical heating element around the bimetallic strip – interrupts the circuit to the solenoid to avoid flooding the engine cylinders with petrol.

The mixture control unit is of course the key component of the system. Essentially, it compromises an air-flow sensor, similar to that in Fig. 10.20, and a fuel metering system. A disc mounted on the end of an arm extending out from a pivot in the mixture control unit is suspended by that arm in a funnel-shaped air intake. If the air flow is zero, the disc falls into the narrowest section of the funnel. When the engine is started, the air pressure differential across the disc lifts it until the air flow through the annular gap between its periphery and the walls of the funnel has increased sufficiently to reduce the pressure differential to a level at which it exactly balances the force tending to pull the disc downwards. This force is independent of gravity, because the masses of the arm and plate assembly are accurately balanced about its pivot, by means of counterweights. The force tending to lower the disc is therefore solely that applied to the arm by the control plunger, Fig. 10.19, the upper end of which is acted upon by fuel pressure in what is termed the *control circuit*.

This circuit goes, through a restrictor, from the primary circuit and thence, through another restrictor, into the upper end of the cylinder in which the control plunger slides. A connection is taken from a point between these two restrictors to the warm-up regulator, which is a valve that maintains the control pressure at a constant 3.7 bar once the engine is warm. During warm-up, however, this regulator reduces the pressure to as low as 0.5 bar, so that the control plunger can rise higher and deliver extra fuel to the injectors, and thus increase the power, by compensating for the loss due to condensation from the mixture onto the cold walls of the combustion chambers and induction passages.

The function of the first-mentioned restrictor, therefore, is to prevent the passage of too much fuel from the primary circuit through the control circuit and thence back to the tank by way of the warm-up regulator valve. That of the second restrictor – at the entrance to the upper end of the plunger cylinder – is to damp the motion of the plunger under conditions of pulsating air flow, for instance at low engine speeds and high load. Under such conditions, the air flow sensor disc would tend to move too far from its base position. This restrictor also damps the transient motion of the plunger during sudden acceleration when, again, the air sensor disc would tend to swing too far upwards.

A bimetallic strip controls the setting of the warm-up regulator valve, Fig. 10.17. When hot, its end lifts, allowing the return spring to close the valve to the point at which it maintains the pressure at 3.7 bar. On the other hand, when cold the bimetallic strip acts against the return spring, thus lowering the pressure in the control circuit. The unit is mounted on the engine cylinder block so that, once the engine is warm, the valve remains under the sole control of its spring. Because of the time lag between the warming of the engine to the point at which no enrichment is necessary and the heat soaking right through

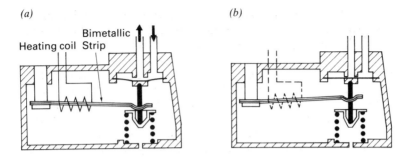

Fig. 10.17 Warm-up regulator, (a) with cold engine, (b) with warm engine

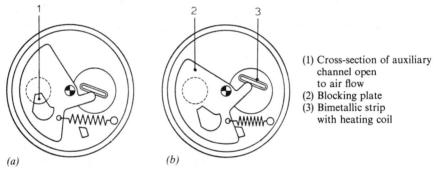

(1) Cross-section of auxiliary
    channel open
    to air flow
(2) Blocking plate
(3) Bimetallic strip
    with heating coil

Fig. 10.18 Auxiliary air device, positions of blocking plate. (a) Partly open, (b) closed

to the warm-up regulator unit, the bimetallic strip is also heated electrically by means of a coil, which is switched on automatically as soon as the engine is started, and switched off again automatically when the engine temperature is high enough to lift the end of the bimetallic strip clear of the return spring.

Compensation is needed also for power lost owing to the high viscosity of the oil when the engine is cold. This is provided by supplying extra air to the engine, by-passing the throttle valve. The auxiliary air device is a rotary plate valve in the by-pass passage. This valve is opened by a bimetallic strip and closed by a return spring, Fig. 10.18. At normal operating temperatures it is closed. When the engine is cold, however, the end of the bimetallic strip, bearing against a lever integral with the rotary plate valve, opens it. As in the case of the warm-up regulator, this valve unit is mounted in a position such that the bimetallic strip is subject to engine temperature, and it is also electrically heated automatically as soon as the engine is started. When the engine is warm, and the bimetallic strip lifts clear of the end of the lever, the current to the heating coil is cut off.

10.13 The fuel distributor

When the fuel enters the mixture control unit, at the 4.7-bar pressure of the primary circuit, it passes into a series of chambers, which are each divided into an upper and lower portion by a metal diaphragm. There is one of these chambers for each cylinder of the engine. Their lower portions are all connected to an annular groove around the control plunger, so all are at the same pressure. When the plunger is lifted by the air-flow sensor lever, fuel passes from this groove radially outwards through vertical metering slits into the upper portions of the chambers and thence out through pipelines to the injectors.

The metering of the fuel to each of the injectors is accomplished by controlling the length of slit uncovered by the plunger and maintaining the pressure difference across the metering slit constant. Obviously the length of slit uncovered is dependent on the height to which the control valve is lifted, which in turn depends upon the air flow past the sensor plate. The pressure difference is determined by the position of the diaphragms separating the top and bottom halves of the chambers relative to the ends of the discharge ports through which the fuel passes into the pipelines to the injectors, Fig. 10.19. A coil spring pushes this diaphragm downwards against the opposing force exerted on its underside by the 4.7-bar pressure in the primary circuit. The compression in this spring is such that the diaphragm is maintained in equilibrium with a pressure differential of 0.1 bar between the upper and lower chambers. If the control plunger is lifted to allow more fuel to flow through the metering slit into the upper chamber, the pressure differential tends to be reduced, allowing the spring to push the diaphragm down. This uncovers the discharge port and allows fuel to flow at a greater rate through the pipeline to the injector. However, the equilibrium, at a differential pressure of 0.1 bar – 4.6 bar in the upper chamber – is still maintained since a higher pressure would open the port wider, allowing the fuel to flow through at an even higher rate, and a lower one would close it. The deflection of the diaphragm is in fact only a few hundredths of a millimetre.

From this it can be seen that the injection valve has no metering function. It is closed by a spring and opens automatically when the pressure in the delivery pipe rises above 3.3 bar. At this pressure the fuel is finely atomised as it passes through the discharge nozzle into the engine inlet valve port.

The only adjustments that can be made to this system in service are those of the engine idling speed and mixture. For adjustment of idling speed, there is a screw that restricts the flow of air through a passage that by-passes the throttle valve when it is closed. The greater the degree of restriction of course, the slower is the idling speed. To increase idling speed, therefore, the screw should be turned anti-clockwise.

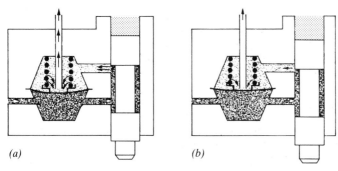

(a)   *(b)*

Fig. 10.19 Metering slit and diaphragm. (a) At high rate of fuel flow, (b) at low rate of fuel flow. The control plunger is that on the right in each diaphragm

The idling mixture strength is adjusted by another screw, which acts on the arm by means of which the motion of the air sensor plate is transmitted to the control plunger. Access can be gained for this adjustment, using a screwdriver, without any dismantling of the mixture-control unit. As can be seen from Fig. 10.11, the idling mixture screw is on the end of a lever swinging about the same pivot as, and approximately parallel to, the arm that carries the air sensor plate. The end of the screw seats on that arm so, when it is screwed clockwise, it increases the angle subtended between the smaller lever and the arm. This raises the control plunger slightly, thus supplying more fuel to the injectors and therefore enriching the mixture.

10.14 Bosch KE-Jetronic system

The KE-Jetronic system, Fig. 10.20, is similar to the K-Jetronic except that it has a simple diaphragm-type fuel pressure regulator, in place of the warm-up regulator, to maintain a constant primary pressure above the control plunger in the fuel distributor, Fig. 10.17. Another change is the flange-mounting of an electro-hydraulic pressure actuator on the fuel distributor, to regulate the pressure in the supply to the lower chambers of the differential pressure valves. Thus, whereas in the K-Jetronic the mixture enrichment is effected by regulating the pressure above the control plunger, this function is performed in

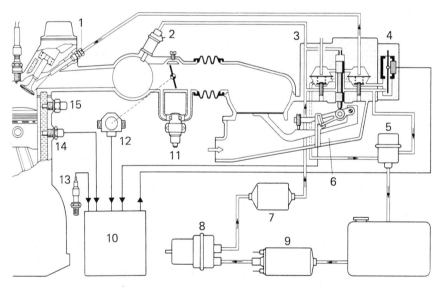

1	Injector	9	Electric fuel pump
2	Cold start injector	10	Electronic control unit
3	Fuel distributor	11	Idle by-pass valve actuator
4	Electro-hydraulic pressure actuator	12	Throttle position switch
5	Fuel pressure regulator	13	Lambda sensor
6	Air-flow meter	14	Engine temperature sensor
7	Filter	15	Thermo time switch
8	Fuel accumulator		

Fig. 10.20 Bosch KE-Jetronic system

the KE by regulating the pressure input to the lower chambers and, moreover, not only for cold starting and warm-up but also for all other situations.

Yet another addition is a potentiometer on the air-flow sensor lever actuating the control plunger. Its function is to signal to the electronic control the rate of air flow into the engine. Additional input signals to the electronic control include engine temperature, engine speed (from the ignition system), idle, overrun and full-throttle signals (from the throttle position switch), exhaust content (from the lambda sensor), atmospheric pressure, and an engine starting signal from the ignition switch. The output from the electronic control goes to the electro-hydraulic pressure actuator.

Fuel is drawn from the tank by the pump and delivered through the hydraulic accumulator to the filter and thence to the electro-hydraulic pressure actuator, in which it is directed through a nozzle on to a plate. The clearance between the mouth of the nozzle and the plate is varied by an axial force exerted by an electro-magnet on a pole-piece on its centre. This clearance is therefore determined by the magnitude of the electric current passing through the windings of the electro-magnet, which in turn is regulated by the electronic control. During overrun, it can totally cut off the supply of fuel.

10.15   Bosch L-Jetronic system

In the L-Jetronic system, Fig. 10.21, the electronic control unit performs the same function as the mixture control unit of the K system. It does this, however, by controlling the duration of opening of the solenoid-actuated valves in the injectors. The advantage of electronic control is that there are fewer mechanical components liable to wear or to stick and thus to malfunction. Moreover, ultimately, more accurate control is possible because the system can be made more easily to respond to a wider range of variables than when a mechanical system is used.

Because the injectors are solenoid actuated, Fig. 10.22, lower delivery pressures are possible than those needed to open the pressure-actuated delivery valves of the K system. Another advantage is that delivery can be made through all the valves simultaneously, which means that the injection system can be simpler than if each had to be opened individually. Actually, to ensure that the distribution of fuel is uniform, to all cylinders, half the required amount of fuel is injected into each port twice, over two separate intervals, during each four-stroke cycle – that is, for each 360° rotation of the camshaft, Fig. 10.23.

The start of each injection pulse is signalled to the electronic control unit by the contacts in the ignition distributor. However, the control unit has to respond to only every second signal from the contact breaker in a four-cylinder engine and every third in a six-cylinder unit – since they have four and six sparks per cycle whereas only two injections are required in each case.

Although the basic signal determining the duration of injection is that from the swinging gate type air flow sensor, Section 10.9, it has to be modified by a number of other signals received by the electronic control. One is engine speed, which is signalled by the frequency of operation of the ignition contact breaker. A throttle valve-actuated switch indicates whether enrichment is needed, for either full load or idling. As in the K system, there is a temperature sensor, but it influences the duration of injection instead of the pressure. It is

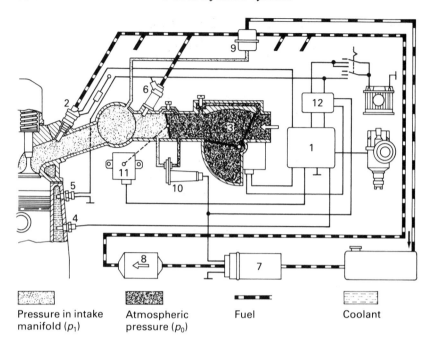

Pressure in intake manifold (p_1) Atmospheric pressure (p_0) Fuel Coolant

1 Electronic control unit	5 Thermo-time switch	9 Pressure regulator valve	
2 Injection valve	6 Start valve	10 Auxiliary-air device	
3 Air-flow sensor	7 Electric pump	11 Throttle-valve switch	
4 Temperature sensor	8 Fuel filter	12 Relay set	

Fig. 10.21 Bosch L-Jetronic system

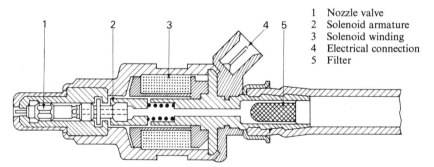

1 Nozzle valve
2 Solenoid armature
3 Solenoid winding
4 Electrical connection
5 Filter

Fig. 10.22 Cross-section of the injection valve

necessary because the density and therefore mass of air drawn into the cylinders is greater in cold than in hot conditions. An enrichment device for acceleration is unnecessary in either system since the air flow sensor gives its signal in advance of acceleration.

A solenoid-actuated start valve comes into operation for cold starting. This is the same as in the K system, as also is the auxiliary air device for by-passing the throttle valve to compensate for the high friction losses. Here, however, there is a relay set. When the ignition is switched on, this relay set switches the

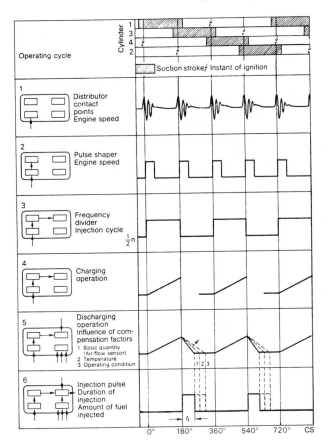

Fig. 10.23   Pulse diagram

battery voltage to the electric fuel pump, start valve, thermo-time switch for switching off the start valve, and auxiliary air device. When the engine starts, the power supply for the pump and auxiliary air device is maintained through a contact actuated by the air sensor. If, on the other hand, the engine fails to start, a thermo-time switch interrupts the circuit to the solenoid-actuated start valve, to avoid flooding the cylinders.

With no control-valve unit, the fuel supply is simpler than before: it passes from the pump through a filter to a pressure-regulating valve and thence directly to the injector and start valve. With such a simple and direct supply a fuel accumulator is unnecessary.

As can be seen from Fig. 10.25, the fuel pressure regulator valve is of conventional design. The fuel flows radially in one side and out the other, while fuel in excess of requirements passes out through the connection at the top of the unit and thence back to the tank. With the L-Jetronic system, the fuel delivery pressure is either 2.5 or 3 bar, according to the type of engine. It is maintained at its set value by a spring-loaded diaphragm, which causes a valve to tend to seat on the port for the return line to the tank. Any increase in pressure pushes the diaphragm down and opens this port. To avoid variations

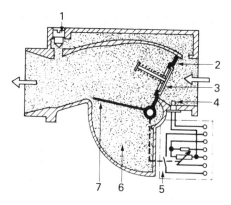

1 Mixture adjustment screw
   for the idle range
2 Air-flow sensor flap
3 Non-return valve
4 Air-temperature sensor
5 Electrical connections
6 Damping chamber
7 Compensation flap

Fig. 10.24 Cross-section of the air-flow sensor

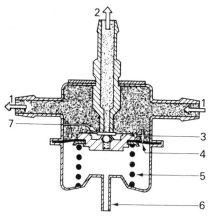

1 Fuel through-flow
2 Return line to fuel tank
3 Valve support
4 Diaphragm
5 Pressure spring
6 Connection to intake manifold
7 Valve

Fig. 10.25 Cross-section of the fuel-pressure regulator

in the back-pressure on the nozzles due to changing induction manifold pressure, a pipeline connection is taken from the manifold to the chamber below the diaphragm.

Bosch also produce a petrol injection system in which the engine load signal is obtained by sensing the depression in the inlet manifold. This is the D-Jetronic system. It will not be described here, however, since the L system is the more advanced one and therefore of greater importance. In any case, wear of the engine causes manifold depression characteristics to change.

10.16   Bosch LH-Jetronic system

All components of the LH are virtually identical to those of the L-Jetronic system, except for the electronic control unit and the substitution of a hot-wire, air-mass-flow meter (Section 10.10) for the volume-flow meter. A major advance made more easily feasible by the use of the hot-wire system is the use of a digital instead of analogue electronic control system, the former being potentially much more flexible. Other advantages of the LH system are negligible resistance to air flow and the absence of moving parts.

The incoming air flows past an electrically heated platinum wire, the temperature of which is maintained constant by using the wire as one arm of a Wheatstone bridge circuit and varying a resistance to balance the bridge. Since the quantity of heat removed from the wire is a function of not only the velocity but also the density of the air flowing past it, the increase in current needed to maintain a constant temperature is a measure of the mass flow of the air. A voltage signal taken from across a resistance through which this current is passed is transmitted to the control unit. Among the other advantages is that the hot wire compensates automatically for changes in altitude.

Other signals required include engine speed, throttle position, and ambient air temperature. An engine load–speed map is stored in the memory of the electronic control system which, taking into account all the other input signals, can in all circumstances accurately regulate the air:fuel ratio for optimum power output, fuel consumption, and exhaust emissions.

During idling the air mass flow is small so, in this condition, the air:fuel ratio is set by a potentiometer. Another factor is that the surface of the wire can become contaminated when it is cold and when the engine is idling. Therefore, to cleanse it and avoid subsequent inaccuracy of the output signals the wire is automatically heated to a high temperature for one minute every time the engine is switched off.

10.17   Bosch Motronic system

Given that a microcomputer is used for regulating fuel injection, it can be even more cost-effective to employ it for other control functions too. Primarily the Bosch Mono-Jetronic system integrates injection and ignition controls, but it can also be adapted for controlling other parameters such as exhaust gas recirculation and evaporative emission canister purging, which are explained in Sections 12.16 and 12.17. Comprehensive information on the electronic components of the Motronic and other injection systems is given in the Bosch publication *Automotive Electric/Electronic Systems*, and in their yellow book entitled *Motronic Engine Management*.

Its intermittent injection system is basically identical to that of the L-Jetronic, but all its signal-processing functions are done digitally. Among the advantages of digital systems is that all the data processing can be done directly by the computer, and much of the electronics can be common to a wide range of applications. Also, operating data can be stored on maps, which can be updated automatically by the computer to take into account changes such as might occur as a result of, for example, wear of the engine in service.

10.18   The electronic ignition control

Control of the ignition system is based on a spark advance characteristic map stored in the memory of the Motronic control unit. The spark advance is continuously changed to correspond to the setting on the map, taking into account throttle position, and engine-coolant and air-intake temperatures. When a spark is required, the electronic controller momentarily opens the circuit to earth, whereupon the collapse of the field around the primary generates the spark voltage in the secondary coil. The resultant high-voltage current is passed through the distributor to the sparking plug. Generally, there

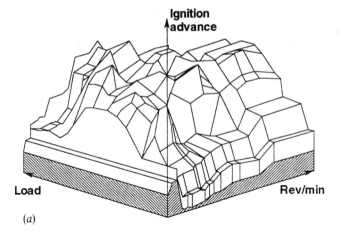

(a)

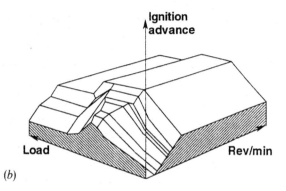

(b)

Fig. 10.26 At (a) is a three-dimensional map showing the degree of accuracy with which the ignition timing can be controlled electronically as compared with, at (b), the best that can be obtained with a mechanical and vacuum advance and retard mechanism

is neither a mechanical contact breaker nor a centrifugal and pneumatic advance and retard system, though some high-speed six-cylinder engines retain the centrifugal mechanism.

Obviously, a conventional mechanically actuated system could not vary the ignition advance to satisfy the complex requirements that are registered on the map, Fig. 10.26, and stored in the memory of the ECU. To obtain the data points on the map, the engine is run on a dynamometer, the ignition advance being optimised in respect of fuel consumption, emissions and driveability. The data thus obtained are recorded electronically and transferred into the memory of the ECU. By virtue of digital recording, the ignition point for each condition of operation can be set independently of all the others.

In operation, the microcomputer first reads from the map the point at which, on the basis of the instantaneous engine speed and load, the next spark should be triggered and then modifies it in relation to throttle position and coolant and air temperatures. An inductive engine speed sensor signals directly from the crankshaft. This is more precise than using a Hall-effect sender in the distributor. Consequently, the spark advance can be optimised while avoiding all risk of detonation, and both fuel utilisation and torque are therefore improved.

Actually, there are two inductive pulse senders on the flywheel. One senses the passage of teeth past its permanent magnet core, for translation into engine speed. The other, for indicating crank angle, senses the passage of a either a pin or hole in the flywheel. These two signals are processed in the control unit to make them compatible with the computer.

The parameters on the basis of which the ignition points are set include fuel consumption, torque, exhaust emissions, tendency to knock and driveability, the weighting given to each differing according to the type of operation. For example, for idling, the priorities are low emissions, smoothness and fuel economy; for part-load operation, they are driveability and economy; and for full-load they are maximum torque and absence of detonation.

For all types of operation other than part-throttle light load, correction factors are applied to the map values. Also included in the control unit is a switch which is actuated automatically during operation in the high-load range, to cater for different fuels and grades of fuels. For starting, there is even a correction routine for adjusting spark timing in relation to cranking speed.

After the generation of each spark, a finite time is required for the re-establishment of the current in the coil to its nominal value, ready for the next firing. The higher the engine speed, and therefore frequency of sparking, the longer is the dwell time needed to allow the current to build up in the coil. Consequently, the relationship between current flow time in the coil and supply voltage has to be regulated, by reference to a dwell angle characteristic map similar to that in Fig. 10.27. As soon as the current has risen to the appropriate level ready for the next ignition point, it is held there by the output stage so that, as the dwell time shortens during acceleration from low engine speeds, the appropriate current can be maintained throughout.

An indication of how the electronic control unit regulates injection and ignition simultaneously can be gleaned from Fig. 10.28. To keep the break-away and release times of the injection valves as short as possible without using current-limiting resistors, the current to them is limited by a special integrated circuit in the electronic control unit. For a six-cylinder engine, for

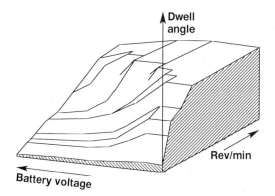

Fig. 10.27 Three-dimensional plot showing how the dwell angle has to be varied relative to the supply voltage and engine speed, to allow the current enough time to build up in the primary winding

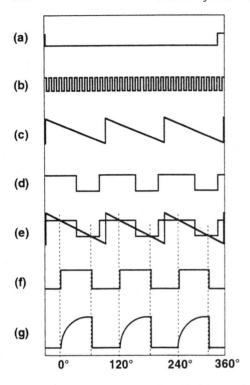

Fig. 10.28 Stages in the production of ignition sparks for a six-cylinder engine, by means of an electronic control such as that in the Bosch Motronic system. (*a*) The reference signal for the crankshaft angle; (*b*) an indication of the degrees of rotation following the occurrence of the pulse signal; (*c*) the saw-tooth-shaped signal of the angle counter; (*d*) characteristic of the instantaneous operating condition, as calculated from the ignition and dwell angle signals and entered in the intermediate memory; (*e*) when the values of the signals from the counter and the intermediate memory are identical, signals are sent to the ignition output stage to switch the ignition coil on or off; (*f*) low-tension signal for ignition; (*g*) current through the coil

instance, the valve opening current is 7.5 amp and, at the end of the injection period, reduced to a holding current of 3.5 amp.

10.19 Fuel supply

As can be seen from Fig. 10.29, the roller-cell type pump delivers fuel at a pressure of 2.5 to 3 bar, through a filter removing particles down to 10 μm, directly to one end of a fuel rail. At the other end is the pressure regulator, Fig.10.30, from which the return flow to the tank passes through a pulsation damper, Fig. 10.31, to the tank. This, by reducing fluctuations in the pressure in the return line, suppresses noises arising from both the operation of the pressure regulator and the injector valves.

By virtue of its large volume relative to the quantities of fuel injected per cycle, the fuel rail acts as a hydraulic damper and ensures that all the injectors connected to it are equally supplied with fuel. Injection occurs once per revolution (twice per cycle) and is directed into the ports.

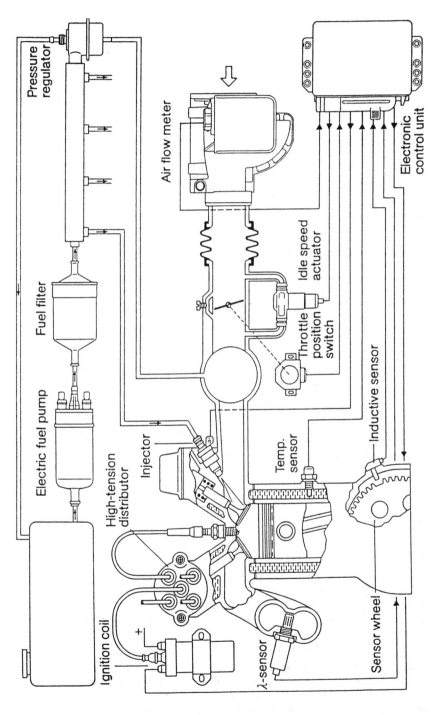

Pressure regulator

Air flow meter

Electronic control unit

Idle speed actuator

Throttle position switch

Fuel filter

Inductive sensor

Electric fuel pump

High-tension distributor

Injector

Temp. sensor

Ignition coil

Sensor wheel

λ-sensor

+

Fig.10.29 Diagram issued by Bosch to represent their Motronic system for a four-cylinder engine, which has a swinging-gate-type air flow sensor

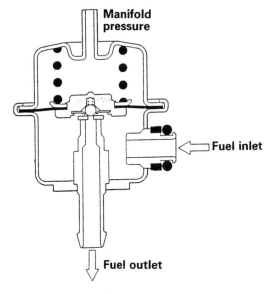

Fig. 10.30 A Bosch diagrammatic representation of the pressure regulator for their Motronic injection and ignition control system

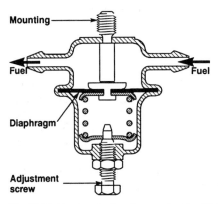

Fig. 10.31 Bosch pulsation damper for the Motronic system

10.20   Overall principle of operation

A swinging-gate-type air flow sensor, described in Section 10.9, is employed in the Motronic system, Figs 10.12 and 10.29. The duration of injection required for maintaining the λ value at 0.85 to 0.95, as needed for engines equipped with three-way catalytic converters, is assessed per piston stroke and in relation to engine speed, instead of per unit of time. Corrections are applied, as required, in response to signals received from detonation, temperature, time and other sensors, and in accordance with plotted values on engine performance maps. The sensors are as previously described for the L- and LH-Jetronic systems, as also are the principles of operation of the ancillary devices such those for

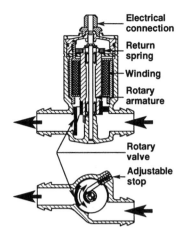

- Electrical connection
- Return spring
- Winding
- Rotary armature
- Rotary valve
- Adjustable stop

Fig. 10.32 When the engine is idling, the Bosch Motronic control system sends signals for this actuator to open a rotary valve to allow extra air to by-pass the throttle and, at the same time, to increase the rate of fuel supply to maintain the appropriate air:fuel ratio needed to cater for low temperature or the switching on of air conditioning or other ancillary loads

regulating air flow through the throttle by-pass. Consequently, it is unnecessary now to present the features other than in the form of a table and footnotes, Table 10.1.

10.21 Other variables

Intake air temperature. A sensor positioned in the air intake, Fig. 10.12, signals to the electronic control unit which, at low temperatures, reduces the fuelling to compensate for the low density, or mass, of charge and, at high temperatures, reduces ignition advance to avoid detonation, especially for turbo-charged engines, in which the higher temperature may not be offset by a lower density of charge.

High altitude. A sensor signalling the reduction in pressure with altitude can be incorporated in the electronic control unit, for reduction of the rate of fuelling with increasing altitude.

Battery voltage. Battery voltage falls off with not only rapidly increasing load but also decreasing temperature and age. Self-induction causes a lag in both the opening and closing of the electro-magnetic injection valves. Breakaway time depends to a significant degree on battery voltage, though little on release time. Therefore, a fall in battery voltage causes a decrease in injection time.

Fuel quality. To cater for premium and low grade fuel, some Motronic electronic control units contain two maps for ignition advance relative to load and speed, and the driver can switch from one to the other. Generally, the program retards the ignition only at high loads

Speed limiting. As in the L-Jetronic system, if required, injection can be inhibited to limit maximum engine speed, the device cuts in and out respectively at speeds of 80 rev/min above and below the required limiting value.

Table 10.1—ADJUSTMENTS FOR VARIOUS OPERATING CONDITIONS

The following symbols are used to indicate the various sensors: TTS, thermal time switch; TVS, throttle valve switch; ET, AT and K, engine and air temperature and knock sensors respectively. Since all control functions call for signals from the air flow, intake temperature and engine speed sensors (reflecting load), these are omitted to avoid repetition.

Operational variables	Fuel supply	Additional sensors	Ignition timing	Additional sensors	Relevant notes
Cold start	Enriched	TTS, TVS	Retarded	TTS, ET	1, 2, 3
Post start	Enriched	TTS, ET	Advanced	TTS	4
Warm-up	Enriched	ET	Advanced	ET	5, 6
Hot idling	Normal	ET	Normal	ET	7, 8
Full load	Enriched	TVS, ET	Controlled	ET, AT, K	9
Acceleration	Enriched	TVS, ET	Controlled	TVS, ET, K	10
Overrun	Cut-off	TVS	Retarded	TVS	11

(1) *Cold start*. Extra fuel delivered either by increasing the duration of opening of the injection valves, or through the cold start valve (not shown in Fig. 10.29) which is in the manifold upstream of the injectors, or both. For most engines, there is no need for a cold start valve; instead, the number of injections per revolution may be increased and, since at very low speeds the quantity of air inducted is constant, the duration of the injections is regulated on the basis of cranking speed, starting temperature and number of revolutions since starting began. At higher cranking speeds throttling occurs, so the duration of injection is reduced.

(2) Rapid variations in speed during starting would lead to inaccurate air flow signals, so a fixed load signal, weighted by engine temperature, is utilised by the control unit.

(3) The lower the cranking speeds and the higher the engine temperature the further must the timing be retarded. In a cold engine, timing earlier than 10° btdc can induce reverse torques, damaging the starter. On the other hand, if the spark is retarded too much with high compression engines, knocking can occur at high intake temperatures. At high cranking speeds, starting is improved if the ignition is advanced.

(4) *Engine firing*. At low temperatures and fast idle speeds, the ignition is advanced to improve both performance and fuel economy. After a short time, it is progressively reduced to normal as the engine temperature rises.

(5) *Warm-up*. Both enrichment and spark advance are progressively reduced as engine temperature rises but, if a cold start valve is incorporated, its cut-off point must be compensated for, by increasing the flow through the injectors. To improve driveability, the spark is further advanced during part load operation. In general, after start-up, the ignition timing is adjusted (on the basis of engine temperature) for idling, overrun, part and full load.

(6) To overcome oil drag, either an auxiliary air device or a thermostatically controlled rotary actuator, Fig. 10.32, by-passes the throttle, and the electronic control supplies the extra fuel needed to maintain the appropriate air : fuel ratio. This control operates on the basis of not only engine temperature but also speed, load and an additional map similar to Fig. 10.33. For lean-burn engines, this is especially important since, in situations in which driveability and good throttle response are critical, extra enrichment can be applied and, in others, the fuelling reduced.

(7) *Hot idling*. By virtue of the application of the Motronic ignition control, enrichment is unnecessary except during overrun when, if no overrun fuel cut-off is incorporated, a slight degree of speed-related enrichment can improve driveability and reduce emissions.

(8) Ideally, the ignition timings for starting and idling should be different. By virtue of electronic control, the spark can be advanced as speed is reduced, to increase torque during idling: consequently, the idling speed does not have to be set to cater for the highest loads from the ancillaries, so both fuel consumption and emissions are reduced.

(9) *Full load*. With Motronic, the degree of enrichment is related to the engine speed, but modified by the map in the memory to cater for pulsations in flow and avoidance of knock. Maximum torque is obtained with $\lambda = 0.9$ to 0.95. The ignition point is set on the basis of the map but, to obtain maximum torque without knock, modified in response to signals from the knock sensor. Thus high power output is obtained with good fuel economy.

(10) *Acceleration*. The degree of initial enrichment is based on the signal from the throttle valve switch and lambda sensor, the electronic control calling for a mixture at $\lambda = 0.9$, for maximum torque and avoidance of a flat spot. For acceleration during warm-up, further enrichment is applied in response to engine temperature signals. Ignition timing is adjusted in response to engine load and speed signals and, if a preset rate of change of load is exceeded, the ignition timing is slightly retarded, to avoid knocking and generation of NO_x.

(11) *Overrun*. After an initial lag, the ignition is first retarded (for a smooth transition) and then, a few cycles later, the fuel is cut off completely. It cuts in again, over a few cycles, at an engine speed slightly higher than idling and, once more, the ignition is first retarded and then, as the fuel begins to flow again, progressively advanced back to normal. All this happens during stop–start situations in city traffic and normal braking and down-hill operation, thus saving fuel and reducing emissions.

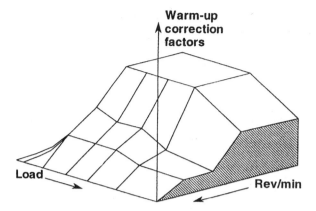

Fig. 10.33 This map is an approximate representation of the correction factors needed for increasing the rate of fuel supply needed, as indicated by the engine temperature, when the engine is cold

Engine stopped. To obviate danger of fire after an accident, a power transistor in the control unit controls an external relay in a manner such that the pump can operate only if the circuit between the starter and the battery is closed or the engine speed is higher than a preset minimum relative to the throttle position. Furthermore, to prevent the coil from overheating if the engine is left switched on after it has been stopped or stalled, the microcomputer turns off the ignition if the speed is less than, say, 30 rev/min.

Stop–start. To save fuel in heavy traffic, an additional controller can be installed to signal either *stop* or *start* to the electronic control unit. To stop the engine, the driver depresses the clutch pedal, and to start it again, he depresses both the clutch and accelerator pedals simultaneously. However, the engine will stop only if the speed is less than 2 km/h and start only when the throttle is less than one-third open. A function of the additional controller is to assess whether, in the light of the fact that each start uses extra fuel, economy is in fact obtainable: if it is not, the engine will not stop.

Computer-aided transmission control. Fuel economy, gear shift quality, and transmission torque capacity and life expectancy can be improved by adapting the Motronic electronic control unit for use also in automatic transmission control. Additional signals needed by the unit, for controlling the transmission's hydraulic pressure regulator, its solenoid valves and malfunction warning, include transmission output speed, kick-down switch, and program (economy or sporting performance, and manual shift). Gear shift performance curves in an electronic memory are much more effective than their hydraulic counterparts for controlling gear changing. Furthermore, the torque of the engine can be regulated during shifts, by momentarily retarding the ignition during a shift, to obtain a part-load feel with a full-load shift.

Exhaust gas recirculation. Exhaust gas recirculation can adversely affect driveability, especially at low speeds and light loads. By employing the electronic control unit for regulation of exhaust gas recirculation (EGR) in relation to the engine performance map, these difficulties can be overcome. The electronic control unit, through the medium of a pneumatic valve, regulates the quantity of exhaust gas recirculated so that NO_x is reduced at high loads, and good driveability retained at light loads and low speeds. Further information on EGR is given in Section 12.20.

Evaporative emissions. Canister purge, as described in Section 12.17, can also be controlled by the ECU.

Boost pressure control. With turbocharged engines, the onset of knock can be delayed by either reduction of boost or retardation of the ignition. However, reduction of boost reduces performance, and retardation of the ignition can cause overheating of the turbocharger. On the other hand if, as soon as knock is detected, both are effected simultaneously in an interrelated manner by the electronic control unit, these drawbacks can be largely avoided. This is done by first retarding the ignition and then, during the lag before the boost falls, advancing it progressively to its optimum value.

Cylinder cut-out. Where, in the interests of economy, there is a requirement for one or more cylinders to be cut out of operation, the Motronic electronic control unit can do so by cutting the fuel supply to the cylinder or cylinders and, as more power is required, restoring it either to one at a time or to groups of cylinders. Moreover, it can control a valve to direct hot exhaust gas through the inactive cylinders, to keep them at normal operating temperature, and therefore with normal values of friction between their moving parts. Another significant advantage is that the working cylinders operate with the throttle opened wider, and there is no throttling of the idle cylinders.

10.22   The Weber electronic control system

The Weber multi-point injection system is in many respects similar to the Bosch systems already described, but its electronic control is totally different and also exercises control over the ignition timing. Injection timing is calculated on the basis of the throttle position, absolute temperature and pressure of the air in the induction manifold, the instantaneous speed of the engine and rotational position of the crankshaft. All these inputs are taken into account together with other data in the memory of the electronic control unit, including an engine performance map and the variation of volumetric efficiency with speed. Compensation is effected for variations in the voltage of the battery output. A more detailed description can be found in *Automotive Fuels and Fuel Systems*, Vol. 1, by T. K. Garrett, Wiley.

10.23   Bosch Mono-Jetronic system

The single injector of the Mono-Jetronic system, Fig. 10.34, like that of the GM Rochester TBI system, Fig. 10.36, injects intermittently into the air intake, just above the throttle valve. A rotary, virtually pulse-free, electric pump in the fuel tank delivers at a pressure of 1 kN/m^2 through a fine filter to the injector. By virtue of its low output pressure, this pump is both light and very economical to manufacture, many of its components being of plastics.

The spray pattern is such that two jets are delivered, Fig. 10.35, one into each of the crescent-shaped gaps between the edges of the throttle valve and its cylindrical housing. Fine atomisation ensures that, even with the throttle wide open, the mixture distribution is homogeneous. Fuel in excess of requirements is returned to the tank, the continuity of supply preventing formation of vapour locks. Each injection is triggered synchronously by the ignition system

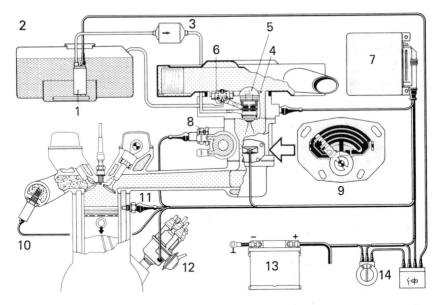

1   Electric fuel pump
2   Tank
3   Filter
4   Air temperature sensor
5   Single point injector
6   Pressure regulator
7   Electronic control unit

8   Throttle valve actuator
9   Throttle valve potentiometer
10   Lambda sensor
11   Engine temperature sensor
12   Ignition distributor
13   Battery
14   Ignition and starter switch

Fig. 10.34 The Diagram issued by Bosch to illustrate their Mono-Jetronic single-point injection system

and is timed to continue for periods of from 1 ms upwards, according to the quantity of fuel needed.

The input to the electronic controller includes signals of engine speed (from the ignition distributor), throttle valve position, and engine air intake temperatures. Stored in its memory is the information it needs for use, in association with the input signals, for calculating the time the injector is required to remain open for supplying the quantity of fuel appropriate for efficient operation of the engine.

The controller is programmed to enrich the mixture for cold starting, warm-up and acceleration. In response to signals from the various electric circuits, the engine temperature and speed sensors, an electric motor adjusts the position of the throttle stop to set the idle speed at an appropriately low level, regardless of what loads are switched in or out. As a contribution to fuel economy and reduction of emissions a fuel cut-off operates the idling system in the overrun condition and, if required in any particular application, at maximum engine speed. Compensation is effected for variations in the voltage of the output from the battery.

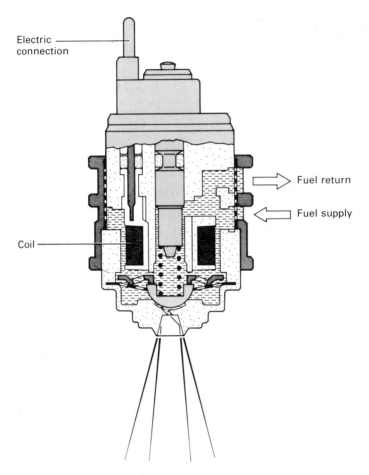

Electric connection

Fuel return

Fuel supply

Coil

Fig. 10.35 Bosch low-pressure injector with twin jets for injecting the spray into the two crescent-shaped openings on each side of the throttle valve as it opens

10.24 The GM Multec single-point system

Many features of the GM Multec single- (or throttle body) and multi-point injection systems are similar to those of the Bosch Mono-Motronic and Motronic systems respectively. A single-barrel, single-point, or TBI, system is illustrated in Fig. 10.36, though twin-barrel versions are also available if required.

From the submerged twin-turbine-type pump, fuel is delivered at a pressure of 0.83 bar and rates from 19 to 26 g/sec, through a 15 μm filter to the throttle body unit. A water separator/fuel strainer is attached to the fuel pick-up beneath the base of the pump.

Fuel injection rates are regulated by an electronic control module (ECM), and a separate electronic ignition module (EIM) controls the spark timing. Air

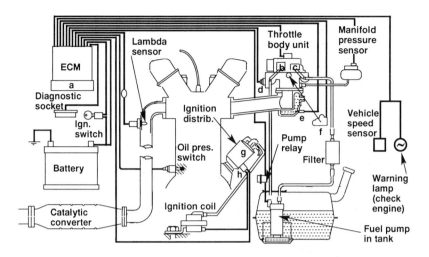

a Plug-in calibration software EPROM   d Idle air control valve
b Fuel pressure regulator   e Coolant temperature sensor
c Injector   f Throttle position sensor

Fig. 10.36 The GM Rochester Multec single-point injection system

flow is metered by a conventional throttle valve. Sensors signal to the ECM the throttle position, and temperature and absolute pressure in the manifold. These and the other sensors are shown in Fig. 10.36.

Illustrated diagramatically in Fig. 10.37 is the throttle body unit. This is mounted on a riser, which is water jacketed to help to vaporise the fuel and prevent icing in cold and damp ambient conditions. A coolant-temperature sensor is screwed into the base of this jacket.

The fuel passes from the inlet lower than the chamber housing the injector, through a fine mesh screen and up to the pressure regulator at the top, so any bubbles of vapour developing will rise and be returned, together with the fuel in excess of engine requirements, to the tank. Metering the fuel injected is a solenoid-actuated ball valve. This valve is closed by a coil spring. When it is open, the constant pressure maintained by the regulator projects a conical spray into the bore upstream of the throttle valve. Regulation of the total rate of delivery of fuel through the valve can be effected by varying either the period open or the frequency of fixed-duration pulses.

The diaphragm-type pressure regulator valve is opened, against the resistance of a calibrated spring, by the pump delivery pressure. It reduces the injection pressure to 0.76 bar. As previously indicated, its primary function is to maintain a constant pressure across the metering jet. The maximum recirculation rate is 27 g/s.

Enrichment for cold starting, warm-up, acceleration and maximum power are effected by the ECM, as also is idling speed. When the throttle is closed on to its stop, extra air by-passes it through a duct in which is a tapered pintle type idle air control valve. This valve is actuated by a stepper motor controlled by the ECM in response to signals from the engine-speed sensor.

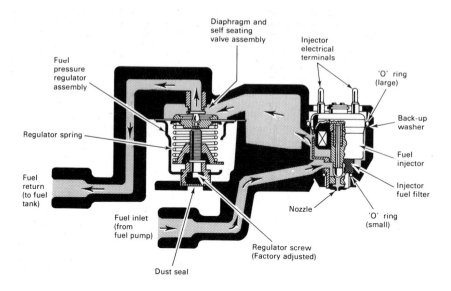

Fig. 10.37 The GM Rochester Multec single-point injector is similar to that of the Bosch Mono-Jetronic

10.25   The Multec multi-point system

In principle, the GM Multec multi-point system resembles the Bosch Motronic, Sections 10.17 to 10.21. To avoid repetition, therefore, only a few brief comments will be made here. It is a complete engine-management system, Fig 10.38, regulating EGR, ignition, fuelling, overrun cut-off, air flow control, including during idling, and open, or closed-loop control over emissions. However, the ECM, is served by signals from the throttle position indicator and manifold pressure and temperature sensors, and meters the air flow on the speed–density principle, so the rate of fuelling is regulated in relation to the computed mass flow. On the other hand, mass flow air metering with a hot wire anemometer is an optional alternative. Among the features of the system are on-board diagnostics, back-up fuel and ignition circuits, and an assembly line diagnostic link. Another option is either direct or distributor ignition timing.

  Direct ignition of course obviates the need for a distributor and separate coil. It can provide 35 kV, though typically it produces a 1700 µs spark at 18 kV. For a four-cylinder engine, a dual tower, twin-spark, epoxy-filled coil is used, the current of which is closed-loop controlled, and there is back-up control over ignition timing.

  The injectors are of GM design, with alternative ratings of 12 or 2 ohm at 3 bar, and flow rates ranging up to 15 g/s. They are a push fit in bosses on an extruded aluminium fuel rail, where they are retained by spring clips and sealed by O-rings. The fuel pump impeller is of the two-stage vane and roller type: the vanes remove centrifugally any vapour bubbles present and prime the roller-type pump of the second stage. A pulsation damper is mounted on top

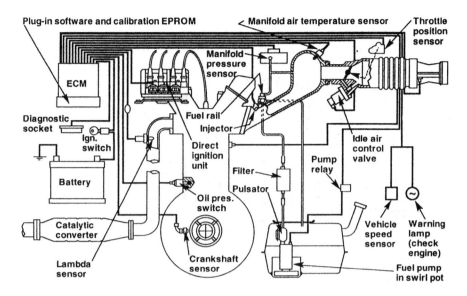

Fig. 10.38 GM Rochester Multec multi-point injection system

and a fuel strainer/water separator sleeve is fitted to the inlet in the base of the pump. At 3.5 bar, a delivery rate of 19 g/s is typical.

The oil pressure switch that can be seen in Fig. 10.38 controls a parallel circuit to drive the fuel pump in the event of failure of either the pump relay or the electronic control. Rated at 1000°C exhaust temperature, the oxygen sensor has a zirconium element. Signals from the vehicle speed sensor indicate to the ECM when overrun cut-off and idle-speed control should be brought into operation.

Incorporated in the ECM are an 8-bit microprocessor, a co-processor, AC/DC converters to enable the digital microprocessor to read the analogue signals from the sensors, and the drivers for the actuators and back-up hardware. Software and calibration are programmed into a 16-kbyte EPROM customised to meet the requirements for specific applications. The co-processor relieves the microprocessor of interruption by the engine timing functions.

Software alogarithms continuously check the status of the ECM outputs and the validity of its inputs. If a fault is detected, a code is stored in the memory and a warning lamp on the instrument panel illuminated. Service technicians can then read the code indicating the nature of the problem and, if necessary, connect to the system a diagnostic facility to obtain further details.

10.26 Rover throttle body injection and ignition control

As previously indicated, single-point throttle body injection offers significant cost economies as compared with multi-point injection. For this reason, Rover retained it in the early 1990s for the models in the lower end of its range, to satisfy the US Federal regulations until at least 1996. At the same time, the old A-Series engine in the Mini Cooper was similarly equipped, mainly for

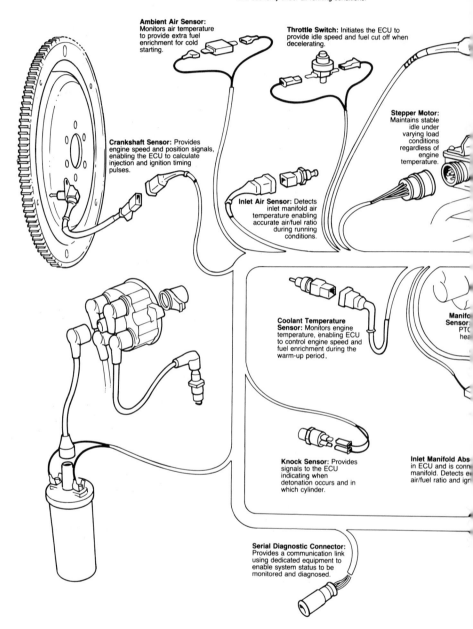

General Description: The Single Point Injection system consists of a numbe components accurately maintaining the precise fuelling and ignition requirem

Electronic sensors monitor operation of the engine ensuring optimum perforr and economy under all running conditions.

Ambient Air Sensor: Monitors air temperature to provide extra fuel enrichment for cold starting.

Throttle Switch: Initiates the ECU to provide idle speed and fuel cut off when decelerating.

Stepper Motor: Maintains stable idle under varying load conditions regardless of engine temperature.

Crankshaft Sensor: Provides engine speed and position signals, enabling the ECU to calculate injection and ignition timing pulses.

Inlet Air Sensor: Detects inlet manifold air temperature enabling accurate air/fuel ratio during running conditions.

Coolant Temperature Sensor: Monitors engine temperature, enabling ECU to control engine speed and fuel enrichment during the warm-up period.

Manifo Sensor PTC hea

Knock Sensor: Provides signals to the ECU indicating when detonation occurs and in which cylinder.

Inlet Manifold Abs in ECU and is conn manifold. Detects e air/fuel ratio and igr

Serial Diagnostic Connector: Provides a communication link using dedicated equipment to enable system status to be monitored and diagnosed.

BH1431/Motor vehicle/ × 65%

Fig. 10.39   The single-point version of the Rover modular engine-management system (MEMS) that has been applied for injection on, among other vehicles in their range, the Mini Cooper

diagnosis is accessed initially via a serial diagnostic connector, which is also for electronically setting CO levels.

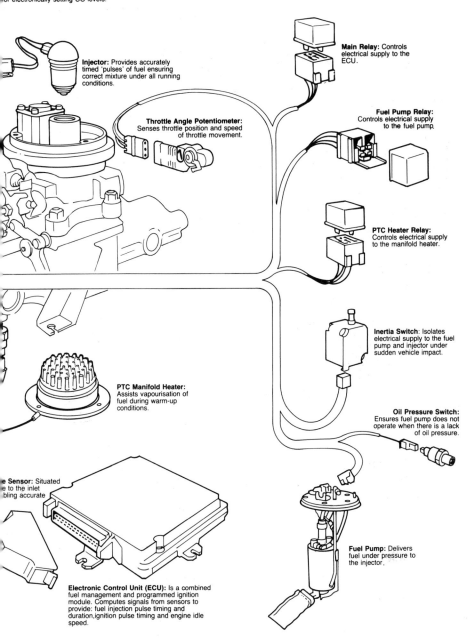

Injector: Provides accurately timed 'pulses' of fuel ensuring correct mixture under all running conditions.

Main Relay: Controls electrical supply to the ECU.

Throttle Angle Potentiometer: Senses throttle position and speed of throttle movement.

Fuel Pump Relay: Controls electrical supply to the fuel pump.

PTC Heater Relay: Controls electrical supply to the manifold heater.

Inertia Switch: Isolates electrical supply to the fuel pump and injector under sudden vehicle impact.

PTC Manifold Heater: Assists vapourisation of fuel during warm-up conditions.

Oil Pressure Switch: Ensures fuel pump does not operate when there is a lack of oil pressure.

e Sensor: Situated e to the inlet bling accurate

Fuel Pump: Delivers fuel under pressure to the injector.

Electronic Control Unit (ECU): Is a combined fuel management and programmed ignition module. Computes signals from sensors to provide: fuel injection pulse timing and duration, ignition pulse timing and engine idle speed.

satisfying the Japanese emissions regulations. Interestingly, because of the reduced breathing capacity of the manifold when TBI is substituted for either twin-barrel or twin carburettors, there is almost invariably a slight loss of power.

For their TBI system for the 1.4-litre engine, Rover use a Bosch injector with solenoid actuated valve, Fig. 10.39. It is installed in a throttle body unit designed by Rover and produced by Hobourn-SU Automotive, and the whole assembly is mounted on the riser of a manifold that is also designed by Rover. Bosch offer injectors with either ball- or pintle-type nozzles, but Rover use the pintle type because experience has shown it to be more durable.

Rover design their own electronic hardware and software, for injection-control and engine-management systems, specifically to match their engines. An indication of the success of this policy is that, while the European average power output per litre for 1.3- to 1.6-litre engines with TBI at that time was 56 bhp/litre, the Rover 1.4 litre developed 68 bhp/litre. Incidentally, the corresponding average for multi-point injection was 72 bhp/litre.

The single-point Modular Engine Management System (MEMS), Fig. 10.40, is a second-generation system, succeeding Rover's earlier electronically regulated ignition and carburation control (ERIC). Manufacture is undertaken by Motorola AIEG, who have specialist production facilities.

An ignition timing map is programmed into the memory of the electronic control unit (ECU), which regulates both the ignition and fuel metering. Short circuit protection is incorporated, and powerful diagnostic facilities store intermittent fault data. Either the Rover Microcheck or Cobest hand-held diagnostic units can be plugged into a separate connector, without disturbing the ECUs main connector. The ECU also controls both an electric heater at the base of the manifold riser and a stepper motor for regulating idle speed.

10.27   Ignition control

Signals required by the ECU for ignition control include crankshaft angle and engine speed (both from the crankshaft sensor), engine coolant temperature and throttle closure, the latter indicated by the throttle switch. Additionally, the manifold absolute pressure sensor converts the pressure to an electrical signal to indicate engine load. To prevent fuel from entering the pressure sensor, a vapour trap is inserted into the pipeline between it and the manifold.

The crankshaft sensor comprises an armature projecting into an annular slot near the periphery of a reluctor disc bolted and spigoted to the flywheel. Spaced 10° apart in the slot are 34 poles, which continuously update the ECU as regards the crankshaft angle and engine speed. The two missing poles, 180° apart, identify the TDC positions.

Basic ignition timing requirements are stored, as a two-dimensional map, in the memory of the ECU, and various adjustments are signalled by the sensors. For example, when the throttle and, with it, the switch contacts are closed, an idle ignition setting is implemented and is further modified by signals from the engine-temperature sensor. This system is so sensitive that the idle ignition timing is continually varying.

The ignition coil has a low primary winding resistance (0.71 to 0.81 ohm at 20 °C), so that the high tension voltage will peak both rapidly and consistently

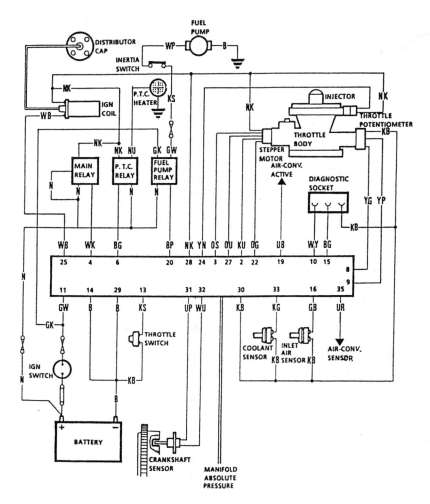

Inputs from (sensors, etc.):

		Outputs to:
Crankshaft angle	Diagnostic unit	Injector
Ignition coil	Power earth	Stepper motor
Manifold absolute pressure	Sensor earth	Fuel pump relay
Coolant temperature	Air-to-converter request	Manifold heater relay
Inlet air temperature	Ignition supply	Main relay
Throttle potentiometer	Battery supply	Diagnostic output
Throttle pedal position (switch)		Air-to-converter clutch

Fig. 10.40 Diagrammatic representation of the Rover electronic control system

throughout the engine-speed range. Since the ignition timing is controlled by the ECU, a simple distributor rotor and cap is used, the rotor being bolted to the D-section rear end of the inlet camshaft.

10.28 The air-intake system

As can be seen from Fig. 10.41, the air is drawn first over a resonator, which reduces noise output, Section 11.20, then through an intake temperature control valve, and on past the air intake temperature sensor to the filter mounted on the combined throttle body and injector housing. The air temperature control valve is a hinged flap over the end of a duct taking heated air (from a shroud around the exhaust manifold) to mix with the incoming cold air. The flap, moved by a diaphragm-type actuator controlled by a Thermac valve, is closed by a return spring and opened by manifold depression.

When the engine is started in an ambient temperature below 35°C, the flap is immediately opened by manifold depression, to allow warm air to pass into the induction system. As the engine warms up and the temperature of the incoming air rises above 35°C, the Thermac valve opens, to vent the depression to the clean side of the air filter. A restrictor in the connection from the manifold to the Thermac valve serves two purposes: first, it damps the motion of the valve so that it does not snap closed and open, but hovers between the two positions, holding the temperature of the air delivered to the cleaner at around 35°C; secondly, it ensures that opening the Thermac valve has a negligible effect on the manifold depression.

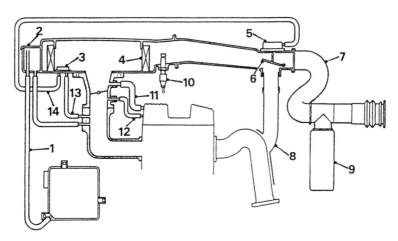

1 ECU pipe	8 Hot air intake
2 Vapour trap	9 Resonator
3 Thermac valve	10 Inlet air sensor
4 Air filter	11 Breather pipe
5 Temperature control diaphragm	12 Breather pipe (restricted)
6 Temperature control flap	13 Thermac valve (manifold)
7 Cold air intake duct	14 Thermac valve (diaphragm)

Fig. 10.41 Schematic representation of the Rover single-point injection system

From Fig. 10.41, it can be seen also that a hose is connected between a fuel trap incorporated on the left of the air filter and the ECU, and there is a second hose between the fuel trap and a point downstream of the throttle. Thus, absolute pressure in the manifold is transmitted to the ECU, in which is a pressure sensor. Incidentally, there is also a throttle by-pass passage, but this is not shown in the illustration. At low engine speeds, fine adjustment to the air flow, and therefore mixture strength, can be made by means of a screw projecting into this passage.

10.29   Throttle body assembly

The throttle body is water jacketed and houses the injector, throttle valve and its potentiometer, a stepper motor and the fuel pressure regulator. Beneath it, in the base of the riser, is the electrically heated hot spot, described in detail in Section 11.3. The water jacket supplies enough heat to ensure complete vaporisation of the fuel under cruising conditions when the engine is warm.

10.30   Stepper motor operation

A screw on the end of the throttle valve actuation lever serves as an adjustable throttle stop. It closes on to the end of a cam-actuated push rod. The cam, driven through reduction gearing by a stepper motor, can be turned through 150°, Fig. 10.42. This motor, which rotates 3.75 revolutions in 180 steps of 7.5°, is set in motion by the ECU, but only when the throttle is closed on to its stop and the engine speed has fallen below a predetermined rate. It is indexed to set the idle speed appropriate to the engine temperature and changes in load due to switching on or off of ancillary equipment, on the basis of signals from the following—

(1) Crankshaft sensor (engine speed).
(2) Manifold absolute pressure (engine load).

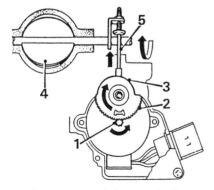

| 1 | Stepper motor pinion | 3 | Cam | 5 | Push rod |
| 2 | Reduction gear | 4 | Throttle disc | | |

Fig. 10.42 In the Rover system a stepper motor sets, in relation to engine temperature and ancillary loading, the degree of throttle opening when the engine is idling

(3) Throttle switch (indicating when throttle is closed).
(4) Battery voltage (state of charge).
(5) Ignition switch (on or off).

To prevent stalling when the throttle is closed suddenly on to its stop, the ECU initially sets a slightly higher than normal idling speed. When the engine is being cranked, the stepper motor is indexed to an idle speed appropriate for starting at the then current engine temperature. During overrun, the motor is indexed to open the throttle far enough for good driveability and low hydrocarbon emissions. If the voltage is down, indicating that the battery is in a low state of charge, the ECU increases the idle speed and therefore the alternator output. When the engine is switched off, the ECU keeps the main relay closed for 30 s, so that it has time to index the stepper motor to set the throttle stop to its a reference position.

Adjustment of the throttle stop screw is done in the factory and should not be altered in service. The previously mentioned throttle by-pass screw should be adjusted only as indicated by the Microcheck or Cobest diagnostic equipment and with the stepper motor indexed a predetermined number of steps. This number varies from model to model and, because the ECU is programmed to learn the changes in engine characteristics occurring in service, it varies also with condition of the engine.

10.31   Fuel metering

The single jet is directed on to the upstream face of the throttle butterfly valve. A submerged Gerotor electric gear-type pump in a swirl pot in the tank, delivers the fuel to a filter mounted in the engine compartment, and thence to the throttle body unit, in which is housed the injector. A proportion of the fuel is then metered by the injector into the air stream, that in excess of the instantaneous requirement being returned through the pressure regulator valve, Fig. 10.43, into the swirl pot.

The swirl pot holds a reserve of fuel when the tank is nearly empty, so that air will not be drawn into the pump when, under the influence of acceleration,

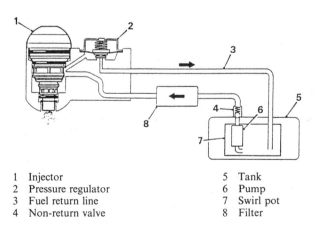

1	Injector	5	Tank
2	Pressure regulator	6	Pump
3	Fuel return line	7	Swirl pot
4	Non-return valve	8	Filter

Fig. 10.43 Fuel supply arrangement for the Rover single-point injection system

it swills around the bottom of the tank. A pressure regulator keeps the delivery pressure to the injector at a constant value, of between 1.0 and 1.2 bar. If the vehicle is subject to an impact, a resettable inertia switch breaks the fuel pump circuit, and a non-return valve in the outlet from the pump keeps the system primed ready for starting again.

Programmed into the ECU is a two-dimensional map of the cruise air : fuel ratio for 10 different engine speeds and nine inlet manifold air density conditions, for continuously matching the requirements as changes are indicated by the load and inlet air temperature signals. There is also an idling air : fuel ratio map, to which the ECU turns when the throttle pedal switch is closed and the engine speed falls below a predetermined value. Idling exhaust CO content is adjustable through the diagnostic port with either Microcheck or Cobest.

Under normal driving conditions, the ECU earths the solenoid of the injector twice per revolution of the crankshaft. The basic pulse width, varying between 1.1 and 8 ms, is determined by the quantity of fuel required to be injected, as indicated by the air : fuel ratio map and the inputs from the crankshaft sensor (engine speed), manifold absolute pressure (load), and inlet air temperature (density). Further refinement, or compensation, of the pulse width is effected on the basis of signals indicating battery voltage, cranking signal, coolant temperature, throttle potentiometer (acceleration) and throttle pedal switch. Below 1500 rev/min, however, for accuracy of metering the very small quantities of fuel, there is only one pulse per revolution. Under all conditions, the start of injection is timed relative to the crankshaft position.

Injector performance is of course influenced by battery voltage, hence the need for compensation by the ECU. Cranking is indicated if the speed is below about 390 rev/min. In this condition, the ECU enriches the mixture in accordance with signals from the coolant sensor but, to prevent flooding, this enrichment is done cyclically, 30 pulses on followed by 30 off. Once the engine starts, this on–off sequencing stops, and the pulse widths are reduced progressively as engine temperature rises.

Signals from the throttle pedal potentiometer indicating increasing output voltage, coupled with rising manifold absolute pressure, trigger acceleration enrichment. This is effected not only by increasing pulse width but also introducing additional pulses at 80° intervals of crankshaft rotation. Overrun fuel cut-off is effected by the ECU on closure of the throttle switch, but only if the coolant temperature is below about 80°C and the engine speed above about 1800 rev/min. In the event of the engine speed subsequently falling below 1500 rev/min, or if the throttle is opened again, injection is resumed to avoid impairing driveability.

To inhibit over-speeding, the ECU breaks the earth circuit from the injectors to cut off the fuel metering supply if the engine speed rises above 6860 rev/min. It is reinstated, again to avoid impairing driveability, when the speed falls to 6820 rev/min.

In the event of failure of inputs from the coolant, inlet air temperature or manifold absolute air pressure sensors, a back-up facility comes automatically into effect. The back-up values are respectively 60 and 35°C for the first two while, in the event of the manifold absolute pressure sensor failure, the ECU regulates air : fuel ratio solely on the basis of engine speed and throttle position.

10.32   The Mechadyne Pijet 90 system

Of particular interest because it would appear to be the first example of
air-shrouded injection to go into service, albeit not on a road vehicle, is the
system originated by Piper F.M. Ltd, and further developed by Mechadyne, as
the Pijet 90 low-cost system for small engines. Injection system cost is not a
function of the maximum quantity of fuel delivered per cylinder, and therefore
is a much higher proportion of the total cost of a small rather than a large
engine.

A thorough analysis by Mechadyne of the performance and other characteristics of the conventional systems revealed a number of shortcomings, as
shown in Table 10.2, in which the relevant characteristics of the conventional
and Pijet 90 digital systems are compared. Basically, Pijet 90 differs from the
other systems in three ways. It has a primary metering slot valve, in principle
similar to that of the Bosch K-Jetronic system but actuated by the throttle
spindle instead of an air-metering flap. This is complemented by a solenoid-
actuated pulser, under the control of an ECU, which delivers into a fuel
accumulator. From this accumulator, the fuel goes to a set of extremely simple
nozzles, one for each cylinder.

10.33   Principle of operation

As can be seen from Fig. 10.44, the principle of operation of the system is as
follows. Fuel is taken from the tank by an electric pump, which first delivers it
through a filter and then past a pressure-regulating valve. This valve holds the
pressure in the delivery line to the primary metering valve, at the relatively low
value of 2 bar, which is adequate for obviating vapour lock and rendering the
system insensitive to the influence of acceleration forces, yet low enough not to
require costly high-pressure pipes and fittings. The primary metering is
effected by a rotary slot valve mechanically interlinked with the throttle valve
and its associated status sensor. It therefore regulates the rate of delivery of fuel
in relation to throttle angle (torque demand) before it passes through the
pulser.

The functions of the pulser, which is a solenoid-actuated pintle valve, are to
trim the rate of delivery of fuel accurately through a fuel accumulator to the
engine, and to regulate the quantity of fuel delivered per injection, in response
to the output from the ECU. Fuel delivered to the accumulator dwells there
until an inlet valve opens, creating a depression pulse that draws it progressively through an adjacent nozzle and porting into the cylinder. Injection is
stopped by the compression wave generated by the closing of the engine inlet
valve.

Each nozzle, Fig. 10.45, in effect comprises a closed cylindrical body with a
pair of diametrically opposed apertures in its sides (through which air enters)
and a small hole in its lower end, through which both the air and fuel pass out
together into the induction manifold. At the upper end of the body is the
connection through which the fuel enters and passes directly into a capillary
tube through which it flows, to be discharged, just above the previously
mentioned hole, into the manifold. The gap between the lower end of the tube
and the upper edge of the hole is set during manufacture as follows. The lower
end of the nozzle is subjected to a calibration depression; the tube is slid up into

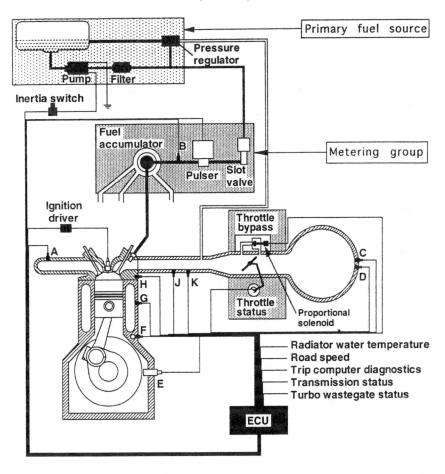

Engine sensors

A	Exhaust gas content	F	Oil pressure
B	Fuel temperature	G	Engine temperature
C	Intake air temperature	H	Knock detector
D	Intake air temperature	J	Manifold air pressure
E	Crankshaft sensor	K	Manifold air temperature

Fig. 10.44 In the Pijet system, a slot valve is mechanically interconnected with the throttle. The primary inputs are for engine speed and throttle angle only. Those for compensation for changes in ambient conditions and for cold-start enrichment are: crankshaft angle, induction manifold pressure, induction manifold air temperature and engine temperature. All the others such as variable valve timing and continuously variable transmission controllers are optional, to suit the vehicle designer's requirements

its holder until the specified depression is measured in the upper end of the capillary. This is an extremely accurate way of calibrating the nozzles.

The fuel is discharged as a fine spray from the lower end of the tube into the air stream passing through the hole into the manifold, so mixture preparation is good. Atomisation is further enhanced by the fact that the capillary tube

Table 10.2—CHARACTERISTICS OF CURRENT AND PIJET 90 DIGITAL SYSTEMS COMPARED

Existing systems	Reasons for deficiencies	The Pijet 90 system
Increasingly, sequential injection is favoured, to obviate transient emissions, but cylinder-to-cylinder timing tends to be inconsistent	Sequential phasing and pulse lengths dependent on accuracy of complex software programs and rapidity of response	Injection timing determined directly by instantaneous conditions in the individual cylinders, not by ECU
Under transient conditions, delay occurring between accelerator movement and injection calls for post-processing compensation	Delays between signal initiating demand for acceleration, the processor looking up reference table, and the injection response	Real time delivery of fuel at optimum timing relative to the induction phase of the Otto cycle
Metering inaccuracies in engines of capacities lower than about 1.5 litres	Electromagnetic injectors both difficult and costly to produce accurately enough for delivery of such small quantities per shot	System adapts well to cylinder sizes down to 0.5 litres
At low-speed, light-load operation, emission control and economy are inferior, and even worse with twin jet nozzles	Poor atomisation of fuel under these conditions, even with low-inertia armatures and pintles	Homogeneity of mixture actually improves as speed and load are reduced, thus also improving knock tolerance and lean burn stability
Cost disproportionately high for application to small engines	Cost of injection system is not a function of size of engine	Only one solenoid needed regardless of number of cylinders, and each nozzle contains only one high-precision component

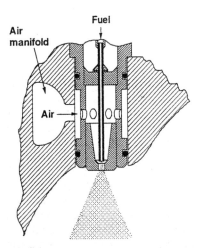

Fig. 10.45 The nozzles, containing neither solenoids nor pintles, are simple and contain no high-precision components other than the capillary tube, and therefore cost little to manufacture

vibrates at its natural frequencies, which are mostly forced at the speed of rotation of the engine.

The delivery pressure in the pipelines to the nozzles is determined by the capacity of the accumulator, the rate of its spring, the duration of opening of the pulse valve and the rate of injection, as determined by the depression in the manifold. Because the pulser is timed to open a few milliseconds before the engine inlet valve, the pressure in the accumulator does not have time to build up to too high a value before that valve opens. This prevents discharge of fuel initially through all nozzles, instead of just the one.

During idling, the fuel delivery pressure is so low as to be incapable of overcoming the surface tension of the fuel at the end of the capillary tubes. So no fuel can be delivered until the force tending to drive it along the capillary tube of one of the nozzles is complemented by the strong depression pulse in the manifold, generated by the opening of the adjacent inlet valve. Once flow has started, it will continue to do so until it is stopped by the pressure pulse generated by the closing of the inlet valve.

With increasing load, and therefore decreasing manifold depression, the rate of delivery from the primary metering valve rises and so also therefore does the pressure in the fuel delivered to the nozzles. Over most of the range, the fuel is injected preferentially through the nozzle adjacent to the inlet valve that is open. However, the magnitude of the flow through the other nozzles does increase marginally as manifold pressure rises, peaking at a maximum torque. Mechadyne state that the incidental flows into the other ports under this condition do not matter, since injection has become virtually continuous and mixture distribution between cylinders is therefore less critical.

When injection ceases, surface tension holds the fuel in the capillary, so there can be no dribble. Because the depression pulses are strongest at low speeds and light loads, metering is particularly accurate in these circumstances, and injection is far more precise, as regards quantity, inter-cylinder balance and timing, than with the conventional systems, especially those dependent upon mass-air-flow meters. Moreover, because the quantity of fuel delivered by each nozzle is a function of the volumetric efficiency of its associated cylinder, compensation is made automatically for wear of the engine in service.

10.34 Idling and the electronic control unit

An additional feature of the Pijet system is automatic control over idling speed, to compensate for variations in load and engine temperature. This is effected by the ECU controlling an electric motor regulating a tapered needle valve in the throttle by-pass duct. In general, because the nozzles do not contain solenoid valves, the demands on the ECU are lighter than those associated with conventional systems, so it can be less complex and costly, its programming is easier and its reaction time potentially faster.

However, the simplicity and low cost of the electronic control system arises not only from this but also because, by virtue of the application of statistical production control throughout the industry, the volumetric efficiency characteristics, Fig. 10.46, of any family of engines in relation to throttle angle and speed vary very little from engine to engine. Indeed, it has been found that these maps can be taken as common to all engines of the same type produced by different companies. The reason for this is that all manufacturers have

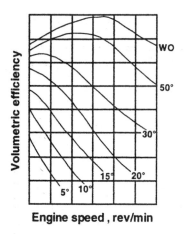

Fig. 10.46 Typical engine volumetric efficiency characteristics

similar design aims, which include low emissions, good fuel economy and approximately 40 to 45 kW/litre power output at about 5500 rev/min. Consequently, with only minor trimming and adjustments, a common volumetric efficiency, and therefore fuelling map, Fig. 10.47, and computer program can be utilised for virtually all engines of the same type.

10.35   Comment

The Mechadyne Pijet 90 system offers the advantage of low cost, coupled with good fuel economy, for cars powered by engines up to about 1.5 litres capacity. It would appear to be a practicable alternative to single-point injection, with its inherent problems of inferior mixture distribution and deposition of fuel in the manifold. Moreover, as a multi-point system, it would be more suitable for satisfying the emissions regulations.

From Fig. 10.50, it can be seen that the sequential firing of the nozzles is not just wishful thinking. The explanation is as follows: at small throttle openings, the depression in the porting in which the open valve is situated is so much greater than that in the other ports that there is a pressure differential favouring discharge through its own nozzle rather than those serving the other cylinders. In these conditions, too, the delivery pressure from the accumulator is so low that the manifold depression is the decisive factor in extraction of fuel through the jet. Although the depression falls as throttle angle increases, this does not matter because simultaneously the injection frequency is progressively increasing until it becomes virtually continuous. The wide open throttle fuel economy is demonstrated in Fig. 10.48.

Under wide open throttle conditions, this system is clearly at a major advantage, as regards fuel consumption between 1500 and 3000 rev/min, Fig. 10.49, even allowing for possible experimental errors. Indeed, it is claimed to be at least as good as the Bosch L-Jetronic system as the maximum of 6000 rev/min is approached. It has the further advantage of superior atomisation,

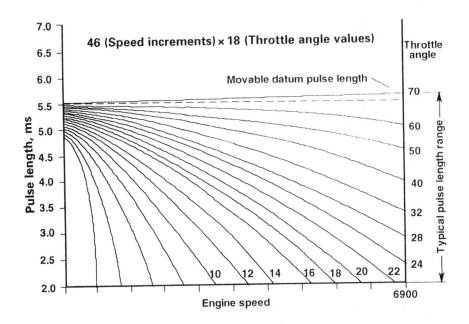

Fig. 10.47 This typical fuelling map with 46 speed increments and 18 throttle-angle values requires 828 byte of memory

and therefore mixture preparation, owing not only to the vibration of the capillary tubes in the nozzles, but also to the fact that the jet of fuel is delivered initially into the jet of air flowing out through the orifice in the inner end of the nozzle, Fig. 10.45.

Mechadyne state that this excellence of mixture preparation has been reflected in better knock tolerance and lower emissions of the engines with which it has been equipped. Another advantage of the Mechadyne system is that there is no need to direct the jets on to the inlet valves, so all risk of corrosion and possible distortion owing to local cooling of the valve can be eliminated.

On the other hand, the depression beneath the nozzle is a function not only of the downward motion of the piston but also of the cross-sectional area and contour of the induction tract and porting between it and the adjacent inlet valve seat. Consequently, if uniform distribution is to be ensured, it would appear that all the ports ought to be dimensionally and physically virtually identical.

In conclusion, it might be said that the Pijet system is an ingenious combination of carburation and sequential and continuous injection, having the advantages of all three but with none of the disadvantages. In common with the carburettor, it is characterised by the use of manifold depression to draw off the fuel, so it is largely self-regulating at low to medium speeds, thus simplifying the electronic controller. Moreover, because it functions in this mode over only the lower portion of the speed and load range, it does not suffer the disadvantage of needing to be compensated hydraulically for the

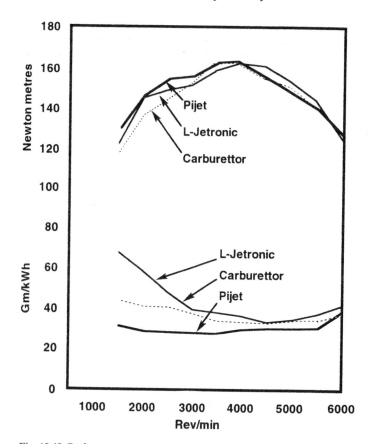

Fig. 10.48 Performance curves at wide-open throttle of an engine equipped in turn with the Pijet 90 and Bosch L-Jetronic injection systems with respectively 40 and 54 mm diameter throttle barrels, and a Weber 2V carburettor with a 25/27 mm twin venturi

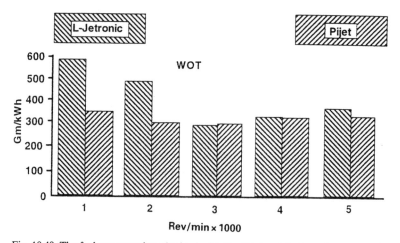

Fig. 10.49 The fuel consumption obtained with the Pijet and L-Jetronic systems during the engine testing in Fig. 10.46

differential rates of increase in flow of fuel and air with increasing depression. As the speed rises through the middle range, there is a progressive transition into the sequential injection mode and, as the top of the range is approached, injection becomes virtually continuous.

Chapter 11

Induction manifold design

For single-cylinder engines one simple pipe is of course all that is needed for transferring the mixture from the carburettor to the inlet port on the cylinder head. To provide for a multi-cylinder engine, however, what is termed an *induction manifold* is generally needed. It generally comprises a main trunk from which branch pipes are taken, one to each cylinder, and on which is mounted the carburettor or injection equipment. By appropriate design, as explained in Sections 11.5 to 11.25, it is possible also to adapt it for enhancing the performance of the engine.

In general, manifolds should be designed to offer a minimum of resistance to flow, and to be of light weight, low cost and easy to manufacture. Minimum resistance to flow is obtained partly by keeping the ducts as straight as possible. Where curves have to be incorporated, their radii should be as large as practicable, though right-angle bends are sometimes introduced to shatter droplets of fuel, and thus promote their evaporation. For example, the base of the riser is usually a T-junction, an arrangement that has been found to provide better uniformity of distribution of flow than if the riser were bifurcated and the two passages thus formed curved round towards the cylinders they served. With sidedraught or downdraught carburettors, long risers tend to encourage smoothness and uniformity of flow. On the other hand, with a downdraft carburettor, a short riser is needed for cold starting, otherwise the droplets of fuel tend to fall back again before they reach the induction manifold.

Any or all of the following may cause the air to be deflected asymmetrically from the fuel metering device into the riser: an air intake tube on the filter, the filter itself, the throttle, a spray tube in the carburettor, a bracket carrying a single point injector, or the injector itself. Without uniform distribution, detonation may occur in the cylinders receiving the weakest mixtures. Consequently, with carburation or single-point injection, fuel consumption is adversely affected since the whole of the mixture has to be enriched until the detonation in those cylinders ceases.

Of major importance is accuracy of alignment of the passages in the manifold and continuity of their inner faces at the junctions between them and both the cylinder head and the throttle barrel. Discontinuities here can cause considerable turbulence and choking of the flow so, to ensure accuracy of alignment, dowel sleeves are sometimes inserted in counterbores in the ends of the ports and pipes.

Smooth interior walls assist flow by reducing viscous friction. On the other hand, a layer of micro-turbulence in the layers hugging the walls helps to promote evaporation. Indeed, if rapid transition from cold to warm running is of greater importance than maximum power output, it may even be desirable to roughen these surfaces. Another consideration may also be relevant: to reduce the speed of flow adjacent to the inner radii of bends, it may in some instances be helpful to roughen the surfaces on the insides of the bends and to polish those on their outsides. This will encourage planar cross-sectional flow round the bend and thus reduce the tendency for vortices to be generated around the sharp radii, which could partially choke the flow.

11.1 Mixture distribution and manifold pressure

Distribution of the mixture uniformly as regards both quantity and quality to each of the cylinders was the overriding design consideration so long as either carburettors or throttle barrel injection were used. However, on diesel and port injected spark ignition engines, the fuel distribution is taken care of by the injection equipment, so the manifold is left with only the air distribution function.

With single-barrel carburettors on four-cylinder engines, manifold layouts such as those in Fig. 11.1 have been used. Multi-barrel carburettors call for manifold layouts. Where the throttles are synchronised, any of the arrangements illustrated in Figs 11.2 and 11.3, and for V engines in Figs 11.16 to 11.18, are commonly adopted. Uniform distribution is more difficult to obtain with sequentially operated throttles. As can be seen from the top right-hand diagram in Fig. 11.4, mixture issuing from the smaller diameter barrel would tend to be distributed preferentially to the lower pair of induction pipes and that from the large barrel to the two upper ones, as viewed in the illustration. With sequentially operated throttles, therefore, a mixing box is generally interposed between the throttle barrels and the riser or main manifold tract.

From the illustrations it can be seen that each barrel of the carburettor serves a group of either two or three cylinders. This is because the opening of an inlet valve in one cylinder generates a suction wave, which passes at the speed of sound back along its induction pipe. When this wave issues from the end of the pipe it tends, if the inlet valve in an adjacent cylinder is still open, to suck mixture from the latter's pipe. As a result, overall volumetric efficiency can be significantly reduced. Hence, unless the ends of the pipes whose induction phases overlap and therefore are liable to rob each other of charge can be widely separated, they should be served through different manifolds.

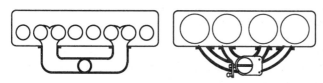

Fig. 11.1 Two possible manifold layouts for four-cylinder engines

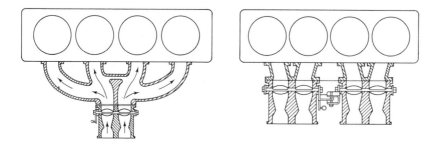

Fig. 11.2 Some manifold layouts for four-cylinder engines with twin-barrel sidedraft carburettors and synchronised throttles

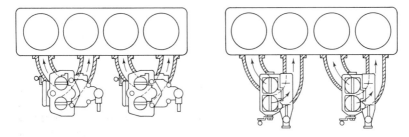

Fig. 11.3 Some manifold layouts for four-cylinder engines with downdraft carburettors and synchronised throttles

With a four-cylinder engine, even if there are only two pipes per manifold, some overlap of induction phases is inevitable. Six cylinders, however, can be grouped in threes between which, given an appropriate firing order (1 3 2), significant overlap can be avoided. In other words, the solution to the problem for a six-cylinder engine is to arrange the manifolding as if it were two three-cylinder engines.

Pressures in manifolds are of concern in relation to gasket design. They mostly range from a maximum of about 150 mmHg under overrun conditions, to 720 mmHg with the throttle wide open at maximum rev/min. These figures translate respectively into about 5.5 and 81 kN/m² absolute (0.8 and 11.8 lbf/in²) below atmospheric pressure. Air leaks into the manifold can adversely affect engine performance.

11.2   Mixture transport problems

When a warm engine is idling, the walls of the manifold are almost dry and the air inside may be virtually saturated with fuel vapour. If the throttle is suddenly opened, the pressure, and therefore density, of the air rises. This, in effect, suddenly squeezes some of the fuel out of the air and deposits it on the

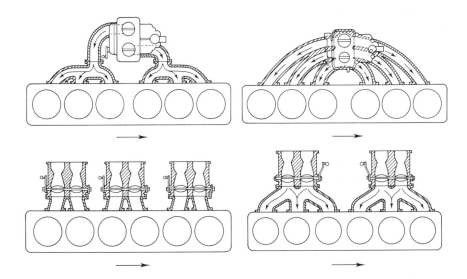

Fig. 11.4 Three induction manifold layouts for six-cylinder engines with synchronised throttles and, top right, a fourth with sequentially opening throttles

walls of the manifold, which of course is one reason why provision has to be made for enrichment of the mixture for acceleration. The other is generally the inertia of the fuel in the jets or injectors. Automatic transmissions embodying torque converters, however, provide a degree of compensation, since the drive is not fully taken up until the engine speed has risen to about 1800 to 1900 rev/min.

During cold starting, the droplets of fuel issuing from the carburettor or single-point injector may not evaporate fully, so there is a strong tendency for pools of petrol to settle on the floor of the manifold. In fact, during cranking as little as about 5% of the fuel may evaporate on leaving the jets so, to facilitate evaporation, there may be a well beneath the riser. Additionally, to make the surface area of the pool in the manifold as large as practicable, it may be of D-section rotated 90° anti-clockwise, to provide a flat floor. Engine installation angle must be taken into consideration, otherwise the liquid fuel will flow to one end of the manifold.

On the other hand, a rectangular section may offer just as large a base for evaporation and, downstream, it can merge first into a square and then a circular section to blend into the valve ports. Alternatively, the progressive change in section may be accommodated within the port in the head casting, provided it is long enough. Enclosing the greatest volume of air within walls of the smallest surface area, a circular section is favoured for port-injected systems, in which condensation does not pose a significant problem.

In some designs, the manifold branch pipes slope down from the inlet ports to the centre of the manifold, to facilitate the collection and subsequent evaporation of the fuel droplets and to prevent them from running into the cylinder, where they would dilute the lubricant and perhaps cause the sparking plugs to misfire. Until about the 1940s there were still vehicles on the road having vertical pipes connected to the centre of the manifold and extending down to a point below the sump, for draining excess fuel away to atmosphere. The aim was to avoid the serious fires that used to be caused by blow-back of the burning gases from the cylinder during the valve overlap period when the ignition was retarded for starting. Generally a ball-type non-return valve in the lower end of the pipe prevented air from bleeding into the induction manifold.

On the other hand, some engines with well-heated manifolds may have branch pipes sloping down towards the cylinders, so that if any fuel remains unevaporated and is deposited out from the air stream during cold starting, it will run on through the inlet ports and subsequently be evaporated by the heat of compression. The aim is to facilitate cold starting, but a fine balance has to be struck between adequate and over-enrichment as well as under- and over-heating. An example is commented upon in Section 3.48.

Straight rake-type manifolds generally have buffer ends, as in Fig. 11.5. This is so that fuel droplets carried into these ends by their inertia will be deposited there and tend to flow down into the evaporation pool, rather than horizontally around into the adjacent cylinders. These buffers can also be tuned to damp out the pulsations caused by the opening and closing of the inlet valves.

As a general rule it can be taken that, to ensure that the fuel will not be deposited on the walls of the manifold, its passages should be of a size such that the maximum velocity of flow through them is no less than about 70 m/s. If, however, the heat coming from carburettor jacket and hot spot is adequate to ensure good evaporation, the volumetric efficiency will be higher at lower velocities.

Risers are generally the same diameters as those of their associated throttle barrels, which range from between 1.2 and 1.3 times those of their venturis. The volumes of the main tracts in simple manifolds for six-cylinder engines range from about 0.65 to 1.1 times those of all the cylinders, while the corresponding ratios for four-cylinder units tend to be higher, up to a maximum of 1.9:1.

11.3   Manifold heating

Manifold and carburettor heating systems are usually integral. There are two separate requirements: one is to warm the carburettor venturi, spray tube, jets and throttle valve, so that ice will not build up on them; the other is to offset the loss of heat due to evaporation, and thus to facilitate vaporisation of the fuel. A jacket through which engine coolant flows can be used both to prevent

Fig. 11.5 A straight rake-type induction manifold for a four-cylinder engine. The buffer ends trap droplets of fuel travelling at high velocity along the main induction tract. The liquid fuel then runs back, evaporating, along the floor of the tract, which slopes downwards towards the centre. In some instances the lengths of the buffers are tuned to damp out resonant pulsations and the tendency towards inter-cylinder robbery of charge

carburettor icing while the engine is running and to provide, immediately below the riser, a hot spot for evaporating the droplets of fuel from the carburettor above impinging on it. This arrangement is generally adopted for aluminium-alloy manifolds.

Among the disadvantages of water heating is that deflecting the flow away from the jacket when the engine is warm is difficult: as the temperature of the coolant varies, there is a possibility of wide variations in volumetric efficiency. For example, the manifold heating may become ineffective when the vehicle is driven down hill over long distances in low ambient temperatures, or become too effective up hill in high temperatures.

Warm-up need not be too slow provided the thermostat in the coolant system confines the circulation to the cylinders, head and carburettor jacket until the coolant temperature has risen to a reasonably high level, before it is released to the radiator. Taking the coolant from the engine can, however, complicate the installation, and there is of course a risk of leakage.

An exhaust heated hot spot is perhaps better, since it warms up rapidly after the engine has started. Exhaust gas can be directed either at a thin plate clamped in an aperture in the floor of the induction manifold casting or, as in Fig. 3.59, at the floor of the casting. A thin plate of course warms up more rapidly than the heavier section of a casting. If the inlet and exhaust valves are on opposite sides of the head, however, taking the exhaust gases over to the inlet manifold may be impracticable.

In the example illustrated in Fig. 3.64, the weight W on an arm on the spindle tends to cause the deflector to turn the gas away from the hot spot but, when cold, a bimetal coil rotates the spindle in the opposite sense and thus causes the deflector to direct the exhaust gas at the hot spot. As the engine becomes warm the bimetal strip relaxes and allows the weight to rotate the deflector back again, thus avoiding over-heating of the mixture.

With the advent of engine-management systems, electrically heated hot spots have begun to take over. They have several advantages. One is that the electronic control system can be programmed to regulate the current through the heater in relation to both atmospheric and engine temperatures, so that volumetric efficiency is not adversely affected.

On the other hand, if a heater element having a positive temperature coefficient is employed, it will tend to be self-regulating. Rover, for instance, in the engines equipped with their TBI injection system, have a water jacket around the throttle barrel but, additionally, an electrically heated hot spot is installed in the base of the riser. The size of the water jacket is adequate only for complete vaporisation of the fuel under cruising conditions when the engine is warm.

A so-called 'hedgehog-type' electrical heater, Fig. 11.6, is employed. Positive-temperature-coefficient (PTC) barium titanate pills, produced by Texas Instruments, are attached to the underside of the cast aluminium base plate from which the spines project upwards into the air stream. The underside of the base plate, and the pills, are exposed to the flow of hot engine coolant passing through a jacket incorporated beneath the riser. When the engine is started from cold, the ECU switches the heater on, and the heat generated in the pills flows up the spines to evaporate any fuel droplets that fall from the injector on to them. The PTC of the barium titanate means that its electrical resistance increases, and the current therefore drops significantly, as the

Fig. 11.6 Top, an electrically heated hedgehog-type hot spot with its back plate beside it; below, the interior within which the black plate, retained by a circlip, covers a set of positive-temperature-coefficient barium titanate pills, two of which can be seen in front of the hedgehog casting in each illustration. The lead passes along the channel pressed in the edge of the back plate and through the hole in the centre of the internal plate, where it is plugged on to a spider-like connector, the arms of which extend radially outwards to contact the centres of the barium titanate pills, which are bonded to the hedgehog casting

temperature rises through the Curie value. Above this value, evaporation of the fuel droplets cools the heater pills again. Consequently, their temperature oscillates about the Curie value, which is the point at which the magnetic properties of the material change. Ultimately, when the engine coolant temperature rises above 75°C, the ECU opens the relay to switch the heater off.

Curves of temperature and current plotted against time, with a constant rate of cooling, are shown in Fig. 11.7. The temperature coefficient is α in the expression—

$$R = R_0[1 + \alpha(T - T_0)]$$

where R is the resistance at a final temperature T, and R_0 and T_0 are the starting temperatures.

Other electric heating elements, with either positive or negative temperature coefficients, are available enclosed or embedded in thick discs of diameters such that they can be clamped beneath the riser. Electrical heating of the throttle barrel is also a possibility, though it tends to be costly and there is considerable opposition to any additions, unless absolutely necessary, to the demands imposed on the electrical system.

11.4 Materials

Until relatively recently, manifolds have been of either cast iron or aluminium, but now various types of plastics are used as alternatives. Cast iron is the cheapest, but heavy. Aluminium is light, offers good thermal conductivity and therefore rapid warm-up, and can be cast accurately. However, it has some disadvantages. For example, because of its high thermal conductivity, the rate of heat transfer through it to the injectors can adversely affect hot starting. Production of cast aluminium manifolds of any complexity is difficult, requiring multiple cores and introducing problems of porosity. Plastics are very light, but are costly. However, they can be produced to much closer tolerances and, therefore, utilised more economically than metals. Another important consideration is that their tooling costs are much lower.

When Ford pioneered plastics manifolds in production for their 1.8-litre diesel engine in 1989, they used a polyester thermoset, dough-moulding compound. This was adequate for such a simple and compact manifold, but would not have been so for the more complex component required for the Zeta engine, which is larger and, embodying exhaust-gas recirculation equipment, more complex. For such a component, this material would have been too brittle, leading to too short a life and, possibly, damage during handling in both production and service.

Consequently, Ford turned to a thermoplastic manifold, which is manufactured at a rate of about 300 000 per annum, by Dunlop Automotive Composites (UK) Ltd. Only two operators man the moulding line, one

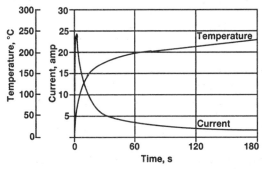

Fig. 11.7–Typical curves of temperature and current plotted against time, for a positive temperature coefficient hedgehog-type induction manifold heater

removing the flash from the cores, and a further nine cope with the finishing, assembly and inspection. The material used is Dupont Zytel, 70G35, a heat-stabilised Nylon 6.6 with 35% short-fibre-glass filler. By virtue of its low viscosity, it easily fills the mould completely, and without moving the cores. Consequently, although this material is more costly than polyester, production cycle times are much shorter.

The primary reason why Ford changed to plastics manifolds was a desire for light weight, and thus improved fuel economy. Indeed, their thermoplastic manifold, complete with brass inserts and other metal components, weighs only 2.5 kg, the net weight of the Zytel being 2.2 kg. The total weight saving relative to aluminium is about 60%, and the overall cost saving, varying with the fluctuating costs of the raw materials, ranges from about 5 to 20%.

11.5   Manifold tuning

Induction manifolds can be designed to improve engine performance by avoiding or taking advantage of four different phenomena. They are as follows—

(1) Inter-cylinder robbery of charge.
(2) Inertia of the flow in the individual branch pipes.
(3) Resonance of the air masses in the pipes.
(4) The Helmholtz effect.

Failure to take the first item into consideration will usually lead to low specific power output. On the other hand, if the engine is run without a manifold, power output is significantly reduced, air intake filtration becomes difficult and engine noise is increased. Therefore, if realistic results are to be obtained, engine bench testing must be done with the standard manifold fitted.

While the first two items listed are simply cyclical, the last two are resonance phenomena. The strongest effects are items (2) and (4), the latter being effective mainly at the lower end of the speed range and the former increasing with engine speed. All four items are superimposed on the flow simultaneously, though not necessarily synchronously.

11.6   Valve timing and inter-cylinder charge robbery

Since the velocity of motion of the piston alternates between acceleration and deceleration according to an approximately sinusoidal law, so also does the depression in the cylinder. The magnitude of this pressure fluctuation is of course affected by both the timing of, and the pressure waves generated by, the opening and closing of the valves. As regards the inlet valve, its opening generates a suction wave, as previously mentioned, and when it closes a pressure wave is generated in its induction pipe.

On a four-stroke engine, the total open and closed periods of the inlet valves are approximately one-third and two-thirds respectively of the 720° of the cycle. During the closed periods, the mixture tends to stagnate in the relevant ports and parts of the manifold. The cyclic rhythm of these effects depends on the number and layout of cylinders and the layout of the crankshaft.

Designers take these effects into account when deciding upon crankshaft layout, though the overriding considerations are dynamic balance and torsional vibration, as explained in Chapters 2 to 4. The timing of the valves, (Section 3.54) on the other hand depends on parameters such as engine speed range, the speeds at which maximum torque and power are required, the degree of flexibility needed and the rate of combustion which, in turn, is a function of combustion chamber design, fuel and engine cycle (two- or four-stroke, spark or compression ignition).

11.7   Crankshaft and cylinder layout in relation to valve timing

As previously indicated, an interference effect, *charge robbery*, arises because the inertia of, and depression in, the air flowing into two adjacent cylinders during overlapping induction phases can cause one to draw in, from the manifold or plenum chamber, some of the air that should be flowing into the other. Consequently, the filling of both cylinders is adversely affected.

Charge-robbing potential decreases with engine speed but, because all cylinders have to fire within two revolutions of the crankshaft, it increases with the number of cylinders. This is the reason for the troughs that appear in torque curves, where manufacturers have not smoothed them for publication. Typical troughs can be seen in Fig. 3.71(*b*). The reason for this tendency to decrease is that, as engine speed increases, less time becomes available for the flow to reverse from one pipe into the other. Consider a typical engine in which the inlet valves are open from 12° BTDC to 52° ATDC, making a total of 242°. With the appropriate length of inlet pipe, this could produce something like the following lower limits below which inter-cylinder robbery could occur—

No. of cycles	Overlap deg.	Lower limit rev/min
3	4	150
4	64	3900
6	124	4500

For induction-system design purposes, the maximum, or limiting, velocity through the throat of the inlet valve is generally taken to be 100.6 m/s (330 ft/s). The mean velocity through the induction pipe is lower, mostly between about $\frac{1}{3}$ and $\frac{2}{3}$ of this value, though instantaneous velocities at points along the pipe may be as high as 140 m/s.

To a limited degree, robbery of the charge between adjacent cylinders can be ameliorated by connecting the ends of the manifold branch pipes to a plenum chamber, where their points of entry into the chamber should be kept as far apart as practicable. However, it is far better to arrange the branch pipes in a manner such that those from any two cylinders whose induction strokes overlap are not situated next to each other.

Consider this in relation to a four-stroke engines with the inlet valves opening at, say, 10° BTDC and closing at 50° ABDC. A twin-cylinder unit with crank throws at 360°, that is, with two connecting rods on a common crankpin, the stagnant period is 120°. Consequently, there is no overlap of the adjacent inlet valve open periods, so the ports can be siamesed. On the other hand, with the cranks at 180°, the stagnation period is 300°, and the induction strokes occur at intervals of 180° and 540°, giving an overlap of 60°. It follows that, unless each cylinder is served separately, the first will have a smaller charge than the second. Moreover, the inertia of the ingoing charge will not be high enough to help significantly to charge the second cylinder.

A 90° V engine fits conveniently into a motor-cycle frame. Generally a single throw crankshaft is employed. In this case, the interval between the induction strokes of cylinders Nos 1 and 2 is 210°, but between Nos 2 and 1 it is only 30°. During the long period of valve closure a higher proportion of the fuel is deposited on the walls of the manifold than during the shorter period. Consequently, unless the surface area of the tract serving cylinder No. 2 is reduced by shortening it, this cylinder will run significantly richer than No. 1.

When designing manifolds, it is important to bear in mind that it is virtually impossible to obtain equal distribution of mixture down the two arms of Y-shaped induction piping. The maldistribution is due to both manufacturing tolerances and because, by the time it reaches the junction, the inflowing mixture has been deflected slightly to one side by the upstream components of the system. Throttle valves, in particular, should be arranged with the axes of their pivot pins at right angles to the division at the junction of the Y shape. Better arrangements for manifolds are either a rectangular twin-pronged fork or a shape more like a T, the arms of which are slightly curved to sweep gently towards the inlet ports.

A horizontally opposed twin-cylinder engine poses similar problems, but exacerbated by the long distance between the inlet ports. An alternative is to accept the cost penalty of one carburettor per cylinder. The stagnation periods are the same as those for side-by-side cylinders with the same crankshaft layouts. An inverted T-shaped manifold is generally employed, the leg of the T being the riser carrying the carburettor; the ends of the arms curve downwards to join the inlet ports. Condensation in the long passages during cold weather renders starting difficult. A generously proportioned hot spot helps, but the penalty is of course lower volumetric efficiency when the engine is hot.

11.8   Three-cylinder engines

Although three-cylinder engines are not of great significance for automotive applications, six-cylinder power units are, and these are generally treated as if they were two three-cylinder units. If the two-stroke engine were to make a comeback, however, the three-cylinder layout could become popular for small cars.

With a four-stroke three-cylinder engine having either a plane crankshaft (cranks at 180°) or with cranks at 120° intervals, the firing order can be either 1 2 3 or 1 3 2. With the former, there is an overlap between the inlet phases of some adjacent cylinders, Fig. 11.8. Generally, a single downdraft carburettor is mounted on a riser over the centre of a straight rake-type manifold having three branch-pipes, one to each cylinder.

On the other hand, with the cranks at 120° and a firing order of 1 3 2, and the inlet valve of cylinder No. 3 opening during the second revolution of the four-stroke cycle, the firing intervals are more uniform. As can be seen from Fig. 11.9(*b*), no overlap occurs between any of the adjacent cylinders, is not of so much concern.

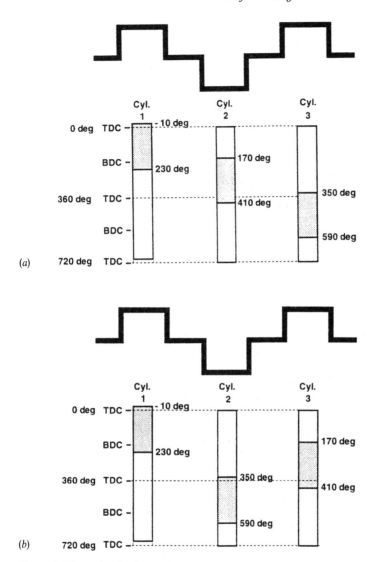

Fig. 11.8 Charts showing inlet valve open periods for a three-cylinder four-stroke engine with a plane crankshaft. Firing orders: (*a*) 1 2 3; (*b*) 1 3 2. There is some overlap between inlet valve opening times in both cases, but in (*b*) the firing impulses of the cylinders are more evenly spaced

11.9 Four-cylinder in-line engines

With four cylinders in-line and crank throws at 180°, although the firing order is generally 1 3 4 2, as in Fig. 11.10, it can be 1 2 4 3. The diagram for the latter firing order can be produced from Fig. 11.10 simply by interchanging the shaded rectangles in the vertical bars for cylinders 2 and 3. A slight advantage of the 1 3 4 2 order is that, when the open periods of the inlet valves in two

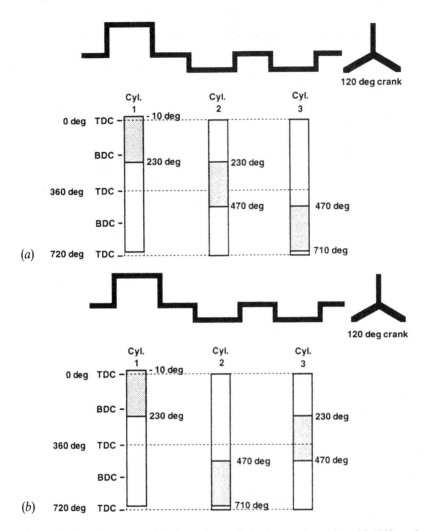

Fig. 11.9 Inlet valve open periods for a three-cylinder four-stroke engine with 120° crank throws. Firing orders: (*a*) 1 2 3; (*b*) 1 3 2. There is no overlap with either, but 1 3 2 is preferred because the firing impulses between cylinders 1 and 3 are more widely spaced

adjacent cylinders overlap, the consequent suction impulses are directed outwards from the centre of the manifold, that is, 2 1 and 3 4.

To avoid the effects of this overlap, the manifold can be cast with twin induction tracts, one serving the outer and the other the inner pair of cylinders, rather like two twin-cylinder engines with cranks at 180° and their shafts joined together. Alternative arrangements, however, are two carburettors on risers situated between each pair of cylinders or, even better, the Weber twin sidedraught carburettor arrangement illustrated in the top left diagram in Fig. 11.2. The balance orifice in the web that separates the pairs of branch pipes

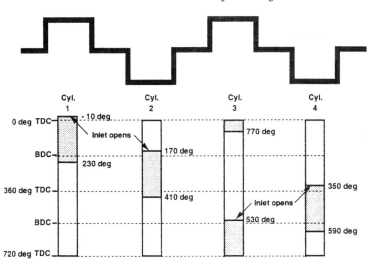

Fig. 11.10 Valve opening periods for a four-cylinder engine with the firing order of 1 2 4 3. The diagram for the more common 1 3 4 2 order can be drawn by transposing the pattern in No. 2 column to column 4, so that Nos 3 and 4 become 2 and 3

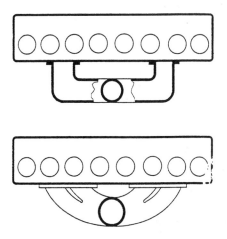

Fig. 11.11 More induction manifold layouts for four-cylinder engines: each branch pipe serves an adjacent pair of inlet valves

ensures that the pressures in each half of the manifold are equal. Two other possibilities are shown in Fig. 11.13.

11.10 Six and eight cylinders in-line

With either of the two 120° crank layouts shown in Fig. 11.12, the six-cylinder engine is dynamically in balance. The upper of these two crankshafts is associated with a firing order of 1 5 3 6 2 4, and the lower with 1 4 2 6 3 5. With

Fig. 11.12 With either of these two crankshaft layouts, a six-cylinder engine is dynamically in balance

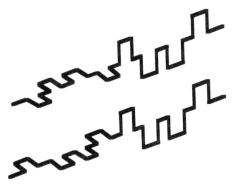

Fig. 11.13 For straight-eight engines, these resembling two four-cylinder plane crankshafts joined coaxially but with their cranks at 90° to each other, are the most common layouts. Their firing orders are 1 5 2 6 4 8 3 7 and 1 6 2 8 4 3 7 5

both, the suction impulses occur alternately in each half of the manifold. Since this is, in effect, two three-cylinder engines, there is some overlap of the suction impulses at intervals of 240°. The first-mentioned firing order gives the more favourable sequence of suction pulses. Again, a balance orifice between the front and rear halves of the manifold is desirable.

The straight-eight engine is uncommon in automotive practice, so it will be dealt with only briefly here. However, readers who are interested in more detail are advised to refer to *Automotive Fuels and Fuel Systems,* Vol. 1, by T. K. Garrett, Wiley. From the two diagrams in Fig. 11.13, it can be seen that the upper crankshaft layout is associated with the firing order of 1 5 2 6 4 8 3 7 5. This is analogous to the 1 2 4 3 firing order of a four-cylinder engine, but firing alternately in first the front and then the rear set of four cylinders instead of in sequence from front to rear. With the lower crankshaft arrangement, the firing order is 1 6 2 8 4 3 7 5. An alternative layout is the equivalent of dividing one four-cylinder engine into two, and placing one half in front of and the other behind a second four-cylinder unit, Fig. 11.14, with a firing order of 1 6 2 5 8 3 7 4. Two straight-eight manifold layouts are illustrated in Fig. 11.15.

11.11   V-layouts

The V-4, V-6 and V-8 layouts were introduced to reduce the length of the

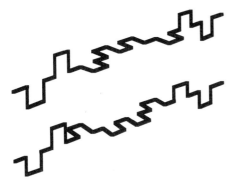

Fig. 11.14 Two alternative layouts for straight-eight engines. Each is the result that would be obtained if one four-cylinder engine were to be divided in two, and the halves attached one to the front and the other to the rear end of a whole crankshaft

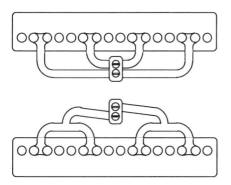

Fig. 11.15 Two induction manifold layouts for straight-eight engines with synchronised throttles. It would be possible, by joining them together at their centres, to use a single-barrel carburettor on each. However, such a large engine generally needs the twin barrels

engine, especially for cars in which it has to be installed transversely. By the early 1960s, the V-4 and V-6 became commonplace even in Europe, and increasingly for transverse installation.

For the V-4, the firing order is normally 1L 3R 4R 2L. In other words, it is the same as that for an in-line four but, in this particular instance, with cylinders 1 and 2 in the left-hand and 3 and 4 in the right-hand banks. The V-angle is usually 60° and the cranks are offset 30° from the 180° positions of those of an in-line four, in a manner such that the pairs of cranks 1 4 and 3 2 are 180° apart. This gives equally spaced firing intervals, the inlet valves of cylinders 1, 2, 3 and 4 opening respectively at −10°, 530, 170 and 350°. An X-shaped manifold with one arm for each cylinder is the norm. A down-draught carburettor is generally mounted on a riser above its centre.

On V-6 engines, the cylinders are usually numbered from the front, but with cylinders 1 and 3 on the left- and 2 and 6 on the right-hand bank. The angle between the banks is 60°, and the crank throws are set at 60° apart relative to their neighbours. Two manifold arrangements are illustrated in Fig. 11.16, the

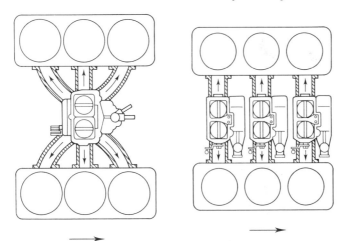

Fig. 11.16 Two induction manifold layouts for V-6 engines with synchronised throttles. V-4 engines have similar manifolds, but the central branch pipes on each side are of course omitted

preferred layout being that on the left. The firing order is generally 1R 4L 2R 3L 5R 6L.

On V-8 engines the banks are 90° apart, and many crankshaft layouts and firing orders are practicable. If cylinders 1 and 4 are on the left and 5 and 8 on the right, a firing order of 1L 8R 4L 3L 6R 5R 7R 2L is generally adopted and, if numbered alternately left and right, it may be 1L 8R 7L 3L 6R 5L 4R 2R, Fig. 11.18. In either case, the manifolds might be described as being in the form of two twin-legged Ys (or asymmetric Xs) mounted on top of each other and each serving two outer and two inner cylinders on opposite banks.

For the alternative manifold layout in Fig. 11.17, the most widely used firing orders are 1R 3R 2R 4R 8L 5L 7L 6L or 1R 3R 7L 4R 8L 5L 2R 6L respectively, according to whether the sequential in-line or alternate left–right numbering is adopted. However, as can be seen from the illustration, the two S-shaped manifolds crossing over each other embody long and clumsy induction tracts and are awkward to fit.

Virtually all V-12 engines have 120° crank throws and a 60° included angle between the banks of cylinders. By fitting a separate manifold or pair of manifolds for each bank, with a twin downdraught carburettor between the banks, they can be treated as two six-cylinder engines. Each barrel of the carburettor discharges into the manifold system serving one bank. An alternative, but one that increases the width of the engine, is one or more carburettors on one or more manifolds on the outer face of each bank. The firing order is either 1 5 3 6 2 4 or 1 4 2 6 3 5 for each bank: in other words, the same as for in-line six-cylinder engines.

11.12   Pipe tuning – the inertia wave

The sequence of the pressure changes initiating the wave due to the closure of the inlet valve is as follows. First, as the gas is drawn into the cylinder, there is a depression in the inlet pipe. When the inlet valve suddenly closes, it therefore

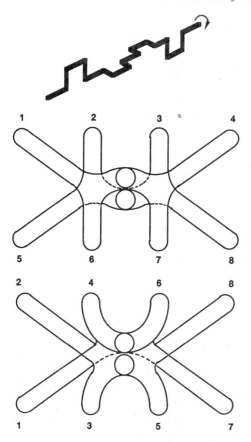

Fig. 11.17 Crankshaft and induction manifold layouts for 90° V-8 engines. The preferred firing order for the upper manifold is 1 5 4 8 6 3 7 2 and, for the lower one, 1 8 7 3 6 5 4 2

reflects a depression wave to the open end of the pipe, whence it is reflected back towards the valve, but as a compression wave. To take advantage of this phenomenon, the length of the induction pipe must be such that the pressure wave arrives back while the valve is open, and therefore the flow into the cylinder is boosted.

The amplitude of the pulse increases directly with engine speed. Obviously, the larger the volume, or mass, of air in the pipe, the greater will be the inertia effect and, therefore, the increase in pressure during the downward motion of the piston. Consequently, the ratio of the induction pipe volume to piston-swept volume is an important criterion. Piston-swept volume is of course determined by other considerations.

Since the desirable length of pipe is dependent on the engine speed at which optimum filling of the cylinder is required, the area is the only independent variable. Two considerations limit the maximum diameter: one is the amount of space available and the other is the practicability of tapering the tract down to the diameter of the inlet valve throat without incurring either increasing manufacturing cost or causing breakdown of the predominantly laminar into

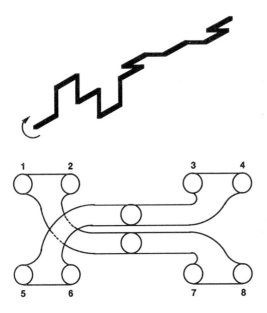

Fig. 11.18 Crankshaft and alternative manifold for a 90° V-8 engine. The most suitable firing orders are 1 3 7 4 8 5 2 6 or 1 3 2 4 8 5 7 6

turbulent flow. Rectangular sections enclose larger volumes of air, but similar limitations apply to them, so circular sections are mostly preferred. An alternative is to taper the pipes, as already mentioned in connection with elimination of pipe end-effects. Tapered pipes do not necessarily have to be installed with their axes parallel, so they can be arranged relatively compactly alongside the head.

11.13   Tuning the pipe to optimise the inertia wave effect

Because it is not sinusoidal, the inertia wave is longer than the standing waves, Fig. 11.19. The time taken for it to travel from the inlet valve and back again is of course twice the length of the pipe divided by the velocity of sound ($2L/c$). Illustrated in Fig. 11.19(a) is a characteristic curve of depression in the valve throat attributable solely to the motion of the piston during one complete revolution. The slight rise in pressure around TDC is caused by exhaust back-pressure during the overlap period. That around BDC is due to the inertia effect.

Optimum benefit is obtained if the length of the induction pipe is such that that the inertia wave arrives back at the inlet valve slightly earlier than the mid-point of its open period, Fig. 11.19(b): if the pressure pulse were to arrive earlier there could be, shortly before the valve closes, a fall in pressure sufficient to cause reverse flow. The optimum timing is such that the pressure difference across the valve port is increased towards the end of the induction period, when it otherwise would be falling. It therefore not only improves the filling of the cylinder, but also tends to prevent reversal of flow at the end of the induction stroke. If, on the other hand, the pulse pressure generated by this

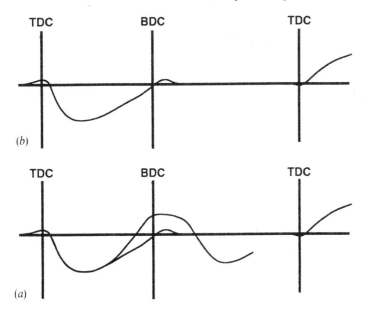

Fig. 11.19 A typical curve of manifold depression at an inlet valve throat is shown at (a) and, at (b), a corresponding inertia wave is imposed on it

wave is timed to persist over the valve overlap period, scavenging of the residual gases will be improved.

Clearly, the reflection time requirement varies with engine speed, so it is more conveniently expressed in terms of degrees of crankshaft rotation than in milliseconds. Another point is that we are concerned with timing relative to the period expressed in degrees of crankshaft rotation during which the valve is open. Therefore, in the calculations in Section 11.17, we shall be using the symbol t to represent time in milliseconds, and θ to represent time in degrees of crankshaft rotation.

11.14 Resonant, or standing, waves

When the air in a pipe is transiently displaced axially, the wave that this displacement generates in it tends to bounce repeatedly from end to end, as indicated in the middle diagrams in Figs 11.20 to 11.22. Each of these diagrams shows the form of the resonant waves in a pipe of a different configuration. Their fundamental frequencies depend on the length of the pipe. Resonance can be initiated by, for example, an external force causing vibration of the pipe, or by some disturbance of the actual air in the pipe at any point along it or at its ends.

In an engine, the principal relevant disturbance is caused by the opening of the inlet valve which, as it lifts off its seat, creates a suction pulse. Because the air is elastic and has mass, it responds by surging forward to restore the pressure, thus initiating an alternating set of depression and compression pulses, which are in fact sound waves travelling along the pipe at the speed of sound.

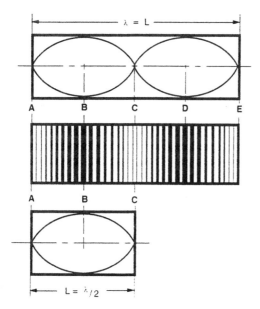

Fig. 11.20 Fundamental and first overtone modes of vibration of air in a pipe both ends of which are closed

At low speeds and with valve overlap, there can also be a slight puff-back of gas from the combustion chamber into the inlet port, but this comes after the initiation of the sound wave. It is less forceful and does not necessarily significantly interfere with the resonance, though the larger the overlap, the longer is the period available for such pulses and others from the exhaust system to have an effect. The kinetic energy in the waves and momentum of the flow increase with engine speed. This is because of the consequently increasing depression in the cylinders, and therefore the pressure ratio across the valve throat, with speed.

Resonant vibration phenomena are associated with mass–spring systems. The mass is that of the column of air in the pipe and the spring element the compressibility of that air. One wavelength λ is a complete cycle, or 2π radians, and therefore is equal to L in the top diagrams in Figs. 11.20 and 11.22, and $4L/3$ in Fig. 11.21. The phase difference between the displacement and pressure waves is always $\pi/4$, or 90°.

At this point, some clarification as to what exactly happens at the open ends of a pipe is necessary. On reaching the open end remote from the valve, a negative pressure wave sucks a slug of air in, and a positive pressure wave propels a slug out. In both instances these effects take place against the influence of atmospheric pressure, so there is an inertia-driven over-swing followed by a bounce-back accompanied by a phase change.

If we plot the axial vibrations in the pipe to a scale such that the maximum amplitude of displacement in each direction equals the radius of the pipe section, they can be illustrated as shown in the top and bottom diagrams in these illustrations. In each, the upper diagram represents the second-order, and the lower one the first-order, or fundamental, mode of vibration.

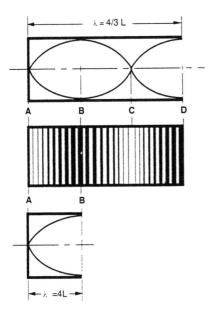

Fig. 11.21 Fundamental and first overtone modes of vibration of air in a pipe one end of which is closed and the other open

From the upper diagram in Fig. 11.20, it can be seen that axial motion of the air is positively stopped by the closed ends, A and E, of the pipe. These ends are therefore displacement nodes. Mid-way between them is a third displacement node, while B and D are displacement anti-nodes. Because the air alternately moves towards and is bounced back from the displacement nodes, A and C and E are pressure anti-nodes. In other words, while the pressure remains constant at B and D, it fluctuates cyclically at A, C and E. This condition can occur in an induction pipe only when both a throttle and inlet valve are closed so, as regards manifold tuning, it is not of practical significance but it is relevant for automotive engineers concerned with body, cab or saloon noise.

If one end of the pipe is open, Fig. 11.21, the air at that end is free to be displaced, so it becomes a displacement anti-node, which accounts for the different arrangement of the displacement curves for the fundamental mode of vibration and the overtones. This condition can arise when the inlet valve is closed and the opposite end of the inlet pipe open.

For a pipe open at both ends, the fundamental and first overtone harmonics are shown in Fig. 11.22. The third harmonic is illustrated in Fig. 11.23. Since this is a condition that arises only when both the inlet valve and pipe end are open, it is of significance in relation to resonance effects initiated by the sudden pening of the inlet valve.

Clearly there must be some displacement beyond the open end before a reflection can occur, so a correction factor has to be applied to the length of the pipe. In fact, the effective length of an open end is L plus about 0.6 times its radius r so, for one open and one closed end, the correction factor is $L(1 + 0.6r)$, and for a pipe with both ends open it is $L(1 + 1.2r)$. The time t taken for the completion of one wavelength is called the periodic time, or the period

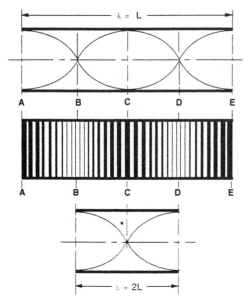

Fig. 11.22 Fundamental and first overtone modes of vibration of air in a pipe both ends of which are open

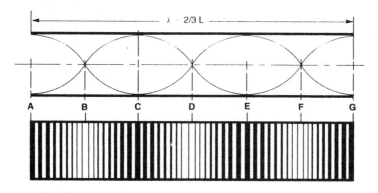

Fig. 11.23 Third harmonic mode of vibration of air in a pipe with both ends open

of the vibration, and the time required for the pulse to return to the inlet valve is $2L/c$, where c is the velocity of sound in the induction pipe. For several reasons, however, this is rarely directly applicable in the context of induction-system tuning. First, the configurations of the ends of the passages are not those of a plain pipe end; secondly, there are other influencing factors such as air temperature and diameter of pipe; thirdly, and perhaps more important, c is not constant for large-amplitude waves such as occur in induction pipes. More accurate results can be obtained if a cyclical mean value of c is used.

11.15 Pipe end-effects

Movement of the air into a pipe in general, and its displacement due to the vibrations, tend to cause turbulence around its open end, reducing the efficiency of flow. This adverse effect can be considerably reduced by flaring the open end of the pipe to form a trumpet of approximately hyperbolic section, so that it guides the air flow smoothly in and thus increases the coefficient of inflow by up to about 2%. The effective length of a pipe with such an end fitting is that of the parallel portion plus about 0.3 to 0.5 of the length of the flare. If the outer ends of the pipes terminate in apertures in a flat plate, or in the wall of a plenum chamber, their flares should not only extend well clear of the flat surfaces but also be clear of any adjacent walls, to ensure that the approach velocity is well below that within the pipe.

Tapering the pipe, increasing its diameter from the inlet port to its open end, also reduces the end-effect. This is sometimes done on very high-speed engines, for example in racing cars. The aim is to reduce the velocity of flow into the open end, and therefore the tendency for turbulence to be generated there. However, it is not conducive to the generation of powerful standing waves. Incidentally, any reduction in the velocity of flow will also reduce the viscous drag between the air stream and the walls of the tube.

11.16 Frequencies, wavelengths and lengths of pipes

From the four illustrations, it is easy to see that the harmonic frequencies for pipes closed or open at both ends are $f_1, f_2, f_3, f_4, \ldots, f_n$, while those of the pipe closed at one end and open at the other are the odd numbers, $f_1, f_3, f_5, f_7, \ldots, f_n$. The formula from which these frequencies can be obtained is $f = c / \lambda$, where c is the velocity of sound in air and λ is the wavelength. The frequencies of the first three modes of vibration in each case therefore are as follows—

Pipes with closed ends	$f_1 = c/2L$	$f_2 = C/L$	$f_3 = 3c/L$
One end open	$f_1 = c/4L$	$f_2 = 3c/4L$	$f_3 = 5c/4L$
Both ends open	$f_1 = c/2L$	$f_2 = c/L$	$f_3 = 3c/2L$

For waves of small amplitude the velocity of sound in dry air is $\sqrt{\gamma p/\rho}$, where p is the gas pressure, ρ is the density, and γ is the ratio of the specific heats of the gas. At the standard temperature and pressure in free air, this velocity becomes 331.4 m/s. Standard temperature and pressure is 298.15 K and 10^5 Pa (1 bar). Potential for some slight confusion arises, however, when referring back to data predating the universal introduction of SI units because, at the latter point, it became 273.15 K (0°C) and 101.325 Pa. At velocities of more than Mach 0.25, viscous friction losses impair performance.

Which ever version of the speed of sound in free air is taken, it is independent of frequency and, because pressure divided by density is constant, it can be considered also to be independent of variations of pressure, certainly of the magnitudes experienced in inlet manifolds. The velocity of sound varies with temperature according to the following relationship—

$$c_\theta = c_0 \sqrt{(1 + \alpha\theta)}$$

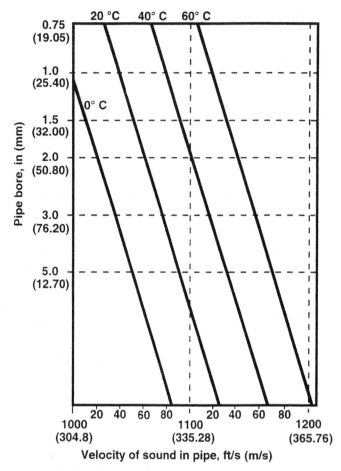

Fig. 11.24 Variation of the velocity of sound with diameter of pipe

where c_θ and c_0 are the velocities of sound at θ and 0 °C respectively, and α is the coefficient of expansion of the gas. While the local velocity of sound is dependent only on the temperature and composition of the gas, in induction pipes it is influenced also by diameter, Fig. 11. 24. This is because of the effect of viscous friction between the gas and the walls of the pipe. Frequency is also affected, but relatively slightly, by the length: diameter ratio and internal smoothness of the pipe, both of which influence the degree of damping of the flow.

Since γ is dependent on the nature of the gas, the presence of fuel vapour, as in carburetted or throttle body injected spark ignition engines, will also affect the speed of sound in the manifold. Even so, because extreme accuracy of calculation is generally unattainable, except possibly where the system comprises a set of straight tubes, this is not of much practical significance. Indeed, induction systems have to be optimised experimentally, for example by the use of telescopic elements, during development.

The amplitudes of the resonant pressure pulsations too are modified by damping. This can be due to roughness of the inner faces of the walls of the

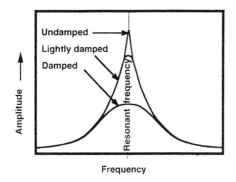

Fig. 11.25 Curves showing the effect of two different degrees of damping on the amplitudes of vibration around the resonant frequency. Without any damping the curve rises to a sharp peak at the point of resonance

induction tract, the presence of bends, and obstructions such as throttle valves and inlet valve stems and guide ends. From damped and undamped resonance curves in Fig. 11.25, it can be seen that the effect of damping is not only a reduction in maximum amplitude but also it rounds off the peak, and spreads the resonance over a significantly wider range of frequencies.

In general, any bends in the pipes should be as close as practicable to the inlet valve ports, blended smoothly into the straight sections, and their radii should not be less than four times that of the bore of the pipe. This arrangement leads to a minimum of both viscous losses and interference with the tendency for the air in the pipe to resonate freely.

11.17   Tuning the pipe to optimise standing-wave effects

The time δ_t, expressed in terms of degrees rotation, required for a single standing wave to be reflected back to its point of origin (the inlet valve) is twice the length of the pipe divided by the velocity of sound $(2L/c)$. From the lower diagram in Fig. 11.21, it can be seen that the wavelength of the fundamental frequency is $4L$, so δ_t is in fact half the periodic time.

During the time δ_t, the crankshaft rotates through an angle $\theta_t = 360N\delta_t/60$. If we substitute for δ_t, this becomes $\theta_t = NL/c$, where the suffix t refers to the time of the reflection, to distinguish it from θ_d, which is the time the valve is open, again expressed in degrees. It follows that if it were practicable for the single wave to be an exact fit in the induction period, it would occur when $\theta_t = \theta_d = 720/2n$, where n is the number of the harmonic or overtone.

If, in our calculations, we substitute the actual velocity of sound in the pipe for that of sound in free air, we have what might be termed an *induction wavefront velocity*. Then perhaps the simplest way to exemplify the time for the wavefront to travel one pipe length is to assume a wavefront velocity of 330 m/s and a pipe length of 330 mm which, of course, will give a time of 1 ms.

11.18   Harmonics of standing waves

In addition to the standing wave at the fundamental frequency, harmonics are generated too by the initial impulse, Fig. 11.26, so a number of modes of vibration, superimposed on each other, occur simultaneously. Consequently,

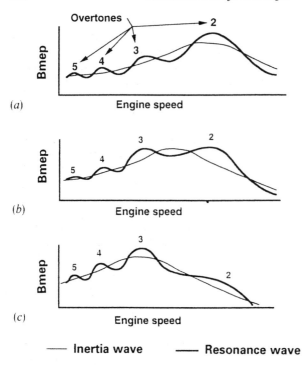

Fig. 11.26 The combined effects of the inertia and resonant standing waves. At (*a*) the system is tuned for maximum power, at (*b*) to obtain a flat torque curve; and at (*c*) for good torque back-up

the initial reflection at the fundamental frequency is accompanied by a ripple of reflections at the smaller wavelengths of the overtone frequencies The successive reflected pulses are of progressively smaller amplitudes owing to attenuation by viscous friction and out-of-phase reflections from bends and other obstructions in their paths. Consequently, no more than one, or possibly two, of the overtones are of significance, depending on whether the valve timing is late or early. Long pipes and high speeds of flow increase both the flow losses and the degree of attenuation of pulses.

The actual timing, relative to the depression wave, of the appearance of the succession of waves at the inlet port can be adjusted by advancing or retarding the opening of the valve. Neither the timing of valve opening nor the duration of overlap, however, have any significant effect on inertia ram, as distinct from resonance (or standing wave) ram, but they do affect exhaust assisted scavenge. Clearly, considerable advantage could be gained by combining induction system tuning with variable valve timing, Section 3.36.

To fit the waves due to resonance into the valve-open period, the following condition must be met—

$$n = \theta_t = \theta_d = 720/2n$$

where *n* is the periodic time of the fundamental standing wave. If θ_t is less than $720/2n$, ripples will be superimposed on the depression pulse; if it is more, they may or may not affect it at all.

Clearly, the inertia effect will be predominant at high speeds. This is because not only do the magnitudes of these pulses increase with speed, but also, as the speed falls, the time available to fit more harmonics into the valve open period increases and, as previously mentioned, each successive wave reflection is weaker than its predecessor. Maximum amplitude of the standing wave occurs when the pipe length is such as to contain a single wave, which occurs when $L = \theta_t = \theta_d = 120°$, and maximum overall amplitude is obtained when both the inertia and the standing-wave effects coincide.

Only the basic information has been given here. In practice, the situation is further complicated by end-effects due to the presence of throttle valves, bends and the progressive motion of the closure of valves and by other factors. For more comprehensive and detailed information, the reader is advised to refer to articles by D. Broome, of Ricardo Consulting Engineers Ltd, and papers by K. G. Hall of Bruntel Ltd. The former is a series in *Automobile Engineer*, Vol. 59, pp. 130, 180 and 262, while the latter were papers presented to the *IMechE and AutoTech 89*, Ref. C399/20. The last mentioned contains a design chart presenting the graphical parameters in a manner such as to facilitate conceptualisation of an optimum induction-system geometry.

11.19 Some practical applications of pipe tuning

The obvious way to vary the length of the induction pipes to vary their resonant frequencies and the timing of the arrival of the reflection of the inertia wave back at the inlet valve is to have telescopic pipes, the lengths of which are controlled by the engine electronic management system. This was in fact done by Mazda on their le Mans winning, Wankel powered racing car. However, whether infinitely variable or a two-position pipe control is used, as in the le Mans car, the mechanism is complex and the whole system bulky and awkward to accommodate in a car. A more practicable alternative is the Tickford rotary manifold, Fig. 11.27, in which the central portion rotates to vary the effective length of the inlet pipe.

A commendably simple system was introduced in 1990 for some of the GM Vauxhall *Carlton*/Opel *Senator* models, and a similar principle has been applied to the Toyota 7M-GE engine. The GM system will be described here. As previously stated, the larger the number of cylinders that have to fire during the two revolutions of the Otto cycle, the more difficult it is to avoid overlap of valve open periods, and therefore inter-cylinder robbery. This problem has

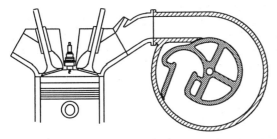

Fig. 11.27 In the Tickford manifold, a central casting, distinguished by closer hatching, can be rotated to vary the effective length of the induction pipes. This portion, extending the whole length of the cylinder block, serves also as a plenum chamber

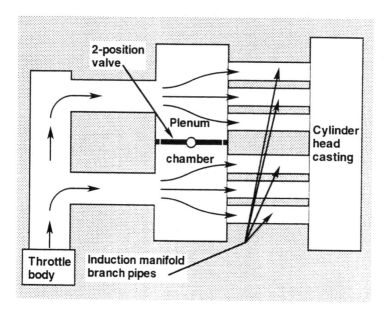

Fig. 11.28 The Dual Ram system, with the two-position valve closed for operation in the 2 × three-cylinder mode. When it is open, the plenum chamber, then unobstructed from end to end, breaks the continuity of the tracts so that only the six short pipes resonate

been avoided in the GM system, called Dual Ram, by controlling the flow through the induction manifold so that at low speeds it has long pipes functioning like those in an in-line six and, at high speeds, it becomes in effect two integrated three-cylinder engines with short induction pipes. How this is accomplished can be seen from Fig. 11.28.

Two tuned pipes take the air from throttle body and inlet plenum to a second, or intermediate, plenum chamber. This chamber is divided by a flap valve so that, when the valve is open it is in effect one, and when closed, two chambers. From the intermediate chamber, the incoming air passes through three short pipes to the six inlet ports in the cylinder head. When the flap valve is closed, which is at the lower end of the speed range, each of the two sets of one long and three short pipes, together with the half plenum between them, form a single tuned duct. At higher speeds, however, the flap valve is open, so that the intermediate plenum, now double the volume, isolates the six short inlet pipes, which of course resonate at a higher frequency. The flap valve is opened at the speed corresponding to the cross-over point of the two torque curves in Fig. 11.29. This valve is actuated by manifold depression and controlled by the ECU. The six 60-mm diameter short pipes are 400 mm long and the length from the inlet valves to the plenum chamber next to the throttle barrel is 700 mm. A smooth transition between the resonant speeds of 4400 and 3300 rev/min respectively is the outcome of this arrangement.

Another tuned induction pipe system of interest is that of the Volvo 2 litre 850 GLT engine, Fig. 11.30. Each induction pipe comprises a pair of siamesed ducts, a section through the top of the pair resembling a figure-of-eight. The diameter of the upper loop of the eight is slightly smaller and its length about twice that of the lower one over which, at its end nearest the head, is a steel flap

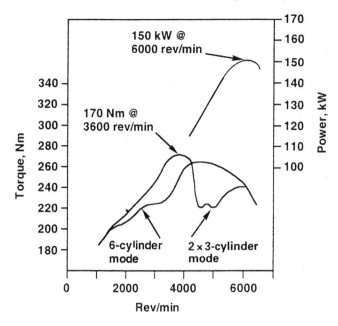

Fig. 11.29 Power and torque curves for the *Carlton* GSi 300 24V engine equipped with the Dual Ram system

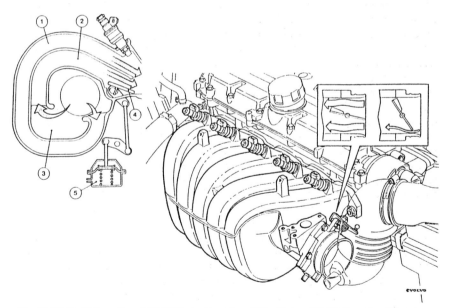

Fig. 11.30 (*Left*) A sectioned V-VIS induction pipe of the Volvo 850 GLT engine. The pipe (1) is about twice the length of pipe (2), and (3) is a plenum chamber. Flap valves (4), one in each pair of pipes, are all moved simultaneously by a single manifold depression actuator (5). (*Right*) The complete system with, inset, a diagram showing how, by thickening one edge of the throttle valve, two-stage opening is obtained to provide a smooth take up of drive from the closed throttle condition. See also Fig. 9.10

valve. Under the control of the ECU, this valve is initially held fully open by its return spring, but moves towards the fully closed position as the manifold depression increases. Each valve is fitted with a rubber seal to obviate the need for machined seats and, when open, is parked in a recess in the pipe so that it does not interfere with the air flow.

At speeds below 1800 rev/min, both ducts are open, providing capacity for acceleration, though whether this adversely affects transient response is open to question. Between 1800 and 4200 rev/min, but only so long as the throttle is 80° or more open, the shorter duct is closed. Above 4200 rev/min, both ducts are open again to afford maximum flow potential. In this condition, because one pipe is half the length of the other, the air is both resonate simultaneously but in different modes. Calculated volumetric efficiencies are shown in Fig. 11.31 and the actual power and torque curves in Fig. 11.32.

11.20   The Helmholtz resonator

Another device that is being applied increasingly to induction systems is the Helmholtz resonator, Fig. 11.33, which, because a larger mass of air may be displaced by it, can be more powerful in its effect than pipe tuning. Because it is effective over only a very narrow band of frequencies, its use has been confined in the past to generating what has now become known as *anti-sound*, to eliminate induction pipe roar and exhaust boom. Anti-sound is of course a sound of the same frequency but opposite in phase to that which has to be eliminated. More recently, the principle of its application to boost the low speed performance of turbocharged engines has been described in two papers presented before the IMechE in May 1990, at the Fourth International Conference on Turbocharging and Turbochargers. One is Paper C405/013, by G. Cser, of Autokut, Budapest, and the other is Paper C405/034, by K. Bsanisoleiman and L. Smith, of Lloyd's Register, and B. A. French, of the Ford

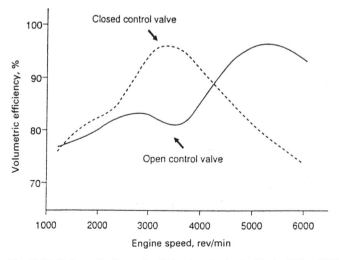

Fig. 11.31 Estimated volumetric efficiencies obtained with the Volvo V-VIS system

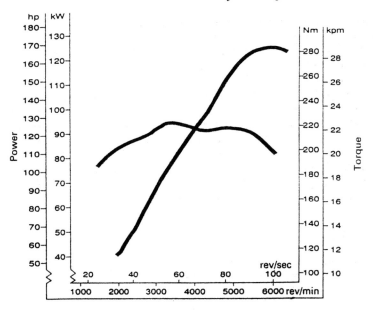

Fig. 11.32 Power and torque curves of the Volvo 850 GLT engine

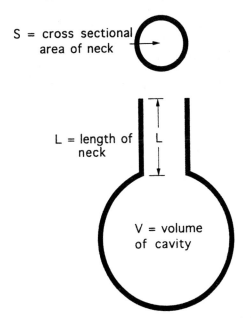

S = cross sectional area of neck

L = length of neck

V = volume of cavity

Fig. 11.33 Diagrammatic representation of the Helmholtz resonator

Motor Company. An earlier and equally interesting paper on this subject by Cser was C64/78, presented at the 1978 conference.

In general, Helmholtz resonators have been used also to detect extremely faint noise signals. Another application is the damping of resonant vibrations, the damping effect being increased by, for example, placing porous material in the neck of the resonator. Also, it can be used to increase the sound pressure in an acoustic field at a particular frequency. This is of interest because of its potential for enhancing the effectiveness of a tuned manifold. By the late 1980s, Helmholtz resonator principle began to be widely applied also as a primary engine-tuning device. Although it is effective at only one frequency, it is particularly useful for improving volumetric efficiency at relatively low engine speeds.

For influencing induction-pipe resonances, either of two locations for the open end of the neck of the resonator are effective. If it is positioned at a displacement anti-node in the induction tract, it is in phase and therefore increases the amplitude of displacement of air in the tract. On the other hand, if placed at a displacement node, it tends to counteract the resonant vibration of the air in the tract, because it is $\pi/2$ out of phase.

The Helmholtz resonator generally comprises a short tube connected to an otherwise totally enclosed cavity. This cavity can be of any shape, though a bulbous form may be preferred because it is less likely than almost any other to have natural modes of vibration that could influence the system as a whole. The air in the neck is assumed to act like a piston, alternately compressing and expanding that in the cavity. In other words, the air in the neck constitutes the mass, while the compressibility of that in the cavity forms the spring of a spring–mass system.

The wavelength of the vibrations it generates is large relative to the dimensions of the cavity. Its natural frequency f corresponds to the value of the angular frequency at which the reactance term disappears, and is therefore given by—

$$2\pi f = c\sqrt{(S/LV)}$$

i.e.

$$f = (c/2\pi)\sqrt{(S/LV)}$$

where c is the speed of sound, L the length of the neck, S the area of the neck and V the volume of the cavity. From the last term in the equation, it can be seen that the natural, or resonant, frequency increases as the square root of the area of the neck, and decreases as both the square root of the length of the neck and of the volume of the cavity, or resonant chamber, are increased. Incidentally, provided the length of the tube is small relative to the wavelength of the sound at the resonant frequency, the effective length of the neck numerically is approximately the actual length plus 0.8 times S. As the cross-sectional area of the neck is increased, the mass of air in it increases, but the relative viscous damping effect falls rapidly. Clearly, however, both the mass of the air and the viscous friction in the neck increase linearly with its length, so the main consideration is the area:length ratio. As regards the volume of the resonator cavity, the smaller it is the higher is its spring rate and therefore also both the amplitude and frequency of oscillation.

An important consideration is the energy content, or what might be termed the 'power' of the resonator, which is a function of the mass of the air in the neck. Therefore, the larger the volume of the neck, the greater is the effectiveness of the system. In acoustical applications, the Helmholtz resonator is most effective at the lower end of the audible frequency range, down to about 20 Hz which, expressed in terms of incidence of inlet valve closure, is from about 600 rev/min upwards.

11.21 Helmholtz resonators in automotive practice

In automotive applications, however, things are not at all simple. For instance, it has been suggested that the Helmholtz system may comprise the induction pipes with their cylinders acting as resonant cavities, but the volumes of the cylinders are of course varying continuously. The suggestion is that the resonator volume can be taken to be that when the piston is at mid-stroke, which is half the piston displacement plus the clearance volume. At this point, when the downward velocity of the piston is at its maximum, a rarefaction wave transmitted from the inlet valve to the open end of the pipe is reflected back as a pressure wave into the cylinder. Optimum tuning is obtained when this wave arrives in the cylinder just before the inlet valve closes. Since the resonance does not continue after valve closure, this type of resonator acts independently of engine speed and therefore can be effective over the whole speed range but, as previously indicated, decreasing in effectiveness as frequency increases. Peak effectiveness occurs when the resonator frequency is approximately double that of the piston reciprocation. Application of the Helmholtz resonator has been investigated in detail and reported by Thompson and Engleman, in ASME publication 69-GDP-11, and a good summary of the situation is presented by Tabaczinski, in SAE Paper 821577.

11.22 Alternative Helmholtz arrangements

In some instances, though mainly in the past, plenum chambers have been designed into the system simply to smooth out pulsations in the flow, or as a means of terminating, or isolating, the ends of tuned inlet pipes. However, as a Helmholtz resonance cavity, it may be a separate component introduced into almost any part of the induction system. For instance, a plenum chamber or the filter housing with its inlet, or zip, tube may be utilised for this purpose.

In most instances, the pressures and densities (and therefore the masses) of air in the pipes will be lower than that of the air in the plenum, and this will affect the resonant frequency. Moreover, the effective volume of the plenum and therefore the resonant frequency and effectiveness of the system may vary according to whether the throttle valve is open or closed. In the latter condition, the incoming air will be passing the edges of the throttle at or near sonic velocity. Other factors come into play too, such as the damping effect of various features of the induction system, including the throttle valve. Damping can be actually helpful, in that it reduces the peakiness of the resonance curve and spreads the response of the resonator over a broader speed, or frequency, range.

With the advent of computer modelling, the introduction of Helmholtz resonators into induction systems no longer involves tedious and repetitive calculations. One such model is the Merlin Model for the Diesel Engine Cycle, information on which is available from Dr Les Smith, Performance Technology Department, Lloyd's Register, Croydon CR0 2AJ.

11.23   Examples of the application of the Helmholtz principle

Perhaps the most common practical application of the Helmholtz principle in the 1960s and 1970s was the suppression of unwanted frequencies in the noise spectrum issuing from the air intake. For this purpose, the air passes through a tuned length of pipe into the filter housing, which serves as the resonator. Any damping provided by the presence of the filter element broadens the band of frequencies over which the noise suppression system is effective.

More recently, it has been used on, for example, turbocharged diesel engines. As the engine speed falls, so also does the torque but at a disproportionately high rate, owing to the square-law performance characteristic of the turbocharger. There is also a tendency for black smoke to be generated. In these circumstances, the low frequency effectiveness of the Helmholtz resonator can be put to good use. A paper on this subject, No. 790069, by M. C. Brands, was presented at the February 1979 SAE Congress.

Similar conditions can arise in naturally aspirated engines with tuned induction pipes. The energy content, or power, of the inertia wave of a tuned induction pipe falls with engine speed. More significantly, however, not only is the tuning of the pipes invariably optimised for the upper speed range, but also, at some speeds, the pulse can actually be in a negative phase when the inlet valve is open, thus reducing the mass of air entering the cylinders below that which would occur even without a tuned system. Consequently, a Helmholtz resonator tuned to oppose the negative inertia wave can significantly increase volumetric efficiency at low speeds. Although it may tend to detract from the volumetric efficiency at high speeds, this effect is not necessarily of much significance owing to the characteristically weak performance of such resonators in the high-frequency range. Incidentally, the latter phenomenon is accentuated by the fact that the spring rate of the resonator is inherently low, while the pressure and therefore the mass of the air in the pipe tends to increase with load, and the resistance to flow increases as the square of its velocity.

11.24   Application to V engines

An objection to the Helmholtz resonator is that it tends to be bulky and therefore difficult to accommodate beside the cylinder head. On the other hand, it can be fitted conveniently between the banks of cylinders in an engine of V layout, especially a V-8 unit. The alternative of fitting long induction pipes to improve volumetric efficiency at low speeds is, by comparison, relatively unattractive. Application to six-cylinder and V-8 engines has been discussed by Watson, in IMechE Paper C40/82.

Induction-system layout for a six-cylinder engine is relatively simple, since resonator cavities can be allocated to groups of three cylinders, each group being isolated by a plenum chamber, as in Fig. 11.34. For a four- or eight-cylinder engine, the situation is more complicated because of the uneven firing intervals. However, in the case of a V-8 with intervals between the passage of the spark in each bank of 0, 180, 270 and 450°, and with valve open periods of 240°, the problem can be overcome by installing two resonators and linking the inlet pipes of the central pair of cylinders in each bank across to the resonator on the opposite side of the V, as in Fig. 11.34(*b*). In this way, the inlet

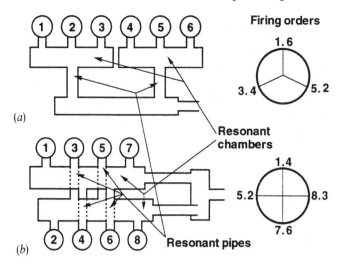

(a)

Firing orders

(b)

Fig. 11.34 Diagrammatic representation of induction-system layouts embodying Helmholtz resonator chambers

events can be equally spaced at 180° intervals although they do not coincide precisely with the corressponding 240° valve periods.

11.25 The Helmholtz resonator in combination with tuned pipes

A naturally aspirated application has been explored in detail, by Shimamoto *et al.*, in Vol. 30, No. 259, 1987, of the *JSME*. A diagrammatic representation of the induction system used for this work is shown in Fig. 11.35. The effects of varying the geometry of the resonator are shown in Fig. 11.36, which illustrates the calculated results obtained by these authors when applying the Helmholtz principle to a single-cylinder engine. Diagram (*a*) shows that increasing the diameter of the neck of the resonator causes the peak of the volumetric efficiency curve to approach the speed corresponding to the tuned frequency of the resonator. Progressively reducing it increases the resistance to flow in the neck and thus increases the time needed for energy accumulation in the resonator. Consequently, as the engine speed is increased, the phase of the resonant vibration lags, causing it to become progressively less effective.

Increasing the length of the neck, diagram (*b*), has a similar effect. Since, in

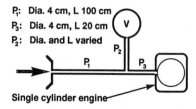

P₁: Dia. 4 cm, L 100 cm
P₃: Dia. 4 cm, L 20 cm
P₂: Dia. and L varied

Single cylinder engine

Fig. 11.35 Diagram used by Shimamoto to explain the effects of varying the geometry of a Helmholtz-tuned induction system

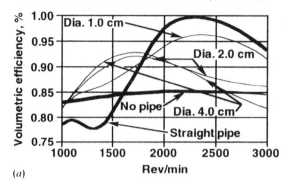

(a)

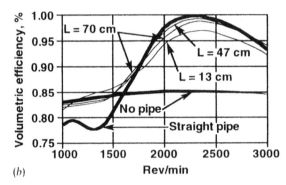

(b)

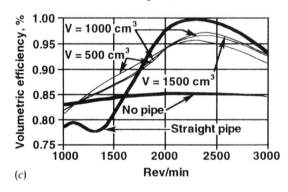

(c)

Fig. 11.36 Curves showing the effects of changes in geometry of the Helmholtz system depicted in Fig. 11.35: (a) varying the diameter of the resonator tube P_2, (b) its length, and (c) the volume of the resonator

this instance, the diameter of the neck was kept constant at 1.0 cm, the tuned frequency of the resonator is high. On the other hand, increasing the volume of the resonator chamber, diagram (c), appears also to increase the turbulence in the flow and therefore the energy loss in it. This effect, however, is slight relative to those of varying the diameter and length of the neck. Although this seems to point to a need to keep the volume of the reservoir low, the generally accepted minimum is about 1.5 times the swept volume of the engine. The results obtained with multi-cylinder engines are similar. However, owing to phase interference effects, they are not so pronounced.

Chapter 12

Emission control

Attention was first directed to atmospheric pollution in Los Angeles in 1947. Subsequently, in 1952, Dr Arie J. Haagen-Smit asserted on the basis of his research that, at least locally, it was due mainly to automotive exhaust emissions. It was subseaquently said, however, that it would have cost the USA less to have moved Los Angeles than to have converted all their vehicles to reduce the obnoxious emissions to the levels now required by law! Japan was close behind the USA with emission control laws, and Europe has practically caught up.

Given complete combustion, each kg of hydrocarbon fuel when completely burnt produces mainly 3.1 kg of CO_2 and 1.3 kg of H_2O. Most of the undesirable exhaust emissions are produced in minute quantities (parts per million), and these are: oxides of nitrogen, generally termed NO_x, unburnt hydrocarbons (HC), carbon monoxide (CO), carbon dioxide (CO_2), lead salts, polyaromatics, soots, aldehydes ketones and nitro-olefins. Of these, only the first three are of major significance in the quantities produced. However, concentrations in general could become heavier as increasing numbers of vehicles come onto our roads. By the end of the 1980s, CO_2 was beginning to cause concern, not because it is toxic but because it was suspected of facilitating the penetration of our atmosphere by ultra-violet rays emitted by the sun. Controversy has raged over lead salts, but no proof has been found that, in the quantities in which they are present in the atmosphere, they are harmful. For many years, manufacturers of catalytic converters pressed for unleaded petrol because lead deposits rapidly rendered their converters ineffective.

Carbon monoxide is toxic because it is absorbed by the red corpuscles of the blood, inhibiting absorption of the oxygen necessary for sustaining life. The toxicity of hydrocarbons and oxides of nitrogen, on the other hand, arises indirectly as a result of photochemical reactions between the two in sunlight, leading to the production of other chemicals.

There are two main oxides of nitrogen: nitric acid and nitrogen dioxide, NO and NO_2, of which the latter is of greatest significance as regards toxic photochemical effects. Under the influence of solar radiation, the NO_2 breaks down into NO + O, the highly reactive oxygen atom then combining with O_2 to make O_3, which of course is ozone. Normally, this would then rapidly recombine with the NO to form NO_2 again, but the presence of hydrocarbons inhibits this reaction and causes the concentration of ozone to rise. The ozone then goes on, in a complex manner, to combine with the other substances present to form chemicals which, in combination with moisture in an atmospheric haze, produce what has been described as the obnoxious smoky fog now known as smog.

Unburnt hydrocarbons can come from evaporation from the carburettor float chamber and fuel tank vent as well as from inefficient combustion due in different instances to faulty ignition, inadequate turbulence, poor carburation, an over-rich mixture, or long flame paths from the point of ignition. The relationships between emissions and the air:fuel ratio are illustrated in Fig. 12.1. Other factors are over-cooling, large quench areas in the combustion chamber, the unavoidable presence of a quench layer of gas a few hundredths of a millimetre thick clinging to the walls of the combustion chamber, and quenching in crevices such as the clearance between the top land of the piston and the cylinder bore.

12.1 Early measures for controlling emissions

A basic essential for spark ignition engine emission control is a carburettor or injection system capable of extreme accuracy in metering the fuel supply relative to the air entering the engine. Diesel engine emissions will be covered from Section 12.17 to the end of the chapter. Some of the emission control measures taken in carburation are described in Sections 9.5, 9.9, 9.12 and 9.14, while all modern fuel injection systems have been developed from the outset specifically for accuracy of metering, for minimal emissions and best fuel economy. Irregular combustion must be avoided during idling and, on the overrun, the mixture must either be totally combustible or the fuel supply totally cut off. In the latter event a smooth return to normal combustion when the throttle is opened again is essential. Idling speeds are typically 750 rev/min with automatic and 550 rev/min with manual transmission.

A capsule sensitive to manifold depression could be used to retard the

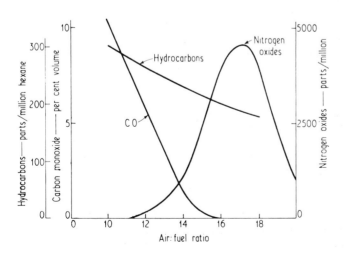

Fig. 12.1 Effect of air:fuel ratio on exhaust emissions

ignition in the slow-running condition, the manifold depression tapping being taken from a position immediately downstream from the edge of the throttle when it is closed. A centrifugal mechanism may retard the ignition from about 5 to 15°, while the depression capsule can further retard it by perhaps another 15°.

Sudden closing of the throttle, and the consequent rapid increase in the depression over the slow-running orifices, may draw off extra fuel that cannot be burned completely. To overcome this, a gulp valve has been developed for admitting extra air into the induction manifold in these circumstances.

Coolant thermostats have been set to open at higher temperatures for improved combustion in cold conditions. Also, thermostatically controlled air-intake valves deflect some air from over the exhaust manifold to mix with the remainder to maintain the overall temperature at about 40–45°C, thus assisting evaporation in cold weather.

In the nineteen-seventies much effort was devoted to the development of various stratified charge engines, some of which are described in Chapter 18. By the nineteen-eighties, however, high-compression lean-burn systems such as that described in Section 4.18 had been the main practical outcome. With increasing pressures for fuel economy as a means of reducing CO_2 output, interest in stratified charge began to surface again in the early 1990s.

Positive crankcase ventilation, PCV, totally eliminated pollution originating from crankcase fumes, and at a modest cost, Section 12.3. By 1968 weakening the air:fuel ratio, retarding the spark timing, preheating the air passing into the engine intake and, on some models, the installation of a pump to inject air for oxidising the HC and CO in the exhaust system reduced the total emissions about 39–41% by comparison with the 1960 cars. In absolute terms, emissions from GM cars, for example, had been reduced to 6.3 and 51 gm/mile respectively for HC and CO. However, there were still no controls on NO_x. New developments then being investigated included carbon canister systems for the temporary storage and subsequent combustion of evaporative emissions of fuel and catalytic converters for controlling exhaust emissions.

12.2 Evolution of the US Federal test procedures

In 1970 the US Congress had adopted regulations requiring by 1975 a reduction of 90% on the then current emissions requirements. The Federal Environmental Protection Agency was formed and introduced a better method of sampling. Previously all the exhaust gas had been collected in one huge bag and then analysed. This had the disadvantage that it gave absolutely no indication of how the engine behaved under the different conditions of operation during the test; moreover, in some circumstances, some of the gases interacted in the bag, giving misleading results.

The new requirement was for collection into three bags, one for each main stage of the test, Fig. 12.2. The first, termed the *cold-transient stage*, comprised cycles 1 to 5 of the test, which represented the beginning of a journey starting from cold. Next came 13 cycles, representing the remainder of the journey with the engine warm and including some operation at high temperature. After this, the engine was shut down for 10 min, to represent a hot soak, before starting up again and repeating the first five cycles, for what is termed the *hot transient stage* of a journey started with a hot engine.

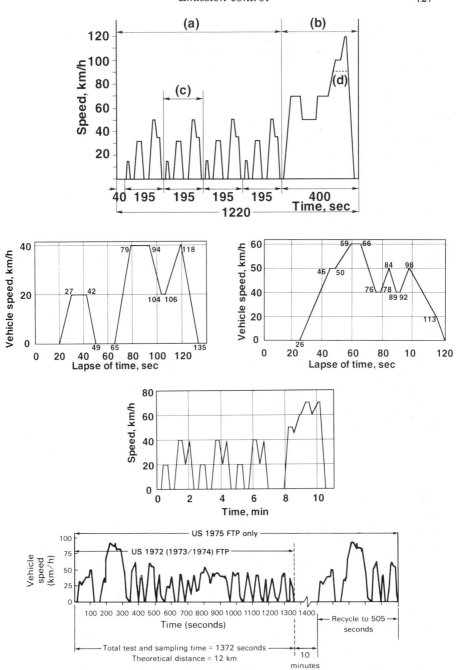

Fig. 12.2 Top, in 1992, a hot transient mode, (b) was added to the earlier EEC test cycle, (a). Sampling begins after 40 s. The lower limit, (d) is for vehicles incapable of attaining the higher maximum speed. Mid-left, the Japanese 10 mode cycle, which has to be run through 6 times. Mid-right, the Japanese 11 mode cycle to be run through 4 times. Below, Japanese hot transient test introduced in the early 1990s, comprising 24 s idle, the first three modes of its predecessor and a 15 mode high speed, or hot, test. Bottom, US Federal Test Program (FTP)

By applying weighting factors to alter the relative effects of the three bag analyses on the totals the results of the test are easily adjusted to represent different characteristic types of operation. Obviously HC emissions are high for the period following starting from cold, while NO_x emissions are of little significance except under hot running conditions.

Conditions that encourage the generation of NO_x in the combustion chamber are principally temperatures above about 1350°C in the gas at high pressures and the length of dwell at those temperatures. Exhaust gas recirculation, EGR, was introduced to lower the temperatures of combustion in cars for California in 1972, and extended nationwide in 1973, when legal limits, at 3.1 gm/mile, were first imposed on NO_x emissions. Subsequently, the overall requirements were progressively tightened as shown in Table 12.1. The 1981 regulations were so tight that, for diesel engines and innovative power units, a delay of 4 years had to be allowed for NO_x and up to 2 years on CO. Since then the regulations have been tightened periodically and clearly this process will continue.

12.3   Catalytic conversion

At this point while some other manufacturers were promoting the lean-burn concept as the way forward, GM engineers, accepting the penalty of low Octane Number, opted for unleaded fuel and catalytic conversion for meeting regulations on both emissions and fuel economy, while avoiding adverse effects on engine durability. As a first step, all their car engines for 1971 were designed for fuel rated at 91 Motor Octane No., mainly by reducing compression ratios and modifying the valves and their seats.

They argued that unleaded fuel offers several benefits: first, the major source of particulate emissions, lead oxyhalide salts, is eliminated; secondly, there is a consequent reduction in combustion chamber deposits, which have the effect of thickening the boundary layers in the gas in the combustion chambers and this, by quenching them, encourages the formation of HC; thirdly, a further reduction in HC is obtained because of the additional oxidation that ocurs in the exhaust system owing to the absence of lead additives and also because the lead salt deposits tend to cause deterioration of the NO_x control system by adversely affecting the flow characteristics of the EGR orifices; fourthly, maintenance of spark plugs, exhaust systems and the frequency of changing lubricating oil are all reduced by the elimination of the lead salts, as also of course is the generation of acids by the halide scavengers that have to be used with them; finally, because catalytic converters call for unleaded fuels controversy over the alleged toxic effects of lead salts in the environment was neatly side-stepped.

Table 12.1

Year	HC, CO and NO_x respectively gm/mile
1975	1.5, 15.0 and 3.1
1977–9	1.5, 15.0 and 2.0
1980	0.41, 7.0 and 2.0
1981 and beyond	0.41, 3.4 and 1.0

12.4 Two-way catalytic conversion

The emissions regulations for the 1975 model year required reductions of 87% in HC, 82% in CO and 24% in NO_x by comparison with 1960 levels. GM concluded that to meet these requirements while simultaneously improving not only economy but also driveability, both of which had deteriorated severely as a result of emission control by engine modifications, two-way catalytic converters were needed. The term two-way conversion implies oxidation of the two constituents in the exhaust, HC and CO, to form CO_2 and H_2O. Such a converter therefore contains only oxidation catalysts and, moreover, without oxygen in the exhaust cannot function. Consequently, the air–fuel mixture supplied to the engine must be at least stoichiometric or, better still, lean. Incidentally, the earlier practice of feeding air into the exhaust was intended primarily for burning the excess hydrocarbons during the first five cycles of the test after a cold start with engines equipped with carburettors. It is unnecessary with the accurate regulation of air:fuel ratio by computer-controlled injection.

If a spark plug were to fail, air–fuel mixture would enter the two-way catalytic converter and burn, seriously overheating the unit. Consequently, high energy ignition systems became a necessary adjunct for the 1975 models, and copper-cored spark plugs were fitted to obviate cold fouling. The overall result of all these measures on the GM models was a reduction in fuel consumption of 28% by comparison with that of their 1974 cars. By 1977 this figure had been further improved by 48% and, by 1982, owing to the stimulus of the Corporate Average Fuel Economy (CAFE) legislation, by 103%. Incidentally, under the CAFE legislation the average fuel consumption of all cars marketed by each corporation in the USA had to improve in stages, from 18 mpg in 1978, by 1 mpg each year to 1980, then 2 mpg annually to 1983 and again by 1 mpg for 1984, and then to 0.5 mpg, to 27.5 mpg, for 1985.

12.5 The converter

Two-way catalytic converters comprise a container, usually of chromium stainless steel, and the catalysts and their supports, all enclosed in an aluminised steel heat shield, Fig. 12.3. Initially, the alumina pellet type of support for the catalyst was the most favoured because it had been developed to an advanced stage in other industries. The monolithic type (one-piece) did not go into regular production until 1977.

Either platinum (Pt) alone or platinum and palladium (Pl) are used as catalysts. The cost of this noble metal content is of the order of 15 to 20 times that of the stainless steel shell that houses them, so other catalysts such as copper and chromium have been tried, with some success, but have not come into general use because they are prone to deterioration owing to attack by the sulphuric acid formed by combustion of impurities in the fuel. A typical two-way converter for an American car contains about 1.6 g of noble metals in the Pt:Pl ratio of 5:2.

12.6 Catalyst support

Considerable development effort has been devoted to monolithic catalyst supports in the form of one-piece extruded ceramic honeycomb structures

Fig. 12.3 The AC-Delco stainless steel housing for monolithic catalyst carriers is enclosed in an aluminised steel outer casing. Sandwiched between top halves of the outer and inner shells is heat insulation material. The perforations in the lower half of the outer shell, termed the grass shield, facilitate local cooling

having large surface areas on to which the noble metal catalysts are deposited. Gas flow paths through them are well defined, and their mass is smaller than that of the pellet type, warming up more rapidly to their working temperature of about 550°C. In some applications, for ease of manufacture two such monoliths are installed in tandem in a single chamber, Fig. 12.4.

Pellet systems are, nevertheless, widely employed in the USA for trucks, where compactness is not an overriding requirement but durability under extremely adverse conditions is. The pellets are relatively insensitive to thermal stress because they can move to relieve it. Moreover, the hottest part of such a bed is near its centre, whereas that of a ceramic monolith is about 25 mm from its leading edge, accentuating thermal stress problems. Packaging for pellets, on the other hand, is more complex, so both assembly and servicing of the monolithic type are easier.

12.7   Metallic monoliths for catalytic converters

Another important aim is of course durability at both very high and rapidly changing temperatures. Ceramics do not satisfy all these requirements, so efforts were directed at producing acceptable metallic matrices. These offer advantages of compactness, minimum back-pressure in the exhaust system, rapid warm-up to the minimum effective operating temperature (widely termed the *light-off* temperature) which, for this type of monolith, is claimed to be about 250°C.

Two obstacles had to be overcome. First was the difficulty of obtaining adequate corrosion resistance with the very thin sections needed for both compactness and acceptably low back-pressure. Secondly, it was difficult to

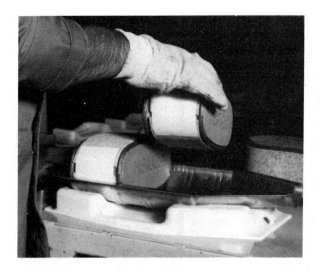

Fig. 12.4 Two monolithic catalyst carriers being assembled in series into their casing in the AC-Delco factory at Southampton

join the very thin sections while retaining the robustness necessary to withstand the severe thermal loading and fatigue.

By 1989, these problems had been solved by Emitec, a GKN-Unicardan company in Germany. They had developed a special stainless steel alloy, called Emicat, which they used in foil strips only 0.04 mm thick to construct the catalyst carriers in monolithic form. These are now made up into matrices comprising alternate plain and corrugated strips, wound in an S-form, as shown in Fig. 12.5. The matrices are inserted into steel casings and the whole assembly joined by a patented high temperature brazing process. S-form matrices proved to be more durable than spirally wound cylindrical units.

Emicat is an Fe, 20% Cr, 5% Al, 0.05% Y alloy. Ytrium, chemical symbol Y, has a melting point of 1250°C. It is a metal but with a strong chemical resemblance to rare earths, with which it therefore is usually classified. Its oxide, Y_2O_3, forms on the surface of the foil and protects the substrate from further oxidation. At a content of 0.05% ytrium is very effective in enabling the alloy to withstand not only temperatures of up to 1100°C over long periods, but also the higher peaks that can be attained in catalytic converters in the event of a malfunction of the ignition system. Even better protection, however, can be had by increasing it up to 0.3%, though at higher cost.

The advantages obtained with the monoliths made of Emicat include: rapid warm-up; resistance to both thermal shock and rapid cyclical temperature changes up to well over 1300°C (both due to the good thermal conductivity of the material and low heat capacity of the assembly); minimal back-pressure, by virtue of the thin sections of the catalyst carrier foil, Fig. 12.5; compactness due to thinness of the sections and the absence of the mat needed around a ceramic monolith (to absorb its thermal expansion); large area of the catalyst exposed to the gas flow (owing to the high surface : volume ratio of the foil); and avoidance of local overheating, by virtue of both the compactness of the unit and the good thermal conductivity of the metal as compared with that of

Fig. 12.5 Comparison between ceramic and metallic monoliths. The thermal stresses in monoliths comprising alternate strips of plain and corrugated metal foil in looped S-form are considerably lower than those that are simply spirally wound

ceramic; and, finally, because the complete unit is directly welded into the exhaust system, the costs of assembling ceramic monoliths and their wire mesh or fibre mat elastic supports into their cans are avoided. The properties of the two types of converter are set out in Table 12.2.

12.8   Ford EGI system for preheating catalysts

Generally, catalysts on ceramic monoliths do not become reasonably effective until they have attained a temperature of approximately 350°C, and are not fully so until a temperature of 450°C is attained. On average, two-thirds of all car journeys undertaken are less than 5 miles in length. Indeed, on a short journey, as much of 80% of the total emissions after starting with a cold engine are produced during the first 2 min, and the situation is even worse in cold climates and in urban conditions.

Ford have obtained catalyst light-off consistently in a few seconds by briefly burning a measured mixture of fuel and air in an afterburner just upstream of the catalyst. They have termed their system Exhaust Gas Ignition (EGI).

Table 12.2—METAL AND CERAMIC MONOLITH MATERIALS COMPARED

Property	Metal	Ceramic
Wall thickness, mm	0.04	0.15–0.2
Cell density, cells/in^2	400	400
Clear cross-section, %	91.6	67.1
Specific surface area, m^2/l	3.2	2.4
Thermal conductivity, W/m K	14–22	1–1.08
Heat capacity, kJ/kg K	0.5	1.05
Density, g/cm^3	7.4	2.2–2.7
Thermal expansion, $\Delta L/L$, 10^{-6} K	15	1

Note: Thicknesses and cross-sections are of metals uncoated with catalyst

Immediately after start-up from cold, three actions are initiated by the electronic control unit (ECU): the engine is run on a rich mixture; air is delivered to the afterburner by a pump which is electrically driven, so that it too can be controlled by the ECU; and sparks are continuously fired across the points of a plug situated in the afterburner. Because the mixture is rich, the exhaust gases contain not only unburned hydrocarbons and carbon monoxide, but also hydrogen. The hydrogen gas is highly inflammable, and therefore is utilised to light up the other constituents. Once alight, the mixture will continue to burn and generate the heat needed to bring the temperature of the catalyst up to its light-off value.

Hydrogen is one of the products of combustion of rich mixtures at high temperatures. The chemical processes, which are similar to those occurring during the production of water gas, are as follows—

$$CO + H_2O = CO_2 + H_2$$
$$CO_2 + H_2 = H_2O + CO$$

These two reactions alternate in the combustion chamber which is why, with the rich mixture, there is always some hydrogen in the exhaust, especially since the some of the products of combustion are frozen by the cold walls of the chamber.

Alternative methods of expediting warm-up can be less satisfactory. For example, placing the catalytic converter close to the engine entails a risk of degradation of the catalyst due to overheating when the car is driven at high speed and load. The alternative of electric heating requires a current of about 500 A at 12 V, and therefore calls for a significant and costly uprating of the battery charging system.

12.9 Three-way conversion

By 1978 GM had developed a three-way converter, the term implying the conversion of a third component, namely the NO$_x$. Whereas two-way conversion is done in a single stage, three-way conversion calls for two stages. By 1980 it became necessary for meeting the stringent requirements for the control of NO$_x$ in California and, by 1981, in the rest of the USA.

The additional catalytic bed contains Rhodium (Rh) for reduction of the oxides of nitrogen. An outcome was an increase to about 3 g in the total noble metal content. In practice, with a 0.1% rich mixture, about 95% of the NO_x can be removed by such a catalytic converter. A reducing atmosphere is essential so the mixture must not be lean, and therefore the conversion of NO_x has to precede the oxidation of the HC and CO.

Oxygen released in the initial reduction process, in the Rh bed, immediately starts the second stage of the overall process while the exhaust gas is still in the first stage. The oxygen that remains unused then passes on into the Pt or Pt-Pl second stage of the converter, in a separate housing downstream of the first. Here extra air is supplied for completion of the oxidation. On the other hand, if what is termed a *dual-bed converter* is used, both stages are in a single housing, though in separate compartments between which is sandwiched a third chamber into which the extra air is pumped, to join the gas stream before it enters the Pt, or Pt-Pl, stage.

With three-way conversion, a closed-loop control system is essential, for regulating the supply of fuel accurately in relation to the mass air-flow into the engine. This entails installing an oxygen sensor in the exhaust, and an on-board microprocessor to exercise control, both to correct continuously for divergencies from the stoichiometric ratio and to ensure good driveability.

12.10   The electronic control system

In practice, the electronic control system has to be more complex than might be assumed from the preceding paragraph. When the engine is being cranked for starting it has to switch automatically from a closed- to an open-loop system, to provide a rich mixture. In this condition the air supply for the second converter bed is diverted to the exhaust manifold to oxidise the inevitable HC and CO content, thus avoiding a rapid rise in temperature in, and overloading of, the second stage of the converter. Owing to the low temperatures in the combustion chambers of the engine, NO_x production is minimal or even zero, so no conversion is required in the first stage.

During warm-up the mixture strength has to be modified for the transition from rich to stoichiometric mixture. However, to cater for heavy loading, such as acceleration uphill, it may again have to be enriched, perhaps with exhaust gas recirculation in this condition to inhibit the formation of NO_x. The on-board microprocessor capabilities, therefore, must include control over idle speed, spark timing, exhaust gas recirculation, purge of hydrocarbons from carbon canister vapour-traps, early evaporation of fuel by air intake heating, torque converter lock-up, and a fault-diagnosis system.

12.11   Warm-air intake systems

Apart from setting the coolant thermostat to open at higher temperatures to improve combustion in cold conditions, several manufacturers have introduced automatic control of the temperature of the air drawn into the carburettor. The General Motors system is built into a conventional air cleaner. There are two air valves, operated by vacuum-actuated diaphragm mechanisms, and controlled by a thermostat. One valve lets warm and the other cold air into the intake.

The thermostat, mounted on the air cleaner, senses the temperature inside, maintaining it at 40 to 45°C. This thermostat operates a two-way control valve directing either induction system depression or atmospheric pressure to the actuator, according to whether cool or warm air is required. The warm air is taken from a jacket around the exhaust manifold.

Ford have developed a similar system. In this case, however, the thermostat senses under-bonnet temperature, and provision is made, by means of a vacuum-actuated override, to enable maximum power output to be obtained during warm-up.

The Austin-Rover design is outstanding for its simplicity. It is a banjo-shaped pressed steel box assembly, the handle of which is represented by the air intake duct to the air cleaner. In both the upper and lower faces of the box is a large diameter port, and a flap valve is poised between them. This flap valve is mounted on a bimetal strip which, when hot, deflects to close one port and, when cold, to close the other. The latter port simply lets air at the ambient temperature into the intake, while the former is connected by a duct to a metal shroud over the exhaust manifold and therefore passes hot air into the intake. It follows that the temperature of the incoming mixture from both ports is regulated by its effect on the bimetal strip, which deflects the flap valve towards the hot or cold port, as necessary.

12.12 Evaporative emissions

The evaporative emissions are mostly hydrocarbons though, with some special fuels and those that have been modified to increase octane number, alcohols may also be present. In general, the vapour comes from four sources—

(1) Fuel tank venting system.
(2) Permeation through the walls of plastics tanks.
(3) The carburettor venting system.
(4) Through the crankcase breather.

Fumes from the fuel tank venting system are absorbed in carbon canisters which are periodically purged, Section 12.17. Permeation through the walls of plastics tanks is controlled by one of four methods. These are—

(1) Fluorine treatment.
(2) Sulphur trioxide treatment.
(3) Du Pont one-shot injection moulding (a laminar barrier treatment).
(4) Premier Fuel Systems method of lamination.

Plastics tanks are generally moulded by extrusion of what is termed a *parison* (a large-diameter tube) which is suspended in a female mould into which it is then blow-moulded radially outwards. The chemical treatments are applied internally, either in the parison or in the blow-moulding. With either procedure, problems arise owing to the toxicity of the barrier chemicals and in the disposal of the chemical waste.

For the Du Pont laminar barrier technology, a modified one-shot extruder is used for producing the parison. It automatically injects into the high density polyethylene (HDPE), which forms the walls of the tank, a barrier resin called Selar RB. This resin forms within the wall an impermeable layer of platelets, in the form of a layer about 4 to 5% of its total thickness. Full details have been published in *Automotive Fuels and Fuel Systems,* Vol. 1, by T. K. Garrett, Wiley.

Item (4) in the second list is a patented method of laminating a plastics tank, which first went into series production in 1994 for the Jaguar XJ 220. It is produced by vacuum moulding, and is designed to meet the Californian requirements which limit the evaporative emissions to 12 parts per million from the whole car in one hour, during the SHED test, Fig. 12.6. There are three laminations. The outer layer is a fabric impregnated with a high nitrile polyvinyl chloride (PVC), while the inner wall is of unreinforced high nitrile PVC. Sandwiched between them is the layer that forms the impermeable barrier. This is of fluorinated ethane propane (FTP, or Teflon), both faces of which are etched to facilitate bonding to the outer layers.

12.13 Crankcase emission control

About 55% of the hydrocarbon pollution is in the exhaust, crankcase emissions account for a further 25% and the fuel tank and carburettor evaporation makes up the other 20%. These figures, of course, vary slightly according to the ambient temperature. In general unburnt hydrocarbons from these two sources amount to no more than about 4 to 10% of the total pollutants.

Crankcase fumes are drawn into the induction manifold by a closed circuit, positive ventilation system. One pipe is generally taken from the interior of the air filter to the rocker cover, and another from the crankcase to the induction manifold. Thus, air that has passed through a filter is drawn past the rocker gear into the crankcase and thence to the manifold, whence it is delivered into

Fig. 12.6 (*Left*) Sealed Housing for Evaporative Determination (SHED) installed alongside a chassis dynamometer (*right*) in a thermostatically controlled chamber at the Shell Thornton Research laboratories

the cylinders, where any hydrocarbon fumes picked up from the crankcase are burnt.

There are three requirements for such a system: first, the flow must be restricted, to avoid upsetting the slow running condition; secondly, there must be some safeguard to prevent blow-back in the event of a backfire and, thirdly, the suction in the crankcase has to be limited. AC-Delco produce a valve for insertion in the suction line to meet these requirements. It comprises a spring-loaded disc valve in a cylindrical housing. When there is no suction – engine off, or backfire condition – the valve seats on a port at one end, completely closing it. With high depression in the manifold, slow running or overrun, the valve seats on a larger diameter port at the other end, and a limited flow passes through the holes which, because they are near its periphery, are covered when it seats on the smaller diameter port. Flow through the larger port is restricted by the valve stem projecting into it. In normal driving, the valve floats in equilibrium between the two seats, and air can pass through the clearance around its periphery as well as through the holes, Fig. 12.7.

12.14 Air injection and gulp valve

A version of the AC-Delco air injection system made in the UK comprises an engine-driven air pump delivering into an air manifold, and thence through a nozzle or nozzles into either the exhaust ports or, latterly, between the reducing and oxidising catalytic beds. At the junction between the delivery pipe and the manifold, there is a check valve, to prevent back-flow of exhaust gas into the pump and thence to the engine compartment: this could happen in the event of failure of the pump or its drive.

A pipe is also taken back from the check valve to a gulp valve. When the throttle is closed suddenly for rapid deceleration, this admits extra air to the induction tract to ensure that there is enough air passing into the cylinders to burn the consequent momentary surge of rich mixture and to prevent explosions in the exhaust system. The gulp valve comprises two chambers. Of these, the bottom one is divided into two by a spring-loaded flexible diaphragm, from the centre of which a stem projects upwards to actuate a valve in the upper chamber. Air from the pump enters at the top and, when the valve is open, passes out through a port in the side, to the induction manifold, 12.8.

Manifold depression, introduced through a smaller port in the side of the housing, is passed into the chamber above the diaphragm. Therefore, a large

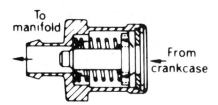

Fig. 12.7 AC-Delco crankcase ventilation control valve. With zero depression in the manifold, the valve seats on the right-hand orifice and with maximum depression on the left-hand one. In normal running it floats between the two

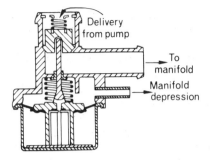

Delivery
from pump

To
manifold

Manifold
depression

Fig. 12.8 AC-Delco gulp valve, for admitting extra air to the induction tract when the
throttle is suddenly closed

transient increase in depression lifts the diaphragm, which subsequently
returns under the influence of the spring. The duration of the lift is determined
by the size of a balance orifice in the centre of the diaphragm. A period of
opening of about 1 and 4 seconds is generally adopted. Slower variations in
depression, which occur in normal driving, are absorbed by flow through the
balance orifice.

In the AC-Delco system, there is also a simple spring-loaded pressure relief
valve, which can be equipped with a small air silencer. This limits the pump
delivery pressure and prevents excess air being injected into the exhaust ports
under high speed conditions of operation.

Another system is the Lucas-Smiths Man-Air-Ox. This also comprises an
air pump, gulp valve and check valve. As with the AC-Delco unit, there is an
alternative to the gulp valve: a dump, or by-pass, valve can be used. This,
instead of admitting a gulp of air to the induction tract, opens to atmosphere a
delivery pipe from the pump. Thus, it effectively stops the supply through the
nozzles, and prevents the possibility of explosions in the exhaust system. It,
too, is suction controlled, but is a two-way valve. The sudden depression
produced by the overrun condition lifts the valve, venting the air delivery to
atmosphere, and at the same time seating the valve on the port through which
the air was formerly being delivered to the manifold.

If air injection is employed for emission control it might also be utilised for
cooling the exhaust valves. However, this is not favoured, because it entails the
use of valves of special steel for avoiding oxidation. Afterburning in the
exhaust gases may also be encouraged by the incorporation of ribs or other
forms of hot spot in the exhaust manifold.

12.15   Air management valves

For engines equipped with catalytic converters something more than a simple
gulp valve is needed. Consequently comprehensive air management systems
are employed to shut off completely the supply of air to the exhaust system, or
to divert it from the catalytic converter into the exhaust manifold, during
phases of operation in which a rich mixture is supplied to the engine.
Otherwise, the excessive quantities of unburnt hydrocarbons passing through
may either cause explosions in the exhaust system or overload, and thus
overheat, the oxidising catalytic converters. Moreover, with carburetted

engines the very high depression arising on sudden closure of the throttle tends to draw off excess fuel, which can similarly cause damage to the exhaust system, including the catalytic converter.

One of the simpler of the wide range of air management valves available is the Rochester Products Standardisation Diverter Valve, Fig. 12.9. During normal operation of the engine this valve is held open by a spring, so that the pump delivers air through it to the exhaust manifold to burn off HC and CO. To counter the effects of over-fuelling due to the very high depression on sudden closure of the throttle, that depression is transmitted to the chamber above the diaphragm. This lifts the diaphragm against its return spring and thus closes a valve in the port through which the pump delivers air to the exhaust manifold, diverting the flow into the dirty side of the air cleaner, to silence the discharge. An orifice in the central plate of the diaphragm assembly allows the depression to bleed away from the diaphragm chamber, the duration of diversion of the air to the cleaner being therefore a function of the sizes of the bleed and depression signal orifices and the volumes of the upper and lower parts of the diaphragm chamber. At high rotational speeds of the engine the rate of supply of air exceeds requirements, so it is this time diverted by the opening of a pressure relief valve, to the air cleaner.

A variant of this system is the Air Intake Control Valve, Fig. 12.10. During sudden deceleration this valve, instead of closing the outlet to the exhaust manifold, simply opens a port to divert some of the air to the induction manifold instead of all of it into the air cleaner. This, by weakening the mixture, not only helps to reduce the HC and CO content of the exhaust gases but also reduces the rate of deceleration. At high engine speeds, again, air in excess of requirements is diverted by the pressure relief valve into the air cleaner.

12.16 Some more complex valve arrangements

Where a computer-controlled catalytic system is installed, a solenoid-actuated air-switching valve is interposed between the output from the normal diverter valve and the delivery line to the exhaust manifold. When the system is

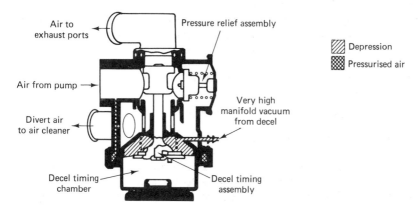

Fig. 12.9 Rochester Standardised Diverter Valve, with high manifold depression applied to the diaphragm to divert air from its normal route, which is to the exhaust system, into the air cleaner

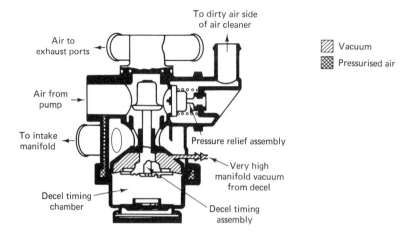

Fig. 12.10 Air intake control valve. This functions similarly to that illustrated in Fig. 11.7, but some of the air is also diverted into the induction manifold to reduce CO and HC output in the exhaust

operating with a cold engine, on the open-loop principle, the Standardised Diverter Valve functions exactly as described in the previous section. However, when the engine is warm and the computer control switches to closed-loop operation, the solenoid-actuated valve diverts the output of air from the exhaust manifold to the second catalytic bed, Fig. 12.11, to enable the reducing catalyst in the first bed to function efficiently while, at the same time, to further the oxidation process in the second bed.

The functions of air management valves so far described have been—

(1) To direct air from the pump to the exhaust manifold during normal operation.
(2) To divert either all the air from the exhaust manifold to the air intake cleaner, or some of it to the induction manifold (the gulp-valve function) during deceleration.
(3) To divert the air excess to requirements, via a blow-off valve, back to the air cleaner.
(4) Where control is exercised by computer, the solenoid-actuated air switching valve diverts air from the exhaust to the second stage catalytic converter for closed-loop operation or back again to the exhaust manifold for open-loop operation.

There are also, however, even more complex air management valves. The additional functions they perform are—

(5) In response to a low or zero depression signal from the induction manifold, to divert the air pump output to the air cleaner when the engine is operating under heavy load from nearly to fully open throttle. This is to avoid the overheating of the catalytic converter that would occur if the excess hydrocarbon required to obtain maximum power output were to be oxidised in it.
(6) In response to a high depression signal from the induction manifold, to

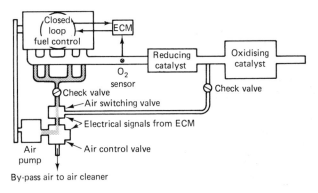

(a)

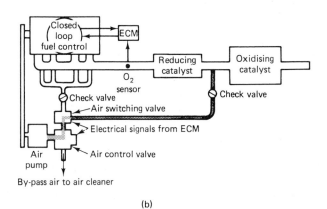

(b)

Fig. 12.11 System for switching air from the manifold (*a*) for open-loop operation to the oxidising catalytic converter, (*b*) for closed-loop operation

divert the air pump output to the air cleaner during normal road load conditions. This improves fuel consumption by decreasing exhaust back-pressure and, to a lesser degree, by reducing the power required to operate the pump. The air flow reverts to the exhaust manifold when the load on the engine increases and therefore the converter is needed to come back into operation to control the hydrocarbon emissions.

(7) By means of a solenoid-actuated valve, to enable air to be diverted by electronic control during any driving mode. With this arrangement, the diaphragm valve is actuated by high depression and a spring, in the normal way, except when the electronic control opens the solenoid valve to introduce air pump delivery pressure beneath the diaphragm to override its spring return mode.

There are other variants of these valves, but not enough space here to describe them all. Their modes of operation can be deduced from a study of Figs 12.12 to 12.14. Which type of valve is selected depends mainly on the engine and emission-control system characteristics.

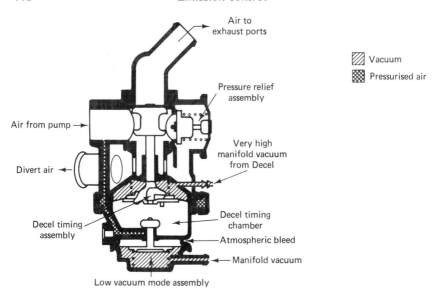

Fig. 12.12 The Rochester Normal Diverter Valve diverts air from the oxidising converter to the air cleaner for operation at heavy load

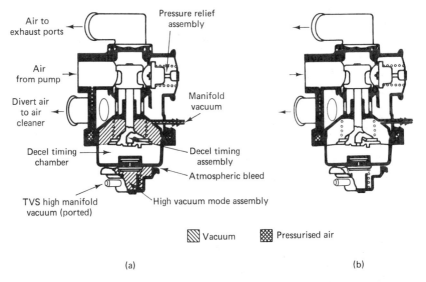

(a)   (b)

Fig. 12.13 This is what Rochester term their High Vacuum Air-control Valve: (a) in the high vacuum mode; (b) in the low vacuum mode. On sudden closure of the throttle valve with the engine cold it operates in the same way as the Standardised Diverter Valve in Fig. 12.8

12.17 Vapour collection and canister purge systems

Carbon-filled canisters for storing vapours emitted from fuel tanks and float chambers are of either the open- or closed-bottom type. The filter in the base of the open-bottom type, Fig. 12.15, has to be changed regularly, generally every

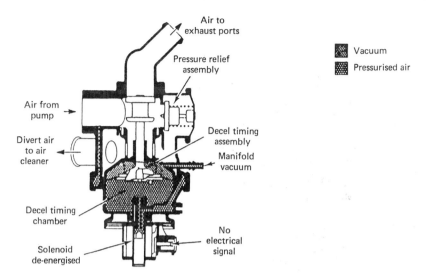

Fig. 12.14 The electrically actuated air-control valve, with solenoid energised

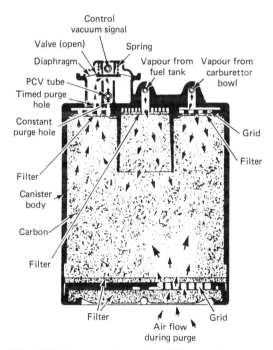

Fig. 12.15 A purge valve canister of the open-bottom type

24 months or 30 000 miles. On the other hand, the purge air for the closed-bottom type, Fig. 12.16, is drawn through the engine air intake before being conducted through a tube to the canister, so its filters do not need to be changed. A major advantage of the closed-bottom canister is that water or

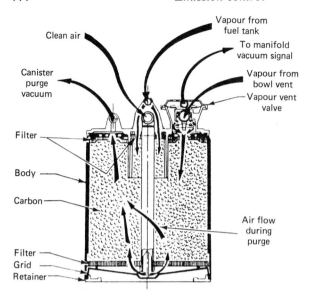

Fig. 12.16 Canisters of the closed-bottom type will not let water in at the base, where it could freeze and inhibit the purge action

condensation cannot enter its base and, in cold weather, freeze in the filter, restricting the entry of purge air. Even so, canisters of the closed bottom type, but also drawing air from atmosphere, are not unknown.

Either closed- or open-bottom canisters may have one inlet tube, through which both the fuel tank and carburettor float chamber fumes enter, and one purge tube. Alternatively there may be separate inlet tubes for these two sources of fumes, making a total of three, as in Fig. 12.16. Generally, the carburettor float chamber vent tube, as in the illustration, is fitted with a diaphragm valve which is normally held open by a spring. However, when the engine is started, closure of the throttle uncovers a hole just downstream of it and transmits manifold depression to close the diaphragm valve, the venting of the float chamber then being effected directly to the carburettor air intake. Yet another option is to take the float chamber vent to a valve on the purge tube, for both the purging and diversion of the float chamber venting in the *timed* manner just described.

There may or may not be a central cylindrical or conical screen around the inlet from the tank vent. The function of such a screen is to ensure that the fumes entering from the tank are initially spread into the carbon particles, instead of being short circuited directly into the induction manifold, which could cause driveability problems when the vehicle is being operated slowly under very light load.

The canister purge tube may be fitted with either of several types of valve as alternatives to that in Fig. 12.15. Purging takes place through both the timed and constant purge orifices. When the depression over the diaphragm falls and the spring lowers the valve on its seat again, the flow continues, but at a greatly reduced rate, through the constant purge orifice.

One of the alternatives is the Rochester Products valve illustrated in Fig. 12.17. When the engine is not running, the spring holds the valve open,

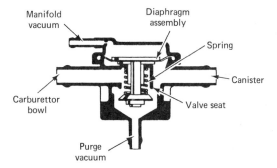

Fig. 12.17 The Rochester Type 1 Canister Control Valve functions as a simple vapour vent valve

allowing the carburettor float chamber to vent into the canister. As soon as it is started, the *timed* manifold depression seats the valve so that the float chamber is then vented internally through ducts in the carburettor body to the air intake.

The valve in Fig. 12.18 performs two functions: it acts as both a carburettor vent valve, as described in the previous paragraph, and a purge valve, as in Fig. 12.15. When the engine is started, the lower diaphragm is lifted by the purge control system depression and this closes the valve serving the float chamber venting system. As the throttle is opened further, the depression signal, again *timed* by the passage of the throttle valve over a hole in the wall of the manifold, lifts the upper diaphragm valve to open it and purge the canister.

When a car is parked for long periods in hot sunshine there could be a tendency for large volumes of vapour to pass from the fuel tank into the canister. This would call for the use of an otherwise unnecessarily large capacity canister so, to avoid this, a fuel tank pressure control valve, Fig. 12.19, may be fitted. This is a diaphragm valve the port of which is normally closed by a spring. To allow for variations in volume of air and vapour in the tank with changes in ambient temperature, the tank is then vented through the restricted orifice alongside the port. Thus, when the engine is not running the major part

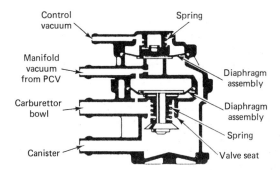

Fig. 12.18 Canister Control Valve Type 2 performs the functions not only of the Vapour Vent Valve but also of the Purge Valve in Fig. 12.15

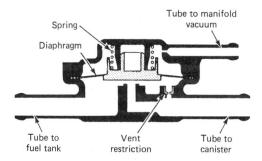

Fig. 12.19 Fuel tank pressure control valve. When the engine is not running, the valve is closed by the spring and the tank is then vented only through the restriction

of the fumes tends to be kept in the tank. When the engine is started the manifold depression opens the valve port, to vent the fuel tank to the canister, which of course will be cleared when the purge valve opens.

Yet another method of controlling the purge is available if an electronic control module is installed. A solenoid-controlled valve can be used and its opening and closing regulated by the computer.

12.18   Diesel engine emissions

Diesel and spark ignition engines produce the same emissions. On the other hand, owing to the low volatility of diesel fuel relative to that of gasoline and the fact that carburettors are not employed, evaporative emissions are not so significant. Crankcase emissions, too, are of less importance, since only pure air is compressed in the cylinder and blow-by constitutes only a minute proportion of the total combustion gases produced during the expansion stroke.

Sulphur, which plays a major part in the production of particulates and smoke emissions is present in larger proportions in diesel fuel than in petrol. This is one of the reasons why the combustion of diesel fuel produces between 5 and 10 times more solid particles than that of petrol.

Because diesel power output is governed by regulating the supply of fuel without throttling the air supply, there is excess air and therefore virtually zero CO in the exhaust under normal cruising conditions. Reduction of NO_x on the other hand can be done only in an oxygen-free atmosphere, so a three-way catalytic converter is impracticable.

As a diesel engine is opened up towards maximum power and torque, NO_x output increases because of the higher combustion temperatures and pressures. At the same time HC, CO and sooty particulates also increase. However, because of the relatively low volatility of diesel fuel and the extremely short time available for evaporation, problems arise also when the engine is very cold. In the latter circumstances, under idling or light load conditions, the situation is aggravated by the fact that the minute volumes of fuel injected per stroke are not so well atomised as when larger volumes are delivered through the injector holes.

12.19   Reduction of emissions: conflicting requirements

Measures taken to reduce NO_x tend to increase the quantity of particulates and HC in the exhaust, Fig. 12.20. This is primarily because, while NO_x is reduced by lowering the combustion temperature, both soot and HC are burned off by increasing it. In consequence, some of the regulations introduced in Europe have placed limits on the total output of both NO_x and HC, instead of on each separately, leaving manufacturers free to obtain the best compromise between the two.

The problem of emission control, however, is not so severe as might be inferred from the last paragraph. Both NO_x output and heat to exhaust become significant only as maximum torque and power are approached. At lighter loads, the gases tend to be cooled because of both their excess air content and the large expansion ratio of the diesel engine. Since the proportion of excess air falls as the load increases, oxidising catalysts can be used without risk of overheating, even at maximum power output.

Fuel blending and quality has a profound effect on emissions. Since fuel properties and qualities are interrelated, it is generally unsatisfactory to vary one property unilaterally. Indeed, efforts to reduce one exhaust pollutant can increase others and adversely affect other properties.

12.20   Oxides of nitrogen, NO_x

To understand the effects of fuel properties on NO_x output, certain basic facts must be borne in mind. First, it depends not only on the peak temperature of combustion but also on the rate of rise and fall to and from it. Secondly, the combustion temperatures depend on primarily the quantity and, to a lesser degree, the cetane number of the fuel injected.

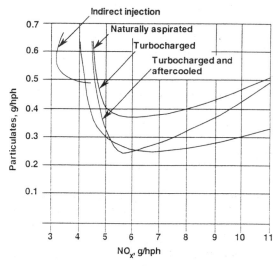

Fig. 12.20 Tests done in the late 1980s to demonstrate how NO_x increases when measures are taken to reduce particulates in different types of engine

Increasing the cetane number reduces the delay period, so the fuel starts to burn earlier, so higher temperatures and therefore more NO_x are generated while the burning gas is still being compressed, Fig. 12.21. However, a smaller quantity of fuel is injected before combustion begins, and this, by reducing the amount of fuel burning at around TDC, reduces the peak combustion temperature. The net result of the two effects is relatively little or even no change in NO_x output. An interesting feature in Fig. 12.22 is the enormous difference between the NO_x outputs from direct and indirect injection systems.

The popular concept that increasing fuel volatility reduces NO_x is an illusion: what happens in reality is that the weight of fuel injected is reduced, and the engine is therefore de-rated. Consequently, combustion temperatures are lowered. This is explained in more detail in Section 12.26, in connection with black smoke.

In the early 1990s, the overall output of NO_x from the diesel engine was, on average, between 5 and 10 times that of an equivalent gasoline power unit with a catalytic converter, but this differential will be reduced as diesel combustion control techniques improve. Efforts are being made to develop catalysts

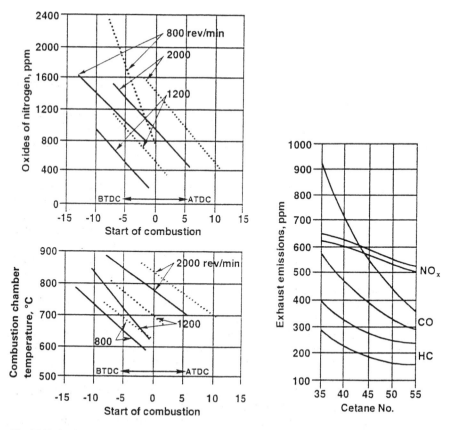

Fig. 12.21 (*Left*) Tests by BP showing how injection timing influences combustion and therefore NO_x output; (*right*) the influence of cetane number on the principal emissions

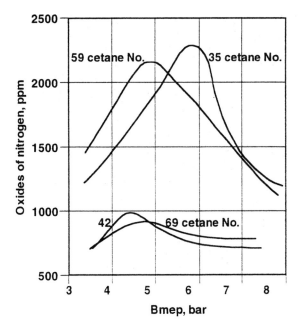

Fig. 12.22 NO$_x$ emissions with direct and indirect injection

suitable for diesel application, but at the time of writing no satisfactory solution has been found.

Unfortunately, most of the currently conventional methods of reducing NO$_x$ also impair efficiency and therefore increase fuel consumption and therefore the output of CO$_2$. The relationship between NO$_x$ output and fuel consumption is illustrated in Fig. 12.23. In general, NO$_x$ tends to form most readily in fuel-lean zones around the injection spray.

Exhaust gas recirculation displaces oxygen that otherwise would be available for combustion and thus reduces the maximum temperature. However, it also heats the incoming charge, reduces power output, causes both corrosion and wear, and leads to smoke emission at high loads. For these reasons it has to be confined to operation at moderate loads. Electronic control of EGR is therefore desirable. Fortunately, heavy commercial vehicles are driven most of the time in the economical cruising range, maximum power and torque being needed mostly for brief periods.

Reduction of the rate of swirl is another way of reducing the output of NO$_x$. It increases the time required for the fuel to mix with the air, and therefore reduces the concentration of oxygen around the fuel droplets. Consequently, the temperature of combustion does not rise to such a high peak. Again, however, it also reduces thermal efficiency. Moreover, unless measures, such as increasing the number of holes in the injector nozzle and reducing their diameter, are taken to shorten the lengths of the sprays, more fuel tends to be deposited on the combustion chamber walls.

Delaying the start of injection has the effect of reducing peak temperatures, and therefore NO$_x$. This is because the combustion process builds up to its peak later in the cycle, when the piston is on its downward stroke and the gas is

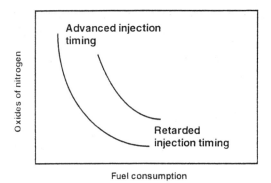

Fig. 12.23 Relationship between fuel consumption and NO_x emissions with (*left*) and without (*right*) charge cooling

therefore being cooled by expansion. However, to get a full charge of fuel into the cylinder in the time remaining for it to be completely burnt, higher injection pressures are needed. Therefore, to avoid increasing the proportion of fuel sprayed on to the combustion chamber walls, the holes in the injector must again be smaller in diameter and larger in number.

Turbocharging increases the temperature of combustion by increasing both the temperature and quantity of air entering the cylinder. After-cooling, however, can help by removing the heat generated by both compression of the gas and conduction from the turbine. It also increases the density of the charge, and therefore thermal efficiency and power output. The net outcome of turbocharging with charge cooling, therefore, is generally an increase or, at worst, no reduction in thermal efficiency.

12.21 Unburnt hydrocarbons

Hydrocarbons (HCs) in the exhaust are the principal cause of the unpleasant smell of a diesel engine, though the lubricating oil also makes a small contribution. There are three main reasons for this. First, at low temperatures and light loads, the mixture may be too lean for efficient burning so the pre-combustion processes during the ignition delay period are partially inhibited. This is why some of the mixture subsequently fails to burn.

Secondly, because of the low volatility of diesel fuel relative to petrol, and the short period of time available for it to evaporate before combustion begins, HCs are generated during starting and warming up from cold. In these circumstances, fuel droplets, together with water vapour produced by the burning of the hydrogen content of the remainder of the fuel, issue from the cold exhaust pipe in the form of what is generally termed *white smoke*, but which is in fact largely a mixture of fuel and water vapours. At about 10% load and rated speed, both HC and CO output are especially sensitive to fuel quality and, in particular, cetane number.

Thirdly, after cold starting and during warm-up, a higher than normal proportion of the injected fuel, failing to evaporate, is deposited on the combustion chamber walls. This further reduces the rate of evaporation of the fuel, so that it fails to be ignited before the contents of the chamber have been cooled, by expansion of the gases, to a level such that ignition can no longer occur. Similarly, the cooling effect of the expansion stroke when the engine is operating at or near full load can quench combustion in fuel-rich zones of the mixture. This is the fourth potential cause of HC emissions.

Unburnt HCs tend to become a problem also at maximum power output, owing to the difficulty under these conditions of providing enough oxygen to burn all the fuel. As fuel delivery is increased, a critical limit is reached above which first the CO and then the HC output rise steeply. Injection systems are normally set so that fuelling does not rise up to this limit, though the CO can be removed subsequently by a catalytic converter in the exhaust system.

Another potential cause of HCs is the fuel contained in the volume between the pintle needle seat and the spray hole or holes (the *sac volume*). After the injector needle has seated and combustion has ceased, some of the trapped fuel may evaporate into the cylinder. Finally, the crevice areas, for example between the piston and cylinder walls above the top ring, also contain unburnt or quenched fractions of semi-burnt mixture. Expanding under the influence of the high temperatures due to combustion and falling pressures during the expansion stroke, and forced out by the motions of the piston and rings, these vapours and gases find their way into the exhaust.

In general, therefore, the engine designer can reduce HC emissions in three ways. One is by increasing the compression ratio; secondly, the specific loading can be increased by installing a smaller, more highly rated, engine for a given type of operation; and, thirdly, by increasing the rate of swirl both to evaporate the fuel more rapidly and to bring more oxygen into intimate contact with it.

Reduction of lubricating oil consumption is another important aim as regards not only control of HCs but also, and more importantly, particulate emissions. Whereas oil consumption at a rate of 1% of that of fuel was, until the mid-1980s, been regarded as the norm, the aim now is generally nearer to 0.2%. Using a lubricant containing a low proportion of volatile constituents helps too.

Avoidance of cylinder-bore distortion can play a significant part in the reduction of oil consumption. The piston rings tend to ride clear over and therefore fail to sweep the oil out of the pools that collect in the hollows formed by distortion of the bores, thus reducing the effectiveness of oil control. Other means of reducing contamination by lubricating oil include improving the sealing around the inlet valve stems, the use of piston rings designed to exercise better control over the thickness of the oil film on the cylinder walls and, if the engine is turbocharged, reduction of leakage of oil from the turbocharger bearings into the incoming air.

12.22 Carbon monoxide

Even at maximum power output, there is as much as 38% of excess air in the combustion chamber. However, although carbon monoxide (CO) should not be formed, it may in fact be found in small quantities in the exhaust. The reason is partly that, in local areas of the combustion chamber, most of the oxygen has been consumed before injection ceases and, therefore, fuel injected into these areas cannot burn completely to CO_2.

12.23   Particulates

Regulations define particulates as anything that is retained, at an exhaust gas temperature of 52°C, by a filter having certain specified properties. They therefore include liquids as well as solids. Particle sizes range from 0.01 to 10 μm, the majority being well under 1.0 μm. While black smoke comprises mainly carbon, the heavier particulates comprise ash and other substances, some combined with carbon. The proportions, however, depend on types of engine, fuel and lubricant.

Measures appropriate for reducing the fuel and oil content of the particulates are the same as those already mentioned in connection with HC emissions, Section 12.21. The overall quantity of particulates can be reduced by increasing the injection pressure and reducing the size of the injector holes, to atomise the fuel better. This, however, tends to increase the NO_x content. Increasing the combustion temperature helps to burn the loose soot deposited on the combustion chamber walls. Various measures have been taken to increase the temperature of these particulates, though mostly only experimentally. They include insulation by introducing an air gap, or some other form of thermal barrier, between the chamber and the remainder of the piston, and the incorporation of ceramic combustion chambers in the piston crowns.

Reduction of the sulphur content of the fuel also reduces particulates. Although the proportion of sulphate + water is shown in Table 12.3 as being only 2% of the total, if the insoluble sulphur compounds are added, this total becomes more like 25%. Because most measures taken to reduce NO_x increase particulates, the most appropriate solution is to use fuels of high quality, primarily having low sulphur and aromatic contents and high cetane number. The relationship between fuel quality and particulate and NO_x output has been demonstrated by Volvo, Fig. 12.24.

A small proportion of the particulates is ash, most of which comes from burning the lubricating oil. Reduction of sulphur in the fuel reduces the need for including, in lubricating oils, additives that neutralise the acid products of combustion; these additives that are responsible for a significant proportion of the ash content.

Incidentally, sulphur compounds can also reduce the efficiency of catalytic converters for the oxidation of CO and HC. In so doing, they form hydrogen disulphide, which accounts for the unpleasant smell of the exhaust when fuels

Table 12.3—ANALYSES, EXPRESSED IN PERCENTAGES, OF PARTICULATES FROM DIFFERENT TYPES OF DIESEL ENGINE

Engine type	Fuel-derived HC	Oil-derived HC	Insoluble ash	Sulphates + water
Ford 1.8 DI	15	13	70	2
Ford 1.8 IDI	48	20	30	2
Average DI HD turbocharged after-cooled engine	14	7	25	4

Note: Horrocks (Ford Motor Co.) differentiated between the carbon and other ash (at 41% and 13% respectively), making the total 44%

with high sulphur content are burnt in an engine having an exhaust system equipped with a catalytic converter.

An ingenious method of reducing visible particulates emitted from a turbocharged engine in a bus has been investigated by MAN. Compressed air from the vehicle braking system is injected in a controlled manner into the combustion chambers to burn off the carbon. This increases the exhaust gas energy content, and therefore compensates for turbocharger performance falling off under light load, including initially during acceleration and while gear changes are being made.

12.24 Particle traps

Basically, particle traps are filters, mostly catalytically coated to facilitate their regeneration (burning off of the particles that have collected). Many of the filters are extruded ceramic honeycomb type monoliths, though some are foamed ceramic tubes. The honeycomb ceramic monoliths generally differ from those used for catalytic conversion of the exhaust in gasoline engines, in that the passages through the honeycomb section are sealed alternately along their lengths with ceramic plugs, Fig. 12.25, so that the gas passes through their porous walls. These walls may be less than 0.5 mm thick.

Ceramic fibre wound on to perforated stainless steel tubes has also been used. These are sometimes termed candle or deep *bed*-type filters. Their pores are larger than those of the honeycomb type, and their wall thicknesses greater.

If catalyst assisted, regeneration is done mostly at temperatures around

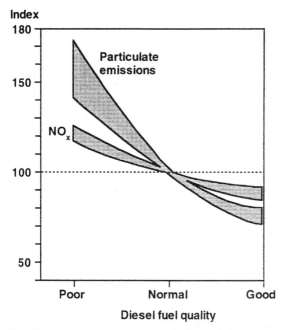

Fig. 12.24 Relationship between particulate emissions and fuel quality, as established by Volvo

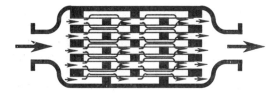

Fig. 12.25 Diagram showing how the gas flow is diverted in a particulate trap

600°C, and if not, at 900°C. Alternatively, special catalysts may be used to lower further the ignition temperature of the particles. These are mostly platinum and palladium, which are useful also for burning HCs and CO. Sulphate deposition adversely affects the platinum catalyst, but the formation of sulphates is largely inhibited by palladium. Copper has been used too, because it reduces the carbon particle ignition temperature to about 350°C, but it suffers the disadvantage of a limited life. Better results have been obtained with alloys of copper with silver, vanadium and titanium.

It is important to regenerate the filters reliably at the appropriate intervals, otherwise they will become overloaded with soot, which can then ignite and burn uncontrollably, developing excessive heat and destroying the filter. Even with normal burn off, however, local temperatures near the centre of the beds can be as high as 1200°C.

Some particulate filters rely solely on excess air in the exhaust for burning off the particulates. Others have extra air injected into them, generally at timed intervals, either from the brake system or by a compressor or blower. With air injection, burn-off can continue when the engine is operating at or near maximum power and therefore with little excess oxygen in the exhaust. At lighter engine loads and therefore with excess air present, once ignition has been initiated, combustion can continue without the burner.

With some systems the engine must be stopped while regeneration is in progress but with others, which are regulated by electronic control, either the particulates are burnt off intermittently while the vehicle is running or there are two filters in parallel, with deflector valves directing the exhaust gases first through one filter, while the other is being regenerated, and then *vice versa*. It is also possible to program the electronic controller to bring both filters into operation simultaneously as maximum power and torque output is approached, to cater for the increased exhaust flow under these conditions. This enables smaller filters to be used.

Claimed efficiencies of particulate removal range from about 70 to 95%, but some of these claims are suspect. However, since the carbon content is easiest to burn off, up to 99% of it can be removed. Currently, the useful lives of the catalytically coated filters vary enormously from type to type.

Since we are in the early stages of development, the situation is in a continuous state of flux; some systems no doubt will soon fall by the wayside, while others will be developed further. Clearly, however, if particulates can be removed in the combustion chamber by using better quality fuel or taking measures to improve combustion, it will be to the advantage of all concerned, except of course the manufacturers of particulate filters!

All filter systems are bulky and present burn-off problems, including high thermal loading. None produced so far can be regarded as entirely satisfactory and, at the time of writing, none has been used on cars. Interestingly, VW have been investigating the possibility of using an iron-based additive to reduce the oxidation temperature of the deposits down to that of the exhaust gas. A Corning monolithic ceramic filter is used, and the additive, developed jointly with Pluto GmbH and Veba Oel, is carried under pressure in a special container. An electronic control releases additive automatically, as needed, into the fuel supply line, so that regeneration proceeds at temperatures as low as 200°C.

Of the systems currently available for commercial vehicles, those requiring the engine to be stopped during burn-off are unsuitable for any automotive applications other than city buses and large local delivery vehicles operated on regular schedules. The continuous and cyclic burn-off filters are generally even more bulky. On the other hand, many of these systems are claimed to serve also as a silencer. There is not enough space here to give details of proprietary filter systems, but most are described in *Automotive Fuels and Fuel Systems, Vol. 2*, by T. K. Garrett, Wiley.

12.25   Influence of fuel quality on diesel exhaust emissions

How individual emissions are influenced by different fuel properties have been summarised by the UK Petroleum Industry Association as follows—

NO_x Increases slightly with cetane number.
Decreases as aromatic content is lowered.

CO   No significant effects.

HC   Decreases slightly as cetane number increases.
Decreases with density.
Relationship with volatility inconsistent.

Black smoke Increases with fuel density and decreases with aromatic content.
Is not significantly affected by volatility.
Increases with injection retard (e.g. for reducing NO_x).

Particulates Reduced as volatility is lowered.
Reduced as cetane number is lowered, though inconsistently.
Unaffected by aromatic content.
Reduced as sulphur content is lowered.

A good quality fuel is generally regarded as one having a cetane number of 50 and a sulphur content of no greater than 0.05%.

In Sweden, Volvo have shown that by bringing the sulphur content down from 0.2 to 0.05% particulate emissions can be reduced by up to 20% and NO_x is also reduced. Furthermore, ignoring the effect of the tax on fuel prices, the cost of such a reduction is only about 2 pence per gallon whereas to obtain a commensurate improvement by reducing the aromatic content and increasing the cetane number would cost about 22 pence per gallon.

12.26   Black smoke

The effect of sulphur content on the formation of particulates has been covered in Section 12.23. Other factors include volatility and cetane number. As regards visibility, however, the carbon content is much more significant. Suggestions that volatility *per se* influences black smoke are without foundation. Smoke is reduced with increasing volatility for two reasons: the first is the correspondingly falling viscosity, and the second the associated rising API gravity of the fuel. A consequence of the first is increased leakage of fuel through the clearances around both the pumping elements and the injector needles and, of the second, the weight of the fuel injected falls. Therefore, for any given fuel pump delivery setting, the power output decreases with increasing volatility. In fact, the real influence of volatility depends on an extremely complex combination of circumstances, and varies with factors such as speed, load and type of engine.

The reason is that each engine is designed to operate at maximum efficiency over a given range of speeds and loads with a given grade of fuel. Therefore, at any given speed and load, a change of fuel might increase the combustion efficiency, yet at another speed and load the same change might reduce it. This is because a certain weight of fuel is required to produce a given engine power output so, if the API gravity is increased, a commensurately larger volume of fuel must be supplied, and this entails injection for a longer period which, for any given engine operating condition could have either a beneficial or detrimental effect on combustion efficiency. Similarly, the resultant change in droplet size and fuel penetration relative to the air swirl could have either a beneficial or detrimental effect.

The reason why the cetane number does not have a significant effect on the output of black smoke is simple. It is that smoke density is largely determined during the burning of the last few drops of fuel to be injected into the combustion chamber.

12.27   White smoke

White smoke is a mixture of partially vaporised droplets of water and fuel, the former being products of combustion and the latter arising because the temperature of the droplets fails to rise to that needed for ignition. It can be measured by passing the exhaust through a box, one side of which is transparent and the other painted matt black. A beam of light is directed through the transparent wall on to the matt black surface. If there is no white smoke, no light is reflected back to a sensor alongside the light source; the degree of reflection therefore is a function of the density of the white smoke. For testing fuels, the criterion is the time taken, after starting from a specified low temperature, for the smoke level to reduce to an acceptable level. After starting at $0°C$, satisfactory smoke levels are generally obtainable with a Diesel Index of 57 and a cetane number of 53.5.

Chapter 13

Fuel pumps and engine intake air conditioning

Some early vehicles had fuel tanks on the dash with gravity feed of the fuel to the carburettor. Where the tank was lower, manifold depression in a device called the Autovac, which tapped manifold depression for lifting fuel from the tank to a small reservoir inside the vehicle. In this reservoir, the fuel was maintained at a constant level while, again, being gravity fed to the carburettor. By the beginning of the Second World War, however, bonnet profiles had been lowered to the extent that dash-mounted tanks were no longer practicable, and Autovac reservoirs inside the vehicle unacceptable, so fuel pumps became universal. With the fuel tank beneath the floor of the boot in cars, and either in a similar position or mounted on the side of the chassis frame on commercial vehicles, the fuel generally has to be lifted against a head of 0.6 m or more to the carburettor or fuel injection equipment.

Fuel tanks are of either pressed steel or plastics, the former having the advantages of impermeability and lower cost. Some information on plastics tanks is given in Section 12.13, and much more can be obtained from *Automotive Fuels and Fuel Systems,* Vol. 1, by T. K. Garrett, Wiley. Steel tanks are given an anti-corrosion treatment internally, as well as externally, since the sulphur impurities combined with the water contamination of fuel can damage them if unprotected.

13.1 Roller-cell positive displacement type pump

Bosch produce a roller-cell type, high pressure fuel pump, driven by a permanent magnet d.c. motor. The output of such a pump does not pulsate as much as most of those of the other positive displacement types. Its pumping element is in the armature housing and the fuel flows right through, from one end to the other, to cool and lubricate all the rotating parts, Fig. 13.1. A permanent-magnet d.c. electric motor drives a radially slotted disc rotating in a housing, the inner cylindrical wall of which is eccentric relative to the disc and shaft. There are five radial slots, each containing a roller. As the disc rotates, the slots revolve past an outlet port, the eccentricity of the housing simultaneously forcing the rollers to move radially inwards, against centrifugal force; during further rotation the rollers move out again, while their slots

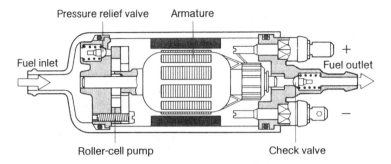

Fig. 13.1

revolve past an inlet port. Thus fuel is alternately forced out to one side and drawn in from the opposite side of the pump. A relief valve between the armature housing and the fuel inlet limits the pressure to 5 bar, and at the opposite end the main fuel delivery port houses a check valve to ensure that the pump remains primed after it is switched off.

A major advantage of an electrically driven pump is that it can be located virtually anywhere, including where the suction head is small and the temperature low, thus enabling vapour lock to be avoided. Indeed, many are installed inside the petrol tank. Another advantage is that, because it is wired into the ignition circuit, it supplies fuel immediately the engine is switched on, thus facilitating starting, and subsequently ceases to do so the moment it is switched off. However, unless either an oil-pressure actuated or inertia switch is incorporated in the electrical circuit, such a pump could become a fire hazard following an accident by continuing to supply fuel even though the engine had stalled.

Pumps submerged in tanks are usually accommodated in a *swirl pot*. This is a cylindrical housing of a diameter such that there is a clearance between it and the pump housing. It has an open top and a small hole or holes in its sides. The sizes of these holes are limited to that needed to allow entry of only enough fuel to supply the engine at maximum power output or, preferably, they are fitted with non-return flap valves. Consequently, when the tank is nearly empty and the fuel swilling around its floor, enough fuel will be trapped in the swirl pot to ensure that the pump will not prematurely suck in air.

13.2   Mechanical diaphragm type pump

The diaphragm type positive displacement pump has the advantage of simplicity and low cost. Moreover, in the presence of water or dirt it is more reliable than an electrical pump. Its principal disadvantages are the pulsating nature of its output and, to provide it economically with a mechanical drive, it generally has to be mounted on the engine. In the latter position it not only has to pump against a large suction head but also may be affected by conducted and radiated heat, especially from the exhaust manifold. Both the suction head and the heat tend to cause vapour lock on the suction side of the pump.

In the early days either metal bellows or fabric diaphragms were employed as the pumping elements but, owing to problems with the development of fatigue cracks in the metal components, the bellows types, originally largely developed under the trade name Sylphon, fell out of favour. Although fabric diaphragms can puncture, they are now remarkably reliable as a result of the development of extremely tough and fatigue-resistant fibre reinforcement fabrics. Gear and reciprocating plunger type pumps are too costly for automative applications.

A successful and reliable mechanical pump which is very widely used is the AC, one form of which is illustrated in Fig. 13.2. The particular model shown is adapted for high level application to a V-eight engine, where it is operated from the camshaft by a vertical pushrod engaging with the inverted cup at the end of the lever L_1. This lever is maintained in close contact with the pushrod by the compression spring shown.

In the majority of applications the pump is mounted on the side of the crankcase at camshaft level, and the operating lever L_1 takes the form shown by the broken lines, acting as a flat cam follower. This lower position of the pump is preferable as it reduces the suction lift. In faulty installations trouble may arise through vapour lock caused by too close proximity of the petrol pipe to the hot exhaust pipe. The high temperature combined with reduced pressure due to excessive suction lift leads to vaporisation of the petrol and interference with the flow. A typical installation can be seen in Fig. 4.6.

To return to Fig. 13.2, the inner portion L_2 of the operating lever fits between the jaws of L_1 and is free to rotate on the bush B. The cross spindle R forms the pivot on which the combined lever turns. L_2 will be driven in a counter-clockwise direction by the rise of the pushrod if there is contact at the heel H of L_2.

This will draw down the diaphragm A, compressing the spring S and providing the suction stroke of the pump. Petrol enters at I and, passing to the dome D through the circular filter gauze F which surrounds the central turret, enters the pump chamber through the inlet valve V_1. This valve with the outlet valve V_0 is mounted in a brass plate P screwed to the under-side of the turret. The valves are small octagonal-shaped discs of reinforced plastics, which allow

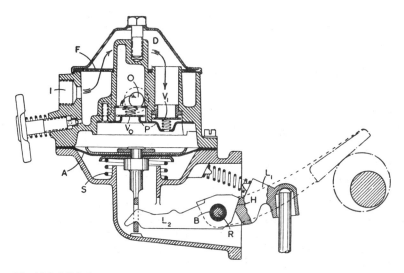

Fig. 13.2 AC fuel pump

for passage of the petrol while keeping the valve central. The outlet to the carburettor is indicated at O which leads to a union on the outside of the pump body.

The illustration shows the diaphragm in the lowest position. If the float chamber control of the carburettor permits delivery, the spring S will raise the diaphragm as the pushrod falls. If not, the diaphragm and L_2 will remain at rest and lost motion will arise at H.

13.3   SU pump

Among the early and most successful electrically actuated diaphragm type pumps was the SU unit, which superseded their Petrolift series. The diaphragm comprises several layers of elastomerically impregnated fabric, and its centre is clamped to the armature A. As can be seen from Fig. 13.3, discs having rounded- instead of square-faced peripheries are interposed between the grooved periphery of the armature and the counterbore in the end of the magnet pot M. These not only centre the armature as it reciprocates axially but also back up the diaphragm, helping to prevent it from ballooning under the load due to the pressure it generates in the fuel. The shoulders in the counterbore in the end of the magnet pot are so proportioned that as the lines of force passing across vary in density and direction to find the shortest path, so the component axial pull on the armature remains approximately constant throughout the travel.

The magnetic circuit is completed through the core C over which is threaded the winding spool W. A fibre disc is inserted in the central recess of the armature to prevent metal-to-metal contact of armature and core, which, as a result of residual magnetism, might prevent the return of the armature under the action of the delivery spring. The suction stroke occurs when the armature and diaphragm are moved to the left by the magnetic pull, and petrol then enters the pump chamber through the filter F and the lower of the two plate

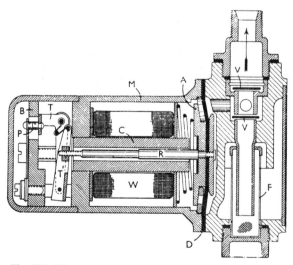

Fig. 13.3 SU pump

valves V. On the breaking of the circuit by the flick-over mechanism the diaphragm is returned by the compression spring, and petrol is delivered to the float chamber of the carburettor through the upper plate valve. Should the float chamber be so full as to close the needle valve, the diaphragm will remain at rest (with the electric circuit broken) until further delivery is required.

The float mechanism of the carburettor must be matched with the spring of the pump so that flooding cannot occur.

The stroke of the pump is about 3 mm and the maximum delivery pressure about 6.98 kN/m². A wide margin of delivery capacity is provided.

The flick-over mechanism for the make-and-break of the current is operated by the rod R, the outer end of which is pivoted to the primary rocker T′. This controls the outer twin rocker T, which carries on a cross yoke one of the contact-breaker points P. The bottom ends of both rockers are pivoted on a pin passing through two of the legs of the plastics moulding B. The second contact point is carried on a spring blade which rests against this moulding when the contacts are open. The auxiliary diagram Fig. 13.4 should make clear how the movement of T′ causes T to flick to the right and open the contacts under the action of the coil spring S mounted on a floating pin. This spring acts as a compression spring between the two rockers.

13.4 Rotary electric fuel pumps

Although the output from a rotary pump is relatively free from pulsations, in most instances it is not entirely so because there are low amplitude, high frequency periodic disturbances due to the passage of parts, such as vanes, of the rotor past the ports. In general, except as regards the rotors, which may take any of several forms, the overall layouts of the electrically driven rotor type pumps are nearly all similar to that of the positive displacement type described in Section 13.1.

Particularly interesting and unusual, however, because it has a two-stage rotor, is the Bosch fuel pump for delivery at pressures below 1 bar. As can be seen from Fig. 13.5, the first stage is a side-channel pump, in which the fuel enters on the left-hand side and has energy imparted to it by radial vanes, to ensure that it is under pressure, and is therefore vapour-free, as it passes radially out to the second stage. The radial vanes on the periphery comprise the second stage of the pump. In this stage the energy content of the flow is further boosted, before the fuel is delivered, now at a higher pressure, into the armature housing. Passing along this housing, it cools the armature, before

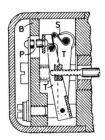

Fig. 13.4

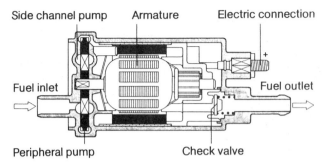

Fig. 13.5

leaving through a check valve in the outlet port. Since this pump is fairly sensitive to vapour and might fail to deliver if any were to form in its inlet, it must be installed under a head of fuel, in the base of the tank.

13.5   Air filters and silencers

There are three types of air cleaner: dry, centrifugal and wet, or oil bath. For cars, dry filters, with pleated resin impregnated paper or, in a very few instances, felt elements are employed. These elements are mostly of cylindrical shape with deep vertical pleats, so that at least some of the dust stopped by them tends to fall to the base, leaving the surface reasonably clear. Deep pleated paper, as in Fig. 13.6, offers the advantage of a very large surface area, and therefore not only long intervals between the need for replacement of the element, but also minimum obstruction to the incoming air flow.

This type of filter costs less, is simpler to maintain and is of higher efficiency than most of the others. To reduce the headroom required under the bonnet, they can be of pancake form and their under surfaces recessed to contain the carburettor or throttle body. Alternatively, they can be separately mounted beside the engine. Filtration areas are generally about 1370 m² per m³/min air flow. However, replacement requirements, which really determine areas, are normally timed to match the service intervals of the vehicle in which the unit is installed.

Silencing is effected by incorporating an air inlet tube of appropriately tuned length, as in Fig. 13.6, to form a Helmholtz resonator, described in Section 11.20. Such a device will, by resonating in opposite phase to a noise at its resonant frequency, take out that frequency from the spectrum of intake noise. Even if damped, it is effective over only a very narrow range of frequencies. In the air-intake application, a small degree of damping is provided by the filter element.

For heavy commercial vehicles, especially if operating in dust-laden atmospheres, for example on civil engineering sites, a centrifugal filter may be incorporated on the inlet side of either an oil bath filter or a paper or felt element. Centrifugal filters are constructed so that the air entering is guided by vanes or a volute to follow a spiral path in the housing, so that heavy particles in the air are centrifuged out to the walls. On striking the walls, the particles lose momentum and drop into the base. Used alone, they would not remove

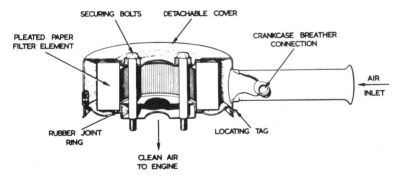

Fig. 13.6 Paper element air cleaner

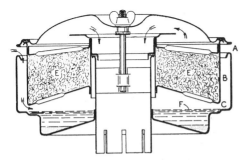

Fig. 13.7 AC oil bath cleaner

the finer articles. Over unmade roads and tracks, the dust content of the atmosphere can be as high as about $45\,g/m^3$, so over a distance of 600 miles an unfiltered engine might induct as much as 50 g of dust.

An example of an oil bath filter is illustrated in Fig. 13.7. Air entering through the peripheral gap at A passes down the annular space B and impinges on the surface of the oil F as it passes through the gap C. The heavier particles break through the surface of the oil and sink to the bottom as sludge. At the same time, the air passing through the narrow gap C whips up an oil mist which entrains the finer particles and deposits them in the oil wetted mesh of steel wool, or fine wire, at E. The air, thus cleaned, continues on through the induction system into the engine.

Whereas the pleated paper elements are simply replaced, usually between 16 000 and 19 000 km, oil bath filters have to be emptied and washed with paraffin or petrol and refilled with oil to the correct level. Although the oil bath type can trap a larger volume of heavy particles before it needs to be cleaned, the paper element type has a higher filtration efficiency, generally between 91 and 99.7% when tested to BS 1701. It is also lighter and more compact than the oil bath type, and in many instances can be installed directly on the air intake without extra support brackets.

Chapter 14

Turbocharging and supercharging

Basically, the power output of an engine depends on the amount of energy, in the form of a fuel, with which its cylinders can be charged, but the quantity of fuel that can be burned in the cylinder is limited by the mass of air that can be introduced. The latter is therefore a critical factor. In a naturally aspirated engine, atmospheric pressure forces the air in but, because of viscous drag in the induction system, and throttling by components such as venturis, bends in pipes and valve throats, the pressure of the air ultimately in the cylinder is lower than atmospheric. So also, therefore is its density. Consequently, any measure such as supercharging, turbocharging or cooling the charge to increase its density must be the key to increasing power output per unit of size and weight of the engine.

Supercharging can be used also to modify the torque characteristic, generally to help to increase the torque over a broad speed range as the throttle is closed, so that fewer gear changes are needed. The most widely used method of increasing the density of the charge, however, is *turbocharging*, though *supercharging*, which has the advantages outlined in Section 14.21, is employed in some applications.

A turbocharger is a compressor driven by a turbine powered by exhaust gas energy. This energy would of course otherwise be wasted. Similarly, a supercharger is a compressor, but mechanically driven and therefore consuming energy taken from the crankshaft. Both can be referred to as *pressure charging*.

14.1 Pressure charging the spark ignition engine

With the spark ignition engine, boosting the pressure in the induction system has the effect of increasing the overall compression ratio. Therefore, to avoid detonation as the boost pressure increases, the engine compression ratio must be reduced relative to that required for natural aspiration. Pressure charging increases the temperature of the ingoing air, and this tends to detract from the potential increase in density.

For every charge temperature rise of 100°C due to pressure charging a spark ignition engine, the increase in absolute temperature at the end of compression will be approximately 200°C. In a diesel engine, with its higher compression ratio, the temperature at the end of compression will be of the order of 300°C higher. In both instances, these effects increase the rate of heat generation, and therefore temperature rise, during combustion.

Consequently, heat disposal problems arise and temperatures may therefore become higher than the components can withstand. This problem is avoided by using high temperature resistant materials for engine components such as pistons, rings, gaskets and exhaust valves and their seats. Cool running sparking plugs may be required too. Moreover, owing to high temperatures of combustion, pressure charging the spark ignition engine tends to increase NO_x emissions. Applied to the diesel engine, however, it can actually reduce hydrocarbon (HC) and CO. Pressure charging can be utilised to compensate for variations in atmospheric pressure, and therefore is of major significance for piston engined aircraft, and can also be an advantage for road vehicles based in mountainous countries.

Maximum temperatures in the combustion chambers of spark ignition engines can be reduced by retarding the ignition but, to avoid thus unduly increasing overall specific fuel consumption, the retardation should be effected progressively as the boost pressure is increased. This can be done by means of an electronic control unit served by a boost pressure sensor. Alternatively, protection can be had by installing a detonation sensor. A cruder, but simpler, measure is to use a manifold depression-actuated mechanical control. However, retardation of the ignition causes late burning and increases the exhaust gas temperatures, which can adversely effect the life of the turbine rotor.

Charge cooling, the other method of increasing density, entails a considerable increase in cost and complexity. Moreover, pressure losses through the cooler, together with the increased volume of the induction manifold system, lead to a reduced rate of response of the engine to the accelerator pedal. Even so, because it increases overall thermal efficiency and helps to reduce output of NO_x, it is tending to become accepted, though more for diesel than spark ignition engines.

14.2   Carburetted engines

With increasing application of electronically controlled fuel injection, choice of the overall layout of the induction system for pressure charging is becoming largely a matter of convenience. Where a carburettor is used, however, placing it before the compressor means that the float chamber can be vented to atmosphere. A disadvantage is that the increased wetted area of manifolding between it and the cylinders can lead to increased condensation and therefore difficulties when starting from cold. On the other hand, as soon as the load and engine speed increase, mixing and generation of heat in the turbocharger offset this tendency. Another problem that arises is that, under conditions of low manifold pressure at light throttle openings, there is a strong tendency for oil to be drawn along the compressor shaft into the induction system. Consequently, good sealing is necessary.

Even more severe problems arise, however, if the carburettor is positioned after the turbocharger, though this arrangement has the advantage that the induction manifold can be common to both the turbocharged and naturally aspirated versions of the engine. To maintain the appropriate pressure differential for drawing the fuel through the jets, the float chamber has to be subjected to boost pressure. This entails the use of a more powerful fuel lift pump and provision of good seals around the choke and throttle controls. Moreover, because of increases in air temperature through the compressor and, usually, wider variations in under-bonnet temperature, the density of the air passing through the venturi will vary more widely than with natural aspiration. Therefore, the fuel metering tends to be less accurate. Additionally, while sudden closure of the throttle instantaneously reduces the rate of flow, the compressor takes significantly longer to accommodate to the changed conditions. The outcome may be compressor surge.

In general, an appropriately designed carburettor installed upstream of the turbocharger meters the quantity of fuel supplied relative to that of the air and thus adjusts the mixture strength automatically throughout the speed and load range. With fuel injection, on the other hand, the electronic, mechanical or hydraulic control system is designed to match the increasing pressure, and therefore density, of the air as the throttle is opened. The higher temperature due to compression of the ingoing charge and other factors has, of course, to be taken into account.

Turbochargers are effective over a relatively narrow speed range, and therefore have to be carefully matched to the engine. How this is done depends on the type of operation. For sports and racing cars, high performance at the upper end of the speed range is generally required. Otherwise, the aim may be at improving torque back-up as well as increasing overall power output. Alternatively, however, the emphasis for a family saloon might be more on turbocharging lower down the speed range and thus increasing overall flexibility, to facilitate driving in urban traffic.

In applying turbocharging to quantity produced spark ignition engines, the aim is mostly at extending upwards the range of power outputs from a single basic design, and thus avoiding the cost of having to design, develop and produce a larger engine. For application to diesel engines, it offers significant additional advantages, including reduction of undesirable exhaust emissions. Consequently, what follows is concerned primarily with diesel practice.

14.3   The diesel engine

Because only air is compressed in the cylinder of a diesel engine, problems associated with detonation do not arise. Moreover, increasing the density of the air reduces the delay period, thus actually producing a more favourable rate of rise of pressure during combustion. Consequently, only about a 30% increase in maximum combustion pressure is required to double the torque, and the engine runs more smoothly than in the naturally aspirated state. Since the air flow is not throttled, the turbine tends to spin faster at light load than on a spark ignition engine. By suitably matching the turbocharger to the engine, a useful increase in torque can be obtained lower down the speed range than would otherwise be possible, and both mechanical and thermal overloading is avoided at maximum power. At the other extreme, if the aim is at increasing maximum power, the engine of course has to be designed appropriately to take the extra loading.

14.4 The two-stroke engine

With two-stroke engines, up to about 40% of the total air delivery is passed straight through to the exhaust, for scavenging. An incidental benefit is a reduction in thermal loading of components, owing to its cooling effect. Superchargers are normally used in preference to turbochargers, since the latter are ineffective at light loads. At higher speeds, however, turbocharging tends to come into its own, because it increases the exhaust back-pressure and thus obviates excessive waste of blown air.

14.5 Turbocharging in general

Turbochargers have been commercially available since well before the Second World War. In fact, it was in 1925 that Büchi first patented an exhaust pulse-driven turbine to run a compressor for boosting the pressure in the induction manifold of an internal combustion engine. However, it is only since about the mid-1950s that compact, high efficiency, reliable units have been available at costs low enough for automotive applications.

Two major advantages render turbocharging preferable to supercharging for automotive engines. First, it does not require a mechanical drive from the engine, which reduces the cost and bulk of the installation. Secondly, because the energy required to drive a turbocharger is not taken from the crankshaft and, thirdly, given good design and matching of turbocharger to the engine, turbocharging can certainly increase overall thermal efficiency by about 3 to 5%, and gains up to 10% have been claimed. The losses due to back-pressure generated in the exhaust system are more than offset by the effect of the higher induction pressures in reducing specific fuel consumption and increasing power.

A disadvantage of turbocharging is what is termed *turbocharger lag*. This arises because of the need, as the demand for torque fluctuates, to accelerate and decelerate the rotor to and from extremely high speeds. Owing to the inertia of the rotating assembly, of which about 60% is generally attributable to the steel rotor of the turbine, it can take as long as several seconds to respond to such changes. With modern designs, however, much shorter response times have been obtained.

On a diesel engine, an additional factor also affects response. Whereas, in a naturally aspirated unit, all the air needed for combustion is present in the cylinder before the driver depresses accelerator pedal, a turbocharged version requires a supply of extra air as well as fuel for producing the extra power. Consequently, if emission of black smoke from the exhaust is to be avoided, the fuel supply has to be increased progressively with the air supply, instead of being injected almost instantaneously with pedal angle change.

14.6 Automotive turbocharger construction

Automotive turbochargers, as can be seen from Figs 14.9 and 14.10, almost invariably comprise a shaft, on one end of which is a radial inflow turbine and, on the other, a radial outflow compressor. Between the two, to keep the unit as compact as possible and to form part of the barrier against thermal exchange between the extremely hot turbine (around 700°C) and the compressor, is a pair of plain bearings. These are generally of the floating sleeve type, to keep rubbing velocities to about half of what they would be with simple plain bearings. Rotational speeds rise up to, or even higher than, 120 000 rev/min.

Because oil temperatures vary widely from start-up to operation at maximum power over extended periods, large bearing clearances (of the order of 20 μm or more) are necessary, to facilitate the flow of oil for cooling the bearings. As speed is reduced at maximum power, the losses (between 2 and 7%) in the bearings may become a significant proportion of the total and can reduce the smoke-limited power of the engine. Rolling-element-type bearings have been used, for example, by Eberspächer, but are are costly, noisy, bulky and, at such very high speeds, tend to wear too rapidly.

The thermal barrier comprises the air gap between both the bearings and between them and the rotors and, in many instances, actual thermal shields too. There may also be an oil thrower just inboard of the compressor, to prevent lubricating oil from being sucked into the air stream. Oil for lubricating and cooling the bearings is usually taken from the engine pressure system. Generous provision must be made for that leaving the bearings to be drained away to the engine sump, otherwise severe thermal deterioration of the lubricant may occur owing to overheating. Iron-ring-type seals, sprung radially outwards, carried in grooves in the shaft immediately inboard of both the compressor and turbine, are generally employed. With spark ignition engines, however, high manifold depression may call for carbon-faced seals for keeping oil out of compressors.

The peripheries of the turbine and compressor wheels are each surrounded by a tapered volute casing. On the turbine, the volute serves to collect the hot gas from the exhaust manifold, convert their pressure into velocity energy, and distribute them around the nozzle directing it into the rotor. They then flow radially inwards, to be discharged around the hub of the rotor into the exhaust downpipe.

Air entering the eye, or *inducer*, of the compressor passes over the hub and radially outwards into the volute which, in this instance, is termed the *diffuser*. This collects the air from the periphery of the rotor, and converts its velocity into pressure energy while guiding it into the induction manifold. Water cooling of the casing, despite the fact that it complicates both the castings and overall installation, is sometimes deemed necessary.

There may or may not be guide vanes in the turbine nozzle and compressor diffuser. Although vanes increase maximum efficiency of the compressor, they also increase complexity and cost and reduce the effective operating range. The reduced range arises because of an increase in the tendency for surge to occur in the compressor rotor as flow rate falls and as the speed of rotation deviates from that appropriate for the angle at which the guide vanes have been set. At the other extreme, as the flow increases, the limit is imposed by choking as the flow exceeds the capacity of the passages in the compressor to accommodate it, Figs 14.1 and 14.2. Without guide vanes, the gases passing between turbine nozzle and wheel and the compressor rotor and diffuser can self-adapt more freely to the variations in mass and velocity of flow. Consequently, although maximum turbine efficiency may not be so high, a good level of efficiency is maintained over a wider range of mass flows.

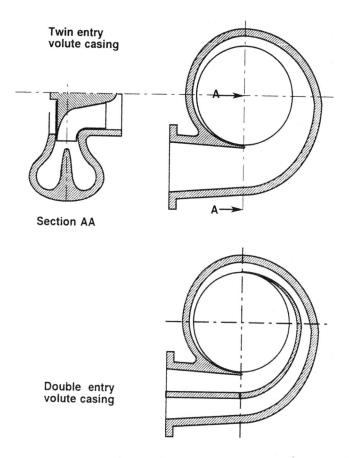

**Twin entry
volute casing**

Section AA

**Double entry
volute casing**

Fig. 14.1 Diagrammatic representations of twin and double entry turbochargers. For the former, the entries are side by side and, for the latter, one above the other

14.7 Operating range and characteristics

The operating range can be improved by varying the geometry of either the ring of vanes or the nozzles as mass flows change, but the mechanisms for so doing are complex and costly and introduce a potential for reduced overall reliability. Even so, because of the current demand for greater efficiencies and higher compressor pressure ratios, much work is now being undertaken on the development of suitable variable geometry systems.

Turbine pressure ratios are low so, to obtain high mass flow rates and thus to convert as much energy as possible into torque for delivery to the compressor, speeds well over 100 000 rev/min are necessary. Gas entry temperatures are about 1000 K on diesel engines, and up to 1200 K on spark ignition engines. Torque output from the turbine must always balance the demand from the compressor, whether running at constant load or accelerating. In automotive applications, torques generally range up to not much more than about 10 Nm.

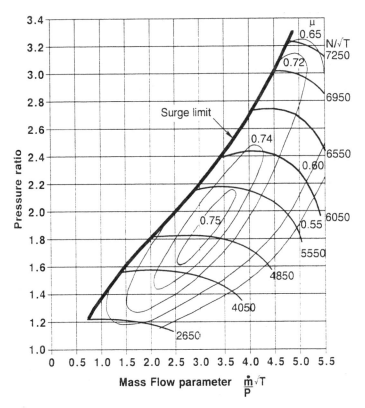

Fig. 14.2 Typical performance map (pressure ratio plotted against mass flow parameter, where N is the rotational speed of the rotor and T and P the absolute temperature and pressure) for a compressor without a vaned diffuser. The surge limit is shown clearly on the left, and the limit on the right is that set by choking of the flow. Between the two are the loops of constant efficiency. The horizontal arcs represent constant-speed conditions

Most turbochargers currently in use on diesel engines operate at compressor pressure ratios of the order of 2 to 2.5 : 1, though the trend is upwards. On spark ignition engines, lower ratios are generally required because detonation has to be avoided. Ratios as high as about 3.5 : 1 are not too difficult to achieve and manufacturers are now looking at ratios of up to 4.5 : 1. The latter require compressor rotors of materials stronger than aluminium alloy. A limitation is the onset of local supersonic gas flows in the compressor. Another is that, as pressure ratio increases, the useful range of operation tends to become progressively narrower, Figs 14.2 and 14.3.

Superimposed on the compressor map in Fig. 14.3 are engine-operating (air requirement) lines at both constant speed and constant load. At constant load the lines rise steeply as engine speed is increased. The air requirement for constant speed, on the other hand, rises much more slowly, because it increases only with the rate of increase of fuelling, instead of with engine speed. Matching the turbocharger to the engine is dealt with in detail in Section 14.15.

For simplicity, suffixes have been omitted in Figs 14.1 to 14.3, but normally lines of constant speed parameter $N/\sqrt{T_{01}}$ and efficiency are plotted on a graph

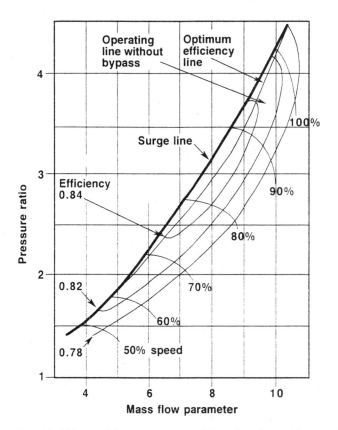

Fig. 14.3 Diffusers with vanes are more efficient than those without, and higher pressure ratios can be attained, but vanes and high pressure ratios mean narrower performance maps. Here, the constant-speed lines are plotted for percentage increments of speed

the co-ordinates of which are P_{02}/P_{01} and the mass flow parameter $m(\sqrt{T_{01}})/P_{01}$, where N is the speed of the compressor rotor in rev/min, $\dot{m}$ is the mass flow rate, and P and T respectively are the pressures and temperatures that would exist in the gases if their flows were suddenly brought to rest under isentropic conditions. Suffixes 01 and 02 indicate inlet and delivery conditions in the compressor and 03 and 04 in the turbine.

14.8 Compressor surge and stall

Two compressor phenomena liable to be confused are surge and stall. Surge is initiated if the energy imparted to the gas by the compressor falls below that needed to overcome the adverse pressure gradient between inlet and outlet: the flow suddenly collapses, as a result of which the output pressure drops to a level at which it can then be re-established. This process is repeated cyclically.

Stall increases the resistance to flow, and therefore may or may not help to initiate surge. It occurs when the streamline flow through the compressor breaks away from the boundary layers over the surfaces of components such as the radial blades, or diffuser walls or vanes. Breakaway occurs when the velocity, and therefore energy, of the streamline flow becomes inadequate either to sweep the boundary layers along with it or to maintain the Bernouilli depression at a level high enough to hold the streamlines down on the surface. The latter condition can arise if the angle of attack between the flow and, for example, a diffuser vane increases beyond a critical level. Stall may or may not become a cyclic phenomenon: if it does, it is termed *rotating stall*.

14.9   Axial or radial flow?

Axial flow turbines and compressors, although capable of operating at pressure ratios up to about 12:1, suffer the disadvantage that, as size is reduced, so also is the available annular space for accommodating the blades, and clearances between blade tips and casing become increasingly a greater proportion of the total cross-sectional area of flow. Consequently, manufacture to very close tolerances is necessary. Additionally, boundary layer flow becomes increasingly significant, tending to choke the flow increasingly as passage sizes are reduced. Furthermore, the number of components needed is very large and, because of the extremely high temperatures, costly materials have to be used.

Turbochargers with radial compressors and turbines are much lighter, simpler, more compact and less costly. However, as the size increases, heat transfer from the turbine rotor becomes a problem and, even more significant, it is difficult to cast satisfactorily large one-piece rotors of materials affording high strength at elevated temperatures.

14.10   The two methods of turbocharging

There are two basic approaches: constant-pressure and pulse turbocharging. In his early attempts, starting in 1909 with a multi-cylinder four-stroke diesel engine, Büchi applied the constant pressure principle, which, however, is now used only on some large industrial engines. Pulse turbocharging, despite the fact that turbine efficiency is reduced by pulsating flow, is virtually universal for automotive applications. The comparisons between the two systems, Sections 14.11 and 14.12, are based mainly on diesel applications.

14.11   Constant-pressure turbocharging

Basically, the constant pressure system entails discharging the exhaust gas into a large manifold, in effect a plenum, to reduce substantially the magnitude of the pulsations before it is delivered to the turbine. Although the efficiency of energy conversion within the turbine in this way is significantly higher than that of pulse turbocharging, the penalty is a loss of much of the kinetic energy of the gas as it leaves the exhaust ports. This arises because of the losses due to throttling and turbulent mixing as the gas discharges, initially at sonic velocity, from the partly opened exhaust valve into the relatively large volume of gas in the plenum. Moreover, unless the large manifold is thermally insulated, the heat losses from it may be considerable.

There are several other disadvantages. Because of the length of time needed to reduce or increase the pressure in the large manifold, response to sudden changes in load is poor, which is one reason why it is unsuitable for automotive applications. Of especial significance for diesel engines is that, under light load, the pressure of the air delivered to the inlet valves may be even lower than that in the exhaust manifold. Consequently, scavenging may be adversely affected and excessive blow-back into the inlet manifold may occur during the overlap period.

Conversely, at high loads, the air delivery pressure may be so much higher than that in the exhaust manifold that scavenging will be too thorough to the point at which air is wasted, and therefore overall efficiency reduced. Air delivery of about 10 to 15% in excess of requirements for complete combustion is generally needed for scavenging, though this does depend on the degree of valve overlap. In spark ignition engines, delivery of excess mixture and its loss into the exhaust system is of course not only wasteful but also increases hydrocarbon emissions.

On the other hand, where compressor pressure ratios, and therefore boost pressures, are high, the consequent high pressures in the exhaust manifold reduce exhaust valve throttling losses. In these conditions, the constant pressure system may show up to advantage. However, with pressure ratios below about 2 : 1 for single or two cylinder groups, and 3 : 1 for sets of three cylinders, pulse turbocharging is positively more suitable.

14.12   Pulse turbocharging

As previously indicated, the principal disadvantage of pulse turbocharging is that, because the flow into the turbine is pulsating, with alternating periods of zero flow from individual cylinders, its efficiency is inherently lower than that of the constant-pressure system. In multi-cylinder engines, however, this disadvantage can be largely obviated by appropriate exhaust manifold design. Indeed it can be more than offset by the availability of a high proportion of kinetic, in addition to the pressure, energy, for conversion into work in the turbine. To take maximum advantage of the pulse energy, the turbocharger should be sited as close as possible to the engine exhaust ports, so that the volumes of the passages from the valve ports to the turbocharger are small. This helps to ensure rapid response in transient conditions. It can also reduce pumping losses, because the back-pressure in the port falls off more quickly.

To avoid excessive throttling losses, it is important to reduce the mean velocity of the gas through the exhaust valve port to sub-sonic levels as rapidly as possible. Therefore, rapid lifting of the valve is desirable. Keeping the cross-sectional area of the manifold small helps too, by minimising the throttling losses between the valve throat and the manifold. Moreover, as previously explained, engine pumping losses are also reduced, because the pressure falls more rapidly towards the end of the exhaust phase.

14.13   Exhaust manifold layouts for turbocharging

Efficiency of pulse turbocharging can be considerably enhanced by appropriate matching of the exhaust manifold layout to the engine. Consider a single cylinder: as the exhaust valve cracks open, the pressure in the cylinder may be of the order of 90 to 120 bar, while that in the manifold will be little, if any, higher than atmospheric. After a short pause for the inertia of the gas to be overcome, the pressure in the manifold will rise steeply to a peak at around BDC, Fig. 14.4. During this period, the pressure in the cylinder will be falling steeply and will continue to do so for about a further 60°. From this point on, the rate of fall will be progressively reduced as the valve closes, leaving the pressure in the cylinder slightly higher than atmospheric, while that in the manifold falls increasingly rapidly as the gas is discharged through the turbine.

If the single-cylinder engine is connected by a relatively small diameter pipe to the turbine, the result of this pattern of pressure variations will be as follows. The initial expansion of the gas at supersonic velocity into the manifold will cause a pressure wave to be transmitted along the pipe to the turbine. The throttling effect of the turbine nozzle will reflect the wave, though at reduced energy level, back to the exhaust valve port.

Given that the pipe is very short, this wave will be virtually superimposed on the initial wave, thus reinforcing its effect on the turbine. With a slightly longer pipe, however, the returning wave would arrive during the blow-down period of the exhaust process, and therefore both reduce the efficiency of scavenging and perform negative work on the piston. Indeed, two successive waves could

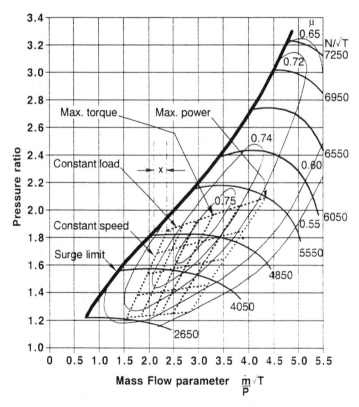

Fig. 14.4 The straight dotted lines superimposed on this compressor performance map comprise a set of engine running lines, or air-requirement plot. In practice, the constant-load lines would be slightly curved. The surge margin is the horizontal distance X between the surge line and the nearest point to it on the engine running line plot

arrive at the exhaust valve during the scavenge period, having an even greater adverse effect on four-stroke operation and perhaps actually stop a two-stroke engine. With a long pipe, however, the initial wave could arrive back after the exhaust valve has closed and thus have no effect on scavenging: in general, it is only the initial wave that is of significance as regards engine performance. It should be noted that we have been considering pressure variations in time. More relevant, because the time actually available depends on engine speed, is pressure variation per degree of crankshaft rotation.

From Fig. 14.4 it can be seen that the availability of pulse energy extends over only a short period before and after BDC, and then the pressure energy available declines increasingly rapidly until the exhaust valve closes. For the remainder of the 720° of the four-stroke cycle, the turbine idles and loses speed. The illustration also shows that relatively little of the kinetic energy available during the initial period, during which the cylinder pressure is falling from its highest point, can be utilised by the turbine.

The ideal number of cylinders is three, all discharging through a single nozzle into the turbine. With this arrangement, and exhaust valves opening at about 240° intervals, the idling periods of the turbine are almost eliminated; moreover, when the exhaust valve of one cylinder opens, all the others are closed so, given appropriate pipe lengths, there need be no obstruction of the exhaust processes of individual cylinders due to pressure waves reflected from others. Even so, because the flow is still pulsating slightly, the overall efficiency of the turbine cannot be so high as that obtainable with a well-matched constant-pressure system, though overall performance is generally better.

The exhaust manifold for a three-cylinder engine usually comprises three pipes joined to a single pipe, which delivers the gas into the turbine nozzle. At the point where each branch joins the main pipe, the pressure pulse from its cylinder generates a secondary pressure pulse travelling back to the other exhaust valves. Therefore, unless the branch pipes are of lengths such that the secondary pulses reach their valves while they are closed, scavenging efficiency will be impaired. This effect is of course additional to that previously described, due to reflections back from the nozzle of the turbine. All these reflections pass back and forth in the manifold, dividing into two as they pass each pipe junction in turn, and therefore with successively decreasing amplitudes.

For larger multiples of three cylinders, each group of three delivers into a common trunk which takes the gas on through a single, double or twin entry into the turbine volute and nozzle, Fig. 14.1. With a single-entry turbine, the tapered volute distributes the flow around the rotor. Given a *double entry*, the gas from one is distributed around half of the periphery of the rotor, and that from the other entry around the other half. The *twin-entry* layout takes the gas from each entry around the whole of the periphery. In this case, to avoid both thermal cracking of the dividing wall and restriction of the nozzle, the dividing wall does not extend right into the nozzle.

A disadvantage of the double-entry system is that, owing to the pulsation of the flow from four or eight cylinders, the pressures through some of the passages between the vanes on the turbine rotor may be lower than those in adjacent ones. Consequently, in those in which the pressures are lower, the flow may periodically cease or even reverse. This can excite vibrations of the shaft and other elements of the structure. The twin-entry system is therefore

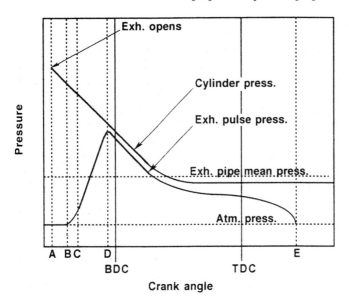

Fig. 14.5 Pulse effect for a single-cylinder engine with a short pipe between the exhaust valve and the turbine. As the exhaust valve cracks open, at A, the pressure drop across it is above the choking, or critical, value so the flow is sonic and no pressure rise is apparent up to B because all the energy due to expansion is lost in turbulent mixing. Between B and C, the pressure rises increasingly steeply, though has no effect on turbine speed. The main rise, between C and D, occurs because the flow of gas into the pipe is greater than that escaping through the turbine. From D to E, the pressure falls down to atmospheric as all the exhausted gas is discharged through the turbine

mostly employed for multi-cylinder automotive applications.

Clearly, the best arrangement for a six-cylinder engine is either a turbocharger for each set of three cylinders as on, for example, some Toyota models, or a single turbocharger with a twin-entry turbine. A 12-cylinder engine will normally require two turbochargers. Engines of the V-layout present a more difficult problem owing to the phase difference between the cycles of the cylinders in different banks. The aim in such cases is, so far as is practicable, to interconnect the pipes from cylinders whose firing sequences are as close as possible to 240° apart.

This principle applies also to, for example, in-line four-cylinder engines. Given a firing order of 1 3 4 2, pairs of cylinders (1 and 2, and 3 and 4) would be connected and the exhaust gas from each pair directed into a different nozzle of a twin-entry turbine. The resultant loss of turbine efficiency due to unsteady flow and a slight degree of pulse interference across the edge of the dividing wall in the nozzle is generally more than offset by the additional energy (kinetic) available to the turbine.

With the pulse system, it is generally possible to arrange for the exhaust manifold pressure during the valve overlap period to be significantly lower than that in both the cylinder and the induction manifold. Consequently, even at light load, when the boost pressure is low, good scavenging can be obtained. Furthermore, scavenging will not be significantly impaired if the turbine becomes fouled with carbon during service.

14.14 Pulse converters

By incorporating pulse converters in exhaust manifolds, it is possible to reduce the adverse effects on scavenging of inter-cylinder interference by reflected pressure pulses. However, this involves increasing the bulk, complexity and cost of the manifold, so such devices are not generally used in road vehicles. If it were not for the increased bulk, however, pulse converters might be worth considering for V-8 engines, owing to their unfavourable firing orders.

Pulse converters for larger engines are based on one of two principles. For twin-cylinder engines, eddies are generated in the mouth of the pipe adjacent to that exhausting impede the reverse flow due to the pressure pulse in the latter, Fig. 14.6. The second method was proposed in 1946, by Birmann, for exhaust manifolds having short branch pipes. Each of these short pipes is constricted as it approaches its junction with the main trunk that delivers the gas to the turbine nozzle. This constriction increases the velocity of flow at the junction and reduces the pressure locally, thus forming a barrier to impede the passage of the pressure pulses back to its exhaust valve. An example of a further development of this system is shown in Fig. 14.7.

In 1957, Brown Boverie and Sulzer went further by feeding pairs of pipes from cylinders whose exhaust phases do not overlap, unrestricted into single pipes. However, the junctions of the last-mentioned pipes with the main delivery pipe to the turbine nozzle are restricted to prevent the outputs from the pairs from interfering with each other.

14.15 Matching the turbocharger to the engine

As previously indicated, the performance of a turbocharger is high only around the design speed, when the angles of approach of the gas flow to the guide vanes and blades of the compressor rotor and turbine approximate to the angles at which those components are set. As these angles diverge, stalling

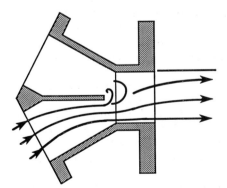

Fig. 14.6 Pulse converters are sometimes used on two-, four- and eight-cylinder engines to reduce or prevent back-flow into cylinders owing to pulsations in the manifold. Each exhaust valve discharges into a short tapered branch pipe which increases the velocity and reduces the pressure of the outgoing gas up to the point at which it enters the main trunk of the manifold. The low pressure thus created in the branch pipe tends to block back-flowing pressure pulses

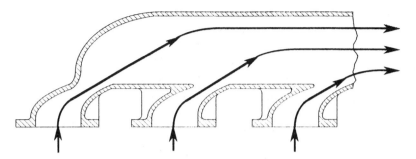

Fig. 14.7 In an alternative type of pulse converter, the narrowing passage is combined with a turbulence generator which spills turbulence over into the mouth of the adjacent branch pipe. This turbulence helps both to interfere with entry of pressure pulses and to prevent them from being significantly diffused as they move along the expanding section

and turbulence tend to occur, ultimately leading to poor efficiency.

Moreover, whereas the pressure drop across the turbine (the expansion ratio) is a function of the square of mass flow, turbine power output is a function of both the mass flow and that pressure drop. Therefore, the speed at which the turbine drives the compressor rises faster than the mass flow through the turbine. Furthermore, the boosting effect of the compressor rises as the square of its rotational speed. The fact that the turbocharger has little or no influence over power output except over a narrow range poses severe problems as regards matching it to the engine, particularly with high-speed spark ignition engines.

If the turbocharger is matched to a diesel engine at the speed at which it develops maximum power, it will not deliver enough air when the speed falls to that at which maximum torque is obtained. The outcome will be black smoke from the exhaust and inadequate torque back-up. On the other hand, if it is matched at the speed at which maximum torque is developed, the maximum power potential will be lower, high peak cylinder pressures will be incurred and pumping losses may increase. For commercial-vehicle operation, good torque back-up is generally the overriding requirement. At light load, a diesel engine is in any case operating with excess air, so the fact that the turbocharger is ineffective in this situation is irrelevant, though a fall off in speed at maximum torque can be accompanied by black smoke from the exhaust.

Matching is done on the following basis. For any given air intake pressure, the mass of air inducted into the cylinder is approximately directly proportional to its speed. We can therefore superimpose a set of engine-running lines at constant speed over the required bmep range on compressor maps for different frame sizes, as has been shown in Fig. 14.3. In many instances, turbine manufacturers issue simplified diagrams such as that in Fig. 14.8, from which the engine builders can make their initial selections of turbocharger sizes.

The left-hand edge of the engine-operating-line map must be kept clear of the surge line by a margin of about 10% mass flow for a multi-cylinder and 20% for a twin-cylinder engine. The location of the right-hand line depends on what minimum efficiency is acceptable. Once the selection has been made, it is necessary to run engine bench tests to ensure not only that the frame size and the compressor and turbine trims are appropriate, but also for making

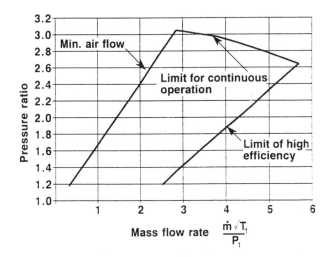

Fig. 14.8 Some turbocharger manufacturers issue compressor matching charts like this, generally for an atmospheric temperature of 15°C and pressure of 1 bar. It is necessary to know, however, what is the surge margin (the distance between the minimum-air-flow line and the surge line), at what efficiency the right-hand limit line is set and whether it is adequately clear of the choking limit

adjustments to the rates of fuelling and timing of injection and of the inlet and exhaust valves, to optimise engine performance.

14.16 Extending turbocharger speed range

Because the turbine does not have to function against an adverse pressure gradient, its range of operation is inherently broader than that of the compressor. The most important requirement for the turbine is that it be capable of generating the torque needed to drive the compressor when delivering at its maximum pressure ratio. Thus, it is the performance of the compressor that is critical. As previously stated, the upper limit is imposed by the sizes of the passages through it since, ultimately, they will choke the flow. Surge imposes the lower limit.

From Fig. 14.9 it can be seen that the volutes for both the turbine and the compressor can be easily removed from what is termed the *main frame*, which is the central assembly comprising mainly the shaft and bearing assemblies in their casing. The effective sizes of the compressor and turbine rotors can be varied by, in effect, machining the edges of their blades back towards the hub and rotor disc, and fitting to the main frame casings appropriately smaller sizes to maintain the same blade edge clearances. These variations, termed *wheel trims*, are utilised for matching the turbocharger to the size and requirements of the engine. There may be as many as 12 trims of compressor but turbines, because of their broader ranges, are generally offered in fewer trims. Adjustment to flows may be also made by changing the nozzle throat size. Any or all of the adjustments explained in this paragraph can be effected to match the performance of the turbine to the requirements of the engine.

Clearly, an ability to vary the effective size of the turbocharger while the

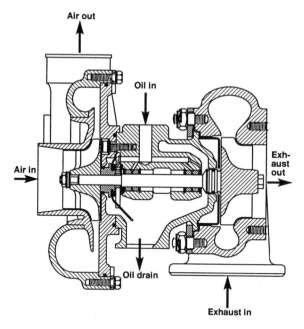

Fig. 14.9 This is a KKK (Künle, Kausch und Kopp) turbocharger with a single-entry turbine. For matching the turbocharger to the engine, it is easy to unbolt the turbine and compressor housings from the frame carrying the shaft, rotor and bearing assembly, and replace them with components of different sizes

engine is running would be advantageous, so that its flow capacity could be made adequate for supplying the air needed for operation under maximum load at high speed while, by reducing its size, the velocity of the air flow through it at lower speeds would remain adequate to prevent the onset of surge. Measures to prevent or delay the onset of stall would be helpful too. Although continuous variation of the size is impracticable, the use of variable geometry can have a similar effect.

14.17   Variable geometry

By either incorporating an axially movable element in the sidewall of the nozzle, as for example in Fig. 14.10, or pivoting each guide vane simultaneously about one of its ends, Fig. 14.11, the effective size of the turbine can be varied. The latter will of course vary the angle of approach of the air flow to the turbine blades. Furthermore, as the vanes are pivoted about one end, the gaps between their other ends tend to open or close, according to the direction of rotation, thus varying the area available for the flow through the nozzle. An alternative is to rotate the vanes about their centres, as shown in Fig. 14.12. The effect on induction manifold pressure of so doing can be seen in Fig. 14.13. Shifting an axially movable element in the nozzle has a similar effect. Varying the turbine nozzle area enables the exhaust gas energy to be utilised to best advantage at low speeds and, by restriction of the flow, it avoids over-boosting the engine at high speeds.

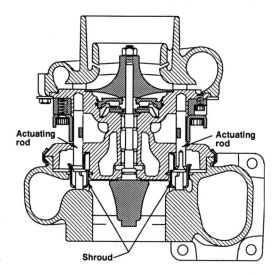

Fig. 14.10 This is a Holset single-entry, variable-geometry turbocharger. A circular shroud, in which are slots that are a close fit around the vanes, is moved axially in and out of the turbine nozzle to vary its effective size. The movement is controlled by either a diaphragm-type actuator subject to manifold depression or an electronic engine management system

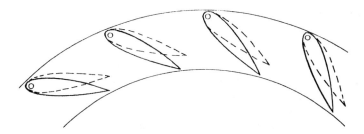

Fig. 14.11 Another way of varying the geometry of a turbine is to pivot simultaneously all the vanes in the nozzle or diffuser about their outer ends. Various actuation mechanisms can be employed, including rotating rings with cam slots into which project pegs on the ends opposite to the pivot pins, or linkages either fixed directly to the pivot pins or connected to pins projecting from the opposite ends of the vanes. For a compressor, the streamlined vanes would be inverted and pivoted about their inner ends, though curved plate vanes are more commonly used than the costly streamlined sections. Note that the throat area would be at a maximum if the principal axes of the vane sections were radially orientated and a minimum if they were in line

Similarly, to vary the geometry of the compressor, the angle of the diffuser vanes or the width of the throat can be varied. In addition to varying the area between diffuser vanes, if incorporated, varying their angle to align them with the outflowing stream of air will reduce the tendency of the flow to stall on entering between them. With fixed vanes, it is primarily the misalignment of the air flow as the rotor departs from the design speed that reduces the width of the compressor map relative to that of a vaneless equivalent. Varying the geometry of a compressor is aimed at obtaining a broad surge-free range of

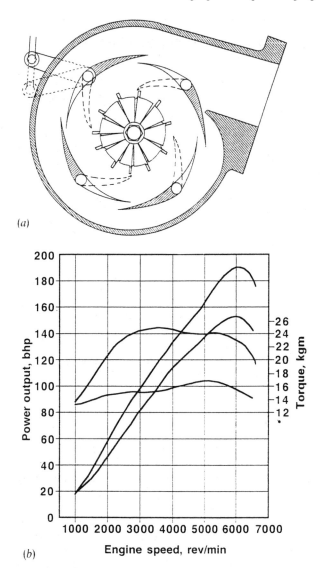

(a)

(b)   Engine speed, rev/min

Fig. 14.12 (a) The Honda Wing turbine with the wings shown unhatched in the closed, and by dotted lines in the fully open, position; (b) performance curves of the Wing turbocharged and naturally aspirated versions of the Honda 2-litre 24-valve + PMG engine for Formula 1 racing

operation, by increasing the separation between the surge and choking lines on its performance map.

A simple measure that can be taken to improve efficiency is to incorporate in the eye, or *inducer*, of the compressor, vanes designed to pre-swirl the incoming air and thus to reduce the tendency to turbulence due to shock as the air stream enters the blades of the rapidly rotating rotor. Another method is to turn the edges of the vanes forward in the area enclosed by the eye of the

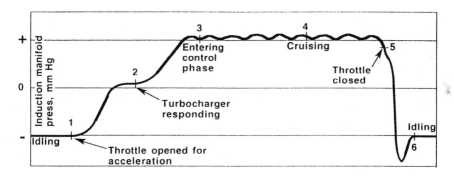

Fig. 14.13 This is a plot of the engine induction manifold pressure, in mmHg, starting at (1) where the Honda turbo wings are in the position indicated by the dotted lines in Fig. 14.12 and ending at (6) where they are returned to the same position. Stages (2) to (5) represent the depression as the throttle is opened and closed again and the turbo wings moved appropriately between the closed and open positions and back again

compressor, in effect causing them to induce initially an axial flow, which is subsequently turned through 90° by the hub, to become radial flow. Holset have carried this a stage further by mounting, on a forward extension of the shaft within the eye of the radial flow rotor, a separate axial flow rotor. This is, in effect, a two-stage compressor, the first stage serving also to pre-swirl the air entering the radial flow section. The axial flow section delivers a pressure ratio of 1.25 : 1, and the potential overall ratio is in excess of 3.5 : 1.

14.18   By-passing the gas flow

The most widely used method of varying the output of the turbocharger to suit the engine characteristics is wastegating, Fig. 14.14. A compressor large enough to deliver the air needed at slow engine speeds is fitted, and its output at higher speeds reduced by opening what is, in effect, a pressure relief valve by passing some of the exhaust gas around the turbine into the downpipe. Thus the over-boosting of the engine at the higher speeds, as a result of fitting the larger compressor, is avoided. The wastegate can be controlled by, for example, a diaphragm type actuator subject to boost pressure, or electronically by a control the functioning of which is based on an input from either a boost pressure sensor or, as an alternative in the case of a spark ignition engine, a detonation sensor.

Another interesting concept, but one that is suitable for use only with constant-pressure turbocharging, is that of regulating the output from the compressor by by-passing some of it to the turbine inlet. This by passed air adds to the mass flow through the turbine so, at least in theory, a smaller compressor might be used. Also, under light load, the engine operating line is moved further from the surge line, so any subsequent increase in boost pressure is less likely to induce surge.

Other methods have been used, but have few attractions and some severe disadvantages. One is the fitting of a restrictor in the outlet pipe from the turbine. Turbine power output is a function of the square of the mass flow, and since this increases with engine speed, it is limited by the rising back-pressure

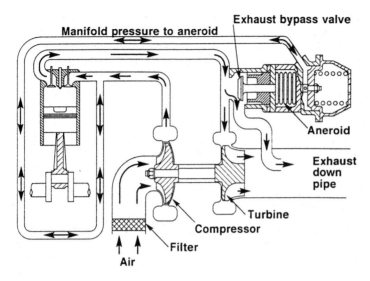

Fig. 14.14 Diagrammatic representation of an exhaust wastegate system

due to the presence of the restrictor. Although this is a simple and reliable method, the consequent high pressures and temperatures in the exhaust manifold, increase the potential for leakage. Fitting a restrictor in either the inlet or outlet of the compressor has a similar effect but has the disadvantages of constant throttling and increasing the charge temperature. It can also, under certain conditions, cause surge.

With another system, the KKK Turbobox, two identical turbochargers are installed in series and controlled electronically. At low speeds, one is by-passed until, as it approaches its strangulation point, the second comes into operation and the air then passes through both. The change-over point is at maximum torque at a medium speed. Alternatively, a large and a small turbine are mounted in series. First, the smallest is brought into in operation, then the largest and, finally, both simultaneously. Because two turbochargers are required, the Turbobox system is of course both costly and bulky.

14.19   Cooling the charge

Turbocharging increases the temperature, and therefore reduces the density of the charge. It does so in three ways. The first is simply by the addition of the energy of compression. Secondly, turbulence in the flow through the compressor also adds heat, doing so increasingly as compressor efficiency falls off. Thirdly, some heat is transferred from the turbine to the compressor. After-cooling helps to compensate for the consequent loss of density. Incidentally, the term *after-cooling* is most appropriate for use with single-stage compression: the alternative, *inter-cooling*, is used for charge cooling between two-stages of compression, a costly system almost unknown in automotive applications.

Cooling the charge after compression brings the following benefits. By virtue of the increase in density of the gas delivered, a higher power output, potentially between about 20 and 25%, is obtainable. Lower temperatures in the cylinders reduce the thermal loading on both the engine and the turbine and, because friction losses are not significantly higher, bmep is increased, and specific fuel consumption is improved by between 3 and 5% though, because charge cooling is greatest when the mass air flow is low, these benefits are obtained mainly at low engine speeds. With a cooler charge, too, the output of NO_x will be reduced. Also, the engine operating lines at constant speed swing over to run more nearly parallel to the surge line, and the constant load lines move away from the surge line on the compressor map. Consequently, all the operating lines tend to fall in the areas of high efficiency, Fig. 14.15, and matching may be easier.

14.20 The heat exchanger

Air-to-water cooling has the advantage that the rate of heat transfer to water is higher than with air-to-air cooling, so the cooler can be more compact. The water can be taken from the engine coolant system. In this case, however, charge heating, instead of cooling, may in some circumstances occur at low

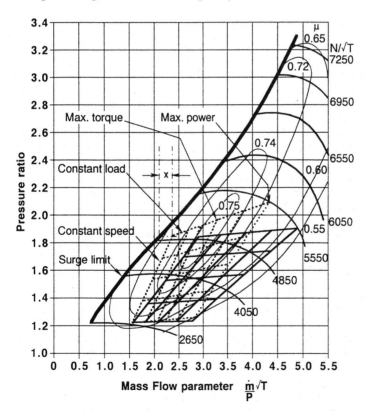

Fig. 14.15 After-cooled engine running lines (full) superimposed on the running lines without inter-cooling (dotted). It can be seen that the engine can be matched more easily to the compressor, by moving the after-cooled lines to the left, mainly because the maximum load line is more nearly parallel to the surge line

speeds. Better, but more costly and bulky, is an entirely independent closed-circuit water system, with a heat exchanger in front of the engine radiator. Such a system has the additional advantage of being less likely to become fouled internally.

Because of a lower heat transfer coefficient, air-to-air charge cooling requires approximately 10 times the heat exchange area, but it does obviate complex plumbing and potentially serious leakage of coolant. The power needed to overcome fluid friction increases as the cube of the velocity of flow so, despite the fact that the rate of heat transfer increases with velocity, low rates of flow as well as large heat transfer areas are desirable. Fins of large area are an advantage on all exchange surfaces exposed to air flow, though this of course applies also on air-to-water coolers.

With vehicles operating under dirty conditions, fouling of air cooled fins may occur, ultimately blocking the passages between them. This can cause thermal over-loading of the engine. On the engine air intake side, the fouling problem, though much less severe owing to the presence of the engine air filter, may be exacerbated by the presence of oil mist in the air. Progressive blocking of the air filter in service may cause compressor surge.

Careful layout of both the inlet and outlet ducting for the cooler is necessary, since poor distribution of flow can have serious effects on cooling efficiency. The length of induction manifolding between turbocharger and engine has to be a compromise. If it is too small, pressure fluctuations in the manifold may trigger surge in the compressor as also can pressure wave reflections; if too large, response to the throttle may be slow.

Until recently, air-to-water after-cooling was generally favoured, because of its compactness. Now, however, the situation has changed, an example of an outstandingly good design and a compact installation being the air-to-air charge cooler on the Ford Mondeo, Fig. 14.16.

For further reading and more detailed information on all aspects of turbocharging, the reader is referred to *Turbocharging the Internal Combustion Engine*, by Watson and Janota, Blackie. A good summary of turbocharger design problems can be found in Paper C405/024, presented by Flaxington and Mahbod, of Allied-Signal Garrett Automotive Group, in the proceedings of the IMechE International Conference on Turbocharging and Turbochargers, May 1990. Optimisation of turbocharger performance, especially under low-speed and transient conditions, has been dealt with in Paper C405/O36, presented by B. E. Walsham, of Holset Engineering, at the same conference.

14.21   Supercharging

Supercharging is the alternative to turbocharging as a method of helping to offset the tendency of the power and torque fall off, Figs 1.7 and 1.8, over the upper end of the speed range of a naturally aspirated engine. The effect is illustrated in Fig. 14.17, which are performance curves for a GM 7.37-litre spark ignition engine for installation in coaches. This power unit was pressure charged by a belt-driven centrifugal supercharger geared to run at 10 times engine speed.

An advantage of supercharging relative to turbocharging is that the mass of air delivered increases approximately in direct proportion to the speed, and therefore air requirement, of the engine. Moreover, its effective range of

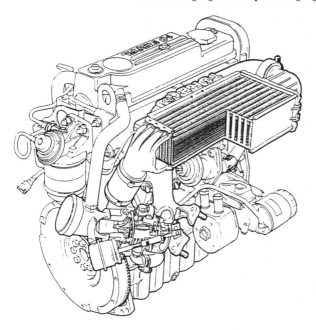

Fig. 14.16 The extremely compact air-to-air charge cooler on the Ford Mondeo 1.8-litre turbo diesel engine

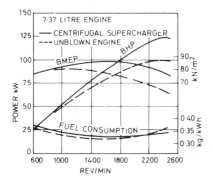

Fig. 14.17 Comparison between the performances of a 7.37-litre spark ignition engine with and without a centrifugal supercharger driven at ten times engine speed

operation is not limited by surge, so a much broader compressor map is obtainable, and its response to sudden demands for extra torque is more rapid.

Figs 14.18 and 14.19 are curves for a 7.68-litre spark ignition engine with a roots-type blower driven through a magnetic clutch engaged only at full throttle. From Fig. 14.18, it can be seen that the height of the bmep, or torque, curve was thereby increased over the upper end of the speed range. Indeed, pressure charging increased the power output to equal that of a 10.1-litre engine, but with an overall road fuel consumption 20% lower. Obviously,

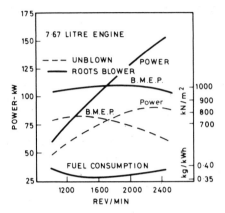

Fig. 14.18 Performance curves for a 7.67-litre spark ignition engine with and without a roots-type blower driven through a magnetic clutch, which was engaged only at full throttle

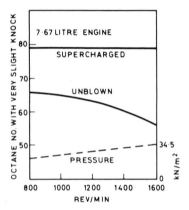

Fig. 14.19 Octane requirement for the 7.67-litre engine the performance curves of which are illustrated in Fig. 14.18

however, maximum temperatures of combustion must have been higher too.

This engine, as can be seen from Fig. 14.17, was found to operate satisfactorily on 79 octane fuel throughout the speed range, whereas in the unblown state, the octane number requirement increased as the throttle was closed. Presumably, this increase was due to a deterioration in fuel metering, or inadequate mixing in the cylinder towards the lower end of the throttle range.

14.22   Two main categories of supercharger

Basically, mechanically driven superchargers fall into two categories. One is the air displacement type in which the air is not compressed in the machine: this type simply carries a fixed volume of air, trapped between lobes or vanes, round from the inlet to the outlet and therefore such machines are widely known as *blowers*, as distinct from *compressors*, which comprise the second category.

Inevitably, the overall fuel consumptions of mechanically supercharged engines are higher than those of their naturally aspirated equivalents. Discharging the air from a blower against a back-pressure generates noise. Moreover, shock waves and pulsations in the flow reduce efficiency. Among the widely used categories of blower is the roots type, Fig. 14.20. This was invented in 1865 in the USA, by F. M. and P. H. Roots, primarily for ventilation applications. Another is the vane type.

The Bendix, Shorrocks and Centric units, Fig. 14.21, are among the vane types. These, however, compress the charge, as their vane carrier rotors are carried on shafts that are eccentric relative to the bores in which they rotate. Because of this eccentricity, the radial distance from rotor hub to housing wall varies cyclically as the shaft rotates and so also therefore do the volumes of the spaces between the rotors. Only the Centric blower will be

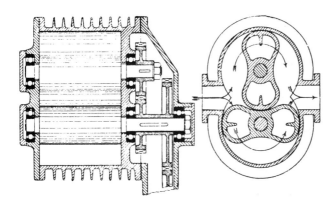

Fig. 14.20 This roots-type blower has twin-lobe rotors, but some have three or four lobes

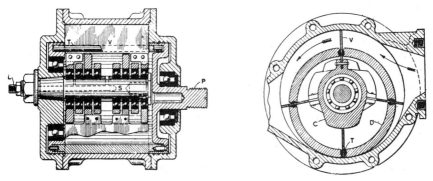

Fig. 14.21 Among the vane-type blowers is this Centric machine. The vanes, the inner ends of which are carried in slots in lugs on a hub freely rotating on a shaft concentric with the housing, are free to slide in trunnions in the eccentric ring that drives the assembly. Thus each vane is continuously aligned at a constant angle relative to the instantaneous tangent at its point of contact with the cylindrical walls of the housing

described here since its rotor is not eccentric and therefore its construction is not so easy to visualise as those with eccentric rotors.

The roots type may have three or even four lobes, though this arrangement is uncommon except in the larger sizes Two intermeshing gears drive the rotors, and manufacturing tolerances on the casings, gears and rotors are very tight, so that the clearances between both the lobes and the casing can be kept as small as possible. The aim is at keeping back-leakage to a minimum to reduce the inevitable decline in efficiency as speed falls. At the higher end of the speed range, efficiency is impaired by turbulence, mainly at the delivery port but also at the inlet. If pressure ratios higher than about 1.6 : 1 are required, the vane type is more satisfactory. A performance map typical of a roots-type blower is shown in Fig. 14.22. From this it can be seen that the best efficiency is likely to be obtained around approximately half engine speed.

In the vane type, provision of adequate lubrication for the edges of the blades, passing at very high velocity over the bore of the casing, is important unless a small clearance can be introduced between these components. Avoidance of entry of lubricant into the engine tends to be difficult. The shapes of the inlet and outlet ports can be such that higher aerodynamic efficiency is obtained than in the roots type and, owing to the absence of clearances between vanes and casing, the fall off in efficiency as speed is reduced is less apparent.

14.23   Vane type with tip clearance

The vane-type supercharger illustrated in Fig. 14.21 is the original version of the Centric unit. Vane carriers C run on ball bearings on the fixed shaft S, which is coaxial with the cylindrical casing. Pairs of these carriers are mounted symmetrically about the centre of the shaft and interlinked. The vanes are riveted to their carriers, and extend radially outwards through plastics trunnions in slots of cylindrical section between the segments of an eccentric drum assembly D.

This drum is rotated by the primary shaft P, and carries round with it the vanes on their free-running carriers. As can be seen from the illustration, the vanes remain constantly radially disposed relative to the cylindrical housing

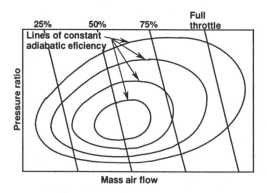

Fig. 14.22 Plots showing how the volume of air delivered by a roots-type blower (straight lines) decreases as the pressure ratio increases

and main shaft D, though at varying pitch. By virtue of the fact that the vanes are constrained by their carriers, an extremely small clearance is maintained between both their tips, their ends and the casing. Thus, both tip-lubrication problems and friction due to high-speed sliding contact are avoided. The only significant losses therefore are those due to windage and friction in the bearings, trunnions and gear drive.

The original main shaft was found to be inadequate for application at high speeds and powers, owing to deflection under the loads applied by the angular acceleration and retardation of the vanes. Therefore, a different layout was subsequently adopted. The main shaft was extended and carried in bearings in the end walls of the casing and became also the shaft driving the drum, through a pinion meshing with an internally toothed ring gear. The reduction ratio adopted depended on the eccentricity of the drum.

14.24   Advantages of blowing

An advantage of both the vane- and roots-type machines is rapid response under transient conditions. An inherent disadvantage of the latter is that, in contrast to machines that compress the air internally, the power absorbed increases continuously as the pressure in the induction manifold pressure rises. However, by incorporating a butterfly valve, electronically controlled on the basis of load and speed inputs, to return back to the inlet a proportion of the air delivered, parasitic losses can be reduced and the proportion of power required made to fall linearly with speed. With such a control, the air flow can be made to match the requirements of the engine, which of course rise linearly with speed. Thus, the torque characteristics obtainable, including in the low speed range, are good.

With the addition of a clutch that engages automatically at about 80% of its naturally aspirated torque and remains engaged up to maximum power, the engine can be smaller for any given power output, and therefore with lower parasitic losses. A fuel-consumption map simulated by Freeze and Nightingale, of Ricardo, for a 1.7-litre diesel engine with a clutched mechanically driven supercharger is shown in Fig. 14.23. Their work (*A Practical and Theoretical Investigation into the Application of Mechanical Superchargers to Diesel Cars*) has also shown that, on a 1.8-litre DI diesel car operated under the LA4 US Federal Test Procedure, the supercharger would be operating for only 6% and, during the Highway Cycle, only 5.5% of the total time. Therefore, durability of the supercharger should not be a problem, though the same comment would not necessarily apply to the clutch. Fuel economies of the order of 13% relative to that of a naturally aspirated engine developing the same maximum power should be obtainable.

14.25   Screw-type compressors

The first screw-type compressor was developed by Krigar in 1878. In the early nineteen forties, two compressors of similar type were introduced for supercharging the internal combustion enginee. One was the invention of A. J. R. Lysholm, of Ljungstrøms Angturbin, Sweden, and the other the unit produced by the Saurer company in Switzerland. In each, spiral lobes on a male rotor meshed with grooves in a female rotor, as shown in Figs 14.24 and 14.25.

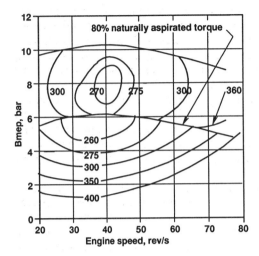

Fig. 14.23 Plots of estimated bsfc of a 1.7-litre diesel engine supercharged by a roots-type blower cut out by a clutch as the load falls below 80% of maximum torque (reported by Freese and Nightingale of Ricardo Consulting Engineers Ltd)

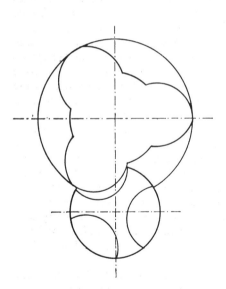

Fig. 14.24 General section of Saurer rotors

Fig. 14.25 Saurer rotors

The spiral on one rotor is left- and that on the other right-handed, the two rotating in opposite directions.

During the rotation, air is drawn through the inlet port, in one end of the casing, where it enters the spaces between, one after the other in turn, the spiral grooves on the female and their meshing lobes on the male rotor. This is the induction phase of the cycle, because the lobe that is beginning to uncover the port is also beginning to withdraw from its groove, so the space between each meshing pair is increasing, allowing the air at atmospheric pressure to enter. As rotation continues, this air is carried around by the grooves in the female rotor until the inlet port is covered by its leading edge. This is the beginning of the compression phase, because the lobe is beginning to mesh into the groove, at the inlet end, and thereafter progressively compresses and displaces the air towards the far end of the casing. Finally, the delivery phase begins as the trailing edge of the groove uncovers the port in the opposite end, whence the air is delivered into the induction manifold. Because the air is compressed before delivery, there is no back-flow so the only significant noise is a high pitched whine due to meshing of the rotors. By virtue of the incorporation of several lobes and grooves, and progressive compression from one end to the other, delivery is practically pulse free.

The Sprintex unit, which can operate at pressure ratios of up to about 2.2 : 1, is a good example of a modern screw-type compressor. Its rotors are coated with PTFE and, by virtue of extremely close tolerances and a tip clearance of only 100 μm, the lobes touch neither the grooves nor the casing. Their rotation is accurately synchronised because they are interlinked by a pair of gears in a housing at one end of the main casing. The range currently in production provides for maximum flow rates of 130 to 330 l/s.

In order to obtain good helical (or axial) sealing, a low addendum (distance between pitch circle and periphery of rotor) is necessary. To achieve this with male and female rotors having equal numbers of lobes and grooves, the female would have to be very small. This condition has been avoided in the Sprintex units by having two more grooves than lobes, which provides a high displacement for any given speed. Sprintex machines operate at higher speeds than most other superchargers. Driving the female rotor limits the crankshaft to supercharger drive ratio and thus reduces the space needed for accommodating the crankshaft pulleys. Volumetric efficiencies of these units peak at a rotor tip speed of about 100 m/s and thereafter fall gradually, as compared with, for example, the roots type, which peak at about 50 m/s and then fall sharply.

Matching to the engine is relatively simple since the output of the supercharger and air consumption of a four-stroke engine are both linearly proportional to speed. The first step therefore is to pick the size of supercharger the air throughput of which matches the consumption rate of the engine. However, some adjustment may then have to be made because the volumetric efficiencies of the engine and supercharger vary differently with speed and pressure. The two-stroke engine calls for a slightly different approach, since pressures and flows have to be related to the additional air requirements of scavenging.

With internal compression, a by-pass valve does not offer the same advantage as with the displacement type, since the work done in compressing the air would be wasted. However, the efficiency of compression is higher, and some economy could be obtained by incorporating a clutch in the drive. Figures quoted by FTD, who developed the Sprintex compressor, indicate that the adiabatic compression efficiencies of the roots type, Sprintex unit and a turbocharger approximate to respectively 30 to 50%, 70 to 75% and 60 to 65%. An adiabatic efficiency of 80% has been claimed for the Saurer unit operating at a pressure ratio of 1.5 : 1 at 5000 rev/min, and pressure ratios of up to 2.0 : 1 are said to be obtainable at up to 10 000 rev/min. Some performance curves for this compressor are shown in Fig. 14.26.

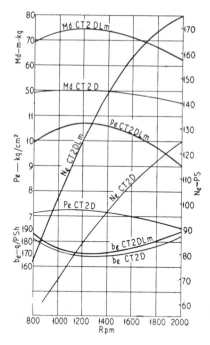

Fig. 14.26 Performance curves for the Saurer supercharger

14.26   Other methods of supercharging

A method that, prior to the Second World War, was fairly widely used for two-stroke engines was to have double the number of cylinders needed for generating the power required. The additional cylinders were used for compressing the air for scavenging and pressure charging those adjacent to them. Well-known examples were the Trojan engines, Section 7.4. This method, however, considerably adds to the bulk, weight and cost of the engine.

In the nineteen-eighties, KKK developed their Ro-Charger, Fig. 14.27. It has two rotors, one inside the other. The inner one is of a figure-of-eight, or dumb-bell, section and is mounted on the driven shaft. This shaft is carried in a pair of sealed-for-life bearings at the driven end and a single bearing at the opposite end. The outer rotor, mounted eccentrically relative to the inner one, is of cylindrical form but with internal lobes meshing with those of the inner rotor. It rotates in two bearings, carried one at each end on eccentric spigots projecting inwards from the end covers.

At one end, a multi-V belt pulley drives the shaft on which, immediately inboard of the bearing at the driven end, is a pinion driving a ring gear mounted in the adjacent end of the rotor. The gearing is such that the ratio of the speeds of the inner to the outer rotor is 3:2, the maximum speed of the outer one being 10000 rev/min.

ZF introduced at about the same time their mechanically driven Turmat centrifugal compressor for engines up to about 3.5 litres swept volume. Relative to the turbocharger, it has the advantage of the absence of a hot and

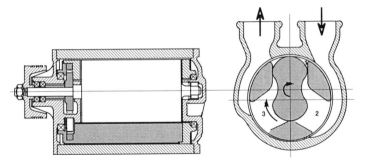

Fig. 14.27 The Ro-Charger has been developed under licence from Felix Wankel. Turning in the same direction about different axes, the two rotors are geared together in the ratio 3:2, inner:outer. During rotation, the lobe of the inner rotor in chamber 1 retreats, drawing air in while rotating past the inlet port. At the same time, the lobe in chamber 3, is advancing to discharge compressed air into the delivery port. Next, the lobe in chamber 2 discharges into the delivery port while that in chamber 3 is inducting air

costly turbine. On the other hand, it is certainly no simpler, since it has a Variator V-belt drive (operating on the principle of the Van Doorne transmission, Section 23.12, which is driven from the crankshaft, either directly or through gears, Fig. 14.28.

The Variator, the ratio of which can be varied between about 1:2 and 1:1.08, drives a 1:15 ratio planetary gear system which, in turn drives the centrifugal compressor. An optional feature is a clutch interposed between the Variator and the planetary gear system. The transmission ratio of the Variator is controlled automatically by a flyweight actuated mechanism, which draws the flanges of the secondary pulley closer together as the speed increases and allows them to be forced apart by the V-belt as the speed decreases. This

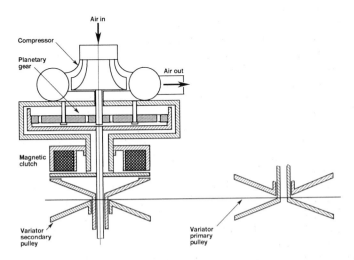

Fig. 14.28 The ZF Turmat supercharger comprises a centrifugal compressor driven by a Variator infinitely variable pulley-type transmission and a step-up planetary gear set. Its magnetic clutch is optional

control can be supplemented or supplanted by a pressure- or depression-actuated control which similarly moves the flanges of the primary pulley. An alternative is electronic control.

14.27 The pressure-wave supercharger

The pressure-wave, or Comprex, supercharger has been developed by Brown Boverie. Basically it takes energy, in the form of pressure pulses, directly from the exhaust gas, in contrast to the turbocharger, which does so mechanically. Consequently, the losses are small and pressure ratios of up to 3 : 1 are said to be attainable.

The principle is illustrated in Fig. 14.29. Energy interchange between the exhaust and inlet gases occurs in a set of straight tubular cells of approximately trapezoidal section within the drum B. These cells, the ends of which are open, are arranged around and parallel to the shaft on which the drum rotates. The shaft is driven by a belt from the crankshaft. Because the unit does not have to compress the gas it absorbs no more than between 1 and 2% of the power output of the engine. Moreover, with such a large number of cells, the compression process is virtually continuous, so synchronisation of rotation with that of the crankshaft is unnecessary.

The sequence of operations is as follows. During each revolution of the drum, one end of each cell in turn passes the end of the exhaust passage A. This allows the exhaust gas, at the pressure in that passage, to flow along the cell, compressing the air that it already contains against the closed far end. Further rotation opens a port at the latter end, which allows the air thus compressed to flow into the inlet passage F, which is then closed again when the cell passes it. Closure of the port reflects a pressure pulse back to the other end, which is now

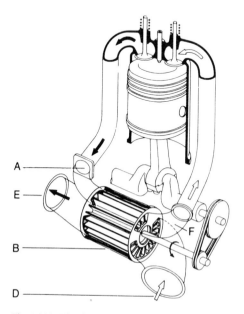

Fig. 14.29 The Comprex pressure-wave supercharger

open to the exhaust downpipe through E, Consequently, the exhaust gas is discharged to atmosphere, at the same time generating a suction wave which travels along the cell. When this reaches the opposite end, the port to the inlet pipe D is open, so a fresh charge of air is drawn into the cell, and the cycle begins again.

With spark ignition engines, unless the Comprex machine is installed upstream of the carburettor, the exhaust gases will come into direct contact with the fuel–air mixture, so it is principally of value only for diesel engines. Since the exhaust gas pressure pulses travel at the speed of sound the Comprex machine has the advantage of virtually instantaneous response. Another advantage is that it is equally effective over the whole of the useful operating range of the engine. Moreover, since the energy utilised is taken from the exhaust gas, significant fuel economy is obtainable. The timing can be arranged to provide a degree of exhaust gas-recirculation during the valve overlap period, and this is claimed to reduce emissions of NO_x by between 20 and 30%.

Chapter 15

Fuels and their combustion

Because the needs of spark and compression ignition engines differ widely we shall deal with them separately in this chapter, taking spark ignition engines first. Petroleum, more widely termed crude oil, is found in natural reservoirs underground. It is the fossilised remains of minute fauna, as opposed to the flora matter, mainly trees, from which coal is derived. This crude oil is distilled in what is called a *fractionating tower*, to separate out its many constituents, or *fractions*, among which are those for blending into petrol. These boil off at temperatures ranging from about 25 to 220°C, Fig. 15.1. They comprise mainly organic compounds, among which three chemical groups predominate.

One group is the *alkanes*, alternatively known as *paraffins*. Typical arrangements of the straight-chain normal molecules characteristic of alkanes can be seen from Fig. 15.2. The longer the chain, the heavier is the molecule and the higher the boiling point of the liquid. Variants termed *alkenes* and *alkynes*, under the generic name of *olefins*, can be produced from the alkanes by cracking processes, Section 15.2. Their molecules are like those of the alkanes, but with some of the hydrogen atoms removed.

Alternative arrangements of hydrocarbon molecules are possible, and are called *isomers*. Of these, perhaps the most widely known is *iso*-octane. For assessing the octane number of a fuel that is being tested, a mixture of *iso*-octane, defined as having an octane number of 100, and normal- or *n*-heptane, defined as having an octane number of zero, is employed. The octane number is the percentage of *iso*-octane in a mixture which has precisely the same tendency to detonate as the fuel being tested. Incidentally, heptane (C_7H_{16}) has nine isomers. Note that the general formula for alkanes is C_nH_{nx+2}, where n is the number of hydrogen atoms. The molecular arrangements of *n*-octane and *iso*-octane are illustrated in Fig. 15.3.

Another of the three main groups referred to in the opening paragraph of this chapter comprises the cyclo-alkanes, so called because of their ring-like molecular structures. Their general formula is C_nH_{2n}. They are present mainly as cyclo-pentane and cyclo-hexane, Fig. 15.4. An alternative name for these products is *naphthenes*, not to be confused with *naphtha*, which is a rather loose term for a mixture of the light hydrocarbons.

The third series, the *aromatics*, also have ring like structures but, as can be seen from Fig. 15.5, the arrangement differs slightly in that each carbon atom is associated with only one hydrogen atom. The aromatics have high octane

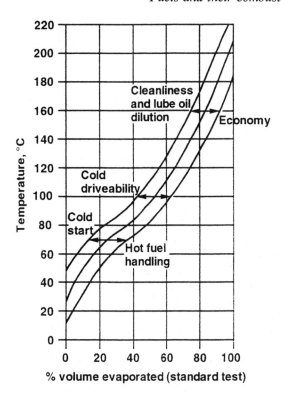

Fig. 15.1 Ease of cold starting is dependent on the percentage of fuel evaporating below 70°C: too high a percentage, however, can lead to vapour formation in the fuel system when the engine is hot

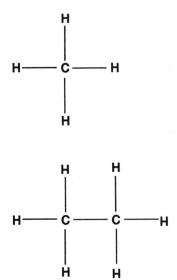

Fig. 15.2 The two smallest and lightest alkane molecules are those of methane (*top*) and ethane (*bottom*). Both are gases at atmospheric temperature

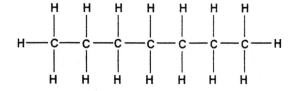

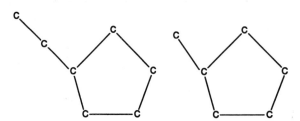

Fig. 15.3 Examples of the long chain heavier alkanes are *iso*-octane (*top*) and *n*-octane (*below*)

Fig. 15.4 Molecules of methyl cyclo-pentane and ethyl cyclo-pentane. This is a simplified diagram, in which only the carbon molecules are shown. Such simplifications are often used and are justified because we can take it that, at least in hydrocarbon fuels, each carbon atom has four arms to which either carbon or hydrogen atoms must be attached

numbers and therefore have been blended into unleaded petrol for improving octane numbers. Some of the heavier among these compounds are suspected of being carcinogens, but this has not been actually proved.

15.1 Distillation and blending

As can be seen from Table. 15.1, the contents of crude oils from different oil fields differ widely. The fractions condense out of the distillation tower progressively into a series of trays, one above the other in the tower, Fig. 15.6: the heaviest fuels (those having the lowest boiling point) condense into the lowest and the others into each tray in turn, until the lightest come out into the

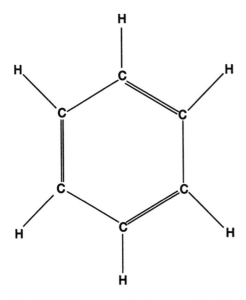

Fig. 15.5 This is a molecule of an aromatic, C_6H_6 benzene. The inner arms of the carbon atoms are linked in pairs instead of to hydrogen atoms

Table 15.1—PERCENTAGE PRODUCTS BY WEIGHT IN CRUDE OIL FROM VARIOUS SOURCES

	N. Africa	N. Sea	Mid. East	N. America	S. America
Sulphur	0.1	0.3	2.5	1.0	5.5
Wax	3	9	6	7	2
Light gasoline, 0–70°C	8.9	5.8	4.7	2.4	0.1
Octane No.	73	76	72	75	70
Naphtha, 70–140°C	16.0	11.0	7.9	6.5	1.1
Kerosine, 180–250°C	26.3	18.6	16.4	15.6	4.4
Diesel oil: 250–350°C	18.2	19.1	15.3	19.6	9.6
Cetane No.	55	53	58	45	30
Residue ⩾350°C	27.5	36.2	47.2	47.9	76.9

topmost tray. A gaseous residue, mainly propane with small quantities of *iso*-butane and *n*-butane issues from the top, and the heavy residue, containing mainly bitumen, is drawn off from the base of the tower.

For two main reasons the distilled fractions have to be refined and blended to render them suitable for use as petrol. First, they may contain impurities such as sulphur that have to be removed and, secondly, there may be high

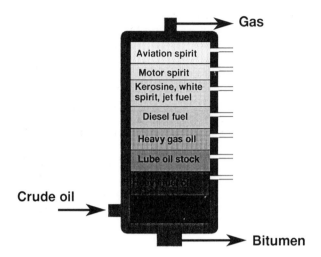

Fig. 15.6 Diagram representing the distillation process. Heat is applied to the crude oil, causing the vaporous constituents to rise to the top, whence the gaseous content is taken away and, in some instances, cooled to separate out the very light constituents. On the way up the tower, the various liquid fractions condense out into a series of trays from each of which they are drawn off through pipes, as indicated, and then put through various refining processes

proportions of some constituents and shortages of others needed for producing fuels suitable for use in road vehicles. The finished product must comprise an appropriate mix of light factions for cold starting, and heavier fractions for normal operation. Blending is done also to provide fuels with the overall properties required, including high octane number for petrol and high cetane number for diesel fuels, Sections 15.5 and 15.16.

15.2 The principal refining processes

The principal refining processes for the production of fuels are thermal, catalytic and hydrocracking, and catalytic reforming. There are many others, such as alkylation, isomerisation and polymerisation, for producing high-octane fuels, and the finishing processes such as caustic washing for removal of certain chemical contaminants, the Merox sweetening and extraction processes for the removal of others, and hydro-desulphurisation. However, we do not have space here for going into details. For further information, readers could consult the *Automotive Fuels Handbook*, by Owen and Coley, published by the SAE.

The cracking processes convert heavy molecules into lighter ones, having lower boiling points. Thermal cracking entails heating the heavy distillate to between 450 and 550°C. Alkanes crack most easily, followed in turn by the cyclo-alkanes and aromatics. This process is more suitable for producing diesel fuel than petrol.

Hydrocracking entails the use of a catalyst and addition of hydrogen, at pressures of about 170 to 180 bar and temperatures of 450°C. The output is mainly olefins. The process, by reducing the ratio of carbon to hydrogen atoms in the molecules, reduces boiling point and increases octane number, though some low octane number constituents may also remain.

Catalytic cracking calls for high temperatures and pressures, and therefore costly equipment. However, it gives a much better yield of high octane fuels than either thermal or hydrocracking.

Catalytic reforming is used for increasing the octane number of naphtha (mixture of light hydrocarbons). During this process several entirely different chemical changes occur. One is the removal of hydrogen from aromatics, to convert them into cyclo-alkanes. A second is isomerisation of the straight-chain molecules of alkanes to produce higher octane branched-chain mol-ecules. The third is dehydrocyclisation, in which alkanes are first cyclised, forming naphthenes and then dehydrogenated to produce aromatics. Other processes occur concurrently, including hydrocracking which forms smaller naphthenic molecules, and dealkylation in which the branch chains of higher aromatics are removed to form lower aromatics.

15.3 Properties required for petrol

Boiling points are a good indication of how easily the fuel will vaporise, though of course some evaporation will occur at almost any temperature. The more volatile the fuel, the easier will it be to start the engine from cold. Because neither condensation in the manifold nor evaporation from a float chamber presents a problem with engines equipped with port injection, these can be run on fractions having relatively low boiling points. However, if the lighter fractions are too volatile, vapour lock might be experienced when the engine is hot, though this is more likely to be experienced with carburetted engines having fuel-lift pumps installed relatively high under the bonnet. A proportion of heavier, high-boiling-point fractions are necessary for operation at normal and high temperatures, with good volatility in the middle range in order that enrichment for cold running can be cut out as soon as possible after starting. If it is not, droplets of low-volatility fractions may be drawn into the cylinders, diluting and even washing away the lubricant, and burning incompletely in the cold combustion chambers, leaving carbon deposits.

Calorific value is an important property, though the difference between the various hydrocarbon, including petrol and diesel, fuels is not of much significance. It becomes a serious factor, however, with alternatives such as the alcohols and, even more so, the gaseous fuels. All the alternatives entail penalties of either large fuel tanks or short ranges of operation. In fact, the energy density of the conventional hydrocarbon fuels is so high that it has few competitors that are economical and safe.

Latent heat of vaporisation is also of some significance. The higher it is the greater will be the cooling effect as the fuel evaporates into the air. This is an advantage with fuel injection, since the charge becomes denser, and therefore its energy content higher, as it cools. On the other hand, it can be a disadvantage for single-point injected or carburetted engines, because it encourages ice formation in the throttle barrel which, as mentioned in Chapter 11, can even stop the engine.

Cleanliness and purity of the fuel is essential. If substances other than hydrocarbons are present, a wide range of problems can arise. Both water and solid impurities, for instance, can cause corrosion and block or reduce the efficiency of carburettor jets and injector nozzles. Sulphur is present in all crude oils and therefore has to be removed during the refining of the fuel, otherwise it will not only cause corrosion but also reduce the effectiveness of the catalytic converters in exhaust systems. Other substances may leave deposits of ash in the combustion chambers after they have burnt.

Perhaps the most important property of a fuel for spark ignition engines is its octane number. If too low, the fuel will burn explosively, instead of progressively, in the combustion chambers, causing over-heating and mechanical damage, especially to components the strengths of which have been reduced by local high temperatures. Piston crowns, rings and exhaust valves are the most vulnerable.

15.4   Fuel-performance requirements

Two combustion phenomena, which should not be confused with each other, have to be avoided: one is pre-ignition and the other detonation. Pre-ignition is the ignition of the charge before the spark has passed. It causes rough running and, in extreme cases, can do damage to the engine. In most instances it arises because the fuel–air mixture is ignited by incandescent particles of carbon in the combustion chamber. On the other hand, it can also be initiated by incandescence of inadequately cooled metal parts in the chamber: sharp corners and thin sections are the most likely to over-heat.

Detonation, on the other hand, sometimes referred to as 'knocking' or 'pinking' because of its characteristic noise, is a spontaneous explosion, instead of the normal progressive spread of the flame throughout the mixture. In normal combustion, the flame initiated by the spark travels across the chamber, heating and expanding the gases that it has consumed and therefore compressing the so far unburned mixture in front of it. If the rising temperature of the unburned gases, due to both their compression and radiation from the flame front, exceeds that of spontaneous ignition, they explode before the flame front reaches them.

There are two potential causes of detonation. One is a combustion chamber in which the distance that the flame has to travel is unduly long. Ideally the spark should occur in the centre of a spherical chamber. This, however, is impracticable for two reasons. First, with a chamber comprising a hemispherical roof complemented by a hemispherical depression in the piston, the compression ratio would be too low. Secondly, a long-reach spark plug placing the electrodes in the centre of such a chamber would overheat and give rise to pre-ignition.

The other cause of detonation is the use of a fuel the octane number of which is too low. Each engine will detonate when run on a fuel of a certain octane number. This number differs slightly from engine to engine. The octane number applicable to an individual engine is termed the co-operative fuel research (CFR) octane number, Section 15.5.

Detonation can occur in two distinctly different circumstances. One, as already implied, is the use of a fuel of too low an octane number. The second is that, even with a fuel of reasonable octane number, a sudden opening of the throttle will cause the pressure in the manifold to rise and the high-boiling-point fractions to condense out on its walls. Those with the highest boiling points tend to be the aromatics, which also have the highest octane numbers, so the overall octane number of the fuel actually being burnt in the combustion chambers suddenly drops. Thirdly, there is high-speed knock. This occurs if the temperatures and pressures in the cylinder become too high at speeds between that of maximum torque and maximum power output. It can be insidious because, under these conditions, the noise of detonation may be masked by the loud mechanical and normal combustion noises in the background. Consequently, the driver may be unaware of it until serious mechanical damage has been done to the engine.

It is sometimes difficult to distinguish between the noises arising from detonation and other causes. Two other such noises are *wild ping* and *crankshaft rumble*. Wild ping is intermittent detonation triggered by incandescence in the combustion chamber ahead of the flame front, under conditions in which detonation would not otherwise occur. Crankshaft rumble is due to resonant vibration of the crankshaft, usually in a bending mode, and it may even alter the mode of rotation of the crankshaft in its bearings, by causing it to orbit around within them. It can be originated by forces generated by either regular combustion or detonation.

15.5 Octane number and anti-knock index

As previously indicated, the octane number is defined as the lowest percentage of *iso*-octane mixed with *n*-heptane that detonates at the same compression ratio as does the fuel under test. *iso*-Octane (C_8H_{18}) is taken as having an octane number of 100 and *n*-heptane (C_7H_{16}) of zero.

Two different fuel octane numbers are in general use. One is the research octane number (RON) and the other the motor octane number (MON). There is also a third, applicable to the vehicle rather than the fuel, and this is termed the road octane number (RON). Yet another indicator of fuel quality is the anti-knock index (AKI), which is the average of the RON and MON. This is a slightly more reliable indication of the detonation resistance of a fuel.

The RON is established on a standard single-cylinder research engine having an accurately adjustable compression ratio. It is therefore a good basis for comparison of different fuels. However, because mixture distribution is not involved, it differs from the results obtained in the multi-cylinder engines in road vehicles, which is why the MON was introduced. This is based on the ASTM D2700 test, run with a representative multi-cylinder engine. The RON of a fuel is significantly higher than the MON: indeed, a typical automotive fuel may have an RON of 98 and a MON as low as 88.

There is also a third indicator of detonation resistance, which is termed the CFR road octane number, the initials CFR standing for co-operative fuel research. This is a criterion of the engine instead of the fuel. It is determined for individual engines on a co-operative basis by the major national and international oil companies, who test a number of cars of each model and pool their results to produce a comprehensive set of data indicating the quality of fuel needed to satisfy the needs in their markets. This work is necessary because, owing to differences between fuel metering systems, engine and manifold layouts and sizes, and fuel distribution between cylinders, each type of engine tends to perform differently with a fuel of any given RON.

15.6 Boiling point, vapour lock and ice formation in induction systems

Vapour lock occurs when pipes or fuel pumps become over-heated. The low pressure both inside the pump and in the pipeline between it and the tank reduces the boiling point of the fuel, so heat transmitted to these components by, for example, the exhaust manifold causes it to vaporise. Because pumps are designed to deliver liquids, they cannot cope with vapour; consequently, the supply of fuel to the engine is interrupted, causing it, at best, to run roughly and, at worst, to stop.

Vapour lock is most likely to occur after the vehicle has stopped, especially after a slow climb to the top of a steep hill in hot weather. In these circumstances, both the forward speed of the vehicle and the rotational speed of the engine are generally low, so also therefore are the rates of flow from both a mechanically driven fan and the water pump, and the engine therefore over-heats. This heat is then conducted out to the surrounding parts, such as the carburettor, fuel pump and pipe lines, in which some or all of their contents vaporise. Consequently, the fuel pump may cease to function efficiently, if at all, and even vapour, instead of fuel, may be delivered to the carburettor float chamber or injection system and, with a carburettor, the fuel may have boiled out of the float chamber. In either event, the engine cannot be restarted. Indeed, if ambient temperatures are high and the float chamber or a pipeline unshielded from and is too close to the exhaust manifold, the engine may even stop while the vehicle is ascending a long steep hill.

Fuel-lift pumps usually deliver the fuel over a weir which, when the engine has been switched off, retains some fuel within the pump to keep it primed ready for restarting. However, this helps only marginally if the suction line is full of vapour, and a long time may elapse before the fuel vapour has condensed and the engine can be started again. The process can be speeded considerably by pouring cold water over the pump and suction line. Vapour can be difficult to clear from fuel-injection equipment so, on modern cars, fuel pumps are usually electrically driven and installed inside the fuel tank, so that they do not have to overcome suction heads.

Ice formation in both carburettors and with single-point injection is due to moisture in the atmosphere wetting throttle valves and barrels and freezing on them. This happens because of a drop in temperature of these parts, arising from the latent heat of vaporisation of the fuel. It tends to occur when the ambient temperature is slightly above freezing point and the relative humidity high. If the moisture is already frozen before it enters the air intake, it is more likely to bounce past the throttle and on into the cylinders. In severe cases, the engine can be stopped by this build up of ice. Subsequently, after the ice has been melted by heat conducted from the surrounding hot parts, the engine can be restarted.

15.7 Composition of fuel for spark ignition engines

In the early nineteen-twenties, the term 'petrol' was originally introduced as a trade name by an English company, Carless, Capel and Leonard, which still exists. The American term 'gasoline' is, however, beginning to gain ground in the British motor industry. Other terms that have been used include 'motor spirit' and 'petroleum spirit', the latter perhaps being the most appropriate for what is basically a complex mixture of distillate from petroleum (crude oil).

Because the contents of crude oils differ widely according to the part of the world from which they come, the major oil companies often have to refine crude stocks from different geographical sources and blend the distillates to produce a petrol suitable for use in motor vehicles. The actual blend depends also on the season in which it is to be used, and a fuel for a hot country must of course be less volatile than one for use in a cold climate. A typical leaded petrol for use in the UK would have a range of volatilities similar to that in Table 15.2. With the introduction of unleaded petrols, there has been a trend towards increasing the proportions of lighter fractions, most of which have higher octane numbers. Such fuels might contain between 24 and 45% aromatics, and from zero up to 26% olefins. The balance would be made up of naphthenes and alkane saturates.

Because benzene, an aromatic having a high octane number, has been said to be a carcinogen, legal restrictions in some countries limit it to about 5% by volume. In general, emissions regulations are now so stringent that most can be satisfied only by fuel injection and closed-loop control. With injection, fuel-delivery pressures are higher than with carburation, and of course there is no evaporation from float chambers. All this has strengthened the trend towards fuels with a higher proportion of light fractions and therefore providing good cold starting.

15.8 Additives

Additives are substances introduced, in small proportions, into fuels to enhance their performance or to offset the effects of certain undesirable properties. It all started in the early nineteen-twenties when the demand for fuel was expanding rapidly and could no longer be satisfied by straight distilled hydrocarbons. Oil companies started to crack the heavier fractions, to break down their molecules into lighter ones and thus increase the supply of petrol. Cracked products of that time tended to be unstable, reacting with oxygen to form gummy deposits causing problems such as blocked carburettor jets and filters. Consequently the first additives were anti-oxidants.

Because of the impending increasingly strict legal requirements regarding

Table 15.2—PROPERTIES OF A TYPICAL PREMIUM GASOLINE FUEL FOR THE UK

Property	Summer	Winter
Specific gravity	0.734	0.732
Octane No.	97	97
Reid vapour pressure, kN/m_2	13.5	7.7
Initial boiling point, °C	34	30
10% fraction boils off at, °C	55	51
25%	74.4	63.5
50%	104.8	92.8
75%	139.2	129.4
Final boiling point	184	185

exhaust emissions and fuel economy, additive technology is now being taken more seriously than hitherto. Even so, at the time of writing, only three oil companies in the UK, handling no more than 30% of the fuel sold there, are marketing additive fuels.

15.9   Lead compounds

Also in the early 1920s, Midgley, in the USA, discovered that adding tetraethyl lead, $Pb(C_2H_5)_4$, in small quantities to the fuel would inhibit detonation. Subsequently it was found that compounds called scavengers, mainly 1,2-dibromoethane and 1,2-dichloroethane, mixed with the lead, would prevent it from forming hard deposits in combustion chambers and on valve seats.

By 1930, mixed in at a rate of about 0.6 g/l, tetraethyl lead (TEL) was widely used to increase octane number. Today, any engine that is not designed for running on unleaded fuel will suffer rapid wear of its exhaust valve seats with a fuel having less than about 0.3 g/l of TEL. The reason is that the combustion process leaves a coating of lead bromide compounds on the seats and these inhibit welding of the peaks of the surface texture of the seats to those of their mating faces on the valves.

By about 1960, tetramethyl lead (TML) began to be used. This has a lower boiling point than TEL, so it evaporates with the lighter fractions of fuel and therefore is drawn preferentially together with them into the cylinders. At one time it was not uncommon for a mixture of TEL and TML to be used as an anti-knock compound. During combustion, the lead additives form a cloud of metal oxide particles. These, because the lead molecules are heavy and the oxides chemically active, interrupt the chain-branching reactions that lead to detonation. Incidentally, sulphur in the fuel reduces the effectiveness of lead additives.

By the 1950s, huge resources were being poured by interested parties into research to prove that burning lead additives in fuel produces toxic exhaust fumes. True, in large concentrations over extended periods, it can adversely affect brain development but, so far, no one has proved it can do so in the concentrations that enter the atmosphere from automotive exhausts, even if deposited on food crops. The reason for the abandonment of lead additives has been that they adversely affect the performance of the catalysts in the converters incorporated in vehicle exhaust systems. Lead additives, though obsolescent, remain the most economical way of increasing octane number.

15.10   Lead-free fuels

One way of producing satisfactory lead-free fuels is to use oxygenate additives, either alcohols or ethers, though these are costly. Alcohols include ethanol, methanol, tertiary butyl alcohol (TBA), methyl tertiary butyl ether (MTBE), tertiary amyl methyl ether (TAME) and ethyl tertiary butyl ether (ETBE). Their octane numbers range from from 104 to 136, and the octane numbers of the fuels in which they have been blended range from 111 to 123. Even so, they have tended to fall out of favour because, under certain conditions, they break down and form hydroperoxides, which are corrosive and, combined with other substances in the fuel, can produce other corrosive compounds.

Mathanol contains 49.9% oxygen, but MTBE and TAME contain only 18.2 and 15.7% respectively. The ethers, whose oxygen contents are lower than those of the alcohols, are nevertheless an attractive alternative.

A widely used method of producing high-octane hydrocarbon fuel is to isomerise the distillate to form mostly light, high-octane derivatives. Such processing, however, is not only costly but also consumes energy that has to be taken from the oil being processed. It therefore increases emissions of CO_2, NO_x and SO_2 into the atmosphere. Moreover, there is a limit to the proportions of light components that can be blended into a motor fuel.

15.11 Detergent additives

Detergents were introduced initially in the early 1960s, in response to driveability problems arising from the formation of deposits in carburettors. 'Driveability' is a term used mainly for describing the smoothness of the response of the engine to movements of the accelerator pedal.

By the late 1960s and early 1970s, the introduction first of positive crankcase ventilation and then exhaust gas recirculation led to the appearance of deposits in all the passages from air filter to inlet valves and even on the valves themselves. Again, the result was poor driveability. Consequently, it became necessary to develop detergents that would be effective not only in the carburettor but also throughout the system. Shell, with its ASD (Additive Super Detergent) fuel, was first in the field, in the late 1960s and early 1970s, and was using these second generation detergents in higher concentrations than hitherto. A significant new feature was the use of carrier fluids, mainly mineral oils, or polymers such as polybutene or polyetheramine, to take the additives right through the induction system.

With both carburettors and throttle body injection, deposits are particularly likely to be formed on hot spots in the induction manifold and any other area in which heat soak increases local temperatures after the engine has stopped. The oily additives partially or completely dissolve these deposits, which are subsequently swept away and burnt in the combustion chambers. Too high a content of such additives, however, can cause valve sticking and increase combustion-chamber deposits, leading to higher octane requirements.

From approximately 1970 to 1980, induction-system temperatures increased significantly, partly as a result of induction air heating and the other measures for overcoming emissions problems. The situation was exacerbated by the trends towards use of leaner air : fuel ratios and higher temperatures, for improving thermal efficiency and hence fuel economy. An outcome was that, because problems such as injector nozzle fouling arose as a result of heat soak, oily additives became no longer adequate alone and therefore detergents had to be used with them. However, the high temperatures involved called for a different type of detergent additive, so polymeric dispersants and amine detergents were then introduced.

In general, detergent additive molecules comprise an oleofilic chain-like tail with a polar-type head, Fig. 15.7. The free arms of the head attach to the particulate deposit and carry it away in the liquid fuel in which these detergent molecules are dissolved.

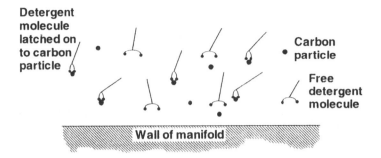

Fig. 15.7 Showing how detergent additive molecules latch on to the dirt particles to carry them away in solution

15.12   Corrosion inhibitors

These additives are particularly desirable with injection systems since, without them, malfunction be caused by corrosion debris blocking the fine filters used and the injector nozzles. Corrosion can also cause fuel tanks to leak, even though they are protected internally by a corrosion-resistant coating. Most corrosion inhibitors react with the acids that form in the fuels and some, like the detergents, have polar heads and oleofilic tails, but the heads latch on to the molecules of the metal surfaces, over which their tails form a protective coating.

The fuel itself can also oxidise, causing the formation of gums that can lead to difficulties both in storage and in the engine. Fuels containing high proportions of cracked products are, as previously mentioned, particularly susceptible to gum formation. Additives inhibiting the oxidation of the fuel are therefore used, but mainly in storage.

15.13   Spark-aider additives

To satisfy emission control regulations, engines have to be operated on weak mixtures, so good driveability can be difficult to achieve. As has been indicated previously, cleanliness can help, but more important are the rapidity with which the engine warms up and the consistency, from cycle to cycle, with which the flame develops and spreads through the combustion chamber.

If the flame kernel around the spark does not expand rapidly to a certain critical size, either the mixture will subsequently burn inefficiently or the flame will die. Even if the engine is cold, the nominally rich mixture supplied can still be weak in the region of the spark plug. This is partly because, on the way to the cylinder, the lighter fractions can condense out and be deposited on cold metal surfaces.

Consistency of combustion can be improved by the use of spark-aider additives. However, if they are used together with lead additives containing halogen compounds, they could lead to sticking and deterioration of the inlet valves. They function by coating the electrodes with a compound facilitating the passage of the spark and thus allowing more energy to be applied to ignite the mixture.

15.14 Diesel fuels

Whereas for the spark ignition engine the fuel and air are supplied pre-mixed to the cylinders, in a diesel engine the fuel is not injected into the air until shortly before TDC. Consequently, there is considerably less time for completion of the mixing and evaporation processes. Furthermore, the diesel engine, having no throttle, is controlled by regulating the quantity of fuel injected per induction stroke. Add to this the fact that ignition cannot occur until the temperature generated by compression is high enough, and it becomes obvious that fuel quality is even more important for the diesel than the spark ignition engine.

Whereas in Europe there is one grade of diesel fuel for road vehicles, in the USA there are two, ASTM D1 and D2. The EEC defines a diesel fuel as containing a maximum of 65% distilled off at 250°C and a minimum of 85% distilled off at 350°C. A UK diesel fuel might have the following properties—

Specific gravity	0.85	Cloud point	− 5.5°C
Sulphur	0.22%	Initial boiling point	180°C
Cetane No.	51	50% vaporisation	280°C
Cold filter plugging point	− 18°C	Final boiling point	360°C

Hydrocracking and catalytic cracking are used to convert fractions having even higher boiling points into hydrocarbons suitable for use as diesel fuels. However, both hydrocracked and catalytically cracked fuels tend to have cetane numbers, Section 15.16, in the region of only 10 to 30. Catalytically cracked fuels, moreover, tend to be slightly unstable in storage.

15.15 Properties required for diesel fuel

A few of the properties required, such as high calorific value (energy content), are common to both gasoline and diesel power units, but most are much different. Diesel fuel mostly comprises fractions boiling off from approximately 150 to 355°C, Fig. 15.8, as compared with about 15 to 210°C for gasoline. As delivered from the fractionating tower, these higher boiling point fractions contain about 20 times more sulphur than those from which gasoline is derived, so extra attention has to be devoted to removing it during refinement.

The following are the properties that must be controlled when diesel fuels are blended—

Volatility	High volatility helps with cold starting and obtaining complete combustion.
Flashpoint	The lower the flashpoint the greater is the safety in handling and storage.
Cetane number	This is a measure of ignitability. The higher the cetane number the more complete is the combustion and the cleaner the exhaust.
Viscosity	Low viscosity leads to good atomisation.
Sulphur	Low sulphur content means low wear and a smaller particulate content in the exhaust.

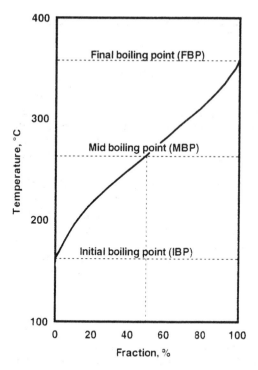

Fig. 15.8 This characteristic distillation curve for diesel fuel is similar in form to that illustrated in Fig. 15.1, but of course the initial and final boiling points are higher

Density	The higher the density the greater is the energy content of the fuel.
Waxing tendency	Wax precipitation can render cold starting difficult and subsequently stop the engine.

As in the case of petrol, properties of a diesel fuel depend in the first instance on the source of the crude oil from which it is distilled. These vary as follows—

UK and Norway	Mainly paraffinic and therefore of high cetane number. Calorific value relatively low and cloud point high. Sulphur content low to medium.
Middle East	Similar, but high sulphur content. Middle East crude oils are a particularly good source for diesel fuel, because they contain a high proportion of alkanes and a small proportion of aromatics.
Nigeria	Naphthenic. Low cetane number, cloud point and sulphur content. Calorific value medium.
Venezuela and Mexico	Naphthenic and aromatic. Low cloud point and very low cetane number, but low to medium sulphur. Calorific value high.

Each of the properties in the lists above influences engine performance, so we need to study them in more detail.

15.16 Cetane number, cetane index and diesel index

Basically, the *cetane number* is the percentage of cetane in a mixture of cetane (*n*-hexadecane) and heptamethylnonane (the latter is sometimes referred to as α-methylnaphthalene) that has the same ignition delay, generally expressed in terms of degrees of rotation of the crankshaft, as the fuel under test. There is a more precise definition but, before we come to it, a brief note on ignition delay is necessary.

Ignition delay, Section 15.32, is important because, if it is too long, the whole charge in the cylinder tends to fire simultaneously, causing violent combustion. With a short delay, ignition is initiated at several points, and the flame subsequently spreads progressively throughout the charge. On the other hand, the injection must be timed appropriately relative to the cetane number of the fuel that will be used: a higher cetane number than that for which the timing was set can lead to ignition before adequate mixing has occurred and thus increase emissions.

Cetane number is defined precisely as the percentage of *n*-cetane + 0.15 times the percentage of heptamethylnonane contents of the blend of reference fuel having the same ignition quality as the fuel under test. Ignition quality is determined by varying the compression ratio to give the same ignition delay period for the test fuel and two blends of reference fuels. One blend has to be of better and the other of poorer ignition quality than the test fuel, but the difference between the two has to be no more than five cetane numbers. The cetane number is obtained by interpolation between the results obtained at the highest and lowest compression ratios.

Carrying out these laboratory engine tests, however, is not at all convenient, so two other criteria are widely used. One is the diesel index and the other the cetane index. The diesel index, which is obtained mathematically, is computed by multiplying the aniline point of the fuel by its API gravity/100. The aniline point is the lowest temperature in degrees fahrenheit at which the fuel is completely miscible with an equal volume of freshly distilled aniline, which is phenylamine aminobenzene. API, measured with a hydrometer, stands for American Petroleum Institution, and degrees API = (141.5/Specific gravity at 60°F) − 131.5. It is a measure of density for liquids lighter than water.

The cetane index is calculated from API gravity and volatility, the latter originally taken as represented by its mid-volatility, or mid-boiling point (50% recovery temperature, T_{50}). Since its introduction, the formula has been modified from time to time, to keep up with advancing fuel technology, and is now based on an extremely complex formula embracing the density and volatility of three fractions of the fuel (those at the 10, 50 and 90% distillation temperatures T_{10}, T_{50} and T_{90}, respectively). This formula can be found in *Automotive Fuels and Fuel Systems,* Vol. 2, T. K. Garrett, Wiley.

The cetane index is, in general, better than the diesel index as an indication of what the cetane number of a fuel would be if tested in a CFR engine in a laboratory, and it is much less costly and time-consuming to obtain. In general, alkanes have high, aromatics low, and naphthenes intermediate, cetane and diesel indices.

Values of 50 or above for either the diesel or cetane index indicate good combustion and ignition characteristics, below 40 are totally unacceptable and even below 45 undesirable. Low values mean difficult cold starting, the generation of white smoke, and the engine will be noisy.

BS 2869: Part 1: 1988 prescribes minimum limits of 48 for the cetane number and 46 for the cetane index. In Europe and Japan the minimum cetane number requirement is 45, and in the USA it is 40, the latter possibly being because a high proportion of their crude oil comes from Mexico and Venezuela. A reduction in cetane number from 50 to 40 leads to an increase in the ignition delay period of about 2° crankshaft angle in a direct and about half that angle in an indirect injection engine.

15.17   Tendency to deposit wax

In cold weather, a surprisingly small wax content, even as little as 2%, can crystallise out and partially gel a fuel. These crystals can block the fuel filters interposed between the tank and injection equipment on the engine, and ultimately cause the engine to stall. In very severe conditions, the pipelines can become blocked and a thick layer of wax may sink to the bottom of the fuel tank. Paraffins are the most likely constituents to deposit wax which, because they have high cetane numbers, is unfortunate.

The various measures of the tendency of a fuel to precipitate wax include *cloud point*, which is the temperature at which the wax, coming out of solution, first becomes visible as the fuel is cooled (ASTM test D 2500). Another test (ASTM 3117) is for the *wax appearance point*, which is that at which the wax crystals become visible in a swirling sample of fuel.

Then there is the *pour point*, which is the temperature at which the quantity of wax in the fuel is such as to cause it to begin to gel (ASTM D 97). To establish the pour point, checks on the condition of the fuel are made a 3 °C intervals by removing the test vessel from the cooling bath and tilting it to see if the fuel flows. Another criterion, the *gel point*, is that at which the fuel will not flow out when the vessel is held horizontal. In practical terms, this translates roughly into a temperature 3° above that at which it becomes no longer possible to pour the fuel out of a test tube.

Another way of estimating operational performance of a fuel is to combine the cloud point, CP, and the difference between it and the pour point, PP, to obtain a *wax precipitation index*, WPI. The formula for doing this is as follows—

$$WPI = CP - 1.3(CP - PP - 1.1)^{\frac{1}{2}}$$

Other tests include the *cold filter plugging point* (CFPP) *of distillate fuels*, IP 309/80 and the CEN European Standard EN116: 1981. This the lowest temperature at which 20 ml of the fuel will pass through a 45 μm fine wire-mesh screen in less than 60 s. However, since paper element filters are now widely used, the relevance of wire-mesh filters is open to question so, at the time of writing, the CEN is debating the desirability of introducing a test called the *simulated filter plugging point*. In the USA, the *low temperature fuel test* (LTFT) is preferred to the CFPP test. The LTFT is the temperature at which 180 ml of fuel passes through a 17 μm screen in less than 60 s. All these and other tests are described in detail in *Automotive Fuels and Fuel Systems*, Vol. 2, T. K. Garrett, Wiley.

15.18   Density

Because injection equipment meters the fuel on a volume basis, any variations in density, because it is related to energy content, will affect the power output. The greater the density of the fuel the higher will be both the power and smoke. Fortunately, however, hydrocarbon fuels differ relatively little as regards their densities.

Density is measured by the use of a hydrometer with scales indicating specific gravity or g/m^3. The sample should be tested at 15°C or the appropriate correction applied. Figures for density differ from those for API gravity, or deg API, in that the higher the number in deg API gravity, the lighter is the fuel.

15.19   Volatility

The volatility of the fuel influences many other properties, including density, auto-ignition temperature, flashpoint, viscosity and cetane number. Obviously, the higher the volatility the more easily does the fuel vaporise in the combustion chamber. Low-volatility components may not burn completely and therefore could leave deposits and increase smoke. Within the range 350 to 400°C, however, the effects of low volatility on exhaust emissions are relatively small. The mix of volatilities is important: high-volatility components at the lower end of the curve in Fig. 15.8 improve cold starting and warm-up while, at the upper end components having volatilities that are too low increase deposits, smoke and wear.

15.20   Viscosity

The unit of kinematic viscosity is the stoke, which is the time taken for a certain volume of fuel at a prescribed temperature and a constant head to flow under the influence of gravity through a capillary tube of a prescribed diameter. That of absolute velocity is the poise, which is the force required to move an area of 1 cm^2 at a speed of 1 cm/s past two parallel surfaces that are separated by the fluid. For convenience, the figures are usually expressed in centipoises and centistokes (cP and cSt). The two are related in that cP = cSt × Density of the fluid. The SI units are m^2/s, and the CGS or stoke's units are cm^2/s (see also Section 16.4).

Increasing viscosity reduces the cone angle of the injected spray and the distribution and penetration of the fuel, while increasing the size of the droplets. It will therefore affect optimum injection timing. An upper limit for viscosity has to be specified to ensure adequate fuel flow for cold starting. Lucas Diesel Systems quote a figure of 48 cSt at −20°C as the upper and, to guard against loss of power at high temperatures, 1.6 cSt at 70°C as the lower limit. BS 2869 calls for a maximum value of 5 cSt and a minimum of 2.5 cSt at 40°C. In Fig. 15.9, all these points are plotted and a viscosity tolerance band established. The viscosity of the average fuel lies approximately mid-way between the upper and lower limits.

Too high a viscosity can cause excessive heat generation in the injection equipment, owing to viscous shear in the clearances between the pump plungers and their bores. On the other hand, if it is too low, the leakage through those clearances, especially at low speeds, can be so high that

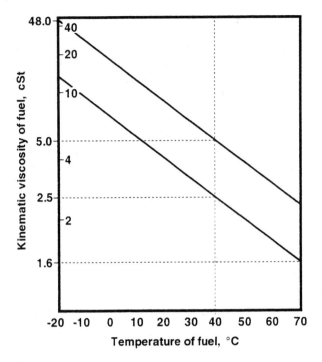

Fig. 15.9 A typical tolerance band for fuel viscosity is illustrated here. At 40°C, the limits are from 2.5 to 5.0 cSt

restarting a hot engine can become impossible. This, however, is because the increase in temperature of the fuel locally due to conduction of heat to the injection system during a short shut-down period further reduces the viscosity of the fuel in the pump prior to starting.

15.21 Smoke

When the engine is started from cold, white smoke comprising tiny droplets of liquid, mainly fuel and water, increases as cetane number and volatility of the fuel are reduced. It persists until the temperature has risen to the point at which the droplets are vaporised in the engine and remain so until well after they have issued from the exhaust tail pipe. The reason why the fuel droplets, although surrounded in the combustion chamber by excess oxygen, remain unburnt is that, in the cold environment, not only do they not evaporate but also their temperature never rises to that of auto-ignition.

Cetane number increases with the density and volatility, and varies with composition of the fuel. Fuels having high cetane numbers are principally the paraffinic straight-run distillates. However, because these have both high cloud points and low volatility, a compromise has to be struck between good ignition quality and suitability for cold weather operation.

Another product of combustion can be black smoke. This is formed because the hydrogen molecules are oxidised preferentially so, if there is insufficient oxygen in their vicinity, it is the carbon atoms that remain unburnt. High aromatic content, viscosity and density and low volatility all increase the tendency to produce black smoke.

Although aromatics tend to produce smoke, they also make a major contribution to the lubricity of the fuel. Consequently, their removal can give rise to abnormally high rates of wear of injection pumps, especially in distributor-type pumps in which all the work is done by only one or two plungers and, perhaps, a single delivery passage might be subject to severe erosion.

As previously explained, the volatility factor is misleading. Because the more volatile fuels have high API gravities (indicating low specific gravities), and their low viscosities allow more to escape back past the injection pumping elements, the weight delivered to the cylinder per injection is smaller. For any given power, a certain weight of fuel must be injected, so a greater volume of the more volatile fuel and a longer injection period are therefore required, and this might affect droplet size. Depending on the air movement and other conditions in the combustion chamber, such changes could have either a beneficial or adverse effect on combustion and therefore smoke generation.

15.22 Particulates

Even though more than 90% of the total mass of particulates in the exhaust, Fig. 15.10, is carbon, sulphur compounds are a problem too. Moreover, as advances in engine and injection equipment design lead to reduced carbon particulates from both the fuel and lubricating oil, sulphur could become a significantly higher proportion of the total. Removal of sulphur from the fuel is a costly, though necessary, process.

15.23 Additives

Until the mid-nineteen-seventies, the diesel fuel generally available in the UK and Europe was of a very high quality. Subsequently, on account of a shortage

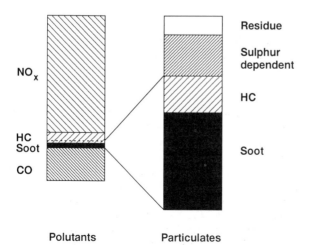

Fig. 15.10 Typical analysis of the pollutants in the exhaust from a diesel engine

of appropriate crude oils, the quality world-wide showed a tendency to fall, though not as rapidly as had been expected. Since fuel reserves are being consumed at an increasing rate, the trend inevitably will remain downwards unless some economical way of synthesising high quality diesel fuel in the enormous quantities required can be developed. The trend has already led to the introduction of additives many of which hitherto had been entirely unnecessary and therefore not even under serious consideration.

Even so, there is no incentive for the oil companies to use additives unless they are cost-effective. Some are sold in the after-market but, unless the purchaser knows in detail the content of his fuel, he could be either paying for products that do not suit it or simply adding to what is already present at saturation level, in which event the extra will make little or no difference. A possible exception is where a commercial vehicle operator has a large residual stock of fuel bought in bulk during the summer, which he needs to convert by adding anti-wax additives for winter use.

15.24   The effects of additives on combustion and performance

One of the early entrants, in 1988, into the multi-additive diesel fuels market was Shell's Advanced Diesel. This contains several additives, one of which has raised its cetane number from the 48 required by British Standards, and the minimum of 50 for Shell's base fuel, to typically between 54 and 56. Among the others are a corrosion inhibitor, anti-foam, cold flow and re-odorant additives. The benefits claimed include lower noise, 3% better fuel economy, Fig. 15.11, 8.4% less black and white smoke, Fig. 15.12, a general improvement in overall engine performance and durability, and a reduction in vehicle downtime.

Of all the additives available, the most obviously important to the operator are the cetane improvers and those that help to overcome the tendency to wax precipitation in winter. Others used include anti-oxidants, combustion improvers, cold flow improvers, corrosion inhibitors, detergents, re-odorants, anti-foamants and, less commonly, stabilisers, dehazers, metal deactivators,

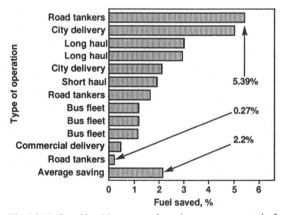

Fig. 15.11 Considerable scatter of results were apparent in fuel consumption trials instituted by Shell on several different fleets and types of operation. Among the variables of course are driver habits

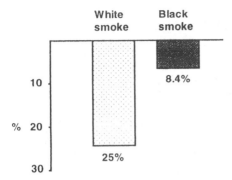

Fig. 15.12 Results of tests carried out by Shell to show that their Advanced Diesel fuel offered a reduction of 25% in white and 8.4% in black smoke as compared with another commercially available fuel

biocides, anti-icers, and demulsifiers. Anti-static additives are used too, but mainly to benefit the blenders by facilitating storage, handling and distribution.

15.25   Cetane number and cetane improvers

Because cetane number is a measure of the ignitability of the fuel, Fig. 15.13, a low value may render starting in cold weather difficult and increase the tendency towards the generation of white smoke, Fig. 15.14. It also increases the ignition delay (interval between injection and ignition). As a result, fuel must be injected into the combustion chamber earlier. Consequently, more fuel is injected before ignition occurs, so the ultimate rate of pressure rise is more rapid, and the engine therefore noisy. Also, because the fuel has less time to burn before the exhaust valve opens, the hydrocarbon emissions are increased.

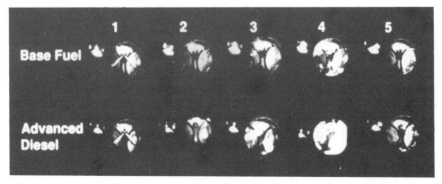

Fig. 15.13 Combustion sequences photographed through a quartz window in the piston crown of an engine running on (*top*) a commercially available alternative fuel and (*bottom*) Shell Advanced Diesel. The two part circles are the inlet and exhaust valve heads and the bright spot to the left of the cylinder is an illuminated pointer over degree markings on the flywheel, which are too small to be visually identifiable here

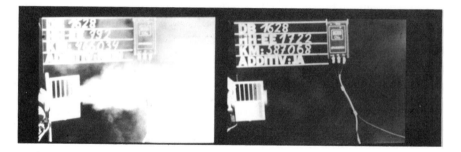

Fig. 15.14 Two photographs showing white smoke in the exhaust gases of an engine running on (*left*) a commercially available fuel and (*right*) Shell Advanced Diesel

On the other hand, if the cetane number is higher than that for which the injection system is timed, power will be lost because a high proportion of the pressure rise will occur when the piston is at or near TDC. Furthermore, the fuel might ignite before it has mixed adequately with the air, so smoke and hydrocarbon emissions may be experienced. Consequently, fuels having high cetane numbers perform best when the injection is retarded. Conversely, with too much retard, there will not be enough time for complete combustion, so smoke and hydrocarbon emissions will again be the outcome. Since high cetane numbers are difficult to achieve, the regulations in most countries specify only the lowest limit that is acceptable.

Cetane improvers, mainly alkyl nitrates, are substances that decompose easily and form free radicals at the high temperatures in combustion chambers. Unfortunately, however, it is the fuels that have the lowest cetane numbers that tend to respond least to cetane improvers.

Combustion improvers differ from cetane improvers in that they are mainly organic compounds of metals such as barium, calcium, manganese or iron and are catalytic in action. Barium compounds could be toxic, so interest now centres mainly on the others. Although these compounds produce metal-based particulates, they also lower the auto-ignition temperature of the carbon-based deposits in both the cylinders and particle traps, causing them to be more easily ignited.

15.26 Cold weather problems

A surprisingly small wax content can partially gel a fuel. Different countries specify cold filter plugging points ranging from about $-5°C$ in the Mediterranean region to $-32°C$ in the far north. In practical terms, the basic consideration is that the engine shall start at the lowest over-night soaking temperature likely to be experienced in service. Once it has started, if the filter becomes partially blocked, the rate of flow could be such that the return flow of warm fuel to the tank is reduced, and therefore also the rate at which the temperature of the fuel in the tank is raised.

15.27 Cold weather additives

All additives for cold weather operation modify the shapes of the wax crystals, which otherwise tend to adhere to each other. They therefore come under the general heading of wax crystal modifiers (WCMs). From Stoke's law we deduce that the rate of settling of crystals is directly proportional to the square of their diameter times the difference between their density and that of the fluid, and is inversely proportional to the viscosity of the fluid. Consequently, small crystal size is the overriding need.

There are three types of modifier: pour-point depressants, flow improvers and cloud-point depressants. Which of these is used depends basically on local requirements and the type of wax to be treated. The last mentioned is mainly a function of the boiling range of the distillates and the country of origin of crude oil. In general, cracked products contain high proportions of aromatics, which have low cetane numbers, but they also contain less wax and, moreover, dissolve what wax there is more readily than do straight-distilled fuels. Fuels having narrow boiling ranges form large wax crystals that are less susceptible to treatment by additives than the smaller and more regular-shaped crystals formed in fuels having wider boiling ranges.

The earliest WCM additives were the pour-point depressants (PPDs) introduced in the 1950s. These modify the shape and reduce the size of the wax crystals. The flat plates of naturally formed wax crystals tend to overlap and interlock, and thus to gel the fuel. Those formed in the presence of a PPD, however, are thicker and smaller, and some multi-axial needle crystals are introduced between the platelets, all of which makes it more difficult for them to interlock.

More effective are the flow improvers. These cause small multi-axial needle crystals to form, instead of larger platelets. Moreover, by virtue of the presence of some additive molecules between the crystals, they tend not to adhere to each other. Although these crystals will pass through wire-gauze strainers, they are stopped by the finer filters used to protect the injection pumps and injectors. Even so, by virtue of their small size and the multi-axial arrangement of the needles, the unaffected liquid fuel still tends to pass between them.

The small compact wax crystals tend to settle in the bottoms of tanks. This is more of a problem in storage than vehicles, but wax anti-settling additives (WASAs) can nevertheless play a useful part in the avoidance of wax enrichment as vehicle fuel tanks become empty, especially in very cold climates. These, too, modify the crystal formation, by both forming nuclei and arresting growth. As the temperature of the fuel falls through the cloud point, their nuclei form centres on which small wax crystals grow and, subsequently, other additive molecules attach to their surfaces and block further growth. Primarily, these additives improve cold filterability, but they also lower the pour point.

Although a cloud-point depressant by itself lowers the cold filter-plugging point of a base fuel, the opposite effect may be obtained by using it in a fuel containing also a flow improver. The improvement obtainable by cloud-point depressants is generally only small, of the order of 3°C, and they are costly. Therefore they are unattractive, except where the cloud point is included as part of a diesel-fuel specification and the blender therefore wishes to lower it.

15.28 Dispersants and corrosion inhibitors

The primary function of dispersant additives is to restrict the size of the particles formed within the fuel at the high temperatures in the engine and, additionally, to remove them from the metal surfaces. However, they must be used continuously, otherwise gum deposits that form when they are not present are dislodged when they are, and tend to block filters.

There are also dispersant modifiers, or detergents, which keep the surfaces of the combustion chambers and injection nozzles clean. However, if used to excess, some can actually cause gums to form.

Finally, there are corrosion inhibitors for protecting fuel-system components and also bulk storage tanks and barrels.

Fuel surfaces can be oxidised in contact with air, causing the formation of gums, sludges and sediments. Surface-active additives can help to prevent this, but must be added immediately after refining the fuel and while it is still warm.

15.29   Detergents and anti-corrosion additives

Detergents are used mainly to remove carbonaceous and gummy deposits from the fuel-injection system. Gum can cause sticking of injector needles, while lacquer and carbon deposited on the needles can restrict the flow of fuel, distort the spray and even totally block one or more of the holes in a multi-hole injector, Fig. 15.15. The outcome can be misfiring, loss of power, increased noise, fuel consumption, and hydrocarbon, CO, smoke and particulate emissions in general, Fig. 15.16. Furthermore, starting may become difficult, because the fuel droplets have become too large owing to the reduction in flow rate.

Detergent molecules are characterised by, at one end, a head comprising a polar group and, at the other, an oleofilic tail. The arms of the polar group latch on to the metal and particulate molecules, Fig. 15.17. Those attached to the metal form barrier films inhibiting deposition and, incidentally, offering a degree of protection against corrosion, while those latched on to the particulates are swept away with the fuel because their oleofilic tails carry them into solution in it.

Anti-corrosion additives (perhaps about 5 ppm) are mainly used to protect pipelines in which diesel fuel is transported, but no more than trace proportions are likely to remain by the time it reaches the vehicle. Therefore, if vehicle fuel-system protection is required too, the treatment must be heavier. As in the case of the detergents, the polar heads attach themselves to the metal, but the water repellent tails form an oily coating over the surface to protect it

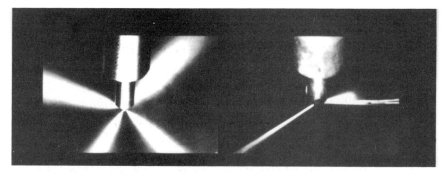

Fig. 15.15 Effects of carbon deposits on diesel injector sprays: (*left*) a new and (*right*) a sooty injector

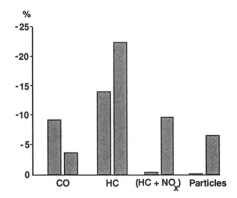

Fig. 15.16 The left- and right-hand columns represent respectively the percentage exhaust clean-up and actual output levels measured after 4000 and 15 000 miles respectively during road tests with Shell Advanced Diesel fuel. The exhaust emissions measurements were taken while running with the vehicle on a chassis dynamometer immediately after it had completed its road mileage

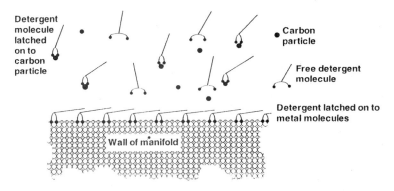

Fig. 15.17 Molecules of detergent additives in fuels are held in solution by their tails, while their polar heads latch on to the molecules of the contaminant. In application to lubricants, some additives have molecules that also latch on to those of the metal surfaces, leaving the tails to protect the latter from corrosion. Alternatively, different additives, functioning in a similar manner but latching only on to the metal, may be used for protecting it against corrosion

against corrosive attack. In Fig. 15.18, test samples that have been left over a long period in the base fuel are compared with those left in Shell Advanced Diesel.

15.30 Anti-foamants and re-odorants

To allow the fuel tank to be completely filled more rapidly, and to avoid splashing, surfactant anti-foamant additives are used. The inclusion of re-odorant additives has been an outcome of the increasing numbers of diesel cars on the road. Total elimination of the smell is undesirable because leaks would be difficult to detect as also would identification of the fuel, so the aim is at modifying it by partially masking it with a more acceptable odour.

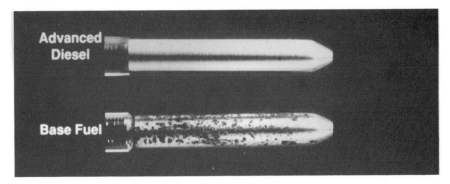

Fig. 15.18 Specimens subjected to the ASTM D.665A corrosion test: (*top*) with Shell Advanced Diesel and (*below*) in a commercially available alternative fuel

15.31 Diesel combustion

Diesel combustion has been largely covered in Chapter 6, so all that remains here is to fill in some details. Combustion is initiated at a number of points throughout the charge. This is because air alone is compressed until its temperature exceeds that necessary for spontaneous combustion of the fuel, which is only then injected into the cylinder as a very fine spray and evaporated. The auto-ignition point of a well-prepared diesel fuel–air mixture can be as low as 220°C, as compared with the ultimate compression temperature of about 600°C. Owing to heat losses, however, temperatures adjacent to the cylinder walls are significantly lower than the latter figure, which is one reason why hydrocarbon emissions can be a problem with a cold engine. Compression ratios range from about 14:1 to 24:1, typical values being 18:1 for direct-injection engines and 22:1 for indirect-injection engines.

Since the droplets in the spray have to mix thoroughly with the air, evaporate, and then burn completely in the very short space of time between injection and the opening of the exhaust valve, the primary requirement is that the fuel be consistently of a high quality, and suitable specifically for diesel engines. This is especially relevant for starting in very cold conditions, when the temperature on completion of compression can be as low as or even lower than 400°C, and the auto-ignition temperature of the cold, and therefore poorly prepared, mixture is as high as about 450 to 500°C. This is why some diesel engines require glow plugs or flame starter systems for use in cold conditions. Immediately following auto-ignition in the combustion chamber, small flames may be alternately initiated and quenched, the temperature at which combustion begins to spread being between approximately 500 and 600°C.

Because the hydrogen content burns preferentially, the maximum power output has to be set by limiting the maximum quantity of fuel supplied per stroke to a level just below that at which black smoke is emitted. Consequently, unlike petrol engines, diesel engines never run with anything approaching a stoichiometric mixture.

Detonation, previously described in connection with spark ignition engines, is not a problem. However, because of the excess air in the cylinders under idling and light-load conditions, an explosive combustion termed *diesel knock* may be heard. The causes of this phenomenon, and why in this context cetane number is important, will be dealt with in the next section.

15.32 Ignition delay

The first point to note is that liquid diesel fuel will not burn: it has to be evaporated first. Obviously therefore some delay occurs from the outset, but this is not what is meant by the term 'ignition delay', which occurs even though the temperature of the air in the combustion chamber is above the auto-ignition temperature of the vaporised fuel–air mixture. Ignition delay is the interval between the evaporation and mixing of the fuel in the air at or above auto-ignition temperature and the initiation of combustion. It is generally of the order of a thousandth of a second, but of course varies according to the properties of the fuel, and the temperature. In principle, the delay can be reduced by increasing turbulence, though of course there are practical limits to the degree of turbulence that can be accepted and, indeed, is effective.

During the delay period, mixing continues and pre-flame reactions take place, in which free radicals and aldehides are formed. Until the visible flame appears, the curve of pressure against crank angle follows the line that it would

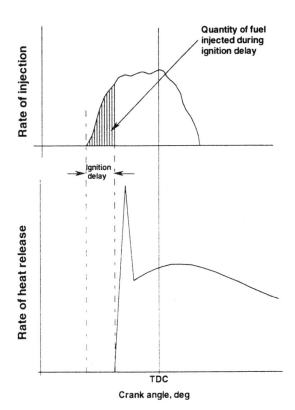

Fig. 15.19 Rate of injection and heat release after the ignition delay

have taken if the engine had been motored without fuel injection. Subsequently, the heat of combustion causes the pressure to rise rapidly, Fig. 15.19, and then falls again as the piston begins to descend on the power stroke.

For any given engine, fuel quality and compression temperature, the delay period is constant in terms of time so, as the speed of the engine increases, it becomes longer in terms of crank angle. Consequently, it is desirable, though not always practicable, to advance the injection timing as engine speed increases.

The longer the delay period, the steeper will be the subsequent pressure rise and the louder the diesel knock. It is always loudest when the engine is running under very light load or idling, especially after starting from cold. Increasing the volatility of the fuel and reducing the cetane number increases the noise.

The main reason for this knock when starting from cold is that, under these conditions, temperatures in the combustion chamber are low so that the ignition delay is long, with the result that a higher proportion of the total fuel charge is injected before combustion starts. Also, only minute quantities of fuel are being injected, so it may not be well atomised, and the quantity of air available for combustion is, by comparison with that of the fuel, huge. The outcome is a sudden late release, at or near top dead centre, of a high proportion of the total energy supplied and, therefore, explosive combustion.

Chapter 16

Friction, lubricants and lubrication

Friction between two mutually contacting sliding surfaces may fall into any of three categories: dry, boundary or viscous friction. Lubrication serves three purposes—

(1) Reduction of the resistance to relative motion between the surfaces.
(2) Prevention of seizure.
(3) Reduction of the rate of wear of the sliding surfaces.
(4) To carry away the heat generated by the friction.

16.1 Dry friction

Dry friction occurs when there is no lubricant between the surfaces. In this case, the resistance to sliding motion is independent of the area of contact and dependent only on the coefficient of friction of the materials and load applied normal to the contacting surfaces. The coefficient of friction is a characteristic of the material and is defined as the limiting friction, which is the load applied along the plane of contact of the surfaces that will just initiate relative movement between them divided by the load normal to them. Static limiting friction is that opposing motion before it starts and dynamic limiting friction is that experienced during the sliding of the surfaces. The latter, which is always lower than the former, falls slightly as the speed of sliding increases. Continuous sliding at high speed under load generates a great deal of heat.

16.2 Boundary friction

Some components, such as cams and their followers, operate under conditions of *boundary friction* for all or at least most of the time and, under extreme loading, even main and big-end bearings may do so. In such circumstances, some lubricant remains between the surfaces but as molecules bonded or keyed to the surface, rather than as a continuous film, so some metal-to-metal contact occurs. The effectiveness of the lubricant under such heavy loading depends on two properties: its film strength, which delays the breakdown of the continuous film and, secondly, its *lubricity*, which helps to ensure that some lubricant remains even where the film has been squeezed out. Good lubricity is

527

obtained when some of the molecules of the lubricant bond either physically or chemically to those of the metal surfaces so strongly that they cannot be removed other than by drastic mechanical or chemical measures.

16.3 Viscous friction

If pressure lubricated, the bearing surfaces are normally separated by a film of oil. In these circumstances, the resistance to motion is solely that due to the shearing of the oil, and is therefore a function of its viscosity. Viscosity tends to increase with pressure and to decrease with temperature. The resistance to motion is proportional to the area being sheared, the velocity of shear and the coefficient of viscosity, and is inversely proportional to the thickness of the oil film.

Where there is no radial loading, the resistance to motion is actually increased by the interposition of lubricating oil between the surfaces, but falls as the clearance is increased. Exhaust valves run hot, so the viscosity of the oil between them and their guides tends to be low and, therefore, for any given clearance, they will slide more freely in their guides than the cooler running inlet valves. To reduce the resistance to motion of pistons, the areas of their skirts in contact with their cylinder bores must be kept to a minimum. This can be done in several ways, as outlined in Sections 3.5 and 3.6. In racing engines, the clearances between pistons and bores are generally larger than in those of road vehicles. The significance of some of the various sources of friction in an engine can be seen from Fig. 16.1.

In general, low viscosity is desirable to facilitate starting from cold, to ensure that the pump delivers oil rapidly to the remote parts of the lubrication system, and to keep the viscous drag to a minimum under all conditions of operation. On the other hand, it has to be high enough to ensure that oil, when hot, will be

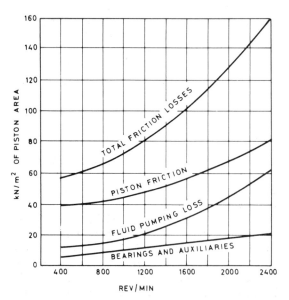

Fig. 16.1 Sources of engine friction

neither too easily squeezed from between the loaded areas nor flow away too quickly from the ends of the bearings, and that the quantities of oil being drawn past the piston rings are kept to a minimum, especially at high speeds and small throttle openings.

From the foregoing paragraphs it can be seen that lubrication has to be a compromise. To keep friction as low as possible, the oil film should be thick and the viscosity low. On the other hand, the clearances in the bearings must not be so large that all the oil is rapidly forced out by the pressure under which it is fed into them, and the viscosity must be high enough to prevent the oil from being squeezed out of the bearings under heavy loading. As can be seen from Sections 16.6 and 16.11, however, lubricity and viscosity are not the only properties that determine the performance of the lubricant in the bearings.

16.4   Measurement of viscosity

Newton has defined absolute viscosity as the ratio of the shear stress between two planes of equal area to the rate of shear. If the thickness of the film of liquid between the two plates is T, their area A, and the shearing force F, the rate of shear is the relative velocity V of the plates divided by their separation distance T. As indicated in Section 15.20, the absolute viscosity of the lubricant is F/A divided by V/T, and its unit is the centipoise, which is the SI unit. However, other units such as pascal seconds, where one $Pa\,s = 10^3$ cP, are used too.

Various methods are employed for measuring viscosity. One is the use of a rotary viscometer, with which the viscosity is a function of the torque required to rotate the rotor at constant velocity in the liquid. Another indicator of viscosity is the *kinematic viscosity*, which is defined as the absolute viscosity divided by the density of the liquid at a given temperature. This can be determined with greater precision than absolute viscosity, by measuring the rate of flow of the liquid through a capillary tube under the influence of gravity. Its units are m^2/s, though more commonly expressed as centistokes, where 1 $cSt = 10^{-6}\ m^2/s$. The relevant standards are ASTM D445 and IP 71. Some empirical scales are also in use, including Saybolt Universal Seconds (SUS) and Redwood seconds. Each is based on flow through a standard orifice under specified conditions. The SAE standard viscosity gradings for engine lubricating oil are shown in Table 16.1.

16.4   Change of viscosity with temperature – viscosity index

To meet all the conflicting requirements as regards its viscosity, a lubricant should have a high *viscosity index* (VI), which means that its viscosity changes little with variations in temperature. The higher the VI the less is the effect of temperature on the viscosity of the lubricant.

Methods of determining VI are set out in ASTM D2270 and IP 226. Originally two reference oils were used, one derived from Pennsylvania crude and defined as having a VI of 0 at 100°C and the other, derived from a Texas Gulf crude, defined as having a VI of 100 at the same temperature. The ASTM method of establishing the VI of a sample of oil is illustrated in Fig. 16.2. At 100°C, the two reference oils have the same viscosity as the sample, but one has a VI of 0 and the other 100. If we call the viscosities at 40°C of the three oils V_0,

Table 16.1—THE SAE VISCOSITY GRADING SYSTEM FOR ENGINE OILS

SAE viscosity grade	Viscosity (cP) at max. temperature (°C)	Viscosity at 100°C, cSt	
		Min.	Max.
0W	3250 at − 30	3.8	–
5W	3500 at −25	3.8	–
10W	3500 at −20	4.1	–
15W	3500 at −15	5.6	–
20W	4500 at −10	5.6	–
25W	6000 at −5	9.3	–
20	–	5.6	9.3
30	–	9.3	12.5
40	–	12.5	16.3
50	–	16.3	21.9
60	–	21.9	26.0

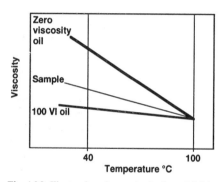

Fig. 16.2 Illustrating the method of establishing the viscosity index of a sample of oil

V_{100} and V_{s}, the viscosity index of the sample is—

$$\mathrm{VI_s} = \frac{V_0 - V_s}{V_0 - V_{100}} \times 100$$

Extrapolation beyond the limits of 40° and 100°C can give misleading results, especially since wax may form at the lower temperatures. The viscosity index of distillate base oils can be improved by the use of additives, Section 16.11.

Base oils in the high viscosity index (HVI) category (produced simply by physical separation from the crude oil) have viscosity indexes ranging from 95 to 105. Very high viscosity index (VHVI) base oils are all synthetic, Section 16.6, and have viscosity indexes greater than 120. They not only have higher viscosity indexes but also are of lower volatility and respond better to anti-oxidants than the VHI oils. Incidentally, some manufacturers use the term XHVI instead of VHVI.

16.5   Types of oil

Vegetable oils have been used in the past, especially for racing-car engines. In the latter application, it did not matter that oil change intervals had to be short, owing to the relatively rapid rate of deterioration characteristic of this type of lubricant. Their main advantages are their high film strength and, owing to good lubricity, excellent protection against wear. However, they are available in only a limited range of viscosities. Moreover, they readily produce gums and lacquers when under stress and hot, which is why oil changes have to be made frequently and the engines have to be stripped regularly and cleaned. They have now been supplanted by modern mineral oils specially formulated for specific categories of these high-performance engines.

Mineral oils are more readily available and cost-effective. They readily respond to additives, and can be produced in a wide range of viscosities. A disadvantage is that they contain wax, which adversely affects cold performance and can clog filters. They, too, degrade to form gums and lacquers, but not so readily as the vegetable oils.

Two authorities set the standards for engine oil standards. One is the American Petroleum Institute (API) and the other is the Association des Constructeurs Européen des Automobiles (ACEA). In 1995, the latter replaced the Comité des Constructeurs Automobiles du Marché Common (CCMC). At the time of writing, the ACEA has not issued its oil performance standards, but they are expected to be ready in a few months.

API standard categories contain the letters SH and CD, S being for petrol and C for diesel engines. The second letters stand for severity of operation, the further down the alphabetical list the letter is the more severe the duty. In the CCMC classification, G is for petrol (gasoline), D for diesel and PD for small diesel engine oils, and these letters are followed by numbers indicating the severity of the duty. When the classifications are finalised, it is probable that standards will be G5 plus for petrol and D6 plus for diesel engine lubricants.

16.6   Synthetic lubricants

Synthetic oils can be produced from several different base stocks. One is ethene gas, $(CH_2)_2$, from which are synthesised the long-chain hydrocarbon poly-alphaolefin (PAO) molecules. This type of synthetic lubricant has a low pour point, and therefore can be used without pour-point-depressant additives. Two more types of synthetic lubricant are obtained from agrichemicals (alcohols) and petrochemicals (acids). A summary of the processes involved is set out in the Table 16.2, and the standard gradings introduced by the SAE in 1994 for API/SH (petrol engine) lubricants can be seen in Table 16.3.

Polyalphaolefins, or more correctly, hydrogenated olefin oligomers, are manufactured by oligormerisation (breaking down into smaller and simpler molecules) of 1-decene, which is derived from the ethene gas. The lubricants derived from the agrichemicals (polyolesters) are biodegradable, and therefore widely used for the lubrication of two-stroke engines in motor boats operated on inland waters. Dibasic acid esters, also extremely suitable for automotive applications, retain their properties at very high temperatures, so they used to be employed mainly for aviation and other gas turbines. However, for

Table 16.2—THE THREE MAIN SOURCES OF SYNTHETIC BASE OILS

Source	Chemical process	Product
Ethene gas	Decene trimer	Polyalphaolefin
Agrichemicals (alcohols)	Alcohol + dibasic acid	Polyolester
Petrochemicals	Organic acid + pentaerythritol	Dibasic acid ester

Table 16.3—SAE BASE STOCK CATEGORIES

Group	Base	% Saturates	% Sulphur	Viscosity index
I	Mineral	<90	>0.03	≥80 to <120
II	Mineral	≤90	≤0.03	≥80 to <120
III	Hydrocarbon*	≥90	≤0.03	≥120
IV	Polyalphaolefins (POA)			
V	Base stocks not included in groups I, II, III or IV			

*Including hydro-isomersed oils.

heavy-duty applications, polyolesters have gained ground because of their superior lubricating properties and resistance to oxidation and aging.

Another category widely regarded as being synthetic comprises the oils derived from the complex very-long-chain molecules of slack wax. These include several different compounds, all unwanted fractions present in the lubricating oil so, to make good use of them, they are catalytically isomerised to produce oligomers. The base oils thus created include alkylated aromatic hydrocarbons and polyglycols. The latter, however, are mostly formulated into industrial oils. Alkylated aromatics are sulphur free and, because they have very low pour points, can be used at temperatures down to −40°C. This renders them particularly suitable for some military applications and for the lubrication of refrigeration equipment.

In general, however, virtually all synthetic oils are available over a wide viscosity range. They are thermally stable, have good natural film strength, excellent low-temperature characteristics and low volatilities, and can be readily engineered to suit specific requirements. A summary of the most important properties of the different types of oil has been produced and is reproduced in Table 16.4.

If the cost of straight mineral oil is taken as 1, those of oils produced from PAO and slack wax are respectively 6 and 3. The reason for the higher cost of the synthetics is that they are produced by chemical processing rather than straight distillation. They are, however, more attractive for several reasons—

(1) They burn more cleanly than mineral oils containing additives, many of which produce ash when burned. Consequently, deposits in the combustion chamber and generally throughout the engine, are reduced.

(2) Their shear strengths are high, so they tend to stay in grade during service.

Table 16.4—COMPARISON OF THE FEATURES OF DIFFERENT BASE OILS

Base-oil type*	Light volatile components	Waxy thickening components	Range of chemical types	Precision of chemical structure	Purity
(1) Mineral	Yes	Yes	Very wide	Very low	Very low
(2) XHVI	Some	Some	Wide	Low/medium	Medium
(3) PAO	None	None	Narrow	Very narrow	High
(4) Ester	None	None	Very narrow	Very high	Pure

*(1) Solvent refined mineral oils; (2) hydro-isomerised base stocks; (3) polyalphaolefin; (4) synthetic diester or polyolester.

(3) They are unlikely to promote cylinder-bore polishing, which can arise as a result of the deposition of hard carbon on and around the top piston ring. This deposit wears the asperities of the honed bores, polishing their surfaces and thus preventing the oil from being retained uniformly over them.

(4) Because their volatilities are low, so also are the associated exhaust and evaporative emissions and engine-oil consumption. In contrast, mineral oils always contain some light fractions, which tend to boil off when hot.

(5) By virtue of their low pour points, they facilitate cold starting and reduce wear during the subsequent warm-up period. Incidentally, in urban operation, wear during the warm-up period may account for more than 80% of the total wear in the engine.

(6) Perhaps the significant advantage of synthetic lubricants is their oxidation stability. This reduces deposition of sludge, varnish and other compounds and, consequently, piston ring sticking.

(7) Because their viscosity indexes are high, they do not need so much VI improver additive which, because it may degrade in service, calls for reduced intervals between oil changes.

16.7   Semi-synthetic lubricants

There is no standard definition of semi-synthetic lubricants. However, they are widely regarded as any blend of at least 10% true synthetic base stock oil with the mineral oil. In practice, most oils advertised as semi-synthetic have a significantly higher synthetic content. The reason why it is important to be specific about the synthetic content is that the quality of the lubricant is largely dependent upon it.

Film strength, not viscosity, determines whether the films will remain intact in bearings under load and enable lubricants of low viscosity to be used. The lower the viscosity the thinner are the films of oil between the loaded rubbing surfaces such as cams and their followers, and the steeper therefore are the stress gradients through these films. Consequently a valuable characteristic of the synthetic lubricants is their high film strengths. Incidentally, older engines, with the larger bearing clearances needed for heavier grades of oil may exhibit high oil consumption if run with synthetic lubricants.

16.8   The wear process and lubrication

The mechanism of wear is simple. Machined or ground metal surfaces of
engine components can never be perfectly polished. Therefore, when a pair are
nominally in contact, they are in fact held apart by the peaks of their
microscopic roughness. A proportion of these peaks pressure-weld together,
while others key into valleys opposite them. Both contribute to the resistance
to sliding and, because metal has to be sheared before the motion can begin,
wear ensues. While the surfaces continue to slide, the peaks ride over each
other generating heat and melting locally, so the wearing process continues.
The microscopic molten peaks are swept down into the valleys, thus further
polishing the surface. This is why the kinetic is lower than the static limiting
friction.

Since the coefficient of friction is a function of the materials in contact, it
makes sense to interpose between their rubbing surfaces a material that has a
lower coefficient. Suitable materials include both solid and liquid lubricants.
Among the solids are PTFE, graphite, molybdenum disulphide, zinc oxide and
certain chlorine and sulphur compounds. Some can be applied as dry coatings,
some are used as additives to the lubricating oil, while others are suitable for
mixing with grease. They should not be used, however, without approval of the
engine manufacturer, since some can deposit in and block oil ways and, in any
case, may over-load and render ineffective any detergent additives present.

As previously indicated, the liquid lubricants may be of either mineral or
vegetable origin, the latter including sperm, rape and castor oils. Lacquers
produced by thermal degradation of vegetable oils tend to fill the microscopic grinding marks on cylinder bores, causing these surfaces to become
polished and thus to lose their capacity to retain the oil after the engine has
stopped. In extreme cases, the oil fails to wet all the polished areas even while
the engine is running, so lubrication is impaired and mechanical damage can
occur. These lacquers deposit on all metal surfaces and cause piston rings,
which run very hot, to stick in their grooves. The alternatives are petroleum
distillates or synthetic oils, used either alone or blended with distillates.
Although the latter are coming into general use, the distillates are still widely
specified.

16.9   Corrosive wear

The reason why about 80% of the total wear in an engine occurs during the
first 90 s following start-up from cold is that water vapour in which corrosive
substances are dissolved condense on components such as the cylinder walls
while the engine is not running. Incidentally, condensation in cylinders can
still occur at any temperature below 70°C, so rapid warm-up is highly
desirable. Components that can be particularly seriously affected include cams
and overhead valve actuation gear, because the oil may take the full 90 s after
start-up to get up to the cylinder head.

When the engine is started from cold, the piston rings may be running
initially in bores that are virtually dry. If these bores are corroded, the debris
thus created increases the abrasive effect. Countermeasures include the use of
chromium-plated and alloy cylinder liners, Section 3.12, though by far the
simplest is to use a lubricant which, by virtue of a high viscosity index, and

therefore relatively low viscosity when cold, will distribute rapidly throughout the engine.

16.10 The lubricant as a coolant

The temperature due to friction in a bearing rises until the rate of generation of heat in it equals that of conduction and radiation from that bearing. If the equilibrium temperature is too high, the viscosity of the oil will be reduced and the rate of flow out of the bearing increased, until first boundary lubrication occurs and then the film breaks down completely, leading to serious metal-to-metal contact. Clearly, therefore, the rate of flow of oil through the bearings in the first instance must be adequate for taking the heat rapidly away to the sump, the air flow around which will dissipate it.

16.11 Oil additives

As can be seen from Table 16.5, a wide range of additives is used. However, some when burnt leave an undesirable ash residue. For example, phosphoric ash can inhibit the action of exhaust gas converter catalysts, and metallic ash in the combustion chamber may act as a catalyst encouraging pre-ignition. In the latter event, engines equipped with knock sensors automatically retard the

Table 16.5—ADDITIVES FOR ENGINE LUBRICATING OILS

Category	Chemical
Detergents	Metallic soaps, sulphonates, phosphonates, thiophosphonates, phenates and phenate sulfides. Ashless types: polymerised olefins, acrylated amines, pyrridones, carboxylic acids and polyglycols
Dispersants	Esters: methacrylates, acrylates, alcoholates and polycarboxylic acid
Friction modifiers	Fatty oils, lard, organic fatty acids and their esters, oleic acid and amines
Anti-wear/EP agents	Zinc dithiophosphate, tricreasyl phosphate, lead soaps, organic sulphur and chlorine compounds and sulphurised fats such as sulphurised sperm oil
Adhesive additives	Latex, linear polymers and isobutylene
Corrosion inhibitors	Zinc dithiophosphate, sulphurised olefins and metallic phenolates
Rust inhibitors	Metallic sulfonates, amines, fatty acids, phosphates, oxidised wax acids and silicone polymers
Oxidation inhibitors	Sulphurised esters, terrapins, olefins and phenols, and aromatic sulphides
Anti-foam agents	Silicone polymers
Pour-point depressants	High-molecular-weight polymers such as polymethacrylate and wax acrylated, or polyalkyl, naphthalines
Viscosity modifiers	Long-chain polymers such as polyacrylate and polymethacrylate, and shorter chain ones such as polyisobutylenes

Notes:
(1) Organochlorine can form extremely toxic compounds when burnt so, as a precaution, such compounds are now used in concentrations of about 50 ppm instead of the former 300 ppm.
(2) Phosphor compounds can coat the catalysts in engine exhaust systems and render them partially ineffective, so there is a tendency to phase them out.

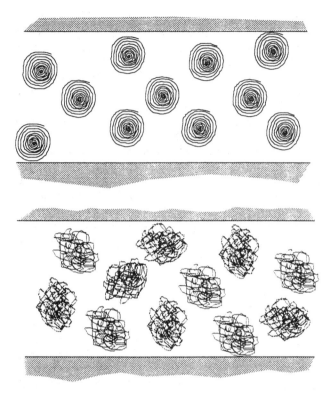

Fig. 16.3 In the cold condition, the molecules of viscosity improvers can be represented, as illustrated, as either tightly coiled long chains or more complex random structures contracted into tight masses. As the oil becomes warmer, they expand and progressively fill the whole volume of the lubricant, thus increasing its viscosity

ignition, which increases both fuel consumption and exhaust emissions. Consequently, the ashless types are preferred for modern engines.

The molecules of detergent additives have polar heads, which latch on to metallic surfaces, and oleofilic tails, which hold them in solution in the lubricant. Once attracted to the surfaces, they form a barrier, as in Fig. 15.17, to inhibit corrosion arising from, for example, sulphur in the base stock or nitrogen fixation during combustion. Their principal function, however, is to prevent the build up of deposits such as lacquer and varnish on the metal surfaces, thus ensuring that deposit precursors and dirt particles remain held in suspension in the oil. They are of particular benefit in diesel-engine lubricants, where piston crowns and rings tend to become extremely hot.

Dispersants serve a purpose similar to that of detergents, but operate in a different way. Their molecules too have polar heads and oleofilic tails, Fig. 15.17: the heads latch on to particles of solid contamination (not on to the metal surfaces, as in Fig. 15.17) and the tails hold them in solution. This prevents agglomeration and holds the sludge in suspension so that it can be removed by filtration. Such additives are especially valuable in lubricants for low temperature stop–start operation and also for diesel engines running at high temperatures in vehicles operating on motorways.

Friction modifier additives, by loosely bonding in multi-layer films on to the oil-wetted metal surfaces, reduce the friction between them. They have mild anti-wear properties but their main function is to reduce friction under boundary, or near boundary, conditions of lubrication. Friction modifier additives are especially desirable for the lubrication of highly rated cams and final drive gears. In addition to the compounds listed in Table 16.5 are some chromium compounds which reduce friction simply by aiding flow. Synthetic lubricant molecules engineered specifically to provide good flow and anti-wear properties generally do not need such additive treatment.

Additives that are held in suspension, such as molybdenum disulphide, colloidal graphite and, more recently, PTFE have been used to reduce friction in boundary lubrication conditions. However, any detergent or dispersant additives present are then engaged in keeping these colloidal materials in suspension, instead of performing their proper function of keeping the engine clean. PTFE is an exception but it deposits on bearing surfaces, including piston rings and grooves where, it has been suggested, it can break down at high temperatures, not only causing sticking but also producing hydrofluoric acid, which is extremely corrosive.

Anti-wear and extreme pressure (EP) additives minimise wear in boundary lubrication conditions, such as when starting from cold. The anti-welding properties of the EP compounds are more pronounced than those of the anti-friction additives. In all cases, heat generated at the point of any metal-to-metal contact causes the additive to react chemically with the metal. The compounds thus formed on the surfaces have shear strengths lower than those of the metal substrates. Therefore shear takes place preferentially in these substances, rather than through the metal-to-metal contact points. Such additives are especially valuable in the lubrication of cams, valve trains and final drive gears.

Anti-wear additives form protective films of relatively low shear strength on the metal parts, but they also inhibit friction welding of the rubbing surfaces in the same way as do the EP additives. Although EP additives are mainly the organic compounds of sulphur and chlorine, a recent trend for gear oils has been to move away from the chlorinated compounds towards adhesive additives: refer to note (1) below Table 16.5.

Adhesive compounds are additives for specialist application. They render the lubricant more cohesive, to prevent splashing and throw by centrifugal force, impact or turbulence, and they can even be made to render the lubricant stringy. These additives are of high molecular weight, and take up space between adjacent molecules of lubricant to resist any tendency for them to be separated. Materials used include latex and linear polymer isobutane.

Corrosion and rust inhibitors and anti-oxidant additives are distinctly different categories. The corrosion inhibitors neutralise the acids produced by oxidation of constituents of the oil and form passive films on all metal surfaces, to prevent corrosion. Rust inhibitors coat, or are preferentially absorbed by, the steel surfaces exposed to moisture both in the oil and in the atmosphere during cold operation. Phosphates tend to deactivate exhaust-cleansing catalysts, so the trend now is towards using the activated zinc compounds: see note (2) below Table 16.5. Anti-oxidant additives inhibit oxidation of the lubricant.

Oxidation inhibitors retard chemical decomposition of the oils. They act as

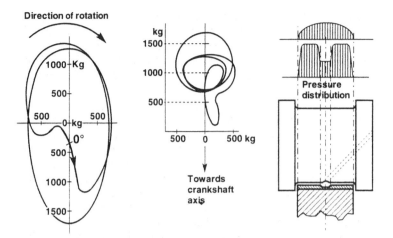

Fig. 16.4 (*Left*) Typical polar diagrams of the forces acting on main bearings. (*Centre*) Forces on big-end bearings. (*Right*) Two diagrams showing the distribution of pressure across the width of a bearing under static (*top*) and dynamic (*bottom*) conditions with the shaft rotating under load

passifiers, deactivating the metal surfaces that could catalytically encourage oxidation of the lubricant. In addition, they preferentially oxidise contaminants in the oil that otherwise might cause it to decompose and form black sludge, for example beneath rocker covers. Insoluble contaminants can cause the oil to gel overnight and therefore, when the engine is started in the morning, fail to flow towards the pump pick-up, as already described in connection with wax. As can be seen from the Table 16.5, phosphorus compounds can be used but, again because they tend to deactivate exhaust gas catalysts, they are falling out of favour.

Foaming of the oil is another phenomenon that can cause partial failure of the pump to deliver enough lubricant to the bearings. Anti-foam agents reduce surface tension, and thus encourage the agglomeration of small bubbles into large ones, which burst easily. Such additives can be especially useful in engines having hydraulic tappets which, if any air were to enter, could become spongy in operation. They are extremely costly but, fortunately, are effective in quantities of only a few parts per million.

Pour-point depressants interfere with the formation, at low temperatures, of crystals of any wax that might remain after the refining of the oil. By coating the crystals or modifying their form they stop their growth and prevent their forming an interlocking matrix. Such additives are much less likely to be needed in good synthetic oils, since these do not contain significant quantities of wax.

Particularly important are the viscosity index (VI) improvers, or viscosity modifiers, which reduce the rate of change of viscosity with temperature. These are polymers, mostly of a rubber-like nature, of high viscosity and molecular weight. When cold, their curved long-chain molecules curl up tightly and thus compact, leaving the remainder of the lubricant free-flowing, Fig. 16.3. With increasing temperature, however, they open to form a matrix which expands throughout the volume of the oil and drags against the other molecules. This, by maintaining its viscosity almost constant, offsets its natural tendency to flow more freely when hot. Some, however, have shorter chain molecules and function by virtue of their increasing solubility with rise in temperature. As previously indicated, synthetics, having naturally high viscosity indexes, need little additive treatment. Incidentally, the first multi-grade oil was Castrolite, introduced soon after 1910.

VI additives have tended to be unstable at high rates of shear, and therefore have to be employed in large concentrations. Stable ones that perform satisfactorily in small doses have now been developed, but are more costly. Incidentally, extreme stresses can arise in the oil as it passes through gear-type pumps, where the individual VI molecules may be sheared, causing the viscosity of the lubricant to revert to that of the base oil. Shearing can also occur in bearings approaching boundary lubrication conditions, and curled molecules can temporarily straighten and align parallel to each other within the film of lubricant, again reducing the effective viscosity.

To produce lubricants of the golden colour attractive to customers on the forecourt, fluorescent dyes may be added. These are especially desirable for synthetic oils which, under static conditions, can be virtually indistinguishable from water. Tracer additives, usually tungsten, or cobalt compounds, are sometimes used to enable the oil producer to check whether cans of pirate products are being sold under his brand name.

Among the additives for specialist application are the emulsifiers, reducing the surface tension so that oil and water can together form emulsions. They are used mainly for marine applications, including steam cylinders, but also for machine tool cutting oils. Another specialist product is the seal expander for hydraulic oils. By chemically modifying their elastomers, seals can be caused to swell by the oil they are retaining and thus seal more effectively. Such additives are generally aromatics, halogenated hydrocarbons or organic phosphates.

16.12 Lubrication systems

Early engines had only elementary lubrication systems. Up to the 1920s, some had manual pumps that were actuated intermittently. For others, up to the 1930s, oil was fed by gravity from a tank through a pipe to the crankcase, where it was either piped directly, or passed through ducts in the casting, to the main bearings, the flow generally being manually regulated by a needle valve. Alternatively, it was delivered into the base of the casting, whence it was whipped up by the rotating cranks into a fine mist to lubricate all the other moving parts.

Ultimately the lubricant issued from the crankcase breather or ends of rotating shafts, to be dumped to atmosphere and drop on to the road. Some was also lost past the pistons and out through the exhaust system.

The next evolutionary stage was the incorporation of a sump, into which the big ends dipped and splashed the oil up into the cylinders. In some engines, scoops were formed on, or fitted to, the big-end bearing caps to actually lift the oil from the sump. In others, the oil splashing around was drained down the walls of the crankcase, to be caught in troughs from which ducts were drilled to take it down to the main bearings. This principle is still applied in some final drive and gearbox units.

16.13   Pressure lubrication

In the meantime, from 1905, a Lanchester car was introduced having a pump for delivering oil to the main bearings, and pressure lubrication systems were soon widely applied. Examples of modern pressure lubrication systems are illustrated in Chapters 3 and 4. The principles are similar to that illustrated in Fig. 16.5. In the early days, pipes were widely employed to take oil to the main bearings and up to the valve gear. However, because they add to installation costs and are vulnerable to damage and leaks, they have now been supplanted, in automotive and most other engines, by drilled and cored ducts in the crankcase and cylinder-head castings.

In modern engines, a pump, gear driven from the crankshaft, draws oil from the sump, through a single pick-up fitted with a strainer to prevent debris from entering it. In the delivery duct from the pump is a pressure-relief valve, which discharges oil in excess of requirements into the inlet side of the pump, thus limiting the maximum pressure. In some instances the excess oil is discharged directly into the sump. This, however, can be a complicated arrangement and, if not carefully arranged, could lead to aeration of the oil in the sump. The actual delivery pressure varies between about 35 and $500 \, kN/m^2$, according to the speed of the engine. Passing up from the relief valve, the oil is taken through a fine filter before being delivered into a large duct which, called the *main oil gallery*, is drilled parallel to the crankshaft in the crankcase casting.

From this hole, it is distributed through smaller ducts to the main bearings whence other holes, drilled through the adjacent crank webs, deliver it to the crankpins and big-end bearings. The pressure in the big ends is attributable partly to that at which it is delivered and partly to centrifugal forces. In many engines, lubrication of the small-end bearings is dependent on oil thrown up to them, by centrifugal force, from the big ends. More highly rated engines, including most diesel units, have small ends lubricated by oil forced through ducts drilled, from the big ends, up the stems of the connecting rods. In some instances, a jet in the top of each small end directs oil up to the underside of the piston crown to cool it. The pressure of the oil delivery into the small-end bearings fluctuates because it is largely influenced by centrifugal force.

In the distant past, some engines had low camshafts rotating in troughs from which they picked up their lubricating oil. The oil either drained from the crankcase wall into the trough or was piped through a restrictor orifice into it. Now, however, a single duct from the main gallery delivers oil up through a vertical hole in the cylinder block and head gasket, into a duct drilled vertically in the cylinder head casting, and usually on into one end of the camshaft, whence it then passes to the rocker shaft, if any. Both of these shafts are hollow, so the lubricant then passes through radial holes in them into their bearings. In some highly rated engines, bleed holes deliver oil from the hollow shaft and cams to points immediately in advance of the most heavily loaded part of their surfaces, which is where they begin to lift the valve.

Issuing from the rocker bearings, some of the oil may run along the downwardly inclined upper faces of the rocker arms, in some instances guided by grooves, to lubricate the contact areas between their hardened end-pads and valve-stem tips. In highly rated engines, ducts are drilled from these bearings to deliver the lubricant to the ends of the rockers. At the other extreme, a lightly loaded valve gear may be lubricated simply by oil mist.

Generally, the supply of lubricant to the valve gear is restricted, to prevent over-lubrication of the valve stems, which require relatively little lubrication, especially where they are tappet actuated, so that there is virtually no side loading. If too much lubricant is drawn, by the depression in the manifold, down the inlet valve stems, oil consumption may be excessive and, in spark ignition engines, the sparking plugs tend to become fouled. Consequently, valve guide ends, especially those of the inlet valves, are generally fitted with seals or shrouds to limit the quantities of oil entering them.

The supply of oil to the head can be either intermittent or limited by a restrictor in the delivery duct. An intermittent supply can be effected by delivering the oil into a flat on the periphery of the rotating overhead camshaft. Once per revolution, when this flat uncovers the delivery duct drilled in the cylinder head, it delivers oil into the flat and on through a hole drilled radially inwards from it into the hollow shaft. The rocker shaft, if there is one, can be lubricated either by oil mist or, more commonly, through the hole drilled from a groove in the camshaft bearing into which the intermittent or restricted supply is directed.

In virtually all systems, the oil drains back to the sump through a large capacity duct cored in the cylinder head and block. With pushrod actuated valve gear, there is a temptation to drain it back through the tubes that house the rods. This, however, may not be satisfactory since blow-by gases pass up these tubes: in some circumstances, the upward flow of gas can impede the downward flow of oil, and thus lead to flooding of the valve gear.

Other components, including the cylinder bores, are lubricated by oil splashing out of the main and big-end bearings and whipped up into a mist by the rotating and reciprocating components. As previously indicated, all the oil eventually drains back to the sump, where it cools and is recirculated.

16.14 Dry sump lubrication

In engines for vehicles, such as air and marine craft, that have to operate while inclined at steep angles, it is impracticable to have large quantities of oil swilling around in a sump, so what is termed the *dry sump* system of lubrication is adopted. Two pumps are required, one for the pressure lubrication system and the other, termed the *scavenge pump*, for continuously transferring the oil from the sump into a separate reservoir. Such systems are used also in some racing cars, partly to keep the ground clearance to a minimum and partly to circulate it through an oil cooler, either of the oil-to-air type or using the engine coolant.

The scavenge pump generally may have two pick-ups, one at the front and the other at the rear of the sump, to keep it dry regardless of the angle of inclination of the engine. Inevitably, some air is drawn through the pump, though not only because of the inclination but also, to keep the sump dry, the scavenge pump must be of a higher capacity than the pressure pump.

16.15 Lubrication of bearings carrying shafts

Rolling element bearings may, where appropriate, be packed with grease and sealed for life. When used in engines and transmission assemblies, however, they are generally lubricated respectively by the engine or gear oil. The supply of lubricant must be limited otherwise excessive drag may arise and the churning of the oil within the bearing will generate heat and reduce overall efficiency.

Heavily loaded plain bearings, on the other hand, must be kept constantly full of oil under pressure, to prevent metal-to-metal contact. Lubricant is generally fed through a hole in the shaft into an annular groove machined in the inner face of the bearing, mid-way between its ends, so that it can then flow axially, in both directions, between the rotating surfaces to lubricate them. As the oil issues from the ends of the bearing, it takes away the heat generated by viscous friction. The local pressure in the oil film is greatest at the edge of its central groove and falls off rapidly at the approaches to its two ends. Superimposed on this pattern, however, is that of the pressure generated by the dynamic loading, which varies continuously. In the main bearing of a single-cylinder engine, for example, the heaviest gas loads are downwards during the firing stroke and upwards during the inlet stroke, with lighter loadings occurring during the compression and scavenging strokes. Superimposed on these are of course the inertia loads. Typical load and pressure distribution diagrams for plain bearings are illustrated in Fig. 16.4.

16.16 Hydrodynamic lubrication

As a circular section shaft rotates in a bearing, it tends to drag the oil film round with it. When the bearing is loaded radially, the shaft is displaced eccentrically relative to the bearing. Consequently, the clearance between the two becomes locally a double wedge section. As periphery of the rotating shaft asses into the leading wedge section, it drags the oil with it. Therefore the

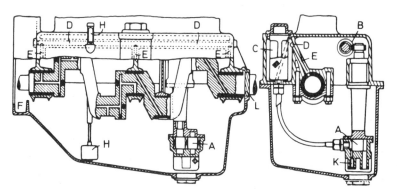

Fig. 16.5 An early pressure lubrication system. Oil is drawn from the sump, through a gauze strainer K by the gear type pump A at the lower end of a vertical shaft driven by a spiral gear B on the camshaft. This pump delivers the lubricant through a filter C into a gallery duct D. From this duct, it is distributed to the main bearings through ducts E and through another duct or ducts to the camshaft, valve gear and any ancillary drive shafts that might be incorporated. In this illustration, two alternative arrangements of ducts for taking the lubricant from the main to big-end bearings are shown. F is a trough for lubricating a chain drive, and L a screw-type oil baffle. The float-type oil level indicator H is not entirely satisfactory, so a dip stick was more commonly used, and generally still is even where electronic instruments indicating oil level are installed. In engines with overhead camshafts, the oil pump is usually gear driven from the crankshaft

hydraulic pressure in the film increases as it approaches the line of heaviest loading, thus continuing to support the load and preventing metal-to-metal contact.

Incidentally, a large clearance, due perhaps to wear, in one bearing may result in the discharge of an excessive proportion of the lubricant through it to the sump. This may starve an adjacent bearing of oil and cause it to fail. It follows that, in the event of a failure, the search for the cause should be confined not only to the bearing that failed, but also to the others.

16.17 Gear-type oil pump

Engine lubrication systems are pressurised mainly by gear-type pumps, though some manufacturers favour the eccentric rotor type. The gear-type pump, Fig. 16.6, comprises a small casting that houses a pair of intermeshing gears, one of which is driven by either an extension of the crankshaft or indirectly from it by a separate shaft. The inlet and outlet (or suction and delivery) ports lead respectively from small chambers on opposite sides of the meshing point. Except at these two areas, the housing is a close fit around the tips of the gear teeth.

As the gears rotate, the oil from the inlet is carried around between their teeth until they begin to mesh. At this point, because the gaps between the teeth are closing, the oil can no longer continue round. Consequently, it has to flow out into the delivery chamber, which is fitted with a pressure relief valve. The delivery pressure rises with engine speed until the relief valve opens, to return the oil in excess of requirements back to the inlet chamber. This prevents the pressure from rising above the maximum desirable level and therefore unnecessarily absorbing too much engine power. The main flow of oil passes from the delivery chamber through a duct to the main oil gallery.

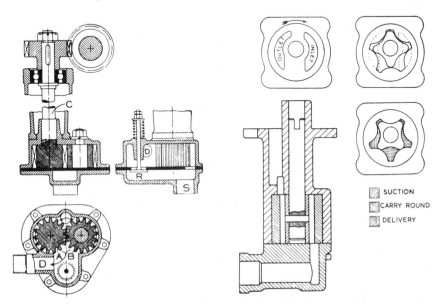

Fig. 16.6 Gear-type pump

Fig. 16.7 Hobourn-Eaton pump

Inescapably, there are clearances around the gears and between their teeth so, even though they are small, such a pump is unable to handle air. Indeed, even when it is pumping oil, there is some back-leakage, so it performs best if submerged in the sump. However, it is generally more practicable to mount it closer to the crankshaft, and this can be satisfactory provided its capacity is large enough. Pumps in this situation should be self-priming: that is, the intake and delivery ports should be high relative to the pump chamber, or weirs should be incorporated so that the oil cannot drain out when the engine is stopped. Air in the pump when the engine is being started, especially from cold, could result in high rates of wear and even mechanical damage anywhere in the engine.

Fitted to the pick-up end of the suction pipe is a strainer to prevent debris from entering the pump. This pick-up is often sited in a miniature sump formed in the main sump so that, if the oil level in the main sump is low, there is less risk of air being drawn into the pump. An alternative is to pivot the pipe freely about its upper end so that the pick-up is always floating on the oil, with the pick-up aperture and strainer submerged.

16.18 Eccentric-rotor pump

Eccentric-rotor-type pumps, such as the Holbourn-Eaton unit in Fig. 16.7, are sometimes used for engine lubrication, though more commonly for powering steering and other auxiliary systems. Invariably, the eccentric star-shaped inner rotor has one fewer lobes than the outer one which it drives. Inner rotors for lubrication systems usually have four lobes, though the larger ones may have five. The inlet and outlet ports are in the walls at the opposite ends of the housing. As the lobes pass over the crescent-shaped inlet port, they are separating and therefore drawing oil into the space between them until they leave the port behind them. As they begin to traverse the similarly shaped delivery port, they are closing together again and therefore forcing the oil between them out through it. A spring-loaded pressure-relief valve, usually of either the ball or the plunger type, is incorporated to by-pass oil in excess of requirements from the delivery to the inlet side of the pump.

16.19 Oil filters

The strainer in the pick-up must not be so fine that it traps the smallest abrasive particles, otherwise it could strangle the flow. Therefore, to protect the bearings, a fine filter is interposed between the pump and the oil gallery. Such a filter has to remove all hard abrasive particles large enough to bridge the oil films in the bearings. When the engine is heavily loaded, these films may become extremely thin. The filter is mounted externally on the crankcase, so that its filtration element can be easily replaced at regular intervals, well before it becomes blocked. Generally, if the element becomes blocked prematurely the oil pressure lifts it off a seat at its lower end to by-pass it.

Originally, relatively thick felt elements were employed. Subsequently, in the 1950s, plastics-impregnated paper element filters were introduced, which trapped so much of the particulate matter that they tended to block more rapidly. The paper is pleated and rolled on a mandrel, to form a star section cylinder presenting a huge surface area to the oil flow. Because these elements

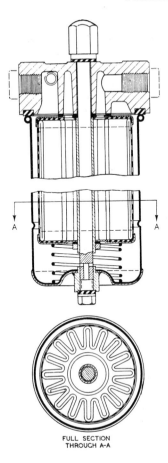

FULL SECTION
THROUGH A-A

Fig. 16.8 Vokes full-flow filter

were so efficient, it became common practice to by-pass some of the oil flow directly into the main gallery, fitting restrictors to limit the proportion filtered. The argument in favour was that, by filtering only a proportion of the oil and by-passing the remainder, an element of the finest mesh could be used without risk of blockage or impeding the flow through the system. Furthermore, the build up of particles in the system was only slow, so not only was full-flow filtration no longer necessary but also actually better.

In fact, in the days of relatively low-rated engines and large bearing clearances, this was good enough though far from ideal, especially where the sump was liable to be refilled or topped up with contaminated oil. As engine performance increased and bearing clearances were reduced, full-flow filtration again became desirable and, ultimately, essential.

Paper elements have some incidental advantages. They are so compact that they can be accommodated in a relatively small-diameter casing. Moreover, they may be self-supporting or reinforced with wire mesh, depending on their overall size and the rate of flow. In the example illustrated in Fig. 16.8, the cup-shaped pressed steel casing is retained by a central bolt passing coaxially though the element and the filter head casting, in which are cored the inlet and outlet ports. Oil is delivered from the pump to the inlet port, on the right in the illustration. Each end of the pleated cylindrical paper element is closed by a plate, that at the upper end being perforated to allow the oil to enter. From the inside, the oil passes radially out into the space between the element and its metal casing, whence it leaves through the delivery duct, on the left, to the main oil gallery.

In the event of severe blockage, the oil pressure, acting against the upward load exerted by the spring in the base, slides the element down the central rod so that the lubricant can pass from the inlet directly into the outlet. The upper end of the element seats on a soft elastomeric sealing ring which accommodates any small particles that might become trapped between the seating faces. Initially the element is by-passed only when the engine is running at high rev/min: at lower speeds full-flow filtration is maintained. As the contamination increases, however, the filter lifts at progressively lower engine speeds.

Formerly, elements alone were replaced; subsequently, they were mostly sealed in their housings, which were screwed on to an adaptor on the crankcase. Replacement therefore was simply a matter of discarding the whole unit, and replacing it with a complete new one. With the trend towards conservation, however, there are now pressures for replacing only the elements.

16.20   Oil circulation and pressure indicators

The only indication that the oil is circulating properly is the reading of a pressure gauge or illumination of a warning lamp if it is not. This is far from satisfactory, but so far nothing better has been found. Oil pressure is of course dependent on viscosity but, in general, any deviation from normal should be regarded with suspicion.

Oil pressure used to be indicated invariably by means of a bourdon tube-actuated pointer in an instrument on the dash fascia. This called for the oil to be piped under pressure to the instrument, so fracture and therefore leakage was possible. Moreover, it was inaccurate. Following the introduction of electronics to automotive applications, these pressure indicators have been supplanted by a small sensor, sometimes termed a *sender*, which is generally screwed into the main oil gallery on the engine. A cable connects it to either a low-pressure warning lamp, or an electrical instrument calibrated to indicate pressure, on the fascia.

The sensor units contain a spring-loaded diaphragm subjected to the oil pressure on one side and having a sensing element mounted on the other. In the more rudimentary examples, deflection of the diaphragm under the oil pressure, against the force exerted by the spring, either opens or closes a pair of electrical contacts, sending a signal either directly or through the on-board computer, either to drive an instrument or to illuminate a low oil pressure warning lamp on the fascia.

In another type, deflection of the diaphragm compresses a piezo-electric element interposed between it and the housing. Again, this generates a signal but, in this instance, it is generally transmitted to the computer which translates it into an actual pressure indication on an instrument on the fascia. A third system is of the analogue type, based on an electrically actuated

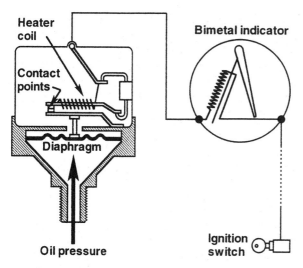

Fig. 16.9 Oil pressure indicator mechanism comprising essentially an electrically heated bimetal strip and vibrator

vibrator, Fig. 16.9.

One arm of the vibrator is a bimetal strip around which is a heater coil. As the temperature increases, the bimetal strip bends, causing the contact on its end to lift clear of a contact mounted on another arm. This opens the circuit to the coil, the bimetal strip cools and the contacts close again, restoring the current to the heater coil. This is a cyclical process at a frequency that is a function of the pressure across the two contacts which, in turn, is determined by the force exerted by the oil on a diaphragm connected to the arm carrying the second contact point. Oil pressure is indicated on the fascia, by the dial of an instrument sensing the average current flow and calibrated in units of pressure.

16.21 Oil level indication

To ensure that the pump is never starved of lubricant, the oil in the sump must be maintained above a certain level. If it is too high, however, the big ends will churn it up, aerating the oil, generating heat, and causing unnecessarily high frictional losses and therefore increased fuel consumption.

In general, the oil level should be checked when the engine is running at its normal operating temperature. This is because the level varies not only with temperature but also with the quantity of oil actually in circulation in the lubrication system.

A dip stick, on which the recommended high and low levels are marked, gives the simplest and most reliable indication. Failure to use it regularly could lead to disaster. Now, however, to satisfy modern emissions regulations, manufacturers have had to reduce oil consumption to such a low level that topping up the sump between regular service operations has become, in most instances, unnecessary. A low-oil-level indication, therefore, generally means something is seriously wrong.

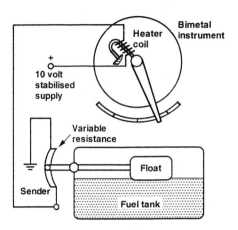

Fig. 16.10 This relatively primitive type of fuel level indicator installed inside the petrol tank comprises a float-actuated lever with a sliding contact brush on the end remote from the float and a variable resistance over which the brush traverses. The needle in the instrument dial is actuated by the current, but is generally calibrated in both gallons and litres

Electronic oil-level detection has been adopted increasingly since the early 1970s. Drivers are alerted to low oil level either audibly, or visually by means of a lamp on the dash fascia. The sensor generally comprises a cup-shaped, oil-tight housing with a diaphragm closing its otherwise open lower end. Inside the cup is a piezo-electric element that vibrates the diaphragm. This element is driven by an oscillator sweeping over a fairly narrow range of frequencies around that of resonance of the diaphragm.

The lower end of the sensor is submerged below the oil in the sump, with its diaphragm parallel to the surface of the oil and in the plane of the minimum permissible oil level. Consequently, the vibration of the diaphragm is damped but, when the oil falls below this level, it is no longer damped. The latter condition is detected by a resonance discriminator, which closes the alarm circuit. Contamination of the diaphragm by oil droplets of oil or dirt has a negligible effect on detection within the range of frequencies swept by the resonance discriminator, so it is much more accurate than the earlier float-type indicators, which in principle functioned in a manner similar to the fuel level indicator illustrated in Fig. 16.10.

Chapter 17

Engine cooling

The high working temperatures of the internal combustion engine which are the cause of its high efficiency are at the same time a source of great practical difficulty in construction and operation. No known materials are capable of enduring continuously the high temperatures to which the cylinder, piston and valves are subjected while at the same time retaining sufficient strength to withstand the high working loads.

The maximum temperature during combustion is approximately the melting point of platinum and the temperature even of the exhaust gases is above that of aluminium. It is thus essential that heat be abstracted from the parts enumerated above at a sufficient rate to prevent a dangerous temperature being reached, for, unfortunately, it is quite impossible to prevent heat entering the metal from the hot gases.

Nevertheless, it must always be borne in mind that such abstraction of heat is a direct thermodynamic loss, and if high thermal efficiency is desired the abstraction of heat should not be more than is necessary to prevent dangerous overheating and consequent distortion of the cylinder, etc. Excessive cooling will also prevent proper vaporisation of the fuel, leading to further waste and also to objectionable dilution of the crankcase oil by unvaporised fuel, which gets down past the piston. On the other hand, too high an operating temperature is bad from the point of view of the power developed, since it results in reduced volumetric efficiency on account of the excessive heating of the incoming charge with the consequent reduction of charge weight and thus of power.

In any given set of conditions there is some particular operating temperature which produces the most satisfactory results, and only experience with a particular engine will determine this temperature. If it were not for the loss of water and the interference with cooling as a result of boiling in an ordinary water-cooled system, the temperature of boiling at atmospheric pressure, namely 100°C or even higher, would not be too high a general operating temperature in the majority of engines. To prevent boiling a lower operating temperature is necessary, as local overheating must be guarded against. It is usually in the neighbourhood of 75° to 90°C unless a pressurised system is used (see Section 17.3).

To come to the actual cooling means employed, whatever these may be, the ultimate result is the dissipation of the dangerous heat to the surrounding air.

Direct dissipation of heat from the cylinder barrel and head to the surrounding air, or air-cooling as it is called, is feasible with small engines and is practically universal with motor-cycles. The heat is dissipated by radiation and convection from thin cooling fins cast on the cylinder head and barrel.

Natural flow air-cooling is at best very difficult to regulate, and as the size of the cylinder increases the difficulties become very serious, for the heat developed increases as the cube of the linear dimensions, while the radiating surface increases only as the square.

In multi-cylinder engines the flow of the air is controlled by suitable cowlings so as to be distributed equally to all the cylinders. However, in large engines, if air-cooling is used, it is generally in conjunction with a large fan which delivers a considerable volume of air to the cylinders, and this fan besides being bulky requires a considerable expenditure of power to drive it, so that unless other considerations enter into the question the chief gain, which is the elimination of the necessity for a radiator and water jackets to the cylinders, with the consequent saving in weight, is not always realised. Air-cooled engines have the reputation of being noisier than the water-cooled form, and very few instances of large fan-cooled engines are on the market, though in the special application of large power units to armoured fighting vehicles, important developments have recently taken place in the US Army. Extremes of climate, difficulty in replacing water lost by evaporation, and all the complications involved in frost precautions in operational theatres, added to the probable greater vulnerability of liquid cooling, are powerful inducements towards the adoption of direct air-cooling in this field.

For the normal medium and large passenger and commercial vehicle water-cooling is practically universal.

Heat is far more readily transferred from metal to water than from metal to air, while the final transfer of heat from the water to the air, in order that the water may be returned to circulate through the engine jackets, can be readily accomplished in a radiator provided with a large cooling surface in contact on one side with the hot water from the engine and exposed on the other side to a strong current of air arising from the motion of the car, aided by a fan driven from the engine.

A steady operating temperature will be reached when the difference between the temperatures of the air and radiator is sufficient, in conjunction with the area of radiating surface and air flow to dissipate heat at the same rate as it is being received from the jackets. During a long climb with a following wind equilibrium may not be reached and the water will boil.

The radiator is normally placed in front of the engine, but whatever the position may be there are two methods of circulating the water through the engine jackets and radiator, namely—

(1) Thermosyphon system.
(2) Pump system.

In the former the water circulates in virtue of the pressure difference arising from the difference in density between the hot water in the jackets and the cooler water in the radiator. For success in operation the passages through the jackets and radiator should be free and the connection pipes large, while the jackets should be placed as low as possible relatively to the radiator, in order that the 'hot leg' shall have as great a height as possible.

In operation the water level must on no account be allowed to fall below the level of the delivery pipe to the radiator top, otherwise circulation will cease.

Fig. 17.1 shows as simply as possible the essentials of the system, XX showing the critical water level.

In general the thermosyphon system requires a larger radiator and carries a greater body of water than the pump circulation system and a somewhat excessive temperature difference is necessary to produce the requisite circulation. On the other hand to some extent it automatically prevents the engine from being run too cold.

The defects mentioned have resulted in the pump circulation system becoming universal in road vehicles. The pump is either a radial flow centrifugal pump, driven in the larger engines by positive gearing, and forcing the circulation from the bottom of the radiator to the bottom of the jackets, or a simpler axial flow impeller, placed usually at the outlet from the cylinder head, and driven by an all round belt drive including the dynamo. This arrangement is used with some smaller engines, and the fan is usually mounted on the impeller spindle. These impellers have less forcing power than the centrifugal type, but give appreciable assistance to the circulation.

Neither type is a positive pump in the sense of being able to force a fixed rate of flow independently of the hydraulic resistance of the circuit.

Fig. 17.2 shows the construction of a centrifugal pump. It consists simply of a casing inside which an impeller P, provided with vanes V, is rotated. The water enters the casing at the centre of the impeller and, being caught up by the vanes, is whirled round. Centrifugal force causes the water to pass to the periphery of the impeller where it is thrown out into the stationary casing. The kinetic energy imparted to the water by the impeller is converted in the stationary casing into pressure or potential energy so that a pressure difference is established between the inlet and the outlet D of the pump. A seal is provided where the impeller spindle passes through the casing, in order to maintain a water-tight joint. This seal is made in modern designs by means of the face contact of a graphitic carbon ring which may revolve with the impeller or be stationary in the housing, on either the suction or pressure sides, according to convenience in design. The face contact avoids wear of the shaft, and the carbon, bearing on a suitable alloy iron, hardened steel or phosphor bronze surface, develops a hard, wear-free surface which requires no lubrication and provides an effective seal against leakage.

Alternative arrangements of the Morganite Unit Seal, supplied by the Morgan Crucible Company as a proprietary fitting, are shown on the left of

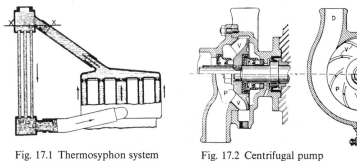

Fig. 17.1 Thermosyphon system   Fig. 17.2 Centrifugal pump

the figure. The carbon ring is bonded to a flexible rubber casing which takes the friction torque. Inside the casing is a light stainless steel spring proportioned to give appropriate pressure between the rubbing surfaces. The whole provides a readily inserted unit assembly requiring renewal only at long intervals.

Pressure relief holes may be provided through the impeller boss to prevent the delivery pressure developing undue end thrust on the back.

A pump circulation system is illustrated diagrammatically in Fig. 17.3. A is the centrifugal pump delivering to the bottom of the jackets and B shows a thermostatically controlled by-pass valve by which the radiator can be short-circuited and the water returned to the pump suction until its temperature reaches a desirable value. In the diagram the valve is in a mid position.

17.1   Temperature control

There are broadly two methods of regulation in use, the first being practically universal in automobile practice—

(1) Control over water flow by means of a thermostat valve.
(2) Control over air flow by means of radiator shutters actuated either manually or automatically by a thermostatic device.

The thermostat consists of a concertina or bellows-like vessel of thin metal, similar to those employed in the aneroid barometer but filled with a suitable liquid. In some cases expansion is magnified by a system of levers to operate the valve. In most modern designs, the large movement necessary for the valve is obtained directly from the bellows by *evaporation* of the liquid, which should be one boiling at about 60 to 80°C. Such liquids are acetone, alcohol, etc.

The valve may control a by-pass as in Fig. 17.3, or it may simply be placed in the main circuit depicted in Fig. 17.4, where (*a*) shows the valve as part of the engine construction and (*b*) illustrates the convenient and simple Motorstat hose-line type, which can be readily installed in any engine in place of the hose connection to the top of the radiator.

Where the thermostatic valve closes the main circuit without a by-pass, which has been the usual practice, it must be capable of closing the circuit

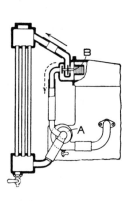

Fig. 17.3 Pump circulation

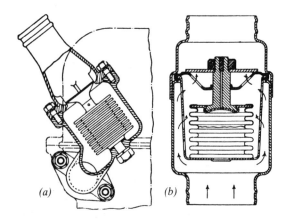

Fig. 17.4 Bellows-type thermostat

against the pressure developed by the centrifugal pump, but no great difficulty is experienced in meeting this condition, since the bellows can be arranged to act as a tension spring, the initial tension being regulated to give any opening temperature desired within the vapour pressure and temperature characteristics of the liquid employed.

When carefully designed and installed these devices are reliable, and their use in various forms has become general in all engine cooling systems.

17.2   Wax-element thermostats

Thermostats of the wax-element type are incorporated in most pressurised cooling systems. Western-Thomson Controls Ltd of Reading were the first to introduce this type of unit into the UK, for the British diesel and gasoline engine manufacturers.

The waxes used are broadly in the microcrystalline group which covers the higher melting point range of paraffin waxes. In many cases the wax is mixed with a very finely divided copper to increase the thermal conductivity of the charge as a whole. This results in a more thermally sensitive element. Over the working range the wax remains in the plastic state, having a coefficient of volumetric expansion of about 0.9%/°C. A maximum lift of 9.525 mm and a thrust of no less than 15.9 kg are given by the standard automotive element.

A tapered piston rod is enclosed in a synthetic rubber boot sealed in the upper end of the capsule (a brass pressing) which contains the wax. The upper end of this rod is screwed into a central hole in a bridge carried by the main seating ring, and after adjustment for calibration, the screw is locked by solder.

As the wax expands on heating, it forces the body of the capsule downwards against the return spring carrying with it the valve which opens against the pump pressure.

To avoid the harmful effects that might arise from a stalled water pump when the valve is shut in a closed circuit, the makers recommend a by-pass system, and when this cannot be arranged without extensive modification, a suitable bleed orifice may suffice.

Failure of the wax capsule would cause the valve to remain closed, with possible disastrous consequences, but the same results would arise with a punctured or cracked bellows. The wax capsule is claimed to be less liable to failure owing to its robust construction and freedom from fatigue effects which may arise in the bellows type due to cyclical distortion.

Fig. 17.5 illustrates a wax element thermostat in one of its simpler forms. The valve, side-shutter type, is pressed on to the outside of the capsule against the swell in diameter, and seats on the main annular pressing which carries the bridge and return spring abutment. The form of the rubber boot and seal will be clear.

The authors are indebted to the makers for information, and a full description will be found in *Automobile Engineer* for March 1962.

The second means of regulation, namely by hand-operated shutters, has both the advantages and disadvantages of being under the direct control of the driver. By the use of these shutters, placed usually in front of the radiator, the air flow may be wholly or partly blanked off in order to obtain the desired working temperature. The shutters and their controls require to be well

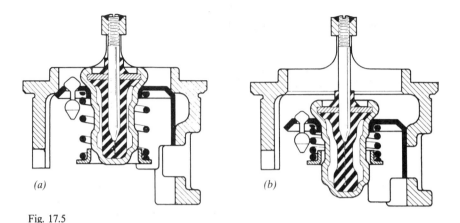

Fig. 17.5

designed and made if they are to be free from rattle. Radiator shutters controlled by a thermostat have been used.

The radiator and bonnet muff so much used in Great Britain, while useful in cold weather to prevent cooling off while the car is standing, is at best a makeshift and unsatisfactory means of control.

17.3   Pressurised cooling system

Experience with evaporative cooling systems in closed water jackets of complicated form appears to have emphasised the danger of local steam pockets forming at critical positions around exhaust valve pockets and spark-plug bosses.

In such locations rapid extraction of heat from the metal is essential, but stagnant or superheated steam which is not rapidly scoured away by fresh supplies of water at boiling point is relatively a very poor agent for the purpose, and local overheating and water hammer action is very likely to take place.

A system which combines the reliability of the more familiar arrangements with some of the advantages of steam cooling is the pressurised system incorporating a light relief valve with an automatic vacuum break valve, the relief valve being spring-loaded to open at a pressure of 20.68 to 27.6 kN/m² above prevailing atmospheric pressure. This is illustrated in Fig. 17.6, which shows a combined pressure and vacuum relief valve.

The pressure inside the radiator block is communicated through the central port H to the pressure disc valve P and the vacuum valve V. These are loaded to pressures of about 24 and 6.9 kN/m² respectively. An escape pipe is provided at E. To guard against spurting on removal of the cap, the spring disc D does not leave its seating until the main body is well clear of the sealing washer S, and the pressure is reduced to that of the atmosphere.

In a closed water system of this type the pressure reached depends upon the temperature attained according to the following table—

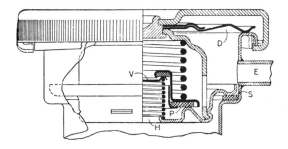

Fig. 17.6 Pressurised cooling system

Temperature (°C)	94	100	107	121
Pressure (kN/m² abs.)	82.7	103.3	131	206.8

The pressure of 82.7 kN/m² corresponds to an altitude of 6000 to 7000 ft, so that boiling would occur at 94°C at this altitude in an open system, whereas in a closed system loaded to 20.68 kN/m² the temperature would be raised to the normal sea level figure before loss occurred. At sea level a margin of 6 or 7° is provided by the pressurised system. These increases may well prove sufficient to avoid the necessity of stopping to cool off in order to prevent loss of water under arduous conditions of hill climbing at high altitudes. The heat-dissipating capacity of the radiator surface would be appreciably increased by such a temperature increase.

A higher loading of the relief valve up to, say, 103.3 kN/m² would give such a gain in temperature that economy in size of radiator might be possible, though the probable necessity of strengthening the block would tend to off-set any reduction in weight. With the lower figure of 20.7 to 27.7 kN/m² above atmosphere special provision for strengthening will not normally be necessary.

17.4 Twin thermostats

With diesel engines, the rate of heat flow to coolant is relatively low. Consequently, in the large engines used in heavy commercial vehicles, the water jackets of which contain a large volume of coolant, the attainment of normal working temperature after starting from cold takes a significant time and, when a correspondingly large thermostat opens, a flood of cold water may pass quickly through the jacket. For vehicles engaged on stop–start delivery, this can happen many times a day, so thermal fatigue – due to alternating stresses induced by repeated local thermal expansion and contraction – can cause cracking around hot areas such as exhaust valve seats, or fretting of joints and gaskets.

To avoid this, twin thermostats are sometimes employed, one opening at say 77°C and the other at perhaps 82°C. This not only reduces the volume of water suddenly flowing through the jackets, but also has other advantages. First, it is easier to pass a large volume of coolant through two open thermostats than through one, since the ratio of total periphery to area of the ports is larger in the former. Secondly, there is a safety factor in that a partial or even a total failure of one will not be so serious if there is a second thermostat.

17.5 Renault R4-L sealed coolant system

An interesting development of the pressurised cooling system is the sealed arrangement first adopted on the R4-L Renault front-engine vehicle.

Described as a *sealed for life* system, both temperature and pressure controls are provided. Communication with a relatively large expansion tank situated under the bonnet is opened by a thermostatic valve when the coolant reaches a temperature of 100°C. The expansion tank is provided with a pressure relief valve set to about $48.26 \, kN/m^2$.

The coolant is a 50:50 mixture of water and glycol charged and sealed permanently subject to abnormal leaks. The air capacity of the expansion tank is sufficient to prevent the pressure (at 100°C) from rising above the relief pressure, and in any case only air would be discharged.

Fig. 17.7 indicates the arrangement diagrammatically. All connections are made with rubber hose including those to the space heater for the passenger compartment.

17.6 Directed cooling

Considerable care is taken in modern designs of cylinder head to provide carefully directed jets of coolant to local hot spots such as valve seats and spark-plug bosses. This may be accomplished by means of a suitably drilled gallery pipe fed directly from the water pump, or by nozzles pressed into the head casting at each of the communicating holes between the cylinder block and head jackets. In the Vauxhall head, shown in section through the exhaust valve in Fig. 17.8, the nozzles take the form of thimbles provided with one or two ports suitably positioned in each case to direct vigorous local streams to the spots desired. A pressed locating snug ensures correct positioning in each case. The six-cylinder head has a total of ten of these jets directed in general towards the exhaust valve pockets.

17.7 Radiator construction

Copper radiators have now evolved into two types of water-tube construction, the early honeycomb or cellular construction shown in Figs 17.9 and 17.10 being obsolete. Though having certain advantages in appearance, ease of repair and smaller proportionate immobilisation by a single obstruction, the earlier types suffer from the important heat transfer defect that the ratio of metal–air to water–metal dissipating surface is practically unity instead of being many times that figure.

In modern tubular constructions, appearance and protection are provided by the ornamental grille which is now universal in passenger vehicles, and the radiator block can be designed simply as an efficient heat exchanger of the lightest possible weight.

Fig. 17.11 shows the vertical water tube type which is to be found on heavier types of commercial vehicles. The construction consists essentially of cast light alloy top and bottom tanks bolted to cast side pillars. The cooling element

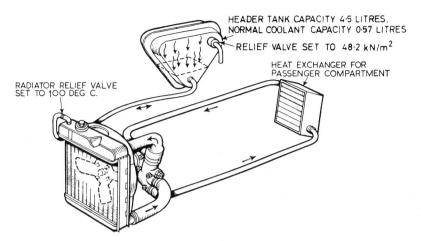

HEADER TANK CAPACITY 4·5 LITRES.
NORMAL COOLANT CAPACITY 0·57 LITRES

RELIEF VALVE SET TO 48·2 kN/m^2

HEAT EXCHANGER FOR
PASSENGER COMPARTMENT

RADIATOR RELIEF VALVE
SET TO 100 DEG C.

Fig. 17.7 Renault sealed system

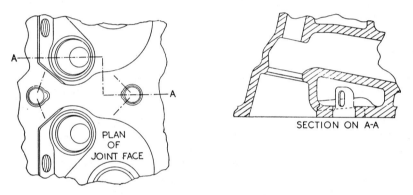

A

A

PLAN
OF
JOINT FACE

SECTION ON A-A

Fig. 17.8 Directed cooling in Vauxhall cylinder head

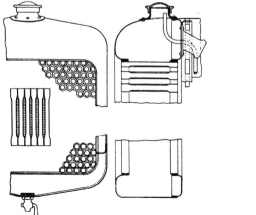

Fig. 17.9 Honeycomb radiator

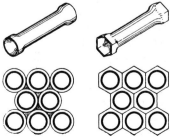

Fig. 17.10 Honeycomb tubes

consists of a nest of copper tubes soldered into brass upper and lower tube plates which are bolted to the top and bottom tanks. This construction is inexpensive and makes for accessibility and ease of repair.

The tubes take various forms, and if of normal circular section, are usually provided with external gills to give additional surface to aid transfer of heat from metal to air. Various forms are illustrated in Fig. 17.12. At (a) is indicated a construction in which a nest of six or eight tubes, running from back to front of the block, is sweated into common gill plates.

Different forms of plate gill applied to individual tubes are shown at (b), while (c) shows the efficient and widely used Still tube which employs a continuous coil of copper wire wound spirally around, and sweated to, the plain tube.

Fig. 17.13 shows a popular current form of flattened tube and plate gill type, which can be assembled and sweated up to the size of block required. The flattened tubes give a more favourable ratio of transfer area to water weight than the circular tube, and they are fabricated from very thin strip metal with a folded and sweated longitudinal joint. The ratio of air surface to water surface can be varied very conveniently by the appropriate choice of the number of gill plates through which the tubes are threaded. The dimples help to ensure the desirable turbulent motion of the air.

Fig. 17.14 illustrates the extreme development of the film-tube type in which the core is built up of units formed from thin embossed copper strip of a width equal to the required thickness of block. This is normally 2 to 3 in thick.

The block is built by first forming airway units of the full height of the block.

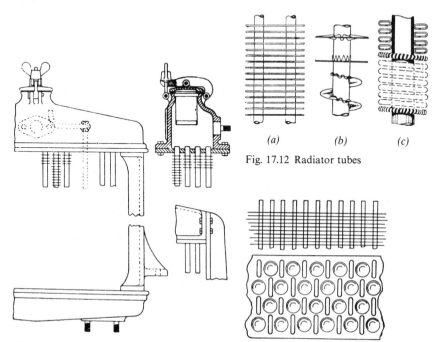

(a)   (b)   (c)

Fig. 17.12 Radiator tubes

Fig. 17.11 Radiator construction for heavy commercial vehicles

Fig. 17.13 Flattened tubes, with plate gills

This is done by folding the embossed strip round serpentine dimpled strip to provide the additional air surface, locating dimples being provided on the embossed strip to position the serpentine strip until the final soldering operation.

These airway units are then assembled in a jig in the required number and sweated together at the corrugated edges, leaving straight vertical waterways of a lateral width equal to the full depth of the edge corrugations. The embossing of the strip provides, in addition, strengthening flutes and spacing supports on the vertical centre line. The actual-size drawing of Fig. 17.14 should make the construction clear. The ratio of air area to water area in the example illustrated is about 5 : 1.

Aluminium radiators have been available for many years but have not been widely used. They are mostly of brazed construction. Because of the high temperatures involved in this method of manufacture, the aluminium becomes annealed, so relatively thick sections have to be used. With the current call for weight reduction to improve fuel consumption, either aluminium or composite radiators – such as plastics tanks with either aluminium or copper tube blocks – are likely to become increasingly attractive.

17.8 Horizontal disposition of copper tubes

The marked tendency towards wider and lower bonnet lines has encouraged designers to test the merits of the horizontally disposed radiator block. As pump or forced circulation is now practically universal, thermosyphon action has little significance, and the high and narrow constructions of block previously called for can no longer be accommodated easily. Increase of width and reduction of height of the block result in a less favourable surface : weight ratio, owing to the greater proportional weight of the top and bottom headers. The larger number of shorter tubes calls for a greater number of soldered joints and increased risk of leakage. Favourable proportions are restored by laying a 'long tube' block on its side, and if desired, introducing baffles so that inlet and outlet may be on the same side of the bonnet, the water traversing the bonnet width twice.

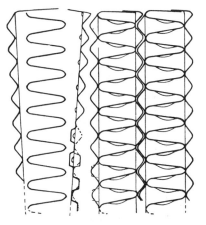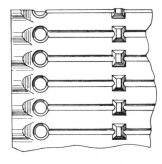

Fig. 17.14 Film tube construction (actual size)

The increased length and smaller total cross-section of the flow area naturally increase the hydraulic resistance, but by careful pump design and choice of speed, the power consumption need not be excessive. The flattened tube type of block shown in Fig. 17.13 is well suited to this horizontal layout.

A comprehensive series of articles on cooling system design, by S. I. G. Taylor, was published in the April, May and June 1970 issues of *Automobile Engineer*.

17.9   Fan drives

With increasing emphasis on fuel economy, the traditional pulley-driven directly-coupled pressed steel fan is tending to give way to more efficient systems. Moulded plastics – nylon or polypropylene – fans are lighter, easier to balance, look better and, above all, can be made more efficient aerodynamically. Incidental benefits are reduced noise and vibration, and less risk of serious injury if the rotating fan blades strike fingers carelessly contacting them.

Basically, there are two approaches. One, by far the simplest, is to incorporate in the hub of the fan a clutch that slips at a certain torque, thus limiting the amount of power that can be transmitted to, and therefore absorbed by, the fan. A good example is the Holset torque-limiting drive, which comprises a disc which is the driving member secured to the V-belt pulley, and a casing which is the driven member within which the disc rotates and on which is mounted the fan. The casing is finned to dissipate heat generated by slip and there is a closely controlled small clearance between it and the front and rear faces of the disc. This clearance is filled with a viscous silicone fluid, the shearing of which allows the drive to slip. Typical characteristics of such a drive are shown in Fig. 17.15.

The second approach is to introduce a thermostatic control either to limit the torque transmitted to the fan, or to cut it out altogether when the coolant temperature falls below a predetermined level. Again, one of the simplest is the Holset air temperature sensing fan illustrated diagrammatically in Fig. 17.16. This is similar to the drive described in the previous paragraph except that its casing is divided by a separator plate into two chambers. Of these, the rear one

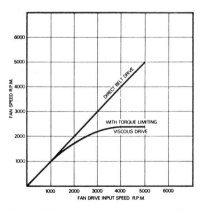
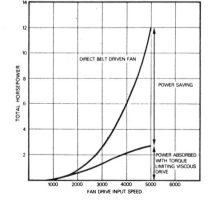

Fig. 17.15 Typical characteristics of a fan drive

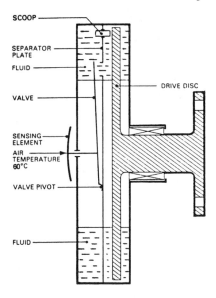

SCOOP

SEPARATOR PLATE

FLUID

DRIVE DISC

VALVE

SENSING ELEMENT

AIR TEMPERATURE 60°C

VALVE PIVOT

FLUID

Fig. 17.16 Holset air temperature sensing fan

houses the driven disc, while the front one is a reservoir. A scoop mounted on the separator plate continuously removes oil from the rear chamber and passes it into the reservoir. Thus, the tendency is for the viscous drive clutch chamber to be emptied so that no drive can be transmitted. However, the oil is allowed to re-enter the clutch chamber through a port nearer than the scoop to the centre of the disc, so that the fluid is therefore distributed over the disc by centrifugal force. This port is controlled by a thermostatically-actuated valve which opens when the engine coolant becomes too hot. The thermostatic control element is a bimetal plate on front of the coupling where it is exposed to the airstream issuing from the radiator.

More elaborate systems are used for some heavy commercial vehicles. These range from thermostatically-controlled hydrostatic drives to the Dynair system, which has a thermostatically-controlled air-actuated cone clutch. With most of these arrangements, the controller is a thermostatic switch, the sensor of which is in the coolant – usually in the radiator outlet. Similar controls are used also in conjunction with electric motor-driven fans – for cars as well as commercial vehicles. The advantages of this and the hydrostatic drive systems is that, because the fan is driven independently of the engine, both it and the radiator can be installed anywhere on the vehicle. Electrically-driven fans are therefore widely employed in cars having transversely-mounted engines.

Chapter 18

Potential for future development

The application of the gas turbine to the smaller types of road vehicle is a revolutionary and conjectural possibility, though the successful enterprise of the Rover Company in producing the first turbine-driven car attracted world-wide interest. Since these post-war developments, research and technical advance have continued, but though the great US corporations and many European firms of high standing are patenting and developing a great variety of appliances, there is no evidence yet of pending large-scale applications in the automobile field. Even under the threat of severe, restrictive anti-pollution legislation, no practical alternative to the internal combustion engine has evolved.

18.1 The gas turbine

The gas turbine, which for many years was the dream of inventors, has inspired a wealth of technical literature and experimental enterprise, but had shown very limited practical and commercial progress until the last four decades, in which we have seen the spectacular development of the turbo-jet propulsion of aircraft. This type of power unit is still considered by some as likely to become a serious competitor of the piston engine for road vehicles.

The following articles, published by the *Automobile Engineer*, should be consulted for further reference: F. R. Bell, *The Gas Turbine Car*, Vol. 43, No. 3, and *Gas Turbine Arrangements*, Vol. 43, No. 8. Additionally, a paper on the Chrysler gas turbine car was presented before the Automobile Division of the Institution of Mechanical Engineers on 13th April 1965. Readers interested in the theory of the design of gas turbines are advised to consult *Jet Propulsion Engines*, edited by O. E. Lancaster and published by the Oxford University Press, London (1959).

18.2 Two fields of successful application

The ultimate objective of automobile and power plant engineers is the replacement of the complicated, noisy, and vibration-prone reciprocating piston engine by a self-contained power unit which shall deliver energy by direct rotative motion.

The valuable features of mechanical simplicity, lightness, and ultimately cheapness, can certainly be achieved, and if fuel economy and reasonable reliability and life can be realised, the piston engine will gradually be superseded.

Unfortunately very great difficulties arise in satisfying these latter conditions, and the two fields in which successful development has taken place represent special applications in favourable circumstances.

The first of these is in the use of the turbine as an adjunct to the piston engine for supercharging purposes, on which some notes are included in Section 18.6, and the second as a means of supplying kinetic energy to the jet used for the propulsion-by-reaction of high speed aircraft.

It should be realised that jet propulsion is quite unsuitable for the efficient propulsion of low speed vehicles, apart from traffic conditions.

The turbo-prop system represents an approach to the self-contained unit delivering shaft power, but in an application where economy and endurance are not factors of overriding importance as in ordinary commercial fields. The system has a higher efficiency than the pure jet system at lower altitudes and speeds.

Fig. 18.1 represents in simple diagram form the turbo-jet and turbo-prop jet systems. The success achieved represents a remarkable advance in overcoming two of the main problems of the rotative unit, the provision of an efficient rotary compressor and the production of turbine blade materials which will give an acceptable endurance in this application.

In Fig. 18.1, a centrifugal compressor is indicated in the pure jet system, feeding multiple combustion chambers which usually fill the available annular space. An enlarged front view of the impeller is shown on the left, and indicates how the blades are curved to provide for axial entry to the eye with the minimum of shock. It should be noted that as the vanes have no curvature in diametral planes, centrifugal force has no tendency to distort their shape.

The lower diagram indicates an axial flow compressor, this type showing

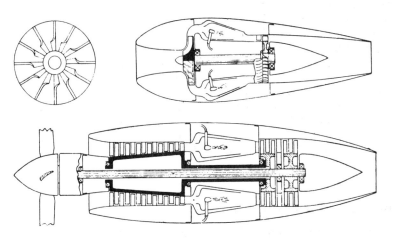

Fig. 18.1

higher efficiency than the centrifugal type. It will be noticed that the two turbines are mechanically quite independent, but utilise the combustion gases in series. The compressors are required to supply a great excess of air over that required for complete combustion, not only to prevent excessive heating of the parts, but also to provide the necessary mass–velocity characteristics of the propulsive jet.

18.3   Essential processes in ic power units

The three essential processes in any internal combustion power unit are compression of the air charge, combustion of the fuel, and expansion of the products as completely as possible to produce mechanical work. The greater the degree of compression and expansion, and the greater the range of temperature effectively utilised, the higher will be the thermal efficiency and the lower the specific fuel consumption. These conditions apply to both piston and turbine engines.

The piston engine can accomplish the first two processes with high efficiency and reliability, but is less successful with the third owing to its unsuitability for handling the large volumes at the low pressure end of the expansion, this being incomplete unless continued in a turbine, which is admirably suited to deal with large volumes at low pressure. Hence, as explained in Chapter 14, the exhaust turbo supercharger forms a mechanically independent but not thermodynamically self-contained high-speed unit which increases the power of the main engine but does not require to be geared to the low-speed power shaft. Moreover, since the gases reach the turbine blades at a manageable temperature, no insoluble temperature–stress–time factor arises with the material of the blading.

18.4   Essential organs in turbine unit

In a self-contained turbine unit separate organs must be provided for all three processes, the compressor as well as the turbine being of the rotary form, and driven by the main or an auxiliary turbine. The combustion chamber or chambers must receive the compressed air, the liquid fuel must be injected, and the combustion be completed before the products are passed to the turbine. This combustion must be done in a light and compact arrangement, capable of withstanding the high temperature of continuous combustion over an acceptable working life.

In the present stage of the art, and with available materials, it is not possible to produce organs which in combination can approach the efficiency of the piston unit, in spite of the ability of the turbine to provide complete expansion.

The most successful rotary compressors yet built achieve at best only about 80% of the efficiency of their piston counterparts, owing to the inherent aerodynamic difficulty of controlling by high-speed rotary means the flow of light fluids. The pressure ratios that can be used are restricted both by the limitations of the compressor and also indirectly by the inability of the turbine blading and rotor to withstand for sufficiently long periods the combination of high temperature and high centrifugal stresses that are involved in the subsequent expansion process.

The net useful output of the plant is the difference between the power developed by the turbine and the power absorbed by the compressor, and it is interesting to note that the net 1489 kW of the Brown-Boveri gas turbine locomotive, corrected for losses in electric transmission, is the difference between the turbine output of about 7457 kW and the compressor consumption of nearly 5970 kW. (See *Proc. I. Mech. E.,* Vol. 141, No. 3, and *Automobile Engineer,* Vol. 32, No. 6.) It will be readily realised that a reduction of 10% in efficiency of both units would eliminate this net difference, and the early days of gas turbine development involved a succession of failures in which the output of the turbine proved insufficient to drive its own compressor. Thus fundamental aerodynamic difficulties in compressor design and material limitations in the turbine have resulted in efficiencies so low as to make the specific fuel consumption of the most successful gas turbine units greater than the average steam plant and far above the best diesel engine practice.

On the other hand, there are many actual and prospective advantages which fully justify expenditure on research and development.

In the case of jet propulsion there are certain favourable factors which are not realised in industrial applications. The low air temperature at high altitudes results in the turbine blading being exposed to less destructive temperatures than would normally arise. The time period during which the blades are so exposed is less, both on account of load factor and the limited endurance that is required or expected, fuel economy is not a first consideration, and finally in the pure jet unit the useful power output is not subject to turbine blading losses as is evident in the case of power delivery by shaft.

18.5   Gas turbines for road transport

Here the situation is much less favourable, in spite of the ultimate advantages of lightness, simplicity and compactness, absence of necessity for a cooling system and negligible oil consumption, all added to the much acclaimed advantage of direct production of rotative effort. This latter may even prove to some extent an embarrassment owing to the extremely high rotational speeds involved relatively to road wheel speeds, necessitating a large gear reduction.

18.6   Essential characteristics of turbine prime movers

When the energy of expansion of high-temperature gases is expanded in giving kinetic jet energy to the gases themselves, gas velocities of several thousand feet per second are generated, depending on the temperature range of the expansion, which in turn is limited by the maximum temperature to which the turbine blades may be continuously exposed.

The efficient extraction and utilisation of this energy involves correspondingly high speeds of movement of the parts operated upon, whether these parts are the blades of a turbine or the vehicle on which a propulsive jet reacts.

As with the Pelton water-wheel, which is the simplest form of impulse turbine, for maximum efficiency the blade speed should be about half the jet speed, and with single-stage impulse wheels peripheral blade speeds of 350 to 365 m/s are now normal practice in gas turbine work.

These speeds may be reduced by using multiple stages, and a gas turbine impulse wheel may be provided with two rows of blades with stationary deflector vanes between them. With this arrangement the gas velocity is extracted in two steps, and the peripheral blade speed may be about one-third of the initial gas velocity. Even so, since the power normally required for automobile work can be obtained from wheels of 152 to 203 mm diameter, rotational speeds of 40 000 to 50 000 rev/min will have to be provided for in design. The high rotational inertia consequent on such speeds leads to troublesome time-lag in response to throttle control.

A turbine is a high-speed, unidirectional prime mover capable of handling very large volumes of fluid flow with high mechanical efficiency – actually of the order of 85 to 90% if correctly designed – and its torque will increase with a decrease of speed for given input; but this will be accompanied by loss of efficiency, since the fixed angles of the blading are correct only for certain relative speeds of jet and blades.

18.7 Automotive power unit

For automobile work is becomes necessary to use either electric transmission, which is expensive and heavy, or a separate work turbine independent of the turbo-compressor shaft. This power turbine is geared to the road wheels, while the compressor unit can be run up to speed with the car stationary. The exhaust from the compressor turbine would thus develop a fluid drive effect on the power turbine and exert on it a starting torque. This arrangement of the power unit is shown in simple diagram form in the upper diagrams of (*a*) and (*b*) in Fig. 18.2. A single combustion chamber CC feeds the compressor turbine CT and the power turbine PT in series. The lower diagrams show two combustion chambers in parallel, so providing independent control for the power turbine. Diagrams (*a*) show the simplest possible layout for each system, while diagrams (*b*) indicate the use of heat exchangers HE which improve

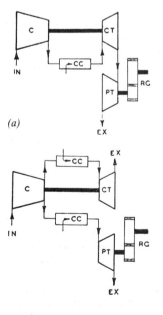
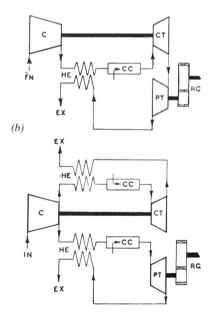

Fig. 18.2

thermal efficiency by transferring some of the waste heat in the exhaust to the combustion air between the compressor and the combustion chamber. RG is the reduction gearing between the turbine and the propellor shaft, a very large ratio being required.

18.8 Fuel consumption

Low thermal efficiency remains the great stumbling-block. Unless a heat exchanger is used, the specific fuel consumption will probably be twice that of the average petrol engine and three times that of the diesel, though possibly of a cheaper fuel. While the commercial user will assess lower capital and maintenance costs as a set-off against higher fuel consumption and cost, this aspect will appeal less to the private motorist, whose accounts rarely include depreciation and interest.

It will, moreover, be many years before a low selling price can be realised, even with reduced production costs, owing to the necessarily very high cost of development work.

All things considered, the gas turbine for the private car can be regarded only as a somewhat remote prospect.

18.9 Heat exchangers

Heat exchangers are essential in large plants to reduce fuel consumption, but their bulk and weight are objections in automobile applications. Much theoretical and practical research is being devoted to improve their construction and performance, and the ultimate position of the turbine power plant in the automobile field will depend vitally on these developments.

Heat exchangers for the application under discussion may be classified into two types. First, the static recuperative type, in which the heat passes by conduction from one fluid stream to the other through plate and tubular surfaces. This type is familiar in a great variety of domestic and industrial applications. In relation to performance the type is bulky and heavy. Secondly, the regenerative, or change-over, type in which the two streams alternately deposit in, and extract the heat from, the refractory material, which in large industrial applications usually consists of a nest of fire bricks, through which the streams are directed in turn.

For light power applications attention is now concentrated on the development of compact revolving units, in which matrices of a suitable material in mesh form are carried in a drum revolving at 20 to 30 rev/min, the two streams passing continuously, each through an appropriate portion of the rotor profile. Each portion of the matrix picks up and deposits heat as it passes from one stream to the other. The Ford research organisation has published information in regard to a rotor construction in which a dual band of stainless steel strip is wound between a 101.6 mm diameter hub and an outer drum about 508 mm in diameter. The drive is through a small pinion engaging a toothed ring on the periphery of the drum.

(See *Engineering Details of Ford Rotary Heat Exchanger for Gas Turbine Engines*, by W. Wai Chao, Section Supervisor, Combustor and Heat Exchanger Section, Scientific Laboratory, Ford Motor Co. Abstract of a Paper presented by the Author at the 17th Annual Meeting of the American Power Conference in Chicago. *Automotive Industries*, 15th May 1955, p. 54.)

The strips, 0.0508 mm thick, are alternately plain and corrugated, giving an xtremely high surface : volume ratio, and a number of flow passages estimated at a quarter of a million. This ensures that the matrix is an extremely efficient means of transfer of heat for its bulk and weight. This is known as the *flame trap system*.

Clearly, means of pressure sealing are required where the rotor profile passes through the boundary wall separating the two streams, and formidable construction difficulties call for solution.

18.10   Turbine developments

The Rover Company achieved striking success in the 1963 and 1965 Le Mans 24 hours race with their Rover–BRM sports car, the first turbine-powered car to compete in this race.

In 1963 no regenerator was fitted, and the fuel consumption was naturally high at 2.445 km/l, but power output and mechanical reliability enabled the car to finish eighth at an average speed of 173 km/h.

In 1964 a heat exchanger of the regenerator type, incorporating ceramic drums, had been developed, but time for reliability tests was limited, and it was decided not to enter for the race.

The race of 1965 saw a remarkable step in fuel economy, achieved by the use of the heat exchanger referred to above, though unfortunately it was necessary to limit performance owing to a mishap to the blading of the compressor turbine early in the race. Nonetheless the car was the leading British entrant, and was placed tenth in a field of 51 of which only 14 finished. At an average speed of 158.3 km/h the consumption was at 4.77 km/l – little more than half the 1963 figure. This was with the 2S/150R gas turbine.

An informative illustrated article will be found in *Automobile Engineer*, Vol. 55, No. 9.

The Firebird II Chassis, which has been developed by General Motors, is powered by the GT-304 Whirlfire turbine unit which incorporates a rotary regenerative-type heat exchanger.

All the basic components were under test in combination, and some reliable forecast of eventual efficiencies should be possible. Improvement in aerodynamic efficiency of the compressor and turbine units will be a slower and less spectacular long-term advance, though since final power output is the net difference between positive and negative powers of the same order, relatively small improvements result in considerable gains.

Metallurgical advance and manufacturing techniques determine cost and endurance rather than thermodynamic efficiency.

The general layout of the GT-304 Whirlfire unit is shown diagrammatically in Fig. 18.3. This specially drawn schematic illustration is based on information given in one of two papers, describing the Firebird II chassis and its power unit, by four members of the research staff of General Motors and published in *J. Soc. Automotive Eng.*, Vol. 64, 1956. The present authors are much indebted to the above Society for permission to make this brief extract.

The general layout consists of an accessories section incorporating a gear-driven hydraulic pump for powering the hydraulic motors which drive

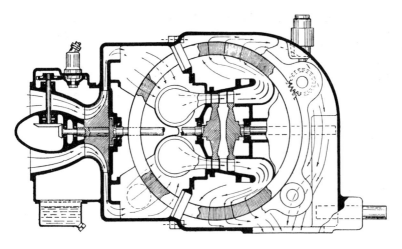

Fig. 18.3 General Motors' GT-304 Whirlfire unit

the rotary regenerators, fuel injection control and lubricating apparatus, and other auxiliary units.

The gasifier section comprises a single-stage centrifugal air compressor delivering to a plenum from which the compressed and warmed air passes radially through a symmetrical pair of revolving-drum-type regenerators. The heat picked up by the matrix of each regenerator as it passes through the hot exhaust stream is imparted to the air as it passes to the four symmetrically-placed 'can'-type combustors. From the combustors the high-temperature gases pass through the nozzle ring mounted in a dividing bulkhead to the turbine chamber. The velocity acquired in the nozzle ring is extracted in the two separate single-stage impulse turbines in series, the power turbine discharging the hot exhaust to a deflector ring which directs the stream to pass radially outwards through the regenerators. The matrices are thereby heated and the cooled exhaust passes out at the bottom of the casing.

The driving pinions, supporting rollers and floating seals where the regenerator drums pass through the dividing bulkhead, are indicated. The construction of the seals is the subject of intensive research and development, and little information is available.

Reduction gearing is provided to the power output shaft, whence a long transmission shaft leads to the final gearing at the rear of the chassis. The normal speeds of the compressor and power turbines are 35 000 and 28 000 rev/min respectively, and the regenerators revolve at 20 to 30 rev/min.

18.11 Ford power unit

Automobile Engineer, Vol. 49, No. 8, and Vol. 59, No. 12, contain descriptions of a gas turbine power unit under development by the Ford Motor Company of America. This unit, known as the 704 Model, exemplifies successful attack on the problem of high fuel consumption by novel arrangement of the various units, giving a somewhat complicated assembly.

Two-stage compression, with intercooling, is employed, the compression ratio of each stage being 4 : 1. The secondary, or high-speed compressor runs at the very high speed of 91 000 rev/min as compared with 46 500 rev/min of the low speed unit. There are also two combustion chambers, the first interposed between the heat-exchanger and the high-speed inward flow radial turbine, from which the gases pass to the secondary combustion chamber where a further supply of fuel is introduced. The exhaust products then pass to the main axial turbine which drives the low speed compressor and the output shaft. The arrangement of the components is shown diagrammatically in Fig. 18.4 with approximate gas temperature at the various stages.

The weight is stated to be 294.84 kg for a power output of about 224 kW, or about one-quarter of the weight of a corresponding diesel engine. The overall dimensions are about 965.2 mm long, 736.6 mm high and 711.2 mm wide. A wide variety of fuels can be used, with a consumption of only 0.294 to 0.353 kg/kWh according to load.

18.12   Chrysler turbine car

In 1963 the Chrysler Corporation achieved limited production of a completely new turbine-driven car which they then supplied to selected customers in order to obtain user experience.

The basic arrangement of the power plant follows the generally accepted system of two mechanically independent turbine wheels of the single-stage impulse type, through which the gas stream passes in series. They are axial flow wheels, the front one driving the single-stage radial compressor with an extension shaft serving the auxiliaries.

Of the two, the rearmost is the power turbine. This is larger than the

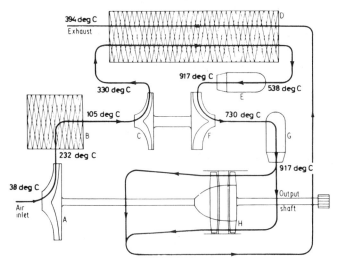

A	Low-speed compressor	E	Primary combustion chamber
B	Intercooler	F	Radial in-flow turbine
C	High-speed compressor	G	Secondary combustion chamber
D	Heat exchanger	H	Axial-flow turbine

Fig. 18.4 Ford 704 Model gas turbine power unit

compressor wheel, and is provided with a novel arrangement of variable angle guide vanes which is one of the many interesting features of the power unit. The direction of the gas stream on to the turbine blading may be varied from maximum power, through economy and idling to a braking position. The operating gear is one of the hydraulic auxiliaries all fed from a single pump.

A heat-exchanger system consisting of a symmetrical pair of rotary regenerators of the multicellular type is employed. These regenerators are about 381 mm in diameter and driven by spur reduction gearing from the compressor shaft at 9 to 22 rev/min. Designed for axial flow, the compactly arranged and well-insulated ducting leads the air flow from the compressor volute through the front halves of the regenerator wheels, where it is pre-heated, to the single combustion chamber mounted underneath the main casing. The housing of the outer diametral seal of the left-hand regenerator may be noted in Fig. 18.5. No details of the sealing strips are available.

As it is injected the fuel burns with excess air, and the combustion products pass first, through fixed guide vanes, to the compressor turbine, and thence through the variable angle guide vanes for the second stage of expansion in the power turbine.

The hot exhaust is ducted to the rear halves of the interchangers where the

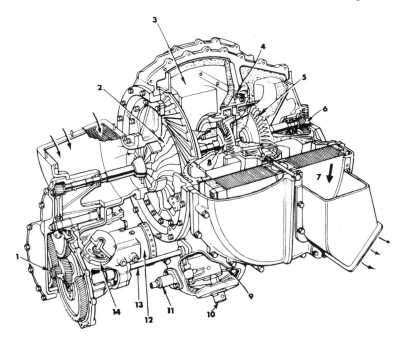

1	Accessory drive	8	Gas generator turbine
2	Compressor	9	Burner
3	R/h regenerator rotor	10	Fuel nozzle
4	Variable nozzle mechanism	11	Igniter
5	Power turbine	12	Starter-generator
6	Reduction gearing	13	Regenerator drive shaft
7	L/h regenerator rotor	14	Ignition unit

Fig. 18.5 Chrysler turbine unit

wheels absorb heat and transfer it to the contrary air stream from the compressor. A pair of aluminium alloy ducts leads the exhaust to the rear of the car.

In the part-sectioned pictorial view, Fig. 18.5, is shown the general arrangement, and many of the details can be seen.

The heat exchangers and ducting are skilfully and compactly arranged, and indeed a great part of the total bulk is due to the auxiliaries and their drives, and the front air intake and cleaner.

Fig. 18.6 gives torque and power plotted against output shaft speed, and the valuable feature of high torque at low speed will be noted.

The power turbine may be stalled to zero speed, but will continue to develop full torque, as the compressor unit will run independently.

An informative article giving particulars of transmission and automatic controls, and further structural details, will be found in *Automobile Engineer*, Vol. 53, No. 12.

18.13   Leyland gas turbine

In March 1967, the Rover team engaged in gas turbine work was regrouped to form Leyland Gas Turbines Ltd. They have developed a new, larger unit developing 260–300 kW intended for commercial vehicles. This engine is the type 2S/350/R and in design it is basically similar to the 2S/150 R, mentioned in Section 18.10. A minimum fuel consumption of 0.2373 kg/kWh at 20°C is quoted. At 26.7°C, the mass flow of air is 1.7 kg/s. The compressor idling speed is 19 000 rev/min, and the output speed range at full load corresponds to a power turbine speed range of 0–32 500 rev/min. This power unit, including ducts and auxiliaries, weighs less than 454 kg.

Its radial-flow impeller is approximately 230 mm in diameter and is produced in two portions. The front part, comprising the inlet guide vanes, is a steel casting, while the remainder is an aluminium alloy forging. Both are shrunk on to the shaft, on the rear end of which is the axial flow turbine. The latter is pressed on to a spigot and its hub bolted to the shoulder on the shaft.

The compression ratio of 4 : 1 is relatively low. At the inlet to the compressor turbine, the maximum temperature of the gas is more than 1000°C. A ratio of

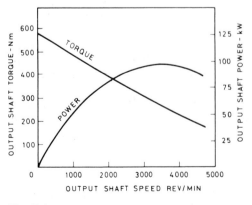

Fig. 18.6

2:1 is quoted for the maximum output torque of the power turbine at stall to that in the maximum power condition.

Provision has been made for a very large exchange of heat between the gas leaving the power turbine and the air flowing from the compressor. The heat exchanger is basically similar to that of the earlier model. It comprises two 711-mm diameter ceramic discs disposed vertically, one each side of the engine, and rotated at 1/1800 of the speed of the impeller shaft, from which they are driven. This heat exchanger is of the rotary regenerator type, and the temperatures quoted are as follows: air inlet 180°C, air outlet 700°C. The pressure loss across each disc is only $2.70 \, kN/m^2$. The exhaust gas is ducted to the rear half of the inner faces of the discs and the air goes through the front halves.

Each disc is driven by a pinion meshing with teeth machined on a peripheral steel ring. The pinion is driven by means of a worm and wormwheel on a transverse shaft. A full description of this turbine was published in *Automobile Engineer*, December 1968.

18.14 Gas turbine prospects

With the maximum permitted loading of vehicles likely to become something of the order of 44 tonne, and a legal requirement of something between 8 and 10 hp/tonne under consideration, engines of about 400 hp (300 kW) will be required for commercial vehicles. For this output, the gas turbine is worthy of serious consideration and, because of its advantages already outlined, it is virtually certain to come into use for heavy commercial vehicles. Another favourable factor is the widespread development throughout the world of motorway networks on which relatively constant speeds of operation for long periods are practicable.

At present, the biggest disadvantage of the gas turbine is the bulk of its heat exchanger. It seems likely, however, that this problem will be overcome when materials better able to withstand loading at high temperatures are developed. There are great hopes for ceramic materials, which currently are being intensively developed for nozzles, blades and perhaps discs. When such materials are available, a higher degree of compression and higher temperatures at the entry to the turbine will be feasible, though the formation of NO_x, polluting the exhaust, might present a problem. Later it might be possible to use multi-stage turbines to get even more work out of the expanding gas, so that a heat exchanger will be unnecessary. This will entail the use of both turbines and compressors of the axial flow type. In other words, the development of the road vehicle gas turbine engine might follow on the lines of that of the aircraft version. In fact, turbine designers have been talking of using rotors of the order of 96 mm diameter. In these circumstances, such an engine could be suitable for the private car.

A good insight into some of the problems associated with the design and development of new materials can be gained from an article on the Ricardo Keramos project, *Automotive Engineer*, Vol. 2, No. 2, April/May 1977.

18.15 Stratified-charge engines

With the rapid increase in crude oil prices over the period 1973–4, attention was drawn to the fact that oil reserves could be exhausted shortly beyond the turn of the century. Suddenly, fuel economy seemed even more important than control of exhaust emissions. In fact, the search for both economy and emission control – especially of unburnt hydrocarbons and carbon monoxide, generally abbreviated to HC and CO respectively – has resulted in intensified efforts to find ways of burning weak mixtures in spark ignition engines. After all, one of the reasons for the relatively high thermal efficiency of the diesel engine at part load is the excess of air over fuel in the charge.

At this stage, it is important to understand the implications of varying the air:fuel ratio. The air:fuel ratio required in theory for complete combustion is approximately 14:1, and this is called the *stoichiometric ratio*, or $\lambda = 1$. Spark ignition engines in general develop maximum power, however, at about 10% air deficiency, $\lambda = 0.9$, but of course the exhaust is then contaminated with HC and CO. The reasons for the higher power output are: first, a slightly higher volumetric efficiency, because of the cooling of the charge by the latent heat of vaporisation of the extra fuel; secondly, the greater amount of heat generated because of the virtual total combustion of all the oxygen in the charge and the lower peak temperatures, and therefore less dissociation – reverse chemical reactions that occur at very high temperatures.

When excess air is supplied, not only are the peak temperatures higher but also the power output is reduced because combustion is slower. When λ rises above about 1.25, or falls below about 0.4, ignition of the charge by the spark can no longer be relied upon. The optimum ratio for producing the cleanest possible exhaust in a conventional engine is about $\lambda = 1.1$. At weaker mixtures, the higher temperatures and slower combustion – and therefore longer dwell at those high temperatures – tend to cause the production of oxides of nitrogen. With very weak mixtures, on the other hand, temperatures fall off again.

Although mixtures weaker than $\lambda = 1.25$ cannot be fired by an ignition source as small as a spark, they can be by a flame. Consequently, if an ignitable mixture can be fed directly to the sparking plug points, it will not only fire locally but also the flame will spread throughout the remainder of the charge even though it is much weaker. The weakest overall fuel ratios that have been used successfully in stratified charge engines are of the order of 64:1, but for best fuel economy and lowest possible emissions, the optimum probably lies between about 20:1 and 25:1.

This has provided the incentive for the development of stratified-charge engines. The earliest work in this field, with the aim of reducing overall fuel consumption, was in 1915 by the late Sir Harry Ricardo. Otto, however, in his Patent No. 532 covering his invention of the four-stroke cycle, proposed the admission of only air at the beginning of the induction stroke, to form a layer next to the piston, for cushioning the effect upon it of the explosion of the subsequently entering combustible mixture.

18.16 Single-chamber versions

Some of the earliest work in this field was done using carburetted mixtures. An example is that of the Société Français du Pétroles, Fig. 18.7. The aim was to feed a rich mixture, through a separate pipe, past the inlet valve and to generate a swirl in the cylinder so that, with the weak mixture entering in the normal way also through the inlet valve, a vortex comprising alternate layers

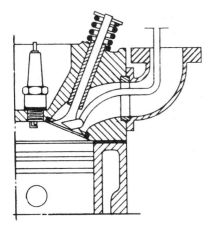

Fig. 18.7

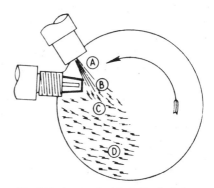

Fig. 18.8 A, atomised fuel; B, air–fuel mixture zone; C, flame front; D, products of combustion

of rich and weak mixtures was formed in the cylinder. Since the rich portion of the mixture was directed initially over the sparking plug points, it could be ignited and, on burning, ignite also the weak mixture.

At much the same time, the Texaco TCCS engine, Fig. 18.8, was being developed on a similar principle, but using direct injection into the cylinder to produce the layer of rich mixture swirling past the sparking plug points. Much later Mitsubishi actually put into production another injected version, but with a bowl-in-piston combustion chamber, Fig. 18.9. Another example of this type is the Ford Proco, Fig. 18.10. These arrangements give better control over the swirling charge, especially since turbulence can be increased by use of the squish effect between the piston crown and cylinder head.

With all these single combustion chamber engines, the principal difficulty is

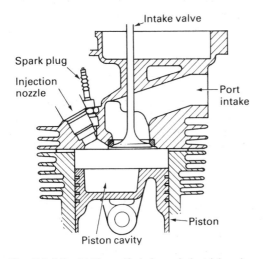

Fig. 18.9 Mitsubishi stratified-charge industrial engine

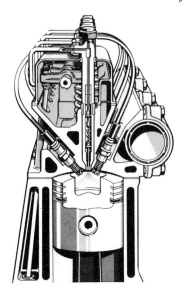

Fig. 18.10 Cross-section of Ford Proco engine showing bowl-in-piston combustion chamber, special injection system and dual ignition

the control over swirl and stratification throughout the range of loads and speeds. Consequently, the Mitsubishi engine is basically an industrial power unit, for running mainly at constant speed, or over a relatively narrow speed range. A major difference between the Texaco and Mitsubishi systems is that the former has early injection, while in the latter it is late. The advantage of late injection in such an engine is that control over stratification is more precise.

By early injection we mean 60° or more before TDC. Late injection is affected as the piston approaches TDC. With early injection, most of the fuel is evaporated before ignition, so combustion is similar to that in a carburetted engine. Late injection, on the other hand, entails combustion on the same principle as that in a diesel engine, though with spark ignition. That is, it occurs as the droplets evaporate. The rate of swirl is such that, when the mixture is ignited, the flame front cannot advance upstream towards the injector nozzle, which it would foul, but spreads only downstream.

The MAN FM engine, originally a diesel unit, was further developed for running as a multi-fuel engine and, later, for stratified charge. As can be seen from Fig. 18.11, fuel is injected directly towards the opposite wall of a bowl-in-piston combustion chamber. Fuel spreading in a thin film over the wall, is evaporated by the air swirling in the chamber to form the stratified charge. This charge is fed progressively to the combustion centre, which of course originates at the sparking plug points. These points are extended down in a groove in the wall of the chamber, so that the spark occurs in the layer of rich mixture. In one version of this engine, a single-electrode plug was used, the spark jumping the gap between it and the piston itself.

18.17 Dual-chamber versions

The main subdivisions of categories of stratified charge engines are those with single combustion chambers, as just described, and those with dual chambers.

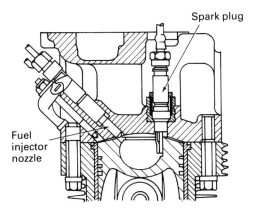

Spark plug

Fuel
injector
nozzle

Fig. 18.11 MAN FM engine as a stratified-charge unit

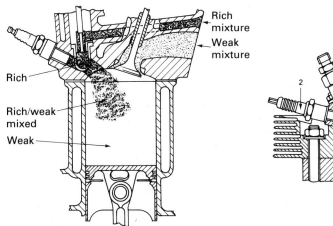

Rich
mixture

Weak
mixture

Rich

Rich/weak
mixed

Weak

Fig. 18.12 Honda CVCC engine,
carburetted, at end of induction stroke

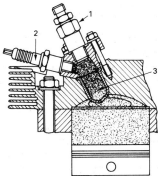

Fig. 18.13 The Porsche SKS

Both can be further divided into carburetted, and early and late injection. The dual-chamber versions are in general similar to indirect injection diesel engines, in that they have a small pre-chamber, generally in the cylinder head, but of course they are spark ignited.

A dual-chamber engine supplied with carburetted mixture is the Honda CVCC unit, Fig. 18.12, which was the first stratified-charge engine to go into production for a car. The Porsche SKS engine, Fig. 18.13, is early injected and the Ricardo experimental engine with a Comet Mk VB, Fig. 18.14, is late injected. A carburetted main chamber and injected pre-chamber engine, the Daimler-Benz DB-TSC, is shown in Fig. 18.15. In each of these examples, a rich mixture is supplied to the pre-chamber where it is ignited, discharging flaming gas into the main chamber to ignite the weak mixture there. Some of these engines have a third poppet valve, for scavenging the pre-chamber, while others do not.

Generally, the pre-chambers are designed for one of two totally different combustion processes. The Honda CVCC, for example, has a pre-chamber

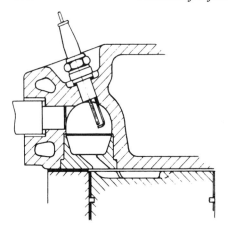

Fig. 18.14 Ricardo Comet Mk VB diesel combustion chamber adapted with special sparking plugs for stratified-charge operation

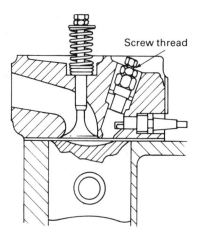

Fig. 18.15 Daimler-Benz DB-TSC engine

with a short but relatively large diameter throat. Consequently, when the combustion and consequent expansion of the rich mixture in the pre-chamber forces the burning gases out, the flame spreads progressively over an ever-widening front through the weak mixture in the main chamber.

With the second system, the throat of the pre-chamber is longer and narrow, generally with sharp edges at both ends. As a result, the stream of burning gases passes through in a turbulent state and at high velocity. In these circumstances, as they pass through the throat, the rate of heat transfer – gases to metal – is so high that they are partially quenched. As they issue into the weak mixture in the main combustion chamber, the stream is torn into a cascade of tiny eddies, each of which becomes a centre of ignition of the weak mixture. This is because much of the ejected rich mixture of partially quenched gases, even though there is no longer any visible flame, is in the ionised state of the initial stages of combustion and therefore extremely unstable and highly chemically active. The result is a rapid spread of combustion throughout the weak mixture in the main chamber.

Similar combustion phenomena appeared to have been observed in the Austin-Rover experimental engine, Fig. 18.16. This is a three-valve carburetted unit, the third valve being for scavenging the pre-chamber. Ciné photographs showed no visible burning in the jet until it was slowed and spread in the proximity of the opposite wall of the cylinder. In this case, however, it was subsequently discovered that, by improving the photographic technique, burning could be observed in the jet.

The effect of sparking plug position is as follows. If it is at the end of the pre-combustion chamber remote from the throat, the combustion kernel, growing from the centre of the spark, rapidly ignites the gas at that end, forcing rich unburnt gases out through the throat ahead of the burning gases. This presumably happens in the Porsche engine.

In the Austin-Rover engine, the plug is at an intermediate point in the pre-chamber and, moreover, directly opposite the throat. Consequently a

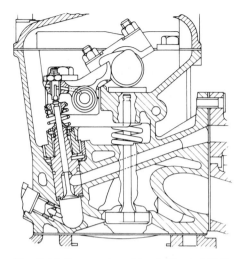

Fig. 18.16 Cross-section of Austin-Rover 1.85-litre stratified-charge cylinder head

much higher proportion of the rich mixture will be ignited before it passes into the throat, so one would expect more active ionised particles to be ejected into the main chamber.

An example of late injection into a pre-chamber, without a scavenge valve, is the Ricardo Comet system. As the piston comes up to TDC compression stroke, air from the main chamber is forced through a tangential throat to generate rapid swirl in the pre-chamber. The fuel is injected against the direction of swirl – tending to reduce its speed locally – but straight at the plug points, which extend to a position somewhere near the centre of the vortex. In this position, the linear velocity of the swirling gases is in any case relatively low, so the flame kernel is able to develop and spread throughout the pre-chamber without risk of its being quenched before it has time to do so.

That there are so many different types of stratified-charge engine is due to the fact that we are still in the early stages of the development of the principle. Ultimately, one or two forms will no doubt emerge as the most suitable, and the others will be dropped. An especially promising unit is that invented by Charles Goodacre, and developed in co-operation between him and Fiat, described in the August/September 1981 issue of *Automotive Engineer*.

An entirely different system, which might be regarded as a compromise between the single- and dual-chamber combustion chamber arrangements, but in which the emphasis is on the use of a high compression ratio, is the high-turbulence, high-compression, lean burn system adopted by Jaguar and described in Section 4.18. Ford, Porsche and Ricardo, too, have done a lot of work on such systems.

18.18   Catalytic ignition

In 1987 a power unit called MCC (Merritt Catalytic Compression-ignition) engine was invented by Dr Dan Merritt, of Coventry (Lanchester) Polytechnic. This is a petrol engine, though with tolerance to a very wide range of volatile fuels and even gases. It is very different from a diesel engine with catalytically assisted ignition, which tends to generate black smoke owing to the rapidity of ignition of fuel contacting the catalyst before it has had time to evaporate fully and mix intimately with the air. This problem does not arise with the MCC engine.

Although a platinum catalyst satisfactorily initiates afterburning in exhaust emission control systems, the problem in the engine itself is principally one of preventing the fuel from coming into contact with the catalyst too early in the cycle. An essential requisite for success, therefore, is application of the following principles—

(1) Segregation in a fuel-management cylinder, i.e. away from the catalyst, of an air:fuel mixture so rich that it cannot be ignited simply by the degree of compression to which it is likely to be subjected in the engine.

(2) Complete purging of a combustion chamber (something like that of an indirect injection diesel engine but cylindrical in shape) during the compression stroke.

(3) The formation of a rapidly rotating vortex of air in the combustion chamber and thorough and rapid dispersion of the very rich mixture into it, by directing two streams of gases in through two slot-shaped ports, one from the main and the other from the fuel-management cylinder, both ports producing ribbon-like jets.

(4) Ignition timing kept constant, by relating it to the position of the piston in the fuel management cylinder.

(5) Initiation of ignition (in an overall mixture whose air:fuel ratio is continuously varying over a wide range) by passing a rich stratum over the catalyst.

(6) This is to be followed by continuous catalytic assistance of combustion.

Dr Merritt applied these principles by constructing a single-cylinder prototype similar to that illustrated in Fig. 18.17, but with pushrod-actuated instead of overhead valves. Had compactness been an overriding requirement, the minor pistons could have been actuated by cams. However, with such an arrangement it would not be possible for the pressure generated by combustion to do useful work on the minor piston.

For production engines, on the other hand, a wide variety of layouts might be preferable. One would be the offset, side-by-side cylinder arrangement with a single crankshaft, Fig. 18.18, or a V-layout with horizontal minor cylinders interposed between the two banks of main cylinders. For large industrial, marine or railway locomotive applications, the delta, Fig. 18.19, or even a square layout, again without any extra crankshafts for the minor cylinders, could be attractive.

During the induction stroke the two pistons move away from TDC but, while that in the main cylinder draws its air past the inlet valves, that in the minor cylinders draws its air through the combustion chamber, thus starting the process of purging it of residual gases. When the piston in the minor cylinders uncovers the hole through which the fuel must be sprayed in the form of a fine mist, through a special nozzle, injection can start. The fuel pressure is between 6 and 8 bar, and the duration of injection is such as to balance the power output against the load, so there is no throttle valve in the main air intake. However, injection must be terminated by the time the minor piston,

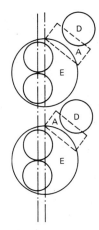

Fig. 18.18 Four-in-line layout
for MCC engine

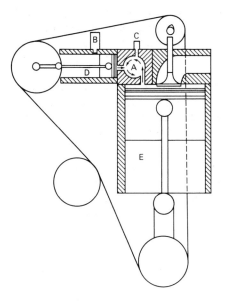

Fig. 18.17 Single-cylinder layout for MCC engine

coming up to TDC compression, once more covers the spray hole. By this time, the mixture in the minor cylinder contains too little oxygen to be ignited by compression at TDC. Heat from the residual burnt gas helps to vaporise the fuel, but even more help could be obtained by directing the spray against the hot wall of the combustion chamber, which closes the end of the minor cylinder.

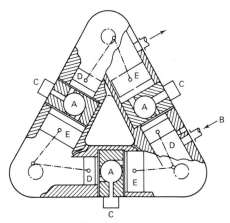

A	Combustion chambers	D	Minor, or fuel-management, cylinders
B	Injectors	E	Main cylinders
C	Glow-plugs for cold starting assistance		

Fig. 18.19 Three-cylinder delta layout for MCC engine

Four design features prevent the minor piston from injecting any of the fuel vapour into the combustion chamber before it approaches TDC. One is appropriate phasing of the crankshafts. Another is arranging for the air to enter the combustion chamber tangentially, thus generating rapid swirl and centrifugally generating a pressure gradient opposing entry of gas from the minor cylinder. Thirdly, the piston-swept volume of the main cylinder is about five times that of the minor one. Finally, the effect of the latent heat of vaporisation is to reduce the pressure in the minor cylinder.

Throughout most of the compression stroke, therefore, air is being forced through the combustion chamber into the minor cylinder, thus completing the purging process. Tests have shown that when, just before TDC, the minor piston ultimately injects the very rich mixture into the combustion chamber, bringing it into contact with the catalyst, and ignition is initiated without any discernible delay even at temperatures and pressures significantly lower than that necessary for spontaneous ignition.

The platinum catalyst, the total cost of which is about \$1 per litre piston-swept volume, can be deposited to a thickness of about 20 μm, directly on the wall of the chamber or on, for example, a stainless steel cylinder, plate or fine wire mesh subsequently to be inserted into the chamber. An advantage of any of the latter arrangements is that the catalyst more easily becomes incandescent, and burns off any carbon that might have become deposited on it. However, the retention of such devices securely in the combustion chamber can be difficult.

The slot, or throat, through which the rich mixture passes may be directed radially at the axis of rotation of the vortex or air or at such other angle that may be found to be most effective. For instance, the mixture might be directed against the direction of swirl to generate turbulence. Alternatively, it could be slipped tangentially, in the direction of swirl, in between the vortex and the catalyst-coated cylindrical wall of the chamber, to form a Swiss-roll type of stratification.

When the fuel contacts the catalyst, combustion proceeds very rapidly, because the rapid rotation of the vortex continually brings any remaining unburnt fuel into contact. Ideally, vaporisation of the fuel should be completed before it is injected into the combustion chamber. In practice, however, it can continue in that chamber. Adjustment to the timing of the initiation of combustion can be effected at the design stage by appropriate phasing of the minor and main crankshafts.

During the power stroke the gases expand into the two cylinders and the work they do on both pistons is translated into a torque on the main crankshaft. Ultimately, the exhaust valves open and the exhaust stroke proceeds until they close again, shortly after TDC.

This engine appears to have remarkably good prospects, for the following reasons. First, it is easy to convert existing engines, so the developing need not be costly. Its advantages by comparison with a diesel engine are compactness and light weight. Fuel economy may be better, because combustion occurs at virtually constant volume and therefore more work can be extracted and less heat lost to coolant than in the diesel cycle. Moreover, since the engine is run on petrol, there is no risk of wax formation in the fuel in cold weather.

The noise and initial and maintenance costs of heavy high-pressure injection equipment are obviated. With catalytic ignition, the compression ratio can be selected solely on the basis of overall efficiency and low noise: it will probably be about 16:1, instead of having to be higher for ease of cold starting, as with a diesel engine. The prototype accelerates smoothly, without any adjustment of ignition timing, from about 300 to 4500 rev/min and, in theory, it ought to be possible by further development to take it on up to perhaps 6000 rev/min. Consequently, the gearbox needed to keep such an engine operating in its economical range would be lighter and less costly than that for a diesel power unit.

For military applications, a two-stroke version could be developed. The main alteration needed would be the delaying of injection until after the inlet valve has closed. A major benefit would be a much higher power output relative to both bulk and weight.

Compared with petrol engines for cars and light commercial vehicles, the major advantages are elimination of electrical ignition equipment and independence of octane number. Indeed, the engine has been run without any trace of knock on not only liquid butane but also a mixture of 10% either in low octane unleaded petrol. Consequently, the geometric compression ratio would not have to be lowered for supercharging such an engine. Since it is not throttled, the efficiency at light load is much higher than that of a conventional spark ignition engine and the overall fuel consumption should be at least as good as, and probably better than, that of a diesel unit. Because very weak mixtures can be burned, it might be possible to avoid the use of catalytic converters in the exhaust system.

In general, compression ratios of the order of 16:1 and the absence of both diesel knock and petrol detonation would imply lighter and therefore less bulky structures. Whether this would be outweighed by the extra mechanical complexity and additional friction remains to be seen. However, the benefits do appear to outweigh the disadvantages.

18.19 Stirling engine

The Stirling engine, invented by Robert Stirling, first built in 1816 and subsequently produced on a small scale, is not an internal combustion power unit. Its working gas is cycled in a closed circuit, passing through a heat exchanger on the way round. Gases such as hydrogen, helium and freon have been used in the closed circuit. Originally, it was a viable alternative to the steam engine, for example in marine propulsion, but it has yet to be proved competitive with the internal combustion engine for road vehicle applications. However, it could become attractive owing to its virtually zero oil consumption and long intervals between oil changes, long service life, relative silence, a thermal efficiency potentially of about 40 to 45% at part load, acceptance of a wide variety of fuels in a continuously burning heater, and a very clean exhaust. Its disadvantages are complexity, bulk and weight. The specific weight of a 10 kW engine is about 10 kW/kW, but becomes lower as the power output increases.

It operates on a four-stage working cycle. One type of Stirling engine is shown diagrammatically in Fig. 18.20, which shows what is termed the *displacer* type. This has the disadvantages that each cylinder contains two pistons, and it can be built up into a multi-cylinder unit only by, in effect, bunching together a group of single-cylinder engines: it cannot be simplified by

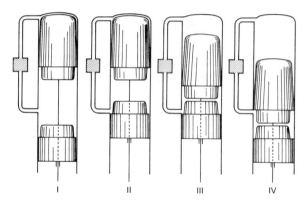

Fig. 18.20 Showing the four stages in the Stirling engine cycle, in a piston displacer-type of unit

integration. There is, however, a less bulky double-acting, or *piston-displacer*, type, which will be explained later.

In the first type, the upper piston is the displacer and the lower one the working piston. The four stages in the working cycle, Fig. 18.20, are—

(1) Both pistons at the outer extremes of their travel, and the working gas in the cold space between them.
(2) The working piston has compressed the gas at a low temperature, while the displacer piston has remained stationary.
(3) The displacer has now descended and pushed the gas through a cooler, regenerator and heater into the hot space above it, while the working piston has remained stationary.
(4) The hot gas has expanded and pushed both the displacer and working pistons down to their lowest positions. Thereafter, the working piston remains stationary while the displacer moves up again, pushing the gas back through the heater, regenerator and cooler into the cold space once more, until stage 1 is repeated and the cycle begins again.

The mechanism by means of which the pistons are moved in this manner is called a *rhombic drive*, Fig. 18.21, in which the two crankshafts are geared together to rotate in opposite directions. Although the motion is complex, the diagram is self-explanatory. With such a mechanism, all the inertia forces can be balanced. The function of the cooler is to reduce the volume of the gas entering the cold chamber which, because of the intrusion of the piston rod, is inherently of smaller volume than the hot chamber. The regenerator takes heat energy from the exhaust and transfers it to the gas on its way to the heater which, of course, supplies additional heat energy to the gas, to increase the total available for conversion into work in the expansion chamber.

There is also a two-cylinder layout, or multiples of two cylinders with single-acting pistons, in the closed circuit of which the gas passes between a cold chamber over one piston to the hot chamber over the piston in the other cylinder, and back again. This of course has a conventional crankshaft, as well as a similar cooler, regenerator and heater arrangement.

However, the double-acting, or piston-displacer, Stirling engine, again with

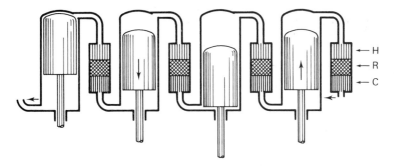

Fig. 18.21 Diagram of the double acting, or piston-displacer, type of engine: H is the heater, R the regenerator and C the cooler

only one piston per cylinder, can be made simpler and more compact than any of the other layouts. With two or more cylinders and double-acting pistons, the cold chamber can be below and the hot chamber above the piston in each cylinder. Again, these chambers are interconnected in a closed circuit, through a cooler, regenerator and heater, in an arrangement represented diagrammatically in Fig. 18.22. The cycle of operations is of course similar to that already described.

Since the bulkiest and most complex component of a Stirling engine is its regenerator, burner and heater, there is much to be gained by placing these in one sub-assembly serving all the cylinders. This can be done by bringing the cylinder heads close together by adopting either a narrow angle V-layout or a swash-plate-type engine having its cylinders arranged around and parallel to a central shaft, Fig. 18.23. As the swash-plate rotates, it moves the pistons alternately up and down.

In general, the sealing of piston rings in Stirling engines is not difficult. This is because there is no combustion in the cylinders, the maximum temperatures are relatively low and, for any given mean pressure, the peak pressure is much lower than in an internal combustion engine. Furthermore, there is no side loading, since this is taken by the crosshead. Sealing of piston rods, however, does pose problems, various seals from a complex sliding seal assembly to a rolling sock type having been used.

A combustor, rotary recuperator and heater unit, developed by Ford for a Stirling engine installed experimentally in its *Torino* car, is illustrated in Fig. 18.24. The use of ceramics or other materials resistant to very high temperatures in this area could lead to an increase in thermal efficiency from its current maximum of about 35% to approximately 40%.

Since a Stirling engine rejects about twice as much heat to coolant as an Otto cycle unit, its cooler has to be correspondingly larger. In all other respects, it is in effect a conventional radiator. Even so, the sheer bulk, weight and cost of such a large radiator is a major disadvantage.

The combustor is not unlike that for a gas turbine engine, comprising a thimble-like casing within which is a cylindrical combustion chamber having large perforations in its peripheral walls. Mounted in its closed end are an electric igniter and a fuel atomiser jet. Air enters the annular space between the casing and the combustion chamber, and passes radially inwards through the

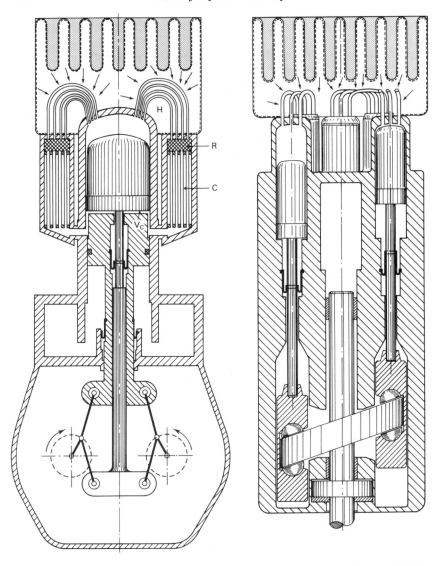

Fig. 18.22 The rhombic drive of a piston-displacer unit having an experimental head operating on the heat-pipe principle. Heat applied to the finned top evaporates liquid sodium from its porous lining, which then condenses on the tubes H containing the working gas. The regenerator is R and the cooler C

Fig. 18.23 A swash-plate-type Stirling engine with heat-tube type head

perforations, to mix with the atomised fuel. The burning gases expand axially along the combustion chamber and pass from its open end into the centre of a cylindrical heat exchanger. This comprises a set of finned parallel tubes arranged in a ring coaxial with that of the burner assembly, but mounted directly on top of the cylinder block. The hot gases from the combustion chamber pass radially outwards between the tubes, inside which the working gas is flowing.

Now let us look at the path of the air from the time that it enters and ultimately, becoming the burned exhaust gas, leaves the unit. The burner assembly is contained coaxially in an even larger cylindrical housing, on opposite sides of which are the air intake and exhaust ports. As can be seen from Fig. 18.24, the incoming air first passes through a rotary heat exchanger, which is actually the regenerator, the principle of which is described in Section 18.9, in connection with gas turbines. The function of the regenerator is to take from the exhaust gases heat that would otherwise go to waste, and to use it to preheat the air immediately before it enters the combustion chamber.

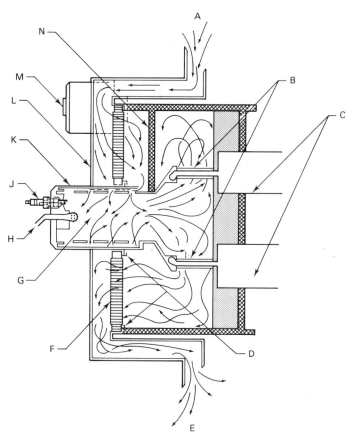

A	Air inlet	H	Fuel atomiser
B	Heater tubes	J	Igniter
C	Cylinder heads	K	Combustion can
D	Regenerator seals	L	Main housing
E	Exhaust	M	Motor
F	Rotary regenerator	N	Baffle
G	Combustion zone		

Fig. 18.24 Combustor and regenerator assembly for the Ford *Torino* engine

After passing through the upper segment of the rotating heat exchanger disc, the air enters the open end of the housing of the burner, through which it proceeds, as previously described, through the burner to the heat exchanger through which the working gas is flowing. From the periphery of this heat exchanger, the gas passes back through the lower segment of the rotary heat exchanger to which it gives up some of its heat. It is then released through the exhaust port.

There are two control systems. One is a closed-loop control over the fuel supply. This is regulated by a cylinder head temperature sensor, and maintains the heater tubes at a constant temperature. The other is for control over power output. This can be done by pumping working gas into and out of the closed circuit. Alternatively, what is termed *dead volume control* can be employed to open the system to one or more fixed volume chambers, Fig. 18.25. For automotive applications a more suitable method, giving more rapid response and better part throttle efficiency, would appear to be to vary the angle of the swash-plate.

A typical indicator diagram for a Stirling engine is illustrated in Fig. 18.26. Note its large area. The performance of such an engine is similar to that of a petrol unit, but its thermal efficiency is higher under all conditions of load and speed, and the torque curve is significantly flatter. Consequently, there might be potential for economy by simplifying the transmission. Inevitably, however, the cost would be of the order of at least 8% above that of an ic engine.

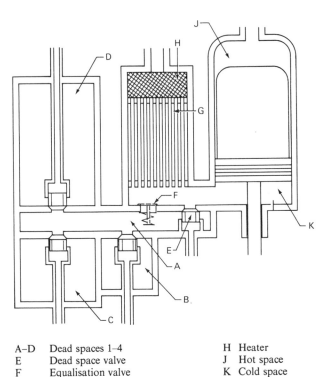

A–D	Dead spaces 1–4	H	Heater
E	Dead space valve	J	Hot space
F	Equalisation valve	K	Cold space
G	Cooler		

Fig. 18.25 The dead volume control system

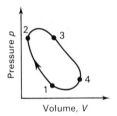

Fig. 18.26 Typical Stirling indicator diagram, with phases from Fig. 18.20

In 1958 General Motors became a licensee of Philips of Einhoven, who had been developing the engine since 1938. Then, in 1968, both United Stirling, of Sweden, and MAN/MWM, of Germany, also became licensees. By 1971, Philips had installed a rhombic drive Stirling engine in a DAF bus, after which Ford, working in conjunction with Philips, installed a double-acting swash-plate type in their *Torino* car. For information on further developments such as the use of a heat accumulator, a heat pipe and a free piston version, the reader is advised to refer to the *CME*, May 1979, and the *Philips Technical Review*, Vol. 31, No. 5/6, 1970.

Chapter 19

Bearings, gearing, chain and belt drives

When a shaft has to revolve it must be supported in bearings that will allow it to rotate. The simplest bearing is just a cylindrical hole formed in a piece of material and in which the shaft is a free fit. The hole is usually lined with a brass or bronze lining, or *bush*, which not only reduces the friction in the bearing but also enables an easy replacement to be made when wear occurs. Bushes are usually a tight fit in the hole in which they fit. To reduce the friction to the lowest amount, ball-and-roller bearings, in which 'rolling' friction replaces the 'sliding' friction that occurs in 'plain' bearings, are used.

Considering the shaft shown in Fig. 19.1, it will be seen that two bearings are provided. This is almost universally done because otherwise the single bearing would have to be of an inordinate length. It will also be observed that the bearings shown are capable of withstanding only loads that are perpendicular to the axis of the shaft. Such loads are called *radial* loads, and the bearings that carry them are called *radial* or *journal bearings*. That part of a shaft that actually lies within a bearing is called a *journal*. To prevent the shaft from moving in the axial direction shoulders or collars may be formed on it or secured to it, as shown in Fig. 19.2, which shows the loose collar secured by a grub screw P. If both collars are made integral with the shaft then obviously in order that the shaft may be put into position the bearings must be split or made in halves, the top half, or *cap*, being put on and bolted in place after the shaft

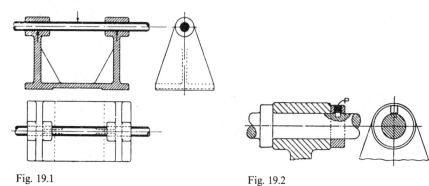

Fig. 19.1   Fig. 19.2

has been put in. The collars or shoulders withstand any end thrusts, and bearings that do this are termed *thrust bearings*.

Figs 19.3 to 19.8 show various forms of ball-and-roller bearings. Fig. 19.3 shows a single-row journal or radial ball bearing. The inner race B is a ring of steel case-hardened (provided with a hardened surface layer but having a soft centre or core) with a groove or track formed on its outer circumference for a number of hardened steel balls to run upon. The outer race A is another ring which has a track on its inner circumference. The balls fit between the two tracks, thus enabling the outer race to turn relatively to the inner one, the balls meanwhile rolling round the tracks. The balls are kept from rubbing against each other by some form of cage. Such a bearing, as its name *radial* implies, is meant to withstand chiefly radial loads; they will usually, however, withstand a considerable amount of axial thrust and are often used as combined journal and thrust bearings. Both the inner and the outer races are then secured in the end direction between nuts and shoulders or by other means. When the bearing is used simply to take radial loads one of the races is left free axially. Some journal-thrust ball bearings can take thrust loads only in one direction; care has therefore to be taken when assembling these to ensure that they are properly placed to take the thrust load.

Fig. 19.4 shows an SKF self-aligning ball bearing. The outer race is ground on the inside to form part of a sphere whose centre is on the axis of the shaft. Hence the axis of the inner race and shaft may be displaced out of parallel with that of the outer race without affecting the action of the bearing. Two rows of balls are used. Such bearings are desirable where the deflection of a shaft or frame cannot be limited to a negligible amount. They are often used in back axles.

An example of a parallel-roller journal bearing is shown in Fig. 19.5. The tracks on the races A and B are now cylinders and cylindrical rollers take the place of the balls. Roller bearings can withstand heavier radial loads than equivalent-sized ball bearings, but obviously the one shown cannot withstand any end thrust. By providing lips or shoulders on both inner and outer races a parallel-roller bearing can be made to withstand small end thrusts and this is sometimes done. If, however, the tracks and rollers are made conical, giving a taper-roller bearing, then large end thrusts can be taken in one direction. A taper-roller bearing is shown in Fig. 19.6. It may be pointed out that the angles

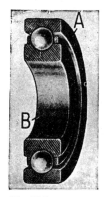

Fig. 19.3

Fig. 19.4

Fig. 19.5

of the cones forming the races and the rollers have to be such that when the bearing is assembled all the cones have a common apex lying on the axis of the shaft.

Roller bearings in which the length of the rollers is equal to several times their diameter are called *needle-roller bearings*. Often these have no cage, the rollers completely filling the space between the inner and outer races. An example is seen at H in Fig. 26.2.

Fig. 19.7 shows a *single thrust bearing*. The tracks are now formed on the inner faces of two discs, A and B, so that an end thrust can be carried. In Fig. 19.8 a *double thrust race* or *bearing* is seen, in which two sets of balls are provided so that thrusts in either direction can be taken; the races A and C abut against shoulders or nuts on the revolving shaft while the central race B is fixed in the stationary housing. The balls may be carried in a cage, but whereas this is almost universal with journal bearings, thrust bearings are often not provided with cages. Thrust bearings are not capable of taking radial loads; these, therefore, have to be taken by journal bearings.

From the point of view of the user the most important point concerning ball-and-roller bearings is the necessity for absolute cleanliness. Dirt, dust, particles of metal and water will quickly destroy them. In machining, the holes in which outer races fit should be truly cylindrical and the outer race should usually be a fairly free fit. The inner race, which is normally the revolving one, should, on the other hand, be a tight fit on its shaft.

Greases containing animal fats must not be used in ball bearings as the acids usually contained in these lubricants destroy the races and balls. Most oil companies put up greases which are suitable for ball bearings and their recommendations should be accepted.

In general, unless bearings are overloaded as a result of bad design or gross overloading of the vehicle, or are exposed to water or dust, they will give no trouble and may easily outlast the rest of the vehicle.

19.1   Types of toothed gearing

Toothed gearing, which is used to transmit motion of rotation from one shaft to another, assumes various forms according to the relative positions of the shafts. When the shafts are parallel the gears are called *spur gears*, and are

Fig. 19.6

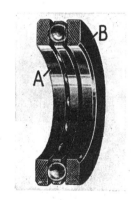

Fig. 19.7

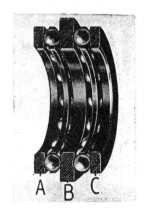

Fig. 19.8

Fig. 19.9   Fig. 19.10

cylinders with teeth cut on the outside or the inside according to whether they are external or internal gears. Examples of such gearing are shown in Figs 19.9 and 19.10 respectively. In these cases the teeth are seen to be cut parallel to the axes of the gears. Such teeth are called *straight* to distinguish them from those shown in Fig. 19.11 which, being parts of helices, are called *helical teeth*. Owing to the inclination of helical teeth there is an axial or end thrust on each gear tending to separate them axially. These end thrusts have to be resisted by suitable thrust bearings. The end thrust is avoided in *double helical gearing,* Fig. 19.12, where the thrust from one half of a tooth is counteracted by that from the other half. In all the forms of spur gearing the forces between the teeth in contact tend to separate the gears radially. These forces have to be taken by the journal bearings of the shafts.

When their axes intersect, shafts are connected by *bevel gearing.* Bevel gears are cones with teeth cut upon them. The teeth may be straight, giving *straight-toothed bevel gears,* as shown in Fig. 19.13, or spiral, giving *spiral bevel gears,* Fig. 19.14. With straight-toothed bevel gears, since the forces between the teeth in contact tend to separate the gears both radially and axially, thrust bearings must be fitted behind each gear. When spiral teeth are used the axial forces on the gears may tend to separate them or to make them mesh more deeply; it depends upon the angle of the spiral, whether it is right handed or left handed, and upon the direction of rotation. With spiral bevel gears, therefore, it is usual to fit double thrust bearings.

When the shafts are neither parallel nor intersect they may be connected by either *skew, worm* or *hypoid gearing.* A pair of skew gears is illustrated in Fig. 19.15. These are usually cylindrical, being then identical with helical toothed spur gears, but one of them is sometimes hollowed out like a 'worm wheel' so that it embraces the other, as this gives a better contact between the teeth and results in an increased efficiency. If the axes of the two gears are at right angles, they are usually termed *spiral gears.* Examples are the distributor drives on most engines having low camshafts.

Fig. 19.11

Fig. 19.12

Fig. 19.13

Fig. 19.14

The ordinary type of worm gearing is shown in Fig. 19.18. The worm (at the top) may be considered as part of a screw having a number of threads. If the wheel is considered to represent a nut it will be seen that if the worm is prevented from moving in the axial direction and then it is rotated, the worm-wheel also will have to revolve. This type is known as *parallel worm gearing*, the worm being a parallel screw. There is another type in which the worm is shaped to embrace the wheel, instead of being parallel. It is known in its general application as the *Hindley, hour-glass* or *globoidal* type, but in its application to motor cars as the *Lanchester worm*, since Dr Lanchester developed it for that application. In both types of worm gearing there are both radial and axial forces acting on the worm and the wheel tending to separate them. The directions of these forces change when the direction of the motion is reversed and, besides journal bearings, double thrust bearings have to be fitted to both worm and wheel. The efficiency of worm gearing depends to a very great extent on the quality of the materials used, the accuracy and degree of

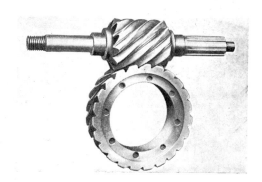

Fig. 19.15   Fig. 19.16

Fig. 19.17

finish and on the lubrication. When these factors are right the efficiency is very high (commonly about 95 to 99%).

The efficiency of good spur gearing is also very high, bevel gears are slightly less efficient and skew gears least efficient of all.

An example of hypoid gearing is shown in Fig. 19.17, from which it will be seen that the chief difference between it and spiral bevel gearing is that the axis of the pinion is placed lower than that of the crown wheel so that the axes of the gears do not intersect. If desirable the pinion axis may be placed above the crown wheel axis. There are two chief advantages claimed for the hypoid gear over the spiral bevel, greater quietness and greater strength; it is possible to make the pinion larger than that of spiral bevel gears having the same gear ratio and same size of crown wheel. If this increased strength is unnecessary then the hypoid gears can be made smaller than the corresponding spiral bevel gears. There is a sliding action in the direction of the length of the teeth of hypoid gears that is absent in bevel gearing and this makes lubrication of the teeth a more difficult matter so that special lubricants frequently have to be used with hypoid gearing.

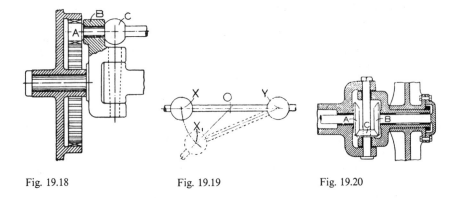

Fig. 19.18   Fig. 19.19   Fig. 19.20

19.2 Gear ratio of toothed gearing

When two shafts are connected by toothed gearing of any of the types described, the ratio between the speeds of those shafts will always be constant. This constant ratio is called the *gear ratio*. In all the types of gearing considered when one wheel drives another the gear ratio between them will be the *inverse* ratio of the numbers of teeth in the wheels. Thus if a wheel A drives a wheel B then—

$$\frac{\text{Speed of A}}{\text{Speed of B}} = \frac{\text{Number of teeth in B}}{\text{Number of teeth in A}}$$

and the bigger wheel goes slower than the smaller one. This will be clear when it is realised that in any given time as many teeth of the wheel A as of the wheel B pass the point P where the teeth mesh together. Now suppose the speeds of the wheels are S_a and S_b rev/min and numbers of teeth N_a and N_b. Then the number of teeth of A that pass the point P in one minute is $N_a \times S_a$ while for B the number is $N_b \times S_b$ and these are equal, thus—

$$N_a \times S_a = N_b \times S_b \quad \text{or} \quad S_a/S_b = N_b/N_a$$

The rule applies to worm gearing, the number of teeth in the worm being equal to the number of threads or starts it has. The number of starts is easily seen by looking at the worm from the end.

19.3 Chain drive

Chain drive is beginning to be reintroduced into transmission lines, mainly as a means of dropping or raising the drive line from one level to another. This requirement is especially liable to arise between the engine crankshaft and gearbox in front-wheel-drive cars and in the transfer box for a four-wheel-drive vehicle.

The advantages include freedom of choice of centre-to-centre distance between the shafts, wider tolerance permissible than for gear drives and, because of the relative flexibility and inherent damping of a chain, it takes up the drive more smoothly than gears. It is also significantly less noisy than a gear drive, especially after wear has taken place in service. A point liable to be overlooked, too, is that simple bearings can be used for chain sprockets whereas, with helical gears, both radial and axial loads have to be reacted by the bearings. Finally, not only are sprockets easier to machine than gears, but also a chain drive is more efficient.

Obviously, the most economical arrangement is a single chain, but if the torque is too high for this to be practicable, a choice has to be made between two or three separate strands and sprockets and duplex or triplex chains. The last two mentioned are more compact but, since there is a clearance between the intermediate plates and the bearing pins, the fatigue life of the plates, and hence of the whole chain, is lower than that of a two- or three-strand drive with simple chain. Two or three strands on staggered teeth also introduce significantly less resonant vibration and noise. The employment of separate strands of chain does not necessarily entail separate chain wheels, since J. Parkinson & Son (Shipley) Ltd, a subsidiary of Renold Ltd, has developed a special technique for shaping as many as three rows of staggered teeth on a single wheel. Chain tensioners are dealt with, in connection with camshaft drives, in Chapter 3.

Another choice that has to be made is between roller and inverted tooth chain. The roller chain is lighter, less expensive and, for a given weight and life, has a higher torque transmission capacity. Maintaining the correct tension with a roller chain is easy, simply by applying a jockey sprocket or simple leaf spring, or a sprung or oil-pressure-loaded shoe to the outer face of the slack run. With an inverted tooth chain, however, these methods cannot be used because it will flex in only one direction, that is, around the sprockets.

An inverted tooth chain, on the other hand, may be the best choice for low speed, high torque applications, because the inverted tooth design has a greater number of plates per unit width, giving increased torque capacity. Moreover, for such applications, the extra weight and inertia can be tolerated. This type of chain has also been used in torque converter transmissions, where, by virtue of the absorption of fluctuations in the torque by the converter, the lower inertia and silence of roller-type chain under this type of loading may not be required. An example is the Saab range of cars. The automatic versions have an inverted tooth chain, while the manual ones are fitted with separate strands of roller chain and a hydraulic tensioner which serves also as a damper.

19.4 Belt drives

An alternative to chain drive in transmission systems is the toothed belt, pioneered by UniRoyal. Originally trapezoidal teeth were used but by the time the original patents had expired, in 1963, UniRoyal had patented a similar belt but with tooth profiles resembling those of gears, and terming it the High Torque Drive (HTD) belt. An even later development was the incorporation of either another set of teeth or, alternatively, a set of multi-V grooves on what would otherwise be the plain outer face of the belt, and then calling it the Twin Power HTD belt. Thus combinations of positively and non-positively driven and contrarotating pulleys can be served by a single belt.

Advantages of the toothed belt drive include high mechanical efficiency, mainly owing to the absence of slip, coupled with its inherent flexibility; quiet vibration-free running; reduced fatigue loading of the structure and wear of the bearings carrying the pulleys; no lubrication is required, obviating maintenance and the need for a sealed cover for retention of lubricant; and, finally, only a light tension is needed in the belt, so the journal bearings and shafts are correspondingly lightly loaded. The Twin Power HTD belt has the additional advantages that extra drives can be taken from its outer face and, at the same time, they can be used to tension the belt, which obviates the need for a separate jockey pulley.

Where silence and positive drive are important, the toothed-belt drive is the obvious choice. It is also used where relative timing – synchronisation – between driving and driven wheels is essential – a V-belt drive of course is not suitable for such applications.

Recent developments with V-belts have increased both their wear resistance and load-carrying capacity. Nevertheless, as a transmission medium, they are suitable only for auxiliary drives. Variants range from the single- to multiple-belt drives with multi-V pulleys, or the integrated banded belt which is, in effect, a number of V-belts joined at their outer edges to form a flat belt with a multi-V section driving face.

On grounds of drive efficiency, the V-belt comes off worst, its principal merits being compactness, simplicity and low cost. Even when properly adjusted its efficiency is likely to be little better than about 94%. Wear and stretch in service, allowing slip to increase, can substantially reduce this value. Largely because of the total absence of slip, the toothed belt efficiency is of the order of 98%, and this is maintained continuously. In general, toothed belt efficiency increases while that of the V-belt decreases with rising torque. The efficiencies of both types, however, decrease as the size of pulleys is reduced. Speed has relatively little effect on efficiency. The efficiency of a roller chain or inverted tooth drive can be about 98% too.

Part 2

Transmission

Chapter 20

Transmission requirements

From the foregoing chapters, it can be seen that the internal combustion engine, as used in road vehicles, has the following requirements and characteristics—

(1) To start it, some form of external energy must be applied.
(2) Its maximum torque is small compared with that of a steam engine or electric traction motor of the same maximum horse power.
(3) Its maximum power is developed at a relatively high speed – ranging from about 1700 rev/min for heavy commercial vehicles to 12 000 rev/min and more in racing cars.

In consequence, it must be used in conjunction with a transmission that differs in many respects from those of either steam- or electrically-powered vehicles.

Various methods have been used for starting engines – for instance, from energy stored in a spring or flywheel, or chemical energy in a cartridge – but the general rule of course is to use a battery-electric starter. To keep the starting system as compact – and therefore as inexpensive – as possible, provision must be made for disconnecting the engine from the drive line during the starting operation. Connecting it again to the drive line, for propelling the vehicle, must be effected as smoothly as possible, both for the sake of passengers and to prevent damage to the vehicle mechanisms.

In passing down the drive live, the torque of the engine is modified, stage by stage, until it becomes the propulsive force, or tractive effort, at the interface between the tyres and the road. If rapid acceleration is required, either when starting from rest or in any other circumstances – for overtaking, for example – that tractive effort must be increased. This is done partly by increasing the torque output of the engine but, since this alone may not be enough, the gear ratios will generally have to be changed too.

In this context, gear-changing can be likened to altering the leverage between the engine and the road wheels, so that the relatively small torque available can be translated into a large tractive effort. A large leverage may also be required for climbing hills or traversing very soft or rough ground.

Since a large leverage implies a correspondingly reduced movement at the output end, this implies a big reduction of rotational speed between the engine and the road wheels. Consequently, the leverage must be reduced as the speed of the vehicle increases, otherwise the engine speed would become too high and the maximum potential speed of the vehicle would be unattainable. Moreover, as explained in Chapter 1, the relationship between torque, power and the speed of rotation of the engine is such that torque falls off as speed increases. For this reason too, therefore, some simple means of varying the leverage – changing gear – is necessary.

Consider a car having road wheels 0.66 m diameter and cruising at 60 mph. Under these conditions, the engine speed would have to be about 3500 rev/min and the road wheels would be rotating at about 800 rev/min. Consequently, the overall ratio of the gearing between the engine and the road wheels would have to be about 4.5 : 1. In practice, this ratio would differ to some extent, depending on the weight of the car and size of the engine. For example, until recently a car with a large engine might have an overall ratio of 3 : 1, a medium weight commercial vehicle 5.5 : 1 and a heavy truck 10 : 1 or even higher. Now, the demand for fuel economy is tending to encourage the use of overdrive gearboxes, and overall gear ratios as high as 6 : 1 could become common on, for instance, cars with five-speed gearboxes.

While the basic principles of transmission remain the same for virtually all classes of road vehicle, the actual arrangements vary – for instance, some may have four-wheel drive and others either front- or rear-wheel drive. Where the engine is installed at the front and the axis of its crankshaft is parallel to, or coincident with, the longitudinal axis of the vehicle, ultimately, the drive must be turned through 90° in order that it may be transmitted out to the wheels the axes of which are of course perpendicular to that longitudinal axis. Such a turn, however, is not necessary if the engine is installed transversely, though other complications, such as a need for dropping the drive line to a level below that of the crankshaft while turning it through 180°, may arise.

Another requirement for the transmission stems from the fact that, when the vehicle is cornering, the outer wheels must roll faster than the inner ones which will be traversing circles of smaller radii, yet their mean speed, and therefore both the rotational speed of the engine and the translational speed of the vehicle, may be required to remain constant.

Then again, to reduce the transmission of vibrations to the chassis frame, the engine is universally mounted on it, while the driving wheels, attached to the frame by the road springs, also have a degree of freedom of movement relative to it. Both these movements must be accommodated by the transmission.

In summary, therefore, the requirements for the transmission are as follows—

(1) To provide for disconnecting the engine from the driving wheels.
(2) When the engine is running, to enable the connection to the driving wheels to be made smoothly and without shock.
(3) To enable the leverage between the engine and driving wheels to be varied.
(4) It must reduce the drive-line speed from that of the engine to that of the driving wheels in a ratio of somewhere between about 3 : 1 and 10 : 1 or more, according to the relative size of engine and weight of vehicle.
(5) Turn the drive, if necessary, through 90° or perhaps otherwise realign it.
(6) Enable the driving wheels to rotate at different speeds.
(7) Provide for relative movement between the engine and driving wheels.

There are several ways in which these requirements can be met, and transmissions fall into three categories—

(1) Mechanical.
(2) Hydraulic:
    (a) hydrostatic,
    (b) hydrodynamic.
(3) Electric and electromagnetic.

While the first of these is the commonest, combined mechanical and hydrodynamic transmissions are becoming increasingly popular, even on certain types of heavy commercial vehicle. Road vehicles with all-hydraulic transmissions have been built but have not gone into large-scale production. On the other hand, such transmissions are not at all uncommon in tractors for agricultural and, more especially, construction and similar industrial equipment, such as diggers, in which transitions from idling to full load may be required to be effected suddenly and repetitively. With a hydraulic pump driving separate motors on each wheel of a pair, there is no need for a mechanical differential gear.

The same advantage can be claimed for some electric transmissions in which a generator or battery supplies power to separate wheel motors. Electric transmissions have been used in tanks and some road vehicles in the past and are still extensively employed in diesel–electric locomotives. With the current revival of interest in electric road vehicles, they are being reconsidered for wider application. Mostly, however, both hydraulic and electric transmissions have some mechanical elements in common. For instance, the hydraulic component generally simply replaces the clutch and gearbox of a mechanical transmission.

Most mechanical transmissions fall into one of the following three categories, each of which has three main elements—

(1) Clutch, gearbox and live axle.
(2) Clutch, gearbox and dead axle.
(3) Clutch, gearbox and axleless transmission.

20.1 Clutch, gearbox and live axle transmission – general arrangement

This system is shown diagrammatically in Fig. 20.1. The engine is at the front, with its crankshaft parallel to the axis of the vehicle. From the engine, the drive is transmitted through a clutch and a short shaft (c) to the gearbox. In cars, this short shaft is almost invariably integral with the primary gear in the gearbox but, in some commercial vehicles, it is a separate component, generally with flexible or universal joints at each end and, in some instances, with a sliding joint at one end. From the gearbox, a 'propeller shaft' or 'cardan shaft' – also with a sliding joint at one end and a universal joint at both ends – takes the drive to a live back axle. A live axle is one through which the drive is transmitted, while a dead axle is one that does not transmit the drive. Bevel or worm gearing (g) within the axle turns the drive through 90°, and *differential gears* divide it equally between the two drive shafts, or *halfshafts* (j), which take it out to the wheels.

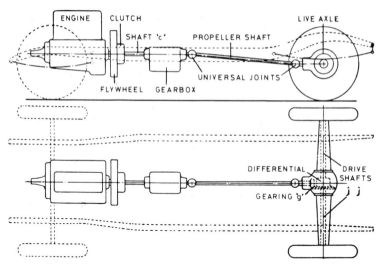

Fig. 20.1

The functions of the components are as follows. A clutch is used for disconnecting the engine from the driving wheels and it must also enable the driver to connect the engine, when it is running, without shock to the driving wheels. Since the clutch is kept in engagement by a spring-loading mechanism and is disengaged by pressure of the foot on a pedal, it cannot be disengaged except when the driver is in the vehicle. Therefore, when the driver wants to leave the vehicle with the engine running – and preferably for starting the engine, too – he has to disconnect the engine from the driving wheels by use of the gear-shift lever, which he sets in a 'neutral', or gears-disengaged, position.

The principal function of the gearbox is to enable the driver to change the leverage between the engine and driving wheels to suit the prevailing conditions – gradient, load, speed required, etc. As the propeller shaft transmits the drive on to the back axle, the universal joints at its ends allow both the engine-and-gearbox assembly and the back axle to move relative to one another, as their spring elements deflect. The sliding joint, usually integral with one of the universal joints, accommodates variations in length of the propeller shaft as its rear end rises and falls vertically with the back axle and its front end pivots about the universal joint just behind the gearbox. Gearing (g), in what is called the *final drive* unit, turns the drive through 90° and reduces the speed in a ratio of about 4:1, since the driving wheels must rotate much more slowly than the engine. Within the final drive unit too is the differential gearing, which shares the driving torque equally between the two road wheels while allowing them, nevertheless, to rotate simultaneously at different speeds while the vehicle is cornering.

In Fig. 20.1, the gearbox is shown as a separate unit, but two other variants of this front engine rear-wheel-drive layout are in use. One is a 'unit construction' – almost universal on cars – in which the gearbox casing is either integral with or bolted rigidly to the clutch 'bellhousing' which, in turn, is similarly secured to the engine crankcase. This has the advantages of cleanliness, lighter weight, neatness of appearance and lower manufacturing costs. Its chief disadvantage is relative inaccessibility of the clutch. In addition to accessibility of both the clutch and gearbox, the layout in Fig. 20.1 enables a shorter, and therefore lighter, final drive propeller shaft to be used, thus obviating potential problems associated with whirling and other vibrations.

The second variant entails incorporation of the gearbox into the back axle, to form what is now widely called a *transaxle* unit. For three main reasons, this arrangement is used only rarely: first, it tends to be significantly more costly than the others; secondly, it entails the use of either a dead axle or an axleless transmission as described in Section 5.3; thirdly, it is extremely difficult to mount a heavy transaxle in such a way as to accommodate the motions, torques, and forces of the input and output driveshafts yet to isolate it to prevent the transmission of noise and vibration to the vehicle structure.

With certain types of gearbox – notably epicyclic – the clutch action is performed within the gearbox itself, to which the drive from the engine is therefore transmitted either directly by a shaft or through a fluid coupling or torque converter. The shaft can be either separate or integral with the gearbox mechanism.

Live axles are built in several forms. That in Fig. 20.1 is called a *single-reduction axle*, because the reduction in speed between the propeller shaft and final drive is effected in one stage, in the final drive at (g). In some heavy trucks, because the reduction ratio may have to be much higher, this reduction is done in two or even three stages, a *double-* or even *triple-reduction axle* being used.

20.2 Layout of rear-engine vehicles with live axles

The rear-engine and live axle arrangement has advantages for buses and coaches, primarily because it allows the floor to be set at a low level and to be flat and clear throughout virtually the whole length of the chassis. In Fig. 20.2 the engine and gearbox are built as a single unit, which is installed transversely behind the rear axle. The clutch is interposed between the engine and gearbox, while at the other end of this box is a bevel gear pair termed the *transfer drive*.

To transmit the drive to the rear axle, the driven gear of the bevel gear pair is coupled by a universal joint to a relatively short propeller shaft, which is similarly coupled at its other end to the pinion shaft of the final drive unit. Obviously, the shorter the propeller shaft, the greater is the angle through which it has to swing to accommodate the relative movements of both the engine on its mountings and the axle on its springs. Therefore, the final drive unit is incorporated at one side of the axle, instead of near its centre. The drive is turned through much less than 90° from the propeller shaft at both its final drive and transfer drive ends, which simplifies the design of both pairs of gears. A difficulty with this layout is the accommodation of the long engine, gearbox and transfer drive within the overall width of the vehicle.

For this reason, several manufacturers have installed their engines longi-tudinally behind the rear axle. This layout has been adopted in Fig. 20.3, where the gearbox is mounted separately, in front of the axle. Because the universal joints on the coupling shaft between the engine and gearbox have to accommodate only relative movements due to deflections of the mountings and vehicle frame or structure – instead of movements of the axles – they can

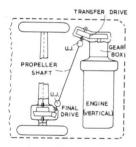

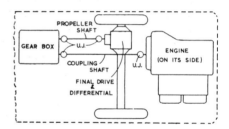

Fig. 20.2   Fig. 20.3

be of a simple type. Constant-velocity joints are needed, however, on the short propeller shaft.

In Fig. 20.4 again, separately mounted transverse engine and gearbox units are employed, but the differential can be nearer to the centre of the axle. Disadvantages of all rear-engine installations include the lengths of the control runs from the driving position and the fact that the driver may not be able to hear the engine and judge its speed, for changing gear, especially in noisy urban traffic. The latter problem does not arise, however, if automatic transmission is used.

Transverse rear-engine installations obviously call for an angle drive, and several transmission manufacturers produce such a unit. It generally consists of a bevel gear pair in a casing that can be bolted on to the gearbox or engine and clutch or torque-converter assembly. In Fig. 20.5, a retarder – transmission brake – is incorporated in the gearbox, which is bolted to the angle-drive casing. A coupling connects the angle driveshaft to the engine and clutch assembly in this MCW Metrobus installation. Some additional comments on front-, rear- and mid-engine installation for cars are given in Section 3.64.

20.3   Dead-axle and axleless transmission arrangements

An advantage of the dead-axle arrangement is a considerable reduction in unsprung weight, which improves both the ride and road holding. Incidentally, the suggestion that a rigid axle always holds its wheels perpendicular to

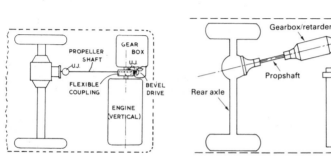

Fig. 20.4   Fig. 20.5

the road is clearly a fallacy: the effect of a bump under one wheel is to incline both wheels equally relative to the road.

Chain drive, Fig. 20.6, now rarely used, is one form of dead-axle transmission. The layouts of vehicles with chain drive are generally similar to those with transaxles, in that the gearbox and final drive unit are in one casing. So, from the engine and clutch, the drive is taken by a propeller shaft to the gearbox and ultimately, in the final drive G, is turned through 90° and divided equally between the two shafts J. At the outer ends of these shafts are the chain sprockets K, around which are the driving chains L for the chain wheels M, which are rigidly secured to the road wheels on the ends of the dead axle D.

Because there is a reduction in the speed of the road wheels owing to the different sizes of the chain sprocket and chain wheel, either a smaller reduction is needed in the final drive gear or, for example in very large tractors, an extra large overall reduction can be obtained. Relative movement between the axle and the frame, as the road springs flex, is accommodated by rotation of the loops of chains about the axes of the chain wheels and sprockets.

Another, and more commonly used, dead-axle layout is based on that employed by the de Dion Company in their early cars, Fig. 20.7. It is most suitable for low cars with firm suspension, and therefore is favoured by some sports car manufacturers. This is partly because of its inherent suspension characteristics, which will be elaborated upon in Section 35.17, and partly because it is more suitable for limited quantity production than the admittedly better, independent rear suspension systems.

The transmission arrangements for the de Dion axle layout and for the axleless systems with independent suspension are similar. From the engine, clutch and gearbox, the drive is taken through a propeller shaft to a final drive unit A, which is mounted on the structure of the vehicle, instead of contained within the axle. Having been turned through 90° and divided equally between

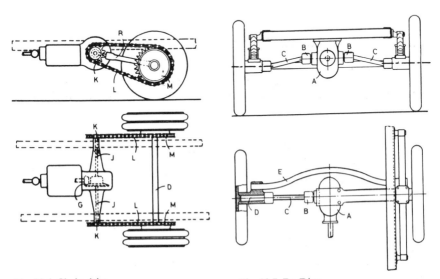

Fig. 20.6 Chain drive   Fig. 20.7 De Dion

two short drive shafts C, it is then transmitted out to the wheels, which rotate on bearings carried in brackets fixed to the outer ends of the dead axle E, or to the suspension mechanism if an independent suspension system is used. Movements of the road wheels relative to the vehicle structure are accommodated by universal joints B at both ends of drive shafts C, and the corresponding variations in lengths – or telescoping – of the shafts C are obtained by making them in two parts joined by some form of splined or sliding coupling. The springs interposed between the dead axle E and the structure of the vehicle seat on the ends of the dead axle, generally on the same bracket that houses the wheel bearings. A de Dion system can also be used with a so-called *transaxle*, the only difference being the interposition of the propeller shaft between the clutch and transaxle unit instead of between the gearbox and final drive unit.

The main advantages of the de Dion layout as compared with a live axle, are that it relieves the axle of the weight of the differential and final drive unit and the wheels remain in a fixed relationship to each other – either parallel or with a slight inwards inclination towards the top, to resist rear-end drift when the vehicle is acted upon by centrifugal force on cornering. At the same time, the troubles involved in the use of chains are avoided.

One disadvantage is the shortness of the drive shafts, and consequently large angles, through which they move, which entails the use of fairly costly universal joints. Another is the fact that, when one wheel alone rises, the contact points between the two tyres and the road move sideways, which momentarily produces a slight, but marked, rear-end steering effect and therefore adversely affects stability and handling.

When there is no axle, the road wheel connections to the frame and the springing arrangements are made in the various ways described in Chapters 36 and 37. With some of these layouts, it is possible to dispense with the outer universal joints and with the sliding coupling on the shafts C.

A front engine and rear-wheel-drive layout with independent suspension is shown in Fig. 20.8. Although the engine, clutch and gearbox unit and the separate final drive unit are all carried on the frame, or basic structure, of the vehicle, universal or flexible joints are still used at the ends of the propeller shaft. This is to accommodate the slight differential movement that can occur between these two units because of deflections of both their flexible mountings and the structure of the vehicle. The forces on the mountings for the final drive unit are by far the greatest problem, owing mainly to reaction from the final drive torque. Loading associated with vibration isolation and supporting the

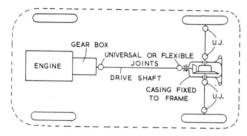

Fig. 20.8

weight of the unit under dynamic conditions is not nearly so severe and therefore calls for softer mountings.

An axleless transmission for a front-wheel-drive vehicle is shown in Fig. 20.9. This represents diagrammatically an installation in a car in which the engine is mounted longitudinally forward of the front wheels and the gearbox to the rear of them, the final drive being interposed between the two. The gearbox in this instance is of the all-indirect type, as described in Section 23.25. Consequently, its output shaft is below its input shaft, instead of the more common arrangement in which the two are in line. The pinion of the hypoid final drive unit is on the end of the gearbox output shaft and meshes with the crown wheel, which is bolted to the cage that carries the differential gears.

On each side the differential gear and crown wheel assembly is carried in bearings in the casing that houses the engine, clutch and gearbox unit. The very short shafts carrying the differential gears and projecting from this casing are coupled by universally-jointed drive shafts and sliding joints to the road wheels.

The transmission layout used in the Austin-Rover *Mini* – as designed by Sir Alec Issigonis – is illustrated diagrammatically in Fig. 20.10. In both the engine and the gearbox, the axes of all the shafts are transverse relative to the longitudinal axis of the car, so spur gears transmit the drive from the driven member of the clutch, downwards to the input shaft of the gearbox. Similarly, spur gears take the drive from the shaft at the output end of the gearbox to the final drive unit, which is positioned virtually on the longitudinal axis of the car. The gearbox is, in effect, in the engine sump, and the final drive unit immediately behind it but in the same casing. All three units share the same lubricant.

Both these front-wheel-drive layouts are suitable for small cars. That of the *Mini* is best of all though, because the whole of the space behind the final drive unit is available for unobstructed accommodation of the occupants. A disadvantage, in addition to the shortness, and therefore potentially wide angles swept by the driveshafts, is the inherently high stiffness of such a short drive line between the clutch and final drive. This may entail the incorporation of extra flexibility in the hub of the driven plate of the clutch, to obviate shock and harshness of take-up of the drive as the clutch is engaged. Another disadvantage is that of the final drive torque, which is that of the engine, multiplied by the gearbox and final drive ratios, has to be reacted by the engine mountings, or special arrangements must be made to react it in some other way.

For cars, the rear engine layout, engine behind or above the rear axle, has

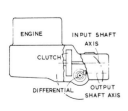

Fig. 20.9

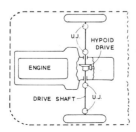

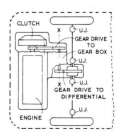

Fig. 20.10

fallen from favour. This is because of the associated instability due to poor weight distribution coupled with a rear suspension that is almost inevitably unfavourable owing to the small space available for it on each side of the engine.

Mid-engine installation is popular for sports cars, because it tends to give a reasonably uniform weight distribution between all four wheels, thus improving stability and, in the case of four-wheel drive, traction too. It entails the use of either of the front-wheel-drive layouts already described, but with the engine in front of, and driving, the rear axle. Major disadvantages are noise inside the vehicle, difficulty of access for maintenance, and encroachment of the engine into space that would otherwise be available for passengers at the rear.

For buses and coaches, the engine is in some instances centrally installed either vertically or horizontally beneath the floor, again for improved weight distribution. In such vehicles, however, because of their high floors, access is not such a problem, especially in the case of large horizontal diesel engines, the rocker covers and other components of which are easily accessible from one side. Access to the top of a vertically installed engine, through a hatch in the floor, tends to be less easy but can be made acceptable. Transmission drive-line layouts similar to those already described for front-engine vehicles in this category are employed.

So far as effectiveness of traction is concerned, front-wheel drive is better than rear-wheel drive, especially on difficult terrain, including ice or snow. This is partly because the weight of the engine on the front wheels enables them to grip the surface better, which of course also applies to rear-engine rear-wheel drive vehicles. Principally, however, the advantage is gained by virtue of the fact that the tractive effort is in all circumstances delivered along the line on which the front wheels are steered. Another factor is that front driven wheels tend to climb out of holes or ruts, whereas rear driven wheels tend to thrust the front wheels deeper down and, in any case, not necessarily in the direction in which they are steered.

20.4   Four-wheel-drive transmission

Typically in a four-wheel-drive transmission layout, Fig. 20.11, a transfer box is interposed between the gearbox and back axle unit. The function of transfer box is to transfer the drive from the main gearbox to both the front and rear axles.

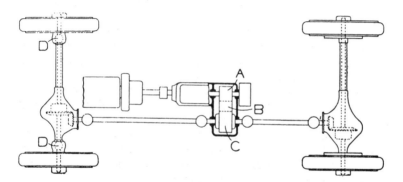

Fig. 20.11 General arrangement of four-wheel driven chassis

In this box is a pinion A, driven by a coupling from the gearbox output shaft. The pinion, through an intermediate gear B, drives a third gear C, mounted on the cage of a differential gear assembly. From the differential gears, one shaft is taken forward to the front axle and the other rearwards to the back axle. Both the axles house their own differentials and final drive gears, but that at the front carries at its outer ends the universal joints D, which are necessary to allow the front wheels to be steered.

The differential at C, in the transfer box, is necessary to distribute the drive equally between the front and rear axles and to allow for the fact that, when the vehicle is driven in a circle, the mean of the speeds of the front wheels is different from that of the rear wheels and therefore the speeds of the two propeller shafts must differ too. Other factors include different rolling radii of tyres owing to, for example, manufacturing tolerances, different degrees of wear and, perhaps, different tyre pressures.

Provision is usually made for locking this differential out of operation, to improve the performance and reliability of traction when the vehicle is driven on slippery ground. For vehicles intended mainly for operation on soft ground, the central differential may be omitted from the drive line, but some means of disengaging four-wheel drive, leaving only one axle to do the driving, is generally provided for use if the vehicle is required to operate on metalled roads.

Since, as can be seen from Fig. 34.2, the steered front wheels always tend to roll further than the wheels on the fixed-geometry rear-axle, because their radius of turn is always the larger, a one-way clutch, or freewheel (see Section 23.21) can be substituted for the inter-axle differential. In practice, this usually takes the form of two freewheels, one on each front hub, on which there are rotary controls by means of which they can be locked by the driver, but, of course, he has to stop and get out to do so. As soon as he again drives his vehicle on firm ground, however, he must remember to unlock the hubs. Should the rear wheels lose traction, on the other hand, and therefore tend to rotate further than the front ones, the drive will automatically be transferred to the front wheels even if they are in the freewheeling mode.

In four-wheel drive, the overall-frictional resistance throughout the transmission is higher than in the two-wheel-drive state, and of course the transmission shafts and gearing will progressively wind up and become highly stressed so long as the wheels on the two axles are rolling different distances. On soft ground, when this wind-up becomes too high the wheels can slip and thus relieve the stresses, but this may not be possible on metalled roads, and a fractured shaft may be the result.

With independently-sprung wheels, the transmission is the same, except that the final drive units are carried on the frame, or structure, of the vehicle. Consequently, universal joints have to be used on both ends of the driveshafts to the road wheels, as with the de Dion layout.

Four-wheel-drive offers two main advantages. First, there is the increased traction obtainable from four driven wheels, which is especially useful on soft or slippery ground. Secondly, if the front wheels drop into a ditch they tend to climb out, whereas with rear-wheel drive they tend to be forced downwards, except when the vehicle is driven in reverse, in which case, of course, the disadvantage of the lower traction of two-wheel drive remains. The principal disadvantages are increased weight, bulk and cost.

Chapter 21

Clutches

A clutch is a releasable coupling connecting the adjacent ends of two coaxial shafts. It is said to be *engaged* or, *in*, when the shafts are coupled, and *disengaged*, or *out*, when they are released. Mechanical clutches fall into two main categories: *positive engagement* and *progressive engagement*.

The former is either positively disengaged, so that no torque can be transmitted from the driving to the driven shaft, or positively engaged, in which case the shafts rotate together, connected by some mechanical devices such as splines, keys or dogs. In contrast, the progressive type is gradually engaged, so that the speed of the driving shaft falls while, simultaneously, that of the driven shaft rises from its initial stationary state until both are rotating at equal speeds.

Positive engagement clutches are unsuitable for connecting the engine to the gearbox, though they are used inside gearboxes. In this chapter, however, we are dealing with the interconnection between the engine and gearbox, so descriptions of this type will be left until later.

For road vehicles, a progressive engagement clutch of the *friction* type is interposed between the engine and the gearbox. As mentioned in Chapter 20, it can be either in an entirely separate housing, with a short shaft in front connecting it to the engine, and another behind to connect it to the gearbox. Alternatively, it can be integral with either unit, or indeed with both. In the latter case it is in either a single or a multi-piece housing, the pieces being bolted together to form a single engine-and-transmission unit. The housing for the clutch can be one of the pieces or, alternatively, it can be integral with either the engine crankcase or the gearbox housing.

21.1 Basic principle of the friction-type clutch

In road vehicles, the simplest friction-type clutch comprises two discs, the more substantial of which is usually the engine flywheel and the other, generally termed the *presser* or *pressure* plate, is lighter. The flywheel is bolted to a flange on the end of the crankshaft, while the other plate slides axially on the output shaft, except in as much as a spring or springs tending to press it against the flywheel. Such a clutch is engaged by its springs or springs, and disengaged by a pedal-actuated linkage under the control of the driver.

To illustrate the basic principles applicable to all progressive engagement clutches, a simple clutch stripped of all complications such as friction linings and actuation mechanism is shown in Fig. 21.1. The two plates E and F are keyed on the ends of shafts A and B, which are carried in bearings C and D. All rotate about a common axis XY.

Because the engine torque has to be transmitted from the engine flywheel to the presser plate by the friction between their two abutting faces, one may be faced with a specially formulated material having a high coefficient of friction, coupled with good wearing properties. However, a much better arrangement, and one that is virtually universal, is the interposition between the driving plates of a third, much lighter, disc, lined on both faces with the high friction material. This disc is free to float coaxially between the other two, and it is carried on a hub splined on to the shaft connecting the clutch to the gearbox. To distinguish it from the presser, or pressure, plates it is termed the *centre plate*, or *friction disc*. It has the advantages of, in effect, doubling the torque capacity of the clutch and halving the speed of rubbing during progressive engagement, thus increasing the life of the friction material.

21.2   Torque transmitted

Consider the progressive clutch engagement operation. Initially the axial force pressing the discs together is too small for the tangential frictional force to overcome the resistances to rotation of the driven shaft, so there is slip at the friction faces, the relative velocity between which progressively reduces until it becomes zero at the instant of full engagement.

If the spring that presses the clutch in Fig. 21.1 into engagement exerts a force P normal to the discs, that is, parallel to their common axis, the frictional force F tending to prevent slip is $\mu \times P$, where μ is the coefficient of friction of the rubbing faces. This frictional force is the sum of all the small forces, each of which is acting at one of the almost infinite number of contact points between the friction faces of the discs.

The resultant friction force may generally be taken as acting tangentially at the mean radius R of the friction faces which, if they are lined, is usually a relatively narrow annular strip, or ring of pads, of the material having a high coefficient of friction. Consequently, the torque Q, that is $F \times R$, about the axis of the plates can be expressed most simply as $\mu \times P \times R$. It must be noted, however, that if the width of the annulus, or ring of pads, is not small relative to its overall diameter, some inaccuracy will be introduced by assuming the mean radius to be the effective moment arm of the force about the axis of the clutch plates. Moreover, if an intermediate plate is employed for carrying the friction

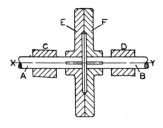

Fig. 21.1

facings, two faces are rubbing, each offering a friction resistance of $\mu \times F \times R$, so the total torque capacity of the clutch becomes $2 \times \mu \times P \times R$.

While the maximum torque a clutch can transmit is dependent on the mean radius of the annulus of friction material, its rate of wear is determined by the area of that annulus. Consequently, although it is desirable to make both the inside and outside diameters of the annulus as large as practicable, if the inside diameter is too large, the area of the annulus will be too small to obtain an acceptable rate of wear.

In conclusion, it can be said that the torque depends on three factors: spring force, coefficient of friction and mean radius of the friction facing. A limit on spring force is set by the magnitude of the effort that a driver may be expected to be able to exert on the clutch control pedal, to compress the clutch engagement spring and thus separate the plates.

Consequently, there is a maximum torque above which a simple pedal-actuated clutch of this type cannot be used. Indeed, for heavy commercial vehicles and some others having very powerful engines, some means of obtaining a higher torque capacity are needed. Any of three measures can be applied—

(1) The use of a cone clutch.
(2) Doubling or trebling the number of intermediate friction plates.
(3) The use of a multiple-plate clutch.

The cone clutch, although sometimes used in the early days, is not now considered suitable as a dry clutch for this application owing to its tendency towards sticking, harshness, and inconsistency in operation. It is, however, employed in transmission systems, generally in oil, and particularly in synchromesh mechanisms, Section 23.15.

21.3 Cone clutch

The cone clutch is so called because each of its two friction faces, equivalent to E and F in Fig. 21.1, is a frustrum of a cone, one male and the other female. That illustrated in Fig. 21.2 is of course a very old design: the female cone surface is ground inside the rim of the flywheel. A, which has a closing plate on

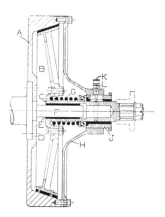

Fig. 21.2

its rear face and thus forms a rotating housing for the clutch. This flywheel and clutch housing assembly is bolted to a flange on the crankshaft.

The male member of the clutch is in two parts. Its disc B, around which is the peripheral conical flange, is of aluminium so that its weight and polar moment of inertia are as low as practicable; it is secured by bolts D to a hub C, which is of steel, not only for strength but also, since it has to be slid axially, good wearing characteristics. A bush-type bearing carries the hub of the clutch on a spigot F, extending rearwards from the tail end of the crankshaft into another bush E in a housing J in the rotating casing. Both these bearings are needed because, although when the clutch is engaged the male component rotates with the crankshaft, it has to be slid axially to the rear for disengagement, and may then be stationary relative to that shaft. A nipple K supplies lubricant directly to the rear and indirectly, through the counterbore in the hub, to the forward bearing bush.

The clutch is held in engagement by a coil spring G around the hub. One end of the coil spring bears against the forward end of the hub and the other on a ball thrust bearing in the housing. This bearing is essential to allow for the previously mentioned continuous rotation of the housing while the composite male member is stationary. Only a short rearward displacement of the male member is needed to disengage the clutch completely.

21.4 Torque capacity of a cone clutch

As mentioned previously, cone clutches have been used where large torques have to be transmitted, but where heavy clutch-engagement springs have been impracticable. From Fig. 21.3, it can be seen that a cone clutch is kept in engagement by the force P, exerted by the spring. The *normal* force Q, uniformly distributed over the conical surfaces, however, is greater than P, since it is augmented by the wedging action of the cones. Consider the resultant of that force as acting mid-way between the ends of the conical surfaces, that is, at c. Draw the line bc, so that its length H represents, to scale, the reaction to the force P; then bc represents the axial and ab the radial component of the resultant of the pressure Q normal to the friction faces.

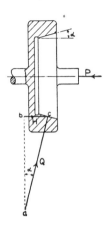

Fig. 21.3

The angle α can be shown to be equal to the cone angle, which is usually such that Q is about five times H. Consequently, the frictional force between the cones is of the order of $5 \times \mu \times P$, and the torsional resistance to slip therefore $= 5 \times \mu \times P \times R$, where R is the mean radius of the annulus of the contacting surfaces between the male and female cones. In an unlubricated or *dry* clutch, α cannot be much less than about 20°, otherwise the cones tend to bind or stick. Moreover, with a more acute angle, even a small amount of wear of the friction faces would introduce a large axial displacement of the cone, which would be difficult to accommodate. In a clutch in which the conical faces are lubricated, however, that is, in *wet* clutches incorporated in transmissions, α is usually of the order of 7°.

21.5   Clutch linings

Many cone- and disc-type clutches in applications other than between engine and gearbox are not faced with friction *linings*. These, however, are usually totally enclosed, so that they can operate in oil to obtain a low rate of wear between the rubbing metal faces and to reduce the fierceness of engagement.

In the early days of the automobile, some cone-type clutches were used, mostly of the dry type, between the engine and transmission. Now, however, cone clutches for this purpose are to be found only on some industrial engines, where they are almost universally of the dry type *lined* with a composite material having a high coefficient of friction. Most cars and commercial vehicles have single-dry-plate clutches. However, with the introduction of vehicles capable of carrying 38 tonne or more, some clutches of the twin- or, less commonly, triple-dry-plate type are being employed.

Linings used universally to be, and in many instances still are, riveted onto the plates or cones, the rivet heads seating in counterbores in the friction material, so that they are recessed well below its working face. Copper or aluminium rivets are employed, because they must not, when the lining is worn, immediately damage the metal surfaces of the discs or cones as they begin to rub against them. To avoid such damage, the linings should be inspected for wear at regular intervals. If this procedure is not strictly observed, the driver may be warned that his linings need to be replaced, by slipping of the clutch and, possibly, the squeaking of the rivets rubbing against the metal. Failure to take prompt action can lead ultimately to total failure of the clutch but, in any event, costly repairs to scored metal faces are likely to be entailed.

With the development of modern high-performance adhesives, linings are now widely bonded to the metal faces. Initially, the main problem was to find an adhesive that would stand up to the high temperatures generated and which would cure rapidly enough for use in high volume production. These and other problems have been solved, however, so we are now able to utilise a larger proportion of the total volume of the friction facing material: with riveted facings, only the thickness of lining down to the rivet heads can be used and, when the linings are replaced, the remainder has to be scrapped.

21.6   Friction materials

There are two types of friction material: woven and moulded. The first is made by spinning asbestos or, more recently, other mineral fibres into a yarn, in some instances on a brass wire core, which is woven into a cloth and then impregnated with a bonding agent. These woven materials are themselves subdivided into two types: solid woven and laminated. In the former, the cloth is woven to the required thickness of the lining before impregnation while, in the latter, several layers of the cloth are placed together and then impregnated. The moulded types may be made by mixing the fibres and bonding agent into a dough and then moulding it under pressure and at elevated temperature in dies.

Originally, brass wire was included in threads because it was impossible to spin asbestos fibres without it. Now, however, the problem of spinning has been overcome, though, alternatively, strength and wear resistance may be improved by including chopped wire or shredded brass in the weave of the cloth, or in the dough prior to moulding.

21.7 Bonding agents for fibres

A wide range of bonding materials has been used as follows—

(1) Asphaltic base with natural gums and oils added.
(2) Vegetable gums.
(3) Rubber-based compounds.
(4) Synthetic resins:
    (a) alcohol soluble,
    (b) oil soluble.

The frictional properties of the materials depend chiefly on the bonding material and the above bonds have broadly the following characteristics—

(1) Coefficient of friction from 0.3 to 0.4 at any temperature up to about 250°C. The coefficient tends to rise at higher temperatures due largely to exudation of the bonding material. Higher coefficients can be obtained by using excessive bonding material but such coefficients are not maintained after the excess of bond has been driven off by a rise of temperature. The wearing properties of materials (usually fabrics) impregnated with this bond are good, particularly when the product is die pressed.
(2) Coefficient of friction from 0.35 to 0.45 up to about 250°C. The wearing properties are rather better than type (1) but otherwise it has similar properties.
(3) The coefficient of friction can be made to have almost any value up to about 0.6 by the incorporation of filling materials, and the vulcanisation of the rubber may be arranged to produce either a flexible or a rigid product. Both types tend to disintegrate under severe conditions, the flexible type being very bad in this respect. The rigid type has a very destructive effect on the surfaces on which it runs. Both types are more affected by water than the other types of bond, and are rarely used.
(4) (a) Coefficient of friction from 0.4 to 0.5 at any temperature up to about 230°C. At higher temperatures, however, it falls off to as low as 0.1, though some materials of this type have been produced in which the problem has been largely overcome. This type of bond is not affected by lubricating oil and can withstand heavy pressures; it consequently finds a field of usefulness in epicyclic gearboxes, etc.

(b) Coefficient of friction from 0.35 to 0.38 which is maintained at quite high temperatures. Vegetable gum and asphaltic bases are often used in conjunction with this type of bond. Both (4a) and (4b) have excellent wearing qualities.

Most modern friction materials, many of which can be used also in brakes, will withstand surface pressures of up to $1380 \, \text{kN/m}^2$ though, in clutches, pressures of about 100 to $200 \, \text{kN/m}^2$ are the usual practice. Fabric-based materials tend to be more porous and to absorb oil which therefore, in small quantities, tends to have only minor detrimental effects on their performances, though any effects that it does have may persist longer than with the moulded materials.

Cotton may be mixed with the other fibres to give a high coefficient of friction, up to 0.6, but it cannot be used at temperatures above about 150°C without becoming charred and ruined, so such mixtures are not suitable for brakes. Cork has been used in clutches, but almost always in oil. It has a coefficient of friction of about 3.0 and will withstand surface pressures of up to approximately $140 \, \text{kN/m}^2$.

21.8   Single-plate clutch

A simple single-plate clutch is shown in Fig. 21.4, from which the principle of operation may be gathered. The flywheel is a simple disc with a thick rim having a flat face, into which are screwed six studs A. These studs carry a thick plate B, which is thus fixed to the flywheel as regards rotation; it is, however, free to slide axially along the studs. In between the plate B and the flywheel is situated the driven plate F; this is riveted to the flange of a hub G, which is connected with the clutch shaft C by splines E. When the parts are disposed as shown there is no tendency for the flywheel to turn the plate F, but if the plate B is moved along the studs so as to squeeze the plate F between itself and the flywheel the frictional forces set up between the surfaces in contact will prevent slip between them.

The necessary axial force is applied to the plate B by a number of springs which are usually arranged to surround the studs A. The latter are then provided with nuts as shown at the bottom of the figure. The face of the flywheel and the inner face of the plate B are sometimes lined with some form of friction fabric or composition, but usually the linings are riveted to the faces of the plate F. The latter arrangement has the advantage that the heat generated when the clutch is slipping during engagement is more easily dissipated, as it can be conducted directly away into the body of the flywheel and outer plate B.

In the former arrangement the heat has to pass through the linings before it can be absorbed and, as the linings are not good conductors of heat, the absorption and dissipation are not so rapid and the clutch will heat up more rapidly. On the other hand, the second arrangement increases the rotational inertia of the driven element which is undesirable; this does not outweigh the advantage obtained.

21.9   Torque transmitted

It will be observed that if slip occurs at one face of the plate F it must also do so at the other face, so that there are two frictional forces tending to prevent slip, each of which will be equal to μP where P, in newtons, is the total spring force and μ the coefficient of friction. Each of the frictional forces acts at the mean radius of the ring of contact, so that if that radius is R metres the torque transmitted will be $2\mu PR$ newton metres (Nm).

To transmit the same torque as a cone clutch of equal overall diameter a single-plate clutch would generally have to be provided with a greater spring force because the wedging action of the cones is lost and also the mean radius of the ring of contact is, owing to constructional exigencies, smaller. The reduction due to these causes is offset by the fact that there are two frictional surfaces in the single-plate clutch as against the single surface in the cone type, but the net result is a reduction necessitating an increased spring pressure.

The spring pressure necessary may, however, be reduced by using more than one driven plate, as is done in the clutch shown in Fig. 21.5.

There are three driven plates which are riveted to hubs that are free to slide on the splined clutch and these are gripped between the flywheel, the pressure plate B and the intermediate plates. Thus there are six slipping surfaces and the clutch is able to transmit three times the torque that a similar single-plate clutch having only two slipping surfaces could so. The intermediate plates are positioned on and are driven by square-headed pegs A A which engage slots formed in their peripheries. The pressure plate B is positioned and driven by three lugs C formed on the cover plate of the clutch and which engage three slots machined in the pressure plate. The force pressing all the plates together so as to transmit the drive is supplied by a single coil spring D; this bears at one end against the pressed steel cup E which seats in a hole in the cover plate, and at the other end against the housing F. Secured to the latter are three plungers G that are free to slide in holes in the cover plate and which engage the ends of the three levers H. These levers are pivoted on pins carried by small brackets J and, at their outer ends, they bear against renewable hardened pads fixed to the pressure plate. The force of the spring is thus increased by the leverage of the levers H. To ensure that all three of the levers press equally on the pressure

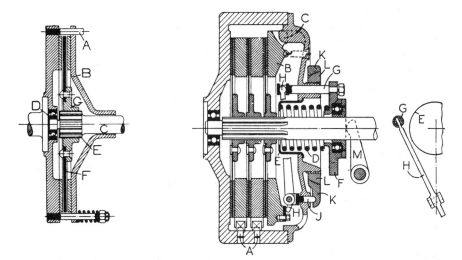

Fig. 21.4 Single-plate clutch   Fig. 21.5 Triple-plate clutch

plate the brackets J are not fixed in the cover plate but their stems are left free to slide in the holes in which they are placed and, at their outside ends, they bear against a disc K which seats on a spherical surface on the nut L. Because the disc K is free to rock in any direction on its spherical seating the forces acting on the brackets J, and hence the forces applied by the levers to the pressure plate, must be equal. The nut L, being screwed on the boss of the cover plate, enables the pressure plate to be adjusted bodily to the left to take up any wear that occurs. The clutch is disengaged by the withdraw fingers M which press against the inner member of the thrust race housed inside the member F, thereby pressing the latter to the left and relieving the pressure plate of the spring force. The pressure plate is pulled back by a number of small coil springs one of which can be seen at the top, and spring loaded levers, not shown, then separate the intermediate plates. The levers H are not arranged radially but as shown in the scrap end view where it will be seen that they pass to the side of the cup E; this enables a longer lever to be used than could otherwise be got in. In the figure the plunger G is shown out of its true position.

The use of a single spring in this manner is, however, no longer current practice.

21.10   Multi-spring single-plate clutch

A typical clutch actuated by a number of coil springs on a pitch circle near its periphery is shown in Fig. 21.6. It is a Borg & Beck clutch, made by Automotive Products Ltd, and is for installation where the engine clutch and gearbox assembly form a single unit. The driven shaft, which normally is a forward extension of the gearbox primary shaft, is supported at its front end in a ball bearing in a hole in the centre of the web of the flywheel, which is spigoted and bolted onto a flange at the rear end of the crankshaft.

A pressed steel cover C, bolted to the rear face of the flywheel, houses the clutch mechanism. In this clutch, nine coil springs D, seating in pockets in the cover C, force the pressure plate A forwards, to clamp the driven plate between it and the rear face of the flywheel. Three lugs B extend rearwards from the periphery of the pressure plate through slots in the cover C, both to locate the pressure plate and to cause it to rotate with the rest of the assembly. The driven plate of course is splined onto the shaft.

When the clutch control pedal is depressed, to release the clutch, it rotates the spindle E, the ends of which are carried in bearings in the bell housing between the engine and gearbox. This causes a pair of fingers F, one each side of the driven shaft, to push the housing G, in which is the graphite-impregnated bearing ring H, forwards into contact with the annular plate J. The interconnection between the fingers F and the thrust bearing housing G is a pair of diametrically-opposite trunnions, which seat in slots in the ends of the fingers and are retained by the springs T, as shown in the scrap view, bottom right in Fig. 21.6. As the pedal is further depressed, plate J presses on the inner ends of the three levers K, causing them to swing about their fulcrum pins L, so that their outer ends push against the small struts M seating on lugs on the pressure plate. This pushes the pressure plate to the rear, against the influence of the nine coil springs, thus releasing the driven plate.

This arrangement of levers and struts, at first sight rather complex, is necessary for reducing to a minimum the friction in the system, which would

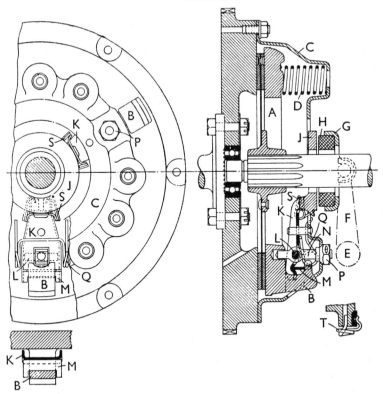

Fig. 21.6 Borg & Beck clutch

otherwise make the clutch control action too stiff. The fulcrum pins L are carried in diametral holes in eyes in the bolts N. Nuts P on the ends of these eyebolts N seat in hemispherical depressions in the cover C, and springs Q press the ends of the levers constantly against the ends of the struts M, to retain them. These struts cannot move laterally, because they are of a channel section into which the projections B register, as shown in the scrap view, bottom left in Fig. 21.6. From Fig. 21.7, it can be seen that lugs R, which project forwards from the annular plate J, in Fig. 21.6, register in slots in the ends of the levers K, where they are retained by the springs S, so these three levers constitute the sole support for the annular plate J.

When the clutch pedal is depressed, the three levers roll round their fulcrum pins, the eyebolts N of which are free to swing about the spherical seats of their retaining nuts P. At the same time, the struts M swing too, to accommodate the small radial motion that this rolling action implies. Obviously the friction in such a system is much less than if pivot pins were employed instead of the rolling contacts. As a further refinement, a bush of bearing bronze is in some instances carried in the holes into which the front ends of the eyebolts N project, to reduce to a minimum the friction between them and the pressure plate as it is pulled clear of the driven plate and flywheel.

On assembly in the factory, the nuts P are adjusted until the face of the plate J is truly perpendicular to the axis of the clutch. They are then staked over to

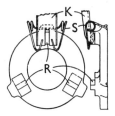

Fig. 21.7

lock the setting. If this adjustment is subsequently disturbed, new eyebolts N and nuts P must be fitted, otherwise faulty operation is virtually a certainty.

Among the various measures that in some instances have been taken by Borg & Beck to reduce friction in the clutch actuation system is the support of the eyebolts N in a three-arm spider bolted to the cover. This spider reacts the centrifugal force on the eyebolts, which would otherwise have to be reacted between their ends and the sides of the holes into which they project in the pressure plate. For greater durability most vehicles now have a ball-thrust bearing instead of the carbon-impregnated ring H.

21.11   The diaphragm-spring clutch

At high rotational speeds, problems can arise with multi-spring clutches owing to the effects of centrifugal force on both the springs themselves and the levers of the release mechanism. These problems are obviated when diaphragm-type springs are used, and a number of other advantages are experienced.

Among these advantages is improved durability, which arises for three reasons. First, by virtue of the compactness of a diaphragm spring in an axial sense, a heavier pressure plate can be accommodated within the same overall length. This increases the thermal capacity of the clutch for absorbing heat generated by friction, so its working temperatures are lower. Secondly, it is easy to design a diaphragm spring so that it will maintain the optimum clamping pressure regardless of the degree of wear on the driven plate facings. Thirdly, the distribution of the clamping load over the facings is more uniform with a diaphragm spring.

Use of the diaphragm spring also reduces pedal loads. This is because of the deflection characteristics of the diaphragm as between the clutch engaged and released positions. Moreover, the total control-travel required for releasing such a spring is generally less than that for a coil spring clutch, and the friction in the moving parts increases less over the life of the clutch.

For a given size of clutch, the diaphragm-spring type has a higher torque capacity. This is partly because of the leverage obtainable due to the geometrical relationship between the diameter of the circle on which lay the ends of the fingers at the centre of the spring, Fig. 21.8(*d*), the diameter of the fulcrum ring G about which the spring pivots, and the diameter of the circle on which the spring applies its load, at J, to the presser plate. Another factor is the compactness of the diaphragm-type spring, as an energy store, and therefore the ability to provide a higher clamping load for any given size of clutch.

The inherent compactness for any given load is a major advantage of the diaphragm-spring clutch. Another is that there are fewer parts and so it is simpler in construction.

In this type of clutch, two examples of which are shown in Figs 21.8 and 21.10, the basic form is the same as that of Fig. 21.4: the driven plate A, in Fig. 21.8(*a*), is clamped between the flywheel B and the pressure plate C, but the clamping is done by the single diaphragm spring D. This is a steel disc which,

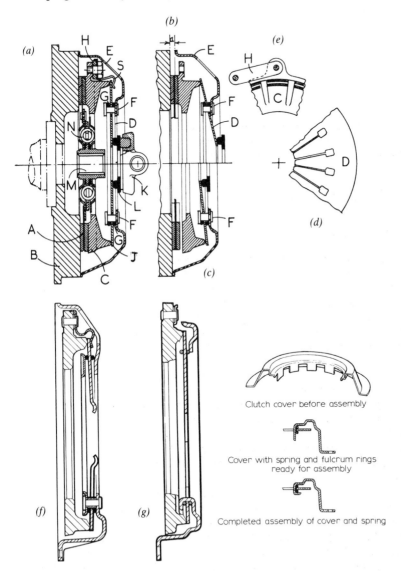

Fig. 21.8 (*a*) When assembled, the diaphragm spring is flattened. (*b*) Prior to assembly, the spring is of conical form, as in the top half here. (*c*) When assembled but disengaged, the apex of the spring is pushed over to the other side, as in the bottom half. (*d*) The spring has radially inward-pointing fingers. (*e*) One of the three straps linking the pressure plate to the cover. (*f*) Supported fulcrum ring in a Borg & Beck clutch. (*g*) Borg & Beck DST clutch

when unloaded, is of conical form as indicated in the upper half at (b). During assembly of the clutch on to the flywheel, the cover pressing E has to be forced axially, to close the gap a, during which the diaphragm is flattened, almost into a simple disc, as seen in the view (a). A number of slots are machined radially from the centre of the diaphragm as indicated at (d), thus forming the 'fingers'.

The spring is held between two circular wire rings G, which are carried on the shouldered pins F (nine in number) and which form a fulcrum for the diaphragm when the clutch is being disengaged. This is done by forcing the centre of the diaphragm to the left, so that it again becomes a cone but now with the apex, as in the lower half of (c). Spring clips S ensure retraction of the pressure plate when the spring pressure is thus removed.

The method of centring and driving the pressure plate in this clutch is different from those used in the clutches described previously. Three strap links are used, one of which is seen at H in the views (a) and (e) of Fig. 21.8. These are pivoted at one end to the pressure plate and at the other to the cover pressing. The principle involved is shown in Fig. 21.9. The small axial movement for disengagement of the clutch is of course easily accommodated by the straps.

Disengagement is effected by the release bearing K which, in the illustration, is a graphite ring but which is more often a ball-thrust bearing. The graphite ring bears against the cast iron ring L, which is carried by three straps in the same manner as the presser plate, and which pushes the inner ends of the fingers of the diaphragm spring forwards to release the clutch. If a ball-thrust bearing is used, the ring L is usually replaced by the inner race of the bearing, the end face of which is appropriately rounded to form a similar section. The latter arrangement eliminates the need for both the ring L and its carrier straps, though the ball-thrust bearing is of course more costly than a graphite bearing.

In recent years, a number of improvements have been made to diaphragm-spring clutches. First, for heavy duty applications, especially for diesel engines, it has been found that the shouldered rivets F serve better if case-hardened. Secondly, these shouldered rivets were then replaced by plain rivets in hardened sleeves which served as distance pieces between the annular support plate and cover, and located the wire rings G and the diaphragm spring, Fig. 21.8(f). This arrangement provided continuous support around the whole periphery of the wire rings, improved the durability of the clutch and increased its efficiency, by reducing the deflections of the fulcrum rings.

Even better support for the fulcrum rings, together with economies in respect of both weight and cost, have been achieved in the Borg & Beck DST clutch. The initials 'DST' stand for *diaphragm spring and turnover*. Basically the clutch is the same as before, but the method of securing the fulcrum rings is as illustrated in Fig. 21.8(g). It has neither rivets nor annular support plate: instead, tags formed in the cover pressing are passed through the diaphragm spring and fulcrum rings, and then bent around them to complete the assembly. This has reduced the number of separate components by up to 19 and has saved about 0.5 kg on a typical clutch for a car, and the durability is even better than that obtainable from the best types of riveted design.

A further refinement has been a minor modification to the ends of the radial fingers of the diaphragm spring, which has had a major effect on their durability. With the arrangement in Fig. 21.8, the radial sliding motion of the fingers over the rounded section of the ring L, or of the bearing race, together with sliding due to any slight eccentricity between the bearing and the drive-shaft, causes wear of the ends of the fingers, ultimately forming a groove in each. Moreover, the eccentricity then causes chatter which is liable to loosen the rivets. To avoid these problems, the face of the ring L, or of the bearing, is now ground flat and the ends of the fingers curved, as shown in Fig. 21.8(*f*). As a result of this simple modification the radial loading is greatly reduced and, because forming the curved ends of the spring fingers work-hardens them, they offer significantly higher resistance to wear.

The Laycock design of the diaphragm clutch is shown in Fig. 21.10. In this the pressure plate A is centred in relation to the flywheel, and is driven by six lugs B which fit into slots formed in the pressing C. The diaphragm spring D is carried in a recess formed inside the lugs and is kept in place by a circlip. The fulcrum for the spring is formed by the lips of the pressings C and E, which are bolted to the face of the flywheel. The diaphragm has 18 radial slots at its centre and the inner ends of the fingers thus formed engage the sleeve F which is pressed to the left by the clutch release bearing when the clutch pedal is depressed.

21.12   Pull-type diaphragm-spring clutch

The rotating components of transmissions for heavy commercial vehicles are themselves heavy, and consequently synchronisation for gear shifting can take an unacceptably long time. To overcome this problem, a small brake is sometimes incorporated on the front end of the gearbox to stop free rotation of the primary shaft when the clutch is disengaged prior to selecting or changing a gear. Obviously it is advantageous if both control motions are in the same direction, so a release mechanism that is pulled instead of pushed to disengage the clutch has been introduced, Fig. 21.11. Then, the initial rearward motion of the clutch release bearing is utilised to disengage the clutch, and its further rearward motion to bring it to bear on a disc brake for stopping the rotation of

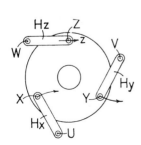

Fig. 21.9

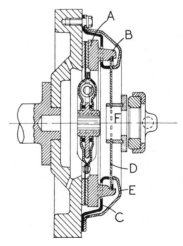

Fig. 21.10

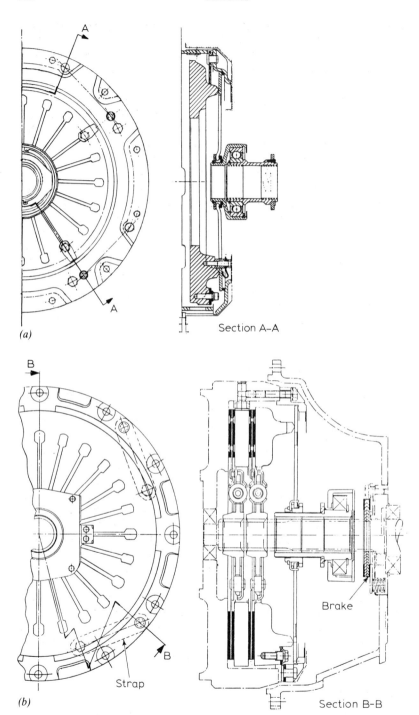

(a)

Section A–A

(b)

Strap

Section B-B

Brake

B

B

Fig. 21.11 Clutches of the pull type. (*a*) AP single-plate, and (*b*) the Fuller twin-plate. In (*b*) the details above the centre line are of the 600 Series and those below it relate to the 900 Series

the primary shaft. This enables the driver to use the brake at his discretion for the initial selection of bottom gear and for subsequent upward changes. Both single- and twin-plate versions of the pull-type clutch are made, the latter being required for some tractors and for the powerful engines needed for the modern, very heavy trucks.

As can be seen from the illustration, the periphery of the diaphragm spring bears on the fulcrum ring, while the diaphragm itself bears against a ring integral with the presser plate to hold the clutch in engagement. This arrangement has three advantages: first, by virtue of the proximity of the fulcrum ring to the rim of the cover of the AP clutch, it is inherently rigidly supported; secondly, less metal is required in the cover, which is therefore lighter and cheaper; thirdly, it is easier to obtain a high mechanical advantage for operation of the diaphragm spring which, for a heavy vehicle, of necessity has to be powerful. From the illustration it can be seen that, in the Fuller twin-plate clutch, the fulcrum ring is carried within the flywheel assembly.

In the AP single-plate clutch, a Belleville, or disc-type, spring is interposed between the release bearing housing and the ends of the fingers of the diaphragm spring. This clamps the outer race carrier firmly enough to the fingers to prevent it from rotating but with a certain degree of resilience for accommodating the angular movement of the fingers. The rearward extension of the tubular component carrying the inner race of the bearing actuates the disc brake on the front end of the gearbox.

21.13 Belleville direct-release clutch

The Borg & Beck direct-release clutch, with a Belleville spring – no fingers extending radially inwards – was first introduced for front-wheel-drive cars with transverse engines, the aim being to reduce to a minimum the space required, beyond the tail end of the crankshaft, for accommodation of the clutch. As can be seen from Fig. 21.12, the release bearing thrusts against a plate attached to the centre of the clutch cover, and both the driven and presser plates of the clutch are on the opposite side of the flywheel. The outer periphery of the Belleville spring seats in the cover, while its inner periphery pushes against a ring formed integrally on the flanged hub of the flywheel.

Movement of the clutch release bearing towards the flywheel thrusts the cover in the same direction, against the influence of the Belleville spring, and carries with it the presser plate on the remote side of the flywheel. This releases the driven plate, which was clamped between the presser plate and flywheel by the Belleville spring.

The advantages of this arrangement are its axial compactness and low rotational inertia. Its disadvantage is that the clamping and release loads are equal. This arises because of the absence of any leverage, which is obtainable only where radially inward extending fingers are incorporated as on the diaphragm-type spring, so that a fulcrum can be provided.

21.14 Driven plate

With the traditional front-engine rear-wheel-drive car, the easiest way to overcome transmission noise problems at the prototype stage is, in most instances, to alter the flexibility of the propeller shift. With front-wheel-drive

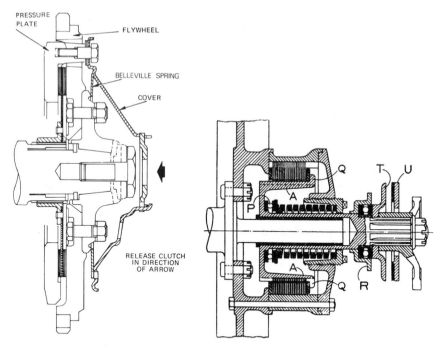

Fig. 21.12 Direct-release clutch with   Fig. 21.13 Multi-plate clutch
Belleville spring

cars, however, which have short, stiff driveshafts, this option is no longer open. As a result, manufacturers have turned to adjustment of the flexibility of the driven plate of the clutch instead.

Originally, the spring centre was incorporated in driven plates solely to absorb shock-loading during engagement of the clutch. Now, however, more complex arrangements are used.

First, to ensure that the drive is taken up progressively, the centre plate, on which the friction facings are mounted, is crimped radially so that as the clamping load is applied to the facings the crimping is progressively squeezed flat. On the release of the clamping load, the plate springs back to its original radially-crimped section. This plate is also slotted so that heat generated does not cause the distortion that would be liable to occur if it were a plain flat plate – this plate of course must be thin, to keep rotational inertia to a minimum.

Secondly, the plate and its hub are entirely separate components, the drive being transmitted from one to the other through coil springs interposed between them. These springs are carried within rectangular holes, or slots, in the hub and plate, and arranged with their axes aligned appropriately for transmitting the drive, as for example in Fig. 21.8. In a simple design, all the springs – perhaps six – may be identical, but in more sophisticated designs they are arranged in diametrically-opposite pairs, each pair having a different rate and different end clearances in the holes that accommodate them. The lowest rate pair have zero end clearance, and therefore are the first to take up the loading. As this pair deflects and the end clearances for the other pairs therefore begin to be taken up, the second pair comes into operation, increasing the overall torsional rate, until the third pair similarly is brought into play, still further reducing the end clearances of any that remain, and increasing the rate, and so on. As many as five stages have been used, though usually there are no more than about three or four. As a further refinement, the flange around the hub is clamped between the two slotted plates that house the springs, sometimes with friction facings interposed between them, to provide a degree of damping for this spring action.

21.15 Multiple-plate clutch

The multiple-plate clutch is now practically obsolete as a main clutch but is still used in epicyclic gearboxes and one example, shown in Fig. 21.13, will be briefly described. Bolted to the flywheel is a drum which on its inner circumference is splined to carry a number of thin metal plates; these must consequently revolve with the drum but are able to slide axially. Interleaved with these outer plates is a number of inner plates that are splined to an inner drum A which is coupled rotationally to the gearbox shaft. This drum is supported on a spigot extension of the crankshaft, a suitable bearing bush being provided. Between the web of the inner drum and a sleeve screwed into the cover plate of the outer drum is a strong coil spring. The inner drum is thus pressed to the left and, being provided with a flange Q, it squeezes the outer and inner plates together so that friction between them transmits the driving torque from the outer drum to the inner one. The clutch is disengaged by pulling the inner drum to the right against the pressure of the spring. The fingers of the withdraw shaft bear against the flange R of the housing of the clutch withdrawal thrust race. A ball bearing P eliminates friction between the spring and the inner drum when the clutch is disengaged. The screwed cup at the right-hand end of the spring enables the spring force to be adjusted. The plates of multiple-plate clutches were at one time made alternately of steel and phosphor bronze but now are all of steel or one set may be lined with a friction material. With metal to metal contact lubrication is essential and so the clutch is made oil-tight and is partly filled with oil. The oil tends to make the plates drag when the clutch is disengaged and so a clutch stop or brake is fitted. It consists of a disc T fixed to the inner drum hub and which comes into contact with a spring loaded disc U carried by any convenient part of the gearbox casing or clutch housing; the disc U has a friction material lining.

21.16 Dry multiple-plate clutch

Multiple-plate clutches are also made to work dry, that is, without any oil. The driving plates are then lined on each side with a friction fabric. In such clutches the driving plates are sometimes carried on a number of studs screwed into the web of the flywheel in the same way as the outer plate of a single-plate clutch is sometimes carried. This construction is inconvenient when oil is used. Several small springs can be used instead of a single spring.

21.17 Clutch release gear

The disengagement of almost all clutches involves the axial movement of a shaft or sleeve that is rotating and this involves the use of a bearing that will permit the motion and transmit the force. A ball bearing is frequently used and is nowadays a dual-purpose bearing capable of supporting both axial and radial loads. In the past bearings which could support only axial loads were used, an example being shown in Fig. 21.13, and the housing R of the bearing had itself to be supported by letting it bear on the clutch shaft. Instead of a ball bearing a graphite block bearing may be used, as has been mentioned, and in this case the housing of the block can itself be supported by the fingers which transmit the axial thrust as is shown in Fig. 21.6. Referring to Fig. 21.13, the shaft E has to be connected to the clutch pedal and this used to be done by means of links and rods, a method that is still sometimes used in commercial vehicles. But the drawbacks to such a linkage are that it would be difficult to arrange it in modern vehicles where the clutch and the clutch pedal are not conveniently near to each other and that, because the engine is carried on the frame or body structure on rubber mountings and the clutch pedal is mounted on a separate part of the body or frame, the relative movements that occur between these two units give rise to poor engagement of the clutch and also to rattles. Hydraulic actuation of the clutch is therefore now used. The units employed are similar to those used for brake actuation and which are described in Chapter 32. The pedal actuates a piston which forces fluid through piping into an actuating cylinder carried on the bellhousing surrounding the clutch and whose piston is connected to a lever carried by the shaft on which the withdraw fingers are fixed. Part or all of the piping can be made flexible and thus both of the difficulties associated with mechanical linkages are obviated.

21.18   Clutch brakes or stops

When a clutch is disengaged while the engine is running the inertia of the driven member will tend to keep it revolving. This inertia is proportional to the weight of the member and to the square of the 'radius of gyration' which is the equivalent radius at which the weight may be considered to act. Since this tendency to continue revolving rendered gear changing difficult, small brakes have been arranged to act on the driven member when the clutch was *fully* withdrawn. The stop member or brake was sometimes backed up by springs so that a gradually increasing pressure acted on the brake. One of the advantages of the single-plate clutch is the small weight and radius of gyration of its driven member.

21.19   Automatic clutch action

Many attempts have been made to produce motor vehicles that can be controlled by the accelerator pedal and brakes only; this can be done in several ways. A centrifugal clutch which automatically disengages itself when the speed falls below and which re-engages when the speed rises above some predetermined values may be used. Alternatively, a fluid coupling, a fluid torque converter or some special form of clutch may be employed. Examples of each of these solutions of the problem will be described in the following sections.

21.20 Centrifugal clutches

The principle of these is shown in the simple arrangement in Fig. 21.14(a) where a single-plate clutch of ordinary construction has its presser plate A actuated by the 'centrifugal' forces acting on masses B formed on the ends of bell-crank levers pivoted on pins in the cover plate C. This arrangement has two principal drawbacks. First, there would be some force acting on the presser plate whenever the clutch was rotating and thus the clutch would never be completely disengaged. Secondly, if the force P due to the centrifugal force CF were sufficient to engage the clutch fully at say 1000 rev/min then it would become nine times as great at 3000 rev/min and it would require nine times the force necessary with an ordinary clutch to produce disengagement at that higher speed by pulling the presser plate back in the ordinary way, and it is desirable to be able to disengage the clutch in that way.

The first drawback can be overcome by putting in springs D (shown dotted) which apply a force Q opposing the force P. The centrifugal forces will then not give rise to any pressure on the driven plate until they have increased sufficiently to overcome the force Q and until then the clutch will be completely disengaged. By choosing the magnitude of Q suitably, the commencement of the engagement can be made to occur at any desired speed. Usually in motor car clutches a speed of about 500 rev/min is chosen. The second drawback can be overcome by modifying the construction as shown at (b) where the bell-crank levers press on a floating plate E between which and the presser plate are placed springs F. These springs transmit the force P from the floating plate to the presser plate. A stop G limits the outward motion of the masses B and thus limits the amount the springs F can be compressed. The force that must be applied to the presser plate in order to pull it back so as to disengage the clutch is now limited to the difference between the force Q and the force exerted by the springs F when the masses (having come against the stops G) have compressed them fully. This difference can be made to have any desired value. The pressure exerted on the driven plate will now be represented by a graph whose shape is the shaded line in Fig. 21.15. The curve OLMH whose ordinates are measured from the line OX as axis shows how the centrifugal force varies with speed. If a new axis O_1X_1 is drawn so that OO_1 equals the force Q exerted by the springs D then the graph O_1LH measured from O_1X_1 gives the pressure on the driven plate when the springs D are fitted. Finally, if the line MN is drawn so that its height above OX is equal to the force exerted

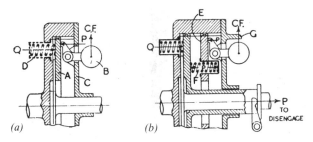

Fig. 21.14 Principle of centrifugal clutch

by the springs F when fully compressed the graph O_1LMN will represent the pressure on the driven plate in the modified arrangement of Fig. 21.14(*b*). This is what is desirable for the variation of pressure with speed, that is, no pressure at speeds below about 500 rev/min, a rapid increase of pressure between 500 rev/min and about 1000 rev/min so that at the latter speed the clutch is fully engaged, and no great increase in pressure at speeds above 1000 rev/min. In Fig. 21.15, S is the force that must be used to disengage the clutch at speeds over 1000 rev/min and R the force required to enable the clutch to transmit full engine torque. If the torque developed by the engine is less than full torque (that is, if the throttle is now wide open) then a force less than R will enable that reduced torque to be transmitted and the clutch will be fully engaged at some speed less than 1000 rev/min.

21.21   Eddy current couplings

The constructional form of one of these is shown in Fig. 21.16. There are two main members, A and B, which are coupled respectively to the engine crankshaft and the input shaft of the gearbox. The member A resembles and, if required, can form the flywheel but its rim is cut away by a number of semicircular recesses as shown in the view (*a*). The metal left between the recesses forms a series of axial bars – connected together at their ends by the continuous portions of the rim. The member A thus resembles the rotor of a squirrel-cage electric motor.

The shape of the outside of the member B is shown in the view (*b*). There is an electrical winding running round the annular space C and the ends of this winding are connected to slip rings D and E on which stationary brushes bear, so that current can be passed through the winding from a battery or generator. When this is done, a magnetic field is built up whose direction will be as shown by the dotted line. The field will take this path rather than pass from the fingers F to the fingers G because the air gap between these fingers is large compared with the gap between the members A and B and the magnetic 'resistance' or reluctance is consequently less through the latter.

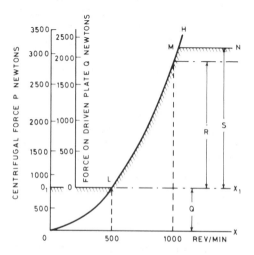

Fig. 21.15 Centrifugal clutch characteristic

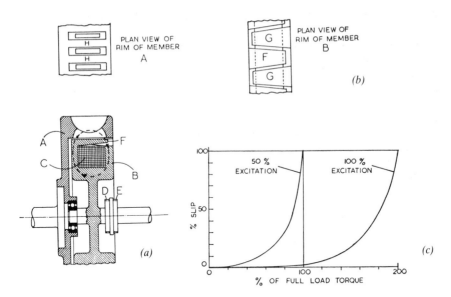

Fig. 21.16

If the member A now rotates relative to B, the magnetic flux will cut the bars of the member A and eddy currents will be generated in that member. There will consequently be a magnetic drag between the members which will tend to make B rotate in unison with A. Some difference of speed will, however, always exist because if there were no such difference, there would be no eddy currents and no drag.

The torque that can be transmitted from A to B depends on the difference in speed that exists between them, on the magnitude of the exciting current (up to that value which will produce saturation of the magnetic circuit) and on the size of the coupling. When transmitting the full designed torque, the difference in speed will usually be about 3% of the speed of the driving member. The variation of the percentage slip (i.e. 100 × (speed of A − speed of B)/(speed of A)) with the torque transmitted and for two values of the magnitude of the excitation current is shown in Fig. 21.16(c). The take-up characteristics of this type of coupling are very similar to those of a fluid coupling (see Section 21.25). Take-up is brought about by increasing the speed of the driving member and by controlling the value of the excitation current.

It should be noted that the output torque from the coupling is for all practical purposes equal to the input torque so that the device is the equivalent of a clutch and not of a gearbox or torque converter. Couplings of this kind are largely used in industrial applications and have been employed in a few automobile vehicles. By fixing the member B relative to the chassis and by driving the member A off the propeller shaft, the device can be used as a brake and this has been done to some extent in heavy lorries.

21.22 The Ferlec electro-magnetic clutch

This is basically a single-plate friction clutch in which the clamping force is provided by an electro-magnet instead of by springs. Its construction is shown in Fig. 21.17, the driven plate A has a spring damper hub which is splined on the output shaft B. The latter is carried in a spigot bearing at the left and in a bearing in the gearbox at the right. The pressure plate C is bolted to an armature disc D which itself is secured to the flywheel by three tangentially disposed links K, as shown in the scrap end view. These links isolate the flywheel from the armature disc magnetically while positioning it and causing it to rotate with the flywheel. The links K also permit the small axial movement required for the engagement of the clutch. There is an air gap of 0.45 mm between the armature and the face of the flywheel when the clutch is disengaged. When the winding F is energised the armature D with the pressure plate C is attracted to the flywheel so that the driven plate is squeezed between the pressure plate and the member G, which is rigidly bolted to the flywheel at three points, as indicated in the scrap end view at the bottom of the figure. The current for energising the winding is led in and out through brushes which bear on slip rings H and I, connected to the ends of the winding.

To provide a smooth engagement the energising current is made to build up gradually, either automatically as the dynamo speed increases or, when the battery supplies the current, by two variable resistances (one for low and one for high gears) which are decreased as the accelerator is depressed. The change from one resistance to the other is made by the selecting movement of the gear lever, the striking motion of which also provides an interruption of the current to the clutch coil and gives the momentary clutch disengagement to enable a gear change to be made. A cam actuated by the accelerator pedal also opens a switch when the pedal is in the idling and starting position, thus ensuring disengagement of the clutch at starting and during idling.

21.23 Fluid flywheel

This consists of two castings (called *rotors*), almost identical in form, one of which is fixed to the crankshaft of the engine and the other to the gearbox shaft. These castings are roughly circular discs in which passages XX, Fig. 21.18, are

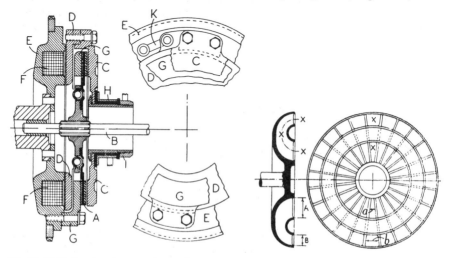

Fig. 21.17 Ferlec electro-magnetic clutch   Fig. 21.18 Fluid flywheel rotor

formed. Since the areas of these passages perpendicular to their centre line (XXX in the sectional view) must be kept approximately constant and since the circumferential width of the opening a is less than that of b the radial size of the opening A is made greater than that of B. In a simplified form the passages may be represented by tubes A and B with right-angled corners as shown in Fig. 21.19.

Imagine these tubes to be full of fluid and suppose them to be rotating about the axis XX with speeds N and n. Then if the outer end of the tube A were closed by a diaphragm C the fluid would be exerting a pressure P_a on that diaphragm, and this pressure can be shown to be proportional to N^2. Similarly, if the tube B were closed by a diaphragm D the fluid would be exerting a pressure P_b on that diaphragm. Now if n is less than N the pressure P_b will be less than the pressure P_a and if the diaphragms were removed the fluid in the tube A would commence to flow in the direction of the arrow E and would force the fluid in the tube B to flow in the direction of the arrow F against its inclination to flow the other way. Thus if the speeds N and n are unequal, the fluid in the tubes will be caused to circulate round and round. If N is greater than n then the circulation will be as indicated by the arrows E and F, that is, clockwise, but if n were greater than N then the circulation would be in the opposite direction and if the speeds of the rotors were equal there would be no circulation at all. Since there will always be some resistance to the flow of the fluid the speed at which the fluid will circulate will reach a steady value which can be shown to be proportional to the difference between N^2 and n^2. Of course, if the speeds of the rotors are not equal then any particular tube A will not always be opposite any particular tube B but there will always be a tube B for the fluid from any tube A to flow into although there may be a certain amount of impinging of the fluid on the walls of the tubes, that is, on the webs of metal between the passages XXX in Fig. 21.18.

Having seen that a difference in the speeds of the two rotors will cause a circulation of the fluid from one rotor to the other it can now be explained how the energy developed by the engine is transmitted to the gearbox. Consider Fig. 21.19. At K is indicated a particle of fluid which is at a distance r from the axis XX. This particle has to rotate in a circle radius r with the angular speed of the tube, that is, N. Its linear speed in the circle is thus $2\pi r N$ and its kinetic energy is—

$$\tfrac{1}{2}\frac{w}{g}(2\pi r)N^2$$

where w is the weight of the particle.

Now if the fluid is circulating as described above (N being assumed greater than n) the particle K will, in a short space of time, arrive at L. It will then be rotating in a circle whose radius is R at the speed N and its kinetic energy will be—

$$\tfrac{1}{2}\frac{w}{g}(2\pi R N)^2$$

Since R is greater than r the kinetic energy of the particle when it arrives at L is greater than it was at K. The increase in the kinetic energy of the particle as it moves from K to L is derived from the energy developed by the engine, the whole of which is utilised in increasing the kinetic energy of the fluid as it flows from the centre to the outside of the tubes A.

Continuing our consideration of the particle we find that a short time after it was at L it will have arrived at M. It will then be rotating in a circle radius R as at L but now at the slower speed n. Its kinetic energy is now—

$$\tfrac{1}{2}\frac{w}{g}(2\pi Rn)^2$$

and since n is less than N the kinetic energy at M is less than it was at L. Some of this difference in the kinetic energy will have been passed to the rotor B but some will have been converted into heat by impact with the webs of metal between the tubes and will for all practical purposes be lost. Finally, a short time after the particle was at M it will have arrived at N where it will be once more rotating in a circle of radius r but now at the speed n. Its kinetic energy will be—

$$\tfrac{1}{2}\frac{w}{g}(2\pi rn)^2$$

and is less than it was at M. Thus, as the particle moves from M to N it loses kinetic energy and the energy it loses is passed to the rotor B and thus to the gearbox.

21.24   Prevention of leakage

It has been tacitly assumed that the fluid cannot escape between the faces of the rotors and it might be thought that those faces would have to be in rubbing contact in order to prevent such leakage. In actual fact, however, there is a gap of about 1.5 mm between the faces and escape of the fluid is prevented by making one rotor with a cover which embraces the other rotor as will be seen in Fig. 21.20.

The relative position of the rotors has been changed in Fig. 21.20 to correspond with the disposition generally adopted in fluid flywheels for motor vehicles. The rotor that is fixed to the flywheel is now the right-hand one A which is bolted to the rim of the flywheel while the left-hand rotor B is fixed to

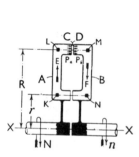

Fig. 21.19

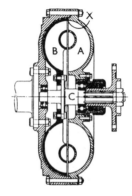

Fig. 21.20

the gearbox shaft C. Fluid does escape between the faces of the rotors and fills the space between the outside of the rotors and the inside of the flywheel. Centrifugal force keeps this fluid in position and maintains a pressure at X which prevents escape of fluid from the insides of the rotors once sufficient fluid has accumulated outside to enable a state of equilibrium to be reached.

21.25   Characteristic of the fluid flywheel

The characteristic which is of chief importance is the way in which the *percentage slip* varies with the speed of rotation. The percentage slip is defined as the quantity $(N - n) \times 100/N$ and is a measure of the difference in the speed of the two rotors. If n were equal to N then the percentage slip would be zero and if n were zero then the percentage slip would be 100. When the percentage slip is plotted against engine speed (N) the resulting graph is the form shown in Fig. 21.21. At any speed less than about 600 rev/min (this speed can be made to have any desired value by suitably modifying the design) the percentage slip is 100, that is, the gearbox shaft is stationary and we have the equivalent of a completely disengaged clutch. As the engine speed increases from 600 to 1000 rev/min the percentage slip falls rapidly to about 12. This corresponds to the period of actual engagement of an ordinary clutch, the speed of the gearbox shaft being rapidly brought up to roughly the same value as the engine speed. From 1000 rev/min up to the maximum speed the percentage slip decreases comparatively slowly from 12 down to possibly as little as 2. The percentage slip at any engine speed depends, however, on the torque being transmitted, the curve above being based on the assumption that the engine exerts full torque at every speed. If the engine torque is reduced below the full torque (open throttle) value then the percentage slip will be reduced. Thus under easy running conditions when the throttle is only slightly opened the percentage slip may be less than 1%. The chain-dotted curve indicates the variation of the percentage slip with the speed for such conditions, that is, level road and no head wind. Whatever the value of the percentage slip, however, it represents a direct loss of energy and thus an increase in the petrol consumption. Thus it is an abuse of the fluid flywheel to allow the engine speed to fall to the region between 1000 and 600 rev/min, full throttle, when the percentage slip becomes considerable and such use is comparable to the slipping of an ordinary clutch,

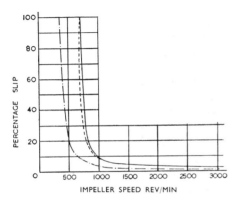

Fig. 21.21

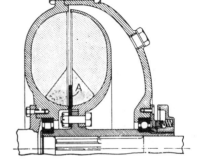

Fig. 21.22

which also increases the petrol consumption, the difference being that whereas an ordinary friction clutch would be damaged by prolonged slipping the fluid flywheel will not suffer any damage although it may become so hot as to burn one's hand if one touches it.

21.26 'Open circuit' fluid coupling

In an alternative design of fluid flywheel the 'torus-ring' that forms the inner wall of the passage round which the fluid circulates is omitted, as is shown in Fig. 21.22. A baffle plate, as shown at A, is also sometimes fitted. The open circuit and baffle ring modify the characteristics of the coupling slightly as may be seen from the dotted curve in Fig. 21.21. The most important difference is a reduction in the drag torque when the coupling is stalled.

Figs 21.23 and 21.24 illustrate the modern forms of seal used with fluid couplings and torque converters. The sealing surfaces are the flat faces of a steel ring C and a bronze or graphite ring D which are pressed together by a spring or springs E. The sealing faces are finished to optical limits of flatness and to ensure proper seating one ring is carried on a flexible diaphragm or bellows F. The design shown in Fig. 21.23 takes up less axial but more radial space than that of Fig. 21.24. In the latter the graphite ring D is left free to float between the metal faces on either side of it.

It can be shown quite easily that the torque exerted by the flywheel on the gearbox shaft is equal to the torque exerted by the engine on the fluid flywheel.

Thus the fluid flywheel does not give any increase of torque and is not the equivalent of a gearbox, the chief function of which is to give an increase of torque when required.

In addition to giving accelerator pedal control the fluid flywheel also reduces the shocks transmitted by the engine to the transmission, and *vice versa*.

The qualities required of the fluid are high density, low viscosity, chemical stability and absence of corrosive action and the fluid that is generally used is a very thin engine oil. A thick, viscous oil would increase the percentage slip, other factors being the same, and would thus reduce the efficiency.

The only important drawback of the fluid flywheel is that even when the percentage slip is 100%, there is a drag on the gearbox shaft which renders gear changing with ordinary types of gearbox very difficult. Consequently, the fluid flywheel is generally used in combination with epicyclic gearboxes which eliminate the difficulty. In the fluid coupling made by MAAG of Zurich and used in the automatic gearbox of their design, the driven member is carried on

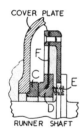

Fig. 21.23

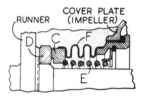

Fig. 21.24

the output shaft on a quick-pitch screw thread so that, when the speed of the driving member falls below that of the driven member, and the latter therefore tends to drive the former, the driven member is itself 'unscrewed' along the thread and moves away from the driving member. This reduces the drag in the coupling and assists in the gear changing.

When it is desired to use an ordinary type of gearbox a friction clutch is sometimes used in conjunction with a fluid flywheel. This clutch is intended to be used only for gear changing and not for taking up the drive during starting and acceleration of the car; it can therefore be made with smaller friction surface areas than a clutch that had to take up the drive but it must, of course, be capable of transmitting the full engine torque. However, if the engine is accelerated with the friction clutch disengaged and then the clutch is engaged it will have to take up the drive. To obviate this misuse of the clutch it may be interconnected with the accelerator pedal so that the latter cannot be depressed while the clutch is disengaged.

21.27   Fluid-friction clutch

The losses that occur in fluid couplings due to the continuous slip at normal running speeds are eliminated in a combination of fluid coupling and centrifugal clutch made by Self-Changing Gears Ltd of Coventry, and called by them the *fluid-friction clutch*. Its construction is shown in Fig. 21.25. The impeller A is bolted to the flywheel and the driven member C is fixed to the output shaft D in the usual way. A spider E is also fixed to the shaft D and carries four shoes F lined with friction fabric. These shoes are pivoted on pins carried in blocks H which are free to slide radially in slots formed in the flange of the spider E. At low speeds the shoes are held out of contact with the inside of the flywheel B by a spring J, which encircles flanges formed on the shoes, but when the speed rises beyond a certain value, the shoes move outwards under the action of centrifugal force and engage the flywheel. The slip between the latter and the driven shaft D is thereby reduced to zero. Thus the smooth take-up of the fluid coupling is retained but the continuous slip associated with it is eliminated. The shoes can be mounted on their pivots so that they either trail or lead (see Section 32.5).

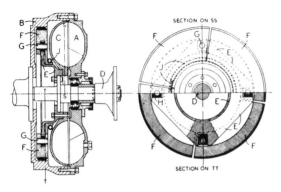

Fig. 21.25

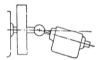

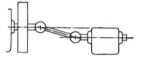

Fig. 21.26   Fig. 21.27

21.28   Connection between the clutch and gearbox

The driven members of the clutches shown in Figs 21.2 and 21.13 are fully supported by the spigot bearings provided within the clutch. The driven members therefore revolve about the same axis as the crankshaft, and they have to be coupled up to the gearbox shaft.

If a *rigid* coupling is used to join two shafts which have to revolve, it is essential that the axes of the shafts shall always coincide. Otherwise very heavy loads will be imposed on the bearings supporting the shafts and on the shafts themselves.

When the engine and the gearbox are independently mounted on the frame it is impossible to ensure that the axes of their shafts, which have to be coupled together, shall always coincide. Although initially the shafts may be *in alignment*, as it is termed, unavoidable flexures of the frame will upset that alignment. A rigid coupling cannot then be used; the coupling must be able to allow for the alignment. The latter may be of two kinds, an angular displacement as in Fig. 21.26, or a bodily displacement as in Fig. 21.27.

The first kind of misalignment can be allowed for by using either a flexible coupling or what is known as a *universal joint*. These will be dealt with in Chapter 26. The second kind of misalignment necessitates the use of two such joints and a short intermediate shaft as shown in Fig. 21.27. When the engine and the gearbox are separate units, therefore, the driven member of the clutch, or the clutch shaft, will be coupled to the gearbox by an intermediate shaft with a flexible or universal joint at each end.

When the gearbox is made integral with, or is rigidly bolted to the crankcase of the engine, so as to form a single unit with the latter, it is possible to ensure the permanent coincidence of the clutch and gearbox shaft axes. Those shafts may then be coupled by a rigid coupling, but usually the clutch shaft will be one with the gearbox shaft. In that case the single clutch gearbox shaft will be supported at one end in the spigot bearing of the clutch and at the other end in the gearbox bearings. The clutches shown in Figs 21.5, 21.8 and 21.10 are intended to be used in this manner.

Chapter 22

Why is a gearbox necessary?

Control over power output, by means of the throttle pedal, simply regulates the rate at which the engine is doing work: at very high speeds, the power output will be correspondingly high but, as has been explained in Chapter 1, the torque output can at the same time be significantly less than at considerably lower speeds. In other words, maximum torque may be available over only a very limited speed range. Consequently, one needs to be able to regulate both the power output and the speed range of the engine relative to the range of speeds over which the vehicle is at any given time likely to be required to operate. Only in this way can the torque at the wheels be balanced against demands for either a steady speed uphill or downhill, or on the level, or for acceleration or deceleration. A gearbox is necessary, therefore, so that the driver can regulate torque by selecting the appropriate speed range or, in other words, the vehicle speed at which maximum torque is obtainable.

When a vehicle is moving at a uniform speed, the driving force, or *tractive effort*, at the wheels must be such as to exactly balance the sum of three categories of variable forces tending to oppose the motion. If it is greater, the car will accelerate, and if it smaller, it will decelerate until a balance is obtained. Such a balance will be established eventually, because two of the forces vary with speed. The three forces are: (1) aerodynamic, or air, resistance; (2) gradient resistance, which can be either positive or negative; and (3) rolling resistance.

22.1 Aerodynamic forces

Air offers a resistance to the passage of bodies through it, as does water or any other fluid. The magnitude of this resistance is dependent directly upon the shape and frontal area of the body exposed to the fluid it is passing through, and to the square of its velocity. Some practical indication of these phenomena can be gained by pushing a small flat plate first edgewise and then frontally in a tank of water, and at a various speeds. With a plate of larger area, similar effects can be felt in air. A graph of air resistance against speed of a vehicle is shown in Fig. 22.1, from which it can be seen that doubling the speed quadruples the air drag.

A practical implication is that for slow vehicles such as farm tractors air resistance is so small that it can be ignored while for high speed vehicles such as racing cars it is of paramount importance. For a family saloon car, it begins to

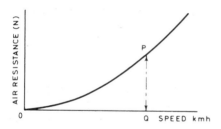

Fig. 22.1

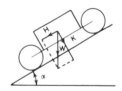

Fig. 22.2

become significant at about 45 mph (72 km/h). The economy of operation of most heavy commercial vehicles, on the other hand, because of the rectangular form of their bodies and therefore almost a flat plate approach to the air, can be significantly affected by increasing speed of operation, especially if they are being driven into the wind.

22.2   Gradient resistance

A diagram representing a car standing on a gradient is shown in Fig. 22.2. The weight of the car, acting vertically downwards, can be resolved into two components: H parallel to the slope and K perpendicular to the slope. To prevent the car from rolling downwards, a force equal and opposite to H has to be applied by the wheels at their contact with the road surface. If the car is to be propelled up the slope, not only that force but also an additional force must be applied, to neutralise the aerodynamic and rolling resistances as well as the force H. It follows that the gradient resistance H is dependent solely on the steepness of the slope and is unaffected by the speed of the car, provided it is constant, either up or down the gradient. It should be noted, however, that during acceleration another force acting, in effect, through the centre of gravity of the car is introduced.

22.3   Rolling resistance

All the remaining forces resisting motion at a constant speed come under the heading of rolling resistance. In some circumstances, they include not only the resistance of the tyres but also the effects of the friction in the transmission system. In general, however, the latter differ according to which gear is engaged so, for calculation purposes, they are normally deducted from the torque available from the engine in that gear.

Rolling resistance is attributable to both the dissipation of energy by the continuous impact of the tyre on the road, and the actual deformation of the tyre and, especially on soft surfaces, of the road too. It should be noted that, on soft surfaces, the wheels are, in effect, being continuously driven up a ramp to lift the vehicle out of the rut into which its wheels have sunk.

It follows that, the softer the tyre or track, and the greater the volume of rubber being deflected, the greater is the rolling resistance, though speed does have a small influence. Impact resistance, on the other hand, is positively influenced by both the speed of the vehicle and its suspension characteristics. However, since impact resistance is a small proportion of the total, it is usual, unless extreme accuracy is required, to take rolling resistance on a good road as being directly proportional to load.

22.4   Total resistance

The total resistance to the motion of a vehicle is the sum of the above three resistances and is thus composed of two parts that are independent of the speed of the vehicle – the rolling resistance and the gradient resistance – and of one part that is dependent on the speed – the air resistance.

A curve of total resistance against speed is therefore obtained by shifting the curve of Fig. 22.1 up vertically by the amount of the rolling and gradient resistances as is shown in Fig. 22.3.

Thus when the speed is OS km/h the total resistance SP is composed of the rolling resistance SR, the gradient resistance RQ and the air resistance QP. If either the gradient resistance or the rolling resistance increases or decreases then the curve would simply shift up or down by the amount of the increase or decrease.

In Fig. 22.4 the curves A, B and C are the curve of Fig. 22.1 shifted up by various amounts, they therefore represent total resistance curves for a given vehicle on roads of either different surfaces or different gradients. Considering the road surface to remain unchanged then curve A might represent the total resistance on the level, curve B the total resistance on a gradient of say 1 in 30, and so on.

22.5   Tractive effort

Having dealt with the resistance to the motion of the vehicle, let us turn to the tractive effort (TE). The source of this is the engine, which turns the clutch shaft with a torque T. This torque is transmitted to the gearbox. Now by applying the principle of the conservation of energy, we shall find that if frictional losses are neglected, and if there is no storing of energy, the whole of the energy put

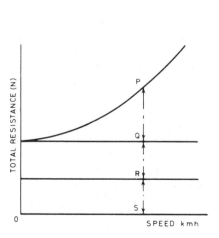

Fig. 22.3

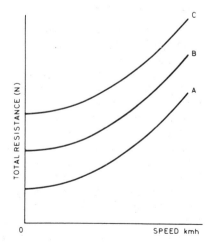

Fig. 22.4

into the gearbox at the engine end must be given out at the propeller shaft end. Since the work done in unit time is measured (when rotations are considered) by the product of the torque and the speed, it follows that the product of the engine torque and the speed of the clutch shaft must equal the product of the torque acting on the propeller shaft and the propeller shaft speed. If, therefore, the propeller shaft speed is $1/n$ the engine speed, then the torque acting on the propeller shaft must be n times the engine torque. We have then, propeller shaft torque $= n \times T$, n being the gearbox gear ratio between the speeds of the engine and the propeller shaft and T being the engine torque.

The propeller shaft drives the road wheels through the final drive where another reduction of speed occurs. If the road wheel speed is $1/n$ of the propeller shaft speed then the torque acting on the road wheel (the two wheels being considered as a single wheel) will be m times the propeller shaft torque, again neglecting frictional losses and energy storage. The torque acting on the road wheel is therefore

$$t = n \times m \times T$$

where n is the gearbox ratio, m the final drive ratio and T the engine torque.

The way in which this torque produces a driving force to propel the car along the road is shown in Fig. 22.5. If the wheel there shown is regarded as being in equilibrium, then the forces that act upon it must be in equilibrium and the couples also. Now at every instant the wheel can be considered as a lever (as shown by the dotted lines) fulcrumed at the point of contact of the wheel and the ground.

Under the action of the torque t the lever will tend to rotate about the point of contact with the ground and the centre of the wheel will tend to go forwards and to take the axle and the vehicle with it. At its centre therefore the wheel is pressing forwards on the axle casing with a force which we will call P_1. The reaction P_2 of this force (P_1) acts backwards on the wheel. Since the wheel is in equilibrium there must be an equal and opposite force P_3 acting on it. This is the adhesive force between the wheel and the road.

The forces P_2 and P_3 constitute a couple tending to turn the wheel in the

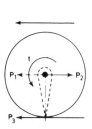

Fig. 22.5

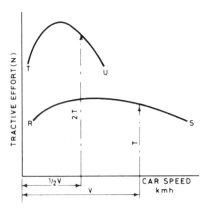

Fig. 22.6

clockwise direction and since the wheel is in equilibrium, and the couples acting on it are therefore also in equilibrium, the couple P_2P_3 must equal the couple t applied to the wheel by the driving shaft. Now the magnitude of the couple P_2P_3 is $P_3 \times R$ where R is the distance between the forces P_2P_3, in this case the radius of the wheel. Hence—

$$t = P_3 \times R$$

and since $t = n \times m \times T$ we have

$$n \times m \times T = P_3 \times R$$

or

$$P_3 = \frac{n \times m \times T}{R}$$

Now the values of the final drive ratio m and the wheel radius R are, for any given vehicle, constants, and for any particular gearbox gear the value of the ratio n is constant. Hence for any particular gearbox gear the value of the fraction $(n \times m)/R$ is a constant which may be called K.

The tractive effort is then given by the equation $P = K \times T$ where T is the engine torque and K is a constant whose value depends upon the road wheel radius, the final drive ratio and upon the gearbox gear ratio.

22.6 Variation of the tractive effort with speed

Now, since the engine is geared to the driving wheels, a particular engine speed corresponds to a particular vehicle speed, and since the tractive effort is proportional to the engine torque, the variation of the tractive effort with the variation of the vehicle speed will depend upon the variation of the engine torque with the variation of engine speed. This last relation has already been considered and a curve showing it is given in Fig. 3.7. The curve showing the relation between tractive effort and vehicle speed will be of the same shape; in fact, the same curve might be used provided that the scales were suitably altered.

Some typical diagrams showing how engine torque varies with speed are illustrated in Figs 1.7 and 1.8 and, since the engine and road wheels are locked together by intermeshing gears, graphs showing the relationship between either torque or tractive effort and vehicle speed are the same shape. Consequently, provided the scales are appropriately altered, the same curves can be used for both. In other words, for a given gear ratio, each engine speed inescapably corresponds to a certain vehicle speed.

A plot of tractive effort, against vehicle speed for a given gearbox ratio, a given final drive ratio and a given road wheel radius is shown at RS in Fig. 22.6. If the gearbox ratio is altered we shall get another curve of tractive effort. Thus if the gearbox ratio is altered so that the total gear ratio between the engine and the back wheels is double what it originally was then the curve RS will become the curve TU, the relation between the curves being that all the horizontal distances of RS are halved and the corresponding vertical distances are doubled to give TU, since for a given engine speed doubling the total gear ratio will halve the vehicle speed but will double the tractive effort.

22.7   Performance curves

Having thus obtained curves showing the variation, with varying vehicle speed, of both the tractive effort and the total resistance to be overcome, let these curves be plotted to the same scales and on the same sheet of paper, as has been done in Fig. 22.7. In that figure the curves A–F are curves of total resistance for a road with a uniform surface but of varying gradient, curve A being the level and curve F the steepest gradient. Curves RS, TU and VW are curves of tractive effort for three different gear ratios, RS being, say, the top gear, TU the next lower gear, etc.

Suppose the vehicle is travelling on the level at a speed represented by OX. Then the resistance to be overcome is XY, while the tractive effort available is XZ. The tractive effort available is therefore greater than the resistance to be overcome and the excess tractive effort YZ will go to increase the speed of the vehicle. Now as the speed increases the resistance also increases but the tractive effort, it will be noticed, falls off. The excess tractive effort, given by the intercepts corresponding to YZ, which is available for acceleration, becomes smaller and smaller as the speed increases until, when the speed is OM, the tractive effort available is only just equal to the resistance to be overcome. There is therefore no excess tractive effort available for acceleration, and the speed cannot be increased further, OM represents the highest speed the vehicle can reach on the level road to which the curve A applies.

The curve RS represents the tractive effort with the engine running with the

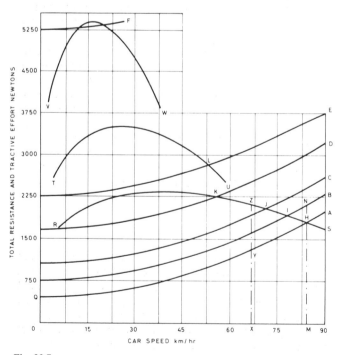

Fig. 22.7

throttle wide open. If it is not desired to increase the speed beyond the value OX then the throttle would be closed until the tractive effort was equal to XY, and the speed would be maintained but not increased.

Now suppose the vehicle is travelling on the level at the maximum speed OM, and that it comes to a gradient to which the curve B applies. At the speed OM on the gradient B the resistance is MN, but the tractive effort available is only MH. Hence the excess resistance HN will slow the vehicle down and the speed will fall to the value given by the point I, at which the tractive effort is equal to the resistance to be overcome.

Now suppose that the gradient becomes steeper and steeper so that we pass in succession from curve B to curve C and so on. The speed that can be maintained gets progressively lower as indicated by the points J, K, etc., and it will be seen that we cannot traverse the gradient E at any speed since the tractive effort curve lies everywhere below the resistance curve.

Now, with ordinary sized engines, if the gear ratio is such as to give a reasonable maximum speed on the level, then quite a medium gradient would bring about the conditions represented by the curve E, and the vehicle would be brought to rest. But if gear ratio between the engine and the driving wheels can be altered we can pass from the tractive effort curve RS to the curve TU and then the gradient E can be traversed at the speed given by the point L.

Thus in order to permit a reasonably high maximum speed on the level and at the same time to be able to climb medium gradients we require to have available two different gear ratios. Similarly in order to permit reasonably high speeds up medium gradients and yet be able to climb steep gradients, we need a third gear ratio, and a fourth or fifth ratio may be desirable.

It should also be clear that at the lower speeds, when quick acceleration is especially desirable, there is a greater excess of tractive effort available on a lower gear than on a higher one.

The above is the complete explanation of the *raison d'être* of the gearbox, and it should be clear that the ideal gearbox would provide an infinite range of gear ratios so that the engine speed could be kept at or near that at which maximum power was developed, whatever the speed of the vehicle. This assumes that maximum speed is the objective. If maximum economy is desired, the engine requires to be run at the highest torque possible for a given power output. Many of the developments in transmission, such as the various hydraulic and electrical mechanisms, have as their principal object the multiplication of the number of ratios available. In general though, a compromise is adopted and three, four or more gear ratios are provided depending on the size of engine fitted and other considerations.

22.8 Clutch action

Fig. 22.7 also serves to show how the clutch enables a car to be started up from rest. When the car is at rest the resistance to be overcome before the car will move is, for the level road A, given by OQ. If the engine were in permanent connection with the driving wheels it would, of course, be at rest and the tractive effort would be zero. The clutch, however, allows the engine to be run at a speed at which a torque giving a greater tractive effort than OQ is developed and enables this torque to be transmitted to the driving wheels, though at the start the latter are at rest.

22.9 Constant power TE speed curve

If the engine of a vehicle could be made to give its maximum power at all speeds then, since Power = TE × Speed, it follows that the TE will be inversely proportional to the road speed. The graph of the TE plotted against the road speed would then be like the full line curve in Fig. 22.8. The TE curves for an actual engine and gearbox combination will touch this constant power curve at one point (corresponding to the engine speed at which the engine gives its maximum power), but will lie everywhere else inside it, as shown by the dotted line curves. The constant power curve is the ideal form of TE curve and its shape is approached by those of a steam engine and of traction electric motors; it could be obtained with any engine if an infinitely variable gearbox of 100% efficiency were available.

22.10 Performance curves on a horsepower basis

Instead of considering the performance as a balance between the tractive effort available and the resistance to be overcome, that is, as a balance of forces, we may consider it as a balance between the rate of producing energy by the engine and the rate of using energy at the road wheels, that is, as a balance of powers. This is shown in Fig. 22.9. The curves A and E represent the horsepower required to overcome the resistances represented by the curves A and E of Fig. 22.7 at corresponding speeds, and the curves RS and TU represent the horsepower available at the road wheels on the two gears corresponding to the curves RS and TU of Fig. 22.7. The power available at any moment is, of course, equal to the power developed by the engine at that moment if frictional losses are neglected, the only effect of a change of gear ratio being to alter the road speed corresponding to a given engine speed.

The greatest speeds possible on the roads to which A and E refer are given by the intersections of the horizontal line representing the maximum horsepower of the engine and the curves A and E, and the best gear ratios for those roads will be such that at those greatest road speeds the engine speed is that at which the maximum power is developed. Any variation of gear ratio from these optimum values will result in a lowering of the maximum speed attainable. The

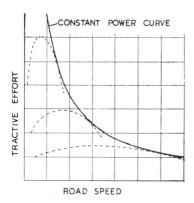

Fig. 22.8

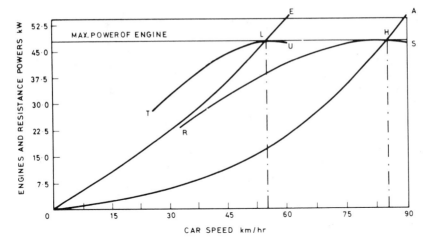

Fig. 22.9

maximum speeds possible with the gear ratios for which the curves RS and TU are drawn are given by the intersections H and L respectively.

The process of settling the gear ratios for a vehicle is ultimately an experimental one, but approximate values which will serve as a starting point may be derived as follows.

A curve corresponding to the curve A of Fig. 22.9 is drawn, and from its intersection with the line of maximum engine horsepower the greatest possible road speed found. The top gear ratio is then made such that this road speed corresponds to the engine speed at which maximum horsepower is obtained. Next the maximum gradient that is to be negotiable must be assumed and the corresponding gradient resistance found. This gives the maximum tractive effort required, and the maximum engine torque being known the gear ratio necessary to enable this tractive effort to be obtained may be determined. The top and bottom gear ratios are thus settled and the remaining ratios may be put in so that they form a geometric progression, this step being justified as follows.

The engine has a certain range of speed within which the power developed is not very much less than the maximum, and the gear ratios should be such that the engine speed can be kept within that range. Let the lower limit of the range be L rev/min and the higher limit be M rev/min. Suppose the car to be moving in lowest gear and the engine to be speeded to the higher limit.

If the engine speed is to be kept within the specified range the gear must now be changed. Immediately after the change has been made the engine speed should be down to the lower limit L while the road speed of the car will be unaltered. If the lower gear ratio is $A:1$ then the road speed corresponding to an engine speed of M rev/min will be equal to $K \times M/A$, where K is a constant depending on the size of the road wheels, and if the next gear ratio is $B:1$ the road speed corresponding to an engine speed of L rev/min will be equal to $K \times L/B$. But these two road speeds are the same; hence $K \times M/A = K \times L/B$ or $B = A \times L/M$, that is, $B = c \times A$ where c is a constant equal to L/M. By similar reasoning it will be found that the next gear ratio $C:1$ is given by $C = c \times B = c^2 \times A$. Hence the gear ratios form a geometric progression.

Now for an engine having a sharply peaked horsepower curve c will be greater than for an engine having a flatter curve, and the number of intermediate gears required to bridge the gap between the top and bottom gears will be greater. Hence a peaked horsepower curve calls for a multiplicity of gear ratios, while a flatter power curve needs but two or three. As an example, suppose the top gear is to be 4:1 and the bottom gear 16:1. Then if $c = 0.5$ the ratios will be 16:1, 8:1 and 4:1; while if $c = 0.63$ the ratios will be 16:1, 10.1:1, 6.35:1 and 4:1, so that in the first case three gears are required and in the second, four.

If performance curves on a tractive-effort basis are now plotted, using the ratios found as above, the performance can be examined and modifications tried.

The number of gears required, however, depends also on the duty for which the vehicle is to be used. For example, a cross-country vehicle obviously will need higher numerical gear ratios (lower gearing) than an equivalent car used only on the road. If the cross-country vehicle is to be used also on the road, and a high top speed is required of it, five speeds will be desirable as compared with, for example, four for the comparable road-only vehicle.

In the heavy range, the requirements vary even more widely according to not only the terrain – gradients, altitudes, etc. – but also the range of loading to be expected. A lightly laden or empty vehicle, for instance, may need only four to six gears, whereas the same vehicle fully laden may need eight to twelve. To meet these requirements we have range-change and splitter gearboxes, as described in Section 23.21. Requirements depend also on the traffic which, if dense, may require that the driver of a heavy vehicle change gear frequently. He will need also more gear ratios so that he can keep with the other traffic without running his engine outside its economical speed range and thus seriously increasing his rate of fuel consumption. With very heavy vehicles – 40 tonne and over – as many as 16 to 17 gear ratios may be necessary.

Chapter 23

Constructional arrangements of gearboxes

The gearboxes which are, or have been, used in motor vehicles may be divided into those in which the drive in every ratio is transmitted indirectly through gear teeth which mesh together, and those in which a direct shaft-drive is provided for one of the ratios while for the others the drive has to be transmitted through gear teeth. Gearboxes may also be classified in the following types—

(1) Sliding-mesh.
(2) Constant-mesh.
(3) Epicyclic.

Gearboxes which are a mixture of two of these types are not uncommon. Spur gearing is used in all three types, the differences between which lie in the manner in which the gears are brought into action.

The sliding-mesh type is the simplest and historically is the oldest; it may conveniently be dealt with first although the constant-mesh type is now the most widely used type.

23.1   Sliding-mesh gearbox

A sliding-mesh gearbox is shown in Fig. 23.1. The engine is coupled, through the clutch and clutch-shaft, to the short shaft A which is integral with the spur pinion B. The shaft A is supported in two journal ball bearings and is fixed axially in a manner that will be described later. The pinion B meshes continuously with the spur wheel D secured to the layshaft E, which is arranged parallel to the shaft A and carried at its ends in the ball bearings shown. Secured to the layshaft (which is splined throughout its length) are four other spur gears, F, G, H and J. The latter is continuously in mesh with a pinion Q which is free to revolve on a pin fixed in the casing. A third shaft (usually called the *mainshaft*) is arranged in line with the shaft A, being supported at one end in a ball bearing housed in the casing C_2 and at the other end by the spigot L which is part of the shaft K and which fits in a bushed hole in the shaft A which, incidentally, is sometimes termed the *primary*, or *first motion*, shaft.

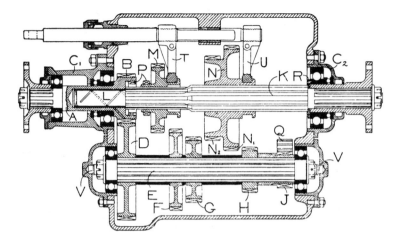

Fig. 23.1 Sliding-mesh gearbox

The shaft K is splined for nearly its whole length and carries the independent members M and N. M is a single spur gear while N is a double gear, being in effect two gears N_1 and N_2 made in one piece. The gears M and N can be slid along the mainshaft when required by selector forks T and U respectively. These forks are secured to rods that can slide in bushed holes in the casing as shown, and their prongs fit into grooves cut in the bosses of the gears so that the latter can revolve freely, but must slide axially when the forks are moved. Teeth P, similar to spur gear teeth, are cut on the boss of the gear M and corresponding internal teeth are cut inside the constant-mesh pinion B and when the gear M is slid along its shaft to the left its teeth P fit into the spaces of the teeth P of the pinion B thus locking those members together. This arrangement is one form of positive clutch; an alternative form is composed of two members, each like that shown in Fig. 23.2. This latter form gives rise to the name *dog-tooth clutch*.

The mainshaft K is connected, through the propeller shaft, to the road wheels. The layshaft is fixed axially by thrust buttons V. The spigot L has a spiral groove cut on it to lead oil to the bearing and at its left-hand end bears, through a thrust button, against the engine shaft A. The mainshaft is thus fixed axially by the thrust button at the left and at the right by the single-thrust ball bearing in the casing C_2. The engine shaft A is fixed axially at the right by the thrust button bearing against the end of the mainshaft and at the left by the

Fig. 23.2 One member of a dog clutch

single-thrust ball bearing in the casing C_1. The proper clearance for the thrust bearings is obtained by grinding the washer R to the proper thickness. Separate ball thrust bearings are not always provided, the journal bearings being arranged to position the shafts axially.

The gearbox shown provides four forward speeds and one reverse, and the operation is as follows.

23.2   First or low gear

The gear M occupies the position shown in Fig. 23.1. The gear N is slid along the mainshaft K until it occupies the position shown in Fig. 23.3(a). It then meshes with the gear H of the layshaft. The drive comes from the engine shaft through the constant-mesh gears B and D to the layshaft, is then transmitted through that shaft to the gear H and thence through the gear N_1 to the mainshaft. The *gear ratio*, or the ratio between the speeds of the engine and mainshafts, is—

$$\frac{\text{Speed of engine shaft}}{\text{Speed of mainshaft}} = \frac{\text{No. teeth in D}}{\text{No. teeth in B}} \times \frac{\text{No. teeth in N}}{\text{No. teeth in H}} = n_1$$

The torque driving the mainshaft is now n_1 times the torque acting on the engine shaft.

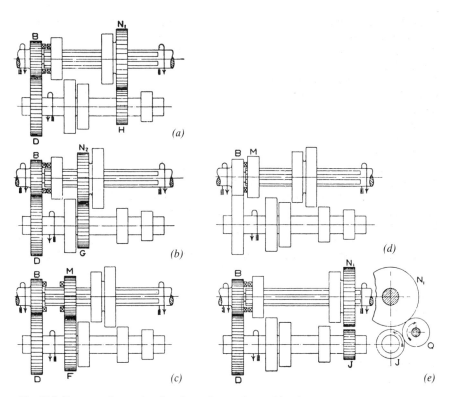

Fig. 23.3 Four-speed gear showing the various ratio combinations

23.3   Second gear

The gear M continues to occupy the same position, but the gear N is slid over to the left as shown in Fig. 23.3(b). It then meshes with the gear G of the layshaft. The drive is from the engine shaft through the constant-mesh gears to the layshaft, through that shaft to the gear G thence to the wheel N_2 and the mainshaft. The gear ratio is—

$$\frac{\text{Speed of engine shaft}}{\text{Speed of mainshaft}} = \frac{\text{No. teeth in D}}{\text{No. teeth in B}} \times \frac{\text{No. teeth in } N_2}{\text{No. teeth in G}}$$

23.4   Third gear

The gear N occupies the position shown in Fig. 23.1. The gear M is slid to the right into the position shown in Fig. 23.3(c), where it meshes with the gear F of the layshaft. The drive is from the engine shaft through the constant-mesh gears to the layshaft, thence to the gear F, through that gear to the gear M and the mainshaft. The gear ratio is—

$$\frac{\text{Speed of engine shaft}}{\text{Speed of mainshaft}} = \frac{\text{No. teeth in D}}{\text{No. teeth in B}} \times \frac{\text{No. teeth in M}}{\text{No. teeth in F}}$$

23.5   Fourth or top gear

The gear N occupies the position shown in Fig. 23.1, while the gear M is slid over to the left as shown in Fig. 23.3(d). The dog teeth on M then engage in the spaces between the dog teeth on B, thus connecting B and M and giving a direct drive between the engine and the mainshaft. The gear ratio is then 1 : 1. The layshaft now revolves idly.

On all the above gears the direction of rotation of the layshaft is the opposite to that of the engine shaft, while the direction of rotation of the mainshaft is opposite to that of the layshaft. The mainshaft rotates in the same direction as the engine shaft.

23.6   Reverse gear

The member M occupies the position shown in Fig. 23.1. The gear N is slid over to the right, but farther than when first gear was being obtained. It reaches the position shown in Fig. 23.3(e); but in this position it does *not* mesh with the pinion J of the layshaft, since that pinion is made small enough to clear the gear N_1. The gear N_1 does, however, mesh with the reverse idler Q which is carried on a shaft on which it is free to revolve. The shaft is fixed in the gearbox casing and the idler Q is constantly in mesh with the pinion J of the layshaft.

The drive is from the engine shaft through the constant-mesh gears and the layshaft to the pinion J, thence to the idler Q, and thence to the wheel N_1 and the mainshaft. The direction of rotation of the idler Q is the opposite to that of the layshaft, and hence is the same as that of the engine shaft. The direction of rotation of the gear N_1 is the opposite to that of the idler Q, and hence is opposite to that of the engine shaft. The gear ratio is—

$$\frac{\text{Speed of engine shaft}}{\text{Speed of mainshaft}} = \frac{\text{No. teeth in D}}{\text{No. teeth in B}} \times \frac{\text{No. teeth in } N_1}{\text{No. teeth in J}}$$

It will be noticed that the number of teeth in the idler does not affect the gear *ratio*, but only the direction of rotation.

A different method of engaging the reverse gear is shown in Fig. 23.4, which shows a gearbox otherwise identical with that just described. The layshaft now carries only four gears, the constant-mesh gear, the third-speed gear, the second-speed gear and the first-speed pinion H. Instead of a simple reverse idler there is now a compound one Q_{1-2}, the two portions of which are of different diameters. This idler is carried on a shaft arranged parallel to the mainshaft, but besides being free to revolve on that shaft it can also be slid axially along it in order to engage the reverse gear.

In the position shown at (*a*) the idler is in the neutral position and there is no drive between the engine shaft and the mainshaft. If, however, the idler is slid to the right, to the position shown at (*b*), the part Q_2 will engage with the pinion H of the layshaft, while the part Q_1 will engage with the gear N_1 on the mainshaft. The drive will then be from the engine shaft through the constant-mesh gears and the layshaft to the pinion H, which then drives the idler Q_2; Q_1 being integral with Q_2 is also driven, and in turn drives the gear N_1 and thus the mainshaft. As before, the mainshaft goes in the opposite direction to the engine shaft, thus giving a reverse drive. The gear ratio is—

$$\frac{\text{Speed of engine shaft}}{\text{Speed of mainshaft}} =$$

$$\frac{\text{No. teeth in D}}{\text{No. teeth in B}} \times \frac{\text{No. teeth in } Q_2}{\text{No. teeth in H}} \times \frac{\text{No. teeth in } N_1}{\text{No. teeth in } Q_1}$$

The chief advantage of this method of obtaining the reverse is that the mainshaft and the layshaft can be made shorter than when the first method is used. The deflection of the shorter shafts under a given load will be less than that of the longer shafts, if the diameters are the same, and one cause of noisy operation is reduced. If the longer shafts are sufficiently stiff then the shorter shafts may be made smaller in diameter. A gearbox of the second type will

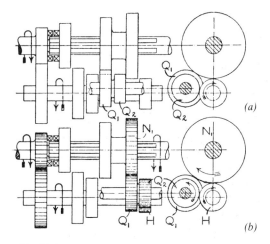

Fig. 23.4

generally be less bulky than a similar one of the first type. The second construction, however, involves the use of three sliding members instead of two, and makes the control mechanism slightly more complicated.

23.7   Control mechanism

The sliding of the members M, N and (when required) Q, is effected by means of selector forks, an example of which is shown in Fig. 23.7. The fork fits into a groove formed in the boss of the gear to be moved, so that although the gear is left free to revolve, it must partake of any sideways movement that is given to the fork. There will be a selector fork for each sliding member in the gearbox. The selector forks either slide on rods fixed in the gearbox casing or are fixed to rods which can slide in that casing, the rods being parallel to the shafts upon which the gears slide. The necessary sliding motions are given to the selector forks by the motion of a gear change lever actuated by the driver, but since there is only one gear change lever and there are two or three selector forks, the driver must be able to select the one belonging to the gear he desires to move. The principal forms of the mechanism that enables this to be done will now be dealt with.

23.8   Sliding-type selector mechanism

This type is now hardly ever used but it has been thought advisable to retain a description of it. In the example shown in Fig. 23.5 (1) is a sectional elevation, the plane of the section being indicated by the line SS in the end view (2). The latter is a section on AB. The third view is a part plan. There are three moving members in the gearbox into which this particular mechanism is fitted so that there are three selector forks, C, D and E. The forks C and D slide on rods F and G fixed in the casing, while E is carried by a pivoted lever Q which is actuated by a member that slides on the third rod H. The forks are moved by a fore and aft rocking motion of the gear lever J which is carried by a shaft L pivoted in the casing and to the inner end of which is secured the striking lever K. The particular fork that is to be moved is selected by a sideways sliding motion of the member JLK. To hold the forks in their various positions spring plungers, one of which is seen in (1), and which spring into grooves cut in the rods FGH, are fitted.

To prevent two forks from being moved at once a locking piece M is provided. This slides on a cross rod N fixed in the casing, and is provided with horns O and P which project into the slots in the sliding members. Between the horns O and P is situated the end of the striking lever K so that the sideways movement of the latter causes the member M to slide on its rod. The gap between the horns O and P is only slightly wider than each of the sliding members, so that the latter can be moved only one at a time.

23.9   Ball-type selector mechanism

In this form of selector mechanism, shown in Fig. 23.6, the control lever is mounted on the transmission casing. Some can have this control on the steering column, as in Fig. 23.10, in which case it is connected by rods and levers to the mechanism on the box, which in principle remains unaltered. The

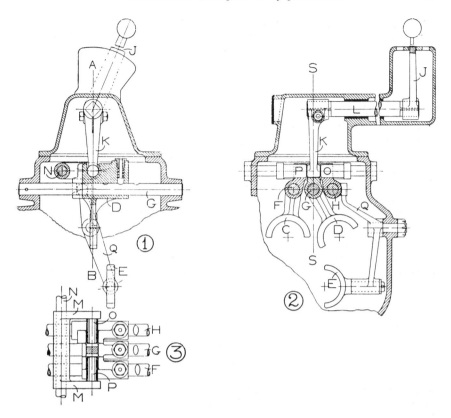

Fig. 23.5 Sliding-type selector mechanism

selector forks A and B slide on rods fixed in the gearbox lid, which in this design carries the whole of the selector mechanism. The shape of the forks is shown by the perspective sketch Fig. 23.7; they are provided with slots C to receive the end D of the striking arm. The latter is the lower end of the gear lever E which is ball jointed in the casing at F. By rocking the lever sideways its end D may be brought into engagement with either of the selector forks, when a fore and aft rocking motion will slide that fork along its rod.

No gate is provided, but small plungers G and H prevent both forks from being moved at once. When both the forks are in the neutral position, and the slots C are opposite each other, the plungers are forced by small springs into holes in the forks, and before either fork can be moved the plunger that locks it must be pressed back into the casing. This is done by the sideways motion of the gear lever. Obviously when one plunger is pressed in to release one of the forks the other plunger is out, and is locking the other fork. These plungers also serve to lock the forks in position when the gears are properly engaged, being arranged to spring into shallow recesses NN in the forks.

Another selector mechanism, suitable for the remote control of the gearbox, is shown in Fig. 23.8. The gear lever A is carried in a ball-and-socket bearing O and its spherical lower end engages a cylindrical hole formed in an arm fixed to

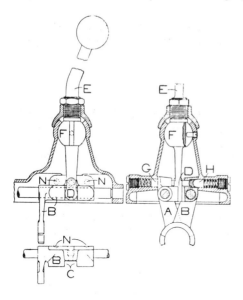

Fig. 23.6 Ball-type selector

Fig. 23.7 A selector fork

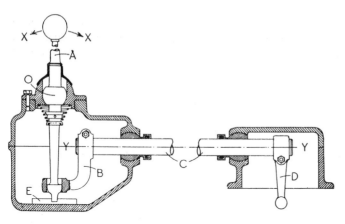

Fig. 23.8

a shaft C that is free both to turn and to slide in bearings. This shaft is coupled rigidly to a shaft carried in the gearbox (see also Fig. 23.14). The striking arm D is fixed to the latter shaft and its lower end engages the selector forks. A sideways rocking motion of the gear lever produces a rotation of the shaft and enables the striking arm to select any desired selector rod and a fore and aft rocking motion of the gear lever then slides the shaft and thus the selector fork that has been engaged, thereby engaging the gears as required. The use of spherical bearings for the intermediate shaft accommodates any relative movement between the gearbox and the bracket that carries the gear lever, alternatively plain cylindrical bearings may be used and universal joints be provided at each end of the intermediate shaft. The lower end of the gear lever engages a grooved plate E which prevents reverse gear being engaged inadvertently. The lever must be lifted up to clear the plate E in order to engage reverse.

23.10   Steering column gear shift control

Gear change levers situated just below the steering wheel are now rarely used, except for some light commercial vehicles, such as forward control vans, in which there is a need for the floor to be free from obstruction so that the driver can move freely from one side of the cab to the other. A typical example of a steering column-mounted control is shown in Figs 23.9 and 23.10. The gearbox concerned is a four-speed and reverse type and there are three striking forks A, B and C, of which A is for the reverse and B and C for the forward gears. The striking forks are slid along their rods by the striking lever D when the lever E is rotated about the axis OO. The appropriate fork is selected by moving the lever F, the lever G then rotates the interlocking member H about the axis XX and pivots the lever D about the pin J, thereby bringing it into engagement with the appropriate fork. The other forks are locked in the neutral position by the fingers K and L of the interlocking member which is shown separately in perspective in the lower left-hand corner of the illustration.

Fig. 23.10 shows the steering column mechanism in which the lever A is pivoted to the end of a rod C and the pin H engages a slot in the bracket D which can rotate about the axis of the rod but cannot move axially. An up-and-down motion of the end of the lever about the pin H raises or lowers

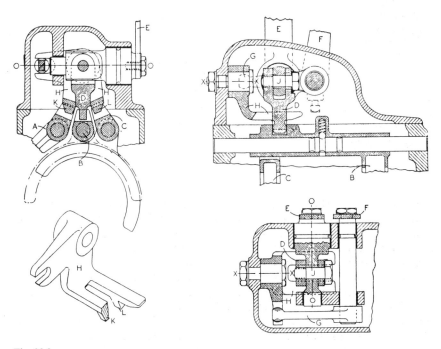

Fig. 23.9

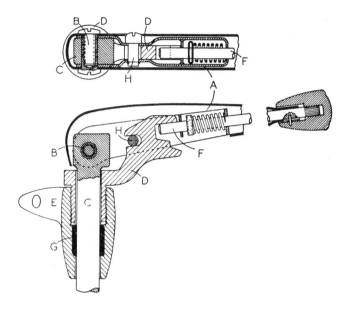

Fig. 23.10

the rod C; this is connected by linkage to the arm F of Fig. 23.10 and this provides the selection. Fore and aft motion of the lever rotates the rod C and an arm at the bottom end of this is coupled to the lever E of Fig. 23.9 to give the striking motion to the striking forks. The selecting motion of the gear lever is normally limited by the engagement of the rod F with a recess in the bracket D so that only forward gears can be selected. By pulling the knob out against the spring the lever can be moved down far enough to select the reverse gear striking fork.

23.11   Constant-mesh gearbox

There are many different forms of constant-mesh gearbox, in some of which the various gears slide axially along their shafts, while in others they have no axial freedom. The characteristic feature of this type of gearbox is, however, that all the pairs of wheels are always in mesh.

The principle of the commonest form of constant-mesh gearbox is shown in Fig. 23.11; the engine shaft A is integral with a pinion B, which meshes with the wheel C on the layshaft. The latter is, therefore, driven by the engine shaft. Wheels E, F and G are fixed to the layshaft just as in a sliding-mesh gearbox, and the mainshaft D is also similarly arranged. The gears E, F and G (the latter through a reverse idler) are, however, in constant mesh with the wheels H, I and J, which are perfectly free to turn on the mainshaft, bronze bushes, or ball or roller bearings, being provided between them and the shaft. The gears H, I and J, therefore, are constantly driven by the engine shaft, but at different speeds, since the wheels E, F and G are of different sizes. The wheel J, being driven through an idler, revolves, of course, in the opposite direction to the engine shaft.

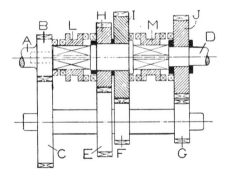

Fig. 23.11 Constant-mesh gearbox

If any one of the gears H, I or J is coupled up to the mainshaft then there will be a driving connection between that shaft and the engine shaft. The coupling is done by means of the dog clutch members L and M, which are carried on squared (or splined) portions of the mainshaft. They are free to slide on those squared portions, but have to revolve with the shaft. If the member M is slid to the left it will couple the wheel I to the mainshaft giving the first gear. The drive is then through the wheels B, C, F and I and the dog clutch M. The other dog clutch is meanwhile in its neutral position as shown. If, with the member M in its neutral position, the member L is slid to the right, it will couple the wheel H to the mainshaft and give second gear, the drive being through the wheels B, C, E and H and the dog clutch L. If the member L is slid to the left it will couple the mainshaft directly to the pinion B and give a direct drive, as in a sliding-mesh gearbox. The reverse gear is engaged by sliding the member M to the right when it will couple the wheel J to the mainshaft. The drive is then through the wheels B, C, G, the idler, J and the dog clutch M.

This type of gearbox has several advantages over the ordinary form of sliding-mesh box. It facilitates the use of helical or double helical gear teeth which are quieter than straight teeth; it lends itself to the incorporation of synchronising devices more readily than the sliding-mesh box; the dog clutch teeth can be made so that they are easier to engage than the teeth of gear wheels, and any damage that results from faulty manipulation occurs to the dog clutch teeth and not to the teeth of the gear wheels. Now, when once the dog clutches are engaged, there is no motion between their teeth, whereas when gear teeth are engaged the power is transmitted through the sliding action of the teeth of one wheel on those of the other. The teeth have to be suitably shaped to be able to transmit the motion properly, and if they are damaged the motion will be imperfect and noise will result. Damage is, however, less likely to occur to the teeth of the dog clutches, since all engage at once, whereas in sliding a pair of gears into mesh the engagement is between two or three teeth.

On the other hand, the wheels on the mainshaft must be free to revolve so that they must either be bushed or be carried on ball or roller bearings. If bushes are used lubrication is difficult, wear will occur and noise will arise. If ordinary ball or roller bearings are used wear is avoided but the gearbox becomes bulky and heavy.

The use of needle-roller bearings overcomes both difficulties.

23.12   A five-speed gearbox

The gearbox shown in Fig. 23.12 is an example of a mixed type of box, the first and second speeds being by sliding-mesh gears and the others by constant-mesh gears. The shaft A is coupled to the driven shaft of the clutch and has the pinion B formed integrally on it. This pinion drives the layshaft to which the gears C, D, E and F are splined. The mainshaft G is supported at the left in a roller spigot bearing housed inside the pinion B, in the centre by a ball bearing carried in an intermediate wall of the gearbox casing and at the right in a combined roller and ball bearing assembly. The ball bearing part of this assembly is used to position the mainshaft axially and to take any thrust that may come on it while the roller bearing takes the greater part of the journal loads. The double gear H is free to slide on splines on the mainshaft and gives first and second speeds when slid to right or left so as to mesh with C or D respectively. The gear J is free to rotate on a bush on the mainshaft and is kept in place by the split ring K which fits in a groove turned in the mainshaft. The split ring itself is kept in place by the overlapping portion of the washer (sectioned in full black) between it and the gear J. The latter is in permanent mesh with the gear E of the layshaft, and when coupled to the mainshaft by sliding the gear L to the right so as to engage the dog clutch teeth M, gives third speed. Fourth speed is a direct drive and is obtained by sliding L to the left so as to engage the dog clutch teeth N. Fifth speed is an overdrive, the ratio being

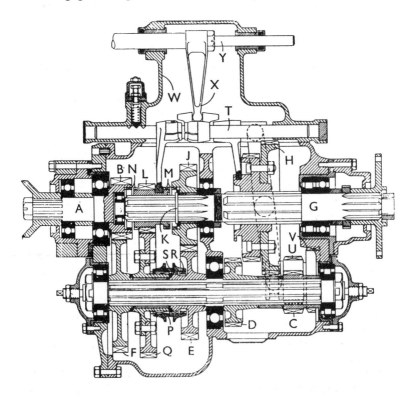

Fig. 23.12 Five-speed gearbox

less than unity, and is given by sliding the sleeve P (which is splined to the boss of the gear Q) to the left. The dog clutch teeth R of the sleeve P then engage the teeth S formed on the boss of the gear E and thus the gear Q is coupled to the layshaft, on which it has hitherto been free to revolve on the sleeve provided for it. The drive is then from B to F, through the teeth S and R to the gear Q, thence to the gear L and the mainshaft. Since the gear Q is larger than the gear F the mainshaft goes faster than the engine shaft A. Reverse is obtained by sliding the reverse idler UV to the left so that the teeth U engage those of the gear H and the teeth V those of the gear C.

The layshaft is carried in a ball bearing in the intermediate wall of the gearbox casing and in roller bearings at the ends. It is positioned axially by the screwed studs shown in the end covers.

The selector forks, by which the various members are slid in order to engage the various gears, are fixed to rods T free to slide in the cover casing W. The selector rods are positioned by spring-loaded plungers as shown. The reverse idler is operated from its selector rod through a rocking lever (shown dotted) pivoted on the wall of the casing. The striking arm X turns about the axis of the shaft Y, to which it is fixed, in order to select the appropriate selector rod and then slides axially in order to slide that rod. The connection between the shaft Y and the change gear lever will be on the lines of that shown in Fig. 23.8. Packing rings are provided in the ends of the bushes carrying the shaft Y and special sealing devices (not shown) are provided where the engine and mainshafts pass through the cover plates in order to prevent the leakage of oil.

23.13 Another example of a constant-mesh gearbox

A modern design of five-speed gearbox in which all the gears except those giving first gear are in constant mesh is shown in Fig. 23.13. The shaft A, which is coupled to the clutch shaft, is supported in two ball bearings the right-hand one of which positions the shaft axially. The pinion B is integral with the shaft A and meshes with the layshaft gear C. The layshaft consists of a splined shaft having an integral pinion E and on which are mounted the gears G, S, I, K and C. It is carried in roller bearings at the ends and a ball bearing in the middle; the left-hand roller bearing positions the shaft axially, both its races being provided with lips for that purpose. The outer races for the roller bearings are mounted in steel shells that are held in the cast aluminium casing of the gearbox by the studs and nuts that secure the end covers.

The gears G, I and K mesh with the gears H, J and N which are free to revolve on the mainshaft D on needle roller bearings, hardened steel sleeves being provided on the shaft for these bearings to run on. The bearings and gears are secured axially by hardened steel washers fixed on the shaft by reason of the nut at the left-hand end. The gears H, J, N and B are coupled to the mainshaft, when required, by the dog clutches L and M; these take the form of sleeves having internal teeth to fit the external teeth of the members O and P which are splined to the mainshaft. The teeth of the sleeves L and M fit the small diameter toothed portions of the gears H, J, N and B to enable the different ratios to be obtained.

Reverse gear is obtained by sliding the gear F to the left to mesh with the portion K of the compound reverse idler; the other toothed portion of this idler meshes permanently with the pinion S of the layshaft.

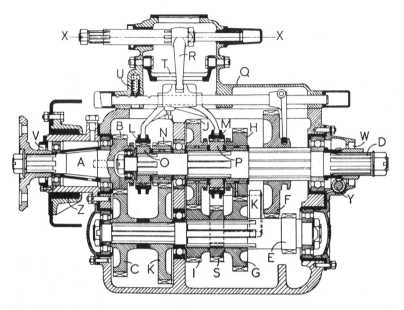

Fig. 23.13

The gear F and the sleeves M and L are actuated by striking forks carried by rods that are free to slide in the cover casting Q and which are moved by the motion of the striking arm R along the axis XX. Selection is done by the rotation of the arm R about the axis XX and a pivoted interlocking member T, similar in principle to that shown in Fig. 23.9 and described previously, prevents two gears from being engaged at once. The selector forks are held in their various positions by the spring-loaded plungers, one of which is seen at U.

Oil seals are housed in the recesses V and W and a speedometer drive is provided at Y.

The box is supported at the front by a rubber bushing Z carried by a cross member of the frame and at the back on two lugs (not shown) which also seat on rubber pads.

23.14 BL cars overdrive, five-ratio gearbox

With fuel economy now a prime requirement, most car manufacturers are turning to five-speed gearboxes with overdrive top gear and, usually, direct-drive fourth speed. Formerly, four-speed boxes were the general rule for all but the more expensive cars, some sports models, and medium weight commercial vehicles, the last-mentioned of course because of the wide range of their weights, from empty to fully laden. A particularly good five-speed design is that developed by the former British Leyland, in the first instance for Triumph and Rover cars. Fig. 23.14 is a diagrammatic illustration clearly demonstrating the basic principles, but it should not be taken as an accurate representation of that box. The main difficulty when five speeds have to be accommodated is

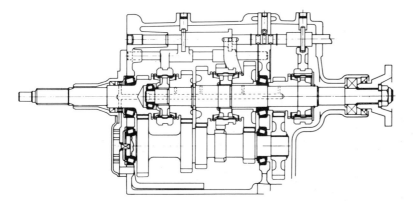

Fig. 23.14

that the shafts tend to become long. This has two main disadvantages: first, the whole structure becomes both costly and difficult to accommodate; secondly, the shafts are more liable to deflect under load and this, by throwing the gear teeth slightly out of mesh, increases the noise, vibration and rate of wear.

These problems have been overcome in this box by carrying the shafts in five taper-roller bearings, and making the overdrive gear pair overhang behind the rear pair of bearings. These two bearings are housed in an intermediate wall clamped between the rear cover and the main portion of the gearbox casing, while the two at the front ends of the main and layshafts are housed in the front wall. The fifth taper-roller bearing is in a counterbore in the primary, or first-motion, shaft and receives the spigot end of the mainshaft.

Taper-roller bearings have many advantages. One is that, because they take both axial and radial loading, no extra provisions have to be made for taking the thrust of the helical gearing. Because of the line contacts between their rolling elements and races, relatively small diameter rollers can be employed and the overall diameter is smaller than that of an equivalent ball bearing, with its point contacts. Alternatively, for a given overall diameter, a larger diameter shaft can be used.

In general, for given size, taper-roller bearings have about double the life of the ball type. Another advantage of the taper-roller is that the clearances in the races can be adjusted after assembly: ideally, this clearance should be zero, since any increase tends to leave only one or two rolling elements carrying all the load, while a pre-load obviously takes up some of the capacity of the bearing for carrying the working load. Assembly of the gears into the box is easy and can be done through one end, so there is no need for side covers – the outer races are inserted into their housing and the caged rollers and gears assembled onto the shafts before they are put in the box.

Other factors ensuring adequate stiffness include the use of a live layshaft. To facilitate the machining of the gears and for ease of assembly into the box, most cars have dead layshafts, with a cluster of gears integral with a sleeve rotating on bearings around it. Live – that is, rotating – layshafts are more widely used on commercial vehicles. Their advantage is that, because no bearings are needed between them and the gears, the shafts can be both larger in diameter and integral with the gears, and therefore much stiffer. In this box, some of the mainshaft gears run directly on the steel shafts, without bearings. This has necessitated the incorporation of an oil pump, driven from the front end of the layshaft, for lubrication of these gears. Pressure lubrication, however, is in any case an advantage and has been used before on some high quality cars.

By overhanging the overdrive gear pair – which of course transmit relatively light torques – the length of shaft between bearings has been kept to only 227 mm – again helping to reduce deflections. The sort of usage pattern that can be expected is: 40% of all running done in fifth gear, 40% in direct fourth gear, 15% in third, 4% in second, and 1% in first. With a four-speed box the averages are the same for first and second, 20% for third and 75% for direct top gear.

23.15 Synchromesh devices

These are used to simplify the operation of changing gear, so that this can be done by unskilled drivers without the occurrence of clashes and consequent damage. The principle of all the devices is that the members which ultimately are to be engaged positively are first brought into frictional contact and then when the friction has equalised their speeds the positive connection is made. The devices can be applied to sliding-mesh boxes, but are almost always used with constant-mesh boxes. The design requirements for this type of synchromesh mechanism are: high resistance to wear of the conical surfaces: a cone angle of the order of 7°; as short as possible a movement into and out of engagement, which entails keeping clearances to a minimum; the inertia of the moving parts should be as small as possible; and the friction in the control mechanism must be as low as possible.

The five constructional arrangements described below cover practically all the different arrangements commonly used.

In Fig. 23.15 the mechanism is shown applied to the direct drive and the next

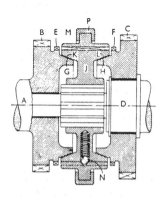

Fig. 23.15

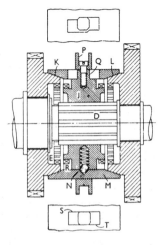

Fig. 23.16

lower gear of an ordinary constant-mesh box whose general arrangement is as shown in Fig. 23.11. Thus A is the engine shaft and the integral gear B meshes with a wheel fixed to the layshaft (not shown) while the gear C is free to rotate on the mainshaft D and is permanently meshed with another wheel fixed to the layshaft. Both B and C are formed with integral dog tooth portions E and F and conical portions G and H. The member J, which is free to slide on splines on the mainshaft, has conical portions K and L to correspond with G and H. Thus if J is slid sideways to the left the cones K and G will come into contact and the friction between them will tend to equalise their speeds. The outside of the member J is formed with teeth which are exactly similar to the teeth E and F and the member M is free to slide on these teeth except that spring-loaded balls N engaging recesses in M tend to prevent such sliding. There are usually six of these balls. If the balls N are overcome, however, and M is slid to the left along the outside of J, the teeth on the inside of M will engage the teeth E and there will be a positive drive between A and D through the teeth E, and the members M and J. The member M is actuated by the selector fork P and thus by the gear lever. In changing gear the gear lever is brought to the neutral position in the ordinary way but is immediately pressed in the direction it has to go to engage the required gear; supposing this to be top gear the effect is to press M to the left. The spring-loaded balls N, however, cause J to move with M and thus the cones K and L are brought into frictional contact. When the speeds of G and J have become equal a slightly greater pressure on the gear lever overcomes the resistance of the balls N, and M slides along J so that its teeth engage the teeth E, thus establishing the positive connection. Unless the gear lever is pressed so as to force the cones G and K together there will be no synchronising action. If the member M is slid along J before synchronisation has been effected a clash will result. This may occur if the gear lever is moved too rapidly and will tend to occur if the springs beneath the balls N are too weak. Synchromesh mechanisms of this type are usually called *constant load* synchromesh units, because an applied load of constant value – the same in all circumstances – effects a gear change.

The construction shown in Fig. 23.16 is that of Fig. 23.15 turned inside out. Thus the cones K and L are formed in the ends of a sleeve M free to slide on a member J which is splined to the mainshaft. The sleeve M must, however, rotate with J because of the projecting pieces Q which engage slots cut in M. The selector fork engages the ring P which is secured to the projecting pieces Q. Spring-loaded balls N again tend to make M move with J. The action is very much as before, thus pressure on the gear lever tends to slide J to the left, say, and the balls N cause M to move with J, thus bringing the cone K into contact. The resulting friction will, given time, produce synchronisation and then an increased pressure on the gear lever will slide J along inside M and the dog teeth E and R will engage, thus establishing the positive connection.

The construction of Fig. 23.15 is rather more convenient where the wheels B and C are small.

When the member M is slowing the wheel B down (as it is in all the early stages of a change up) there will be a pressure between the sides of the projections Q and the sides S of the slots as shown in the part plan. At the moment of synchronisation the pressure will become zero and after synchronisation (supposing the dog clutches are not engaged) the pressure would be between the other sides of the projection Q and the side T of the slot. The friction consequent on these pressures will help in preventing J from being slid along before synchronisation has occurred and this latter occurrence can be almost entirely prevented by shaping the slots as shown in the part view at the top of Fig. 23.16.

23.16   Baulk type of synchromesh

The mechanism used by Vauxhall Motors (and other General Motors concerns) works on a principle different from that used in the types described previously. It is shown as applied to the direct-drive engagement in Fig. 23.17. The cone A is loosely carried on the male cone of the gear B and has three fingers D whose ends engage the spaces between the splines of the mainshaft E but with considerable clearance as shown in the end view and plan. Thus while A has to rotate at the same speed as the shaft E it can move relatively to that shaft through a small angle. The gear B has internal dog clutch F which can engage the external teeth G on the member H, which is splined to the mainshaft and which has three slots J cut in it. These slots are just a little wider than the fingers D. Springs C are carried by the member H as shown.

The action will be described as it occurs when top gear is being engaged. Because of the small amount of friction always present between the cones in A and B there will be a drag on the former member which will keep the fingers D up against one side of the space between the splines of the shaft E, as shown in the plan and end view. This drag will be accentuated as soon as an attempt is made to move the member H along so as to engage the teeth G and F, because firstly the springs C will press against the fingers D and secondly the corners of the slots J will bear against the corner of the fingers D. Thus the cones will be pressed together and the friction between them will tend to bring about synchronism. The clutch being disengaged, synchronism will in due course come about and, if the dog clutch teeth were not then engaged, the speed of B, which has up to now been higher than that of E, would become lower. The

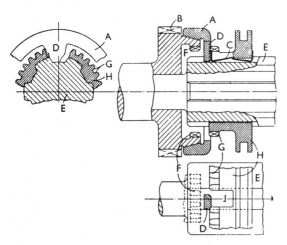

Fig. 23.17

fingers D would then be pressed against the opposite sides of the spaces between the splines of the shaft E.

Consequently at about the moment of synchronisation the fingers are about to leave one side of the spaces between the splines and pass over to the other side. The pressure on the sides of the fingers D is thus, at about the moment of synchronisation, quite small and so the force acting on the member H, in conjunction with the bevelled corners of the fingers D and slots J, can move the fingers slightly so that they come opposite the slots J. The member H can then be slid over so as to engage the teeth C and F, the spring C being slightly depressed as this is done. Clearly until synchronisation is almost effected the greater the force that is applied to the member H the greater will be the force pressing the cones together and the greater will be the friction between them and the sooner will synchronisation be brought about. This is an advantage over the first type above, where the maximum force that can be used to press the cones together is limited by the strength of the springs pressing the retaining balls into contact with the cone members. Because the engagement of the dog clutch is baulked by the fingers D until synchronisation occurs, this synchromesh unit can be said to be of the *baulk* type.

23.17 Baulk-ring synchromesh

A further example of a synchromesh mechanism is shown in Fig. 23.18. The dog clutch sleeve A is free to slide on splines on the hub B, which is fixed to the mainshaft C, and is controlled by the striking fork of the gear change mechanism. When moved to the right its internal splines (a) ultimately engage the dog teeth D to give third gear and when moved to the left they engage the dog teeth G to give fourth gear. The synchronising action is provided by the baulk rings H and J, which are coned internally to engage the cones formed on the gears E and F and which have external teeth (h) and (j) similar in form to the dog teeth D and G and are therefore the counterpart of the internal splines (a) of the sleeve A.

The hub B has three grooves formed in it, as shown in the upper part of the elevation and in the scrap plan and end views, and fitting freely in these grooves are three fingers K. At their ends these fingers engage grooves in the baulk rings H and J but these grooves are wider than the fingers K so that although the baulk rings are constrained to rotate at the same speed as the hub they can rotate slightly relative to that hub. The fingers K are pressed outwards by circlip springs L and M and thus the ridge in the middle of the fingers is kept in engagement with a groove formed in the middle of the splines of the sleeve A.

Consider the action when a change from third to fourth gear is being made. At the beginning of the change the gear F will be rotating at a higher speed than the mainshaft, hub B and the baulk rings. When the gear lever is pushed so as to try to engage fourth gear it tries to press the sleeve A to the left but this merely causes the fingers K to push the baulk ring J against the cone of the gear F. Friction between the cones then tries to speed up the baulk ring and, conversely, to slow down the gear F. The baulk ring is thus forced into the position shown in the scrap plan view (a). In this position the teeth (a) are half opposite the splines (j) of the sleeve A. Thus even if the pressure applied to the sleeve A by the striking fork is sufficient to depress the fingers K against the

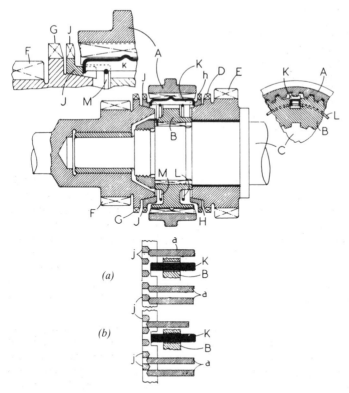

Fig. 23.18

force of the spring M the sleeve cannot move to the left and the pressure will merely be transferred to the baulk ring by the contact between the ends of the splines (a) and the teeth (j). Thus the harder the gear lever is pressed the greater will be the friction, between the baulk ring and the cone of the gear F, which is tending to synchronise those members.

When synchronism is attained the frictional drag between the baulk ring and the gear F is reduced to zero and the pressure applied to the sleeve A is able, in conjunction with the inclined ends of the teeth (j) and splines (a), to rotate the baulk ring slightly to its central position as shown in the lower plan view (b) and the sleeve A can then move across so that the splines engage the dog teeth G thus clutching the gear F to the hub B and mainshaft.

23.18   Multi- and double-cone synchronisers

Smiths Industries Ltd had developed a multi-cone synchromesh mechanism by 1963, but never put it into quantity production. Meanwhile, synchromesh gearboxes have been increasingly demanded for heavy commercial vehicles and ZF have introduced their D-type synchroniser and have applied it to, for example, their Ecolite gearboxes.

The Smiths version is illustrated in Fig. 23.19, in which the parts have been annotated with the same letters as the corresponding parts in Fig. 23.18. The

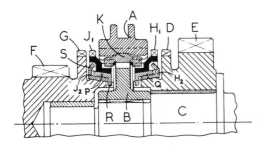

Fig. 23.19

important difference is that the cones J and H are now made in two parts J_1 and J_2 and H_1 and H_2 and between these parts there are intermediate cones P and Q. The outer cone J_1 is driven by the member K as in Fig. 23.18 and the inner one by the projection S which engages a recess formed in J_1. The intermediate member P is driven by the projection R which engages a slot formed in the cone of the member F. When the shaft C is rotating relative to F there are now three slipping surfaces in the cone assembly whereas in Fig. 23.18 there is only one and hence for a given axial force on the sleeve A and with the same dimensions the synchronising torque will be approximately three times as great.

The ZF D-type double-synchroniser, Fig. 23.20, is similar in principle, but its intermediate cup-cone floats between the main cup and cone, and the detent mechanism is loaded by coil instead of leaf spring. Advantages claimed for it include faster synchronisation, less power consumed in effecting synchronisation, easier release and engagement of the cones, increased service life owing to

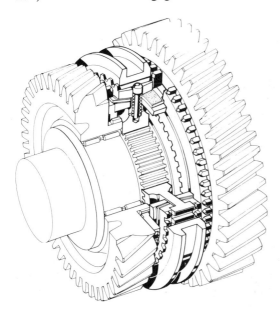

Fig. 23.20 ZF D-series, double-cone synchroniser. A significant attraction of this type is short axial length

the large area of the conical surfaces, more effective control with large and heavy clutches, less effort required for gear shifting and therefore lower loads on the shift mechanism, short axial length and therefore a potentially compact gearbox design.

23.19 Porsche synchromesh

The principle of this will be explained with reference to Fig. 23.21 which shows a design used by Henry Meadows Ltd of Wolverhampton, in some of their gearboxes. The gear G is on the end of the clutch shaft and drives the layshaft; H is the mainshaft of the gearbox. When changing up to the direct drive, the sleeve A, which is splined to the three-armed spider K, is moved to the left by the gear lever, and the conical inner surfaces of its splines make contact with the conical outer surface of the ring B. Friction then causes the latter to move relative to the member C so that the clearance at X is taken up and the end of the ring bears against the key E. The friction on the ring then tends to expand it and this increases the pressure between the ring and the sleeve A and makes it virtually impossible to move the latter any further to the left. When synchronism is achieved, the frictional force on the ring B is reduced to zero and it becomes easy to move the sleeve A over so that its internal splines engage the splines or teeth formed on the member D, which is itself splined to the gear G. This provides a positive drive between the gear G and the shaft H. The ring B is thicker at its centre, opposite to the key E, than at its ends though this is not very apparent from the drawing. This tapering is designed to obtain a uniform pressure between the ring B and the sleeve A, and to accommodate it, the outer surface of the ring C is made eccentric to the inner surface. The key F compels the ring C, and thus the ring B, to rotate with the gear G.

23.20 Lubrication of the gearbox

The lubrication of gearboxes, other than epicyclic types, which will be considered later, is usually effected by putting enough oil into the box to ensure that at least one gear will dip into the lubricant. When the gears are rotating the oil will be thrown about inside the box, thus lubricating the various parts.

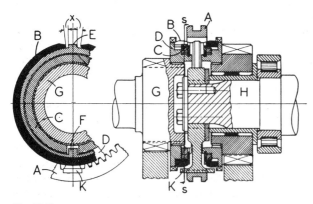

Fig. 23.21

The oil required is different from that suitable for an engine, the conditions being quite different. Temperatures are much lower and carbonisation has not to be considered. On the other hand, the pressures to which the oil films may be subjected may be much heavier than in an engine. The instructions of makers and the advice of reputable oil companies should always be followed.

It should not be thought that filling a gearbox to a higher level than that recommended will reduce the frictional losses in the box. On the contrary, the loss due to the churning of the oil will be greatly increased. However, over-filling is usually obviated by suitable placing of the filler spout.

Special oil seals of various patterns are fitted where the gearbox shafts pass through the casing and these usually give no trouble, but leakages might be caused by the expansion of the air enclosed in the box, and a vent should always be provided. Large washers are sometimes fitted on the inside of the ball bearings supporting the shafts of gearboxes to keep any particles of grit or chips from the gear teeth, from the bearings.

Some of the exterior parts of the selector and gate change mechanisms are best lubricated with a thin machine oil or with engine oil.

23.21   Freewheel devices

These are sometimes termed *one-way clutches*, since they transmit torque in one direction only and are generally used to disengage the drive automatically when torque is applied in the reverse direction. They have been placed in drive lines, usually on the rear end of the gearbox, so that the engine could idle at a controlled speed under overrun conditions, without need for selecting neutral in the gearbox. This can be useful with two-stroke engines, which tend to fire very irregularly under overrun conditions, and it can also facilitate gear changing since it automatically de-clutches the drive when the throttle pedal is released. However, overrun braking by the engine is of course lost, which is why the device is installed so rarely. More common applications are in four-wheel drive vehicles, Section 20.4, and in automatic transmissions, as in Sections 25.1 and 25.3.

There are two main types of one-way clutch, or freewheel. Both are generally contained within the annular space between two hardened steel rings, rather like the inner and outer races of roller bearings. The inner one is keyed or otherwise fixed to the shaft and the outer one to a hub, or sleeve, from which the drive is ultimately taken.

Of the two types, one is the Borg-Warner, or sprag, type, Fig. 23.22, the essential elements of which are a series of rocking tumblers, called *sprags*, held together within a cage, also comprising two rings but with a crimped ribbon-spring in the annular space between them. All three components of the cage are pierced to accommodate the sprags.

The ribbon-spring tends to push the sprags upright, into a radial disposition between the inner and outer rings, or races. However, the radial clearance between these rings is not enough for the sprags to rock right up to their TDC positions, so they jam. Consequently, so long as the drive in the direction tending to keep the sprags thus jammed, the torque can be transmitted from the shaft to the outer ring. If, however, the drive is reversed, the sprags tend to lay down, or trail, against the influence of the ribbon spring, so no torque can then be transmitted.

The second type of freewheel is exemplified by the ZF unit, Fig. 23.23. This is usually called the *roller* type, because it has rollers instead of sprags. These rollers are generally housed in what resemble splines in either the inner or outer hardened steel ring. However, the base of each spline is not tangential to a circle centred on the axis of the shaft but, instead, is inclined, as viewed from the end of the assembly. Consequently, when the shaft is rotated in one direction, the rollers run up the inclines and therefore jam between the inner and outer rings. If the shaft is then rotated in the other direction, they run down the inclines again and are freed in the annular clearances between the two rings. The effects are similar to the jamming and unjamming of the sprags in the Borg-Warner unit. To obviate backlash in the system, the rollers are generally lightly spring loaded up their short inclined tracks.

23.22 Auxiliary gearboxes and overdrives

The gear ratios adopted in all vehicles have to be a compromise since the factors concerned in their choice are conflicting. Thus for good acceleration in top gear, a ratio that is low in comparison with that which would give the best top speed on a good level road would be used. Again, for the best fuel economy higher ratios would be used than if a lively performance was the chief requirement.

The provision of a large number of ratios reduces the element of compromise and so there is a tendency to employ gearboxes giving a large number. This can be done in two ways: one is to provide the large number of ratios in a single box of more or less conventional form and the other is to use an auxiliary gearbox in tandem with a main gearbox giving only three or four ratios. The first method complicates the construction of the box and may make gear changing a difficult operation but gearboxes giving five ratios are common in lorries and are becoming so in cars.

If an auxiliary gearbox is used, it may be an entirely separate unit but it is usually attached to the main box so as to form a single unit; it is, however, sometimes housed in the driving axle. The auxiliary box usually provides only direct drive and one reduction and this may be such that, when in use, all the overall ratios are lower than those when direct drive is used; this is the

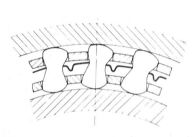

Fig. 23.22 Borg-Warner sprag-type freewheel

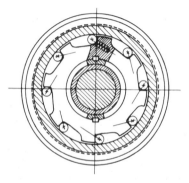

Fig. 23.23 Roller-type freewheel, with rollers retracted against their springs

common practice with 'off-the-road' vehicles and the auxiliary box is kept in direct drive for road work and in reduction ratio for cross-country use. According to the duties for which the vehicle is required, control may be effected by either separate levers for the main and auxiliary boxes, or the auxiliary shift might be controlled by a small lever or switch mounted near the top of the main gear-shift lever.

The ratios in the auxiliary box may, however, be such that the overall ratios in direct drive fall between those of the main box. In this case, the driver changes through all the ratios in sequence, generally using a small finger-actuated lever or switch on top of the main gear lever for changing the auxiliary gear ratio back and forth from direct to indirect drive, as he moves to the main lever from gear to gear. This type is termed a *splitter* gearbox. The sequence, for example, in a box giving an overdrive top gear might be from direct to indirect first, to direct and indirect second and so on, up to indirect overdrive and overdrive top gears, or the auxiliary gear set could be inoperative on overdrive top. Alternatively, in a box giving direct top, the sequence could be from indirect first to direct first and then up to indirect to direct top gear.

Auxiliary boxes may be either simple constant-mesh types with a layshaft or epicyclic. The change-over is often power operated, compressed air being generally used. Sometimes the auxiliary box takes the form of an overdrive unit, either automatically or electro-mechanically controlled, for obtaining a ratio of less than $1:1$ in top gear and thus reducing engine speed during long-distance, high-speed cruising, or to obtain a more suitable third gear ratio for cruising in semi-urban conditions, or for overtaking at higher speeds.

Another possibility is to have an auxiliary ratio such that, after all the gears have been changed in sequence in the main box, selection of the auxiliary ratio and reselection of first gear gives an overall ratio one higher than direct-drive top gear. Again, there may be two separate levers for controlling the main auxiliary gears but, alternatively, with, for example, a four-speed box there may be two H-gates side by side so that after the driver has changed through the first four gears, he moves the lever over into the next H-gate to obtain automatically the next four. In this case the mechanism is designed so that movement of the lever from one H-gate into the next actuates the range-change, which is why it is termed the *range-change* system.

23.23 A Leyland ten-ratio gearbox

This provides a good example of the first of the above methods. It will be seen from Fig. 23.24 that the shaft arrangement is conventional, the mainshaft being in line with the clutch shaft and the layshaft below on a parallel axis. Two pairs of gears AB and CD provide alternative drives to the layshaft, the gears B and D being connected to it as required by muff-type dog clutches. The muff E of the dog clutch for coupling the gear D has gear teeth on its exterior and provides a power take-off, the drive for which goes from A to B and thence to E which will be in its neutral position as shown. As there are two alternative drives to the layshaft the four pinions on it, which mesh directly with gears on the mainshaft, will provide eight ratios; the direct drive between the clutch shaft and the mainshaft gives the ninth ratio while the tenth is an overdrive and is obtained by clutching the gear F to the layshaft. It should perhaps be

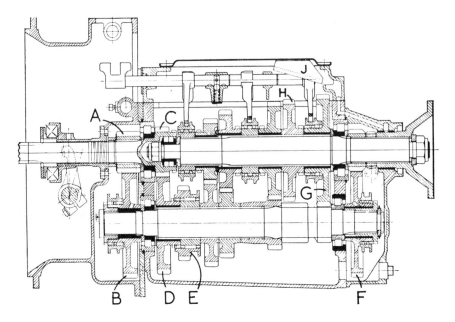

Fig. 23.24

pointed out that the drive from the gear G to the mainshaft is obtained by clutching it to the gear H which is splined to the mainshaft. The reverse ratios are given by sliding a reverse idler into mesh with the gear H. The ratios are—

Low 1st	9.352	1st	6.988
Low 2nd	5.764	2nd	4.308
Low 3rd	3.552	3rd	2.655
Low 4th	2.147	4th	1.605
Top	1.0	Overdrive	0.76

The gearbox is lubricated by splash provided by partly immersing the layshaft and reverse idler in oil. A trough J catches and feeds oil to the plain bearing on the outside of the hub of the output coupling.

23.24   The Fuller twin-countershaft gearbox

This provides an example of the second method of obtaining a multiplicity of ratios but also embodies a principle which enables the size of the box to be reduced. By using two layshafts (or countershafts) the tooth loads are halved and so the gears can be made only a little more than half as wide as they would have to be if only one layshaft were used. This will, however, be successful only if the torque is in fact equally divided between the two layshafts, and the way in which this is ensured is described later.

The gearbox shown in Fig. 23.25 comprises a five-speed and reverse box followed by a two-speed one, the combination thus providing ten forward and two reverse ratios. The input shaft A ends at the right in a narrow splined portion which engages splines inside the pinion B but enough clearance is

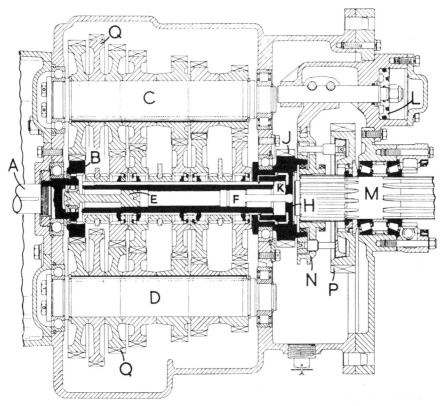

Fig. 23.25

provided between the splines to allow the pinion to float radially by several thousandths of an inch. The pinion meshes with gears on the two layshafts C and D which are identical and are carried in ball and roller bearings in the casing so that they rotate about fixed axes. The gears are driven by keys which ensure that they are assembled in the correct relative positions to enable them to mesh correctly with the mainshaft gears. The latter are free radially of the shaft and are positioned axially in pairs by collars, thinner ones being fixed inside the gears and the outer ones to the shaft and whose faces abut. The mainshaft is supported at each end by relatively flexible dumb-bell shaped members E and F; one end of each of these is fixed inside the mainshaft and the other ends are supported respectively in the spigot bearing inside the shaft A and in the boss H which is, in effect, part of the gear J. The latter is carried in the casing in a large ball bearing. The mainshaft gears are coupled to the shaft when required by sliding dog clutches. Because the mainshaft is free to float radially it will move slightly, if the tooth loads are unbalanced, until the upward tooth load at one side of the gear which is clutched to it is balanced by the downward tooth load at the other side. The torques transmitted by the layshafts are thus equalised. In order not to compromise the freedom of the mainshaft to move in this way it is connected to the output member J through an intermediate member K which is splined on the inside to the mainshaft and on the outside to the member J; both sets of splines are made with sufficient clearance to allow the freedom desired. The reverse ratio pinions on the layshaft (at the right) engage idlers which are carried on roller bearings and which mesh with the reverse gear on the mainshaft. The gears QQ are for a power take-off.

The auxiliary gearbox is housed in the right-hand part of the casing and a direct drive is obtained by clutching J to output shaft M by means of the dog clutch N which is operated by air pressure applied to the piston L. The low gear is obtained by engaging the dog clutch with the gear P, the drive then going from the member J through the two layshafts (not shown in the view because they do not lie in the plane of the drawing) to the gear P. The layshafts are carried in the casing on ball and roller bearings and rotate about fixed axes. The gear P, however, is free radially and can thus float sufficiently to enable the layshaft tooth loads to balance each other. The dog clutch N is provided with synchromesh for both its engagements.

23.25 An all-indirect gearbox

A good example of an all-indirect gearbox is illustrated in Fig. 23.26. This layout is especially suitable for rear-engine rear-wheel-drive or front-engine front-wheel-drive cars in which the engine is installed longitudinally.

Clutch shaft A is coupled to the input or driving shaft B by a splined muff C. Five pinions are either integral or fixed to the shaft B and four of these pinions mesh with gears that are carried on the driven or output shaft D on needle roller bearings N. The driven shaft is made integral with the pinion E of the hypoid final drive and is carried in the casing on taper roller bearings. The input shaft is carried at the right in a needle roller bearing and at the left in a ball bearing. The gears on the driven shaft are coupled to that shaft when required by dog clutches and synchromesh is provided. Reverse is obtained by sliding a reverse idler (not seen in the view) into mesh with the pinion F, which is fixed to the tail of the shaft B, and the corresponding gear fixed on the end of

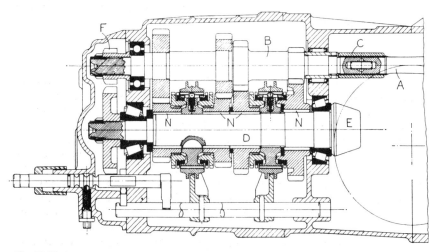

Fig. 23.26

the driven shaft; both these gears are outside the end wall of the gearbox casing and are housed in the end cover. Now that gears can be made with sufficient accuracy to ensure reasonable silence in operation there is no real objection to the absence of a direct drive and the convenience of the all-indirect design is very great.

23.26   Multi-speed splitter gearbox

A multi-speed splitter gearbox termed the 'Twin Splitter' was introduced by Eaton in 1985. Its gear shift lever is in an H-gate and, at each of its four stations, a three-position splitter switch near the top end of that lever can be used so that, in addition to the ratio into which the main box was initially shifted, two more can be obtained. The driver preselects the splitter ratios, and therefore does not need to use the clutch, except for starting and stopping.

It was termed the Twin Splitter unit because it is basically a four-speed twin-countershaft gearbox, in principle similar to that illustrated in Fig. 23.25, but with a three-speed splitter auxiliary gear train added on its rear end to convert it into a twelve-speed transmission, Fig. 23.27.

An essential feature of the transmission is a sensor-blocker unit accommodated inside each of the constant-mesh mainshaft gears, and sliding axially on their splined shaft, Fig. 23.28. Each sensor-blocker comprises a dog clutch, a coil spring compressed between the dog clutch and a large circlip round the shaft, and a conical blocker ring that acts like the cone of a synchroniser. The cup is machined in the boss of the mainshaft gear. Projecting radially inwards from the bore of the blocker ring are three pairs of teeth spaced 120° apart. Their thickness at the pitch circle diameter is less than that of the teeth on the dog clutch. Each pair can rotate back and forth in one of three gaps, also 120° apart, formed by partly machining away three of teeth around the dog clutch to enable the ring to block engagement of the dogs, in a manner similar to that described in Sections 23.16 and 23.17.

The blocker acts like a cup-and-cone synchroniser. However, because it is required to do no more than to sense absence of synchronisation, rather than actually to synchronise the gears, it is so lightly loaded that, say Eaton, it lasts the life of the vehicle.

23.27   Operation

When the driver actuates the splitter switch to initiate a change of ratio an air-actuator moves a selector fork to push one of the splitter gears towards engagement with the dog clutch on the mainshaft. If the gear and shaft are not synchronised, however, the clutch teeth will have been rotated out of alignment by the blocker ring and engagement will be blocked, as shown in Fig. 23.28(*b*). Consequently, all that happens is that the coil spring is compressed between its abutment ring on the shaft and the dog clutch. Synchronisation is effected by the driver using his accelerator control pedal, depressing or releasing it respectively to accelerate or brake the mainshaft assembly so that, when the teeth of the dog clutch rotate into alignment, the clutch is automatically slid into engagement by the spring.

By virtue of this mechanism, ratcheting or crashing of the clutch teeth, and the associated wear, is impossible. Moreover, since the main pedal-actuated

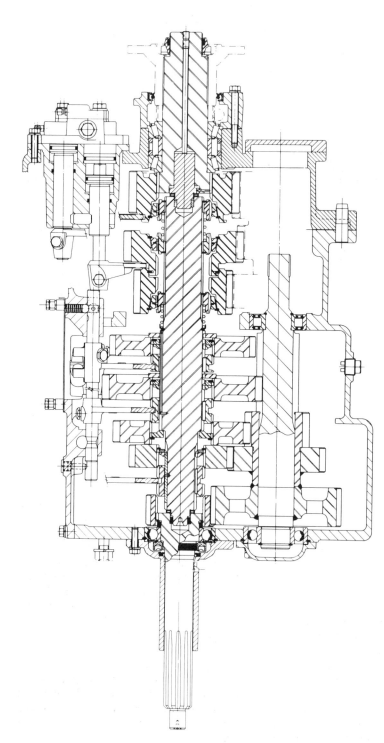

Fig. 23.27 The Eaton Twin Splitter transmission comprises a four-speed twin countershaft (*left*) with a three-speed splitter box mounted on its rear end (*right*)

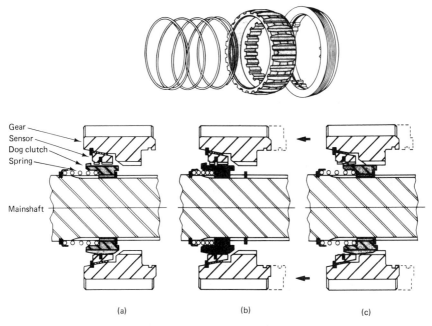

Gear
Sensor
Dog clutch
Spring

Mainshaft

(a)   (b)   (c)

Fig. 23.28 (*Above*) exploded view of the dog clutch (jaw clutch) and blocker ring assembly. (*Below*), the three stages of engagement of the splitter gears: (*a*) neutral; (*b*) preselected; (*c*) splitter gear engaged

clutch is not used for changing the splitter gears, and the gears in the main box, too, can be changed without it, wear of the main clutch is also reduced. In any case, even if the clutch is used every time for engagement of the gears in the main box, it will still be operated only four instead of twelve times in changing up through all the ratios.

23.28 Clutchless changes

The way in which, except for stopping and starting, actuation of the clutch pedal can be avoided is simple. Assume, for example, that top gear is in operation in the main box, with bottom ratio of the splitter box. If the driver wants to change down one ratio he first preselects the highest of the three ratios with his splitter switch. Next, he breaks the drive-line torque by using his throttle to enable him to move his gear shift lever into neutral. At this point, first the bottom ratio splitter gear will engage, and then he will either blip his throttle to synchronise quickly the gears in the main box, or of course he can wait until the speed of his vehicle has fallen adequately, so that he can move his main gear lever on smoothly into engagement. After a little practice, either method becomes simple.

If with this transmission a typical-long distance truck geared to cruise at 87 km/h (54 mph) is driven to keep within the economical engine speed range, of around 1500 rev/min, it will be in direct drive at all speeds above about 40 km/h (25 mph). Consequently, in long-distance operation almost all the gear shifting would normally be done using only the splitter switch, thus greatly reducing the driver's workload.

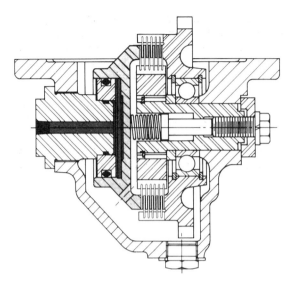

Fig. 23.29 The pneumatically actuated upshift brake

23.29   An upshift brake

One of the further refinements that have been incorporated is an upshift brake, resembling a multi-disc clutch, Fig. 23.29. In some circumstances, for example when a fully laden vehicle is moving away from rest on a steep upgradient, the driver will not want to lose the momentum of his vehicle by wasting the second or so waiting for the engine speed to fall before each splitter gear ratio he preselects engages. Therefore, an arrangement has been made to enable him to speed up his shift by using the upshift brake to reduce the rotational speed of the primary gearing. To actuate the brake all he has to do is to push the clutch pedal down beyond its normal travel, against the resistance offered by a kick-down spring detent. This causes the brake to be applied pneumatically. Since this brake is in constant mesh with one of the gears of the primary gear pair it reduces the rotational speed of the whole gear train through the twin countershafts to the mainshaft.

23.30   Additional features

Semi-automatic operation of the transmission has been obtained by incorporating an electro-pneumatic servo mechanism on the shift-bar housing. The clutch and throttle are both controlled by a microprocessor, and are activated automatically when the drive gives the appropriate signal by means of a switch on the console. This switch is moved forwards to indicate an upshift or backwards for a downshift, and an indicator on the instrument panel shows which gear is engaged. Skip-shifting is done either by multiple movements of the switch or automatically, depending on the load and road conditions. Pedal operation of the clutch is still necessary when starting and stopping the vehicle.

A fully automatic version offers even greater potential for economy. Its control module has forward, reverse and hold positions. Consequently, the driver can hold any gear in anticipation of the road conditions, except in that provision is made for avoiding either under- or over-speeding the engine. Cruise control can also be incorporated, with road-speed governing, and a manual throttle control is available for a power take-off.

The main advantage of this overall transmission concept is that it is based on a system that is simple and has not only been proved to be robust and easy to maintain but also service engineers are familiar with it. Most of the components of all the transmissions, manual, semi- or fully automatic, of any one of the sizes in which it is offered are common throughout the range. Only the add-on pieces such as the electro-pneumatic servo and self-diagnostic microprocessors are special parts.

Chapter 24

Epicyclic and pre-selector gearboxes

Hitherto we have dealt only with what is termed *ordinary gearing*, the characteristic feature of which is that the axes of the various gears are fixed, the motions of all the gears being simply rotations about their own axes. The characteristic feature of *epicyclic gearing* is that at least one gear not only rotates about its own axis but also rotates bodily about some other axis. Example of epicyclic gearing is shown in Fig. 24.1. A spur pinion A, integral with its shaft, is free to rotate about its own axis XX, being carried in bearings in the frame E. A cranked shaft or arm C is also free to rotate about the same axis XX and carries on its crankpin a spur pinion B. The latter is free to rotate about its own axis YY but must also rotate bodily about the axis XX when the arm C is rotated in its bearings. The pinion B meshes with the pinion A and also with an internally toothed ring or annulus D which forms part of the frame E and is therefore fixed. The annulus D is, of course, circular and is concentric with the axis XX. This arrangement is an epicyclic train of gearing providing a definite and fixed speed or gear ratio, between the shaft A and the shaft of the arm C; it is *not* a gearbox.

The action of this gear train will be understood by reference to Figs 24.2(*a*) and (*b*), in which the effective part of the pinion B, the only part which for the

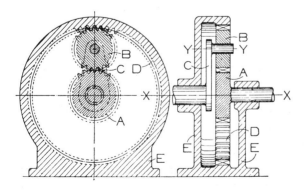

Fig. 24.1

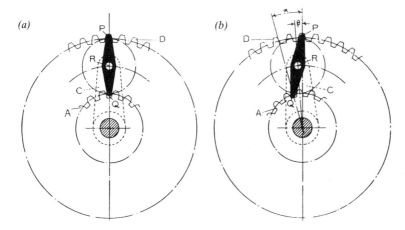

Fig. 24.2

moment is useful, has been blacked in. It will be seen that this portion constitutes a lever, the end P of which fits in a tooth space of the annulus D, the other end fitting in a tooth space of the pinion A. At its centre R the lever is pivoted on the pin of the arm C. Now suppose the pinion A to be turned through a small angle as shown at (b). The tooth space of A moves round to the left slightly, taking with it the end Q of the lever. Since the other end P of the lever is in a tooth space of the fixed annulus D the lever has to pivot about that point P as fulcrum. In so doing the centre R of the lever will move to the left, but since R is nearer than Q to the fulcrum P the point R will not move as far as Q. Now the lever PQ is attached at R to the arm C, hence the end of that arm moves round to the left a smaller distance than did the tooth space of A, and since the point R is farther from the axis XX than is the tooth space of A the angular movement of the arm C will be still smaller than that of the pinion A.

The above action can only be supposed to take place for very small movements of the pinion A because obviously for large movements the lever PQ will go out of engagement with the pinion A and the annulus D. But we have in reality not a lever such as PQ but a complete pinion which acts as a succession of such levers so that the action is kept up, and as the pinion A is turned so the arm C is turned at a slower speed in the same direction.

24.1   Another epicyclic train of gearing

The arrangement of gears just described is only one of many different forms of epicyclic gearing; different, that is, in their mechanical arrangement. Fig. 24.3 shows a second arrangement.

The wheel A, integral with its shaft, is free to rotate about its axis XX in the frame E. It meshes with the teeth B_1 of a compound pinion which is free to revolve on the pin of the arm C. The latter is carried in the frame, being free to revolve about the axis XX. The portion B_2 of the pinion meshes with a wheel whose axis is XX but which, being part of the frame, is fixed. When the arm C is turned the wheel A is driven round in the same direction as the arm but at a lower speed.

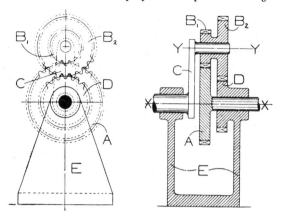

Fig. 24.3

To illustrate the action Figs 24.4(*a*) and (*b*) have been drawn, in which the effective portion of the pinion is blacked and shaded and, again, it can be regarded as a lever, although the resemblance is not so easily seen as in the previous example. The end P of this lever engages a tooth space of the wheel D which is a fixture, the other end Q engages a tooth space of the wheel A, while at R the lever is pivoted to the arm C. Imagine the arm C to be turned through a small angle, as shown at (*b*). The point R of the lever is moved slightly to the left and, as the only motion possible to the lever PQ is to pivot about the point P, it turns about that point as a fulcrum. The end Q will, therefore, move in the same direction as the point R but, being nearer to the fulcrum, it will move a shorter distance. Thus the wheel A will be moved, as shown, in the same direction as the arm C, but through a smaller angle. As there is actually a complete pinion B instead of a lever PQ the action can be sustained.

It will be noticed that the fixed wheel D is smaller in diameter than the

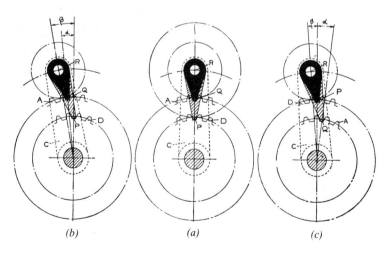

(*b*)   (*a*)   (*c*)

Fig. 24.4

driven wheel A. When the fixed wheel is larger than the driven one the latter will be driven in the opposite direction to the arm. This is shown at (*c*). As before, when the arm C is turned through a small angle the point R of lever PQ is moved a small amount to the left, and the lever turns about the point P as fulcrum. Now the end Q of the lever is on the opposite side of the fulcrum P to the point R which moves to the left. The end Q of the lever, therefore, moves to the right taking with it the wheel A with which it engages. Thus the driven wheel A moves in the opposite direction to the arm C.

The wheels that are carried on the rotating arm in epicyclic gearing are generally called *planet wheels*, while the wheels at the centre about which the planet wheels roll are called *sun wheels*. Epicyclic gearing is sometimes referred to as *planetary gearing*.

24.2   Other forms of epicyclic gearing

There are many other forms of epicyclic gearing using both spur and bevel gears, but the two forms described are the most important as regards motor vehicles. If the working of these is thoroughly understood no difficulty should be experienced in following the working of other forms. An example of an epicyclic train using bevel gears is shown in Fig. 30.7 and the ordinary differential dealt with in Chapter 28 is nothing more than a simple epicyclic train. In the following paragraphs it is shown how the gear ratio of an epicyclic train may be worked out and the method is applicable to all the various forms of epicyclic gearing.

24.3   Gear ratio of epicyclic gearing

Consider the epicyclic train shown in Fig. 24.1. The gear ratio of the train is the ratio between the number of turns made by the arm and the (consequent) number of turns made by the sun wheel. If the arm C is turned once round (keeping the annulus D fixed) then the sun A will be turned a certain definite amount. Now, provided that in the end we arrive at the result that the annulus D has not moved and that the arm C has made one turn, the number of turns made by the sun A will be the same however we arrive at that result. This may be seen by considering as an analogy a wheel rolling on a board, as shown in Fig. 24.5. If the wheel is rolled one half turn (the board being fixed) it will move along by the amount S (half the circumference of the wheel). The result of the

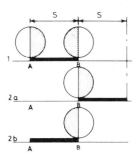

Fig. 24.5

operation is that the wheel has turned half a turn about its axis and has moved bodily the distance *S*, while the board has not moved at all. Suppose now that from the original position the board and the wheel are moved solidly, both together, the distance *S*. The position will then be as at 2*a*. Now, keeping the wheel at B, let the board be moved back under the wheel to its original position (2*b*). The board must, therefore, be moved back the distance *S* under the wheel, and in so doing it will turn the wheel half a turn.

It should be clear that when the board has arrived back in its original position the result is the same as before. The board (having arrived back into its original position) has not moved. The wheel has moved bodily the distance *S* and has made half a turn about its axis. We have arrived at the same result in two steps instead of in a single step. By adopting a similar procedure with epicyclic gearing the motions of the various members are easily found. We arrive at the desired result in two steps instead of in one and we so arrange the steps that in one of them we *turn every member the same amount*, and that in the other step we *keep the arm of the train fixed*, so that for that step the train works as ordinary gearing, which can easily be dealt with. An example will make the process clear, especially if the steps are recorded in tabular form.

Thus, considering the gear shown in Fig. 24.1, suppose the wheels to have the following numbers of teeth: A, 40; B, 20; D, 80. We are concerned with the motions of the various wheels and the arm and our table will have four columns—

	Arm C	Annulus D	Sun wheel A
First step			
Second step			
Desired result	+ 1	0	?

Now when we have arrived at the desired result the arm C must have made one turn and annulus D no turns. These figures may, therefore, be inserted as shown. Secondly, it is required that in the second step the arm of the train shall be fixed, hence we must have zero for the arm in the second step, thus—

	Arm C	Annulus D	Sun wheel A
First step			
Second step	0		
Desired result	+ 1	0	?

It follows that the motion of the arm in the first step must be + 1, and since we want all the members to have the same motion in the first step all the members must receive + 1 turn, thus—

	Arm C	Annulus D	Sun wheel A
First step	+ 1	+ 1	+ 1
Second step	0		
Desired result	+ 1	0	?

Again, it follows that the motion of the annulus D in the second step must be -1 in order that the two steps shall add up to zero. The minus sign indicates that in this step the annulus is to be turned *backwards*. Now, considering the second step, the arm is going to be fixed and the annulus is going to be turned backwards one turn. The wheel B will then turn in the same direction as the annulus, that is, backwards, while the sun wheel A will turn in the opposite direction to B and hence in the opposite direction to the annulus. In the second step, therefore, the sun wheel moves forward so that whatever the amount of the motion the sign is $+$. Care must be taken to get the sign of the motion right. The amount of the motion is given by the rules of ordinary gearing as—

$$1 \times \frac{\text{No. teeth in annulus}}{\text{No. teeth in pinion B}} \times \frac{\text{No. teeth in B}}{\text{No. teeth in A}}$$

This becomes—

$$1 \times \frac{80}{20} \times \frac{20}{40} = 2$$

Our table now appears as under, and on adding up the motions of the sun wheel in the two steps its resultant motion is found—

	Arm C	Annulus D	Sun wheel A
First step	$+1$	$+1$	$+1$
Second step	0	-1	$+2$
Desired result	$+1$	0	$+3$

Since the sign of the resultant motion of the sun wheel A is positive, that wheel moves in the same direction as the arm, so that the gear is a forward one. The gear ratio—

$$\frac{\text{Arm C}}{\text{Wheel A}} = \frac{1}{3}$$

so that the sun wheel A goes three times as fast as the arm. If an epicyclic train of this type is to be used in a motor car it will have to be with the engine driving the sun wheel A and with the arm C coupled to the driving wheels. As will be seen later, a train of this type is used in the Wilson epicyclic gearbox.

A pair of planet pinions meshing with each other is often used to replace a single pinion as is shown in Fig. 24.6(a). One pinion meshes with the sun and the other with the annulus. If the annulus is assumed to be the fixed member and the sun to be the driving member, the ratio of the train becomes—

$$\frac{\text{Sun}}{\text{Planet carrier}} = \frac{S - A}{S}$$

and as S must necessarily be smaller than A the ratio is negative, i.e. a reverse is obtained. One of the pair of pinions may also be made a compound pinion having two sets of teeth as in Fig. 24.6(b), where S meshes with P_1, P_1 with P_2 and P_3 which is integral with P_2 meshes with the annulus A. The ratio of this train with the annulus as the fixed member is—

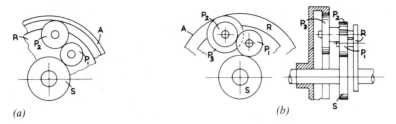

Fig. 24.6 Double planet pinnions

$$\frac{\text{Sun}}{\text{Planet carrier}} = 1 - \frac{A - P_2}{P_3 \times S}$$

24.4 Clutch action of epicyclic gearing

In the train of gearing shown in Figs 24.1 and 24.2, when the pinion A was turned the lever PQ pivoted about the point P and thus caused the arm A to turn. Now suppose that the annulus D is made a separate member from the frame E, free to revolve about the axis XX, then when the pinion A is turned, instead of the lever PQ pivoting about the point P and thus moving the arm C, the latter will remain stationary and the lever PQ will pivot about its centre R and cause the annulus D to revolve in the opposite direction to the pinion A. If now a brake is applied to the annulus D so as to slow it down and finally bring it to rest, then arm C will be gradually speeded up and will finally rotate as it would do if D were part of the frame E. Therefore, by making what will ultimately be the fixed member of the train free, and by gradually bringing that member to rest, the driven member will gradually be speeded up and there is no longer any necessity for a separate clutch such as is required with ordinary gearboxes. The clutch action by which the engine is gradually coupled to the driving wheels is obtained in the epicyclic gearing itself. Hence cars employing epicyclic gearboxes have no clutch in the ordinary sense of the word.

24.5 Epicyclic gearboxes

So far we have dealt only with epicyclic trains giving a single ratio between two shafts. We go on to consider boxes giving a number of ratios but will point out that any single epicyclic train can be arranged to give a 1 : 1 ratio in addition to its epicyclic ratio by providing a clutch to lock any two of its members together. It is then necessary to leave free the member of the train that is fixed during epicyclic action and to fix it when that is required; this is generally done by means of a friction brake, though the more costly multi-plate alternative is employed because it requires less attention in service.

Multi-ratio epicyclic boxes can be built up in three principal ways. In the first, the members of a single train are arranged so that each can be used either as the input, the output or the fixed member; in a simple three element train (e.g. sun, annulus and planet carrier) seven different drives are possible but the unsuitable values of some of the ratios and the complexity of the construction make the method of little practical value. In the second method, several separate trains are arranged in sequence so that each can provide its own ratios when required, i.e. a reduction or a $1:1$ ratio, and by locking all the trains an overall $1:1$ ratio is obtained. The number of ratios obtainable by this method theoretically increases very rapidly with the number of trains used, thus with four trains 15 ratios could be obtained but, in practice, it would be found impossible to use all of these. The trains can be all of the same type or be of different types. Fig. 24.7(*a*) shows two sun and annulus trains in tandem and (*b*) two double-sun trains so arranged. By making some of the members function as part of both trains the constructional problems are simplified but the number of ratios theoretically available is reduced. Thus in (*c*) the two sun gears of (*a*) have been joined together and the planet carrier acts for both the trains, each of which acts as a simple epicyclic train; the arrangement gives three ratios as compared with the four of (*a*). In (*d*) two double-sun trains are arranged with a common planet carrier and one sun, S, and its mating pinion function for both the trains. The brake B_2 when applied gives a reverse gear because the sun S_2 is larger than the driven sun S whereas S_1 which gives a forward gear when the brake B_1 is applied is smaller. This ability to provide a reverse gear easily made the arrangement popular in the past but it is now little used.

The third method of getting a multi-ratio epicyclic gearbox is by compounding a number of simple epicyclic trains. The Wilson boxes, developed more than half a century ago and still widely used, employ this method and one of these will now be described.

One train of epicyclic gearing is used for all the various ratios, its sun S_1, Fig. 24.8(*b*), being secured to a shaft D coupled permanently to the engine and its arm R_1 to the shaft E which is coupled permanently to the driving road wheels and the various ratios are obtained by driving the annulus A_1 at different speeds in relation to the engine speed. Supposing the latter to be constant at say 1000 rev/min, then if the speed of the annulus A is zero (that is, it is held at rest), then the speed of the arm R_1 would be 1000 $S/(A + S)$, where A and S represent the number of teeth in the annulus and sun respectively. Taking a numerical example, suppose $A = 100$ and $S = 25$ then the speed of R will be

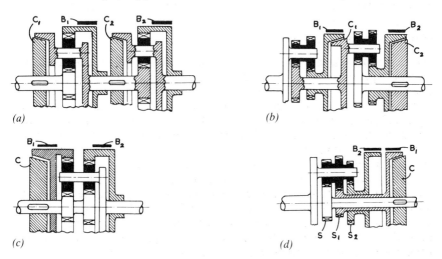

(*a*) (*b*)

(*c*) (*d*)

Fig. 24.7 Epicyclic gearbox arrangements

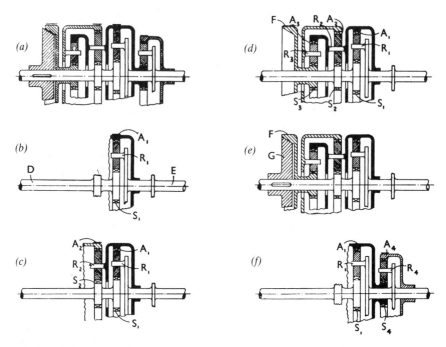

Fig. 24.8 Wilson gear ratios

200 rev/min and the gear ratio is 1000 to 200 or $5:1$. Suppose now that the annulus instead of being held at rest is driven in the same direction as the engine at, say, 100 rev/min then (the engine speed being 1000 rev/min as before), the speed of the arm R will be 280 rev/min, instead of 200 rev/min. Thus a higher gear is obtained. Similarly, if the speed of the annulus was made 200 rev/min, then the speed of the arm would be 360 rev/min. If the annulus were driven in the opposite direction to the engine then the arm would rotate slower than when the annulus was fixed. Thus, if the speed of the annulus were, say, 400 rev/min backwards then the speed of the arm (engine speed still 1000 rev/min) would be -120, that is, backwards, thus giving a reverse gear. Thus, by driving the annulus at different speeds (in relation to engine speed), the driven shaft is driven at different speeds, that is, different gear ratios are obtained.

To enable the annulus of the main train to be driven at these different speeds, auxiliary epicyclic trains are used. Thus, for second gear the train $S_2 R_2 A_2$, Fig. 24.8(c) is utilised. In this train the sun S_2 is fixed to the engine shaft D, the annulus can be held at rest by a brake and the arm R_2 is coupled to the annulus A_1. If the annulus A_2 is held at rest and the engine drives the suns S_1 and S_2 then, considering train $S_2 R_2 A_2$, it should be clear that R_2 will rotate in the same direction as the engine but at a lower speed. But R_2 is secured to A_1 and so A_1 is rotated in the same direction as the engine and R_1 will rotate at a higher speed than it did on first gear when A_1 was fixed.

To get third gear, the arm R_1 has to be made to rotate faster than on second gear and to do this the annulus A_1 and thus the arm R_2 must be made to rotate faster. To do this the annulus A_2 is now caused to rotate in the same direction as the engine. This is brought about by the train $S_3 R_3 A_3$, of which the sun S_3 is held at rest by the brake on its brake drum F, while the arm R_3 is coupled to the annulus A_2 and the annulus A_3 is coupled to the arm R_2. Considering the train $S_3 R_3 A_3$ if S_3 is fixed and A_3 is rotating in the same direction as the engine then the arm R_3 will also be rotating in the same direction as the engine. But R_3 is coupled to A_2 and so in the train $S_2 R_2 A_2$ both the sun S_2 and the annulus A_2 are rotating in the same direction as the engine and thus the speed of the arm R_2 must be greater than when A_2 was fixed. The speed of the annulus A_1 is thus greater than on second gear, that is, a higher gear is obtained.

The direct drive for top gear is obtained by locking S_3 to the engine shaft D by sliding the male cone G, which is splined to D, along so as to engage the female cone formed in the drum F which is fixed to S_3. This has the effect of locking the gear so that it must all revolve 'solid'.

To obtain a reverse gear the auxiliary train $S_4 R_4 A_4$ is brought into action. The sun S_4 of this train is secured to the annulus A_1, the arm R_4 is fixed to the driven shaft D and the annulus A_4 can be held at rest by a brake.

Considering train $S_4 A_4 R_4$ if A_4 is fixed and R_4 is rotating backwards then S_4 will be rotating backwards at a higher speed than the arm and thus in train $S_1 R_1 A_1$, although the sun S_1 is rotating forwards, the annulus A_1 is rotating backwards at such a speed that R_1 goes backwards also. This explanation can only be made convincing by working out the gear ratio as follows.

Suppose the numbers of teeth $S_1 = 25$, $A_1 = 100$, $S_4 = 40$, $A_4 = 80$. Keeping A_4 fixed, let R_4 be rotated backwards one turn, the motions of the members of train $S_4 R_4 A_4$ may be obtained by the method previously described and as shown in the table Fig. 24.9. Now considering train $S_1 R_1 A_1$ is secured to S_1 it has received -3 turns and since R_1 and R_4 are both fixed to the driven shaft D the arm R_1 has received -1 turn the same as R_4. Hence in train $S_1 R_1 A_1$ the motions of R_1 and A_1 are settled and the consequent motion of S_1 may be found as shown in the table Fig. 24.10.

Thus the engine shaft has turned $+7$ turns and the driven shaft has turned -1 turn and a reverse gear ratio of $7:1$ is obtained.

The clutches used to give the direct drive in modern Wilson boxes are multi-plate designs as shown in Fig. 24.11. Compressed air, or occasionally fluid pressure, is used to engage the clutch and acts in a cylinder A to rotate the lever B and thus move the lever C, which is in the form of a ring, about its fulcrum at D. The ring C is pivoted on pins F at both sides of the housing E and through the ball bearing compresses the plates of the clutch between the presser plate G and the hub H. The latter is splined on the input shaft J and carries the inner plates while the outer plates are housed in the drum K which is splined to the hub of the sun S_3.

R_4	A_4	S_4
-1	-1	-1
0	$+1$	-2
-1	0	-3

Fig. 24.9

R_1	A_1	S_1
-1	-1	-1
0	-2	$+8$
-1	-3	$+7$

Fig. 24.10

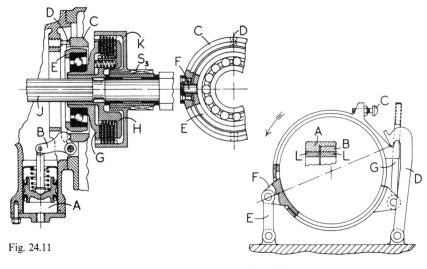

Fig. 24.11

Fig. 24.12

The brakes are in two parts: an outer band A, Fig. 24.12, anchored by the hook link D and the inner band B which is anchored by the lug F that project through a slot cut in the outer band. The two bands are thus anchored at diametrically opposite points thereby making the fairly large anchorage forces cancel each other so that no large force has to be sustained by the bearings supporting the brake drum. The outer band is pulled up at its free end by a rod G in order to apply it while the inner band is brought into action by the overlapping part of the outer band. This obviates the need for two separate actuating mechanisms. An example of the actuating mechanism is shown in Fig. 24.13. Air pressure on the piston A lifts the strut Q and rotates the lever B about its fixed axis O, the roller C then acts on the cam surface of the lever K and lifts it so as to pull up the end of the outer band. The cam surface is made so that initially the lift of the lever K for a given piston travel is large and the clearances in the brake are taken up rapidly while subsequently the lift is reduced so that maximum leverage is obtained and the large operating force is provided with the minimum effort. This action is also promoted by the approach of the point P of the lever B to a dead-centre position above the axis O and similarly by the lever K. In order to keep this mechanism in its proper operating condition as the brake linings wear an automatic adjuster is used; this is seen at the top of the pull rod and its action is illustrated in Fig. 24.14.

24.6 Automatic adjusters

The action of the automatic adjusters will be made clear by reference to Fig. 24.14. Surrounding the nut H on the top of the pull rod is a coil spring (b), one end of which is secured to a pin fixed in a plate (a) which is free to turn round the nut H. The spring is coiled round the nut a few turns, is looped round the pin in the plate (a) again and finally is secured to a pin passing through a slot in the plate (a) and fixed in the knife edge plate J. When the pull rod is moved upwards by the bus-bar its upper end moves over and the plate (a) approaches

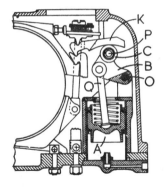

Fig. 24.13

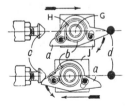

Fig. 24.14

the stop (c) fixed to the brake band (to the gearbox casing for top gear). When the brake is in proper adjustment it comes on fully just as the plate (a) touches the stop (c), but if wear has occurred the plate hits the stop and is rotated about the nut H in an anticlockwise direction. This rotation does not turn the nut, however, because the spring (b) uncoils and becomes free on the nut. When the pull rod is released its upper end moves back and the plate (a) hits the stop pin D, fixed in the casing. This causes the plate to rotate in a clockwise direction and this rotation is communicated to the nut H because for this direction of rotation the spring (b) coils up and grips the nut. The nut being turned slightly the brake is tightened up slightly. The action described occurs every time the brake is applied and released, until the nut has been tightened up so much that when the brake is fully on the plate (a) only just reaches the stop (c); the brakes are then in proper adjustment. Thus, provided the stops (c) are positioned properly and that the automatic adjusters work regularly the brakes will be maintained always in the proper adjustment.

For public service passenger vehicles, in which Wilson boxes are widely used, it is important that the motion of the vehicle when starting and when gear changes are being effected should be smooth and the accelerations should not be unduly high. This is ensured by means of a semi-automatic operation of the gear changes.

These are initiated by the movement of a small lever by the driver but the operation of the solenoid-controlled valves admitting air to the brake and clutch-actuating cylinders is dependent on the prevailing conditions, in particular the road speed and the position of the accelerator pedal. The pressure of the air supply to the actuating cylinders is also modulated in accordance with the gear being engaged, a lower pressure being provided for the lower gears than for the higher ones. The pressure is also varied during the engagement of the brakes and the clutch, being increased towards the end of the engagement to ensure that this is complete.

24.7   Cotal spicyclic gear

This is shown in Fig. 24.15. The wheel A is integral with the engine shaft and meshes with pinions carried by a spider B which is free to slide along the outside of the engine shaft. When the spider B is slid to the left its teeth E mesh

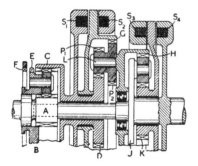

Fig. 24.15

with teeth F of an annulus which is fixed to the gearbox casing. The pins of the spider then form fixed bearings for the pinions, and so the annulus C with which the latter mesh is driven in the opposite direction to the wheel A. This gives the reverse drives. When the spider B is slid to the right its teeth E engage the teeth of the annulus C and then the wheel A, pinions, spider B and annulus C revolve 'solid'. This gives the forward drives.

The four forward ratios are obtained by means of two epicyclic trains arranged in tandem. One consists of the sun D (fixed to the annulus C), the compound planets P_1 P_2 (carried on the pins L of the arm which is integral with the annulus H of the second train) and the annulus G which can be held fixed.

When this is done the annulus H is driven in the same direction as the sun D but at a lower speed. The second train consists of the annulus H, the sun K and the arm J which is fixed on the output shaft. The sun K can be held at rest so that the train gives a reduction between the annulus H and the arm J and it can also be locked to the output shaft so that the train must revolve solid. The annulus G can also be locked to the sun D so that the first train must revolve solid.

The fixing and locking of the members is done by electromagnets whose windings S_1 S_2 S_3 S_4 are energised as may be required. For first gear S_2 and S_3 are energised and both epicyclic trains provide a reduction since both annulus G and sun K are fixed. For second gear, S_2 and S_4 are energised and the second train revolves solid, the only reduction being in the first train. For third gear, S_1 and S_3 are energised, the first train is locked solid and the only reduction occurs in the second train. For fourth gear S_1 and S_4 are energised and both trains revolve solid so that a direct drive is obtained.

The windings S_1 and S_4 are carried by parts that sometimes rotate and so these windings are connected to slip rings on which brushes bear. The current for energising the windings is supplied by the battery or generator of the car and is between 2 and 3 amps.

The control is extremely simple consisting merely of a switch which connects the appropriate winding to the battery. This switch is usually mounted at the centre of the steering wheel.

Chapter 25

Torque converters and automatic gearboxes

A torque converter is a device which performs a function similar to that of a gearbox, namely, to increase the torque while reducing the speed, but whereas a gearbox provides only a small number of fixed ratios the torque converter provides a continuous variation of ratio from the lowest to the highest.

Constructionally, a torque converter is somewhat similar to a fluid flywheel from which it differs in one important aspect, namely, in having *three* principal components instead of only *two*. Torque converters all consist of (a) the driving element (*impeller*) which is connected to the engine, (b) the driven element (*rotor*) which is connected to the propeller shaft, and (c) the fixed element (*reaction member*) which is fixed to the frame. It is the last element which makes it possible to obtain a change of torque between input and output shafts and, as has been seen, the fluid flywheel, which does not have any fixed member, cannot produce any change of torque.

A three-stage torque converter is shown in the several views of Fig. 25.1. The impeller, shown in sectional perspective at (d), is a conical disc provided with blades A the outer ends of which are tied together by a ring a. If this impeller is immersed in fluid and rotated then the fluid between the blades will be flung out more or less tangentially and a flow will be established from the centre or eye of the impeller to the periphery. The velocity of a particle of fluid on leaving the impeller is indicated by the line V_a in the view (a); this velocity may be resolved into a purely tangential component V_t and a purely radial component V_r.

The rotor or driven element is sectioned in black in the views (a), (b) and (c) and is shown in sectional perspective at (f). It consists of a portion similar to the impeller, comprising the disc member b, which carries the blades F, and the hollow annular member g which is carried by the blades F and which in turn carries blades B and D; these latter blades are tied together at their outer ends by rings as shown.

The fixed, or reaction, member consists of a drum-like casing h fitting the shaft portions of the impeller and rotor at the centre and thus enclosing those members. The reaction member carries blades C all round its periphery and blades E project, in a ring, from the right-hand end wall.

The action of the converter is as follows: the fluid flung out at the periphery

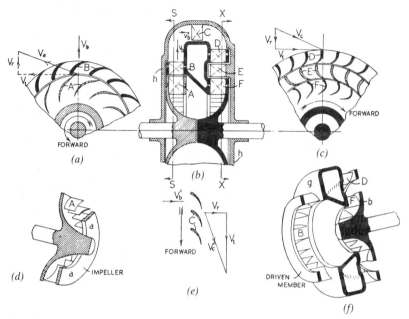

Fig. 25.1

of the impeller impinges on the blades B of the rotor and is deflected by those blades, the tangential component of the velocity of any particle, for example, V_t in view (a), being abstracted, more or less completely, so that the velocity of the particle on leaving the blades B is more or less radial as indicated by the arrow V_b. The particle being considered has therefore lost momentum in the tangential direction and this momentum has been gained by the blades B, that is, by the rotor. In being deflected backwards by the blades the fluid applies a pressure forwards on the blades. On leaving the blades B the fluid is guided round by the fixed casing and enters the blades C. In passing through these blades the velocity of the particle is changed from a more or less purely axial velocity, as V_b^1 in views (b) and (e), into a velocity having a considerable tangential component, as V_t in view (e). In deflecting the fluid in this way the blades C, and thus the fixed casing, receive a backwards thrust and unless the reaction member were fixed these thrusts would make it rotate backwards. The particle of fluid is now guided round by the fixed casing and enters the blades D of the rotor with a velocity V_c in view (c), which again has a considerable tangential component V_t and again on passing through the blades this tangential component is abstracted and the momentum associated with it is acquired by the rotor. The particle of fluid now enters the blades E which restore the tangential component of velocity once more and finally it enters blades F which finally abstract the tangential momentum, which is acquired by the rotor. The particle of fluid has now found its way back to the eye of the impeller and the cycle commences all over again.

The rotor or driven element thus receives three driving impulses, one from the blades B, one from the blades D and one from the blades F, and this converter is consequently called a *three-stage converter*.

The characteristics of a torque converter of this kind are shown by the graphs Fig. 25.2(a) and (b).

The graphs (a) show the manner in which the torque increase and efficiency vary when the rotor speed varies from zero to the maximum value (2700 rev/min), the impeller speed being constant at 3000 rev/min. When the rotor speed is zero (because the resistance opposing its motion is large enough to hold it fixed) the torque tending to rotate it will be nearly $6\frac{1}{2}$ times the torque developed by the engine at its speed of 3000 rev/min. If the resistance to the motion of the rotor now decreases so that the rotor starts to rotate, then as it gathers speed so the driving torque action on it falls off. At a rotor speed of 1000, for example, the driving torque would have fallen off to about three times engine torque, at 1900 rev/min the driving torque would be only just equal to engine torque, while at 2700 rev/min the driving torque would have fallen to zero. The efficiency on the other hand starts at zero when the rotor speed is zero, because although the driving torque acting on the rotor is then large no rotation occurs and the torque does no work; thus no work is being got out of the converter but a lot of work is being put in by the engine and so the efficiency is zero. As the rotor speed increases so does the efficiency and a peak efficiency of 86 to 90% is reached at a rotor speed of about 1000 rev/min. As the rotor speed continues to increase the efficiency falls off again and at 2700

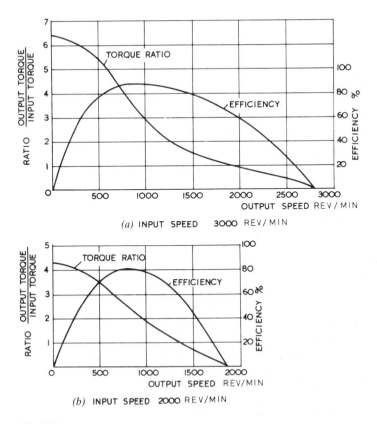

(a) INPUT SPEED   3000 REV/MIN

(b) INPUT SPEED   2000 REV/MIN

Fig. 25.2

rev/min becomes zero once more, this time because although the rotor is revolving rapidly the driving torque on it is zero and no work can be got out of it.

The graphs (*b*) show the same things but for an impeller speed of 2000 rev/min instead of 3000 rev/min. It will be seen that the driving torque acting on the rotor when it is stalled, that is, held at rest, is now only $4\frac{1}{3}$ times engine torque and it falls off to zero at a rotor speed of 1800 rev/min. The efficiency reaches a maximum value of only about 80% instead of 85 to 90%.

It is thus seen that only over a rather narrow range of rotor speeds is the efficiency reasonably good and it must be borne in mind that if the efficiency is, say, 60%, then 40% of the power developed by the engine is wasted, being converted into heat which raises the temperature of the torque converter fluid and which has to be dissipated by some means, commonly a radiator. The fall-off of efficiency at the low speed end of the range can be tolerated because those speeds are normally used only for short periods when starting and climbing severe hills, but the fall-off at high speeds cannot be tolerated and must be circumvented. There are two principal ways in which this can be done, (*a*) by substituting a direct drive for the torque converter at high speeds and (*b*) by making the torque converter function as a fluid flywheel at the higher speeds.

25.1   Torque converter with direct drive

Referring to Fig. 25.3, a double clutch, provided with two separate driven plates A and B, is situated between the engine and the torque converter (TC), only the impeller and part of the rotor of which are shown. The plate A is connected to the shaft C which is permanently coupled to the propeller shaft while the plate B is connected to the impeller of the torque converter. The rotor of the latter is connected through the freewheel D to the shaft C and thus to the output shaft. The intermediate plate E of the clutch can be pressed either to the left or to the right. When pressed to the right it grips the plate B and thus drives the impeller of the torque converter and the drive passes through the torque converter and the freewheel D to the output. If now the plate E is pressed to the left the plate B (and torque converter) will no longer be driven but the drive will pass direct through plate A to the shaft C and output which will override the rollers of the freewheel D; the rotor of the torque converter will thus come to rest. The efficiency of the direct drive is 100% and the combined efficiency curve will be as shown in Fig. 25.4. The change-over from

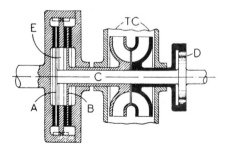

Fig. 25.3

converter to direct drive is done by the operation of a lever or pedal by the driver. Three-stage converters are no longer used in road vehicles because single-stage ones used in conjunction with gearboxes have been found adequate.

25.2   Turbo-Transmitters converter

The second method of obviating the fall-off in efficiency at the higher output speeds is used in the Turbo-Transmitters converter unit shown diagrammatically in Fig. 25.5.

The impeller is seen at A and is permanently connected to the engine crankshaft; it differs from that of the unit shown in Fig. 25.3 in having blades that extend over nearly half of the complete fluid circuit instead of over only about a quarter of that circuit. The impeller thus more nearly resembles the impeller of a fluid flywheel from which it differs chiefly in having blades that are curved in the end view whereas the blades of a fluid flywheel impeller are straight. The driven member, shown sectioned in solid black, has two sets of blades B and C and is fixed to the output shaft D. The reaction member also has two sets of blades E and F. The blade unit E is carried on the unit F on which it is free to rotate in the forwards direction but is prevented from rotating backwards by pawls that engage ratchet teeth of F. The member F in turn is free to rotate on the fixed member G but again only in the forwards direction. Backwards rotation is prevented by a multi-plate clutch, situated at H, which engages and locks F to G whenever the member F tries to rotate backwards but which disengages when F tries to go forwards, which motion is thus allowed. This clutch is shown in Fig. 25.7. The converter is a two-stage one, two driving impulses being given to the driven member, one when the direction of the fluid is changed in the blades B and a second when the direction of the fluid is changed in the blades C. When the torque acting on the driven member BCD is greater than the engine torque applied to the driving member A, there will be a reaction torque acting on the blades E and F; this will be transmitted by the pawls and ratchet teeth from E to F and by the clutch at H to the member G. Whenever the torque acting on the blades B and C tends to fall below the torque applied to A, a forwards torque will be applied to the blades E and F; this will merely cause those blades to rotate forwards and the converter will then function as a fluid flywheel; the percentage slip will then be quite small, say about 5 or 6%, and will decrease as the speed of the driven member increases.

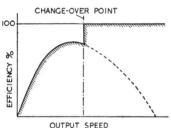

Fig. 25.4

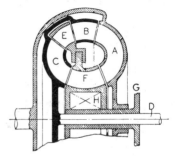

Fig. 25.5

The characteristics of this converter will thus be somewhat as shown in Fig. 25.6, which shows torque increase and efficiency curves for an input speed of 3500 rev/min. The converter being only a two-stage one, the maximum torque is only about four times engine torque as compared with $6\frac{1}{2}$ to seven in the Leyland converter, but otherwise the curves are very similar in general shape. The change-over point, at which the reaction members E and F begin to rotate forwards, and the unit commences to function as a fluid flywheel, is at 1200 rev/min, and from that speed onwards the output torque will be approximately equal to the input torque and the output speed will be only some 5 or 6% less than engine speed.

When the engine speed is less than the maximum 3500 rev/min assumed above, and when the throttle opening is reduced so that the engine torque is less than maximum, then the change-over speed will be lower than the 1200 rev/min corresponding to maximum engine speed and torque.

The fact that the change-over is quite automatic is important and is responsible for the good performance of this type of converter at part throttle loads and medium engine speeds.

25.3   Other arrangements of torque converters

Four arrangements of single-stage converters are shown in Figs 25.8 to 25.11.

In Fig. 25.8 the reaction member R is permanently fixed, which makes it unsuitable for use in motor vehicles unless some form of direct drive is provided and the converter is emptied when the direct drive is engaged.

The design shown in Fig. 25.9 is widely used, being simple constructionally; the one-way clutch S is sometimes placed outside the casing by providing the reaction member with a sleeve which passes through the cover of the impeller member.

In Fig. 25.10 an auxiliary impeller I_2 is provided and is carried on a one-way clutch S_3 on the main impeller I_1; the reaction member is also divided into two portions R_1 and R_2, each of which is anchored separately by the one-way clutches S_1 and S_2. This arrangement, which was introduced on Buick cars

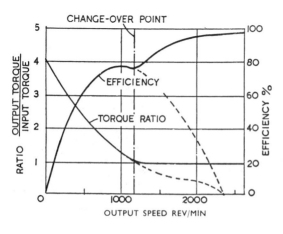

Fig. 25.6

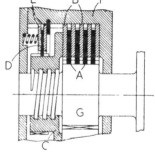

Fig. 25.7

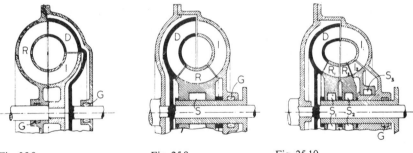

Fig. 25.8   Fig. 25.9   Fig. 25.10

some years ago, is claimed to increase the efficiency of the converter and to make the change-over to coupling action smoother.

In the Borgward converter shown in Fig. 25.11 the whole reaction member RR_1 and the impeller I on which the reaction member is mounted on ball bearings B, is moved to the left when engaging the cone clutch H and is done by the reaction of the pressure that exists inside the coupling at L and which acts on the exposed area of the driven member D. To obtain direct drive, oil under pressure from the gearbox control unit is passed to the space K and moves the reaction member and impeller to the right so as to engage the cone clutch M. The one-way clutch S enables the output member E to drive the impeller, and thus the engine, in the forwards direction and so permits the engine to be used as a brake or to be started by towing the car.

The arrangement of a two-stage converter shown in Fig. 25.12 differs from that of Fig. 25.5 in that only a single reaction member is used and so the second driven member D_2, which is bolted up to the member D_1, discharges direct into the inlet of the impeller. The single-plate clutch C gives a direct drive when it is required; the method of engaging the clutch is not shown but various means are used; a very convenient one when the converter is associated with an automatic gearbox is to make the pressure plate of the clutch function as a piston in a cylinder formed in the flywheel and to engage the clutch by admitting pressure oil to this cylinder.

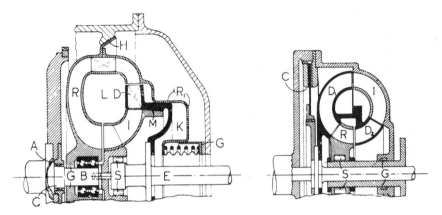

Fig. 25.11   Fig. 25.12

25.4　Chevrolet Turboglide transmission

This is a combination of a converter and an epicyclic gear and is shown in Fig. 25.13. The converter has five elements, the pump P, three turbine or driven elements T_1, T_2 and T_3, and a reaction member R. The latter is free to rotate in the forward direction on the freewheel F_1 and is provided with a set of blades B, whose angles are adjustable; the mechanism for making the adjustment is not indicated.

The first turbine element T_1 is coupled by the shaft D to the sun S_2 of the second epicyclic train; the second turbine T_2 is coupled through the sleeve E to the annulus A_1 of the first epicyclic train and the third turbine T_3 is coupled to the output shaft H by the sleeve G_1, the clutch C_1 (which is always engaged except when neutral and reverse are selected), the sleeve G_2 and the planet carrier R_2. The sun S_1 is normally prevented from rotating backwards by the freewheel F_2, since usually the clutch C_2 is engaged and the member K is fixed so that the sleeve J cannot rotate backwards. The annulus A_2 is also prevented from rotating backwards by the freewheel F_3 which locks it for such rotation to the sleeve J. Engagement of the clutch C_3 fixes the annulus A_2 against forwards or backwards rotation, and this is done when 'low' is selected so as to reduce the load on the freewheel F_3, when the engine is pulling hard under adverse road conditions, and to allow the engine to be used effectively as a brake on down gradients.

At low forward speeds of the output shaft H relative to the engine speed, the sun S_1 and annulus A_2 will be stationary because the torques on them will tend to make them rotate backwards and this motion is prevented by the freewheels F_2 and F_3. Both epicyclic trains then provide speed reductions and torque increases, and all three turbines will be driving.

As the output speed rises, the torque passing through the sun S_2 will fall and at some point will tend to become negative, and then the annulus A_2 will start to rotate forwards and the turbine T_1 will be effectively out of action. At a higher output shaft speed, the sun S_1 will start to rotate forwards and the turbine T_2 will go out of action. The drive will then be through T_3 direct to the output shaft, the only torque magnification then being that due to the torque

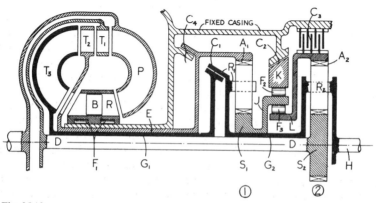

Fig. 25.13

converter itself. Finally, the reaction member R will start to rotate forwards and the torque converter will run as a fluid coupling. The speeds and torques at which these events occur will depend on the angle at which the blades B are set.

Reverse is obtained by engaging the clutch C_4 and disengaging C_1, C_2 and C_3. The trains 1 and 2 are then compounded and give a reverse ratio, the whole of the driving torque being transmitted by the turbine T_1 and sun S_2. Forward motion of S_2 tends to drive R_2 forwards and A_2 backwards; backward motion of A_2, however, results in backward motion of S_1 (through the freewheel F_3 and the sleeve J) and so in train 1, whose annulus is fixed, the sun tends to rotate the planet carrier R_1 backwards. The backward torque on R_1 is greater than the forward torque on R_2 (from S_2), and so R_1 and R_2 will move backwards.

25.5   Automatic transmission in general

Transmission systems that will function purely automatically without any attention from the driver of the vehicle have been sought for many years by designers, but even today such systems have not been fully achieved since present-day automatic transmissions still include control levers or press buttons, which have to be operated by the driver to enable the transmission to cope with the widely varying driving conditions encountered. However, systems which require no attention under ordinary road conditions and which require only the selection of the appropriate operating range when heavy gradients or very difficult terrain are to be traversed are now in extensive use. The transmissions can be divided into those which employ mechanical gearboxes, usually epicyclic, for providing all the required variations of gear ratio, and those which use a torque converter in conjunction with a mechanical gearbox.

It is very difficult to design an automatic control system for an ordinary gearbox that will function as satisfactorily under all conditions as can a really competent driver, but present-day systems will give better results than the average driver can achieve. In general, the control must bring about changes from low to higher gears when the vehicle speed rises and from high to low gears when the vehicle speed falls. However, it is frequently possible to employ the higher gears even at low vehicle speeds, for example on level roads and with following winds, when the resistances to be overcome are low.

The control system must therefore take account of the engine load and, in general, produce changes up when the load is light and changes down when the load is heavy. There are, however, occasions, such as on descending hills, when it is desirable to employ a low gear although the load on the engine may be nil or the engine may be acting as a brake. It is under these diverse conditions that the human element has to be retained in the control.

The automatic systems now in use consequently all utilise the two factors mentioned above, namely vehicle speed and engine load, in their operation. The vehicle speed factor is dealt with by employing some form of speed-sensitive unit which is driven off the output side of the gearbox and is thus responsive to vehicle speed. These units may be mechanical, hydraulic or electrical and examples of each of these are given in the descriptions which follow. The engine load factor is introduced at present only by indirect means; it is assumed that engine load is a function of the depression of the accelerator pedal and the latter supplies the corresponding control, the actual form of which will depend on whether the system is a mechanical, a hydraulic or an electrical one.

It is essential that the vehicle speed, at which any change from a lower gear to a higher one is produced when the vehicle speed is rising and the accelerator pedal position is constant, shall be higher than the speed at which, when the speed is falling and the accelerator pedal position is unchanged, the corresponding change down will occur. If this is not so, a change up is likely to be followed immediately by a change down, and this sequence may go on indefinitely. This phenomenon is known as *hunting*. It is also generally desirable that in traffic, when the accelerator pedal is released and the vehicle comes to rest, the control system shall produce all the changes down from top to bottom, but shall retain the gear that is in use when the accelerator pedal is released until the vehicle speed has fallen nearly to zero, and shall then engage the low gear ready for the ensuing acceleration. It is also desirable that it shall be possible to start off in first or in second gear according to the prevailing road conditions.

The above considerations should be borne in mind when reading the descriptions of the systems that follow since many of their apparent complications are due to the necessity to comply with the requirements outlined above.

25.6   Borg-Warner Models 35, 65 and 66 transmissions

Model 35 is shown in Figs 25.14 and 25.15. It consists of a single-stage torque converter IDR coupled to a three forward and one reverse ratio epicyclic gear. The driven member D of the converter is, in effect, integral with the drums E and G of the clutches C_1 and C_2. When C_1 is engaged the drive goes to the sun S_2 and if the brake B_2 is applied gives the low forward ratio while if the brake B_1 is applied instead of B_2 then the immediate ratio is obtained. The one-way sprag clutch F prevents the planet carrier R from rotating backwards but allows it to rotate forwards. By engaging both clutches C_1 and C_2 simultaneously the gear is locked solid and the direct drive is obtained. To get reverse the clutch C_2 is engaged and the brake B_2 is applied, the drive then goes from S_1 to P_1 and thence to the annulus A, the planet carrier being fixed.

The teeth seen on the outside of the annulus in Fig. 25.15 are engaged by a

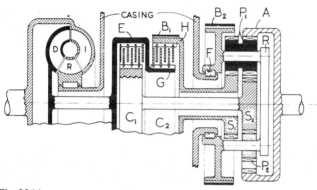

Fig. 25.14

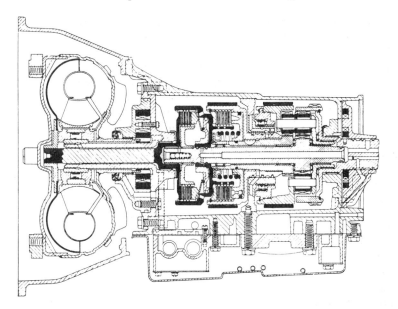

Fig. 25.15

detent when the control lever is put into the parking position and this holds the car stationary. Two oil pumps provide the oil pressures required to engage the clutches and apply the brakes. One is housed in the left-hand end of the box and is driven off the sleeve of the impeller so that it is working whenever the engine is running while the other is seen at the right-hand side and is driven off the output shaft of the box so that it will be running when the car is in motion. The principle underlying the action of the control system is similar to that of the Hydramatic boxes which are described in Sections 25.13 and 25.14.

In principle, the Model 65 is similar to the Model 35, which went out of production about 1974. Their gearsets are virtually identical, though that of the 65 is designed for heavier duty. Detail modifications have improved the quality of operation and, by lowering the brake actuation cylinders to a level below the brake bands and bringing them into the casing, the overall width of the transmission has been significantly reduced.

Further development of the Model 65 resulted in the introduction of the Model 66, which is designed for even heavier duty or for a higher level of durability in cars at the upper end of the market range. The shafting has been strengthened, and the lubrication system improved by fitting a deeper oil pan and enlarging some of the ducts. Externally, however, the dimensions remains the same.

25.7 Borg-Warner Models 45 and 55 transmissions

The demands for increasing fuel economy have tended to expand the market for cars having smaller engines. Consequently it is in this area that manufac-turers of automatic transmissions must look for volume sales. This led to the introduction by Borg-Warner of a transmission specifically designed for cars

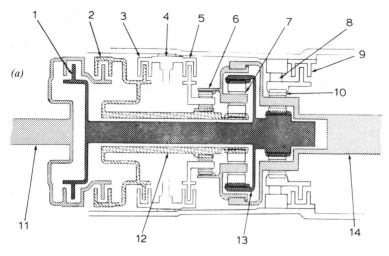

(a)

1 Forward clutch
2 Direct clutch assembly
3 3rd gear brake
4 Centre support (reaction member)
5 2nd gear brake
6 Front planetary train
7 Middle planetary train
8 Sprag clutch
9 Reverse gear brake
10 Rear planetary train
11 Input shaft and forward clutch
   cylinder assembly
12 Front and middle sun gear assembly
13 Rear sun and middle ring gear
   assembly
14 Output shaft and rear carrier
   assembly

Fig. 25.16(*a*) Borg-Warner Model 45

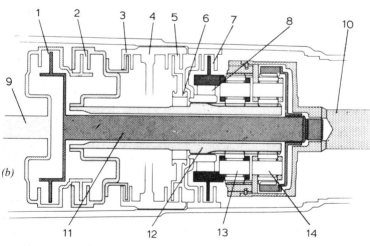

(b)

1 Forward clutch (C.1)
2 Direct clutch (C.2)
3 Brake (B.1)
4 Centre support (reaction member)
5 Brake (B.2)
6 OWC no. 1
7 Brake (B.3)
8 OWC No. 2
9 Input shaft
10 Output shaft
11 Intermediate shaft
12 Sun gear
13 Front planetary train
14 Rear planetary train

Fig. 25.16(*b*) Borg-Warner Model 55

in the 1- to 2-litre categories, as opposed to the 1.5- to 3-litre range associated with the Models 35, 65 and 66.

The Model 45 has four speeds and the ratio spread between top and bottom gears is thus widened without increasing the steps between ratios, so that good acceleration is obtainable with cars of modest power: weight ratios despite the use of a high final-drive ratio for good economy at cruising speeds on motorways. As in the two three-speed models previously described, a torque converter-coupling giving a conversion ratio of about 2:1 is used, but the gearset is of course different and the brakes are of the multi-disc type. Multi-disc brakes have the advantage, over the band type used previously, of not requiring any attention in service; moreover, the actuation pistons and cylinders are all coaxial with the transmission shaft, so the unit is narrower and shallower than the earlier ones. Further reductions in weight and size have been achieved by the ingenious use of steel pressings for many parts of the running gear, which has made possible by the employment of electron beam welding for their fabrication.

The control quadrant for the gear selection, having stations marked PRND32, is similar to those of other automatic transmissions. Manual selection of either third or second gear – the last two of the stations – inhibits automatic shifting to the higher gears, P is for parking, R reverse, N neutral and D is for driving automatically.

There are three planetary trains in the gearset, giving four forward speeds and one reverse. The front train has three planet pinions and the others four. These gears are selected by means of two clutches and three brakes, all of which are applied by hydraulic pressure and released by their return springs. Additionally, there is a one-way sprag-type clutch between the ring gear of the rear train and the transmission case.

Illustrated diagrammatically in Fig. 25.16(*a*) is the arrangement of the gearset, clutches and brakes. When first gear is selected, only the rear planetary train is in operation, the other two rotating freely. The front clutch is engaged, transmitting power to the central shaft and thus the rear sun gear. Since the ring gear is prevented by the sprag clutch from rotating backwards, the planets and their carrier orbit at a speed slower than that of the sun gear – speed ratio 3:1. The planet carrier is connected directly to the output shaft.

Power flow in third speed is equally easily explained because, again, only one gear train is in operation – this time the middle one – while the others rotate freely. The front clutch is engaged and it transmits power to the ring gear of the middle train. At the same time the foremost, or third gear, brake is applied, holding the sun gear and thus causing the planets to orbit at a speed ratio of 1.35:1. This planet carrier too is connected to the output shaft.

For selection of second gear, both the front and middle planetary trains are brought into operation, while the rearmost ring gear rotates freely. This is done by engaging the front clutch, thus conveying power to the middle ring gear, which drives its planets and their carrier in the same direction while tending to drive the sun gear backwards, but at a rate determined by the planetary action of the front train, as follows: the planet carrier of the front train is held by the second gear brake – the middle of the three. Consequently, the backwards rotation of the front sun gear – which is, in effect, integral with the middle sun gear – drives its planets in a manner such as to cause its ring gear to rotate forwards. This ring gear is directly connected to the planet carrier of the middle train, which in turn is connected to the output shaft. The effect of coupling these two trains is a multiplication of the torque, giving a ratio of 1.94 : 1.

Fourth gear is a direct drive, with the two foremost clutches engaged to connect the input shaft to both the sun and ring gears of the middle planetary train simultaneously. In this condition the planet pinions are locked between the two, so the drive is transmitted through their carrier to the output shaft at a ratio of 1 : 1.

In common with other Borg-Warner transmissions, control over the gear changes is exercised by four basic devices. The first is the manual gearshift lever operated by the driver for selecting fully automatic drive, or limiting upward changes to either second or third gear, or for selecting reverse, park, or neutral. In the park position, a pawl locks the output shaft. The second control is a primary valve regulating the line pressure to the system and directing fluid at reduced pressure to the torque converter, the lubrication system and, if fitted, the oil cooler. Thirdly, there is a governor on the output shaft, which actuates a valve to regulate the pressure further in relation to vehicle speed. Finally, there is a valve actuated by the carburettor throttle. This receives fluid at line pressure and directs it, at a pressure related to throttle position, to the shift and other valves. At or beyond about seven-eighths full throttle, not only is maximum pressure attained but also ports are open for directing fluid for the 'kickdown' function, delaying upward gearshifts to obtain maximum acceleration.

The Model 55 transmission is a development of the Model 45. However, since it is intended for cars having engines ranging up to about 3 litres swept volume, it has only three forward speeds and the torque converter. As can be seen from Fig. 25.16(*b*), it has a totally different gearset, with a single long sunwheel machined on the rear end of a hollow shaft within which is the intermediate shaft. The sunwheel has two sets of four plant gears rotating around it. For the Model 55, there is an extra one-way clutch and a different brake arrangement too. The hydraulic control system also is different, and this transmission is also available with an overdrive unit mounted on its rear end.

Incidentally, a major difference between the Model 45–55 gearboxes and the 35, 65 and 66 family is the use of oil-wetted multi-plate brakes in the former to replace the brake bands in the latter. Multi-plate brakes are inherently better able to take heavy loads without showing significant signs of wear or requiring automatic adjustment devices.

25.8   The AP automatic gearbox

This is a four-speed and reverse gearbox in which bevel gears are used as ordinary gearing for some of the ratios and as epicyclic trains for the other ratios. It incorporates a single-stage torque converter whose reaction member is held by a one-way clutch up to the change-over point. The automatic changes are controlled by the speed of the vehicle through a governor driven by the output shaft of the box and by the position of the accelerator pedal. The selector lever has seven positions: reverse, neutral, 1, 2, 3, 4 and D2, this last one giving full automaticity. The selector lever can be used to give manual control at all times, the gear selected being held until the lever position is changed. In automatic operation, kickdown changes are available.

The layout of the gearbox is shown in Fig. 25.17; there are four bevel gears S_1, S_2, S_3 and S_4, the latter being always the driven gear while either S_1 or S_2 can be the input or driving gear according to whether the clutch A or the clutch B is engaged. The clutches are actually multi-plate ones and incorporate a piston to enable oil under pressure to engage them. The plant carrier has two compound planet pinions, only one of which is shown in the diagram. The ratios are obtained as follows—

1st	Clutch A is engaged so that S_1 becomes the driving gear, the sprag F prevents backward motion of the planet carrier which therefore remains stationary because the reaction torque that acts on it is backwards. The drive is from S_1 to P_1 and thence from P_2 to S_4, the planet pinion rotating on its fixed pin.
2nd	Clutch A remains engaged and the brake D is applied to fix the sun S_3. Rotation of the sun S_1 then causes the planet carrier to rotate forwards and so the planet P_2 drives the sun S_4 forwards.
3rd	Clutch A remains engaged but the brake C is applied to fix the sun S_2. Rotation of S_1 again causes the planet carrier to rotate forwards but gives a higher ratio because S_2 is equal in size to S_1 whereas in second gear S_3 was larger.
4th	Both clutches A and B are engaged so that the suns S_1 and S_2 are locked together and the gear has to rotate solid thus giving a $1:1$ ratio.
Reverse	The clutch B is engaged and the brake E is applied to hold the planet carrier stationary. The drive is from S_2 to P_1 and thence from P_2 to S_4.

The oil pressure to apply the brakes and engage the clutches comes, when the engine is running, from a pump that is driven off the cover of the torque converter. To enable a towed start to be obtained a second pump driven off the output shaft is provided. The change-over from one pump to the other is made by the *flow-control and tow-start valve* seen at the bottom left-hand corner in Fig. 25.18, where its position corresponds to that assumed when the engine is

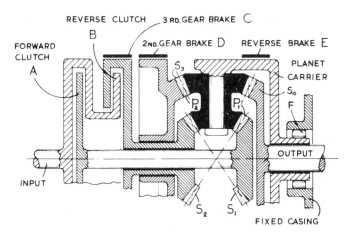

Fig. 25.17 AP automatic gearbox running gear

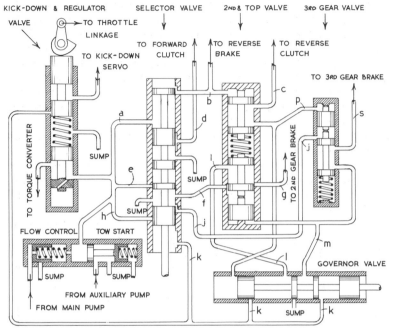

Fig. 25.18 AP Automatic gearbox control system

running and oil is passing from the main pump through the flow-control valve spool to the control system. In this condition the tow-start valve is held in the position shown so that the auxiliary pump can discharge freely back to the sump. When the engine is not running the flow-control valve is moved to the left by its spring and blocks the pipe coming from the main pump, the tow-start valve spool is also moved to the left thereby closing the passage to the sump and connecting the auxiliary pump to the control system so that if the vehicle is towed the auxiliary pump can supply oil to the control system.

The pressure of the oil supply is maintained at the designed value by the regulator valve which is the lower spool in the *kickdown and regulator valve* assembly. The oil from the pump passes through the diagonal hole drilled in the spool and thus sets up a pressure that moves the spool upwards and restricts the entry of oil so that a balance of forces is reached between the upwards oil pressure and the downwards spring force.

The control system. The box can be used either as a manually-controlled box, the gear changes being produced by moving the selector lever to the required position, or the changes, except into reverse, can be obtained automatically by putting the selector lever into the D position. Considering the manual operation Fig. 25.18 shows the selector valve in the position corresponding to N, or neutral, all the brakes being off and the clutches disengaged. If the valve is now moved upwards to reverse then oil will pass from pipe *a* to pipe *b* and so to the reverse brake; it will also pass to the top of the *second and top gears valve* and will move the spool downwards to open pipe *c* and so pass oil to the reverse clutch which, on engagement gives the reverse gear. When *1* is selected the valve is moved downwards and oil passes from *a* to *d* and engages the forward clutch to give first gear; no brake has to be applied because the planet carrier is held by the sprag. Pipe *b* is opened to the sump to release the reverse brake and the spool of the *second and top gears valve* moves back to its top position thus opening pipe *c* to the sump via pipe *p*. The supply to the forward clutch remains open throughout all the subsequent changes so that the forward clutch remains engaged. By moving the selector lever to 2 oil is passed from pipe *e* to *f*; it enters the *second and top gears valve* and raises the middle spool so that oil flows via pipe *g* to the second gear brake, giving the gear. When 3 is selected the oil flows from pipe *h* to *j* and so to the *third gear valve* where it depresses the lower spool so that oil flows to the third gear brake via pipe *s*; at the same time the pipe *f* is opened to the sump so that the second gear brake is released. When 4 is selected oil flows from pipe *h* to *k* and, if the engine speed is not low, from *k* to *p* and thence via *c* to the reverse clutch, thus locking S_1 and S_2 together and giving the 1:1 ratio.

When the selector is put in the *D* position the selector valve reaches its lowest position, pipe *a* remains connected to pipe *d* so that the forward clutch remains engaged while pipe *h* is connected to pipe *k* to give a supply of oil to the *governor valve*. As this valve moves to the left because of the increasing speed of the car it first connects *k* to *l* and thence to the second gear brake via the *second and top gears valve* and pipe *g*, thus giving second gear. Further movement causes the pipe 1 to be connected to the sump, thus releasing the second gear brake and at the same time *k* is coupled to *m* and hence to the third gear brake; at the same time the bottom spool in the *second and top gears valve* is moved upwards to give a free flow of oil from the second gear brake to the sump and take that brake off quickly. When the governor valve moves to its extreme left position the pipe *k* is connected to pipe *p* and hence to pipe *c* so that the reverse clutch is engaged as well as the forward clutch and gives the 1:1 ratio.

The kickdown changes are produced by the operation of the cam on the spool of the kickdown valve which results in oil passing from pipe *k* to *t* and so to the kickdown servo. This is a small ram which bears on an arm of the governor linkage and forces the governor valve spool to the right thereby producing the required downwards change.

25.9   AP hot-shift automatic gearbox

Conventional automatic transmissions suffer two major shortcomings. One is the inevitable energy losses in the torque converter and the other is the complexity of the epicyclic gear arrangements, including clutches and brakes, required if ratio changes are to be made without interrupting the power flow to the road wheels. Moreover, the smaller the engine the greater becomes the significance of the power lost in the torque converter relative to the total power available.

Because of the increasing prices of fuels, the trend for the future must be towards smaller engines driving lightweight cars. This stimulated Automotive Products to re-examine the prospects for developing a simpler and more efficient automatic transmission based on the traditional, highly efficient, manual gearbox. A major advantage was that both manual and automatic could, if necessary, be produced on the same production line, or at least a minimum of investment would be required in new manufacturing facilities and equipment for production of the automatic transmission.

A prime requirement was the development of a *hot-shift system* – changing gear without interrupting the power flow to the wheels. Closing the throttle and re-opening it again for each gearshift is not only a tricky operation, but also it can even be dangerous – for example, potentially initiating skids on ice – and can render emission control difficult. Consequently, a new system, based on the use of two clutches and transferring the drive from one to the other for each shift, was developed.

To understand how it has evolved, look at Fig. 25.19, ignoring the peripheral equipment, including the clutches, and concentrating attention on the mainshaft and the layshaft, top and bottom respectively. The main difference between this and the equivalent manual gearbox is that the gear pairs for second and third speeds have been interchanged – the order, left to right, in the conventional box would be 4, 3, 2, 1. Additionally the mainshaft has been divided, leaving first, third and reverse gears driven conventionally through the main clutch A, on the engine flywheel, while second and fourth gears are integral with a sleeve rotating freely on the mainshaft and driven through a second, smaller clutch B, mounted on the rear end of the gearbox. This second clutch is driven by a quill shaft extending from the engine crankshaft right through the gearbox.

The main clutch is released by the hydraulic actuator, top right in the

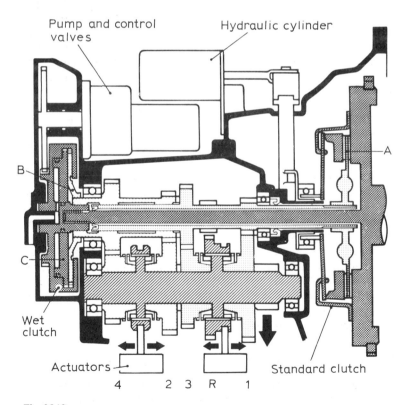

Fig. 25.19

illustration, and engaged by its diaphragm spring, in the usual way. Since the secondary clutch is not used for driving away from rest, but only for gear changing, it is not subject to much slipping and therefore can be smaller and is of the wet type. It is engaged by hydraulic pressure acting on the rear face of the pressure plate C. Hydraulic power is provided by the pump, top left, driven by the continuously rotating portion of the second clutch, and the gears are shifted by hydraulic rams, represented by the two rectangles at the bottom of the illustration.

Overall control is effected electronically by a computer, which could be designed to perform also the functions of engine control and thus optimise engine and transmission operation. So far as the transmission is concerned, the inputs to this control come from a governor, and sensors for indicating which gear is engaged, throttle position and engine and gearbox output speeds – the throttle-position sensor of course is, in effect, a torque-demand sensor. The outputs go to control the valves for the hydraulic actuators of the clutches and gear shift rams. All gears are held in engagement by hydraulic pressure and are sprung out into neutral.

The sequence of operations is as follows: with the engine idling and gearbox in neutral, selection of drive, using the manual control, first causes disengagement of the primary clutch and then a shift into first gear. Depression of the accelerator initiates automatically the gradual engagement of the primary clutch, thus setting the vehicle in motion.

At an appropriate speed, the governor signals a gear change. Since the secondary clutch is already disengaged, second gear can be selected by its actuator operating through its synchroniser. Then, the rear clutch engages while the front one simultaneously disengages, to perform the hot-shift. Finally, the hydraulic pressure to the first gear is released, so it springs out of engagement, and the primary clutch is re-engaged, to keep all parts rotating at engine speed and thus keeping the demands on the synchroniser for the next gear shift to a minimum. Subsequent gear shifts are made similarly, by transferring the drive from one clutch to the other, which is why the second and third gears in the box had to be interchanged for this design.

The claims made for this fully automatic transmission are as follows: its efficiency is as high as that for a manual box except in that a little power is absorbed by the hydraulic pump. Both the weight and cost are certainly no more, and are expected ultimately to be less, than for a conventional automatic box. Other advantages include ease of servicing, minimum of investment required in new tooling and the ease with which the control can be integrated with an electronic engine-management system.

For front-wheel-drive cars, a compact four-speed design is attainable, using a three-shaft layout, Fig. 25.20. Similarly, a six-speed version can be designed on a three-shaft basis, Fig. 25.21, and with so many gears it is possible to approximate fairly closely to the characteristics of an ideal continuously-variable-ratio gearbox, Fig. 25.22. The outcome, AP claim, would be a potential saving of 25% in fuel consumption.

25.10 The Ricardo ALT automatic transmission

The search for a low cost, high efficiency automatic transmission continues and, in late 1987, the Ricardo Automatic Layshaft Transmission (ALT) unit, Fig. 25.23, was announced. It was designed by Peter Windsor-Smith and

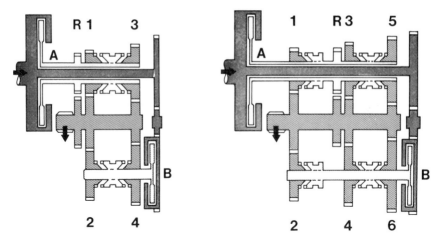

Fig. 25.20   Fig. 25.21

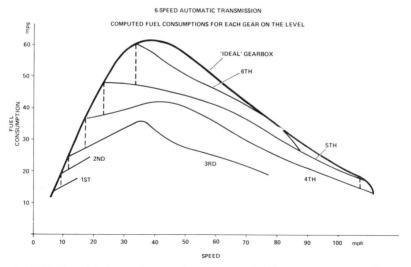

Fig. 25.22 Six-speed automatic transmission; computed fuel consumptions for each gear on the level

developed by Ricardo Consulting Engineers. Peter Windsor-Smith also played a major part in the design and development of the Maxwell transmission for buses, which is described in detail in Section 25.18. Although the ALT transmission is being developed, in the first instance, for cars, it has potential also for heavy duty applications.

Basically, the gearbox is identical to a conventional five-speed synchromesh layshaft type unit but multi-plate wet-clutch units replace the synchronisers. These clutches are hydraulically actuated, automatic control over the hydraulic system being effected electronically. A signal from the automatic control initiates engagement of the clutch in the constant-mesh gear pair appropriate

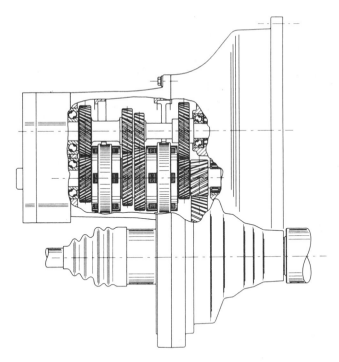

Fig. 25.23 The Ricardo automatic layshaft transmission (ALT), as originally developed for a 1.1-litre car, has multi-plate wet clutches where the synchronisers would otherwise be

to the prevailing engine torque and vehicle speed. As the conditions change, the clutch serving a gear pair that has subsequently become more appropriate automatically takes over from the first. If all clutches are disengaged, no drive can be transmitted. The clutch actuators do not rotate, so there is virtually no leakage past their seals. Two-pedal control is practicable, since the conventional single-dry-plate clutch can be dispensed with, and starting the vehicle from rest is effected by the highly developed multi-plate clutch that engages first gear.

25.11   Van Doorne Variomatic and Transmatic transmissions

The twin-belt Variomatic automatic transmission developed by Van Doorne was first used in 1955 on Daf cars – more recently taken over by Volvo. By 1980, steel belt derivatives were at an advanced stage of development by Fiat, Borg-Warner and others.

The Variomatic transmission system, as originally produced, Fig. 25.24, comprises six main sub-assemblies: the propeller shaft, power divider, two belt-drive units, and two final-drive reduction gear units. Of these, the propeller shaft, with flexible couplings at each end, instead of universal joints, transmits the drive from a centrifugal clutch to the power divider, which is mounted on a sub-frame beneath the rear of the vehicle. The input shaft to the power divider has a bevel gear on its rear end, which meshes with two bevel pinions, one on each side, to turn the drive through 90°.

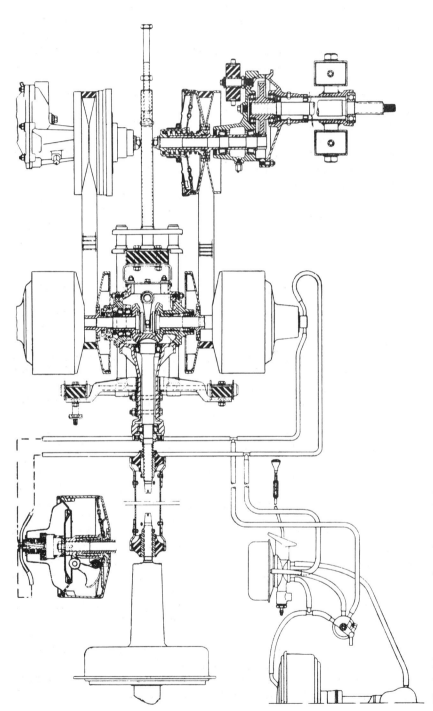

Fig. 25.24 Van Doorne Variomatic transmission

These pinions, therefore, are driven in opposite directions, but they rotate independently of their transverse shaft the outer ends of which carry the two driving, or primary, pulleys. Splined on to the centre of this shaft, and floating axially between the two pinions, is a coupling sleeve with dogs on its ends. When the coupling is in its mid-position, neither of the pinions is engaged, so the transmission is in neutral. If it is moved in one direction along the splines, its dogs engage with slots in the forward-rotating pinion and, if it is slid in the other direction, it engages the pinion rotating in the reverse sense.

One of the flanges of each driving pulley is fixed on the shaft, while the other can be moved axially on splines either to squeeze the belt outwards, if the two are closed together, or to allow it to sink deeper into the groove of the pulley, if they are moved apart. Similarly arranged flanges on the driven pulley move in the opposite sense to those on the driving pulleys, and thus vary the ratio of the drive over a range of slightly more than 4 : 1. The ratio of the bevel drive in the power divider is rather less than 2 : 1 and that of the final drive reduction is about 4.75 : 1. In practice, the maximum overall ratio is of the order of 4 : 1.

The flanges in the driving pulley are moved apart by a combination of manifold vacuum and the centrifugal force on bob-weights, and they are closed together by a diaphragm spring. This spring also transmits the torque from the shaft to the moveable pulley flange, thus obviating the need for a sliding splined joint and consequent problems due to its sticking. A control valve cuts the vacuum assistance out of operation below about quarter throttle and above about three-quarters throttle. In each case, this increases the ratio – for engine braking with the throttle closed, and for maximum acceleration with it wide open.

Unlike the driving pulleys, the driven ones are moved only by springs, but in this case a combination of a diaphragm and coil spring is used, the two together giving an almost constant closing load characteristic. Ratio adjustment is therefore controlled solely on the driving pulleys, the driven ones being automatically self-adjusting to accommodate to them.

Because of the inherent load-sensitivity of the drive, and the use of a separate pair of pulleys for each wheel, the system has a differential drive capability, which can be explained as follows: when the vehicle turns a corner, the acceleration of the outer wheel occasioned by the external forces – i.e. applied to it at its contact with the road – causes the torque transmitted by the belt to be reduced and thus lessens the tension in it. Consequently, the belt is squeezed radially outwards and the effective diameter of the driving pulley automatically increased. The converse effect of course is obtained with the deceleration of the inner wheel, which pulls the belt down deeper into its groove.

Since both the driven pulleys and the spur-type final-drive reduction gears are carried by the swinging arm of the rear suspension, and the axis about which it swings intersects the neutral axis of the tension-loaded belt at the point where it leaves the driving pulley, wheel displacement neither causes misalignment of the pulleys nor varies the tension in the belts. In overrun conditions, when the opposite run of the belt is loaded in tension, the loading is so much lighter that the variations in geometry are of no consequence.

A later version of the same transmission, Fig. 25.25, has a single belt, with a differential gear and final drive reduction gear together forming a transmission unit, which is carried on two transverse members secured by rubber mounting

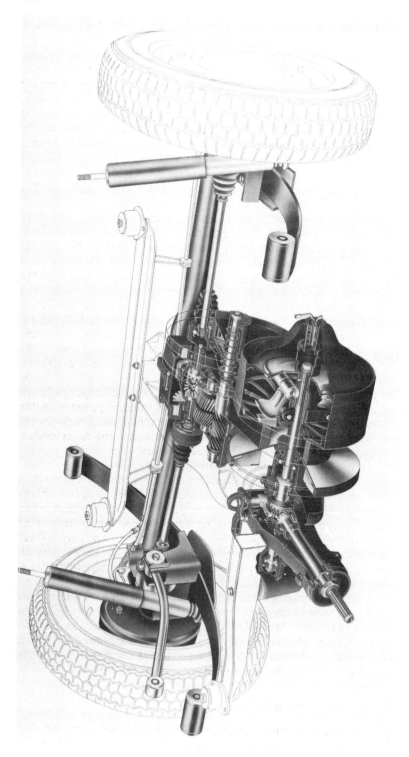

Fig. 25.25 Variomatic transmission with a single belt

beneath the vehicle structure. A de Dion axle with single-leaf springs is employed.

As before, the drive line from the propeller shaft is taken to the two floating bevel gears, with a forward-or-reverse selector coupling sliding between them. Thence, however, it is taken to a single centrifugal- and vacuum-actuated drive-pulley assembly, and through the belt to the spring-loaded driven pulley, which is splined on the end of the final drive reduction pinion spindle. The reduction gear, of the helical spur type, drives a ring gear bolted to the differential cage, the differential gears being connected by constant velocity joints to the swinging driveshafts out to the wheels.

25.12   Van Doorne Transmissive BV steel CVT

The Van Doorne Transmatic continuously variable transmission (CVT), further developed by Fiat and Ford and adapted for the *Uno* and *Fiesta* models in 1987, is similar in principal to the single-belt Variomatic but is much more compact and has a single segmented metal drive-belt, Figs 25.26 and 25.27. Again the pulley ratio spread is about 4:1 but the maximum overall ratio is only about 25:1. This is for an experimental installation in the Fiat *Strada*.

From the illustration it can be seen that it is particularly suitable for front-wheel-drive cars. A centrifugal clutch, top left, transmits the drive to the primary pulley and, through a quill drive, to a hydraulic pump, top right, the adjustable flanges of both the primary and secondary – driving and driven – pulleys slide axially on linear ball bearing splines, that of the primary one being controlled by hydraulic pressure in the cylinder to the left of it, while the secondary one is closed by the coil spring plus hydraulic pressure in the smaller-diameter cylinder on its right. A splined muff coupling between the two gears to the left of the secondary pulley is disengaged in its central position, engages forward drive if moved to the left, and reverse – through the medium of an idler gear – if moved to the right.

Of especial interest is the segmented steel belt. It comprises a set of plates about 2 mm thick by 25 mm wide by about 12 mm deep, with slots in each side to receive the two high-tensile steel bands which hold them together, rather like two strings in a necklace. The sides of the plates slope to match the V-angle of the grooves of the pulleys in which they seat. Unlike a conventional V-belt, however, this one transmits the drive by compression, instead of tension, in the run on one side between the pulleys. The run on the other side of course is unladen and runs free. Buckling of the strand under compression is prevented by tension in the high-tensile steel bands. The principle advantage of this type of transmission is that its torque ratio is steplessly variable. Transmission efficiencies of between 90% at a 1:1 ratio and 86% at the extremes of the ratio range are claimed. The whole unit is totally enclosed in a cast housing and the belt-pulley interface is lubricated by oil passing from the primary pulley actuation cylinder through ducts into the base of the V-groove.

25.13   Hydramatic transmissions

These fully automatic transmissions were developed and are now manufactured in large numbers by General Motors and they have been made under

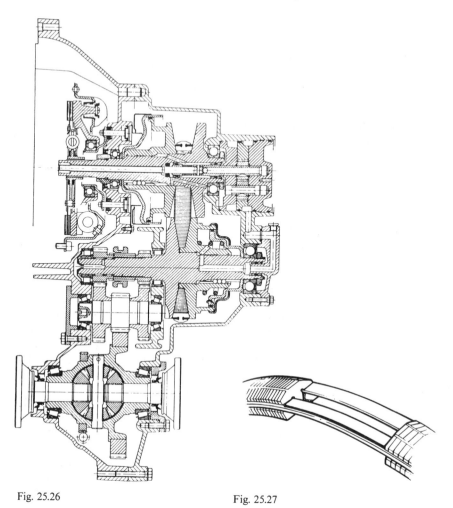

Fig. 25.26   Fig. 25.27

licence by many European firms. They are essentially three- or four-speed and reverse epicyclic gearboxes having brakes and clutches operated by oil pressure and which are controlled by the joint action of a governor, whose speed is proportional to that of the car, and of a valve actuated by the accelerator pedal.

The speeds at which the changes up will occur, supposing the accelerator to be pressed hard down, are, usually, as follows: 1st to 2nd, 19–23 mph; 2nd to 3rd, 37–40 mph; 3rd to 4th, 70–74 mph. At low throttle openings these changes will occur at 7–9, 14–16 and 19–22 mph. The changes down are arranged to occur at lower speeds than the corresponding changes up in order to prevent hunting. Thus at full throttle the down change speeds are 4th to 3rd, 38–17 mph; 3rd to 2nd 13–10 mph; 2nd to 1st, 8–6 mph.

The gearbox was originally produced in 1938 and was redesigned in 1956; the constructional arrangement of the Strato-flight model is indicated in Fig. 25.28. It comprises three sun-and-annulus-type epicyclic trains 1, 2 and 3, two

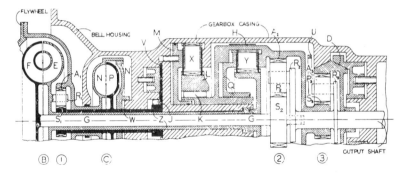

Fig. 25.28 Strato-flight gearbox

multi-plate clutches X and Y, of which the former functions as a brake, a double-cone brake D, two fluid couplings B and C and two one-way clutches or sprags J and K. The coupling B is always filled with oil and transmits power on all the gears but the coupling C can be filled or emptied as required and is used for 2nd and 4th gears only.

The epicyclic train 3 provides the reverse gear only and will be dealt with later on; the trains 1 and 2 together provide the four forward ratios. Each of these trains can be used so as to provide either a reduction or a direct drive; thus train 1 gives a reduction of 1.55 : 1 when its sun S_1 is fixed and the annulus A_1, which is permanently coupled to the flywheel, drives the planet carrier R_1. The latter is integral with the driving member E of the fluid coupling B and thereby drives the member F which is coupled by the shaft G to the sun S_2 of the second epicyclic train. If the annulus of the train 2 is fixed then the sun S_2 will drive the planet carrier R_2, which is the output member of the box, with a reduction of 2.55 : 1. The four forward gear ratios are therefore obtained as shown in Table 25.1.

Because some slip will occur in the coupling B the sun S_2 will be driven at a lower speed than that of the planet carrier R_1 and this slip will modify the numerical values quoted above and will make all the reductions slightly greater.

The epicyclic trains of the gearbox are essentially the same as in the models produced prior to 1956 but the methods used to control the elements of the trains in order to give either a reduction or a direct drive are different. Whereas in the earlier models the sun S_1 was fixed by the application of a band brake it is now fixed by a *sprag*, or one-way roller clutch J which prevents the inner member of the unit, which is integral with the sun S_1, from rotating backwards. Similarly the annulus A_2 of the second train was fixed by a band brake in the earlier models and is now fixed, when necessary, by the sprag K (or the band brake H or by both together) which prevents it from rotating backwards provided the outer member L of the sprag itself is prevented from rotating by the engagement of the multi-plate neutral clutch X which locks it to the casing of the gearbox. This is done by admitting pressure oil behind the piston M. To obtain first gear the clutch X is engaged, the clutch Y is disengaged and the sun S_1 and annulus A_2 are fixed by the sprags J and K respectively. To obtain second gear the epicyclic train 1 has to be put into direct drive. In the earlier models this was done by engaging a multi-plate clutch that was arranged

Table 25.1—EPICYCLIC COMBINATIONS AND GEAR RATIOS IN HYDRAMATIC GEARBOX

Gear	Train No. 1	Coupling C	Sun S_1	Brake V	Train No. 2	Brake X	Annulus A_2	Clutch Y	Brake D	Ratio
1	In reduction Ratio 1.55:1	Empty	Fixed	On	In reduction Ratio 2.55:1	On	Fixed	Disengaged	Off	1.55×2.55 $= 3.966:1$
2	In direct drive Ratio 1:1	Full	Rotating	Off	In reduction Ratio 2.55:1	On	Fixed	Disengaged	Off	1×2.55 $= 2.55:1$
3	In reduction Ratio 1.55:1	Empty	Fixed	On	In direct drive Ratio 1:1	On	Rotating	Engaged	Off	1.55×1 $= 1.55:1$
4	In direct drive Ratio 1:1	Full	Rotating	Off	In direct drive Ratio 1:1	On	Rotating	Engaged	Off	1×1 $= 1:1$
Reverse	In reduction Ratio 1.55:1	Empty	Fixed	On	Compounded with train No. 3	Off	Rotating backwards	Disengaged	On	4.31:1

between the planet carrier member R_1 and the sun S_1, and this method is used in the Dual-Range model, but in the Strato-flight model it is done by filling the coupling C, which in first gear was empty, so that the driving member N, which is permanently coupled to the flywheel, drives the member P which is permanently attached to the sun S_1. If no slip occurred in the coupling C then the train 1 would give a 1:1 ratio since the sun and the annulus would be running at the same speed. Because some slip must occur in the coupling C the sun S_1 will run slightly slower than the annulus A_1 and the train will give a reduction slightly greater than unity. This slip will also modify the ratios quoted above for second and fourth gears. This use of a fluid coupling instead of a multi-plate clutch is to provide a much smoother take-up of the drive than was previously obtainable.

Third gear is obtained by emptying the coupling C so that train 1 goes back into reduction and by engaging the clutch Y so that train 2 goes into direct drive. This is done by admitting pressure oil behind the piston Q; this is the same as in earlier models.

To obtain reverse the clutch Y is disengaged and the coupling C is emptied while the annulus A_3 is fixed by admitting oil behind the piston D. Train 1 then provides a reduction, its sun being fixed by the sprag J, and the trains 2 and 3 function as a compound train to give a reversal of direction and a reduction in speed between the sun S_2 and the output member R_2. This action is precisely as in the reverse gear of the Wilson gearbox (see Section 24.5). During reverse the sun S_3 and hence the annulus A_2 will be rotating backwards and to permit this rotation to take place the sprag K must be made inoperative; this is done by disengaging the clutch X so that the outer member L of the sprag is free to rotate; the band brake H must also be free.

When the train 1 is being used to provide a reduction and the sun S_1 is prevented from rotating backwards by the sprag J the single-disc brake V may be employed to hold the sun against any tendency to rotate *forwards*, which tendency would arise if the car over-ran the engine: this enables the engine to be used as a brake.

The coupling C is filled by means of a large capacity vane-type pump and the emptying is done by opening a valve provided in the cover N of the coupling which will permit the oil to be discharged by the action of centrifugal force. The operation of this 'dumping' valve is by means of oil pressure which is admitted to the space W and thence through passages in the cover N to the valve.

When the clutch Y is engaged the engine torque is divided at train 1; part of the torque is transmitted direct from the arm R_1 through the sleeve Z and the clutch Y to the annulus A_2 while the remainder is transmitted through the coupling B and the shaft G to the sun S_2. Approximately 40% of the engine torque goes through the coupling B.

Because the coupling C runs at engine speed and has to transmit only just over one-third of engine torque it can be made smaller in size than the coupling B which has to transmit slightly over one and a half times engine torque on the first and second gears (when the clutch Y is disengaged) and which runs at a lower speed than the engine.

25.14 Hydramatic Strato-flight gearbox controls

The steering column control lever has six positions, parking (P), neutral (N), normal driving (Δ Dr), fast accelerating, driving (Dr Δ), low range (Lo) and reverse (Rev). Parking differs from neutral in that a positive lock engages the teeth U, Fig. 25.28, to lock the output shaft of the gearbox when (P) is selected.

At the rear end of the gearbox there is a governor which is driven by the output shaft and whose speed is thus proportional to the road speed of the vehicle. The governor is actually two governors in one assembly and is shown diagrammatically in Fig. 25.29. Since both governors function in the same manner only one need be described. Considering the valve B and assuming that the pump that supplies the system with pressure oil is operating, the action is as follows. The valve is shown in the equilibrium position, the upper end of the portion B is just closing the port by which pressure oil is introduced while the lower end of B seals the pipe line 1 from the exhaust port. The pressure in line 1 will be lower than the pump pressure and, as will be seen shortly, will be proportional to the rotational speed of the governor. The position is that the centrifugal force acting on the valve and tending to make it move outwards (downwards in the figure) is just balanced by the force due to the oil pressure acting on the underside of the piston A and which tends to move the valve inwards. Suppose now that the governor speed increases a little, then the centrifugal force will increase and will overcome the oil pressure so that the valve will move outwards; this will open the pressure oil inlet from the pump and the pressure in the line 1 will rise until the increased pressure acting on A is sufficient to balance the centrifugal force when the valve will move into the equilibrium position again. Thus the oil pressure in line 1 will always be proportional (up to the limit of the pump pressure) to the centrifugal force acting on the valve and thus proportional to the square of the governor speed. The reason for using two valves is that a single valve that provided sufficient pressure to control the changes at low car speed would give pressures that would be much too great at the higher speeds while if a valve suitable for the high speed changes were used it would give insufficient pressure for the low speed changes.

The governor valve B thus gives a pressure in line 1 which is dependent on

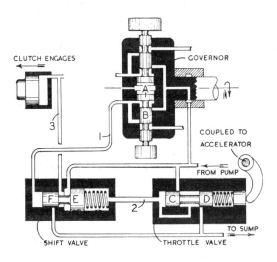

Fig. 25.29

the road speed of the vehicle. This pressure will be referred to as G_1 *governor pressure*, that from the other valve will be called G_2 *governor pressure*.

The throttle valve acts in a manner similar to the governor valve and is also shown in the equilibrium position, the pressure of the spring towards the left being balanced by the oil pressure in line 2 which acts on the left-hand end of the valve and tends to move it to the right. If the accelerator pedal is further depressed the balance will be upset and the valve will move to the left, the part C will then uncover the inlet port from the pump and the pressure in the space between the parts C and D will increase and this increase will be communicated by the by-pass to the left-hand end of the valve and by the pipeline 2 to the right-hand end of the shift valve. The increase of pressure will continue until the force exerted by the oil on the left hand of the throttle valve moves that valve back to the equilibrium position. The spring is now compressed a little more and the pressure in line 2 is a little higher than before. On the other hand, if the accelerator pedal is released a little then the oil pressure on the left-hand end of the valve will move the valve to the right, thus uncovering the exhaust port. The pressure in the line 2 will then fall until the spring force moves the valve back to the equilibrium position again.

The throttle valve thus gives a pressure that is dependent on the throttle opening. This pressure will be referred to as *throttle-valve* (TV) *pressure*.

The shift valve shown in Fig. 25.29 is a simplified version of the actual shift valves but enables the fundamental action to be explained easily. In the position shown the governor pressure in line 1 which acts on the left-hand end F of the valve has overcome the combined force of the spring and the TV pressure acting on the right-hand end E and the valve in its extreme right-hand position in which the pipe coming from the pump is connected to line 3 which goes to an operating member of one of the epicyclic trains, say the clutch Y of Fig. 25.28. That clutch will therefore be engaged so as to give the direct drive in the train and this corresponds to the change up from first to second or from third to fourth gear.

If the speed of the car decreases or the accelerator is depressed further, then the shift valve will be moved to the left and will cut off line 3 from the pump line 1 and open it to exhaust so that the change down would occur.

The motion of the valve is made a snap motion by reason of the difference in diameter of the ends E and F and this also imparts a delay action so that the valve will not move back so as to change down again unless the car speed decreases appreciably or unless the throttle is open considerably. These actions are obtained as follows. In the position shown the forces acting on the valve to the right are due to governor pressure acting on the end F and pump pressure acting on the difference in area of E and F, whereas when the valve started to move to the right the latter pressure was absent. Hence as the part E uncovers the inlet port an additional force urges the valves to the right and makes the valve motion rapid. Again before the valve can move back to the left the force on F due to the governor pressure must fall below the force on E due to the throttle valve (or the latter must rise above the former) by an amount corresponding to pump pressure multiplied by the difference in area of E and F. The valve cannot move back, therefore, unless an appreciable fall of car speed, or an appreciable increase in throttle opening, occurs and thus hunting is prevented. The snap action of the valve when moving to the left occurs because as soon as the part E covers the inlet ports and the part F uncovers the exhaust port the force due to pump pressure acting on the difference in areas of E and F disappears and the valve moves across rapidly to its extreme left position.

The complete control system is shown in Fig. 25.30, in which the manual valve is shown in the Δ Dr (drive left) position, corresponding to normal driving, and the shift and coupling valves are shown in the position corresponding to low gear; the car speed and governor pressure being assumed to be low. The main pump supplies oil to the governor which gives G_1 pressure in line 9 and thus on the left-hand end of the boost valve. The latter acts in a manner similar to the throttle valve to give a pressure in line 12 which is proportional to, but higher than, the governor G_1 pressure. This boosted governor pressure is passed via line 12 to the transition valve and then by line 13 to the left-hand end of the coupling valve. The latter, in the position shown, opens the line 15, which controls the emptying valve of the coupling C, Fig. 25.28, to exhaust so that the emptying valve is open and the coupling C is empty. The epicyclic train 1 is thus in reduction. Since pump pressure via lines 2, 6, 8 and 10 is applied to the neutral clutch (X, Fig. 25.28) the latter is engaged, thereby enabling the freewheel unit, Fig. 25.28, to anchor the annulus of the rear epicyclic which is thus also in reduction, giving first gear. For the moment it will be assumed that the throttle opening is kept constant and at a moderate value, then as the speed of the car rises the G_1 pressure acting on the coupling valve will rise and will eventually move the coupling valve to the right. This will cut off line 15 from exhaust and open it to line 14 so that pump pressure via line 1 will be applied to the coupling-emptying valve so as to close it. At the same time the coupling valve will open line 3 to line 16 so that pump oil will be supplied to the coupling C so as to fill it. As the coupling fills so the first epicyclic will go into direct drive, thus giving second gear. Further increase in car speed and governor pressure, which acts via lines 9 and 52 on the 2–3 shift valve will cause the latter to move to the right and bring about the change up from second to third gear as follows. Pump pressure via line 6 from the manual valve will be passed via lines 8 and 54 to line 17, which has hitherto been opened to exhaust; thus pump pressure will be passed to the rear epicyclic clutch (Y) so as to engage it and via line 19 to the left-hand end of the transition valve. The latter will consequently move to the right, thus cutting off line 12 from line 13 and opening the latter to line 20 and thus to exhaust. The oil pressure on the left-hand end of the coupling valve will thus be released and the coupling valve will move back to the left, thereby opening lines 15 and 16 to exhaust. The coupling C will thus empty and the first epicyclic will go back into reduction. Since the rear epicyclic is now in direct drive, third is obtained. This action is controlled by the orifice which restricts the entry of oil into the transition valve and by the accumulator whose piston will have moved to the right before pressure can build up in the line going to the clutch Y.

The change up from third to fourth will follow if the car speed and governor pressure continue to increase, the action being that governor (G_1 and G_2) pressures which act, via lines 9 and 53, on the 3–4 shift valve, will move that valve to the right, producing the following results. Line 20 will be cut off from exhaust and be opened via lines 7 and 6 to pump pressure from the manual valve, and this pressure will be passed via the transition valve (which is to the right) and line 13 to the coupling valve, thereby moving that valve to the right and opening lines 15 and 16 to pump pressure; the coupling C will thus be filled

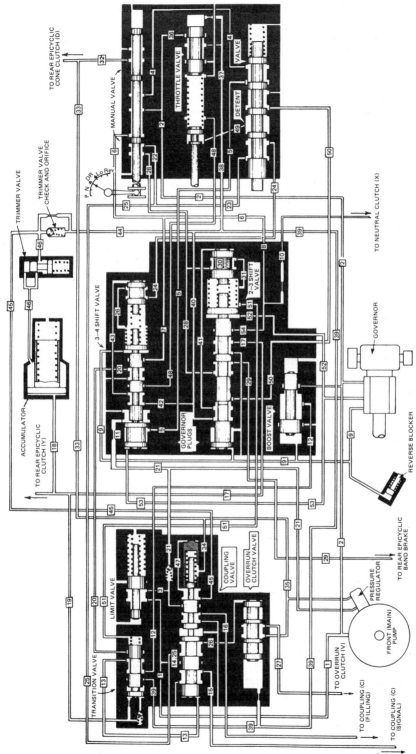

Fig. 25.30

and the first epicyclic will go into direct drive, thus giving fourth gear. At the same time pump pressure from the manual valve via lines 4, 5 will be passed to line 21 and thus to the inner end of the pressure regulator of the pump where it will act so as to reduce the maximum pressure produced by the pump. This is desirable since it reduces the power consumption of the pump, and permissible since at all except large throttle openings, the torques in fourth gear are not heavy.

If the car speed and governor pressure fall the above changes will occur in the reverse order and the changes down will occur successively.

The effect of varying the throttle opening will now be considered. TV pressure via lines 36, 37, 38, 40, 41, 42 and 43 is applied to the right-hand ends of the 2–3 and 3–4 shift valves and via lines 44, 45 and 47 to the right-hand end of the coupling valve, and so increasing TV pressure will necessitate higher governor pressure in order that these valves may be shifted to produce the changes up and these changes will consequently occur at higher car speeds. The TV pressure also acts via line 46 on the annular area of the accumulator piston so as to tend to hold that piston down; this tends to speed up the engagement of the rear epicyclic Y on a 2–3 change as the throttle opening is increased. The TV pressure that can be applied to the accumulator is limited by the trimmer valve to $517 \, kN/m^2$, in order that the clutch engagement shall not be made unduly harsh. It will be noticed that when the coupling and shift valves move to the right the TV pressure acting on the right-hand ends of those valves is cut off; this produces the difference in the car speeds at which the changes down will occur and also provides some of the snap action of the valves.

A part-throttle change down from fourth to third can be produced, provided that the car speed is not too high, by depressing the accelerator about half-way to the bottom position. This opens line 48 to TV pressure which is thus passed via line 43 (the 3–4 shift valve being to the right) to the right-hand end of the 3–4 shift valve regulator plug which is accordingly forced to the left and thereby opens line 26 to TV pressure which is enabled to act on the large area of the shift valve and move it to the left, so producing a change down from fourth to third. A forced change from third to second can be produced by pressing the accelerator to the floor; this moves the detent valve to the right and passes TV pressure via lines 49, 50 and 41 (the 2–3 shift valve being to the right) to the right-hand end of the 2–3 shift valve regulator plug, forcing it to the left so that TV pressure is caused to act via line 31 on the large end of the shift valve and move it to the left to produce the change down. When this occurs the coupling C has to be filled and the rear epicyclic clutch Y has to be released and it is necessary to ensure rapid movement of the transition valve to the left and of the couplings valve to the right as soon as the 2–3 shift valve moves to the left. The transition valve is moved rapidly by passing TV pressure via the detent valve to line 50 and hence as soon as the 2–3 shift valve opens line 40, to line 39 and thus to the right-hand side of the large left-hand end of the transition valve (which is to the right), rapid movement of which is allowed by the lifting of the ball valve and by-passing of the restricting orifice. The moving of the transition valve passes governor boost pressure via lines 12 and 13 to the end of the coupling valve (which in third gear is to the left) and moves it to the right. To speed up this motion a momentary reduction of the TV pressure acting on the right-hand end of the coupling valve is produced as follows. The movement of the 2–3 shift valve to the left opens line 17 and thus lines 18 and 19 to exhaust so that the accumulator piston moves rapidly to the left (in the diagram); this produces a sudden lowering of pressure in line 46 and, because of the trimmer valve check and orifice, in line 45 also and hence via line 47 behind the right-hand end of the coupling valve. As soon as the latter moves to the right the line 45 will be closed so that although the pressure in line 45 will build up slowly through the trimmer valve orifice it will not act to move the coupling valve back to the left again.

When the manual valve is moved to the Dr Δ (drive right) position it opens line 22 to pump pressure and hence pump pressure is applied via lines 23, 24 and 26 to the right-hand end of the 3–4 shift valve; this has the effect of preventing governor pressure via lines 9 and 53 from moving the shift valve to the right except at very high car speeds so that the gearbox is virtually reduced to a three-speed box and the increased acceleration of third gear is retained. Pump oil is also passed via lines 25 and 26 to the overrun clutch valve and moves that valve to the left so that line 26 is opened to line 27 and pump pressure is applied to engage the overrun clutch. The engagement of the overrun clutch is delayed until the sprag has engaged because the volume of oil trapped in the left-hand end of the overrun clutch valve has to be discharged through a restricted orifice. This enables the engine to be used as a brake in third gear if required since the overrun clutch will prevent forward rotation of the sun of the front epicyclic whereas the sprag will not do so. The line 25 is cut off from line 26 whenever the coupling valve moves to the right, as it does for second gear, so that the overrun clutch valve is moved back to the right-hand position and the overrun clutch is released as is necessary to enable the front epicyclic to go into direct drive.

When the manual valve is moved into the Lo position, pump pressure is passed to line 28 and thus via line 30 and 31 to the right-hand end of the 2–3 shift valve, and via line 51 to the right-hand end of the transition valve, so as to prevent that valve from moving to the right except at very high car speeds. The gearbox is thus restricted to first and second gears. This is intended to enable very heavy going, such as in sand, to be negotiated. Since under these conditions torques will be high, oil from line 28 is passed via line 29 to the rear epicyclic band brake which thus assists the rear sprag in holding the annulus of the rear epicyclic and also holds that annulus against forward motion so that the engine can also be used as a brake in second gear if required. If the car speed should increase sufficiently to enable governor pressure to move the 2–3 shift valve to the right so as to produce a change up to third the shift valve will cut off line 28 from line 29 and thus release the band brake. In this position of the manual valve the line 4 is cut off from pump pressure so that pressure is not applied to the pump pressure regulator and the maximum pump pressure is not reduced.

When the manual valve is moved to the Rev position pump pressure is applied direct via line 32 to the rear epicyclic cone clutch (D) and via line 33 and 34 to the inside of the coupling valve regulator plug. The coupling valve is thus prevented from moving to the right and the front epicyclic is kept in reduction. It will be noticed that governor pressure is always to the reverse blocker so that if the forward speed of the car is higher than about 10 mph the blocker will be held out against spring pressure and will prevent the selector and manual valve from being moved to the Rev position. Line 33 is also connected via line 35 to the outer end of the pump pressure regulator so as to increase pump pressure in order to provide ample holding power to the cone clutch D. There is also a connection via line 27 to the overrun clutch so that the engine can be used as a brake in reverse.

The action of the limit valve, which has not yet been dealt with, is that it acts as an excess pressure safety valve since if the pump pressure via line 1 exceeds the maximum safe value the valve will move sufficiently far to the right to open line 1 to exhaust. The valve also acts to prevent the pressure from falling below a value of about $380 \, kN/m^2$, as it might do if a leak developed in coupling C, because at lower pressures it will not move far enough to uncover line 3. Under these conditions the filling of coupling C will be done by oil from the manual valve via lines 4, 5, 11, 21 and the non-return valve into line 3.

25.15   Automatic transmissions for commercial vehicles

Automatic transmissions have been slow to gain ground in the commercial vehicle field. The principal reasons are that, because the numbers of vehicles involved are so much smaller than in the case of cars, the cost per transmission is correspondingly higher; secondly, with most such transmissions there is a fuel consumption penalty which, even though it may be only a few per cent, represents nevertheless a considerable addition to the operating costs of a vehicle travelling very large mileages at a rate of fuel consumption which, in any case, is high.

However, efficiencies have improved over the years and, with only a few very large-scale specialist manufacturers – principally Allison, Voith and ZF – producing increasing numbers of heavy-duty automatic transmissions to meet the demands from over the world, costs per unit have been falling as the quantities sold have continuously increased. These manufacturers, while admitting that a skilful driver can get a better fuel consumption with a manual transmission than with automatic, generally assert that this is only if such a driver is positively aiming at low consumption, and that his performance will in any case fall off as the hours pass during his working day. The main argument for automatic transmission though is unrelated to fuel costs: it is that maintenance costs, including vehicle downtime, are greatly reduced since there are no clutch changes and virtually no other work to be done on the driveline.

Types of operation for which automatic transmission may be economical are as follows—

(1) Buses – not only because of reduced drive-line maintenance but also increased passenger comfort and safety, especially for the elderly.

(2) Stop–start operation with goods vehicles – door-to-door deliveries and special-purpose vehicles such as refuse collectors.

(3) Rear-engine vehicles and vehicles that operate in conditions in which the driver may not be able to hear his engine as he changes gear – e.g. where aircraft or road works are causing high levels of extraneous noises.

(4) Where the driver has other controls to operate while operating the vehicle – e.g. road sweepers, tippers, diggers and dumpers.

(5) When dual control is required and the installation of such controls is more costly than that of automatic transmission.

(6) Where drivers are trained and employed primarily for duties other than driving – e.g. refuse collection, telephone maintenance, etc. – or where they may successively have to take over vehicles of many different types in the course of a day's work.
(7) For emergency services such as fire appliances, ambulances and police.
(8) Operation in confined spaces close to valuable equipment, such as aircraft, that must not be damaged.
(9) For costly special-purpose vehicles where, for operational reasons, downtime for repairs to drive lines would be expensive or unacceptable.
(10) Vehicles used among crowds of pedestrians.

The number of gear ratios required varies according to the type of operation. For instance, a light bus in a small town in a flat country such as Holland might require only a torque converter – ratio perhaps between 2.0 and 2.5:1 – and one extra forward ratio plus reverse. In general, however, a city bus might require four speeds, a coach five speeds and a long distance truck six speeds; all three types of vehicle of course would need a torque converter too, operating perhaps on only the bottom two gears.

25.16 Voith Diwamatic transmission

This is shown in Fig. 25.31 and is an example of a 'split-torque' transmission, the engine power being transmitted partly by a single-stage torque converter and partly by direct mechanical action through an epicyclic gear. A forward and reverse gearbox of conventional form is also employed. The converter is somewhat different from the usual arrangement in that the impeller I discharges directly into the fixed casing R whose passages lead the fluid into the driven member D. The division of the engine torque is done by the epicyclic train ACS in which the annulus is driven by the engine. When the vehicle is at rest and the planet carrier C is therefore stationary (assuming the dog clutch J to be engaged with the gear E) the impeller I will be driven at a relatively high speed in the opposite sense to the rotation of the annulus A. The fluid will then develop a high torque on the driven member D and this will begin to rotate (in the same sense as the annulus) and will apply a torque to the gear E through the gears G and H, the one-way clutch J and the pinion L. A driving torque will also be applied to E direct through the planet carrier C. As the speed of the

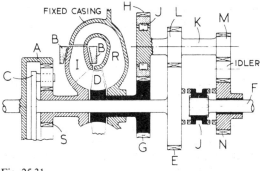

Fig. 25.31

vehicle increases so the speed at which the impeller is driven will fall and the torque acting on the member D will also fall. At a speed determined by a governor driven from the output gear E the brake BB will be applied and the impeller brought to rest. The whole of the engine torque and power will then be transmitted mechanically through the planet carrier C and the dog clutch J. The reverse drive is obtained by engaging the dog clutch with the gear F and as this is done a cam operates a valve which empties the torque converter. The brake BB has therefore to be applied in order to obtain a drive and the application of this brake will also give the clutch action which is normally obtained from the torque converter.

25.17   ZF HP500 fully automatic transmission

As can be seen from Fig. 25.32, the ZF HP500 automatic transmission has a hydraulically-actuated lock-up clutch in front of the torque converter. Its lock-up mechanism is engaged automatically to lock the impeller and turbine wheels together, to give direct drive, at a predetermined ratio of turbine-to-engine speed. However, a kick-down switch actuated by the accelerator pedal unlocks them again to provide extra torque when called for by the driver for climbing a hill, or overtaking on one, in first or second gear. A one-way clutch in the hub of the converter prevents the stator from rotating during torque conversion, but allows it to freewheel when the unit is operating as a fluid coupling, that is at a $1:1$ ratio. A secondary oil pump, driven from the output shaft of the gearbox, can be fitted as an optional extra if a tow-starting facility is likely to be needed. In such circumstances, this pump performs the duty that is otherwise the function of the engine-driven pump.

The torque converter – maximum ratio $2.5:1$ – is used only for starting from rest in first or second gears or reverse, except under kick-down conditions, as previously described – hence the high overall efficiency of this transmission. Behind it is a space in which, if required, a hydraulic retarder can be installed without adding to the overall length. In Fig. 25.32, the retarder, item 4, is shown installed.

For the four-, five- and six-speed versions of this transmission, there are three planetary gear trains, but for the five- and six-speed boxes there is also an alternative internal arrangement of four gear trains and therefore a wider spread of ratios. All have a central multi-plate clutch, just forward of the group of gearsets, the other clutches and the multi-plate brakes being in an annular space between the housing and the gearsets and central clutch.

Simplest of all of course is the four-speed box, which has, in addition to the central clutch, only one annular clutch and three brakes. Then there is a five- or six-speed variant having an extra annular clutch. Finally, there is the five- and six-speed version with an extra gear train behind the others, for which there is another brake.

The fourth gearset – rearmost – is manufactured as a separate module so that it can be fitted or omitted, as required. When it is not fitted, the power flow pattern is the same, except in that 1st, 2nd, 3rd, 4th and 5th speeds are respectively as illustrated for 2nd, 3rd, 4th, 5th and 6th, and the alternative power-flow is adopted for the 6th speed. The power flow for the four-speed version is similar, i.e. 2, 3 and 4 illustrated here become 1, 2 and 3, but the alternative 4th speed power-flow is adopted.

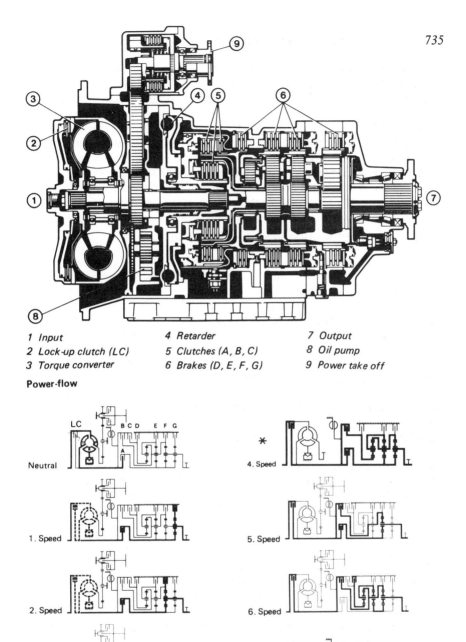

1 Input
2 Lock-up clutch (LC)
3 Torque converter

4 Retarder
5 Clutches (A, B, C)
6 Brakes (D, E, F, G)

7 Output
8 Oil pump
9 Power take off

Power-flow

LC   B C D   E F G

Neutral

1. Speed

2. Speed

3. Speed

4. Speed

* 4. Speed

5. Speed

6. Speed

* 6. Speed

Reverse

* Alternative 4 and 6 speeds

Fig. 25.32 ZF5 and 6 HP 5000 automatic transmissions (version 2, with fourth gearset)

25.18 The Maxwell automatic transmission

Automatic transmissions based on hydraulic torque converters have some serious disadvantages: they are bulky, heavy, complex, costly, their servicing requires a high degree of skill, and losses in the converter entail a fuel consumption penalty. The Maxwell automatic transmission was designed to overcome all of these disadvantages, and is especially useful for rear-engine buses in which the engine is installed transversely. Most of its servicing can be done without removing the engine-transmission unit from the vehicle: with a conventional unit, removal and replacement of the engine and transmission, for replacement of a small component such as a brake band valued at only a few pence, may cost as much as about £1000.

At the front of the Maxwell transmission is a fluid coupling. Splined into the rear end of this and coaxial with it is a quillshaft, which extends rearwards through a short hollow mainshaft and integral gear cluster. On the rear end of the quillshaft is the primary gear.

Arranged round the mainshaft are four short layshafts each carrying a pinion meshing with a layshaft gear, and having, at its rear end, another pinion which meshes with the primary gear. Secured to the rear face of each layshaft pinion is a pneumatically actuated multi-plate wet clutch. Engagement and disengagement of this clutch couples and uncouples the pinion with the primary gear. All the gears and pinions are in constant mesh.

The whole assembly is enclosed in a housing of rectangular shape, the layshafts being positioned one in each of its four corners. Access for attention to the clutches in service is gained by removing four bolted-on circular cover plates from apertures in the end of the housing.

In Fig. 25.33 are two diagrams, one showing the first gear layshaft, the mainshaft and output gear assemblies, and the other showing the corresponding arrangement for the fourth gear layshaft. Note that the spiral bevel output gears are in a bolted-on housing near the front end of the box, so that they can be easily removed in service and, important to the vehicle designer, the

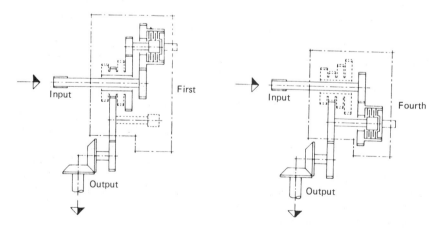

Fig. 25.33 Showing the transmission paths when the clutches for the first and fourth speeds (*left* and *right* respectively) are engaged in the Maxwell transmission

driveshaft can be taken almost straight back to the back axle. This is why this transmission is so suitable for rear-engine buses: with a conventional transmission, having its output shaft at the rear end, the drive has to be taken through an acute angle for connection to the final drive unit in the back axle. This acute angle drive also of course adds to the overall length.

The clutches are actuated by an electro-magnetic valve, which is controlled automatically by an electronic system. Prior to a gear's being selected for starting from rest, all the clutches are disengaged. To select first gear, the clutch on the first gear layshaft is engaged; then, to change upwards again, this clutch is progressively disengaged while that for the next gear is being engaged. For the selection of reverse, the first speed layshaft gear is slid by a pneumatic actuator, against the resistance of a return spring, out of mesh with its mainshaft gear and into mesh with an idler gear in constant mesh with the first gear on the mainshaft, and then the first gear clutch is engaged.

A useful feature of this box is the fact that the first and second speed clutches can be held in partial engagement, to act as retarders when either the third or fourth speeds are engaged. While the retarder is in operation, oil from a pump in the base of the box is delivered at a rate of up to 68 litre/min through ducts in the casing into the hollow layshafts for first and second speeds and then radially out between their clutch plates, to dissipate the heat generated. It is then collected and returned through a separately mounted oil-cooler to the sump. The retarder is brought into operation by the initial movement of the brake pedal, during which it takes up some lost motion before the service brakes are brought into operation.

25.19 Leyland continuously variable transmission

With a transmission the ratio of which can be varied progressively from reverse through zero to the maximum needed it becomes possible to operate the engine virtually continuously at its most economical speed, Fig. 25.34, regardless of changes in load and vehicle speed. Furthermore, the jerk-free ratio changes, Fig. 25.35, and total absence of shock loading due to clumsy gear shifting means a minimum of wear and tear on the transmission line. Consequently, a continuously variable transmission (CVT) has been the aim of automotive designers since the invention of the road vehicle. A major obstacle to such a development has been the fact that the requirements for control over such a transmission are too complex to be exercised satisfactory by a human being. Now that microprocessors can be used for this control, however, not only has this obstacle been removed but also prospects have been opened up for its use in conjunction with regenerative braking, see Section 33.22.

Leyland's CVT is developed from an idea originally patented by W. D. Hoffman in 1899. Further development was begun in 1928 in the USA by F. A. Hayes, and his transmission was installed in about 600 cars in the nineteen-thirties. Yet more work was done in the UK after the Second World War by Forbes Perry, for some of the time with BL. The outcome was first the installation of an experimental transmission of this type in a Triumph *Dolomite* and then, by 1983, the announcement of a similar, but much larger, transmission for installation in the Leyland *Terrier* truck.

In essence, the Leyland CVT, Fig. 25.36, comprises what is termed a *variator* and a planetary gearset. The variator is interposed between the primary gears

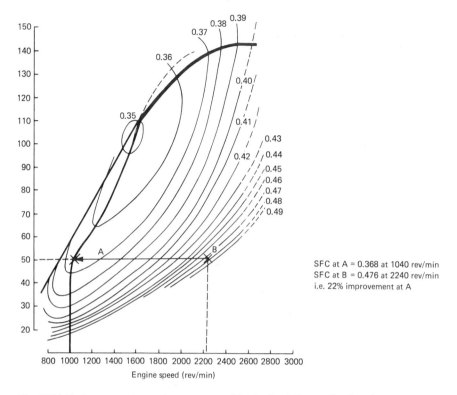

SFC at A = 0.368 at 1040 rev/min
SFC at B = 0.476 at 2240 rev/min
i.e. 22% improvement at A

Fig. 25.34 Fuel consumption and power map of the Leyland *Terrier* diesel engine. By controlling it to operate mostly between about 1400 and 1600 rev/min, optimum fuel consumption can be obtained. The line drawn from 1000 rev/min represents the power output of the engine when operated with the CVT

and planetary gearset, as shown diagrammatically in Fig. 25.37. In a common housing, both are driven in parallel by the engine, via a fluid coupling. Each is coupled by a clutch, one connecting the sun gear, the other the annulus to the output. It should be noted that, for simplicity, the idler gear has been omitted from between the two primary gears in Fig. 25.37, so the direction of rotation of the variator is not the same as in Fig. 25.36. Control is exercised by a microprocessor.

The torque input to the variator passes from the primary gears to a layshaft on which is mounted a pair of discs, with a single disc output member interposed between them. Torque is transmitted from the input to the output discs by what are termed rollers, but which look more like wheels, rather like planetary gears but with friction instead of teeth transmitting the drive. It followed that the output disc rotates in a direction opposite to that of the twin input discs. From the output disc, the torque is transmitted through a large diameter cup-shape component to the sun gear of the gearset. The planet carrier is driven by the mainshaft and primary gears, while the annulus gear is the output from the gearset.

In principle, the variator is simple: the axes about which the three rollers rotate are simultaneously tilted through equal angles so that their peripheries

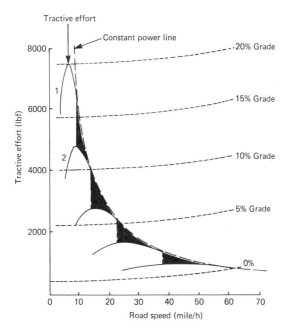

Fig. 25.35 A particularly useful characteristic of the Leyland CVT is its jerk-free ratio changes as compared with, for instance, the five-speed manual shift transmission, as shown here

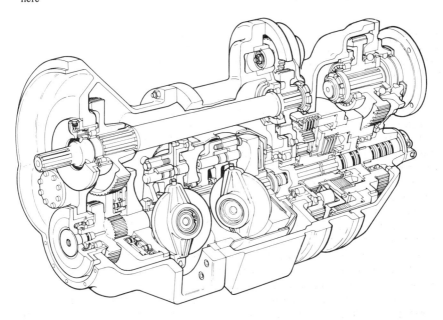

Fig. 25.36 In the Leyland CVT the idler between the primary, or input, gear pinion sets the layshaft rotating in the same direction as the mainshaft. Immediately behind the primary gears, on the layshaft, is the variator. The mainshaft passes above it, straight through this section of the box to the rear section, where it carries a gear meshing with a pinion that drives the planet carrier. The sun gear is coaxial with and driven directly by the variator

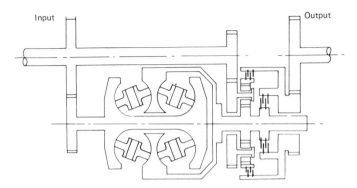

Fig. 25.37 Diagrammatic representation of the variator and planetary gearset, but without the idler between the primary gear pair. The two clutches, in turn, couple and uncouple the sun gear and annulus gear to the output pinion

roll along a track of different diameter on the input disc to that on the output disc. In Fig. 25.37, the track on the input disc is of smaller diameter than that on the output disc, so the transmission is in a geared-down ratio. If the rollers are tilted so that both tracks are of equal diameter, the drive ratio is 1 : 1, while a further swing to make the track on the input larger than that on the output disc takes it into overdrive.

The tracks are of a toroidal section, such that the rollers are always in contact, regardless of the angle through which they are tilted. Moreover, the disc and roller assembly is axially preloaded hydraulically, at loads varying up to about 15 tonne, according to the torque being transmitted. Even so, there is no metal-to-metal contact between the peripheries of the rollers and the discs, since a thin elasto-hydrodynamic film of oil is drawn in between them, as between the rollers and races of bearings. The torque therefore is transmitted by shear in this film, so the rate of wear of the rollers and tracks is extremely low.

In what follows, it is necessary always to bear in mind that the directions of rotation of the input and output from the variators illustrated in Figs 25.36 and 25.37 are different. Reverse, forward and neutral gears are obtained by using the variator to increase or decrease the speed of the sun gear relative to that of the planet carrier which, as previously mentioned, is directly driven from the engine. The overall ratio, Fig. 25.38, is a function of the ratios of the variator and the gearing from the primary gear to the planet carrier.

In this illustration the change from low into high regime occurs at the point when the overall ratio is 0.43 : 1 which, in the conventional transmission of the *Terrier*, is approximately equivalent to second gear. For operation in the low regime, the annulus gear clutch is engaged and the sun gear clutch disengaged. At the change-over point both the sun gear and annulus are rotating at the same speed and in the same direction, so no load is transmitted through their clutches. This is termed the *synchronous speed*, or *synchronous ratio*, of the transmission.

To progress from low to high regime, the sun gear clutch is engaged and the annulus clutch immediately disengaged. In the high regime all the torque is taken through the variator, while the planet carrier and annulus rotate

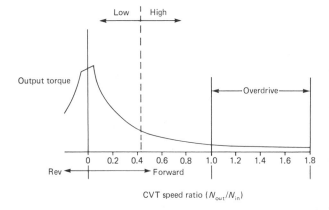

Fig. 25.38 The modes of operation of the Leyland CVT, from reverse, through forward in low, and on up into high regime

unladen. However, should there be a requirement for holding low regime, for instance for engine braking, the electronic control can be called upon to actuate the clutches appropriately.

The really clever part of the whole transmission is the system for controlling the ratio of the variator. Central to this control system is the microprocessor, a significant feature of the demand on which is that it is continuous. In contrast, a microprocessor in a conventional automatic transmission springs into action only when a gear shift is called for. Consequently, the CVT has a 16-bit microprocessor with 8 kbyte of software. It actuates an electro-hydraulic valve controlling a low pressure system for loading the variator axially in proportion to the torque transmitted, as well as for steering the rollers and actuating the clutches. The three main inputs to the electronic control unit are driver demand and the speeds of the engine and of the vehicle.

Ideally, the engine would operate continuously at its most economical speed, while the speed of the vehicle is regulated by the transmission. However, since constant speed implies constant power output, the only way to obtain extra power for acceleration or climbing a hill is to relieve the engine momentarily of the transmission load, so that the power thus made available can be utilised for accelerating the engine to a speed appropriate for its required operation at a higher power output. Then the engine and transmission have to be coupled together again to re-establish a steady-stage level of operation at that output.

Changing the ratio of the variator is effected by steering the rollers on their tracks. Signals from the microprocessor regulate the movements of the hydraulic valves which, in turn, control the hydraulic actuators of the steering mechanism. From Fig. 25.39, it can be seen that the rollers are carried in roller bearings in carriages, vaguely resembling pulley sheaves, which tilt about trunnions at their ends. From the cross-section of the tilting mechanism, Fig. 25.40, it can be seen that the trunnions pivot on the ends of Y-shaped rockers, the legs of which extend radially inwards to register in slots in a control sleeve. This sleeve can be rotated a few degrees in either direction to alter the angle of tilt of the rollers by steering them inwards or outwards until they run along tracks of diameters appropriate to the speed called for by the driver.

Fig. 25.39 Each roller of the variator rotates in a carriage pivoted about a trunnion at each end. These trunnions are carried on the ends of Y-shaped centrally pivoted levers (out of sight here) the legs of which project radially inwards, as shown in Fig. 25.40

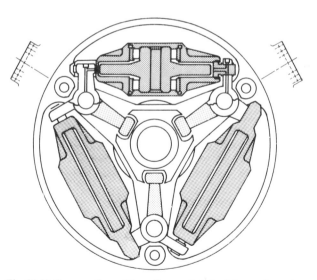

Fig. 25.40 Cross-section of the rollers and their tilting and steering mechanisms

Equalising the share of the load between the rollers in each set of three is also an important requirement. Failure to do so greatly reduces the potential rating of the transmission. So long as the reactions at the ends of the arms of the rockers remain equal, the sleeve remains coaxial within the shaft. If, however, the reactions at the end of one arm or the other increases or decreases, the resultant displacement of the leg of the Y causes the sleeve to deflect to one side or the other, within the clearance between the sleeve and the shaft. This steers the roller, causing it to tilt so that is downshifts or upshifts appropriately until the reactions are again equal. During this process the sleeve progressively returns to the position in which the reactions are again all in equilibrium.

Since the input and output discs rotate in opposite directions, steering the rollers in one direction causes their peripheries to move simultaneously inwards on one and outwards on the other disc. The pivot geometry is such as to give the rollers a castoring tendency, so the unit can run in only one direction, which is why the epicyclic gear has to be utilised for obtaining reverse, as previously described. Such a transmission could lead to fuel savings of as high as about 22% and is intended to replace the conventional transmission in vehicles, such as buses and delivery vans, used for stop–start operation.

Chapter 26

Universal joints and driving steered wheels

A universal joint is a form of connection between two shafts, whose axes intersect, whereby the rotation of one shaft about its own axis results in the rotation of the other shaft about its axis.

The principle on which the Hooke's type of universal joint works is illustrated in Fig. 26.1. The shaft A is formed into a fork at its end and pivoted between the prongs of this fork is a cross-piece C. The cross C can therefore pivot about the axis XX relatively to the shaft A. The other shaft B is similarly forked and the other arms of the cross are pivoted between the prongs of this fork. The shaft B can therefore pivot about the axis YY relatively to the cross C and, since the latter can pivot about the axis XX relatively to the shaft A, the shaft B can assume any angular position relatively to shaft A. Or from another point of view, if the shafts A and B are supported in bearings with their axes at an angle, then when the shaft A is turned about its axis the motion is communicated to the shaft B and it turns about its axis; the arms of the cross meanwhile oscillating in the prongs of the forks.

The axes XX and YY intersect at O and are perpendicular one to the other. The axes of the arms of the cross C are also perpendicular to their respective shafts. The axes of the shafts intersect at O, which point is the 'centre' of the joint.

It does not matter how the pivoting action is obtained; all that is required is that the shaft B shall be able to pivot independently about two intersecting perpendicular axes such as XX and YY, relatively to the shaft A.

26.1 Constructional forms of universal joints

There are several types of universal joint all working on the principle outlined above, but in motor cars and lorries the joints are of two principal forms. These may be called—

(1) Cross type.
(2) Ring type.

Examples of these are shown in Figs 26.2 and 26.3.

A proprietary design of cross-type joint is shown in Fig. 26.2. The yoke

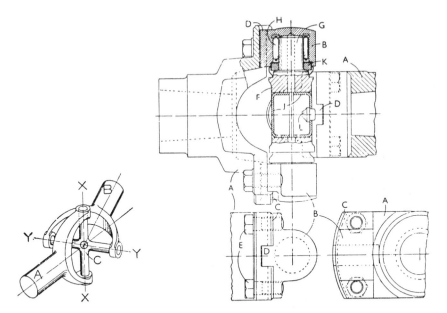

Fig. 26.1 Hooke's universal joint   Fig. 26.2 Cross-type universal joint

members A are secured to the shafts that are connected by the joint and carry bushes B. These are positioned by the projecting lips C of the yokes, which fit machined portions of the bushes and by the keys D integral with the bushes and which fit in keyways formed in the yokes. The keys D transmit the drive and relieve the set screws E of all shearing stresses. The two yokes are coupled by the cross member F which consists of a central ring portion and four integral pins G. The ends of the latter bear on the bottoms of the bushes B thus centring the joint. Between the pins G and the bushes B are needle bearings H. The hole in the centre of the cross member F is closed by two pressings J and forms a reservoir for lubricant which reaches the bearings through holes drilled in the pins G. The cork washers K form a seal at the inner ends of the pins and also serve to retain the needle rollers when the joint is taken apart. A single filling of oil suffices for practically the whole life of the joint. Any excessive pressure in the reservoir which might lead to oil being forced out past the seals K is prevented by the relief valve L.

Fig. 26.3 shows a ring-type joint. The member A is bolted to one shaft by its flange and the fork B is secured to the other shaft by splines. The two members are coupled by the ring C. This ring is made of two steel pressings each forming half the ring and being bolted together by the nuts on the trunnion portions of the four bushes D whose shape is clearly seen in the separate plan view. The pins of the fork member B fit in two of the bushes and the ends of the pin E, which is fixed in member A, fit in the other two bushes. The space inside the ring forms a reservoir for oil which may be introduced through a nipple not shown. The joint between the two halves of the ring is ground to form an oil-tight joint and escape of oil at the points of entry of the pins is prevented by the compressed cork washers F. The shafts are centred relatively to the ring by

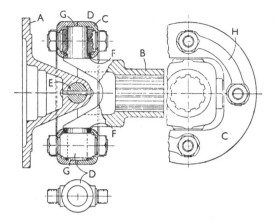

Fig. 26.3 Ring-type universal joint

reason of the fitting of the pins on faces G accurately machined inside the ring, and not by the cork washers. The nuts securing the ring are locked by a pair of tab washers H.

Another universal joint construction is shown in Fig. 26.4. It consists of a ball A having two grooves formed round it at right angles. In these grooves the forked ends of the shafts E and F fit. Obviously when the joint is put together the shaft E can slide round in its groove, thus turning about the axis XX. Similarly the shaft F can slide round in its groove, thus turning about the axis YY. This type of joint was used at one time in front wheel brake linkages. The arrangement of the shaft bearings has to be such that the shafts cannot move away from the centre of the ball, otherwise the joint would come apart. It is not suitable for use in the transmission.

26.2   Flexible-ring joints

A joint that acts by the flexure of a flexible ring is much used to connect shafts between which the angular displacement will not be very great. Such a joint is shown in Fig. 26.5. The shafts are provided with three-armed spiders, the arms of which are bolted to the opposite faces of a flexible ring, the arms of one spider being arranged mid-way between the arms of the other. The flexible ring

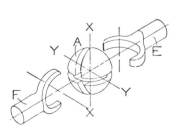

Fig. 26.4

Fig. 26.5

is usually made of one or more rings of rubberised fabric made in a special way so as to provide the necessary strength. A number of thin steel discs are sometimes used instead of the fabric rings. When the shafts are revolving about axes which are not coincident there is a continuous flexing of the ring. This type of joint has several advantages over the universal joints described above, the principal of which are the elimination of the need for lubrication and cheapness of manufacture. The joint cannot cope with such large angular displacements as the universal joints and when the torque to be transmitted is large it becomes very bulky.

26.3   Rubber-bushed flexible joints

Joints of this kind are now widely used and there are several forms of them, three being shown in Figs 26.6 to 26.8. The first of these is the original Layrub; it is basically a ring type of joint. The shafts that are connected by the joint carry the two-armed spiders A and B projecting from which are bolts carrying special rubber bushes C. These bushes are housed inside the coupling ring D which is made of two exactly similar steel pressings bolted together. The rubber bushes by distorting slightly enable any misalignment of the shafts to be accommodated. Angular misalignments up to 15° can be allowed for but

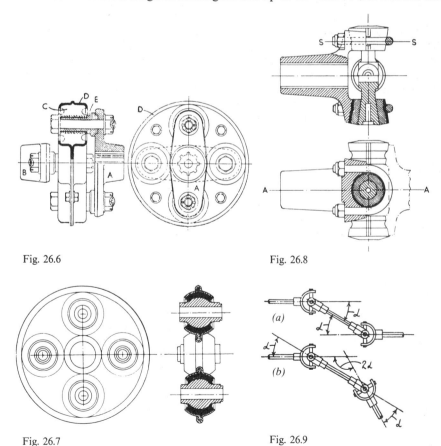

Fig. 26.6   Fig. 26.8

Fig. 26.7   Fig. 26.9

generally the misalignment is limited to about half that amount. The joint can also accommodate a considerable amount (up to 12.5 mm) of axial movement of the one shaft relative to the other and when two of them are used, one at each end of a propeller shaft, it is usually possible to dispense with the sliding joint that is essential when all-metal joints are used.

Since the only connection between the shafts is through the rubber bushes, the joints also assist in smoothing out vibrations; this property has been used to give a flexible clutch plate in a single-plate clutch the driven plate being connected to the clutch centre by four Layrub bushes.

The bushes are made with concave ends as shown in order to keep the internal stresses in them approximately uniform and to increase their flexibility. They are made with a metallic gauze insert on their insides and are forced on to the sleeves E which are made somewhat larger than the holes in the bushes. The outside diameters of the bushes are also greater than the diameters of the pockets in the ring D in which the bushes are housed and so when the coupling is assembled the bushes are compressed to such an extent that although when the joint flexes the distance between the sleeve E and the ring D may increase on one side the rubber remains in compression and is never in tension. The sleeves E have spigots which fit into holes in the spiders so that the bolts are not called upon to transmit the torque and are not subjected to any shearing.

Fig. 26.7 shows the very effective Metalastik unit in which the rubber bushes are bonded to the spherical pins and are compressed when the two metal pressings which form the ring of the joint are assembled together. These pressings are held together by spinning the lips of one of them over those of the other. The design in Fig. 26.8 is basically a cross-type joint and is made by the Moulton company. The rubber bushes are bonded on the inside of the tapered portions of the arms of the cross and on the outside to steel shells. The latter fit into depressions formed in the flanges of the joint and are held in place by stirrups which are bolted up to the flanges.

26.4   Constant-velocity joints

The Hooke's type of universal joint suffers from a disadvantage which is obviated in some other types of joint. It is that supposing one of the shafts connected by a Hooke's joint is revolving at an absolutely constant speed then the other shaft will not revolve at a constant speed but with a speed that is, during two parts of each revolution, slightly greater and, during the other two parts of the revolution, slightly less than the constant speed of the first shaft. The magnitude of this fluctuation in speed depends on the angle between the axes of the two shafts, being zero when that angle is zero but becoming considerable when the angle is large. This disadvantage becomes of practical importance in front wheel driven vehicles and in the drives to independently sprung wheels where the angle between the shafts may be as large as 40°. It can be obviated by using two Hooke's joints arranged as shown in Fig. 26.9(a) and (b), the intermediate shaft being arranged so that it makes equal angles with the first and third shafts and the fork pin axes of the intermediate shaft being placed parallel to each other. The irregularity introduced by one joint is then cancelled out by the equal and opposite irregularity introduced by the second joint. Examples of front wheel drives using this arrangement are shown in Figs

26.16 and 26.17. A slightly different arrangement using the same principle is given in Fig. 37.9.

Constant-velocity joints are joints which do not suffer from the above disadvantage but in which the speeds of the shafts connected by the joint are absolutely equal at every instant throughout each revolution. Although such joints have been known for very many years they have not been used to any extent until relatively recently.

The Tracta joint, manufactured in England by Bendix Ltd, is shown in Fig. 26.10, from which the construction will be clear. The joint is a true constant-velocity joint but the theory of it is beyond the scope of this book and those who are interested in this theory and in those of the joints described below, are referred to an article by one of the authors in *Automobile Engineer*, Vol. 37, No. 1. Another true constant-velocity joint, the Weiss, which is used to a considerable extent in America, where it is manufactured by the Bendix Products Corporation, is shown in Fig. 26.11. It consists of two members each with two fingers or arms in the sides of which are formed semi-circular grooves.

Fig. 26.10 Bendix Tracta universal joint

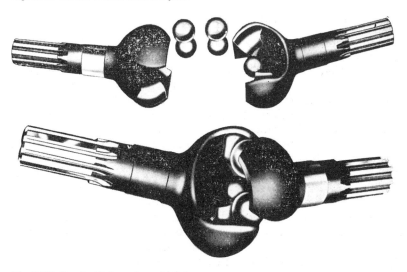

Fig. 26.11 Bendix Weiss universal joint

When the two members are assembled the fingers of the one fit in between the fingers of the other and balls are inserted in the grooves of the fingers and form the driving connection between them. The formation of the grooves is such that the balls lie always in a plane making equal angles with the axes of the shafts connected by the joint, this being a fundamental condition that must be satisfied if the drive is to be a constant-velocity drive. This joint has the property that the shafts connected by it may be moved apart axially slightly without affecting the action of the joint and this axial motion is accommodated by a rolling of the balls along the grooves in the fingers of the joint members and so takes place with the minimum of friction.

A third example is shown in Fig. 26.12. It is the Rzeppa (pronounced Sheppa) and it consists of a cup member A with a number of semi-circular grooves formed inside it and a ball member B with similar grooves formed on the outside. Balls C fit half in the grooves A and half in B and provide the driving connection. For true constant-velocity operation the balls must be arranged to lie always in a plane making equal angles with the axes of rotation of the members A and B. This is ensured by the control link D and the cage E. The former has spherical ends one of which engages a recess in the end of the member B while the other is free to slide along a hole formed inside A; the link is kept in place by the spring F. The spherical enlargement G of the link engages a hole formed in the cage E which has other holes in which the balls C fit. When the shaft B swings through an angle relatively to A the link D causes the cage E and the plane XX of the balls C to swing through half that angle and thus the balls are caused to occupy the required positions for the correct functioning of the joint.

In some designs of this joint, intended for use where the angular deviation of the shafts is small, the control link D is omitted.

A joint developed by Birfield Transmissions Ltd which gives constant-velocity ratio transmission and allows for a plunging motion of one of the shafts relative to the other is shown in Fig. 26.13. The inner member is grooved to carry the balls that transmit the motion and its outer surface is ground to a sphere whose centre is at the point A. The balls are housed in recesses in the cage and this is ground on its inside to fit the outer surface of the inner member while its outer surface is ground to a sphere whose centre is at the point B. The outer member has a cylindrical bore with grooves formed in it to take the balls and the outer spherical surface of the cage fits the cylindrical surface of the

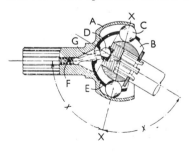

Fig. 26.12 Rzeppa universal joint

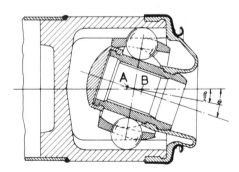

Fig. 26.13 Birfield constant-velocity universal joint

outer member. The inner member can therefore move bodily along the bore of the outer member thus giving the plunging motion required in the drives to most independently sprung wheels and which usually has to be provided by sliding splines. The off-setting of the centres of the spherical surfaces of the cage keeps the plane of the balls at all times in the plane bisecting the angle between the shaft axes as is necessary for the maintenance of a constant-velocity ratio.

26.5 Driving and braking of steered wheels

Various methods of driving a steered wheel are shown in Fig. 26.14. In the examples (*a*) and (*b*) a rigid driven axle is assumed, but in the others independent suspensions are shown. The arrangement at (*a*) is the simplest, a single universal joint U being provided to accommodate the steering motion of the stub axle S. Unless this joint is of the constant-velocity type, there will be an irregularity in the drive to the wheel whenever the stub axle is turned for steering purposes while, if the wheel is given any camber and the wheel shaft A is inclined to the half-shaft H, the irregularity will always be present. This constructional arrangement was adopted by the Four Wheel Drive Company in their lorries, which were among the earliest four-wheel-driven vehicles, and although the details of this arrangement are obsolete the general arrangement still represents current practice. An ordinary live axle is used so far as the final drive, differential and axle casing surrounding those components are concerned, but at their outer ends A, Fig. 26.15, the drive shafts are carried in bushes and are forked to form one member of a Hooke's type universal joint. The other shaft of this joint is seen at B and conveys the drive to the hub cap of the road wheel. The hub is carried on bearings on the stub-axle member, which is made in three pieces D, E and F bolted together as shown in the right-hand view. The inner spherical surfaces of the portions E and F touch the corresponding surfaces of the end of the axle casing in order to make the housing oil tight and to exclude mud and dust, but those surfaces do not carry any loads. E and F are carried on the projecting swivel pins of the axle casing.

In the arrangement shown in Fig. 26.14(*b*), two universal joints are used and are symmetrically disposed relative to the king-pin axis OO. When the stub axle is turned about that axis for steering purposes, the angles between the

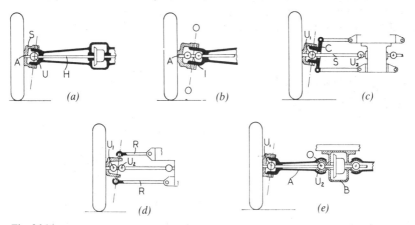

Fig. 26.14

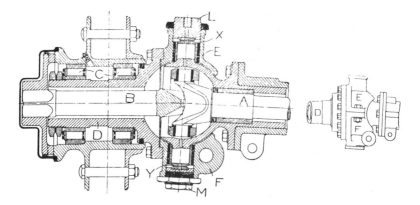

Fig. 26.15 Details of the FWD stub axle

intermediate shaft I and the steel shaft A and half-shaft H respectively will be maintained equal as in Fig. 26.9, and so a constant-velocity drive will be obtained. An example of this construction is shown in detail in Fig. 26.16. It is the design of the Kirkstall Forge Engineering Company and incorporates a second reduction gear which is housed in the wheel hub. This second reduction is between the pinion which is splined to the end of the shaft A and the annulus C which forms the hub cap of the road wheel and is bolted to the hub of the latter. The intermediate pinions B are carried on pins D, which are supported in the member E. The latter fits the cylindrical extension of the stub axle and a key prevents rotation. Because the intermediate member coupling the two

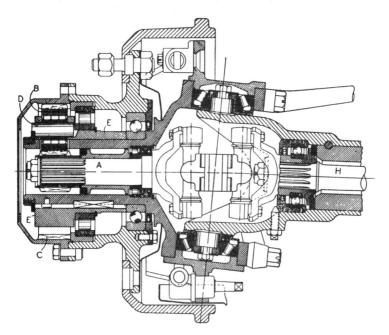

Fig. 26.16 Kirkstall steered axle

universal joints is rigid, and the forks of the joints are rigidly attached to the half-shaft H and wheel shaft A respectively, one of these shafts must be left free to float axially. This will be seen from Fig. 26.19, in which the full lines show the position when the wheels are in the straight-ahead position, and the dotted lines the position when the stub axle is turned for steering. It is clear that the distance X_1Y_1 is less than the distance XY. This variation is accommodated by leaving the shaft A, Fig. 26.16, free to float. It is therefore carried in a parallel roller bearing at the right end and is supported by the contacts with the three pinions B at the left end. The omission of a bearing at the left end ensures equal division of the driving torque between the three pinions.

The example shown in Fig. 26.14(c) is a conventional double-arm type of suspension, in which a stub axle carrier C connects the two arms. The drive shaft S is provided with universal joints U_1 and U_2. The first of these accommodates the steering motion of the stub axle and, in conjunction with the second, allows for the vertical motion of the wheel assembly. Because the distance between the centres of the universal joints cannot be kept constant, the shaft S must be provided with some axial freedom. This is usually done by leaving one of the universal joint forks free to slide on the splines of its shaft. Obviously, U must be a constant-velocity joint.

In the example shown in Fig. 26.14(d), the stub-axial carrier is omitted and the stub axle is carried directly by the arms RR, to which it is connected by ball and socket joints which accommodate the steering motion as well as the vertical motion of the road wheel. The joint U_2 now has to be supported from the stub axle through the joint U_1, and the construction of a joint which provides this support is shown in Fig. 26.17. The joint is made by the Glaenzer Spicer Company, of Poissy, France. The forks A and B, integral with their shafts, are coupled by four-armed spiders and an intermediate member C. The shaft B is supported relative to A by the ball and socket DE. The ball D is free to slide along the spigot shaft F, and the socket E is integral with the spigot G. The connection keeps the two universal joints and the intermediate member in the correct relationships to provide a constant-velocity drive, as described above.

Fig. 26.14(e) shows a swinging-arm type of independent suspension, in which the arm A which carries the stub axle is pivoted to the final drive casing B on the axis O. Two universal joints are necessary; one (U_1) to accommodate

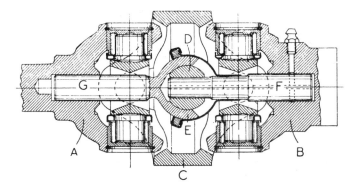

Fig. 26.17 Glaenzer Spicer axle

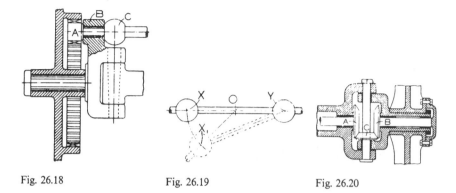

Fig. 26.18   Fig. 26.19   Fig. 26.20

the steering motion and the other to allow for the swinging of the arm. The arrangement does not provide a constant-velocity drive unless both the joints are of the constant-velocity type. The casing B is carried by the frame of the vehicle.

In the arrangement shown in Fig. 26.18 there is a gear reduction between the driveshaft and the road wheel. This makes the speed of rotation of the driveshaft higher than that of the wheel and reduces the torque the universal joint has to transmit. The driveshafts are more exposed and difficult to protect from mud and dust but, being higher than the axle, are more out of the way of damage from the striking of obstacles. The arrangement is only very occasionally used on special types of vehicle.

The use of a universal joint can be avoided by using the arrangement shown in Fig. 26.20, where the half-shaft carries a bevel gear A which drives a second bevel gear B fixed to the road wheel shaft through the intermediate wheel C. The latter is free to rotate on bearings on the swivel pin. The turning of the stub axle for steering purposes is accommodated by the rolling of the wheel B round the wheel C, and although this introduces an epicyclic action which causes an acceleration or deceleration of the road wheel this action only occurs during the time the stub axle is actually being turned. The arrangement can be made somewhat more robust than the universal joint drive, but is rather clumsy and is very little used.

Chapter 27

The differential

The differential is the device that divides the torque input from the propeller shaft equally between the two output shafts to the wheels, regardless of the fact that they may be rotating at different speeds, for instance on rounding a corner. In principle, as can be seen from Fig. 27.1, it is a set of two bevel gears with a bevel pinion between them. The bevel pinion can be likened to a balance beam pivoted at its centre, its ends registering between pairs of teeth on the differential gears.

If a force P is applied at the central pivot, in a direction tangential to the two differential gears, and if the beam either does not swing, or swings at a uniform velocity, it follows that the forces at the ends of the beam will be equal to $P/2$. Hence, equal forces are applied at equal distances from the centres of the differential gears, and therefore the torques they transmit to the halfshafts are equal. Clearly, the force P represents the pressure between the differential pinion and its pin, while the forces $P/2$ represent the pressures between the teeth of the bevel pinion and those with which they mesh on the two differential gears.

In Fig. 27.2 is a typical differential unit in the back axle of a car. It consists of a drum-shaped cage A, which is carried in ball bearings BB in the axle casing and is therefore free to rotate about the axis XX of the road wheels. Fixed to the cage A is a crown wheel C driven by the bevel pinion D. The arrangement is similar to that of Fig. 28.1, except that the crown wheel is fixed to the cage, in which the bevel gears F_1F_2 and their pinions GG rotate on axes mutually at right angles. There are now two 'differential' or 'drive' shafts E_1E_2, the outer ends of which are connected to the road wheels. Their inner ends pass through the bosses of the differential cage A in which they are quite free to turn. Inside the differential cage, their ends are splined into the bevel wheels F_1F_2 with which the bevel pinions GG mesh. The pinions GG are free to turn on the pin H fixed in the differential cage.

It should be clear that if the differential cage is held and the wheel F_1 is turned in the forwards direction at, say, 2 rev/min then the wheel F_2 will be turned backwards at 2 rev/min, since it is equal in size to F_1. Moreover, since these motions are relative to the differential cage, they will not be affected by any motion of that cage. If, therefore, the differential cage is rotating at, say, 200 rev/min in the forwards direction and the wheel F_1 is still turning at 2

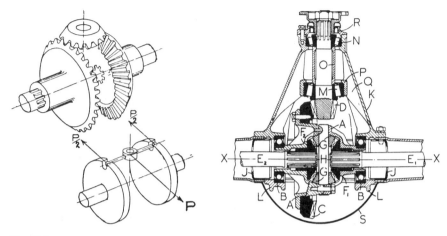

Fig. 27.1   Fig. 27.2

rev/min forwards relatively to it, the wheel F_2 will still be turning at 2 rev/min backwards relatively to it.

The actual speed of the wheel F_1 will then be 202 rev/min, because its forward motion of 2 rev/min relatively to the differential cage is added to the forwards motion of 200 rev/min of that cage. The actual speed of the wheel F_2 will be 198 rev/min because its backwards motion of 2 rev/min relatively to the differential cage is subtracted from the forwards motion of 200 rev/min of that cage.

This is the action that occurs when a car moves in a circle: the road wheels are constrained to move at different speeds and do so by virtue of one wheel going faster than the differential cage while the other goes an equal amount slower than the differential cage. Thus the speed of the differential cage is the mean of the road wheel speeds. When the car moves in a straight line, the road wheels turn at the same speed as the differential cage, and the differential pinions do not have to turn on their pins at all.

The above description should make it clear how the road wheels can turn at different speeds; it remains to show that when so doing they are driven with equal torques. In Fig. 27.1 the bevel wheels are shown replaced by discs having notches in their peripheries. Lying with its ends in these notches is a beam. If a force P is applied to the centre of the beam in a direction tangential to the discs as shown, then if the beam does not turn about the vertical axis, or if it turns about that axis with uniform velocity, the forces at the ends of the beam must be equal and each will be equal to $P/2$. The reactions of the forces acting on the ends of the beam act on the discs, hence equal forces are applied to the discs at equal distances from their axes, and therefore the twisting moments or torques acting on the discs are equal.

It should readily be seen that the bevel pinion acts in a manner precisely similar to the beam. Hence the torques transmitted to the driveshafts are equal and each will be equal to half the torque applied to the differential case by the final drive. In the actual differential the force P appears as a pressure between the bevel pinion and its pin while the forces $P/2$ appear as pressures between the teeth of the bevel pinion and the bevel wheels.

27.1 Another arrangement of the bevel final drive

The bevel final drive is sometimes arranged in a different manner from that described previously, generally because some other difference in axle construction necessitates the change. The principle of this other method is shown in Fig. 27.3. The propeller shaft is coupled to the shaft A which passes right across the centre portion of the axle casing B in which it is supported. At the centre of the axle the shaft A is enlarged and formed into pins PP to carry the differential pinions CC. These mesh with the differential wheels Q_1 Q_2, which are integral with the bevel pinions D_1 D_2, of which D_1 meshes with the large bevel wheel E and D_2 with the smaller bevel wheel F. The bevel wheels E and F are supported in the axle casing and are splined on to the shafts which drive the road wheels.

Of course, the gear ratio between D_1 and E is the same as the ratio between D_2 and F. The action of the differential is just the same as in the conventional axle, but the reduction of speed now occurs between the differential and the road wheels instead of between the propeller shaft and the differential cage. The differential therefore runs at a higher speed than the road wheels, enabling smaller wheels to be used in it.

It should also be clear that it is quite possible to have the axes of the shafts inclined to each other in the end view, and advantage has been taken of this feature to arch the back axle casing and to tilt the rear wheels in order to reduce the overhang on the axle and to bring the road wheels perpendicular to the curved surface of a cambered road. Arched back axles are not now used, and the form of final drive described is uncommon.

27.2 Spur, or planetary type, differential

Most planetary type differentials have spur instead of bevel gears, as in Fig. 27.4, and therefore are often called the spur type. The wheels A, B are splined to the driveshafts that drive the road wheels. Meshing with the wheel A is a spur pinion E_1, whose teeth extend nearly across the gap between the wheels A and B. The spur pinion F_1 meshes in a similar way with the wheel B, and at the centre the two pinions mesh together. The pinions are carried on pins which are supported by the ends of the differential cage, which in this design is formed of the worm wheel C of the final drive and two end cover plates DD. It should be clear that if the differential cage is held fixed and the wheel A is turned in, say, the clockwise direction, the pinion F_1 in a clockwise direction and the wheel B in a counter-clockwise direction. Hence, if one of the differential

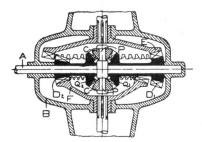

Fig. 27.3

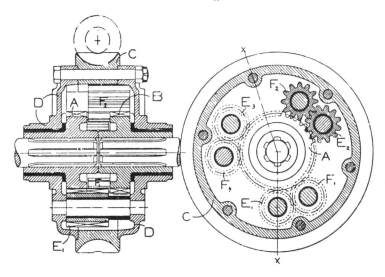

Fig. 27.4 Spur-type differential

wheels goes faster than the differential cage, the other differential wheel will go an equal amount slower, just as in the bevel type.

As regards the equality of the torques, the torque on the wheel A is due to the pressure of the teeth of the pinion E_1. This pressure tends to make the pinion E_1 revolve on its pin, and this tendency is opposed by a pressure between the teeth of the two pinions at the centre. This last pressure tends to make the pinion F_1 revolve on its pin, and this tendency is opposed by the pressure between the teeth of the pinion F_1 and the wheel B. If the pinion E_1 (and therefore the pinion F_1 also) is at rest relatively to, or if it is revolving uniformly on, its pin then the two pressures acting on it must be equal. Hence, all the pressures between the teeth of the wheels A and B and the pinions E_1 and F_1 are the same, and hence the torques acting on the wheels A and B are equal. Three pairs of pinions are provided as shown.

27.3   Traction control differentials

With the introduction of four-wheel-drive cars for use on the road, in countries where snow and ice are prevalent in winter, and for rallying, traction control devices, ranging from the complex electronic systems to simple differential locks, became necessary. The reason is that, if the grip of a driven wheel on one side is reduced, for example by ice, to zero, the total traction that can be transmitted by a conventional differential is also reduced to zero. This follows from the definition of a differential in the opening sentence of this chapter: a simple differential without any form of traction control reduces the torque transmitted to the wheel on the side that is gripping until it equals that to the wheel that is slipping. For similar reasons, traction control may be beneficial when applied to the differentials interposed in the transmission lines between the front and rear axles of four-wheel-drive vehicles, see Section 20.4.

Various forms of differential designed to prevent total loss of traction if one wheel spins freely have been in use since soon after the road vehicle was invented, and a list of those currently available is given in Table 27.1. The simplest is a mechanism by means of which the differential cage can be locked to the inner ends of the halfshafts, to cause the whole differential and halfshaft assembly to rotate in unison with the crownwheel. In most cases, this mechanism is either a dog clutch or a sliding muff coupling. Such systems are commonly used on off-road vehicles, including light as well as the heavy commercial types such as tippers. They may be actuated pneumatically, hydraulically, mechanically by rod or cable, or electrically.

Their main disadvantages are the extra complexity and weight represented by the controls and the fact that, in most instances, the vehicle has to be stationary while the lock is engaged: if, when the vehicle is stationary, the driven wheels bed down into very soft ground, moving off again may be difficult. A serious disadvantage, too, is that if the driver forgets to disengage the lock as he takes the vehicle off soft on to hard ground, transmission wind-up will ensue and cause excessive tyre wear and, ultimately, a mechanical failure in the driveline.

In some electronically-controlled automatic systems a wheelspin sensor, similar to the wheel-lock sensor in anti-skid systems, as described in Section

Table 27.1—SOME TRACTION CONTROL DIFFERENTIAL SYSTEMS

Trade name	Type	Supplier
Sure-Drive	Freewheel	Formerly Borg-Warner, now Auburn Gear Inc.
ZF	Cam-and-pawl	Zahnradfabrik Friedrichshafen (ZF)
Lok-O-Matic	Multi-plate clutch, ramp-actuated	Zahnradfabrik Friedrichshafen (ZF)
Powr-Lok	Multi-plate clutch, ramp-actuated and rampless	GKN Axles Ltd, Light Division, ZF, Spicer Axle Division, Dana Corporation
Trac-Aide	Multi-plate clutch, ramp-actuated, but rampless	GKN Axles Ltd, Light Division, ZF, Spicer Axle Division, Dana Corporation
Trac-Loc	Multi-plate clutch, rampless	GKN Axles Ltd, Light Division, ZF, Spicer Axle Division, Dana Corporation
Traction Lok	Multi-plate clutch, rampless	Ford
Traction Equaliser	Clutch type	Rockwell
Sure-Grip	Cone clutch	Formerly Borg-Warner, now Auburn Gear Inc.
Super Traction	Cone clutch	Formerly Borg-Warner, now Auburn Gear Inc.
No-SPIN	Face dog clutch	Tractech, Dyneer Corporation
Detroit Locker	Face dog clutch	Tractech, Dyneer Corporation
True Trac	Helical gear	Tractech, Dyneer Corporation
Knight	Helical and worm	Knight-Mechadyne Ltd
Torsen	Worm and spur	Gleason, Power Systems Division and Quaife Power Systems Ltd
Max-Trac	Variable leverage gear	Fairfield Manufacturing Co.
Super Max-Trac	Variable leverage gear, with friction	Fairfield Manufacturing Co.
Viscodrive	Viscous coupling	FF Developments, Viscodrive GmbH, Tochigi Fuji Sangyo KK
Mercedes ASD	Electronic	Daimler-Benz

33.15, is used in conjunction with a computer which simply applies a brake to stop the spinning wheel so that the torque can be reacted back through a conventional simple differential to drive the wheel on the other side. There are also systems which, instead of applying the brake, reduce the engine power by cutting the fuel injection to one or more cylinders, which has the advantage of not only reducing the power output but also of using the inactive cylinder or cylinders as compressors to produce engine braking. Alternatively, in diesel engines, the fuel supply can be reduced to all cylinders or cut completely to one or more. An advantage of automatic operation, electronic or otherwise, is the absence of any remote manual controls and of the risk of the driver's forgetting to disengage the lock.

27.4   Vehicle design implications of traction control

A prime requirement, especially for road-going vehicles, is that the traction control device should come both into and out of operation rapidly. It must do so, however, entirely unobtrusively, otherwise the traction can become jerky and the steering can be adversely affected. Other problems arising include dissipation of the heat developed by the slipping clutch and the cam devices, wear, juddering and noise generated, and a loss of overall efficiency.

Ideally, the slip should be limited to shedding only that proportion of torque that exceeds the tractive capacity of the driven wheel on the side on which the wheelspin is tending to occur. This reduces jerk and optimises acceleration potential. Generally, however, the best that can be achieved is either a total lock-up or a torque bias ratio of up to 5 or 6 : 1 for rear axles but, for front axles, the torque bias has to be limited to no more than about 2.5:1, for the avoidance of adverse effects on steering. The effect of both axles is usually understeer while, on a front axle, there is also an increase in effort required at the steering wheel; both effects increase with torque bias ratio.

On the back axle, the steering effect is limited to that due to the asymmetry of the drive, coupled with the influence that is has on the slip angles of the tyres when cornering. At the front, the loads in the track-rod go out of balance instantly when the differential locks. In fact, dependent on the steering geometry and the degree of slip in the differential, the steering may even lock, especially when the vehicle is being turned from a straight-ahead course. This is easy to understand if one envisages a condition in which changing the direction of the steering can be effected only if there is relative rotation between the two front wheels: with the differential locked, this rotation cannot occur.

It follows that only soft-action, low bias-ratio traction control differentials are practicable for front axles. Indeed, most vehicle manufacturers avoid the problem simply by fitting an inter-axle traction control differential or a viscous coupling either alone, as on the rear-engine VW Transport Synchro 4 × 4, or in conjunction with a similar one in the rear final drive gear.

With only an inter-axle traction control unit installed, one axle can take over the drive if traction is lost beneath either one or two wheels on the other. On the other hand, if rear wheel spin occurs where both inter-axle and rear traction control differentials are fitted, the lost traction is transferred to the other wheel on the same axle, the front wheels continuing to transmit their share of the drive. However, if either front wheel spins, all the traction is transferred to the rear axle. This layout is particularly appropriate for vehicles in which, to give characteristics similar to those of rear-wheel-drive, the transmission system is geared so that the proportion of torque transmitted to the rear is in any case higher than that to the front axle.

If a viscous coupling serving as an inter-axle differential has to transmit all the torque from the engine it must be fairly large. Its actual dimensions, however, normally depend on what proportion of the total torque has to be transmitted from one axle to the other, and this in turn depends on factors such as the handling characteristics desired and the effects of the torque transmission on tyre slip angles.

Renault, for their front-wheel-drive cars, advocate a 65:35 torque split in favour of the front end, to retain the typical front-wheel-drive handling characteristics. For the Ford front-engine rear-wheel-drive *Scorpio*, on the other hand, the gearing is such that 66% of the drive goes to the rear axle and only 34% to the front, so the rear wheels are the more likely to spin. If this happens all the torque is transmitted to the front wheels. On this model the viscous coupling is on the rear end of the gearbox, and the torque is transmitted through a multi-strand toothed chain drive to a shaft, extending forwards alongside the engine, to the front axle.

In the VW installation, the inter-axle viscous drive coupling, see Section 27.10, is housed in the front final drive casing so that its weight helps to offset what otherwise would be a preponderance on the rear, owing to the fact that the engine is there. The gearing ratio, front to rear, is 1:1 but, owing to slip in the coupling, all the drive is normally transmitted directly, through the rear wheels. However, if one of these wheels spins, and traction is therefore lost on the rear axle, the coupling takes over and transmits all the torque to the front axle.

In general, the proportions of a viscous coupling incorporated in a driven axle depend on whether it is interposed between the two halfshafts or between the differential cage on one halfshaft.

27.5   Multi-plate clutch-type traction control device

For cars, the most commonly used automatic device is a limited-slip mechanism incorporating some sort of clutch. The aim is at introducing into the differentially geared coupling between the two halfshafts a degree of friction which will not only prevent one wheel from spinning freely, but also, by reacting the input torque to that wheel, provide for the transmission to the other wheel a torque equal to that arising from the frictional resistance.

The multi-plate clutch type is the most popular for cars. Most are of the loading, torque-sensitive type: as the torque differential between the two halfshafts increases, so also does the pressure on the clutch plates, as can be seen from Fig. 27.9, ultimately causing it to lock completely. Some other types, of which the ZF cam-and-pawl type is a good example, are speed sensitive. Only these two types will be described in detail, the former because of its widespread use and the latter because its principle of operation is perhaps the most difficult to understand. For full descriptions of the others listed, the reader is recommended to refer to an article in *Automotive Engineer*, February/March 1987, Vol. 12, No. 1.

The other types listed in Table 10 can be summarised in broad principle as follows. The simplest is a freewheel coupling, see Section 23.21, on each

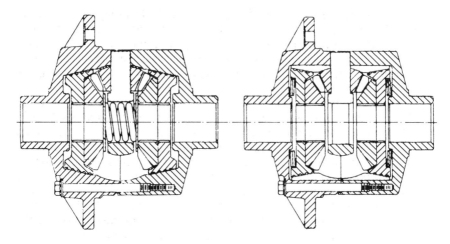

Fig. 27.5 Warner Gear loading (*left*) and unloading (*right*) type traction control differentials

halfshaft so that, for example during a turn, the outer wheel and shaft overruns its coupling and all the drive is transmitted by the inner wheel. This has several disadvantages; first, during differential operation, the total tractive potential is limited to that capable of being transmitted through only one tyre; secondly, the transitions from symmetrical to asymmetrical drive can adversely affect the steering; thirdly, unless the freewheels can be locked, not only engine braking but also driving in reverse are impossible.

27.6   Some other clutch types

Among the cone-clutch types that were first introduced by Borg-Warner, and are now produced by Warner Gear Inc., are both loading and unloading types. The clutch of the unloading type, Fig. 27.5, in contrast to that in the loading type, is engaged by Belleville springs and therefore its arrangement is such that, as the transmitted torque increases, the engagement force progressively unloads it until it disengages altogether. Consequently, the loading type is best for vehicles for operation at high speed and in which, as the speed is reduced, the torque is increased, for example by using lower gears. In contrast, the unloading type is principally for vehicles, such as those employed for earthmoving, operated mainly at low speed and high torque.

Traction control differentials incorporating dog clutches are harsh in action and are suitable mainly for applications in which total torque transfer is required. Consequently they are mainly for off-highway vehicles. Most of the gear type units, on the other hand, depend for their operation on the friction generated in either the irreversible or semi-irreversible gears. The action of the semi-irreversible type is relatively gentle and therefore has minimal effect on steering, so the majority are suitable for incorporation in front axles.

27.7   Gear type traction control devices

A gear type in which friction is not the factor relied upon is the Max-Trac, produced by Fairfield Manufacturing Company. This has gears with wide teeth, the diametral pitch of which varies along their width. As the gears rotate, the contact points traverse to and fro along the teeth, thus varying the effective ratio and pulsating the output. This pulsation of torque, it is claimed, is a measure a driver might take deliberately to get out of difficulty if his vehicle were stuck in the mud. The Super Max-Trac is more effective because friction is introduced in the gearing system, to provide a differential torque in addition to the pulsation effect.

Of the other gear types listed, all are of the planetary type. Both the Dyneer and Quaife units have pairs of helical differential pinions, similar to those in Fig. 27.4 but floating in pockets in the housing that contains the differential gear assembly. The gears of each pair intermesh, one meshing also with the differential gear on one side and the second with that on the other side. Torque bias, up to about 2:5 to 3:1, is generated by the friction due to both the axial and radial thrusts of these gears in their housings. If the torque reactions from the wheels on each side are equal, the axial thrusts arising from the helical teeth of the pinions balance, so these pinions float in their housing pockets; if the torque reactions become unequal, the axial thrusts on the pinions go out of balance, forcing them against the ends of their individual housing pockets and generating friction.

An advantage of this type is that its maximum torque bias (the highest ratio of the torque differential between the wheels) and its load capacity can be adjusted at the design stage by two methods. First, varying the helix angle of the gear teeth effects not only its capacity but also both the gear-pinion separation forces and axial thrust, and thus the friction due to radial pressure between the pinions and their housings. Secondly, both the friction and torque capacity can be increased by varying the number of pinions.

Of similar layout is the Knight Mechadyne unit, but instead of intermeshing, its planetary pinions are interconnected by wormwheel idlers, the axes of which are radially disposed, as in Fig. 27.6, instead of being parallel to those of the gears and pinions. The advantages of this arrangement are first its compactness and secondly, by virtue of the substitution of radially and axially

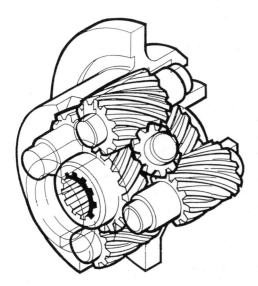

Fig. 27.6

orientated intermeshing gears for parallel ones, the obviation of some high precision machining operations.

The Gleason Torsen type, Fig. 27.7, was originally invented, in 1958, by Vernon Gleasman, who called it the Dual Drive. Subsequently, it was manufactured by Tripple-D, before the production was taken over by Gleason, in 1982. It differs from those already described in that its output gears are wormwheels meshing with pairs of worm type planetary pinions, the two in each pair being interconnected by straight spur gears on their ends. These pinions could not themselves intermesh because they must have teeth the profiles of which are designed for meshing, not with each other but specifically with those of their wormwheels. If the torque differential between the two halfshafts suddenly rises, perhaps due to wheelspin, the irreversibility of its worm and wormwheel elements causes the unit to lock. On the other hand, if the torque is relatively equally divided, the worm and wormwheels are in floating mesh, so differential rotation can occur, for example when the vehicle is steered round a curve.

27.8   ZF limited slip differential

The ZF cam-and-pawl type unit, Fig. 27.8, has the advantages of simplicity, and therefore inherently low cost, and light weight and compactness. On the

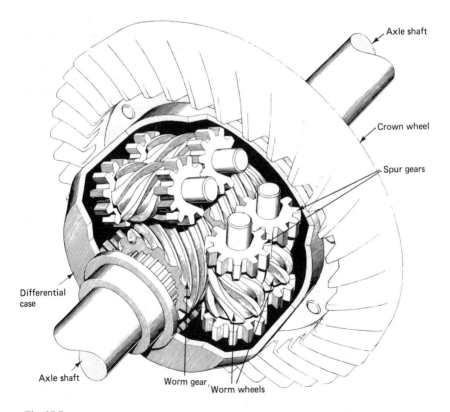

Fig. 27.7

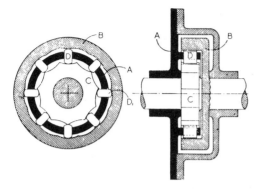

Fig. 27.8

other hand, it is liable to wear fairly rapidly if worked hard. It is particularly suitable for application to racing cars, where rapid wear is immaterial because it is normally not required to last longer than a few races. Moreover, racing drivers are all highly skilled and can cope with the steering and other effects associated with its jerky engagement and disengagement.

It comprises three main parts: a driving member A which is, in effect, integral with the crownwheel and which therefore corresponds to and takes the place of the differential cage of a conventional final drive gear; two opposed cam-rings, B and C, splined one to each halfshaft, and therefore corresponding to the differential gears, which they replace; and a set of pawls, or cam-followers, free to slide, radially between the cam rings, in slots in the previously mentioned driving member. These cam rings can alternatively be in the form of face-cam rings, to reduce the overall diameter, though at the expense of increased width.

In general, the performance of the unit depends to a major extent on the numbers of lobes on its cam-rings. For most applications, about 13 are ground in the outer ring and only 11 around the periphery of the inner one so that, when the rings rotate relative to one another, the radial motions of the plungers are phased, giving a slight gearing effect: when ring B rotates once relative to A, ring C makes 13/11 revs.

Torque from the crownwheel is transmitted through the flanks of the slots in the driving member A to the pawls D. So, if the car were jacked up clear of the ground, both halfshafts would rotate at equal speeds. Imagine now that the driving member A is rotating anti-clockwise and a brake is applied to only the wheel driven by the halfshaft to which ring C is splined: in these circumstances, the cam at C would force the plunger between the letters B and C in the left-hand diagram outwards but, to be able to move out, it would have to force the outer ring B, and its associated road wheel, to rotate anti-clockwise faster than A.

Obviously, there would be a considerable frictional resistance between the plungers and cams, so the anti-clockwise motion of the unbraked wheel, despite its being clear of the ground and therefore otherwise free to rotate, would absorb a great deal of torque. This torque would, of course, be reacted by the brake on the opposite wheel until it was released, when both wheels would again rotate at equal speeds.

Now imagine the brake released and the car's being lowered, one wheel coming down on to wet ice, and therefore still free to spin, and the other descending on to dry road. The circumstances, so far as the drive is concerned, will ultimately be similar to those in the previous paragraph except in that while the wheel on ice will be rotating slowly and laboriously, owing to friction in the differential, the torque reaction on the other wheel, instead of being applied to the brake, will drive the car.

27.9 Multi-plate clutch type

The most widely used clutch type has been developed from the Powr-Lok unit produced originally by the Thornton Axle Co. but now manufactured by the Spicer Axle Division, Dana Corporation, who also make a variant that they call the Trac-Aide. Both have multi-plate clutches interposed between the differential gears and the outer ends of their cages. The difference between the two is as follows. The multi-plate clutches in the Powr-Lok are engaged primarily by axial pressure exerted as a result of movement of the differential spindles up the flanks, or ramps, in V-shape slots, though about a third of the engagement thrust is attributable to the axial components of the meshing forces, tending to move the differential gears outwards relative to their pinions; the similar clutches in the Trac-Aide and the later Trac-Loc devices, on the other hand, are engaged solely by the axial components.

Units of these types are produced also in the UK by GKN Axles Ltd, Light Division, and in Germany under the name Lok-o-Matic by ZF, while others have been produced under names such as Anti-Slip, Traction Lok, Posi-Traction, and Super Traction by various companies including Borg Warner, Ford, GM and at least one company in Japan.

A disadvantage of most of the multi-plate clutch-actuated devices is that, being servo actuated by the transmitted torque, they are useless if there is very little or no tractive resistance. Most, however, have clutches with preload springs, usually of the Belleville type, to ensure that there is always a slight degree of resistance to relative rotation and to cater for dimensional variations arising from manufacturing tolerances and wear. In applications where wheel spin occurs over a high proportion of operating time, however, this arrangement can tend to accelerate the rate of wear of the clutch plates.

The principle of the ZF Lok-o-Matic, ramp-actuated type is illustrated in Fig. 27.9. Around one end of the cylindrical housing for the combined limited slip and differential mechanism is a flange to which the crownwheel is bolted. This housing is carried in two bearings, one each side, in the axle casing. Machined in its bore are four slots equally pitched around, and parallel to, its axis of rotation. Within it is the differential cage, which is split into halves on the longitudinal vertical plane that contains the axes of the differential pinions; each half houses a differential gear. Projecting radially outwards from, and equally pitched around, the periphery of each half of the cage are four lugs, which are a clearance fit in the slots in the bore of the housing so that, although the cage is driven by the rotation of the housing, its halves are allowed a limited degree of differential rotation.

The differential gears are splined to the inner ends of the halfshafts and mesh, in the usual manner, with four differential pinions. However, these pinions, instead of being carried on a one-piece spider, are on the ends of two

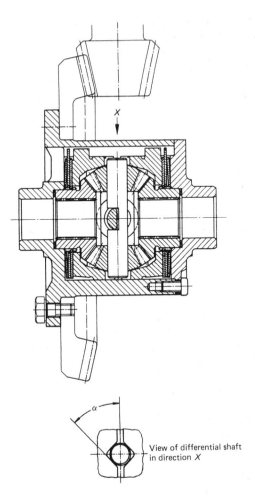

View of differential shaft
in direction *X*

Fig. 27.9

separate pins cranked at their centres, where they cross over to form a
two-piece, four-arm spider. Furthermore, the ends of the pins are not carried in
cylindrical bearings but seat in V-slots in the mating faces of the two-piece
differential cage, as in the scrap view in Fig. 27.9. In some versions, however,
the cylindrical form of these ends is retained.

With either arrangement since the transmission of torque to the halfshafts
tends to rotate the spider relative to the differential cage it causes the ends of
the pins to ride up the flanks of the V-shaped slots in which they seat. This
forces outwards the two parts of the cage and thus applies pressure axially to
the clutch plates, which are splined alternately to the differential cage and the
bosses of the differential gears. Consequently, the greater the applied torque,
the tighter is the engagement of the clutches, and thus any tendency towards
differential rotation of the gears and the halfshafts to which they are splined
becomes increasingly limited.

27.10   The traction control by viscous coupling

Gaining ground rapidly is the viscous coupling, initially introduced as a practicable transmission component by FF Developments Ltd and now manufactured by Viscodrive GmbH, a joint GKN-ZF company. It is installed in, among others, BMW, Ford and VW cars and light commercial vehicles. Its major advantages are simplicity and relative freedom from wear and maintenance. A disadvantage of its being incapable of transmitting a differential torque without a differential speed is that there tends to be a significant time lag before it comes into operation as a limited slip device.

The performance of this type of coupling is to a major extent dependent on the properties of the fluid used, which is the main reason why development, from its pre-FF conception in the nineteen-twenties to its widespread acceptance in the early nineteen-eighties, took such a long time. Although viscous couplings have been used for competition cars they are more suitable for other types and, because they can be designed for soft action, they can be installed in front axles. Since their torque transmission capacity increases with rotational speed, they tend to be less suitable for drive-axle than for inter-axle installation, that is, before the speed has been reduced by the crownwheel and pinion, though they are used for both.

In its simplest form the coupling consists of two sets of plates alternately splined to a housing and a shaft, with the viscous fluid between them, the housing is driven and the torque is transmitted from it by viscous friction between the driving and driven plates to the shaft. The plates are perforated to increase viscous drag by optimising the distribution of fluid between them. These perforations also optimise the hump mode of operation, a phenomenon that will be explained later.

A silicon fluid is employed because its viscosity falls linearly, but only very little, with increase of temperature. The housing is not initially full of fluid, otherwise the expansion due to generation of heat internally as a result of work done on the fluid would burst either it or the seals. Consequently, thermal expansion of the fluid progressively increases the wetted (effective) area of the plates, thus tending to offset the effect of the simultaneous reduction in viscosity owing to the rise in temperature.

As temperature increases further, the housing eventually becomes full of fluid, at which point input of energy causes a rapid rise in temperature, and therefore pressure, locally between the plates. These instantaneous local pressure rises cause the plates to move axially and the fluid between some of them to be squeezed out through the perforations. The outcome is metal-to-metal contact and a rapid rise in transmitted torque: this is called the hump mode of operation. Then, as the speed differential drops precipitately because of the suddenly increased friction, the unit rapidly reverts to the viscous mode.

The hump mode must not, however, be regarded as normal. It is an advantage only as a self-protection effect, considerably increasing the traction at the one wheel, and thus rapidly freeing a vehicle stuck because the other wheel is spinning. Alternatively, if the traction potential of the previously effective wheel is exceeded, the load on the coupling is again relieved because it, too, will spin. If, on the other hand, neither of these reliefs comes into operation, the engine stalls and the load on the coupling is once more effectively shed.

There are two ways of installing this device in a drive line, Fig. 27.10: one is in series, as a viscous transmission (VT), and the other in parallel, as a viscous coupling (VC). Moreover, in a differential there are, again, two ways of installing it, and these are: in the shaft-to-carrier and shaft-to-shaft layouts, Figs 27.11 and 27.12 respectively. Possible four-wheel-drive layouts are illustrated in Fig. 27.13.

In a shaft-to-carrier layout one set of discs is splined to the differential carrier, while the other set, the alternate discs, is splined to the differential gear on one side which, in turn, is of course splined to its shaft. On the other hand, with a shaft-to-shaft arrangement the discs are connected alternately, one set to each of differential gears. With the latter arrangement, although the viscous

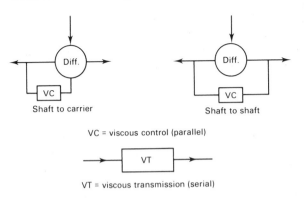

VC = viscous control (parallel)

VT = viscous transmission (serial)

Fig. 27.10

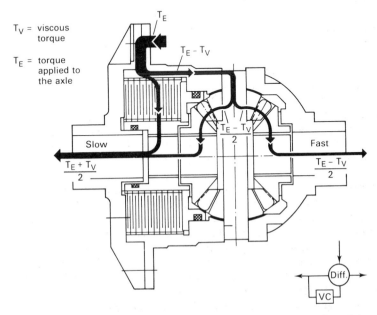

Fig. 27.11 Schematic torque distribution of a shaft-to-carrier axle differential

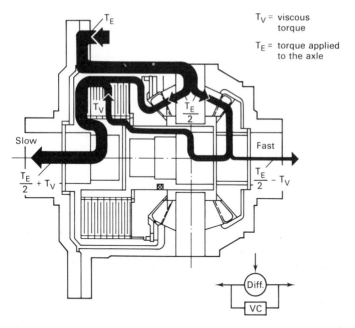

Fig. 27.12 Schematic torque distribution of a shaft-to-shaft axle differential

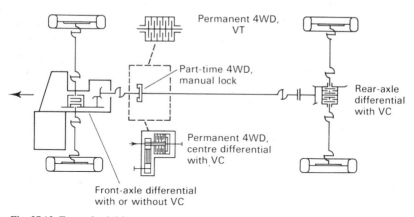

Fig. 27.13 Four-wheel-drive systems

coupling is connected, in effect, in series between the ends of the two halfshafts, the different gear is nevertheless still in parallel with it.

Either arrangement leaves the differential carrier and pinions to function normally except when there is a significant speed difference between the halfshafts, in which case the viscous coupling comes automatically into operation at a limited rate of slip, thus greatly reducing the potential for relative rotation. For any given speed difference, however, the shaft-to-shaft layout has approximately three times the locking torque of the shaft-to-carrier design. It is therefore the preferred layout for applications in which the space available is restricted and high torques are to be transmitted.

Chapter 28

The back axle

Having dealt with the mechanical transmission system between the engine and output from the gearbox, we now turn to the three alternative final stages. As listed in Chapter 20, these are: live axles, dead axles and axleless transmissions.

28.1 Live back axles

A live axle is one that either rotates or houses shafts that rotate, while a dead axle is one that does neither, but simply carries at its ends the stub axles on which the wheels rotate. Live axles perform two functions—

(1) To act as a beam which, through the medium of the springs, carries the loads due to the weight of the carriage unit and its contents, and transmits these loads under dynamic conditions through the road wheels – rotating on its ends – to the ground. The dynamic loading is principally a result of the motions of the wheel and axle assembly over the ground and the reactions due to its mass, the flexibilities of the tyres and road springs and the mass of the carriage unit and its contents.

(2) To house and support the final drive, differential, and shafts to the road wheels, and to react the torques in both the input and output shafts.

Most live axles, therefore, are of hollow or tubular construction and usually, though not necessarily, of circular cross-section outboard of the final drive unit.

28.2 The final drive

The functions of turning the drive from the propeller shaft through 90° to distribute it to the two wheels, and of reducing the speed of rotation – thus increasing the torque – is performed by the gearing carried in the final drive unit, usually housed in the back axle. For relatively small reductions – up to about 7 : 1 – single-stage gearing is used; but for greater reductions, two or even three stages may be required, and the gearing for one or more of these stages may be housed in the wheel hubs. The terms *single-*, *double-* and *triple-reduction axles* are therefore used.

Generally, the first stage is either a bevel pinion and what is termed the *crown wheel*, or a worm and worm-wheel, both of which of course turn the

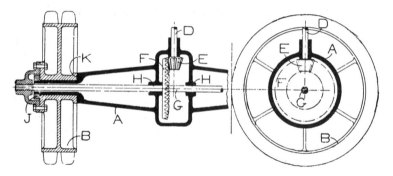

Fig. 28.1 Single-reduction axle

drive through 90°. Worm drives have the advantages of silence, either a low drive line or a high ground clearance – according to whether the worm is underslung or overslung relative to the wheel – ease of providing for a through drive to a second axle in tandem with the first, and the fact that a high single-reduction ratio can be readily provided – even as high as 15:1.

Bevel and hypoid bevel final drives are, however, far more common because they are less costly to manufacture and have a higher efficiency – the sliding action of worm teeth generates a lot of heat, especially if the gear ratio is high, and makes heavy demands on the lubricant. A hypoid bevel gear is one in which the axes of the crown wheel and the pinion are not in the same plane, and in which therefore some sliding action takes place between the meshing teeth. The one advantage is that a low propeller drive line can be obtained, so that the floor, and therefore centre of gravity, of the vehicle can be kept down.

28.3   Single-reduction live axles

An elementary single-reduction live axle – with a differential – is illustrated diagrammatically in Fig. 28.1. It has a hollow casing A, which carries on its ends the road wheels B. The weight of the body and load is supported by the casing A through the springs which are attached to the body and to the axle in a manner which will be described later. The casing in turn is supported at its ends by the road wheels. It therefore acts as a beam and is subjected to a bending action as is shown in Fig. 28.2, where the forces *P* are the supporting forces supplied by the road wheels, and the forces *W* are the body load, applied to the casing through the springs. The casing has to be stiff enough to withstand this bending action without undue flexure.

Supported in bearings in the casing A is a short shaft D integral with which is a bevel pinion E. The shaft D is coupled by means of a universal or flexible

Fig. 28.2

joint, outside the casing, to the propeller shaft and hence to the mainshaft of the gearbox. Inside the casing the bevel pinion E meshes with, and drives, a bevel wheel F which is fixed to a transverse shaft G. This shaft is supported in bearings HH in the casing and is bolted to the hubs of the road wheels B at its outer ends. Obviously, when the pinion shaft D is turned by the propeller shaft the drive is transmitted through the bevel wheel to the transverse shaft G and hence to the road wheels. The road wheels are kept in place on the casing A in the end direction by nuts J and shoulders K of the casing. Although a bevel gear drive is shown, the principle would have been similar – only the gear arrangement different – had a worm drive been used.

28.4 Torque reaction

From Fig. 28.1, it can be seen that the propeller shaft applies to the shaft D a torque which, as it is transmitted through the bevel gearing, is increased in the same ratio as the speed is reduced. This increased torque is then transmitted through the shaft G to the road wheels. From Newton's third law of motion, we know that action and reaction must be equal and opposite, so not only will this torque tend to rotate the wheel, but also the reaction from the wheel will tend to rotate the shaft G in the opposite sense. Therefore, there will be a tendency for the pinion and its shaft D to swing bodily around the crown wheel, and this tendency will be reacted by the axle casing. Some means therefore must be introduced to prevent the axle casing from rotating in the opposite direction. This may be simply the leaf springs themselves, or additional links – torque-reaction or radius rods – may be used and will be essential if coil, instead of leaf, springs are employed.

Similarly, the axle casing will tend to rotate about the axis of the bevel pinion in a direction opposite to that of rotation of the propeller shaft. However, since the torque transmitted by the propeller shaft is less than that in the driveshafts, it can in most circumstances be reacted satisfactorily simply by the suspension springs.

28.5 Driving thrust

Again, according to Newton's third law of motion, the driving thrust, or tractive effort, of the road wheels is reacted by the vehicle structure, the reaction being the inertia of the mass of the vehicle if it is accelerating, or rolling resistance of the other axle plus the wind resistance if it is not – the rolling resistance of the tyres of the driving axle involves of course purely local action and reaction. In effect, therefore, the driving axle has to push the carriage unit along, so it must be connected to the structure of the vehicle in such a way that this forward thrust can be transmitted from one to the other. This connection can be either the leaf springs or some other linkage for locating the axle relative to the carriage unit. The relevant members of this linkage are known as *thrust members*, or *radius rods*.

28.6 Torque and thrust member arrangements

In addition to the torque and thrust, sideways forces also have to be transmitted from the carriage unit to the wheels, and *vice versa*. The connections between the axle and the frame must therefore be capable of dealing with—

(1) The weight of the carriage unit.
(2) Torque reaction – from both drive line and brakes.
(3) Driving thrust.
(4) Brake drag.
(5) Lateral forces.

Although various connection arrangements have been employed, only four are still generally accepted. These are—

(1) The springs reacting all forces.
(2) As in (1) but with separate torque reaction members.
(3) As in (1) but with torques and thrusts reacted by separate members.
(4) The springs transmitting only the weight of the carriage unit, leaving the torque, thrust and drag reactions, and lateral forces to be dealt with by separate members.

These four systems are outlined in more detail in Sections 28.7 to 28.11.

28.7   Springs serving also as torque and thrust members

This system, Fig. 28.3, known as the *Hotchkiss drive*, is the most widely used. The springs A are rigidly bolted to the axle casing B. Their front ends are pivoted in brackets on the frame or vehicle structure, and their rear ends connected to the structure by means of either swinging links, or shackles C, or simply sliding in brackets as in Figs 35.7 and 37.18.

Obviously torque reaction causes the springs to flex, or wind up, as shown exaggeratedly in Fig. 28.4. Brake torque of course would flex them in the opposite direction. Since the front ends of the springs are anchored to the pins on the structure, they will transmit drive thrust and brake drag. The freedom of their rear ends to move fore and aft of course allows for variations in the curvature, or camber, of the spring with vertical deflection.

Wind-up of the springs under brake or drive torque causes the axle to rotate

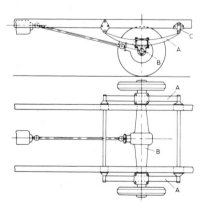

Fig. 28.3 Hotchkiss drive

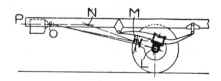

Fig. 28.4 Spring deflection due to drive torque

through a small angle, causing its nose either to lift, as in Fig. 28.4, or to drop. In the illustration, the spring wind-up has shifted the alignment of the final drive bevel pinion shaft from its normal attitude LO to LN, in which circumstances the propeller shaft would be subjected to severe bending loads were it not for the universal joints at O and M.

When the axle moves upwards relative to the carriage unit, it must move in the arc of a circle whose centre is approximately the axis of the pivot pin at the front end of the spring. The propeller shaft, on the other hand, must move on the arc of a circle centred on its front universal joint. Because these two centres are not coincident, the distance between the front universal joint and the forward end of the bevel pinion shaft will vary as the propeller shaft swings up and down. This variation is accommodated by the incorporation of a sliding joint somewhere in the drive line between the gearbox output shaft and bevel pinion in the axle. Usually a sliding splined coupling is formed on a fork of one of the universal joints, but sometimes a universal joint of the pot type, as for example in Fig. 26.12, is used. The example illustrated is the Birfield Rzeppa constant-velocity joint, another would be the very neat and simple universal joint used on the inner ends of the swinging halfshafts of the 1955 Fiat 600 rear-engine car. In the latter instance a rubber joint at the outboard end of each shaft accommodated the cyclic variations in velocity.

Rotation of the axle about a longitudinal axis, for example if one wheel only rises, is accommodated mainly by flexure of the springs, in a torsional sense, of rubber bushes, and by deflections of the shackles or within clearances in sliding end fittings. For cross-country vehicles, however, special forms of connection of the spring ends to the frame are sometimes used to isolate the springs from such twisting effects. Figs 37.13 to 37.15.

28.8 Hotchkiss drive with torque reaction member

With the simple Hotchkiss-drive arrangement, making the springs stiff enough to react the torque adequately can leave them too stiff for giving a good ride. To avoid a compromise, a separate torque reaction member can be introduced, but the penalty is increased complexity. This system is now rarely used.

Ideally, since with such a system the springs do not have to react the torque, their seating pads would be free to pivot on the axle. However, to simplify construction and obviate lubrication points, rigid spring-seatings are sometimes used.

When a torque reaction member has been used, it has been mostly a triangular steel pressing, as in Fig. 28.5. Sometimes one has been employed and sometimes two. With other arrangements, a tubular torque reaction member has enclosed the propeller shaft, in some instances having its forward end carried by a ball bearing on the propeller shaft, adjacent to its front universal joint, which then has to take the vertical force necessary for reacting the torque. Whatever its form, the torque reaction member has to be secured rigidly at its rear end to the axle casing. Its front end, however, may be connected by a shackle to the frame, or structure of the vehicle. This is necessary to allow for the fore-and-aft motion of the axle resulting from the flexure of the semi-elliptic springs about their front pivots. For the avoidance of shocks, for example if the clutch is engaged too rapidly, the front end of the torque member may be sprung, as shown in Fig. 28.5.

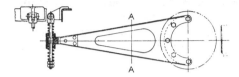

Fig. 28.5

28.9 Single combined torque-thrust reaction member, with springs taking only vertical and lateral loads

This form of construction is shown in Fig. 28.6. Bolted to the axle casing A and surrounding the propeller shaft is a tubular member B, the front end of which, C, is spherical and fits in a cup D bolted to a cross-member of the frame, or to the back of the gearbox. The springs are bolted to seats pivoted on the axle casing, and at each end are shackled to the frame. Clearly the member B will transmit the thrust from the axle to the frame and will also take the torque reaction. Since the centre line of the bevel pinion shaft will always pass through the centre of the spherical cup, if the propeller shaft E is connected to the gearbox shaft F by a universal joint situated exactly at the centre of that cup, neither an additional universal joint nor a sliding joint will be necessary, since both pinion shaft and propeller shaft will move about the same centre, namely that of the spherical cup, when the axle moves up or down. Because the axle is constrained to move about the centre of the spherical cup, the springs of course have to be shackled at each end to allow for the variation of their camber with deflection.

An alternative to the ball-and-cup construction is shown in Fig. 28.7. The tubular member B is again bolted to the axle casing at its rear end, but at the front it has pivoted on it a forked member A which is pivoted on pins C carried by brackets riveted to a cross-member of the frame. By pivoting on the pins C the axle can move about the axis XX, both rear wheels moving up or down together, while by the tube B turning in the bracket A about the axis YY, one rear wheel can move up without the other. The universal joint must have its centre at O, the intersection of the axes XX, YY.

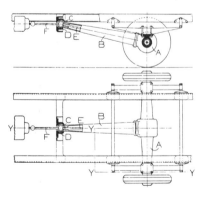

Fig. 28.6

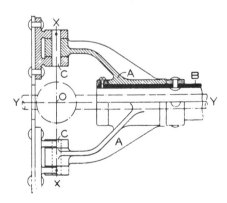

Fig. 28.7

In this system the spring seats are sometimes articulated on spherical bearings on the axle casing, to relieve the springs of twisting stresses. The same advantage was sought in some early designs by attaching the spring shackles to the frame on a pivot whose axis was parallel to the centre line of the frame.

28.10 Transverse radius rods

Where coil, torsion bar or air springs are used, which of course cannot locate the axle, other measures have to be introduced. Two such arrangements are illustrated in Figs 28.8 and 28.9, where transverse radius rods A, usually termed *Panhard rods*, are employed – to take the lateral loads – in conjunction with a single combined torque-thrust member B of the type illustrated in Fig. 28.6. With the arrangement of Fig. 27.8, the Panhard rod is parallel to the axle and therefore can have simple pivots at its ends. The advantage in Fig. 28.9 is that the Panhard rod is longer and therefore has less tendency to pull the axle laterally, as it moves up and down. On the other hand, its pivots must be rubber bushed, unless it can be arranged to lie parallel to the axle.

28.11 Three radius rods

The principle of a system often used is shown in Fig. 28.10. Radius rods A and B are placed parallel to the longitudinal axis of the vehicle and at the ends of the axle, while a wishbone or A-shaped member C is placed at the centre. The rods A and B are provided with ball-and-socket joints at both ends, while the wishbone member is pivoted to the frame on a transverse pin joint and to the axle by a ball-and-socket joint. The wishbone member deals with all the sideways forces, while all three rods between them deal with the driving and braking thrusts and torques. The torques are transmitted to the frame by tension and compression forces in the rods. Thus the driving torque reaction (which would act in a clockwise direction as seen in the end view) produces a tension force in the member C and compressive forces in the rods A and B, while the brake torque produces a compressive force in the wishbone and tension forces in the rods A and B. An approximation to the system of Fig. 28.10 is sometimes made by replacing the triangular upper member C by two separate radius rods arranged at about 45° to the axis of the vehicle and coupled at their ends by rubber-bushed joints to the frame and axle respectively.

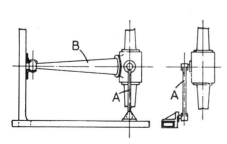

Fig. 28.8

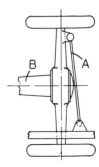

Fig. 28.9

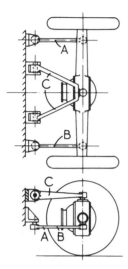

Fig. 28.10

If the upper radius rod of Fig. 28.10 is a simple link with a single pivot connection to the frame, then a transverse radius rod must be provided to position the axle sideways. These arrangements are fairly common when air or coil springs are used. Various methods of locating axles and reacting torques and thrusts are described in Chapters 35 and 36.

Chapter 29

Axle constructions

The pinion shaft of a bevel final drive and the worm shaft of a worm drive must be supported in journal bearings at two, or more, points and must be held axially. The latter can be done by using separate thrust bearings, or by means of the journal bearings.

In Fig. 27.2 is shown a differential and final drive such as might be used in a car or medium-weight lorry. The differential cage A is made in two halves which are bolted together and thus secure the pins H on which the differential pinions G are free to revolve. The tubular extensions of the differential cage carry the inner races of ball bearings B so that the cage is free to rotate about the axis XX. The outer races of the bearings B are in the differential carrier casting K, caps L being provided so that the differential cage, complete with its bearings, may be placed into position. The carrier K bolts up to the front face of the circular centre portion of the axle casing while the rear face is closed by a pressed steel cover S. In cars this cover is frequently welded in place because the petrol tank makes it inaccessible but in lorries it is usually removable so as to give access to the final drive and differential. The centre portion of the axle casing, which is formed of steel pressings, is stiffened by pressings J and the caps L are machined so as to bear against the machined opening in the axle casing in order to give them the maximum support. The bevel pinion D is carried in two taper roller bearings M and N. The inner races are fixed to the pinion shaft by the nut at the outer end, a spacer O being provided, and the outer races are in the differential carrier casting K. The housing P for the outer race of the bearing M is supported by several webs Q that connect it to the body of the casting K. By the proper machining of the spacer O the bearings M and N are 'pre-loaded', that is, there is a load on them when the pinion is free of all load; this enables the axial deflection of the bearings due to the load applied to the pinion when driving, to be reduced in magnitude and helps in maintaining the proper contact between the teeth of the pinion and crown wheel. Similarly the bearings B are also pre-loaded. An oil seal is provided at R.

Ball journal bearings are often used for the pinion shaft instead of taper roller bearings. The outer ball bearing is then commonly used to position the pinion shaft axially, its races being fixed respectively to the shaft and the carrier casting. The bearings are frequently pre-loaded. Taper-roller bearings are frequently used instead of ball bearings to support the differential cage.

A hypoid gear final drive assembly is shown in Fig. 29.1. The pinion A, integral with its shaft, is now supported in two large ball bearings at its outer end and in a parallel roller bearing at its inner end. The ball bearings take the axial loads and position the shaft axially, a selected washer B being used between the flange of the sleeve carrying the bearings and the neck of the casting C in order to give the correct adjustment. The roller bearing is carried in the web C_2 of the casting C and this web also houses two angular contact bearings which support one side of the differential cage, the other side being supported by a parallel roller bearing housed in the web C_1. Both webs are provided with caps. The axial position of the differential cage and crown wheel is determined by a selected washer interposed between the bearings and a flange of the web C_2 and cap D. Tubular arms F are bolted up to the central portion C of the axle to complete the axle structure and oil seals are provided at E.

An underhung worm-drive axle is shown in Fig. 29.2. The worm A, which is integral with its shaft, is supported by parallel roller bearings B and positioned axially by the dual purpose bearing C. As the latter is being used as a thrust bearing its outer race is an easy fit in the bore in which it is housed but is held securely against axial movement. This type of bearing is now regarded as superior to an ordinary thrust bearing of the type shown in Fig. 19.8. The bearings of the worm are carried in the cover plate casting D which bolts up to the underside of the axle casing K. In an overhead worm drive the cover plate would be bolted to the top of the axle casing. An oil seal is provided at J. The differential case is formed of the worm wheel itself and of two end plates that are bolted up to it at each side. Ball bearings which fit on the bosses of the end plates and are carried in the cover plate casting D serve to support the differential cage. Bearing caps E enable the differential assembly to be placed in position and these caps are arranged to bear against machined faces G on the axle casing so as to get the maximum lateral support. Washers F, which are selected of suitable thicknesses during assembly, enable the lateral position of the worm wheel in relation to the worm to be adjusted. A bevel-type differential is used, four pinions being provided and carried on the arms of a cross-shaped spider; these arms fit into four rectangular blocks H that in turn fit into slots formed in the bore of the worm wheel, thus enabling the spider and pinions to be assembled, and driven by the worm wheel. The blocks H are held

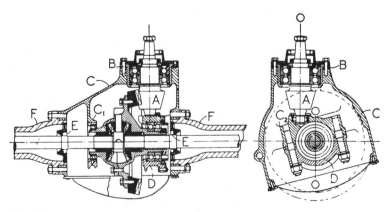

Fig. 29.1

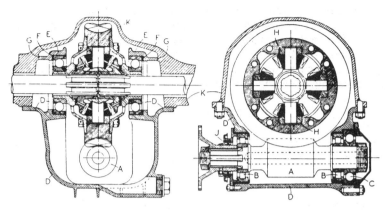

Fig. 29.2

sideways by the end plates of the differential cage. The axle casing is a forging and is like an inverted pot at the centre; it has tubular arms which carry the spring seats and wheel bearings.

29.1 Effects of wheel-bearing layout on axle loading

Some comments on wheel-bearing details are given in Section 34.11. The actual layout has a significant influence on axle loading. For instance, in both the elementary axle of Fig. 28.1, where the road wheels run on plain bearings, and the more advanced design of Fig. 29.5, in which the wheel is fully supported on two bearings on the axle casing, the halfshafts carry no loads other than the driving torque. This is termed a *fully-floating* axle layout.

To reduce the cost, complexity, weight and bulk of the hub, however, other arrangements are often used, especially for the lighter cars and other small vehicles. Again, the basic principles can be demonstrated by the use of a simple diagram, Fig. 29.3, where the road wheel is mounted on the end of the half-shaft which, in turn, is carried in plain bearings in the axle casing. The vertical load applied through the spring to the axle is W, and the equal and opposite reaction is also W and is at the point of contact between the wheel and the ground.

Because of the offset between this point of contact and the outer bearing carrying the halfshaft, a bending load is applied to the shaft, Fig. 29.4(*a*), the couple being reacted by equal and opposite loads P on the inner and outer bearings. There is also a shear load on the halfshaft, which could cause a failure of the type illustrated in Fig. 29.4(*b*). While the shear stress is constant between the wheel and the outer edge of the bearing, the bending moment increases linearly up to this section and so a maximum combined stress will occur here.

If for any reason a side load S is acting on the wheel, this too will apply a bending load, but in the opposite sense, to the halfshaft, Fig. 29.4(*c*). In this instance it will remain constant up to the outer edge of the outer bearing, and will be reacted by equal and opposite loads at the two bearings. These loads, − Q in Fig. 29.3, are of course opposite in sense to the loads P.

For the maintenance of a state of equilibrium of forces, if a side load S is applied at the wheel, there must be an equal and opposite reaction S. Since the halfshaft is placed in compression there must be a thrust bearing in the region of the centre of the axle, to provide this reaction.

Thus the shaft is subjected to five different loads, and the maximum combined stresses arising from them will occur in a plane adjacent to the outer edge of the outermost bearing. The loads are—

(1) Bending under the vertical load at the wheel.
(2) Shearing due to the vertical load at the wheel.
(3) Bending arising from the couple due to the side load on the wheel.
(4) Axial loading – compression or tension – due to this side loading.
(5) Drive torque.

Between the extremes of totally supported and fully-floating axle, there are several intermediate, or *semi-floating*, arrangements in common use. In most, the centre of the single bearing used is either in or as close as possible to the plane in which the vertical load is applied to the wheel. This reduces to a minimum the bending load on the halfshaft and avoids the use of two bearings. The halfshaft, however, still has to share with the bearing the side loading and its couple.

29.2 Some actual bearing arrangements

Some typical road wheel bearings are shown in Figs 29.5 to 29.8. The first of these is a full-floating bearing for a lorry; the hub A of the road wheel is carried on two taper roller bearings B whose inner races fit on the end of the axle casing C. One of these bearings will take any end thrust to the right while the other will take any thrust to the left. The inner races are pulled up against the spacer F by the nut D which is screwed to the end of the axle casing and locked by a bolt whose head enters a hole in the casing. The spacer E prevents excessive loads being applied to the bearing by tightening the nut too much and also enables the left-hand race to be firmly held. The spacer F is bolted up to the flange G, the backplate H of the brake assembly being placed between these two members. The flange G is welded to the axle casing. The casting K which is bolted up to the hub A and which houses the oil seal L serves to keep oil or grease from getting through on to the brake shoes or mud from getting into the wheel bearings. The wheel hub is driven by the halfshaft J whose end is formed into a flange that is bolted to the hub. The brake drum is bolted to the

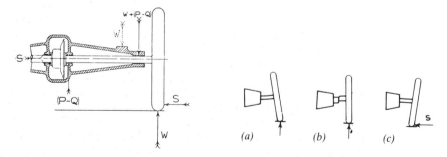

Fig. 29.3   Fig. 29.4

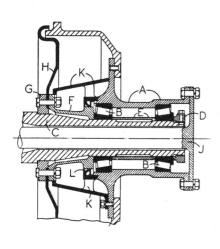

Fig. 29.5

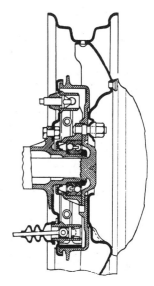

Fig. 29.6

outer side of the flange of the hub and so can be removed without disturbing the bearings.

In the bearing assembly shown in Fig. 29.6 a double row ball bearing is used, its inner race being held on to the outer end of the axle casing by a nut and its outer race being held between the flange of the halfshaft and a member which is bolted to the flange. The pressed steel wheel centre and the brake drum are also bolted to this flange. The bearing thus takes end thrust in both directions while the halfshaft takes only the driving torque although it also assists the bearing in taking the tilting actions due to end loads acting on the wheel. The back plate of the brake assembly is bolted up to a flange that is integral with the end of the axle casing.

In the assembly shown in Fig. 29.7 the halfshaft A is formed with a flange at its end to which is bolted the brake drum E and also the road wheel (not shown). The halfshaft is supported by a single row ball bearing B whose outer race is housed in the cup C which is bolted up to the flanged end of the axle casing D. The backplate F of the brake assembly is also bolted to a flange of the member C. The outer race of the bearing B is thus held so that it can take end thrusts in both directions while the inner race abuts up against the shoulder of the halfshaft on the right-hand side and is held by the collar G, which is a force fit on the shaft, at the left. The bearing thus takes all the end thrusts but the halfshaft is subject to shearing force and bending moments. An oil seal, indicated at G, prevents the egress of oil from the axle casing.

In Fig. 29.8 the end of the halfshaft A is tapered and the brake drum E is bolted to it, a key serving to transmit the drive. A single taper roller bearing B supports the halfshaft, its inner race fitting the taper on that shaft while its outer race fits in the end of the axle casing and is held in by the plate C which, together with the oil seal housing G and the brake backplate F is bolted up to the flange of the axle casing. The bearing B can thus take end loads acting to

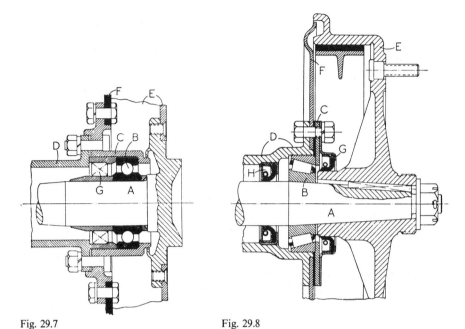

Fig. 29.7   Fig. 29.8

the right but loads to the left have to be carried through the halfshaft to the bearing at the other side of the axle. The halfshafts therefore abut at the centre of the axle. An oil seal H serves to prevent oil from escaping from the axle.

29.3 Axle casing constructions

Space is not available to allow all the axle constructions that are used to be described; only the most important types can be dealt with.

A common type when a worm final drive is used is shown in Fig. 29.9. The central portion A is a hollow iron or steel casting open at the top and having tapering extensions at each side. Bolted to the top of this member is a cover plate B in which are arranged the bearings for the worm and for the differential cage. The latter is carried in ball or roller bearings housed in the cups CC, the upper halves of which are integral with the cover plate, the lower halves being separate caps in order that the differential cage may be got into position. The caps are secured by the nuts and bolts shown, but studs screwed into the cover plate are sometimes used instead of bolts.

By supporting the differential cage in this way the worm and worm wheel can be adjusted to the correct relationship on the bench, and then the whole assembly can be bolted to the axle casing A. The gearing can also be easily withdrawn for inspection.

Pressed into the extensions of the central casting are two steel tubes D, which at their outer ends carry the bearings for the road wheels, ball or roller bearings being arranged between them and the inside of the wheel hubs. The springs are bolted to spring seats pivotally mounted on the cylindrical portions of the extension castings E, the latter being bolted by flanges H to the central casting. The wheels are kept on by nuts J screwed to the ends of the

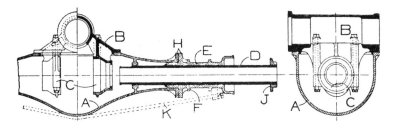

Fig. 29.9 Worm-drive axle casing

tubes D, while side thrusts towards the centre of the axle are taken on the ends of the members E.

This construction provides a strong and rigid axle.

Most axles have a built-up construction, the axle beam that carries the bending stresses being formed of a central member to which tubular side members are welded or bolted on as in Fig. 29.1. On the ends of the tubular sections castings are fixed, or pieces welded, to form the spring seatings and radius rod attachment point and the brake backplate. The central member may be a casting or may consist of a number of pressings welded together; it usually has an approximately circular face at the front and may also have one at the rear. A casting that houses the final drive and differential assembly is bolted to the front circular face and a cover plate to the back one but this may be permanently closed by a cover that is welded on. These faces usually have reinforcement rings and, because the construction resembles a banjo, particularly after only one tube has been assembled to it, it is usually described as a *banjo*-type axle. Axles of this type are illustrated in Figs 27.2 and 28.10. On the other hand, axles of the type illustrated in Fig. 29.9 may be referred to as the *pot* type.

Another term sometimes used to describe axle construction is the *portal* type. This is an axle the tubes of which are cranked to form what looks something like a portal frame. Its central portion may be dropped below the level of the hubs and has a pair of gears at each of its outer ends both to lift the drive line up again to that level and to provide an extra reduction ratio so that the central differential casing can be relatively compact. This is one method of ensuring that the top of the central differential casing is below the level of the top of the chassis side frames, so that the floor of, for example, a bus or van can be as low as practical and flat. Alternatively, by elevating the central portion, the ground clearance beneath the differential can be greatly increased; for example, for off-road vehicles.

In Fig. 29.2 the axle is a single forging consisting of a cylindrical annular central portion with tubular extensions but as this is expensive to produce it is not often employed.

Chapter 30

The double-reduction axle

In this type of axle the permanent reduction of speed between the engine and road wheels is obtained in two separate steps. Double-reduction axles are used chiefly on heavy lorries, buses, etc., for the following reasons: such vehicles run at low speeds in comparison with passenger cars; thus, while cars run at speeds up to 100 mph, lorries do not exceed between 50 and 80 mph; they also have larger diameter wheels. Thus, although lorry engines run at much slower speeds than car engines, the reduction of speed between the engine and road wheels is a good deal larger, being from 5 : 1 up to 10 : 1, as against 3.5 : 1 up to 6 : 1.

If these large reductions were obtained in a single step, using, say, bevel gearing, then either the bevel pinion would have to be made very small with few teeth, when it would be both weak and inefficient, or the crown wheel would have to be made very large, which would result in a heavy and expensive axle, and would reduce the clearance between the axle and the ground too much. Similar conditions are found, to a lesser degree, with worm final drives, and so the double reduction axle is adopted when the final drive ratio has to be large. On some vehicles they are used to enable a very low body position to be obtained.

30.1 Both steps at the centre of the axle

This type of double-reduction final drive is shown in Figs 30.1 to 30.5. Referring to Fig. 30.1, the bevel pinion A is driven by the propeller shaft and gears with a small crown wheel B. The latter is fixed to a layshaft C, to which is also fixed a spur pinion D. The layshaft is carried in ball or roller bearings at its ends in the axle casing, suitable thrust bearings being provided to take the thrust of the bevel gears. The spur pinion D meshes with a large spur wheel E which is bolted to the differential case F just as the crown wheel of a single reduction axle is bolted. The differential cage is carried in ball or roller bearings in the axle casing in the usual way.

The arrangement of the gearing in Fig. 30.2 is slightly different from the above. The differential is here situated between the two halves of the layshaft instead of between the two driveshafts. The propeller shaft drives the bevel pinion A, meshing with the small crown wheel B that is bolted to the differential cage. The latter is supported by the shaft C which is carried in ball

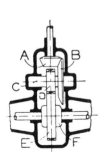

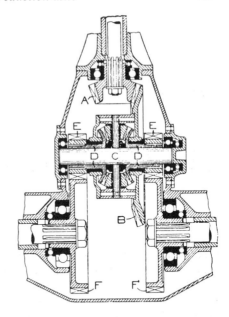

Fig. 30.1   Fig. 30.2

bearings in the axle casing. The differential wheels DD are now integral with short sleeves free to rotate on the shaft C, and which carry spur pinions EE at their outer ends, outside the differential cage. The pinions EE mesh with spur wheels FF fixed to the driveshafts. Thus the differential action is obtained between the pinions EE instead of between the driveshafts themselves. The chief advantage of this arrangement is that the differential now revolves at a higher speed than the road wheels, and so, for a given torque on the latter, the forces on the differential wheel teeth are smaller, enabling smaller wheels to be used. To set against this advantage, two spur pinions and wheels are required for the second reduction, thus increasing the cost.

A third arrangement is shown in Fig. 30.3. The spur pinion A driven by the propeller shaft, drives a wheel B fixed to the layshaft C, to which is also fixed the bevel pinion of the bevel drive. The arrangement of the latter is normal. It will be observed that the axis of the propeller shaft is higher than it would be with a single reduction bevel gear axle, and can thus be brought more nearly into line with the mainshaft of the gearbox, so that the work put upon the universal joints is reduced. This advantage is also obtained with the other types of double-reduction axle and with the overhead worm-driven axle.

With single-reduction bevel axles and with underhung worm drives the engine and gearbox are sometimes inclined in the frame in order to obtain this advantage.

An arrangement in which the second reduction is by an epicyclic gear is used by Scammel on some of their lorries and is shown diagrammatically in Fig. 30.4. A bevel first reduction and a normal bevel-type differential are used, the differential cage being carried in bearings in the end plates A which are bolted up to flanges on the central drum-shaped portion of the axle casing. The differential shafts carry pinions B which form the sun wheels of epicyclic gears

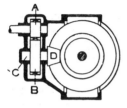

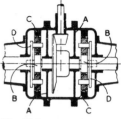

Fig. 30.3   Fig. 30.4

of the sun and annulus type. The annuli C are fixed, being formed integral with the end covers which are bolted up to the central casing on the same flanges as the plates A. The arms D which carry the planet pinions are made integral with the driveshafts and are supported in bearings in the end covers C. The tubular end portions of the axle casing are bolted up to the end covers C on flanges as shown.

30.2   Kirkstall double-reduction axle

An interesting and original design by the Kirkstall Forge Engineering Company of Leeds, who manufacture axles for all kinds of vehicles, is shown in Fig. 30.5.

The first reduction occurs between the worm and worm wheel, the latter being fixed to the annulus member A which is carried in bearings C and D in the axle casing; the second reduction is obtained from the epicyclic action of the train consisting of the annulus A and sun S, in conjunction with the concentric spur train (not epicyclic) consisting of the pinion H and the annulus J. The annulus A meshes with three planet pinions which are carried by the member E and the latter is splined to the left-hand halfshaft F. The pinions mesh at the centre with the sun S, which is integral with the pinion H. The latter drives the annulus J through three pinions K, which are carried on pins L

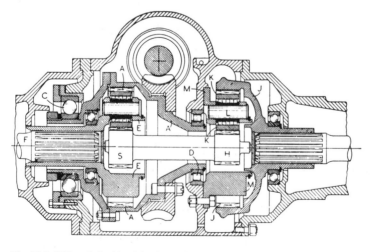

Fig. 30.5 Kirkstall double-reduction axle

supported by the member M which is fixed to the axle casing. The annulus J is splined to the right-hand halfshaft. The left-hand halfshaft is thus driven in the forward direction by forward motion of the halfshaft is thus driven in the forward direction by forward motion of the annulus A, as it would be if the sun were fixed; but the sun, being free, will be driven in the backward direction so that the speed of the planet carrier E and its halfshaft will be lower than if the sun S were fixed. The backward motion of the sun S and pinion H is converted to forward motion of the right-hand halfshaft by the pinions K, and some reduction of speed occurs in this train.

In order that equal torques shall be exerted on the two halfshafts, the gear train H, K, J must have a particular ratio in relation to that of the epicyclic train A, S. This particular ratio is found as follows. Let the torque imparted to the annulus A by the worm and wheel be T, and let the number of teeth in the annulus and sun be denoted respectively by s and a. Then the torques applied to the member E will be $[T \times (a + s)]/a$ and the torque acting on the sun S will be $T \times (s/a)$ (backwards). The torque applied to the annulus J is therefore $T \times (s/a) \times (j/h)$, where j and h are the numbers of teeth in J and H respectively. Hence if the torques applied to the two halfshafts are to be equal, we must have—

$$T \times \frac{a + s}{a} = T \times \frac{s}{a} \times \frac{j}{h}$$

that is $(a + s)/s$ must equal j/h.

The value of the second reduction produced by the trains A,S and H,J may now be found by using the tabular method described in Chapter 24. Let N_A, N_F and N_S be respectively the speeds of the annulus A, the left-hand halfshaft F and the sun S. Then we have the results shown in Table 30.1 and the speed of the right-hand halfshaft will be $-N_S \times (h/j)$ or

$$N_R = \left[N_A \frac{a}{s} - N_F\left(1 + \frac{a}{s}\right) \right] \times \frac{h}{j}$$

the minus sign occurs because the annulus J goes in the opposite direction to the pinion H.

Since j/h is made equal to $(a + s)/s$, as stated above, we get

$$N_R = N_A \times \frac{a}{s} \times \frac{h}{j} - N_F$$

But in straight-ahead running $N_R = N_F$, hence

$$2 N_R = N_A \times \frac{a}{s} \times \frac{h}{j}$$

Table 30.1

Planet carrier E or left-hand halfshaft F	Annulus A	Sun S
N_F	N_F	N_F
O	$N_A - N_F$	$(N_F - N_A)^{as}$
N_F	N_A	$N_S = N_F(1 + {}^{as}) - N_A \cdot {}^{as}$

The second reduction is therefore

$$\frac{N_A}{N_R} = 2 \times \frac{s \times j}{a \times h}$$

It will be noticed that the shaft SH is supported solely by the contacts between the teeth of the gears S and H and the planets and pinions K. This ensures equal division of the torques acting on the pinions S and H between the three pairs of teeth of each pinion which transmit the torques.

30.3   One step at centre of axle, the other at road wheels

Double-reduction axles of this type may be divided into two classes, those in which the second reduction is provided by a simple spur pinion and wheel, and those in which it is given by an epicyclic train. An example of the second type is given in Section 30.2, while an axle of the first type, to a design by Mercedes-Benz, is shown in Fig. 30.6. It will be seen that the parts required to fulfil the two functions of a live axle are to a great extent independent members. A forging A, to the ends of which are bolted the members B, forms the axle beam and the road wheels, the hubs of which are seen at C, revolve on taper roller bearings on the ends of the members B. The springs are bolted to seats formed on the forging and the covers B. A separate casing D houses the first reduction and differential, and light tubular covers protect the halfshafts. At their outer ends the latter are splined to pinions E, which are carried in roller bearings and which mesh with gears F. The latter are also carried in roller bearings and are splined to the shafts G, which are integral with the hub caps and by which the wheel hubs C are driven. Seals H serve to keep lubricant in and mud out.

This arrangement has the advantage that the halfshafts revolve at higher speeds than the road wheels so that the torques they have to transmit are correspondingly smaller than if they drove the road wheels direct. The system gives a very rigid construction, and by arranging the halfshafts below the axle beam, a very low body position can be obtained. Alternatively, by placing the halfshafts above the axle beam, a very high ground clearance could be achieved, but this would introduce certain difficulties and has not become

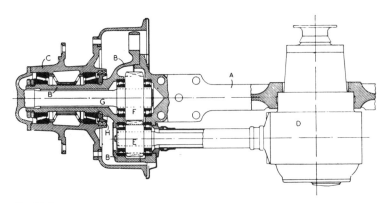

Fig. 30.6

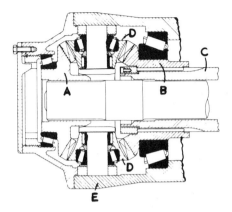

Fig. 30.7 Foden bevel-gear hub reduction

widely used. In the axle shown the halfshafts are situated behind the axle beam and the only gain in road clearance is that, because of the second reduction, the crown wheel of the first reduction, and hence the casing, D, can be smaller than otherwise.

In some early axles of this type the gear F was replaced by an internally toothed ring with which the pinion E meshed. The drawback to this arrangement was the great difficulty in providing satisfactory seals.

30.4 A bevel-gear hub reduction

A design by Foden Ltd that uses a bevel epicyclic train as the second step of a double-reduction axle, is shown in Fig. 30.7. Splined on the end of the halfshaft there is a bevel gear A and a similar gear B is splined to the end of the axle casing C. Between these two gears is a two-armed spider on whose pins the planet pinions DD are carried on taper roller and needle roller bearings. The spider is fixed inside the wheel hub E. The bevel gear B is thus the fixed member of the epicyclic train, the gear A is the driving member and the planet carrier and wheel are thus driven at half the speed of the halfshaft.

Part 3

The Carriage Unit

Chapter 31

The basic structure

So far, we have dealt with the propulsion sub-systems of road vehicles –
engines, transmissions, axles and wheels. Now we come to the main assembly
on which the engine and transmission are mounted, and the interconnections
between it and the axle-and-wheel sub-assemblies. This of course is the
carriage unit, which also includes the body for carrying the occupants in
comfort and safety or, in the case of trucks, the payload safely and without
damage.

31.1 The frame

Until the end of the Second World War, the overwhelming majority of cars
had separate frames similar to, though of course much smaller than, those now
associated with most modern commercial vehicles. The function of the frame is
to carry all the major components or sub-assemblies making up the complete
vehicle – engine, transmission, suspension, body, etc. In Fig. 31.1, a fairly early
ladder-type frame is illustrated, but it has been drawn with dissimilar side
members, to show two different types of layout. These side, or main
longitudinal, members normally would be virtually identical, though opposite
handed. In this illustration, that at A is straight, and therefore easy to
manufacture, and it is inclined inwards towards the front; that at B, on the
other hand, is cranked inwards. In the first instance the taper, as viewed in
plan, and in the second, the cranked formation of the side member, are for
clearing the front wheels when steered on full lock but are also convenient for
taking the engine mountings. As viewed in side elevation, these members in
both instances are cranked upwards near the rear to clear the back axle.

Transverse members C (often confusingly called cross members – nothing to
do with cruciform members) separate the side members, contribute to the
overall reaction to torsional loading, and help to support components
mounted above or slung underneath – the gearbox or a transaxle for instance.
Ideally, if the suspension spring or linkage is laterally offset relative to the
frame side members, one transverse member is fitted between each of the
mounting points, D and E, to translate the torsional loading applied to each
side member – by the offset – into a bending load in the transverse member.
Additional transverse members may be required to carry the body or other
major units such as the engine or final drive unit – if the latter is not in the axle

– and for locally supporting the side members where they are cranked inwards towards the front. The last-mentioned supports are needed because of the tendency of the side members to flex where they are kinked. Such flexure would occur owing to severe local deflections due to the tensile and compressive loads induced in their flanges – by bending in the vertical plane – and would rapidly cause fatigue failures.

Where the centres of the suspension spring attachment points are directly beneath the effective flexural axes of the side members, transverse support is required to take only the side loading at the spring eyes and, in these circumstances, the lateral bending may not be large. It may therefore be taken by extra-stiff front and rear extensions of the side members, at D. These extensions are called *dumb irons*.

In general, the usual reason for the omission of transverse members where one might have thought them to be essential is the need to allow clearance for the propeller shaft, as it moves up and down with the suspension, or the presence of some obstruction such as the engine sump, clutch or gearbox. In some instances, the propeller shaft passes through a vertically elongated hole in the centre of a transverse member.

Where the cranked portion towards the front of a frame is unsupported by a transverse member, substantial gusseting or some other form of reinforcement is necessary. In other instances, for example at the rear spring attachments, the frame side members may be locally reinforced to transfer the loads from the spring support brackets to the nearest transverse member.

On the lowest of the side members shown in Fig. 31.1, two brackets F extend laterally outwards. These are for carrying the running boards fitted to most pre-Second World War cars. Similar, but generally much shorter, brackets carry the body sides and other components such as engine, gearbox, and fuel tank mountings.

Early cars and most modern commercial vehicles traditionally have had channel section frame members, *a* in Fig. 31.1 – either steel pressings or fabricated from steel plate or strip. Where practicable, brackets are attached by bolting, riveting or welding them to the webs of the sections. This avoids introducing stress raisers in the flanges, which are the most heavily stressed

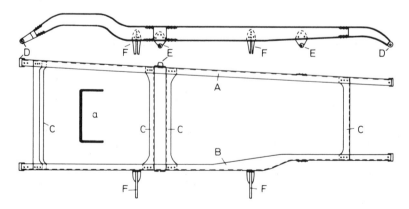

Fig. 31.1 Inset (*a*) is the cross-section of both the longitudinal and transverse members

parts when the whole sections are subject to bending loads. For heavy commercial vehicles, bolts are frequently used in frame construction, including for the attachment of the transverse member, because such frames are usually fairly accessible and maintenance is therefore easier than on car frames. Rivets must be very skilfully fitted, preferably cold, otherwise they are liable to loosen. Welds tend to suffer fatigue failure under the severe racking loads experienced with commercial vehicles. Because of this loading and the need to avoid local high stresses due to sudden changes in overall stiffness along the length of the frame – for example between a cab and the body – torsionally rigid frames of box section are widely regarded as unsuitable for heavy commercial vehicles. Mild steel – easily pressed and welded – used to be the invariable choice for all frames, but modern heavy commercial and even some light vehicles frequently have frames of carbon manganese steel with a yield stress of about 3620 kg/cm^2.

With the introduction of independent front suspension, chassis frames were called upon to take much higher torsional loading. This was because, whereas the centres of semi-elliptic leaf springs on a beam axle have to be well inboard of the front wheels to leave a clearance for steering them, the effective spring base – distance between spring centres – with independent front suspension is approximately equal to the track. In these circumstances, when a wheel on one side only rises over a bump, the upward thrust it exerts on the frame has a much greater leverage about the longitudinal axis of the car.

The outcome is that means have to be sought for increasing the torsional stiffness of the frame. Provided that their ends are closed – by, for instance, welding to them a flat or flanged plate – or are otherwise strongly reinforced so that they cannot lozenge or in any other way distort, tubular sections of any shape – round, oval, triangular, square, rectangular, etc. – are inherently very rigid torsionally. Such sections therefore began to be used for both longitudinal and transverse members on car frames. A selection of sections that have been used is illustrated in Fig. 31.2.

The transverse members most heavily loaded in torsion are of course those that support the independent front suspension. This is partly because of brake-torsion reaction which is applied by the rearward thrust of the road on the tyre contact patch and transmitted through the brake disc or the drum brake backplate to the stub axle, and thence through the suspension links to

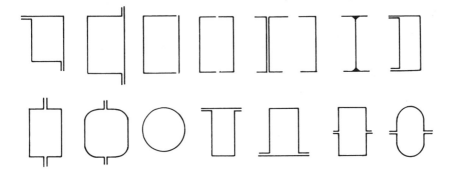

Fig. 31.2 Some chassis-frame sections

the frame. Additionally, an entirely different torsional loading arises in this transverse member as a result of single-wheel bumps – when the wheel on only one side rises. Such a bump, lifting one side of the front end of the frame, leaves the far side and the rear end down in their original positions, thus causing the side members to tend to twist the front transverse member and, incidentally, all the others. Hence, heavy gusseting is needed between the transverse and side members. Sudden local changes in stiffness at or near the junctions between the transverse and side members have to be avoided, otherwise trouble due to fatigue failures will be experienced.

Among the last separate frames to be designed for large cars produced in reasonable quantities in the UK was that of the Humber *Super Snipe*, Fig. 31.3. This has box section members of various cross-sections selected for the ease with which they can be accommodated beneath the floor and secured to each other. The front cross member is a top-hat section of very large proportions, with a closing plate welded to its bottom flanges. Such a section is fairly convenient for the attachment of the gussets and brackets, top and bottom, that carry the suspension link pivots, springs and shock absorber mountings. The frame has also a cruciform bracing member, made from channel sections welded back to back. At the centre, the top flanges of the cruciform member arch over the clearance hole in the web, through which the propeller shaft passes.

There is also extensive gusseting between the ends of the cruciform and the cranked portions of the side members. The torsional stiffening effects of a cruciform bracing member on a frame can be accounted for in either of two ways. First, together with the adjacent side and transverse members of the frame, it forms in effect, a single huge tubular member of flat rectangular section, the outer boundaries of which are, on each side, the webs of the side-members at each side and, top and bottom, the flanges of the side members, transverse members and cruciform brace. From the structural viewpoint this is similar in effect to extending all the flanges to form a continuous web closing the top and bottom of the frame, to form the huge flat rectangular section. Indeed, total enclosure was in fact done successfully to solve a torsional vibration problem that was causing dash shake in the

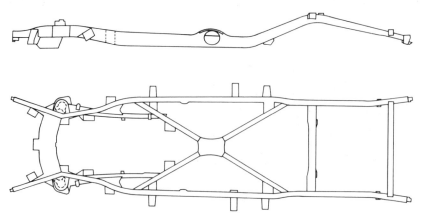

Fig. 31.3 Humber *Super Snipe*. All the main members of the frame are of 12 swg steel sheet

development stage of the Austin A90 *Atlantic*, and was patented in about 1949 by the author. The stiffness of any closed tube of given wall thickness, incidentally, is directly proportional to the cross-sectional area it encloses.

The alternative explanation, which applies particularly to an earlier form of cruciform bracing, is that the cruciform is in effect four tetrahedrons, Fig. 31.4 – a tetrahedron is the simplest basic geometric structure that has torsional stiffness. In most road vehicles, the front and rear tetrahedrons of the cruciform bracing have acute angles and those on each side have obtuse angles at their apices. There can of course be a hole through the central vertical post, to clear the propeller shaft, provided that post is suitably reinforced around the hole.

Backbone-type frames have also been used. The tubular type, Fig. 31.5, has been adopted principally for some Czechoslovakian, German and Austrian designed vehicles – Tatra, Daimler-Benz and Steyr Puch. In some instances the ends of the tubular backbone are bolted directly, at the front, to the gearbox and, at the rear, to the final drive casing, while in others, they have forked extensions, between the arms of which these components are accommodated. An example of a backbone-type frame, but made by in effect bringing the two box section side members together, side by side in the centre, is illustrated in Fig. 31.6. It is that of the 1960 Triumph *Herald*, probably the last quantity-produced small car to be designed in the UK with a separate frame.

The advantages of the backbone frame include high torsional stiffness at low cost, and light weight. A disadvantage is the length of the outrigger arms needed to carry the body sides. These arms tend to introduce torsional vibration problems because of their bending flexibility.

Separate frames, in general, have the merit of being capable of accepting a wide variety of different bodies – hence their attraction even for light commercial vehicles. The principal disadvantage is that, because such a frame has to be accommodated under the floor of the vehicle, its depth is of necessity restricted and therefore its bending stiffness – proportional to the cube of its depth – and its torsional stiffness are limited. This can be overcome by the use of a space frame, of which there are plenty of examples in racing cars, but these are of complex construction and therefore costly. They are suitable

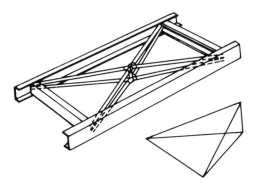

Fig. 31.4 Cruciform bracing and tetrahedron

(a)

(b)

Fig. 31.5 On the Steyr Puch Hafflinger chassis (*a*), the floor frame of the body is mounted on top of the springs, to which some of the vertical and torsional loading is directly applied through the body platform, as can be seen at (*b*)

where exceptionally light and simple bodywork, for instance of reinforced plastics, is to be fitted; but, where the body in any case has to be a fairly substantial structure, considerable weight saving and economy can be effected by designing the body so that it can perform the functions of the frame.

Fig. 31.6

31.2 Sub-frames

Sub-frames are employed for one or more of three basic reasons. The first is to isolate the high frequency vibrations of, for example, an engine or a suspension assembly, from the remainder of the structure. In this case, rubber or other resilient mountings are interposed between the sub-frame and main structure.

Secondly, a sub-frame can isolate an inherently stiff sub-assembly such as the engine or gearbox from the effects of the flexing of the chassis frame. This is done generally by interposing a three-point mounting system between the sub-frame and main frame, one of the mountings being on the longitudinal axis about which the main frame twists, and the others one on each side.

Thirdly, a sub-frame may be used to carry, for instance, the front and rear suspension sub-assemblies, where to utilise the front and rear ends of the body structure for this purpose would increase unacceptably its complexity or cost, or introduce difficulties in either manufacture or servicing, or both. A good example of such sub-frame usage is the BL *Mini*, the front and rear sub-frame assemblies of which have been used by some kit car manufactures because of the ease with which the engine and front suspension, on its sub-frame, can be bolted to the front, and the rear suspension, similarly on its sub-frame, bolted to the back of a different body designed to receive them.

31.3   Integral and chassisless construction

The terms *integral* and *chassisless* construction are often confused, but the difference is simple. Integral construction is that in which a chassis frame is welded to, or integrated with, the body. It was the first stage in the evolution of the chassisless form of construction, in which no chassis frame can be discerned. The first quantity-produced vehicle in the latter category was almost certainly the Austin A30, the design of which was very fully described in *Automobile Engineer*, October and December 1952, and March and April 1954. The first two of these four references describe the vehicle itself, while the second pair elaborate on methods, developed by the author, for use in its structural design.

The details of chassisless construction are much too numerous, varied and complex to be described here. In principle, however, its advantages stem from the facts that beams formed by the body panels may be something like 50 cm deep, whereas a chassis frame for a car is only about 8 to 13 cm deep, and the area enclosed by a complete body is similarly vastly bigger than that enclosed by the cross-section of a frame side or transverse member. Since the strength and stiffness of a beam are proportional respectively to the square and cube of its depth, while both the torsional stress and stiffness of a box section are

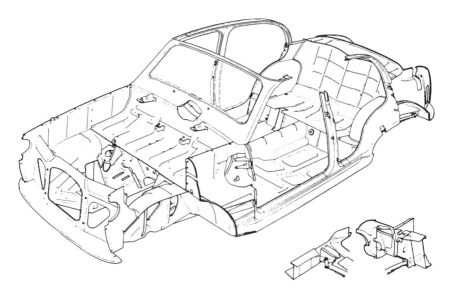

Fig. 31.7 The Austin A30 was almost certainly the first car of truly chassisless construction to go into quantity production anywhere in the world. Virtually all the panels of 0.914 mm thick steel, the principal exceptions being some 1.299, 1.626 and 2.032 mm brackets carrying the front and rear suspension, the 1.626 mm front apron and two inverted channel sections on each side of and parallel to the engine

proportional to the area enclosed by it, it follows that the strength and stiffness of a body shell are potentially much greater than of a chassis frame.

One of the main facts to be borne in mind when designing sheet metal structures is that even a simple flange will carry satisfactorily a surprisingly large load provided that it is stabilised – supported against buckling or other forms of distortion. A useful rule of thumb is that a flange will carry a stress up to the yield strength of the material provided that its width is no more than 16 times its thickness. This is approximately valid provided that the deflection of the beam as a whole is negligible. If not, as would be the case if, for example, an inverted top-hat section beam freely supported at its ends and point loaded by a weight on top of its centre, the component would fail prematurely.

In such a beam, the loading would induce a compressive stress in the top layers and a tensile stress in the bottom layers, that is the flanges, of the section. As the yield stress was approached, the beam as a whole would bow downwards, and the tensile loading in the flanges would cause their outer edges to tend to stretch straight – in other words chordwise – instead of following the bowed form of the remainder of the beam. This, in turn, would reduce the effective second moment of area – moment of inertia – of the section, so the stress in the stiffer portions of the flanges – adjacent to the vertical walls of the section – would be higher than the calculated values, the metal would yield and the section buckle and collapse prematurely.

A typical chassisless body structure is illustrated in Fig. 31.7. Spot welding is used extensively in its construction, though some unstressed panels may be bolted on, for ease of replacement in the event of damage.

Chapter 32

Brakes

The operation performed in braking is the reverse of that carried out in accelerating. In the latter the heat energy of the fuel is converted into the kinetic energy of the car, whereas in the former the kinetic energy of the car is converted into heat. Again, just as when driving the car the torque of the engine produces a tractive effort at the peripheries of the driving wheels, so, when the brakes are applied the braking torque introduced at the brake drums produces a negative tractive effort or retarding effort at the peripheries of the braking wheels. As the acceleration possible is limited by the adhesion available between the driving wheels and the ground, so the deceleration possible is also limited. Even so, when braking from high speed to a halt, the rate of retardation is considerably greater than that of full-throttle acceleration. Consequently, the power dissipated by the brakes, and therefore the heat generated, is correspondingly large.

When a brake is applied to a wheel or a car, a force is immediately introduced between the wheel and the road, tending to make the wheel keep on turning. In Fig. 32.1 this is indicated as the force F; this is the force which opposes the motion of the car and thereby slows it down. The deceleration is proportional to the force F, the limiting value of which depends on the normal force between the wheel and the road, and on the coefficient of friction, or of adhesion, as it is called. Since the force F does not act along a line of action passing through the centre of gravity of the car, there is a tendency for the car to turn so that its back wheels rise into the air. The inertia of the car introduces an internal force F_1 acting at the centre of gravity in the opposite direction to the force F. The magnitude of the inertia force F_1 is equal to that of the force F. The two forces F and F_1 constitute a couple tending to make the back wheels rise as stated. Since actually the back wheels remain on the ground, an equal and opposite couple must act on the car somewhere so as to balance the overturning couple FF_1.

This righting couple is automatically introduced by the perpendicular force W_1 between the front wheels and the ground increasing by a small amount Q while the force W_2 between the back wheels and the ground decreases by an equal amount Q. The forces $+Q$ and $-Q$ constitute a couple which balances the overturning couple FF_1. The magnitude of the latter is $F \times OG$, so that other things being equal the smaller the height OG the less the overturning couple. The magnitude of the righting couple QQ is $Q \times SS$, so that the greater the wheelbase SS the less the force Q, that is, the less the alteration in the perpendicular forces between the wheels and the ground.

When going down a hill the conditions are changed. From Fig. 32.2 it will be seen that the vertical force W, the weight of the car, can be resolved into two components H_1 and K. The component K is the only part of the weight of the car that produces any perpendicular force between the wheels and the ground, and is, therefore, the only part of the weight giving any adhesion. Thus, on a hill, the adhesion available is necessarily less than on the level. The component H_1, however, tends to make the car run down the hill, and if the car is merely to be kept stationary, a force H equal and opposite to H_1 must be introduced by applying the brakes. The forces H and H_1 constitute an overturning couple, which is balanced by an increase L in the perpendicular force between the front wheels and the ground, and an equal decrease in the rear.

If, instead of being merely held stationary, the car has to be slowed down, then an additional force F must be introduced between the wheels and the ground by applying the brakes harder. An equal inertia force F_1 is then introduced by the deceleration of the car. This inertia force acts at the centre of gravity of the car, and together with the force F constitutes an additional overturning couple, which is balanced between the wheels and the ground. The perpendicular force between the front wheels and the ground is thus increased by an amount $L + Q$, and that between the rear wheels and the ground is decreased by the same amount. Thus, on a hill, the deceleration possible is less than on the level for two reasons. First, the maximum perpendicular force between the wheels and the road is reduced from W to K, and secondly, part of the adhesion is neutralised by the component H_1 and is not available for deceleration.

If the rear wheels only are braked, the conditions are still worse, because the force producing adhesion is still further reduced by the amount $L + Q$.

A little consideration will show that the opposite action occurs when the car is being driven forward. The perpendicular force between the front wheels and the ground is then decreased, and that between the rear wheels and the ground is increased, so that from the point of view of adhesion the rear wheels are a better driving point than the front wheels. This is particularly so when accelerating up a hill.

The extent of this alteration in the weight distribution depends directly upon the magnitude of the deceleration, which, in turn, assuming the brakes are applied until the wheels are about to skid, depends upon the coefficient of

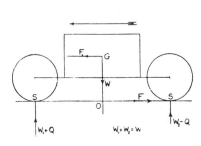

Fig. 32.1

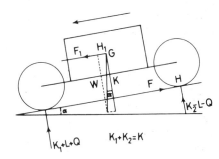

Fig. 32.2

adhesion between the wheels and the road. When that coefficient is low the maximum deceleration is low also, and the weight distribution is altered only slightly. Under these conditions the relative effectiveness of the front and rear wheels is in the ratio (approximately) of the weights carried by these wheels, and if the weight carried by the front wheels is only a small part of the total weight little will be gained by braking them.

The decelerations possible with modern braking systems are, however, high enough to make the braking of all the road wheels desirable and this is a legal requirement in most countries.

32.1   Two functions of brakes

Two distinct demands are made upon the brakes of motor vehicles. First, in emergencies they must bring the vehicle to rest in the shortest possible distance, and secondly, they must enable control of the vehicle to be retained when descending long hills. The first demand calls for brakes which can apply large braking torques to the brake drums, while the second calls for brakes that can dissipate large quantities of heat without large temperature rises. It may be pointed out that the same amount of energy has to be dissipated as heat when a car descends only 400 yards of a 1 : 30 incline, as when the same car is brought to rest from a speed of 35 mph. Thus heat dissipation hardly enters into the braking question when emergency stops are considered, but when descending long hills the problem is almost entirely one of heat dissipation.

32.2   Braking systems

A driving wheel can be braked in two ways: directly, by means of brakes acting on a drum attached to it; or indirectly, through the transmission by a brake acting on a drum on the mainshaft of the gearbox, or on the bevel pinion, or worm, shaft of the final drive. A brake in either of the latter positions, being geared down to the road wheels, can exert a larger braking torque on them than if it acted directly on them. If the final drive ratio is 4 : 1, then the braking torque exerted on each road wheel is twice the braking torque exerted on the brake drum by the brake, that is, the total braking torque is four times the torque on the brake drum. Thus, brakes acting on the engine side of the final drive are much more powerful than those acting on the wheels directly. A transmission brake, however, gives only a single drum to dissipate the heat generated, whereas when acting directly on the road wheels there are two or more drums. Also in many vehicles a transmission brake would be badly placed as regards heat dissipation, but in commercial vehicles it can sometimes be better in this respect than wheel brakes since the latter are generally situated inside the wheels and away from any flow of air. The transmission brake has the advantage that the braking is divided equally between the road wheels by the differential but the torques have to be transmitted through the universal joints and teeth of the final drive and these parts may have to be increased in size if they are not to be overloaded. The transmission brake at the back of the gearbox is fixed relatively to the frame so that its actuation is not affected by movements of the axle due to uneven road surfaces or to changes in the load carried by the vehicle. In vehicles using the de Dion drive or an equivalent, the brakes are sometimes placed at the inner ends of the drive shafts and here again the torques have to be transmitted through universal joints and also through sliding splines which may cause trouble.

In present-day vehicles the wheel brakes are usually operated by a foot pedal and are the ones used on most occasions; they are sometimes referred to as the *service brakes*. The brakes on the rear wheels can generally be operated also by a hand lever and are used chiefly for holding the vehicle when it is parked and are consequently called *parking brakes* but as they can, of course, be used in emergencies they are sometimes called *emergency brakes*.

32.3 Methods of actuating the brakes

Considering manually-operated brakes, the brake pedal or lever may be connected to the actual brake either mechanically, by means of rods or wires, or hydraulically, by means of a fluid in a pipe. Before considering these connections, however, we must deal with the brakes themselves.

32.4 Types of brake

Brakes may be classified into three groups as follows—

(1) Friction brakes.
(2) Fluid brakes.
(3) Electric brakes.

The last two types are, in practice, confined to heavy vehicles and are not used in cars. The principle of the fluid brake is that a chamber has an impeller inside it that is rotated by the motion of the road wheels so that if the chamber is filled with fluid, usually water, a churning action occurs and kinetic energy is converted into heat thereby providing a braking effort. To dissipate the heat the water may be circulated through a radiator.

The construction is somewhat similar to that of a fluid flywheel and the unit is generally placed between the gearbox and the front end of the propeller shaft but it can be incorporated with the gearbox. The chief drawbacks of this type are that it is difficult to control the braking effort precisely and that while it can provide large braking efforts at high vehicle speeds it can supply very little at low speeds and none at all when the road wheels are not rotating. Thus it can be used only to supplement a friction brake and so such devices are often called *retarders* rather than brakes.

The electric brake is, in effect, an electric generator which, being driven by the road wheels, converts kinetic energy into an electric current and thence, by passing the current through a resistance, into heat.

The 'eddy current' brake employs the same principle as the eddy current clutch described in Section 21.21. The rotor is coupled to the road wheels, being often mounted on a shaft that is interposed between the gearbox and the propeller shaft, and the stator is mounted on the frame of the vehicle. The heat generated is dissipated chiefly by convection but this may be augmented by some kind of fan which may be incorporated into the rotor.

This type of brake suffers from the same drawback as the first type of fluid brake, namely, that it cannot provide any effort at zero speed and can be used only to supplement a friction brake. A fairly large number of such brakes are in use at the present time, as retarders, and have been quite successful.

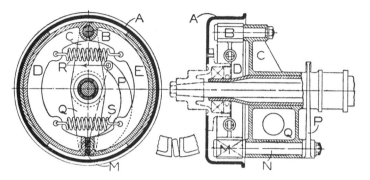

Fig. 32.3 Internal expanding rigid-shoe brake

The vast majority of brakes are friction brakes and these may be subdivided into: (1) drum brakes and (2) disc brakes, according to whether the braked member is a drum or a disc. Drum brakes are still widely used and are invariably expanding brakes in which the brake shoes are brought into contact with the inside of the brake drum by means of an expanding mechanism. External contracting brakes are now used only in epicyclic gearboxes.

The principle of the internal expanding rigid-shoe brake is shown in Fig. 32.3. The brake drum A is fixed to the hub of the road wheel (shown in chain dotted lines) by bolts which pass through its flange. The inner side of the drum is open, and a pin B projects into it. This pin is carried in an arm C which is either integral with, or secured to, the axle casing, a rear wheel brake being shown. The brake shoes D and E are free to pivot on the pin B. They are roughly semicircular, and in between their lower ends is a cam M. The latter is integral with, or is fixed to, a spindle N free to turn in the arm Q of the axle casing. A lever P is fixed to the end of the cam spindle, and when this lever is pulled upon by a rod which is coupled to its end, the cam spindle and cam are turned round slightly, thus moving the ends of the brake shoes apart. The shoes are thus pressed against the inside of the brake drum, and frictional forces act between them, tending to prevent any relative motion. This frictional force thus tends to slow down the drum, but it also tends to make the shoes revolve with the drum. The latter action is prevented by the pin B and the cam M. The pin B is therefore called the *anchorage pin*. The magnitude of the frictional force, multiplied by the radius of the drum, gives the torque tending to stop the drum, that is, the braking torque.

The reaction of this braking torque is the tendency for the shoes to rotate with the drum, so that this reaction is taken by the pin B and cam M, and ultimately by the axle casing and the members which prevent the axle casing from revolving, that is, the torque-reaction system. Most modern car brakes do not have actual pins for the shoe anchorages but, instead, have simple abutments against which the rounded ends of the webs of the shoes bear and are kept in contact by springs, but in lorries a separate anchorage pin is often provided for each shoe, as indicated in Fig. 32.4 which shows a design of the Kirkstall Forge Engineering Company. The anchorage pins are seen at A and B and are carried in the projecting arm C of the brake anchorage bracket. The latter is a force fit on the end G of the axle case, and a key is provided to prevent any rotation. The actuating cam D is now of S shape, which provides a greater

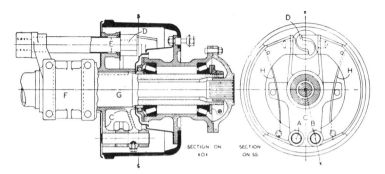

Fig. 32.4

amount of expansion of the shoes and a more constant leverage than is provided by the simple cam shown in Fig. 32.3. The cam D is integral with its shaft and is supported in needle roller bearings, one of which is seen at E. The pull-off springs H are now single-leaf springs, which are easier to remove and replace than coil springs. The seats F, to which the road springs are bolted, are formed integral with the brackets C.

The cam expanding mechanism described above is simple in construction and fairly satisfactory in action but there are others and two are shown in Figs 32.5 and 32.6. The first, Fig. 32.5, is used in heavy lorries and is a variation of the S-cam described above, it is actually a double crank and connecting rod mechanism, it provides a greater movement with less friction than the ordinary cam; when the S-cam is used friction is often reduced by employing rollers at the shoe ends against which the cam surfaces bear. In the second example, Fig. 32.6, a wedge T is used and is pulled inwards towards the centre of the vehicle by the rod R in order to apply the brake. The wedge operates through rollers that reduce friction and forces the plungers or tappets U and V apart. The body W that houses the tappets may be fixed to the backplate of the brake assembly in which case the forces applied by the wedge to the shoes may be unequal or it may be free to slide and then forces will be equalised. Equalisation of the shoe forces can be obtained although the housing is fixed by leaving the wedge free to rock or slide sideways and an example of the latter is shown in Fig. 32.31.

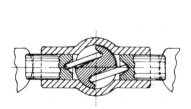

Fig. 32.5 Brake shoe expanding mechanism

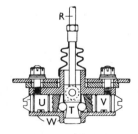

Fig. 32.6 Wedge-type actuator

32.5   Elementary theory of the shoe brake

Consider the simple shoe shown in Fig. 32.7. An actuating force W will give rise to a normal force P between the shoe and the drum (this force is shown as it acts on the shoe) and this normal force will produce a frictional force μP if the drum is rotating as shown. Now the shoe is in equilibrium under the action of the forces shown, together with the forces acting at the pivot, but the latter have no moment about the pivot and consequently the clockwise moments due to the forces P and μP must be balanced by the anti-clockwise moment due to W. Hence we get the relation—

$$W \times L = P \times M + \mu P \times R$$

and hence that—
$$P = \frac{WL}{M + \mu R}$$

Now the braking torque acting on the drum is due entirely to the frictional force μP and is equal to $\mu P \times R$, or, substituting the expression obtained above for P, we get—

$$\text{Brake torque, } T_t = \frac{\mu WLR}{M + \mu R}$$

Considering the show shown in Fig. 32.8, the balance of moments about the pivot gives—

$$WL + \mu QR = QM$$

and hence—
$$Q = \frac{WL}{M - \mu R}$$

so that the expression for the braking torque is—

$$T_1 = \frac{\mu WLR}{M - \mu R}$$

It is now easily seen that T_1 is greater than T_t, the other factors being equal. Let $\mu = 0.4$, $L = 0.15\,\text{m}$, $M = 0.075\,\text{m}$, $R = 0.1\,\text{m}$ and $W = 500\,\text{N}$. Then—

$$T_1 = \frac{0.4 \times 500 \times 0.15 \times 0.1}{0.075 - 0.04} = \frac{3}{0.035} = 86\,\text{Nm}$$

while—

$$T_t = \frac{3}{0.075 + 0.04} = 26\,\text{Nm}$$

Thus T_1 is 3.3 times T_t.

The shoe shown in Fig. 32.7 is called a *trailing shoe* while that shown in Fig. 32.8 is called a *leading shoe*. It should be clear, however, that in a conventional brake the leading shoe will become the trailing one if the direction of rotation of the brake drum is reversed, and *vice versa*.

An actual brake shoe acts in a similar manner to the simple one considered above, the only difference being that the frictional force μP will act at a larger

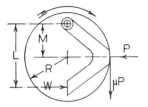

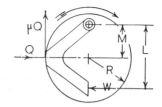

Fig. 32.7   Fig. 32.8

radius than the radius of the brake drum and this will accentuate the difference between the torques developed by the shoes.

In a brake of the type shown in Fig. 32.3, however, the expanding cam will not apply *equal* forces to the shoes but will apply a greater force to the trailing shoe. Taking the data assumed above and supposing that a total actuating force of 1000 N is available then the cam would apply a force of 767 N to the trailing shoe and only 233 N to the leading shoe. The total braking torque would then be 8000 Nm. If, however, the whole 1000 N available for actuation had been applied to the leading shoe alone then the brake torque would have been 17 144 Nm, that is, more than twice as great and this result can be obtained by making both shoes leading shoes and applying 500 N to each.

If the actuating mechanism were of the type that applies equal forces to the shoes then each actuating force would be 500 N and the total brake torque developed by the shoes would be 8571 + 2608 = 11 179 Nm. Thus a floating or equalising actuating mechanism gives an increase in brake torque for a given actuating force but it has the disadvantage that the wear of the leading shoe (assuming the shoes to have linings of the same material) would be 3.29 times that of the trailing shoe. The brake with two leading shoes would not suffer from this drawback and, as has been seen, gives an even greater brake torque. Such brakes are, therefore, widely used, particularly for front wheels. When hydraulic actuation is used it is a simple matter to make both shoes leading ones for the forward direction of rotation; the brake is arranged as shown in Fig. 32.9, two actuating cylinders, connected by a pipe, being used instead of one cylinder. For the reverse direction of rotation both shoes would be trailing shoes and the brake would be rather weak. For this reason it is usual to employ the two-leading-shoe brake in the front wheels only, the rear brakes being the conventional leading and trailing shoe type.

When the brake actuation is mechanical it is not so simple to make both shoes leading ones but a relatively simple mechanism has been developed by Girling and the principle of this is shown in Fig. 32.10. The expanding mechanism does not act directly on the shoe but on one arm of a bell-crank which is freely pivoted on a pin carried by the shoe. The other arm of this bell-crank bears against a fixed anchorage (shown cross-hatched) and the shoe itself can bear on this anchorage and on another one at the top, as shown. It should be clear that, supposing the arms of the bell-cranks to be all equal in length, the force in the strut will be equal to the actuating force W and this force will act on the bell-cranks as shown; also that the bell-crank at the bottom will press on the anchorage with a force W and the anchorage will press back equally on the bell-crank as shown. The resultant force on each bell-crank will thus be a force R as shown and these forces will press the shoe into contact with

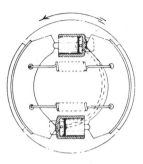

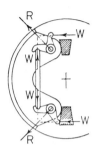

Fig. 32.9   Fig. 32.10

the drum. If the drum is turning clockwise the shoe will now move round clockwise very slightly until it bears on the anchorage at the top and is thus a leading shoe while if the drum is turning anti-clockwise the shoe will turn anti-clockwise and will bear on the anchorage at the bottom, being once again a leading shoe. Thus, by employing two shoes, each with the bell-crank and strut mechanism, a brake which is a two-leading shoe brake for either direction of rotation and which requires only one actuating mechanism is obtained.

32.6 Brake shoe adjustments

In order to take up the wear of the brake linings and to enable the clearance between the shoes and the drum to be adjusted the anchorages on which the shoes pivot are frequently made adjustable so that the shoes can be moved outwards. In the example shown in Fig. 32.11, the brake shoes bear on the ends of tappets carried in a housing that is fixed to the backplate of the brake assembly. These tappets can be forced outwards by screwing in the adjusting wedge, thereby reducing the clearance between the brake shoes and drum. The adjusting wedge is roughly conical but the cone is actually a series of flats; this enables the pull-off springs to lock the adjustment positively.

A design by Lockheed suitable for heavy duty brakes is shown in Fig. 32.12. The housing G is bolted to the backplate H and the tappets A and B have screwed inner members D and E that are prevented from rotating by their engagement with the ends F of the brake shoe webs. The outer members of the tappets have teeth K formed on them so that they can be rotated from the

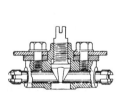

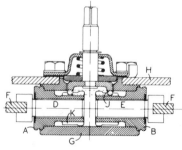

Fig. 32.11 Wedge-shaped
adjustment device

Fig. 32.12 Lockheed heavy-duty brake adjusting
mechanism

outside by means of a simple crown wheel J and thus move the inner portions of the tappets outwards. Other designs of adjusting mechanisms can be seen in Figs 32.14 and 32.15.

An alternative point at which an adjustment can be made is between the actuating mechanism and the ends of the brake shoes.

32.7 A modern rear-wheel brake

The brake shown in Fig. 32.13 is suitable for the rear wheels of a car because it incorporates a handbrake actuation. It is a Girling design. The shoes bear against the flat faces of an abutment carried by the backplate at the bottom and against the ends of the plungers of the hydraulic actuation cylinder at the top. They are held in place by two springs, only the bottom one being shown. Flat strip springs S at the middle of the shoes bear against the webs and hold the shoes against projections (two for each shoe) formed on the backplate.

The handbrake actuation is by means of a bell-crank lever B and a connecting link or strut, made in two parts E and F. The bell-crank lever is pivoted on the upper end of a pillar C and when its long arm is pulled inwards as indicated by the arrow the short arm D applies a force to the right-hand shoe and the reaction of this force moves the upper end of the pillar C to the left, this motion being possible because the pillar is carried at its lower end on a flexible rubber moulding. The motion of the pillar is transferred through the strut EF to the left-hand shoe and so the shoes are applied with equal forces.

The ends of both the parts E and F of the connecting link are flat but the

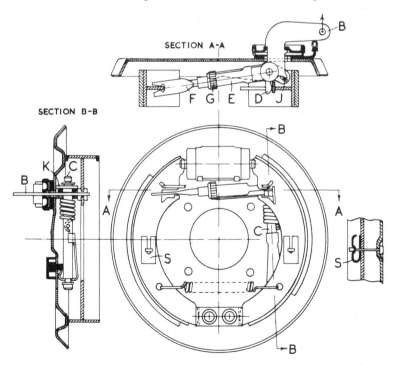

Fig. 32.13 Girling drum brake with auto adjuster

middle parts are cylindrical, E being hollow and F having a screw thread formed on it on which there is a nut G, rotation of which will alter the effective length of the strut. The flat end of the member E pivots on the pillar C and a ratchet lever (shown in dotted line in the upper view and seen below the strut in the main view) is also pivoted there. A slot in a short arm of the ratchet lever is engaged by the end J of the spring that surrounds the pillar so that the spring tends to rotate the ratchet lever in an anti-clockwise direction, this motion is limited by a small shoulder formed on the bell-crank. The long arm of the ratchet lever can engage teeth formed on the outside of the nut G. When the bell-crank is rotated to apply the brake the spring causes the ratchet lever to follow up the motion and if, due to wear of the linings, this is greater than normal the lever will rotate the nut G and take up the wear. When the brake is released, the shoulder on the bell-crank will move the ratchet lever back to its original position. The actuating force applied to the lever B and the force in the strut both tend to pull the pillar up against the backplate and so a roller K is provided to reduce friction when the pillar moves during the brake actuation.

An automatic adjusting mechanism made by the Lockheed Company is shown in Fig. 32.14. The actuating cylinder is single-acting, i.e. it is closed at one end and its piston has an arm A secured to it. This arm carries a pin B that engages a bell-crank lever C which is free to turn on a fixed pin D. The end E of the bell-crank forms a pawl that engages the teeth F of an adjuster sleeve that is free to turn inside the piston but must move axially with it. The tappet G is screwed into the adjuster sleeve but is prevented from rotating by the brake shoe that it engages. When the piston is moved outwards (to the left in the diagram) the pin B rotates the bell-crank and if the movement is greater than normal the pawl E will ride over one or more teeth of the adjuster sleeve F so that when the brake is released and the piston moves back the pin B will rotate the bell-crank and turn the adjuster sleeve so as to take up the wear.

Some designers prefer to use adjusting mechanisms that operate only when the brake is applied and the car is moving backwards and an example of this is shown in Fig. 32.15. The brake is a duo-servo type in which one shoe is used to apply the other through the adjusting turnbuckle whether the brake drum is rotating clockwise or anti-clockwise. The shoes are prevented from rotating very far by a fixed abutment C, against which one of the shoes will abut when the brake is applied. An expanding mechanism operated by the handbrake linkage may also be incorporated between the upper ends of the shoes. For forward motion of the drum, which is assumed to be anti-clockwise, the shoe A is the servo shoe and it will be brought into action by the left-hand piston of the

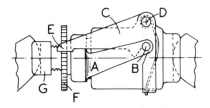

Fig. 32.14 Lockheed auto adjuster

actuating cylinder. The frictional force on the shoe will cause it to rotate slightly with the drum and so acting through the adjusting turnbuckle it will apply the shoe B, causing it to bear against the abutment C at the top. The force Q applied to this shoe will be considerably greater than the force P applied by the piston to the upper end of the servo shoe. For backwards motion the shoes will change their functions and the shoe B will become the servo shoe. The slight rotation that it then gets will cause the lever D to rotate about its pivot E on the shoe because the upper end of the lever is constrained by the wire D which is anchored at its top end. If undue wear has occurred the rotation of the lever D will be sufficient to cause the pawl at its end to rotate the nut of the turnbuckle and so take up the wear.

Another example of a duo-servo brake is shown in Fig. 32.16 (which illustrates the principle), and in Fig. 32.17 which shows the complete brake and all its components. The brake is a Girling design and suitable for heavy lorries. In the diagram, the left-hand side shows the service actuation and the right-hand side the parking actuation but the brake is actually symmetrical. Double-acting hydraulic cylinders at top and bottom provide the service actuation and bring the shoes into contact with the drum when, assuming clockwise rotation, the left-hand shoe will move round until it comes up against the abutment at A, and the right-hand shoe will come up against the adjuster at B. Both are fixed points, A being part of the backplate while the body of the adjuster which carries the tappets B is fixed to the backplate. For anti-clockwise rotation the abutments A and B will be interchanged. The servo action is thus effective for both directions of rotation.

The parking brake actuation is provided by a wedge-type expander which moves the tappets D outwards and through levers C applies forces to the centre of the shoes. The lever C fulcrums on a flat formed on the tappet B and lies in a plane that is slightly offset from the central plane of the shoes. It will be seen from Fig. 32.17 that the wedge of the expander is free to slide in a slot formed in the plunger of the drawlink and so the forces that are applied to the levers C will be equalised. The adjusting mechanism for the tappets B is duplicated and operated by small bevel gears on the same principle as that shown in Fig. 32.12.

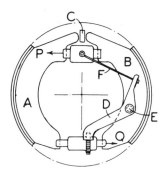

Fig. 32.15 Girling duo-servo
brake with auto adjuster

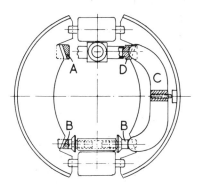

Fig. 32.16 Girling
two-leading-shoe brake

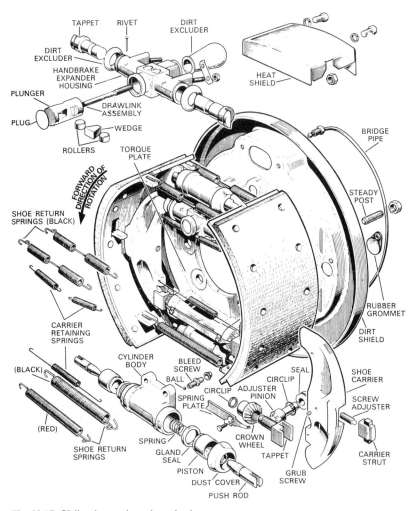

Fig. 32.17 Girling heavy-duty drum brake

32.8 Disc brakes

Brakes using flat discs as the friction surfaces have been used in the past but until the last three decades have not been so successful. They are now almost the commonest type for the front wheels of cars and are often used for the rear wheels and on some light vans. The earliest disc brakes were made on the same lines as a multiple-plate clutch but most present-day designs use a single disc and almost always have sector-shaped friction pads of relatively small area. The great advantages are, first, that despite the small area of the pads compared with the area of lining of a drum brake occupying approximately the same amount of space the rises of temperature are smaller and consequently the linings are less subject to fade and their life is comparable with that of drum brakes. Secondly, the action of the disc brake is unaffected by the occurrence of wear or by expansion due to rises of temperature, both of which are sources of trouble in drum brakes.

Perhaps the simplest construction is the fixed calliper double piston type shown in the diagram Fig. 32.18. The disc A is secured to the wheel hub which rotates on bearings on the axle casing or stub axle depending on whether a rear or a front brake is concerned. A calliper member C is bolted to the member B, which will be referred to as the mounting, and two pistons D and E are carried in cylinders formed in the calliper. These bear on pads consisting of a metal plate to which friction material facings are bonded. The metal backplates fit in recesses in the calliper so that they are prevented from rotating with the disc. The callipers of such brakes usually have to be made in two parts to enable the cylinders to be machined and also must have openings through which the friction pads can be removed for replacement. Their actual form is therefore more complex than as shown in the diagram.

The brake torque provided by a disc brake of this type is given by $T = 2\mu paR$ where μ is the coefficient of friction, p is the fluid pressure, a is the area of one cylinder and R is the distance from the point at which the frictional force acts to the wheel axis, this may be taken to be distance from the wheel axis to the centre of area of the pad. Calculation will show that to obtain the same torque as from a drum brake occupying the same volume the forces applied to the pads of the disc brake must be much higher than those applied to the shoes of the drum brake. This is for two reasons, firstly the radius R is necessarily less than the radius of an equivalent drum brake and secondly there is no servo action in the disc brake because the frictional forces do not help in the application of the brakes as they can do in drum brakes. It is consequently imperative that the axial forces on the disc shall be balanced and this is clearly so in the arrangement shown; it can however be obtained in other ways and these are considered later.

Although the axial forces are balanced there is an unbalanced tangential frictional force on the disc and this will have to be supported by the wheel bearings. By using two callipers placed diametrically opposite each other the brake torque can be doubled and the tangential forces be balanced although, of course, the volume of fluid that has been displaced to apply the brakes will be doubled. This arrangement is seldom used. It is also imperative to provide automatic adjustment for wear in disc brakes and this is usually done very simply. A rubber ring placed near the pad end of the actuating cylinder is carried so that when the piston moves outwards the ring distorts enough to

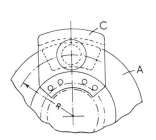
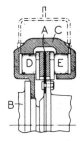

Fig. 32.18

allow the normal clearance to be taken up without any slip occurring between the ring and the piston but if the movement is more than this, slip occurs and when the fluid pressure is released the piston retracts only by the amount of the distortion and thus the normal clearance is restored.

Some methods of balancing the axial forces without using two cylinders are shown in Fig. 32.19. In (*a*) the calliper is carried on two links G pivoted at one end to the calliper and at the other end to the mounting B; it can therefore float sideways so as to equalise the axial forces. The design shown in (*b*) uses a single pivot but has the disadvantage that if the friction facings are initially of uniform thickness they will wear to a wedge shape and there will be a waste of friction material but this can be avoided by making the facings wedge shaped initially. In (*c*), the calliper is allowed to slide along pins fixed in the mounting and in (*d*) the calliper is again carried on a single pivot but this is now placed approximately tangential to the periphery of the disc and in its central plane. If the axis of the pivot is offset from the central plane, then the tangential forces on the pads will have a moment about the axis and the balance of the axial forces will be upset but the magnitude of the effect is not likely to be great. The pads will also wear to a wedge shape when the pivot is offset but again this can be allowed for by making them wedge shaped initially. In all these examples only one actuating cylinder is required and this is one of their chief advantages as it avoids having to put one cylinder inside the road wheel where it cannot always be adequately cooled.

The disc itself can be a simple flat one but this can sometimes lead to trouble because of the stresses that can be set up due to temperature rises; this can be mitigated by using a top-hat shape for the disc – a shape which may sometimes be forced on the designer to enable him to get the working surface into a suitable position. The cooling of the disc can be improved by venting it with radial passages so that a flow of air can be established through it. This is not much done in Europe but is fairly common in America.

A more detailed drawing of a brake in which the calliper is pivoted as in Fig. 32.19(*b*) is given in Fig. 32.20. This is a Girling design which has an unusual indirect method of actuation. The calliper body is in one piece to which one pad is fixed directly, the other pad is carried on the cylindrical outer member assembly E, which slides in a bore in the calliper. It is moved by the piston A acting through the connecting piece D, the lever B and the inner part of the assembly E which is in two parts screwed together. By holding one part and turning the other the effective length of the unit is increased and the wear is

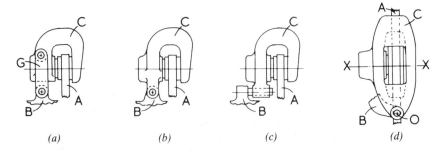

Fig. 32.19 Floating calliper disc brake arrangements

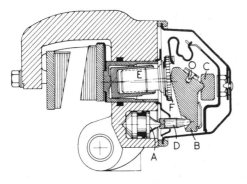
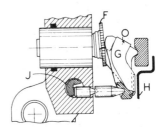

Fig. 32.20 Girling indirect floating calliper disc brake

taken up. This is done by means of ratchet teeth F and a stirrup-shaped member G; when, due to wear, the level B moves farther than usual the stirrup comes up against an abutment H and is caused to rotate about an axis at O, a projecting finger of the stirrup then engages the ratchet teeth and rotates the inner screwed portion of the assembly E. Provision is made for mechanical actuation of the brake so that it can be used as a parking brake; this is shown in the scrap view where J is a rod having a lever at its outer end and shaped at its inner end so that it can also actuate the lever B. This scrap view is a section by a plane parallel to that of the main view but displaced perpendicularly to the paper slightly, the plane of the piston A in the main view is displaced slightly the other way in relation to the assembly E.

Another design by Girling that may be regarded as a sliding calliper type is shown in Fig. 32.21. The cylinder A is open at both ends and is fixed to the mounting K, it is provided with two pistons B and C, the latter bearing directly on the pad D while the former acts on the pad F through the plate E to which the pad is secured. The plate E, which is in effect the calliper, is supported in slots H formed in the cylinder and in a nylon insert J inside the piston B. Springs, not shown, take up any clearance and prevent rattle. The plate is curved as shown in the end view to enable it to clear the disc while contacting the pad F and the piston B at their centres. The chief advantage of the plate over the conventional calliper is that as it is a steel pressing instead of a casting, it can withstand the stresses set up by the forces acting on the pads better and is appreciably lighter.

An American design of a sliding calliper brake, the Delco-Moraine, is shown in Fig. 32.22. The calliper body A is free to slide on chromium-plated bushes B that are fixed to the mounting D by the bolts C. The pad E is fixed to the calliper and the other, F, is supported by, but free to slide on, the bolts C.

A Lockheed single-cylinder disc brake is shown, somewhat simplified, in Fig. 32.23. It is a pivoted calliper type in which the calliper is again a steel plate pressing. It is supported on an angle bracket B that is bolted to the mounting and is pivoted at O on a pin that can be adjusted to a small extent in a direction perpendicular to the disc; this enables initial adjustments to be made during assembly of the vehicle. The actuating cylinder D has slots formed in a flange at its open end and can be slid into an opening formed in the calliper. The cylinder also has small slots at its corners which position its closed end by engaging the calliper plate. A spring clip, not shown, holds the cylinder in

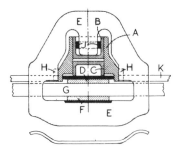

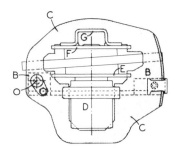

Fig. 32.21 Girling sliding calliper disc brake

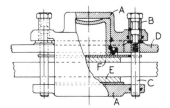

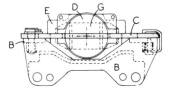

Fig. 32.22 Delco-Moraine sliding calliper
disc brake

Fig. 32.23 Lockheed pivoted
calliper brake

place. The pads are moulded on to backplates E and F and these fit into recesses formed in the calliper so as to transmit the tangential forces. A plate spring and split pins hold the pads in place radially. The pad F is supported by a steel pressing G which has a slot cut in its U-shaped centre part to fit the calliper plate. A spring at the pivot O and a spring-loaded clip at the other end prevent shake or rattle.

In the Tru-Stop brake, shown in Fig. 32.24, equality between the forces acting between the friction linings and the disc is obtained by leaving the link B, which carries the two shoes, free to pivot on the fixed pin C which is carried by a part of the gearbox casing. The shoe A is pivoted on a pin which is fixed in the link B while the shoe D is pivoted on the pin E of a crank or eccentric F, which is carried in bearings in the link B and to which the actuating lever G is fixed. When a pull is applied to the lever G the shoes are drawn together and grip the disc.

32.9 Self-energising disc brakes

The principle of a brake using complete annular discs for the friction elements instead of small pads and which was used on some military vehicles many years ago is shown in Fig. 32.25.

The rotating part is the drum which is made in two parts A and B and the non-rotating part is the inner member F on which the discs D and E are carried. The disc D is prevented from rotating by a key but the other one, E, rotates slightly. Both discs carry friction linings CC. On the inner faces of the discs there are several conical recesses in which steel balls H are placed. The brake is applied by moving the discs apart and in the diagram this is, for simplicity, done by admitting fluid to the annular cylinder J. The frictional

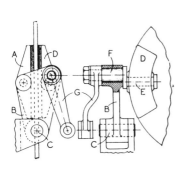

Fig. 32.24

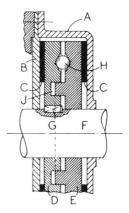

Fig. 32.25

force applied to the disc E as soon as it makes contact with the drum rotates it slightly and causes the balls H to force the discs apart and thus adds to the actuating force. The magnitude of the servo or self-actuating action depends on the included angle of the ball recesses.

The brake torque is given by—

$$T = \frac{2\mu p a R R_b}{R_b - \mu R \tan \alpha}$$

where μ is the coefficient of friction, p the actuating pressure, a the area of the actuating cylinder, R the mean radius of the friction lining, R_b the radius to the centre of the balls and 2α is the included angle of the ball recess. If the angle α is made larger the denominator of the above expression will become smaller and the torque T larger; there is a limiting value for the angle α which makes the denominator zero and the torque theoretically infinite, i.e. the brake locks itself as soon as the discs make contact. This limiting value of α must not be approached too closely because it is dependent on the value of μ and this may vary.

Brakes acting on this principle are used to a considerable extent on agricultural tractors and contractors' vehicles. They may be either 'dry' or 'wet' according as to whether oil is kept away from the friction surfaces or whether they are housed inside a gearbox or axle casing and are exposed to oil. The constructional arrangement of these brakes differs from that described above and is shown in Fig. 32.26.

The rotating members are now the discs A,A which are splined to the shaft B to which the brake torque is to be applied and they are pressed outwards against the fixed casing F to apply the brake. This is done by pulling on the rod H thereby causing the links J and K to pull on the lugs L and M and cause the discs to rotate slightly in opposite directions and bring into action the self-energising forces. Depending therefore on the direction of rotation of the shaft B one or other of the discs will become anchored by its lug P or Q coming up against a stop on the casing. The discs are centred by means of projections G which are in contact with machined surfaces inside the casing. The

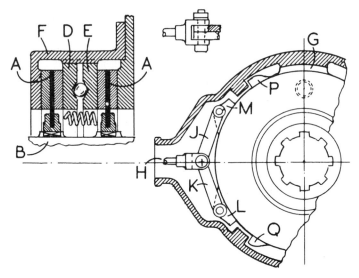

Fig. 32.26 Girling self-energising disc brake

construction of the wet form of this clutch is similar but the friction linings are of course of different material and, because of the lower coefficient of friction obtainable, the single discs of the dry form are replaced by packs of several plates as in a multiple-plate clutch.

32.10 Brake linkages

Hydraulic actuation is now almost universal for the service brakes of vehicles but linkages consisting of levers, rods and wires are often required for the operation of the parking brakes and these may conveniently be considered first. A simple linkage for applying the two rear brakes of a vehicle is shown in Fig. 32.27; it is called an *uncompensated linkage* because there is nothing to ensure that equal forces will be applied to the two brakes. This should be evident if the effect of lengthening one of the rods H or K is considered. In the linkage partly shown in Fig. 32.28, however, the beam C (which is pulled on at its centre by the hand lever) ensures the equality of the forces transmitted by the rods D and E to the brakes and the linkage is consequently called a *balanced* or *compensated* linkage. When a balance beam is employed however it is usually made quite short and may sometimes be arranged to lie in a vertical plane, the rod from one going direct to one brake while the other end is coupled by a cross-shaft, with levers at both ends, to the other brake.

When the application of the brakes is effected by rods that pull inwards towards the centre of the vehicle the compensation between the two sides is most easily effected by one of the arrangements shown in Fig. 32.29. The principle of both of these is that the levers L are free to move sideways and thus enable the forces P and Q to balance each other. The freedom is obtained by pivoting the levers on members S that can rotate about the axes XX. In cars, the compensation is sometimes obtained by using a flexible wire that passes round a pulley and has its ends connected to the brake actuating levers. The pulley is itself pulled on at its centre by a rod coupled to the hand lever.

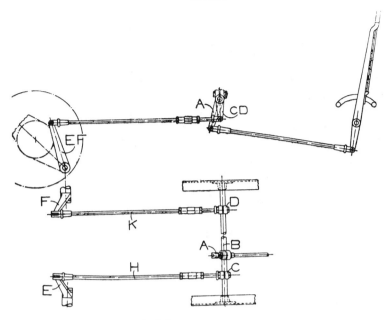

Fig. 32.27 Uncompensated brake linkage

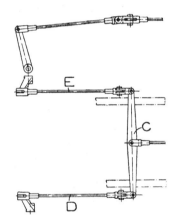

Fig. 32.28 Simple type of balanced linkage

32.11   Leverage and adjustment of the brake linkage

The forces that have to be applied to the brake shoes in order to produce the maximum deceleration of the vehicle in an emergency stop are very large, approximating to the weight of the vehicle, and to enable the driver to produce these forces with an effort which cannot exceed about 700 N and which is normally kept down to about one-third of that amount, the brake linkage has to provide a considerable leverage. The leverage that can be provided, however, is strictly limited by several factors. First, a definite clearance has usually to be maintained between the brake shoes and drum; supposing this to be only 0.25 mm then, with a leverage between the shoes and brake pedal of

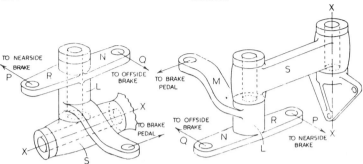

Fig. 32.29

100 to 1, 25 mm of pedal travel would be required merely to take up the clearance. Secondly, the brake linkage cannot be made rigid and so when the brake pedal force is applied the parts of the linkage stretch and give slightly and to take up this stretch may use up another 25 mm of pedal travel.

Now the total travel that can be conveniently accommodated is limited to about 100 mm and so only 50 mm is available to take up the wear of the linings and this corresponds, with the leverage assumed above, to only 0.05 mm. Thus it is important to keep the leverage provided down to the lowest value consistent with reasonable pedal pressures and to keep the stretch of the brake linkage (including deformation of the brake drums themselves) as low as possible. But even when this has been done the brakes will still have to be adjusted at intervals. The best place for this adjustment is as close to the brake shoes as possible and in cars nowadays the adjustment is always either at the anchorage or between the actuating mechanism and the shoes.

32.12 Hydraulic systems

The general arrangement of a hydraulic system is as follows.

The foot pedal actuates the piston in a master cylinder and forces fluid along a piping system to operating cylinders situated in the wheel brakes. The pistons of the operating cylinders are forced out and thus operate the brake shoes. Since all the operating cylinders are connected to the one master cylinder, it follows that the pressure in all the cylinders is the same, and hence compensation is automatically obtained. It is important that all air should be eliminated from the piping system and cylinders, because if air is present then, when the piston of the master cylinder is depressed, instead of fluid being forced along the pipes into the operating cylinders the air in the piping will be compressed, and since air is readily compressible the pressure in the system may not rise sufficiently to actuate the brakes. Of course, if enough travel could be given to the master piston the presence of air would not matter, because the air could be compressed sufficiently to raise the pressure until the brakes were operated; it is not possible to do this, however. Arrangements are therefore made to keep air out of the system.

A Lockheed master cylinder is shown in Fig. 32.30. It has an integral sheet metal reservoir that should always contain a reserve of fluid in order to keep the cylinder replenished and thus prevent air from getting into the system. The piston is sealed by a cup A and a thin, slightly corrugated washer is interposed

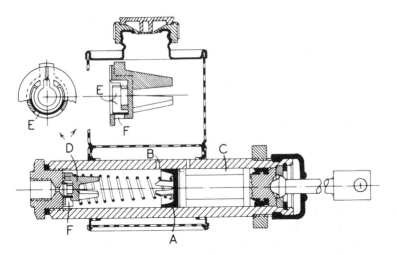

Fig. 32.30 Lockheed master cylinder

between the cup and the piston. In the 'off' position the hole B is uncovered and so fluid can flow from the reservoir to make good any losses that may have occurred. A second hole keeps the annular space C full of fluid and helps in keeping air out of the system. The valve assembly D is held against the end of the cylinder by the spring and it incorporates a second valve consisting of a flexible arcual strip E of copper alloy that will lift inwards when a pressure is set up in the cylinder.

When the brakes are applied, therefore, the fluid flows through the opening F into the piping system but, in order to return when the brakes are released, the whole valve D has to be lifted. This requires a difference in pressure across the valve to overcome the force of the spring and so, when the valve closes, a low pressure remains in the piping system. Such an arrangement is called a *trap valve* and it helps to prevent a sudden reduction of the pressure in the piping system if the brake pedal returns at a faster rate than the brake actuating pistons in the brakes.

32.13 Operating cylinders

Various types of operating cylinder are used according to circumstances; one type is open at both ends and is fitted with two pistons which bear, either directly or through simple struts, on the brake shoes which are operated when fluid is forced in between the pistons. This type must clearly be situated inside the brake drum and this is sometimes a drawback since the space available is very limited and also the temperature may rise considerably and thus lead to vaporisation of the brake fluid.

A second type of operating cylinder is closed at one end and has a single piston; this type may be placed either inside or outside the brake drum and may be used to operate a single brake shoe or a pair of shoes. For the latter purpose the operating cylinder is sometimes pivoted to one of the shoes and has its piston pivoted to the other shoe and is connected to the piping system by a flexible pipe. Alternatively the cylinder may be carried by the backplate of the brake but be free to slide, parallel to the cylinder axis, on that backplate; the piston then bears on one brake shoe and the cylinder on the other shoe. When only one shoe has to be operated the cylinder will be fixed to the backplate and the piston will bear on the shoe. A 'bleeder' screw is usually incorporated in the operating cylinders so that the piping system may be bled until it is free of air. The system is bled as follows.

A piece of rubber pipe is connected to the nipple of the bleeder screw and its end is kept submerged below the level of some brake fluid contained in a glass jar. The reservoir of the master cylinder is filled with fluid and is kept full during the bleeding operation. The bleeder screw is then loosed a turn or two and the brake pedal is pushed down and allowed to come *right* up several times so that fluid is pumped through the piping system and out through the bleeder screw into the glass jar and pumping is continued until the fluid coming out is seen to be free of air. The bleeder screw is then tightened. Each bleeder screw throughout the system is treated in this way.

A single-acting Lockheed operating cylinder that incorporates provision for the mechanical operation of the brake shoes is shown in Fig. 32.31. The body A is carried by the backplate B but is free to slide parallel to the cylinder axis. The piston is made in two parts C and D and bears, through a Micram adjuster, on the leading shoe of the brake. The cylinder body bears on the trailing shoe at E. When fluid enters the cylinder a force is exerted on the piston and thus on the leading shoe; the reaction of this force acts on the cylinder body and is transmitted to the trailing shoe, because, as mentioned above, the body is free to slide on the backplate. Mechanical operation is through the lever F which is pivoted on the pin G carried by the cylinder body. When the lever is pulled inwards (towards the centre of the car or upwards in the figure) it applies a force to the piston member D and thus to the leading shoe. The reaction of this force acts on the pin G and is transmitted by the cylinder body to the trailing shoe. A rubber filling piece is provided in the cylinder in order to reduce the volume of fluid enclosed inside the brake drum where it is subject to any temperature rise that may occur.

Operating cylinders that are situated outside the brake drums are less likely to be affected by temperature rises and so are used to a considerable extent on commercial vehicles. A Lockheed design is shown in Fig. 32.32. The piston is in two parts A and B, the former only being moved when the actuation is mechanical while, when the actuation is hydraulic, the actuating lever itself is not moved. The brake shoes are expanded by the wedge-shaped end of the strut D acting on the plungers or tappets E. These are carried in the housing G and bear on the brake shoes. Rollers are interposed between the wedge and the

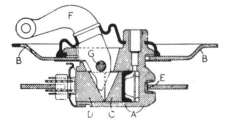

Fig. 32.31 Lockheed single-acting operating cylinder

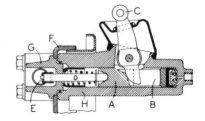

Fig. 32.32 Lockheed external actuating cylinder

plungers and these are kept in place by a cage carried on the end of the wedge. The whole assembly is clamped to the backplate F of the brake between the body H and the housing G.

32.14 Divided and dual brake systems

In most countries it is now obligatory to design the brake system in a manner such that a failure in any part of it cannot cause total loss of braking – usually any two of the four wheel brakes must remain fully effective. Duplication of the complete system would be both costly and unnecessary, so vehicle manufacturers generally install a double master cylinder, as in Fig. 32.33, which has two pistons, each serving a different part of the brake system. The system itself, therefore, is also divided into two independent sections.

With such an arrangement, it would be possible for a failure of part of the system to go unnoticed by the driver. Consequently, the regulations in most countries stipulate that a failure in any of the hydraulic control circuits shall be indicated on the dash fascia. One method of doing this is to apply the pressure from each circuit to opposite ends of a plunger. The plunger has a peripheral groove around its centre in which, provided the pressures on its ends are equal, is seated a spring-loaded pushrod. If, however, the plunger is moved in either direction by differential pressures on its ends, the pushrod is forced out of the groove, closing a switch and thus illuminating the warning lamp.

The simplest way to divide the hydraulic control circuit is to take the output from one part of the master cylinder to the front wheels and the other to the rear ones. However, for cars, even ignoring the effects of weight transfer, if the front–rear axle load ratio is 70 : 30 and the coefficient of friction between road and tyres is 0.8, the maximum efficiency of the rear wheel brakes acting alone cannot be more than about 25% and therefore is inadequate for emergency braking, which is when most sudden failures occur. Should it be the rear circuit that has failed, the residual braking at the front would give about 55% efficiency – the total (front + rear) of course corresponding to the coefficient of

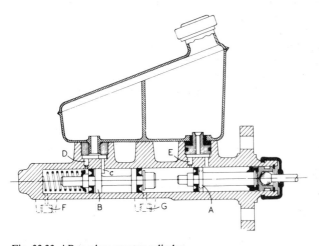

Fig. 32.33 AP tandem master cylinder

friction of 0.8. Such a system may, however, be used on heavy commercial vehicles which, if unladen, still have sufficient power in the front brakes to stop satisfactorily and, if laden, have a suitable front–rear load distribution for either the front or rear brakes to be effective.

A further disadvantage of such a system on cars is the possibility of loss of control as a result of locking of the two wheels being braked. Locked front wheels result in the vehicle continuing in a straight line, even if the steering wheel is turned. The even more dangerous two-rear-wheels-locked condition renders the vehicle unstable, almost invariably causing it to spin. This latter phenomenon occurs because of the difference between the static coefficient of friction between, at the front, the rolling wheels and the ground and, at the rear, the significantly lower coefficient of sliding friction between the locked wheels and the ground, which means that unless the vehicle is braked extremely accurately in a straight line and the road is level and not cambered, the lateral component of the forces on the vehicle will meet with much less resistance at the rear wheels than at the front ones, so the rear end will swing round.

Connection of one part of the master cylinder output to the two wheels on the left and the other to those on the right would be dangerous because of severe pulling of the vehicle to one side in the event of a failure. On the other hand, a diagonal split – each half serving a front wheel on one side and a rear wheel on the other – is frequently used. However, this gives only 40% braking efficiency and, moreover, is unsuitable for vehicles having a significant offset between the centre of the front wheel contact with the ground and the kingpin axis projected to the ground. This is because the couple due to this offset when only one front brake is in operation will not be reacted by an equal and opposite couple on the other wheel so, again, the steering could pull suddenly and unexpectedly to one side on application of the brakes. Such a system, sometimes termed an *X-split*, has two advantages though: premature wheel locking is ruled out because the normal criteria for front-rear wheel braking ratios still apply; and, secondly, there will always be one freely rolling wheel on each side for steering with and for reacting any tendency for the rear end to swing.

There is also a *Y-split*, in which the two front brakes are served separately, but both rear brakes are controlled through a common pipeline, though independently of the front brakes. This is suitable for cars with little or no steering offset, as described above. Even so, in the event of a failure of the front brake on one side, the braking of that on the other side can have a steering effect.

The so-called L-split, in which each part of the divided circuit serves the two front and a different rear brake, has certain limitations. Application of extra pressure to make up for the loss of braking on one rear wheel is impracticable because it would cause locking of that rear wheel and therefore further loss of efficiency. Consequently, the maximum obtainable braking efficiency of one circuit is still only about 40%.

By far the best system would appear to be what is termed the *IH split* – one circuit serving all four wheels and therefore, if viewed in plan, can be likened to the letter H, while the other serves the front brakes only, and therefore can be likened to the letter I. With such an arrangement the circuit serving only the front wheels may be connected to large-diameter cylinders in the disc brake calliper, while that serving both front and rear brakes is connected to smaller diameter cylinders. The outcome is that, by increasing the pedal pressure, a braking efficiency of about 55% can be obtained with the I circuit alone, while that with the H circuit alone a good braking performance – about 45% – is still obtainable even though greatly increased pressure may not be practicable for fear of locking the rear wheels.

In all the systems two master cylinders are required and these are sometimes physically separate, their pistons being actuated by the brake pedal through a balance beam that equalises the forces applied to them. The motion of the balance beam is limited and so if one system fails the other can be applied although the pedal travel has to be increased to do this. The use of tandem cylinders became the most widely used arrangement however and a typical example made by Automotive Products Ltd is illustrated in Fig. 32.33. The two pistons A and B work in a single bore in the body casting, the front one A being operated directly by the brake pedal pushrod while the rear one B floats between the front one and the end of the cylinder.

In the 'off' position shown, the space between the two pistons and that between the rear piston and the end of the cylinder are open to the two parts of the reservoir via the holes D and E and this keeps these spaces and the piping systems connected to them full at all times. In normal action the brake pedal moves the front piston and as pressure is built up in front of it, so the rear piston is moved along and thus equal pressures are built up in both systems. If the rear piping system should fail, the rear piston would move along without generating any pressure but as soon as it came into contact with the end of the cylinder pressure would start to build up between the pistons and the other system would function normally except that the pedal travel would have to be greater. If the piping system connected to the space between the pistons should fail, then the front would move along until it contacted the rear one and the rear system would then function normally. In systems of this kind (including those employing separate master cylinders and balance beams) attention must be paid to the adjustment of the brakes themselves in order to ensure that the available stroke of either piston is not used up before the brakes it operates are applied.

Chapter 33

Servo- and power-operated, and regenerative braking systems

As vehicle weight increases then the force that must be exerted on the pedal of a simple brake system to produce the maximum deceleration permitted by the road conditions also goes up, and when the weight exceeds two to three tons the force required may become greater than a man can exert. The driver must then be given some assistance in applying the brakes. This can be done in two ways: (a) by using a servo mechanism which adds to the driver's effort although that effort remains a considerable part of the total effort applied to the brakes, or (b) by using power operation in which case the effort of the driver is a controlling effort only and is not transmitted to the brakes at all.

Servo systems are usually lower in cost than power-operated systems and can frequently be fitted as an addition to a vehicle having an ordinary brake system. They are widely used on vehicles coming within the medium weight range, say, 2 to 6 ton, whereas power operation is used for heavy vehicles where the weight exceeds six tons or so and for vehicles used with trailers.

There are two essential features of both servo and power brake systems: (1) the time-lag of the system, that is the time interval between the moment when the brake pedal is depressed and the moment when the brakes come on, must be very small and (2) the system must be such that the driver can judge the intensity of application of the brakes fairly accurately. The second feature usually requires that the force applying the brakes shall be closely proportional to the force exerted by the driver on the brake pedal.

The source of the additional effort supplied by a servo system may be (a) the momentum of the vehicle itself, (b) vacuum in a reservoir, obtained by connecting the reservoir through a non-return valve either to the induction manifold of the engine or to a separately driven exhauster, (c) oil under pressure supplied by a pump driven by the engine or some part of the transmission system, or (d) air under pressure supplied by an air compressor driven by the engine.

The first servos to be introduced were purely mechanical in action but had to be very carefully designed and made in order that they should function properly, and this made them costly to manufacture. They became obsolete when cheaper vacuum-operated mechanisms were introduced.

33.1   Vacuum brake operation

Vacuum may be used in two ways in the application of brakes. In each method a cylinder fitted with a piston, or diaphragm, is used and the piston is coupled to the mechanical brake linkage.

In one method the cylinder is permanently open to the atmosphere at one end and the other end also when brakes are off. The brakes are applied by exhausting the latter end of the cylinder to the desired degree thus setting up a force on the piston equal to the pressure difference on its two sides multiplied by its area. In the other system the cylinder is closed at both ends and when the brakes are off the same degree of vacuum exists in both ends. To apply the brakes one end is open to the atmosphere so that the pressure in that end rises and gives a force on the piston which applies the brakes.

The second system is commonly referred to as a *suspended vacuum system*. Its chief advantages over the first system are that it is slightly more rapid in action and that there is a smaller amount of air that has to be exhausted either by the engine through the induction manifold, or by the exhauster, if a separate one is used. Both of these advantages arise from the fact that under the second system vacuum has to be destroyed in a volume equal to the piston area multiplied merely by the piston travel necessary to apply the brakes; whereas in the first system the volume that has to be filled, and subsequently exhausted, is equal to the piston area multiplied by the maximum piston travel possible, plus any clearance which may then remain between the piston and the end of the cylinder. By increasing this clearance volume in a suspended vacuum system, the servo cylinder can be made to function as a vacuum tank and a separate tank can therefore be dispensed with. This is done in many light commercial vehicles.

33.2   Clayton Dewandre master servo unit

The first successful vacuum-operated servo was made by the Clayton Dewandre Company of Lincoln, and as it embodied the basic principle of present-day designs and because large numbers of them are in present-day use it has been thought desirable to retain a description of one. This is shown in Figs 33.1 and 33.2. It consists of a drawn-steel cylinder A fitted with a piston B sealed by a cup-leather C. The piston is connected by a link E to the lower end R of a balance lever F, the upper end of which bears against the valve G. In the position shown, the right-hand end of this valve is clear of the face of the valve J so that the port H through the centre of the valve G opens the passage K and the cylinder A to the atmosphere. The balance lever F is pivoted on the pin O, which is carried by the support levers MM. The latter are pivoted on the fixed pin N. The brake pedal is coupled by the rod D to the balance lever at the point Q, but the connecting pivot pin S is hollow and there is ample clearance between it and the pin Q. The latter serves to couple the support levers MM to the brake linkage through the rod P. The arrangement is shown in perspective in Fig. 33.2.

When the brake pedal is depressed the balance lever is rotated clockwise about the pivot pin O, and its upper end pushes the valve G to the right. This seats on the valve J and thus closes the vent H to the atmosphere. It then

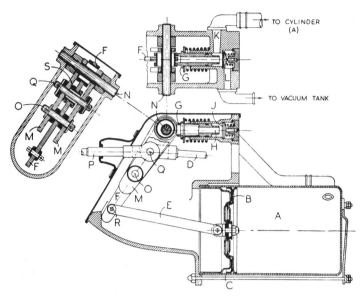

Fig. 33.1

pushes the valve J off its seat, so that the vacuum tank is put into communication with the cylinder C via the passages L and K. The pressure in the cylinder A starts to fall so that a force X is exerted, which tends to pull the lower end of the balance lever to the right. When this force reaches a value sufficient to balance the force applied by the rod D, the balance lever will be rotated slightly anti-clockwise, and this will bring the valve G to the 'lap' or 'brake applied' position, the valve J being seated but the vent H being still closed. In this position the cylinder A is isolated. The force X is now balancing the force Y applied by the brake pedal, and the relation of these forces is determined by the ratio of the lever arms of the balance lever. Thus—

$$X = RO = Y = QO \quad \text{so that} \quad X = Y = \frac{QO}{RO}$$

The resultant of the forces X and Y applied to the balance lever, equal approximately to $Y[1 + (QO/RO)]$, is applied to the pin O and through the support levers and the rod P to the brake linkage. As the brakes come on, the support levers may rotate slightly, but this will not upset the balance of forces since in the lap position the upper end of the balance lever is concentric with the pivot N, about which the movement will be occurring.

The force transmitted to the brake linkage is thus seen to be proportional to the force applied by the driver to the brake pedal, and the degree to which the latter will be augmented by the servo is determined by the ratio of the lever arms of the balance lever.

If for any reason the vacuum formed in the cylinder A is insufficient, the brakes may be applied purely by the force Y, the action being as follows. The depression of the brake lever rotates the balance lever clockwise about the pin O until the clearance between the eye of the balance lever and the pin N is

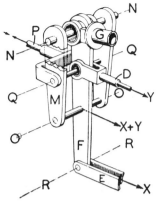

Fig. 33.2

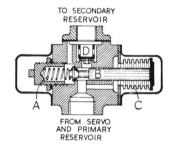

Fig. 33.3

taken up. The pin N then becomes a fulcrum for the balance lever and the force *Y* can be transmitted through the pin O and the levers MM to the rod P. Since the rods D and P are approximately in line, the force in the rod P will then be equal to the force in the rod D.

When hydraulic actuation of the vehicle brakes is adopted, the master cylinder of the system is bolted up to the vacuum servo unit, but the latter is rearranged slightly. The rod D is now coupled to the support levers and operates the piston of the master cylinder, while the rod P is pivoted to the balance lever and connected to the brake pedal. The forces in the rods D and P will now be compressive forces instead of tension forces.

33.3   Reservoirs

The reservoirs used with vacuum brake systems are usually simple cylindrical (sometimes spherical) vessels fitted with a non-return valve in the pipes connecting them with the induction manifold or the exhauster, as the case may be. A drain plug or cock is usually fitted at the lowest point. To give a sufficient reserve to enable several brake applications to be made even though the engine has stopped requires a large reservoir; but a large reservoir takes a long time to evacuate and this is a drawback. To overcome this difficulty dual reservoirs are sometimes used. These comprise a small tank which can be quickly evacuated and a large one which provides a suitable reserve but which does not begin to be evacuated until the smaller one has been completely evacuated. The diverter valve which governs this action is shown in Fig. 33.3. Until the small or primary tank is exhausted the spring A keeps the valve B on its seat and the large or secondary tank is blanked off. As the small tank is evacuated and the pressure in its falls a force acting to the left will be set up on the bellows C, which is subject to atmospheric pressure externally and to the pressure in the small tank internally. When the pressure in the small tank is reduced to the designed value the valve B will be unseated and the large tank will begin to be evacuated. The large tank supplements the small one when necessary because if the pressure in the small tank rises above that in the large tank the valve D will open and put the two tanks into communication.

33.4   Bendix Hydrovac

The Bendix Hydrovac was one of the earliest suspended vacuum systems and is still in extensive use, it combines a vacuum servo with the master cylinder of a hydraulic brake system. In Fig. 33.4, the size of the valve gear has been exaggerated in order to make the details clear. When the brake pedal is in the off position the valve A will be off its seat and so both sides of the piston B will be subject to vacuum and there will be no force acting on the piston. The plunger C, which is integral with the piston rod, will be off its seating in the piston H so that the hydraulic piping system leading to the brakes will be open to the master cylinder D and thus to the brake fluid reservoir. When the brake pedal is depressed a pressure will be set up in the master cylinder and this pressure will act on the plunger E of the vacuum valve. The diaphragm F will be pushed up until it seats on the valve A and then the valve G will be pushed off its seat. The left-hand end of the vacuum cylinder will thus be opened to atmosphere and the pressure in it will rise so that a force will act on the piston B to push it to the right. The movement of the piston B will first seat the plunger C on to the piston H and then will force the latter into its cylinder so as to apply the brakes. The force due to the oil pressure acting on the left-hand side of the piston H will be added to the force due to the piston B so that the system is a servo system. The pressure in the left-hand end of the vacuum cylinder also acts on the top of the diaphragm F and when the pressure reaches a value sufficient to balance the hydraulic force applied to the plunger E the diaphragm will be depressed slightly and the valve G will be seated again. The left-hand end of the vacuum cylinder will then be isolated and the pressure in it will be proportional to the hydraulic pressure acting on the plunger E, that is, proportional to the brake pedal force. If for any reason no vacuum is available then the pressure set up in the master cylinder D will be transmitted direct to the brake cylinders because the plunger C will remain of its seat.

33.5   Direct-acting vacuum servos

In many cars it is convenient to operate the master cylinder piston directly from the brake pedal but the use of servo assistance is required; for such conditions *direct-acting servos* are used.

The design by Automotive Products shown in Fig. 33.5 consists of a pressed steel body or vacuum chamber to the front of which the master cylinder is bolted and which is closed at the rear by a pressed steel cover between which and the chamber a rubber diaphragm is secured. At its centre the diaphragm is sealed to the stem of a piston E that houses the control valve. There is a sliding seal between the cover and the stem of the piston which is bored out to take a small piston H that is in effect integral with the pushrod of the master cylinder and behind this piston is a rubber reaction disc K. A spring holds the servo piston in the 'off' position. The left-hand side of the vacuum chamber is connected to the induction manifold of the engine by a pipe containing a non-return valve so that a vacuum always exists in that side and in the 'off' position the same vacuum will also exist in the right-hand side because the two sides are connected via the passage F, the gap B and the passage G.

When the brake pedal is depressed the valve C is moved to the left and this permits the valve D, which is a reinforced rubber moulding, to close the gap B,

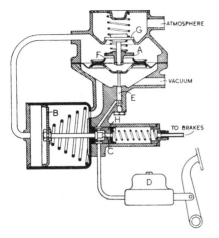

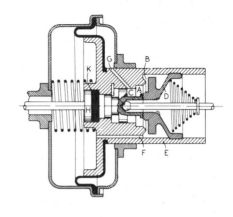

Fig. 33.4   Fig. 33.5 AP direct-acting vacuum servo

thereby closing the connection between the two sides of the vacuum chamber. Further motion of the valve C then opens a gap at A and puts the right-hand end of the chamber into communication with the atmosphere so that the pressure rises and sets up a force to move the servo piston to the left. This force causes the rubber reaction disc K to be compressed and the rubber to deform until its fills up the space on its right and makes contact with the left-hand end of the valve C. The pressure set up in the rubber and therefore the forces acting on the valve C and the piston H will increase until the valve is pushed to the right to the equilibrium position where the gaps at A and B are both closed. In this condition the force acting on the piston H and thus on the master cylinder piston will be nearly proportional to the force applied to the valve C by the brake pedal.

When the brake is released the valve B is pushed off its seat by the valve C and the force exerted by the rubber K and the system returns to the 'off' position. The relative sizes of the parts in the figure have been modified in order to make the construction clear.

33.6 Power-operated brakes

These may employ either oil or air as the working medium and in what follows the term *fluid* may often be used since it covers both media and the systems using them have much in common.

A fluid system that has been widely used is the Lockheed design shown in Fig. 33.6. The pump supplies fluid to the cut-out valve and in normal operation when the pump is running and the accumulators are charged up this fluid flows from A to B and thence from C to D through the driver's foot valve and so back to the supply tank. The pressure in the pipeline BC is then only a little above atmospheric and so the pump is running light. When the foot valve is depressed the flow from C to D is restricted and a pressure builds up in the pipeline BC and as this is connected to the actuating cylinders of the front brakes these are applied. The pressure that is set up will be proportional to the

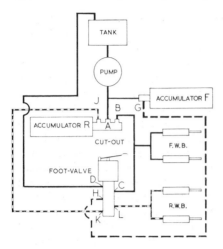

Fig. 33.6

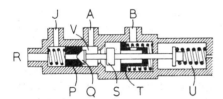

Fig. 33.7

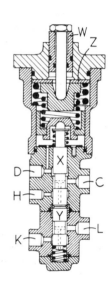

Fig. 33.8

force applied by the driver to the brake pedal. If the pump is not running then the front brakes will be operated by the accumulator F which is coupled by the pipe GH to the foot valve and the flow will then be from H to C. The rear brakes are normally operated direct from the accumulator R via the pipe JK and so through the foot valve to L. Again, the pressure set up in the brake-actuating cylinders will be proportional to the force applied by the driver.

The cut-out valve regulates the pressure in the accumulators by interrupting the supply when the upper limit is reached and restoring it when the lower limit is reached. In Fig. 33.7 the valve is shown in the position it occupies when the accumulator pressure is above the lower limit and is able to keep the sliding seat P seated on the end Q of the valve QS. The pump is assumed to be running and it will keep the valve S off its seat in the casing and will also press back the sliding member T so that fluid can flow from A to B. As the pressure in the accumulator falls so the valve QS and the sliding seat P (still in contact) will move to the left and when the lower limit for the accumulator pressure is reached the valve S will seat on the casing. The pressure in the space V will then force the sliding seat P away from the valve Q and fluid will flow into the accumulator. The pressure in the pipeline connecting the pump and the space V also acts on the non-return valve of the accumulator F, Fig. 33.6, and so if the accumulator pressure is low it also will be charged up.

The construction of the driver's foot valve is shown in Fig. 33.8. Its operation is as follows: in the position shown there is a passage from C to D and the fluid from the pump via the cut-out valve will flow without any great pressure drop. But when the brake pedal is depressed the valve spool X will be depressed through the inner spring and this will reduce the area for the flow of fluid from C to D and so the pressure in the space C and the reduced part of the valve X will rise and, as explained above, the increased pressure will apply the front brakes. The pressure in the space C is transmitted through the holes (shown dotted) in the lower end of the valve X and will act in the space between X and Y. This pressure will tend to push the valve X upwards and that valve will consequently reach an equilibrium position in which the fluid pressure acting on its bottom end balances the force exerted on its upper end by the inner spring and thus by the brake pedal. Thus the pressure in the brake-actuating cylinders is made proportional to the force applied to the brake pedal. Returning now to the lower spool Y, this has been pushed down a little by the valve X but not enough to open the passage from K to L. However, as the pressure in the space between X and Y increases the valve Y will be pushed further down and the passage from K to L will be opened. Fluid will then flow from the accumulator R, Fig. 33.6, to K and L and thus to the rear-brake-actuating cylinders. The pressure set up in the latter will be transmitted through the holes drilled in the lower part of the valve Y and will act in the space at the bottom of that valve. The valve will thus reach an equilibrium position in which the fluid pressure on the bottom end balances that on the top end and as the latter is proportional to the pedal force so the pressure in the rear-brake-actuating cylinders is also kept proportional to the pedal force. Clearly, the front brakes are independent of the rear ones and vice versa.

The purpose of the outer spring at the top of the assembly is to limit the normal vehicle deceleration to a value which will not be objectionable to passengers while still permitting greater decelerations in emergencies. The compression of the inner spring is normally limited by the plunger W coming into contact with the plate Z which is held up by the precompressed outer spring. But extra force applied by the driver in an emergency will compress the outer spring further and thus increase the force applied to the valve spool X and thus will increase the brake-actuating pressures.

33.7   A dual power brake system

The Clayton Dewandre system shown in Fig. 33.9 represents a modern design in which great importance is attached to the elimination of total failure of the system and all the components are duplicated so that one set is available even if the other fails.

A dual pump B driven from the engine has its two cylinders fed from the separate compartments of the reservoir A and delivers fluid through separate pipelines to two accumulators CC and thence to a dual foot-operated valve F that is described later on. Low-pressure warning light switches D and filters are included in the pipelines. A branch pipeline that contains a non-return valve E brings fluid to the hand-brake control valve G which is used to operate the spring brake unit H; this also is described later on.

The dual foot valve is shown in the diagram Fig. 33.10, only one valve being shown, the other being identical. The two valve plungers are depressed by the

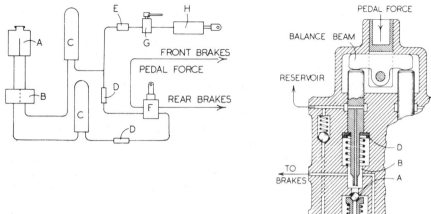

A	Dual reservoir	E	Non-return valve
B	Dual pump	F	Dual foot valve
C	Accumulator	G	Hand valve
D	LP switch	H	Spring brake unit

Fig. 33.9 Clayton Dewandre dual hydraulic brake system

Fig. 33.10 Clayton Dewandre dual pedal valve

foot pedal equally between the two plungers. As only a very small motion of the plungers is needed it is possible to actuate one plunger even if there is no pressure acting in the other valve, the tilting of the beam being sufficient for this.

When the brakes are applied the plungers move downwards and their lower ends contact the balls A thus sealing off the brake pipe from the reservoir, the balls are then pushed off their lower seats so that fluid can pass through the pump or accumulator to the brake pipes. The pressure in the pipelines also acts in the spaces B and upwards on the reaction discs D and so tends to push the plungers upwards. The pressure will increase until the plungers do move upwards to an equilibrium position when the balls A re-seat themselves but still remain in contact with the ends of the plungers and seal off the passage to the reservoir. The plungers are then in equilibrium under the action of the downwards force applied by the brake pedal and the upward force due to the fluid pressure acting on the reaction discs D. The fluid pressure is thus made proportional to the pedal force. When the pedal is released the valve plungers move upwards and open the spaces B and the pipelines to the reservoir through the holes in the plungers.

33.8   Compressed air systems

Most of the compressed air brake systems in use are power systems although there is no reason why compressed air should not be used in servo systems, in fact most of the units described above for vacuum could be modified so as to work equally well with compressed air.

A complete system suitable for a vehicle having a trailer is shown in Fig. 33.11. The compressor A may be driven off the engine or, in electrically-propelled vehicles, by a separate electric motor. The compressor charges a reservoir B to a pressure which is regulated by the governor valve G. When this pressure is reached, the unloading valve U lifts the inlet valves of the compressor so that the latter runs 'light'. The air supply to the brake cylinders DD of the tractor vehicle is supplied direct from the reservoir through the brake valve C, which controls the pressure in the brake cylinders to a value proportional to the force exerted on the brake pedal. The brake cylinders D_1D_1 may be supplied in the same way if the dotted pipeline replaces the units R and K.

When the brake cylinders are at some distance from the brake valve, however, they may be operated indirectly from an auxiliary reservoir K situated close to them. In this case, the air from the brake pedal valve operates a relay valve R which passes air from the reservoir K to the brake cylinders D_1D_1 and regulates the pressure to the same value as determined by the brake valve. The reservoir K is charged from the main reservoir when the brake valve is in the 'off' position.

The trailer brakes are operated by air from the trailer reservoir S by means of an emergency relay valve X. The reservoir S is charged direct from the main reservoir on the tractor, through the emergency pipeline Y, and the brake operation is controlled by the air supply coming from the brake pedal valve via the pipeline H through the medium of the valve X.

A filter F is usually fitted, and sometimes an anti-freeze device as well. The latter feeds small amounts of methanol (methyl alcohol) into the air drawn into the compressor and this lowers the freezing point of any moisture that is in the air and prevents the system from frosting up.

Dual reservoirs are used as in vacuum systems, the smaller reservoir being charged up to full pressure very rapidly and before the larger one. The actuating cylinders employed are similar to those used with vacuum brakes but are smaller in size because of the higher pressures that can be used. Pressures of the order of 550 to $700 \, kN/m^2$ are usual.

Some of the components used with compressed air systems are shown in Figs 33.12 to 33.14. The reservoir valve that controls the pressure in the reservoir is shown in the first of these. The port A is connected to the reservoir

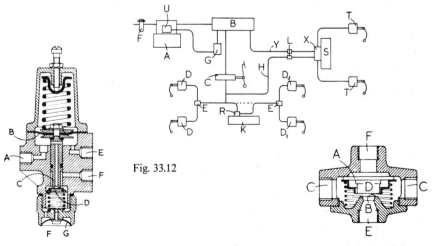

Fig. 33.12

Fig. 33.11

Fig. 33.13

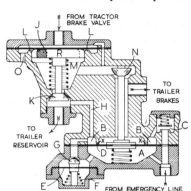

Fig. 33.14

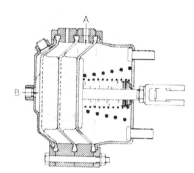

Fig. 33.15

whose pressure thus always acts on the lower side of the diaphragm B. This pressure is also transmitted to the driver's pressure gauge by a pipe connected to the port E. The valve is shown in the position corresponding to the lower limit of the air pressure. The spring has overcome the air pressure acting on the lower side of the diaphragm and has pushed the valve D off its seat, thereby opening the port F to the atmosphere through the dust protecting valve G. The port F is connected to the unloading mechanism of the compressor and with only atmospheric pressure applied to it the compressor will be functioning and passing air to the reservoir so that the pressure in the latter will rise. It will be noted that the hole down the centre of the valve C is closed by the contact with the valve D. The diaphragm is in equilibrium under the action of the spring force downwards and the air pressure upwards. The pressure acts on the area of the diaphragm minus the area of the valve C but plus the area of the hole of the latter. As the reservoir pressure increases the diaphragm will gradually move upwards until when the upper pressure limit is reached the valve D will seat on the casing and contact between it and the end of the valve C will be broken. Immediately this occurs the force tending to push the diaphragm upwards will increase by the amount due to the air pressure acting on the annular area of the valve C and so the diaphragm and valve will move up with a snap action. Air from the reservoir can now pass through the valve C and the port F to the unloading gear and unload the compressor by holding its inlet valve up. As the reservoir pressure gradually falls the valve C will move downwards until it contacts the valve D, thereby shutting off the air passage from the reservoir to the port F and pushing the valve D off its seat. This opens the port F to atmosphere so that the unloading mechanism will bring the compressor into action again. This action will also be a snap action because when the valve D is pushed off its seat the air pressure will no longer act on the annular area of the valve C and so the upward force on the diaphragm will be suddenly decreased.

A quick-release valve is shown in Fig. 33.13. The port F is connected to the brake pedal valve, the port E to atmosphere, and the ports CC to the brake cylinders. In the 'off' position shown, the brake cylinders are open to the atmosphere through the port E. When pressure air enters the port F, the diaphragm A will be moved downwards and its stem D will meet the seat B and seal off the exhaust port E. Pressure air will pass round the periphery of the diaphragm to the ports CC and the brake cylinders. When the pressure in the latter equals that in the port F, the diaphragm will re-seat on the upper seating, but the stem D will remain seated on the seat B. This is the 'brake holding' position of the unit. As soon as the pressure in the port F falls below that in the ports CC, the stem D will lift off the seat B and the air in the brake cylinders will escape to the atmosphere without having to pass along the pipe from F to the brake pedal valve.

The Bendix emergency relay valve is shown in Fig. 33.14. When the trailer reservoir pressure is low, air will pass from the tractor reservoir along the emergency pipeline (Y in Fig. 33.11) and through the non-return valve C and the space B to the trailer reservoir. The emergency line pressure is also exerted on the top of the diaphragm E and the latter will be depressed and the valve G will open, as shown. Air to the trailer reservoir can thus pass also via the space A and the valve G. If, however, the pressure in the emergency pipeline should fall below that in the trailer reservoir, the valves C and G will close and the trailer reservoir pressure will be maintained. The pressure below the diaphragm D will keep the valve N off its seat and the diaphragm itself in contact with the face X so that air from the trailer reservoir can reach the trailer brake cylinders only via the valve K.

In the ordinary way the brakes will be applied by the admission of pressure air from the brake pedal valve to the upper side of the diaphragm J. This will seat the diaphragm on to the seating L and will seal off the brake cylinders and the space M from the exhaust port O. The piston R will also be depressed, and the valve K will be unseated so that air from the trailer reservoir can pass to the trailer brake cylinders. The pressure set up in the latter will be such that the force exerted on the underside of the diaphragm J just balances the force applied to the top side, and so the trailer brake pressure will be equal to the pressure set up in the service pipe by the brake pedal valve, and will be proportional to the force exerted on the brake pedal.

If the trailer is disconnected or breaks away from the tractor, the pressure in the emergency pipeline will fall to atmospheric and the trailer reservoir pressure in the space B will depress the diaphragm D and seat the valve N. Air will then pass from the trailer reservoir via the space B and round the stem of the valve N to the trailer brake cylinders, and the pressure set up in these will be approximately equal to the trailer reservoir pressure, so that a full application of the trailer brakes will result.

Power operation is especially suitable when a large number of brakes have to be operated, as, for example, on eight-wheeled vehicles and when trailers are used, because although the driver's valve proportions the intensity of the braking to the force exerted by the driver, this latter force does not increase with an increase in the number of brakes operated as it would do in a mechanically-operated system. The connection between a tractor and trailer is simpler with air operation, relative movement between the two being allowed for by the use of a flexible hose connection and having no effect on the brake operation, a condition that is not easily obtained with mechanical linkages. The trailer brakes can also be arranged to come on automatically should the trailer break away from the tractor.

33.9   Actuating cylinders for air brakes

Air pressure brakes are usually actuated by diaphragm cylinders in which the diaphragm acts as a piston but has the advantage that no sliding air-tight joints are necessary. To enable a single cylinder to be used for both foot-pedal and hand-lever operation Clayton Dewandre introduced the triple-diaphragm cylinder shown in Fig. 33.15. When the brake pedal is depressed air is admitted to the space A while operation of the handbrake admits air to the space B. The space between A and B is open to atmosphere. For both forms of braking the force applied to the brakes will be regulated by a brake valve so that it is proportional to the force exerted by the driver. Although air pressure is used for the application of the handbrake the brake will be held on by the usual ratchet even if the air pressure should fall after the brakes have been applied.

33.10   Spring brake units and locks

In road vehicles the foot brakes are used chiefly for stopping the vehicle and for controlling its speed on gradients and the handbrake is used almost solely for parking purposes or in conditions where the driver does not have a foot free to operate the foot brake, for example when starting on a gradient.

The two systems are made independent to a considerable extent in order to meet legal requirements and no great difficulty arises in designing satisfactory systems except in very heavy vehicles and in tractor–trailer combinations. The requirements of the foot brake can be met even in heavy vehicles by using servo assistance or power operation but these do not suffice for parking brakes that must often be left on for long periods. To meet these demands springs have been used to apply the brakes and power to release them and this is done in a *spring brake unit*, which may be placed at the driver's end of an ordinary mechanical linkage as in the system shown in Fig. 33.9, or one may be used at each brake.

The system comprises a spring that is sufficiently powerful to apply the brakes fully but which is normally held so as to be inoperative by means of a piston in a cylinder to which air or fluid under pressure is admitted. The construction is shown in Fig. 33.16. Fluid from the handbrake control valve is passed to the port F so as to hold the piston in the position shown, the spring E being then inoperative. To apply the handbrake, the control valve is moved so as to release the pressure in the cylinder and the spring then applies the brakes. As the fluid pressure can be regulated by the control valve the brake can be applied by the spring for normal applications if required. However, a separate diaphragm cylinder C is incorporated and this provides for the service applications, fluid or air being admitted through the port G by the pedal valve so as to operate the piston D. To enable the unit to be dismantled when required a draw bolt K and nut H are provided to hold the spring in compression.

An alternative to the use of a spring unit is to employ a mechanical lock to hold the brakes on after they have been applied by air pressure. This is generally used when direct-acting diaphragm chambers are situated close to, or actually mounted on the backplates of, the brakes. An example is shown in Fig. 33.17.

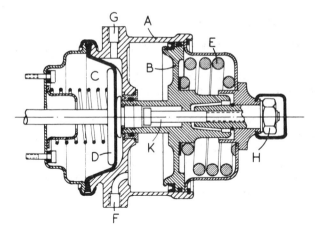

Fig. 33.16 Clayton Dewandre spring brake unit

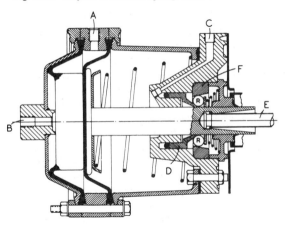

Fig. 33.17 Clayton Dewandre brake cylinder with lock

The service brake application is obtained by admitting air to the port A from the pedal valve and the parking application by admitting air through the port B and this can, if required, be used to supplement the foot brake. To hold the parking brake on, even when the air pressure is released, locking rollers R are provided; these are normally held so as to be inoperative by the piston D to which air is admitted through the port C but to lock the brakes this pressure is released and then the spring pressing on the rollers will force them into the tapered bore of the collar F and they will jam so as to prevent the piston from moving back. To release the brake, air must first be admitted to the port B so as to relieve the pressure on the locking rollers and then these can be moved to the free position by admitting air to port C; this sequence may be provided automatically by the control valve and it prevents the brakes from being released inadvertently because after the lock has been released the parking brake has to be released in the normal way.

33.11   Brake limiting device and anti-slide systems

As the brakes are initially applied, a miniscule degree of slip occurs between the tyres and road, but the slip as well as the brake torque progressively increase with the load applied to the brake pedal, until the wheels actually lock and the tyres slide along the road. During the last phase of brake application the rate of angular deceleration of the wheels becomes considerably greater than the linear rate of deceleration of the vehicle without any wheels sliding.

It is well known that the coefficient of friction between the tyres and the road is reduced with the onset of sliding, which is therefore to be avoided if the shortest possible braking distance is to be achieved.

The point at which wheel-lock will occur is dependent upon the vertical forces between the tyres and the road. Therefore, application of the brakes transfers, in effect, some of the weight from the rear wheels to the front ones, so during hard braking the rear wheels will tend to slip earlier than those at the front. A vehicle having a static front : rear weight distribution of 60 : 40 would generally have adequate adhesion for all its brakes to perform effectively when its weight distribution ratio coincides with a braking ratio of about 70 : 30. At higher decelerations this level may rise to 80 : 20, so if the rear wheels are not progressively relieved of some of the braking effort they are liable to slide. Sliding rear wheels have no capacity for exerting any additional grip on the road to produce any stabilising force. Consequently, the vehicle becomes unstable when braked hard and, given the slightest lateral disturbance, its rear end may slide sideways, causing the vehicle to swing round through as much as 180° or more.

To prevent this unstable condition from arising, a *brake pressure apportioning valve* can be fitted in the hydraulic system. Its function is to reduce the control pressure applied to the rear brakes relative to that to the front ones and, preferably, it should cater for both the fully laden and driver-only conditions. A whole range of such valves is produced by Lucas Girling, the principles of which are explained later.

In the nineteen-sixties it became common practice to insert a simple *pressure-limiting valve* in the line to the rear brakes only. This allowed the applied pressure to be transmitted up to a predetermined level to both front and rear brakes, after which the pressure in the line to the rear brakes was either held constant or its rate of increase was reduced. However, this pressure characteristic was constant regardless of the load carried, so the rear brakes were significantly under-utilised when the vehicle was fully laden.

Hydraulic pressure from the master cylinder is applied at A, in Fig. 33.18, and transmitted through the axial hole in the centre of the shouldered piston and on through the metering valve D, whence it goes through the outlet B to the rear brakes. At the same time it acts upon the crown of the piston. Because the effective area of the crown is greater than that of the shoulder on the other side of the piston, it progressively pushes the piston back, compressing the spring C, until the metering valve is seated. Then, any further increase acting on the smaller diameter of the shoulder of the piston pushes it to the right as viewed in the illustration, again opening the metering valve. Thus, the pressure applied through outlet B to the rear brakes continues to increase, as shown in Fig. 33.22(*a*), the front : rear pressure ratio at the brakes being predetermined

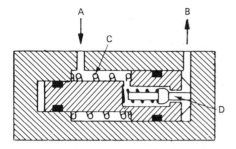

A   From master cylinder
B   To rear brakes
C   Spring
D   Metering valve

Fig. 33.18 Diagram of a pressure-conscious brake apportioning valve

by the ratio of the smaller and larger effective areas of the shouldered piston and the force exerted by the internal spring.

The next development was the introduction of the *deceleration-conscious* valve, Fig. 33.19, generally termed a *G-valve*, which was designed to improve the fully laden performance. In principle it is similar, but the mechanism is different. The metering hole is closed and opened respectively by a ball rolling up and down an inclined base: in practice, instead of incorporating a ramp the whole unit is inclined, so that the centre of the ball will remain at all times axially in line with the hole.

Again, the applied pressure enters at A and is transmitted through the axial hole in the shouldered piston out at B to the rear brakes. At a predetermined deceleration the ball rolls up the slope and closes the hole in the piston, thus preventing the pressure in the pipeline to the rear brake from increasing further until the pressure on the small end of the piston is high enough to

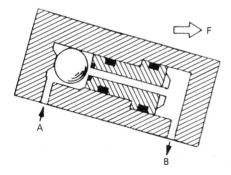

A   From master cylinder
B   To rear brakes
F   Direction of forward travel

Fig. 33.19 A deceleration-conscious brake pressure apportioning valve

overcome the pressure acting on the large end of the piston. The pressure differential between the front and rear brakes is, of course, that across the piston and is fixed by the ratio of the areas of its ends.

From the diagram showing the pressure characteristics of this type of valve, Fig. 33.22(*b*), it can be seen that it can respond to changes in gross vehicle weight but not axle weight. With a front-wheel-drive car a change from the driver-only to the fully laden condition may increase the load on the front axle by perhaps 18% and on the rear axle by about 76%, so a deceleration-conscious valve is only slightly better than a pressure-conscious one. Therefore the next step was to introduce a *load-conscious valve*.

33.12   The load-conscious valve

This valve, Fig. 33.20, has a shouldered piston similar to that of Fig. 33.18 but loaded by an additional spring acting externally on the piston. Since the deflection of the internal spring is no more than that necessary to open and close the metering valve, the force it exerts is almost constant. Its function is to counteract the hydraulic pressure. The degree of compression of the external spring, on the other hand, is dependent upon the height of the sprung mass above the rear axle and therefore the load carried, and reduces to zero for the driver-only condition.

During deceleration this type of load-conscious valve is affected by the weight transfer from rear to front. In Fig. 33.22(*c*), A is the cut-in point of the pressure-reducing valve in the driver-only and B in the fully laden condition, for which the line AC is non-existent.

From Fig. 33.22(*c*) it can be seen also that, during progressive weight transfer, the correspondingly progressive reduction of the external spring force to zero has the effect of inclining the slope of the pressure characteristic down and ultimately, at point C, causing it to join and turn up along the line whose

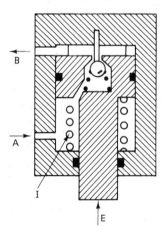

A   From master cylinder
B   To rear brakes
I   Internal spring
E   Force applied by external spring

Fig. 33.20 The simple load-conscious brake pressure apportioning valve

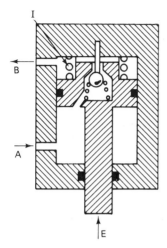

A From master cylinder
B To rear brakes
I Internal spring
E Force applied by external spring

Fig. 33.21 The revised load-conscious brake pressure apportioning valve

slope is determined by the internal spring, that is, for the driver-only condition. This is still far from ideal for the fully laden case, especially on a vehicle equipped with a front-wheel sensed, anti-lock system. During anti-lock operation such a system might operate at a 75% utilisation factor, the effect of which is indicated by the line XX in the illustration. For further information on how to calculate the spring forces needed, the reader is recommended to refer to a paper entitled 'Brake Pressure Apportioning Valves', by G. P. R. Farr, *Proc. I. Mech. E*, Vol. 201, D3.

33.13 Apportioning valves for front-wheel sensed anti-lock systems

To offset the effect on the rear brakes of a fully laden vehicle having front-wheel sensed anti-lock brakes it is necessary to increase the ratio of the effective area of the crown to that of the underside of the piston of Fig. 33.20. However, to avoid too steep a rise in pressure above point A in Fig. 33.22(*c*) this must be accompanied by a reduction of the external spring force throughout the deceleration range in the driver-only condition. This is accomplished by fitting the internal spring on the opposite side of the piston, as in Fig. 33.21, so that it opposes the external spring and therefore the force exerted by this spring can be increased.

With this arrangement the force tending to displace the piston is reduced as the hydraulic pressure, and therefore the deceleration, increases. Even so, the external spring applies a force to the piston under all conditions, even during maximum deceleration with only the driver in the car. This force, however, reduces progressively with increasing deceleration throughout the whole range of loading up to the fully laden condition.

The combination of the positive loading applied by the external spring and

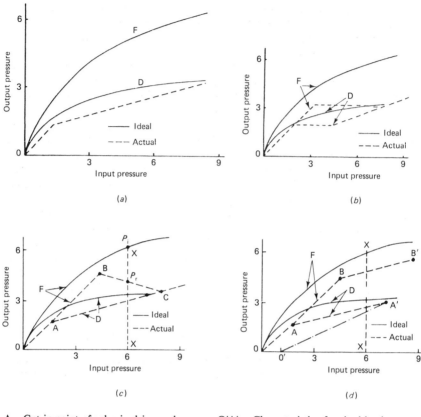

A   Cut-in point of valve in driver-only condition
B   Cut-in point of valve in fully laden condition
C   Cut-out point of valve, fully laden
B'   Cut-out point for the revised load-conscious valve

O'A'   Characteristic of revised load-conscious valve with broken external spring or linkage
XX   Limit of pressure that can be applied to front brakes when anti-lock system is in operation in fully laden condition

Fig. 33.22 Performance characteristics of (*a*) a pressure-conscious, (*b*) a load-conscious, (*c*) a deceleration-conscious, (*d*) a revised load-conscious valve, where curves D are for the driver-only and F are the fully laden condition

the negative effect of the internal spring gives the required cut-in forces. Comparison of the lines XX in Fig. 33.22(*c*) and (*d*) reveals how the rear brake pressure in the fully laden condition is improved by adopting this arrangement. Such a system is fail-safe, since the result of a fracture in the external spring or its linkage system is that the input pressure then follows the line O'A'.

This type of valve was specially designed for X-split anti-lock braking systems. With this combination the pressure to each front wheel brake and diagonally opposite rear brake is modified, as appropriate, through the medium of an anti-lock device, which delivers also to the rear brakes, though at a pressure that is ultimately further modified by a load-sensitive apportioning valve for each wheel. Such an arrangement corrects for the lateral weight transfer during cornering, and thus improves both the vehicle handling and brake effectiveness. With the simpler systems the reduced loading on the inside wheel during a turn can cause it to lock, while the brake on the more heavily loaded outer wheel is under-utilised.

With all apportioning valves that sense the height of the sprung mass above the rear axle, correct front : rear brake distribution is dependent on the accuracy of setting of the valve and freedom from friction in its moving parts. Moreover, if variable-rate road springs are employed the linkage between the sprung mass and the valve, which is mounted on the axle, must be designed to take their variable rate into account. This can be difficult, so an alternative arrangement, termed a *direct-acting valve*, has been devised, Fig. 33.23. It is similar to that illustrated in Fig. 33.21, but its external spring is the rear suspension spring and the internal one is replaced by a bias spring interposed between the upper seat for the suspension spring and the valve housing.

Because of the use of the powerful suspension spring, the piston stem has to be much larger and care has to be taken to keep to a minimum the volume of fluid displaced during the closing of the metering valve. An incidental advantage of this arrangement is that piston seal friction becomes a correspondingly smaller proportion of the total force on the unit. With the types of valve previously described, reduction of seal friction to a minimum is a major requirement.

33.14 Apportioning valves for heavy commercial vehicles

Brake apportioning valves designed for air-actuated brakes for heavy commercial vehicles are based on principles similar to those for cars, but they tend

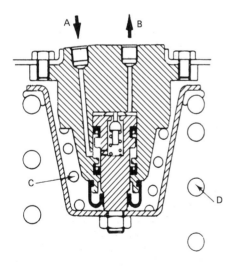

A   Input from master cylinder
B   Output to rear brakes
C   Bias spring
D   Vehicle suspension spring

Fig. 33.23 Direct-acting brake pressure apportioning valve

to be much larger and slightly more complex than those for hydraulic operation. Such a device, made by Clayton Dewandre, is shown in Fig. 33.24. It regulates the pressure set up in the brake pipe B in relation to the pressure at the inlet A and in accordance with the position of the roller D which is controlled by the distance between the axle of the vehicle and the frame. The pressure of the air at the inlet is itself regulated by the brake pedal valve so as to be proportional to the force exerted by the driver. In action, the air entering at A depresses the piston F so that the end E of the dumbell GH contacts its seating and closes the exhaust port and the end H is moved off its seating so that air can pass to the brakes. The pressure that builds up in the brake pipe also acts via the passage J on the piston K and (through the pin L which is now in contact with the beam E because of the downward movement of the piston F) applies a force to the beam and pulls the piston F upwards. An equilibrium position is reached when the valve H contacts its seating but the valve G still remains in contact. The forces at the ends of the beam are then the ratio of the lever arms x,y and the pressures at the ports A and B are then in a ratio that is determined by the ratio $x : y$ modified by the relation between the areas of the two pistons. The ratio between the inlet pressure and the brake pressure can thus be modified by moving the roller D and this is done by any sustained motion of the axle relative to the frame by means of a linkage whose principle is shown by the diagram to the right. Rapid movements of the axle such as are due to road irregularities are accommodated by the springs S and the bell-crank member R is held virtually stationary by the damper T, but any sustained movement of the axle will move the bell-crank and thus alter the position of the roller D.

In the system described above, the magnitude of the forces acting on the roller fulcrum prevents any alteration of the pressure ratio setting during the actual brake actuation. This drawback is obviated in another Clayton Dewandre system shown in Fig. 33.25. In addition to the form shown, the device is made in two simplified forms but all three regulate the ratio between the pressure applied to the brake actuating cylinders and the pressure of the air coming from the brake pedal valve.

In the simplest form, known as the *direct type*, all the components above the line ZZ are omitted and replaced by a simple cover and the double piston T_1T_2 is replaced by a simple one-piece piston. The port A is connected to the brake pedal valve and, via the passage P and the dotted opening shown, air reaches the top of the piston T. The force thereby applied to that piston forces it down against the springs thereby closing the gap C, through which the brake actuating cylinders have been open to the atmosphere and then opening the

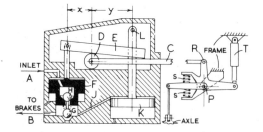

Fig. 33.24 Clayton Dewandre Mk I brake-limiting system

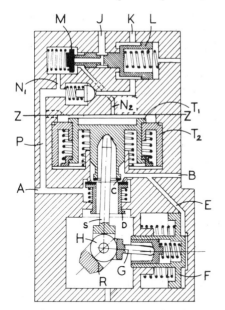

Fig. 33.25 Clayton Dewandre Mk II brake-limiting system

valve D so that air from the port A can pass to the brakes. The pressure that builds up in the port B also acts on the piston F via the passage E. The forces acting on the pistons T and F are made to balance each other by means of the struts S and G which are pivoted together, a ball bearing H being mounted on the pivot pin. This ball bearing is in contact with a ramp R which is pivoted in the casing on an axis that coincides approximately with the axis of the ball bearing. The inclination of the ramp is controlled by the 'height' of the suspension through a linkage similar in principle to the one described earlier. When the pressure on the piston F reaches an appropriate value the ball bearing H moves slightly up the ramp and the strut moves up to allow the valve D to seat so that the air from the brake valve is cut off and the system reaches an equilibrium position. The ratio between the pressures at A and B then depends on the areas of the pistons T and F and the angle at which the ramp is placed by the axle. When the brake pedal is released, the force on the piston T falls to zero and the springs open the gap at C, thus releasing the air pressure from the brake cylinders.

In the second form, known as the *relay type*, the port A is connected to a reservoir and the brake pedal to a port in the cover above the piston T. When the brake pedal is depressed air passes to the top of the piston T and the action proceeds as before except that the air to the brake cylinders comes from the reservoir via the port A instead of through the brake pedal valve.

In the third form, the *relay-emergency type*, shown in Fig. 32.20, the brake pedal valve is coupled to the port J and an emergency pipeline to the port K. Pressure is normally maintained in the emergency pipeline and this holds the valve L to the right, allowing the valve M to seat on the shoulder of the body member and leaving a gap between that valve and the end of the valve L. Thus, when the brake pedal valve is depressed, air passes via port J through the hole in the valve L and through the passage N_2 to the top of the piston assembly T_1 T_2 and the action proceeds as in the second type, the air passing from the reservoir connected to port A to the brakes via port B. The reservoir can be charged via the emergency pipeline air which can overcome the non-return valve so that air passes to the reservoir via the passage P. Both the relay and the relay-emergency types have double piston assemblies T_1 T_2 as shown and the outer piston moves down to pre-load the inner one; this provides a smoother action when the brakes are applied.

If, for any reason, the pressure in the emergency pipeline falls to a low value the valve L will be moved to the left and air from the reservoir will pass via the passage P, N_1 and N_2 to the top of the pistons to give a full application of the brakes.

But even these devices cannot take into account the changes in road conditions that occur due to icy or greasy surfaces and they can be set only for average conditions; the difficulties can, however, be almost completely overcome in the anti-lock or anti-wheel-slide systems now coming into use. The first of such systems was the Dunlop-Maxaret (Section 33.16) but present-day systems all employ the same basic principle and this will now be considered.

33.15   Basic principle of anti-lock systems

For the sake of simplicity a single wheel will be assumed in the following description of the fundamental principles of anti-lock systems. Basically, an anti-lock system comprises a sensor to detect incipient wheel-locking, together with a system for relieving momentarily the hydraulic pressure to the brakes, to prevent locking before it actually occurs. As explained in Section 33.11, when the wheel is slipping only to a small extent, the deceleration will be low in comparison with the value appertaining to the approach of sliding and so, when the deceleration exceeds a certain value, the control releases the brake, the deceleration of which will then fall to a low value and so the brakes will be reapplied. This release and reapplication of the brakes must take place in an extremely short time if the system is to work satisfactorily and, in practice, the cycle will occur up to as high as 15 times per second.

In the early systems the wheel deceleration was measured by purely mechanical means but in present-day systems electronic circuits are used because they give much quicker responses and can be controlled more easily. These circuits are beyond the scope of this book and only an outline of their action can be attempted.

The deceleration sensor usually consists of a toothed disc attached to the hub of the wheel, and a pick-up placed near to the periphery of the disc. The pick-up is essentially a horseshoe magnet with a winding, and the projections of the disc act as a succession of keepers which bridge the poles of the magnet thus momentarily causing an increase in the magnetic flux through the winding and setting up a current in it. The frequency of this current will depend on the speed of the disc and the rate at which that frequency changes will be proportional to the deceleration of the disc. This rate can be measured relatively simply electronically and can then be used to supply a signal for the control of the brakes. The remainder of the system therefore consists of valves actuated by the signal and which control the actuation of the brakes.

A sensor may be provided for each wheel to be controlled but sometimes it is practicable to control the wheels in groups. A common system is to have a sensor for each of the front wheels and a single sensor with its disc on the propeller shaft for the two rear wheels together. It will be appreciated that the incorporation of these anti-lock systems is facilitated by the use of power operated brakes, also that they are somewhat expensive and are used therefore only on the more expensive cars and on certain classes of commercial vehicle.

33.16   Dunlop-Maxaret system

The Dunlop-Maxaret system was developed for application to the driving wheels of the tractor unit of articulated vehicles in order to prevent the vehicles from jack-knifing. The system has been most successful in doing this.

The general layout of the system is shown in Fig. 33.26 and details of the valves appear in Fig. 33.27. When there is no incipient wheel slide there will be

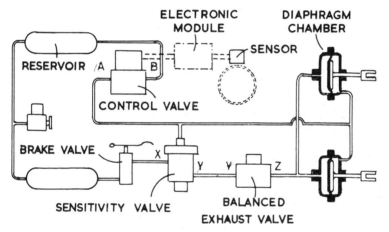

Fig. 33.26 Dunlop-Maxaret anti-slide system

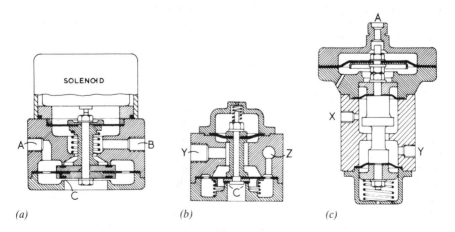

Fig. 33.27 Dunlop-Maxaret anti-slide system valves

no signal current from the electronic module to the control valve and so the port A, Fig. 33.27(*a*), will be open to atmosphere through the gap at C and there will be atmospheric pressure on the right-hand sides of the brake actuator diaphragms. Hence, when the brake pedal is depressed, air will pass freely from the service (lower) reservoir to the port Y of the balanced exhaust valve, Fig. 33.27(*b*). This pressure will deflect the outer portion of the lower diaphragm against the force of the spring so that air will pass to the port Z and thence to the left-hand sides of the brake actuator diaphragms, thereby applying the brakes. Under poor road conditions depression of the brake pedal will again pass air to the brake actuators to apply the brakes but if wheel slide becomes imminent the electric modules will pass current to the solenoid of the control valve, the gap C will be closed and air will pass from the anti-skid (upper) reservoir to the right-hand side of the brake actuator diaphragms and release the brakes.

The pressure air from the control valve will also depress the piston assembly of the sensitivity valve, Fig. 33.27(*c*), and this will restrict the passage of air from the brake pedal valve to the brakes. The pressure that acts on the left-hand side of the brake actuator diaphragms also acts on the underside of the balanced exhaust valve and if it exceeds the pressure acting on the upperside from the port Y the central portion of the diaphragm will be lifted so as to open the port Z to the atmosphere via the gap opened at C. Thus the valve equalises the pressures at Y and Z and the brake actuating pressure will at all times be equal to the pressure determined by the brake pedal valve.

As soon as the anti-skid pressure on the right-hand side of the brake actuators is released by the cessation of the signal from the electronic module the brakes will be reapplied. This action will be repeated with a frequency of several cycles per second as long as the wheel-slide condition continues.

33.17 Lucas-Girling WSP system

The general layout of the Lucas-Girling system, as applied (for the sake of simplicity) to a single wheel, is shown in Fig. 33.28. It is designed for use with brakes employing fluid application. Under normal conditions the brake is applied by the master cylinder in the usual way because the valve A of the actuator unit of the system is open as shown.

The valve is held open by its spring and by fluid pressure on the left-hand side of its piston, this pressure being maintained at a constant value by a pump that is driven by the engine of the vehicle. When a signal is passed by the electronic module to the solenoid of the control valve, oil from the pump will pass to the right-hand side of the actuator piston and as the effective area of this side is greater than that of the left-hand side the piston will move to the left to close the valve. Because of the decrease in the volume of the stem of the valve that projects into the chamber B, the pressure in the brake cylinder will drop and the brake will be released. When the signal to the control valve ceases the right-hand side of the actuator piston will again be opened to the atmosphere in the reservoir and the valve will open. The action will be repeated with a frequency up to some 15 Hz, this frequency being modified to some extent by auxiliary circuits in the electronic module. When several wheels are to be controlled each must have its sensor, electronic module circuit, control valve and actuator but the pump will be common to all. On the other hand, to reduce

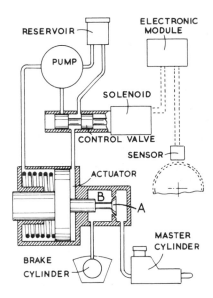

Fig. 33.28 Lucas-Girling WSP system

cost, some vehicle manufacturers elect to sense the occurrence of wheelspin per axle instead of per wheel.

33.18 Ford *Escort* and *Orion* anti-lock systems

For the Ford front-wheel-drive *Escort* and *Orion*, the Lucas-Girling, low cost system is used. It has sensors for detecting wheel-lock on only the front wheels, locking of the rear wheels being initially inhibited by a pressure-limiting valve. These models have an X-split brake control system, as described in Section 33.14 so, when operating, the anti-lock system alternatively relieves and reapplies the pressure to the brake not only on the front wheel that is about to lock but also on the diagonally opposite rear wheel, Fig. 33.29.

Having two instead of three or four wheel-lock sensors of course is an economy, but other measures, including the substitution of a mechanical instead of an electronic sensing and control system and the avoidance of any need for separate, electric or engine-driven hydraulic pump to supply the braking pressure also make major contributions to the overall cost reduction. A flywheel incorporating an overrun device serves as the mechanical sensor. This is driven from the front wheel by a toothed belt which gears it up to 2.8 times the driveshaft speed. The flywheel, a modulator valve unit and a cam-actuated reciprocating pump are all in a common housing, Fig. 33.30.

In normal conditions the flywheel accelerates and decelerates with the road wheel, and the hydraulic modulator valve is functioning as shown in Fig. 33.31, in which the black areas are those in which the hydraulic pressure rises as the brakes are applied. In this condition the pump plunger (11) is held clear of the cam (10) by the plunger spring (12).

If, however, the angular deceleration of the wheel attains a value equivalent

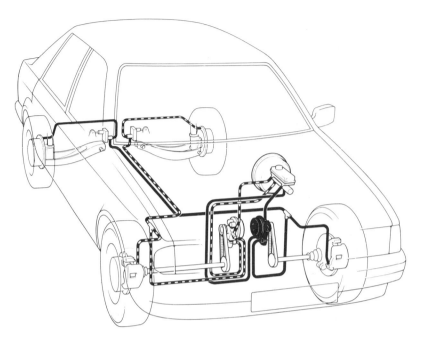

Fig. 33.29 Ford *Escort* and *Orion* anti-lock system. Note that, as compared with Fig. 33.30, the control units are upside down

to a 1.2*g* deceleration of the vehicle, wheel lock is likely to occur and so the overrun torque generated by the flywheel, due to its inertia, rotates it a few degrees relative to the hub. This rotation occurs within a ball-and-ramp mechanism (4), which causes the axial displacement shown in Fig. 33.32. The consequent axial movement displaces the dump-valve lever (9) about its pivot, thus opening the dump valve (7).

The opening of the dump valve releases the pressure above the de-boost piston (15) and consequently also relieves that in the pipeline to the brakes. Since the downwards pressure on the pump plunger (11) has also been released by the opening of the dump valve, the master cylinder pressure acting on the piston forces it into contact with the cam (10). Even so, the consequent reciprocation of the pump plunger cannot generate any hydraulic pressure so long as the dump valve remains open.

Simultaneously, the de-boost piston, under the influence of the hydraulic pressure below it, rises to allow the cut-off valve (13) to close, as in Fig. 33.33, thus cutting off the input from the brake master cylinder and relieving the pressure in the pipelines to the brakes. Therefore, the road wheels accelerate to the speed of the still decelerating flywheel. At this point the flywheel, moving back and contracting its ball-and-ramp mechanism, is accelerated at a rate controlled by the clutch that can be seen in Fig. 33.30. As the lever (9) is released, the dump valve closes.

This allows the reciprocating pump to increase the pressure above the de-boost piston and in the pipeline to the brakes. If it again causes the road

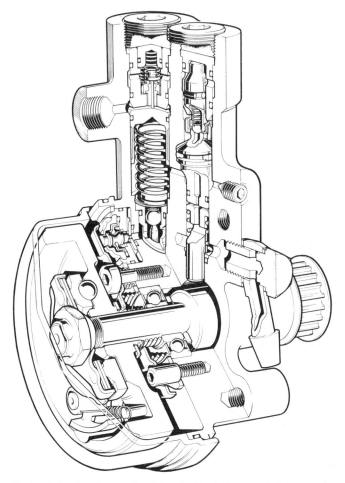

Fig. 33.30 Sectioned control unit for the Ford *Escort* and *Orion*

wheel to lock, the cycle of events is repeated but if, without wheel lock occurring, it rises to equal the pressure applied by the driver's pedal to the master cylinder the cut-off valve (13) is opened again, the pump disengages and the master cylinder is reconnected. The effects of the whole sequence of operations on the input to the brakes and on the wheel spin is illustrated in Fig. 33.34.

33.19 Ford *Granada, Sierra* and *Scorpio* anti-lock systems

The ABS (Anti-lock Braking System) on the rear-wheel-drive *Granada, Sierra* and *Scorpio*, each of which have a Y-split (Section 32.14) brake system, is the outcome of co-operation between Ford and the German brake manufacturers ATE. It has an electronic control, with electro-magnetic sensors on all four wheels, Fig. 33.35, and an electrically driven pump and hydraulic accumulator for maintaining sufficient reserve pressure to enable the anti-lock system to

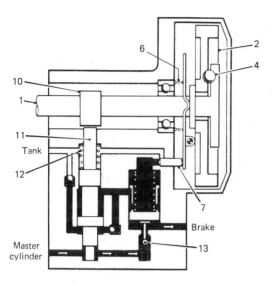

Fig. 33.31 Positions of the valves in the normal brake operating condition

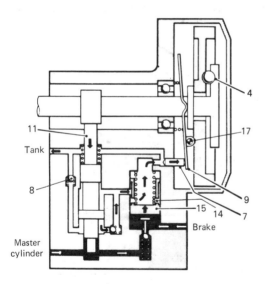

Fig. 33.32 Axial displacement of the flywheel displaces the lever to open the dump valve to the reservoir

release and reapply the brakes repeatedly at rates of up to twelve times per second. The electronic control module has two microprocessors which not only duplicate the processing of the incoming signals but also monitor each other continuously to check that both are functioning properly. In the event of a total system failure the brake control reverts to conventional operation without anti-lock control and an indicator on the dash is illuminated to warn the driver.

Frequency signals from the four wheel sensors are translated by the

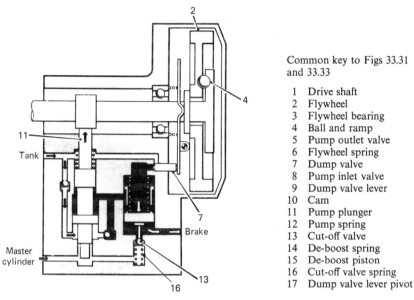

Common key to Figs 33.31 and 33.33

1 Drive shaft
2 Flywheel
3 Flywheel bearing
4 Ball and ramp
5 Pump outlet valve
6 Flywheel spring
7 Dump valve
8 Pump inlet valve
9 Dump valve lever
10 Cam
11 Pump plunger
12 Pump spring
13 Cut-off valve
14 De-boost spring
15 De-boost piston
16 Cut-off valve spring
17 Dump valve lever pivot

Fig. 33.33 Closure of the dump valve allows the pump to build up the brake pressure again

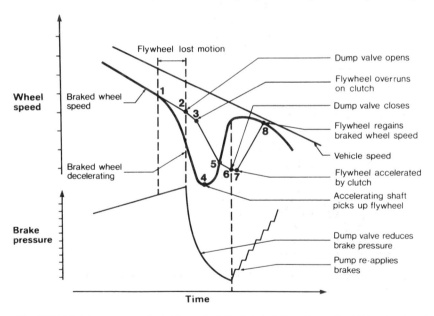

Fig. 33.34 Brake pressure and wheel speed plotted against time throughout the sequence of operations of the Lucas Girling anti-lock brake system

electronic module first into wheel-speed and acceleration values and then into vehiclespeed and wheelslip. When the slip becomes so great that wheel-lock is imminent, the control alternatively energises and de-energises the appropriate hydraulic inlet and outlet solenoid valves in the ABS electro-hydraulic unit,

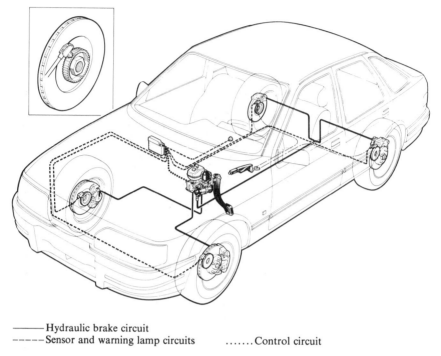

————— Hydraulic brake circuit
– – – – Sensor and warning lamp circuits   . . . . Control circuit

Fig. 33.35 Layout of the anti-lock brake system of the Ford *Sierra* with (*above left*) the electro-magnetic pick-up to a larger scale

Fig. 33.36, to relieve and reinstate the pressures in the lines to the brakes. There are three hydraulic circuits, one for each front wheel and the third for both rear-wheel brakes. The advantage of having a single control over both rear brakes is good stability when cornering under maximum braking and a reduction of oversteer by reducing the brake torque on the most heavily loaded rear wheel. At the same time, the reduction in the overall braking of the vehicle is minimal because of the effect of the apportioning valve in limiting the contribution by the rear wheels to only a fraction of the total.

During operation without anti-lock the front brakes are actuated by the master cylinder with assistance from its integral hydraulic servo, while the rear ones are actuated by pressure from the hydraulic accumulator. This accumulator is maintained at 140 to 180 bar by the electric pump. The control valve linked to the piston rod of the tandem master cylinder, Fig. 33.36, maintains a constant relationship between the hydraulic output pressure from the servo and the input force applied by the brake pedal to the master cylinder.

33.20   Traction control

For the similar Ford models that have also four-wheel-drive based on the use of viscous couplings, Section 27.10, a further development of the ABS electronic control system takes into account also the interactions between the four wheels, through the viscous couplings, under varying engine torques. In

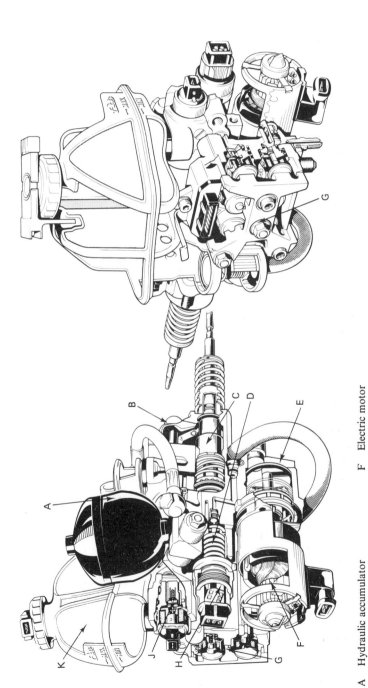

A   Hydraulic accumulator
B   Control valve
C   Hydraulic booster
D   ABS master cylinder
E   High pressure pump

F   Electric motor
G   ABS valve block with six solenoid valves
H   Pressure warning switch
J   Main valve
K   Hydraulic fluid reservoir

Fig. 33.36 Two views of the Teves combined master cylinder and ABS unit. The control valve equalises booster and master cylinder output pressures

other words, what is termed *traction control* is incorporated. This entails automatic alternate application and release of the brake on either driven wheel as soon as the microprocessors detect that it is about to spin. Obviously, therefore, the electronic control module has to differentiate between wheel-lock and wheelspin. With full traction control, as soon as the driving wheel on one side spins the brake is applied on that side, but if the wheel on the other side then spins, the electronic control closes the throttle or reduces the rate of fuel injection to reduce the torque output from the engine. All these control operations are effected within milliseconds.

For starting from rest the traction control system must be much more sensitive to drive slip than for either simple acceleration from one speed to another or deceleration in anti-lock systems. Additionally, it must also differentiate between wheel-speed differential due to the vehicle's being simultaneously driven round a corner. The electronic control unit is virtually identical to that for the Mk IV system, Fig. 33.39, which is described in the last two paragraphs of Section 33.21.

33.21   Teves Mk IV ABS and traction control

In Fig. 33.37 the Teves Mk II system, for the Ford two- and four-wheel-drive cars, Sections 33.19 and 33.20, is compared diagrammatically with the Mk IV system. Cost reduction, to render the equipment suitable also for less upmarket cars, was the primary incentive for the Mk IV development. The principal economies were the substitution of a vacuum for a hydraulic servo, or booster, and the use of a pump of higher output, at extra cost, to obviate the need for an accumulator. With the abandonment of the hydraulic booster J containing the control valve, equalisation of the output pressure from the ABS

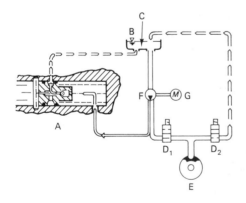
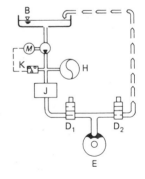

A	Master cylinder	F	Non-return valve
B	Fluid level sensor	G	Electrically driven pump
C	Reservoir	H	Hydraulic accumulator
D_1	Solenoid valve (inlet)	J	Control valve in the hydraulic booster
D_2	Solenoid valve (outlet)	K	Pressure switch
E	Disc brake		

Fig. 33.37 Comparison between the Teves Mk II (*right*) and Mk IV (*left*) anti-lock brake systems. The hydraulic circuit to only one brake is shown since the others are identical to it

pump to the brakes with that from the master cylinder has had to be effected by means of valve A incorporated in the centre of the piston of the master cylinder.

The Mk IV system, too, has acceleration/deceleration sensors on each rear wheel, and the brake pressure to both rear wheels is reduced or increased, as appropriate, to the level necessary to control the locking, or spinning, wheel. All-wheel control gives the sensitivity needed for traction control, and greatly reduces the possibility of rear-end instability in all modes. The electric pump is switched on by the electronic control system only when it is required to build up the brake pressure in the anti-lock or anti-spin modes (ABS or traction control operation), so little energy is consumed.

From Fig. 33.38 it can be seen that the general arrangement of the Mk IV system is similar to that of the Mk II except that, in the Mk IV, the master cylinder and reservoir together have become a separate unit. Also, the system illustrated is designed for an X-split braking system, as compared with the Y-split of the Ford two-wheel drive layout.

If a wheel tends to lock, the electronic controller closes the solenoid-actuated hydraulic inlet valve to its brake and opens an outlet valve in the line back to the reservoir. Simultaneously, it switches on the electric motor. Since the inlet valve to the brake circuit is closed, the fluid delivered from the pump can only force the piston in the master cylinder back until the valve in the centre of its piston, Fig. 33.37, opens to release all fluid in excess of that required for ABS operation back to the reservoir. This ensures that the pressure generated by the pump cannot exceed that induced by the driver through his brake control pedal, and that the driver does not lose the feedback from (the feel of) his brake control.

As the unlocked wheel accelerates back to the appropriate speed, the outlet

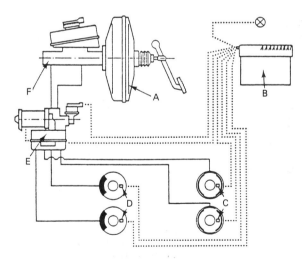

A	Vacuum servo	D	Front brake sensors
B	Electronic controller	E	Hydraulic module
C	Rear brake sensors	F	Tandem master cylinder

Fig. 3.38 Schematic layout of Teves ABS IV anti-lock brake system

valve D_2 in its brake circuit closes and the inlet valve D_1 opens, so the pump brings the brake application pressure back up to the level dictated by the force exerted on the pedal. If the wheel again starts to lock, the sequence is repeated. Incidentally, the non-return valves shown in Fig. 33.37 prevent fluid from flowing back to the reservoir under pressure exerted by either the pump, hydraulic accumulator or master cylinder when the brakes are applied.

For safety, the electronic control circuit, Fig. 33.39, is duplicated. There are two identical microprocessors, each with its own comparator. The comparators check both the internal and external signals from the wheel-speed sensors and to the valves respectively. If they do not correspond the defective circuit is switched off and a warning lamp illuminated on the dash. There is also a continuous monitoring system for checking the performance of the sensors, connections, solenoid valves and hydraulics. In the event of a failure, the brake system reverts to operation without ABS, and again the driver is warned by a lamp on the dash.

This system can be expanded to include traction control. The extra cost is small because the same sensors and valves are used, though some extra valves do have to be introduced. Some expansion of the hardware and software in the electronic controller is necessary, too, since a traction control system may have to apply, instead of release, a brake to prevent wheelspin and, moreover, it has to prevent the wheels from spinning at any vehicle speed. It is also required to intervene in the engine control, as described in Section 33.20. An advantage of the X-split hydraulic system shown in Fig. 33.38 is that the driven wheels can be controlled individually to provide optimum traction, instead of perhaps having to reduce the traction on both wheels to that obtainable from the more lightly loaded wheel when cornering.

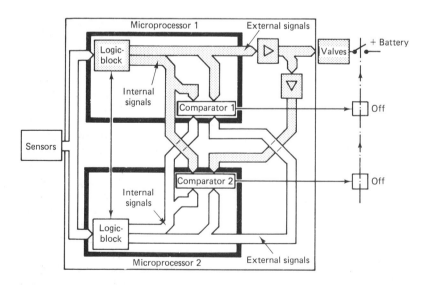

Fig. 33.39 The Teves MK IV electronic control system is duplicated

33.22   Regenerative braking systems

A simple form of regenerative braking system is often employed on electric vehicles. It is necessary because the energy storage capacity of a battery of a weight and size practicable for installation in road vehicles is so small that one cannot expect to get more than 30 to 40 miles (38 to 64 km) out of it, even with regeneration of the energy that would otherwise be dissipated in braking. This type of vehicle has an electric motor and control system such that, when current is passed through it, it drives the vehicle, but generally when its control pedal is released, or more unusually during the initial movement of the brake pedal, the current supply to the motor is cut off and it actually generates current which is utilised to contribute to recharging the batteries. Thus, a braking torque is applied to the road wheels, by virtue of the fact that they are driving a generator.

With the advent of electronically-controlled, constantly variable transmissions it has become practicable to introduce regenerative braking for petrol and diesel vehicles. Leyland has been experimenting with such a system since before 1980, using its CVT with a flywheel for energy storage, while Volvo had a hydraulic accumulator regenerative system installed on acceptance trials in a London bus in 1985. Indeed, regenerative braking is particularly attractive for urban bus operation, since much of the power from the engine is used for acceleration from bus stops, soon after which it is dissipated again in braking for the next stop.

Flywheel storage has some disadvantages. First, in the event of an accident in which excessive shock loading is transmitted to the flywheel, it might burst and cause casualties. Secondly, it adds significantly to the weight of the vehicle, thus offsetting some of the gains as regards fuel economy. Thirdly, it is bulky. Fourthly, it will run down overnight, so the engine has to be started electrically in the morning.

Most of these disadvantages can, to a major extent, be designed out. For example, by using fibre-reinforced material for the flywheel it can be made so that it does not burst into large fragments when ruptured but rather tends to shear along the fibres and to be retained by them. The weight can be reduced by the use of a very dense material as a rim on a very light disc, so that its polar moment of inertia is high relative to its weight. Little can be done about its bulk, since it must have a flywheel of reasonably large diameter, though it can be installed with its axis of rotation vertical. To keep it spinning for a long time it could be housed in a vacuum, but this is hardly practicable; alternatively, the housing can be filled with a very light gas such as hydrogen or helium. Even so, to keep it freewheeling for, say, twelve hours is scarcely a reasonable demand.

With a hydraulic accumulator, on the other hand, overnight storage presents no problem so that, in the event of, for example, a fire in a bus garage all the vehicles could be driven out instantly by drawing on the accumulator for energy, without having to wait for their engines to start and become warm enough to move off, and without generating any exhaust fumes. Disadvantages of high pressure hydraulic drive systems, however, include problems of leakage, and they are inherently noisy under certain conditions, owing to turbulence and very high local velocities of fluid flow. With low pressures and velocities, the system becomes unacceptably bulky.

In trials of a Volvo bus in service in Stockholm the use of hydraulic regeneration has indicated average savings in fuel of between 28 and 30% in urban operation, though in an extreme case a saving of 35% was made. A significant proportion of this economy is attributable to the fact that, under load, the engine can be run virtually continuously at its most economical speed, the stored energy being used for acceleration and assistance in hill climbing. With suitable electronic control it might even be possible to stop the engine during braking and the initial stages of acceleration, though in the Volvo bus it is kept idling under these conditions.

The layout of the control system of what Volvo call their Cumulo system can be seen in Fig. 33.40 and of the hydro-mechanical system in Fig. 33.41. Power is derived from a 180 kW diesel engine installed in conjunction with a fluid flywheel and four-speed gearbox, and with a final drive ratio of 4.87 : 1. A power take-off of the sort used for driving auxiliaries alternately drives and is driven by, according to the mode in which the vehicle is operating, a swashplate type hydraulic pump/motor with a 40° Z-shaft and spherical pistons. When this unit is operating as a pump the fluid is delivered from the reservoir into the hydraulic accumulator. During operation as a motor the flow of fluid is of course reversed. Pressure for regulation of the angle of the swashplate on the Z-shaft has to be obtained from a separate pump: if it were dependent on the pressure in the accumulator, control would be lost if that pressure became too low.

The electronic control, with its 8-bit microprocessor, is programmed not only for normal conditions of operation but also for warming up the engine and charging the accumulators. It also monitors the system continuously at a frequency of 20 Hz to detect malfunction and check that the safety system is operational at all times.

The inputs to microprocessor include a potentiometer coupled to the accelerator pedal for sensing the torque demanded by the driver and another connected to the brake pedal to sense the deceleration required. A position indicator senses whether the gearbox is in a drive ratio or neutral and a pulse pick-up senses the rotational speed of the propeller shaft and thus the vehicle speed. Another position indicator senses whether the power take-off clutch is engaged or disengaged. A potentiometer senses the displacement of the swashplate pump and a position indicator signals the volume of oil in the hydraulic reservoir, which determines when the engine should be brought into operation to take up the drive. Finally, there is a pressure sensor on the accumulator. When the accumulator is full, the incoming oil is diverted to the reservoir via a pressure relief valve.

Energy stored in the accumulator is locked in overnight by the shut-off valve, which is actuated automatically by the electronic control system. Consequently, the vehicle can be driven out of the garage in the morning, using stored energy. When it is in the open air the diesel engine can be started and the bus driven away, still using the hydraulic energy. As the speed rises to 22 mph (35 km/h), or if the hydraulic pressure drops below a predetermined level, the engine is automatically accelerated from idling up to the same speed as the propeller shaft, at which point the engine and gearbox take over from the hydraulic drive. When the brakes are applied, the engine reverts to idling and the hydraulic motor to a pumping mode, to charge up the accumulator. The friction brakes come into operation only if the control pedal is depressed

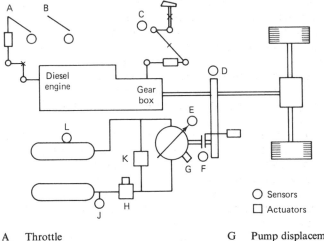

A	Throttle	G	Pump displacement
B	Brake pedal	H	Shut-off valve
C	Drive/neutral sensor	J	Pressure switch
D	Vehicle speed	K	Relief valve
E	Reservoir contents	L	Reservoir contents
F	Clutch engagement		

Fig. 33.40 Diagrammatic layout of the Volvo Cumulo control system, showing the sensors and actuators

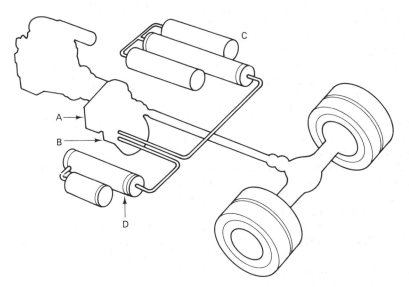

A   Power take-off gearing   B   Pump/motor unit   C   Hydraulic accomulator

Fig. 33.41 Of the three interconnected cylinders in the Cumulo system, the two outer ones contain nitrogen gas, while the central one also contains gas but is separated by a free piston from the hydraulic fluid

beyond a spring-loaded detent. Fully charged, the accumulator stores 0.22 kWh energy, which is adequate for running a half-laden vehicle at a constant slow speed for about three-quarters of a mile (1.2 km) or accelerating it, at a constant rate of 1.8 m/s², up to the cut-in speed of the engine.

The accumulator in Fig. 33.41 comprises three interconnected cylinders, the outer two containing only compressed nitrogen and the inner one both gas and hydraulic fluid separated by a free piston fitted with Teflon seals. When fully discharged, the gas pressure is about 200 bar and, fully charged, about 350 bar. The system is designed to bring a half-laden city bus to rest from 31 mph (50 km/h), the difference between this figure and that of 22 mph (35 km/h) for acceleration from rest being accounted for by the rolling and aerodynamic resistances. While the overall efficiency of the transmission is about 80 to 85%, that of the hydraulic system is 90 to 96%.

33.23   Lucas-Girling Skidchek GX

Following some years' experience with their original Skidchek system for heavy commercial vehicles, Lucas-Girling introduced the GX version. This was virtually a redesign based on an analysis in detail of the performance of the original system in service. The fundamental considerations, they found, were as follows.

To minimise the delay between the driver's observing a need to brake and the actual application, a rapid rise in air pressure is essential. It is this that is liable to cause a skid, because of the tendency for the pressure to overshoot the target pressure level and apply the brakes too hard. Unfortunately, however, it is not practicable to reduce the rate of pressure rise for the initial application without unacceptably increasing the stopping distance, as measured from the instant the driver depresses the foot valve.

The only possibility for reduction of this tendency to overshoot, therefore, is early detection of wheel-lock. One requirement is to have a sensor on each wheel and to trigger the air pressure release sequence following detection of incipient wheel-lock on the slowest wheel of the pair on an axle, or of two pairs on a bogie. Lucas-Girling considered that to economise by using only one sensor and placing it on the propeller shaft could increase the reaction time. This is because vibrations in the drive line can generate spurious signals and the electronic detection system then has to wait a few milliseconds to establish whether these signals are real before it can initiate an anti-lock sequence. Obviously, earlier detection of incipient wheel-lock might also be made by designing the electronic control system to respond to factors other than just wheel acceleration – for instance, to rate of change of acceleration, possibly in relation to wheel and vehicle speeds.

Once the initial brake application has been completed and the anti-lock cycles begin, however, it is possible to modify the rate of pressure build-up for each successive reapplication. This is done in the GX system by the use of a modified relay valve, which memorises the pressure at which the wheels previously locked. The new valve, called the *memory-controlled relay* (MCR) valve, is illustrated in Fig. 33.42. It is similar to a conventional relay valve except that a solenoid, a latch valve and the memory chamber have been added.

When the driver depresses the brake pedal, air from the foot valve enters at

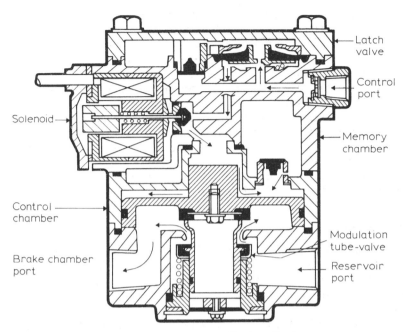

Fig. 33.42 MCR valve operation for rapid reapplication of brakes

the control port. The solenoid is not energised, so this air lifts the latch valve and flows rapidly down, past the solenoid valve – seated on its exhaust port – into the control chamber, pushing the control piston down. This causes the piston to seat on the end of the tube below, which is the exhaust port for the brake application system, and to push this tube downwards against its return spring. The latter action opens the modulation tube valve, by lowering it from its seating, thus allowing air to flow rapidly from the reservoir port, past this valve to the brakes.

When the electronic control unit detects a wheel-lock, it energises the solenoid, lifting the solenoid valve off its exhaust seat and closing it on to the control pressure supply port. This allows the pressure in the control chamber to exhaust past the solenoid to atmosphere, so the control piston rises again, closing the modulation tube valve and allowing the pressure in the brake line to exhaust rapidly to atmosphere through that tube valve.

In the meantime, while the brakes were on, two things have happened: first, air from the control port has leaked through the restricted orifice in the centre of the latch valve and equalised the pressure above and below it; secondly, the non-return valve in the base of the memory chamber has lifted, so the pressure in that chamber is that of the control system, as dictated by the force applied by the driver to his pedal.

Then, when the solenoid was energised and the pressure below the solenoid valve dropped to atmospheric, the valve in the base of the memory chamber dropped on to its seat, so the pressure in that chamber could fall only at the slow rate dictated by the size of the orifice on the right, just below that seat. Consequently, when – in response to a signal from the sensor indicating that the road wheel has regained its appropriate operating speed – the electronic control de-energises the solenoid, the pressure remaining in the memory chamber is dependent on both the original pressure that initiated the wheel-lock and the time that the wheel has taken to recover to its normal speed. The latter time of course is a function of the grip – or lack of it – of the wheel on the road surface.

This de-energisation of the solenoid – closing its exhaust valve and opening the control valve supply port – releases the control pressure into the control chamber. The resultant drop in control pressure pulls the latch valve down on to its seat, leaving only a small passage open past the seat to the control chamber. Despite the small size of this passage, however, the flow is still enough to move the control piston rapidly downwards, so that the rate of application of the brakes is still high. This rapid action is attributable to the fact that the volume of the control chamber is very small when the piston is in its uppermost position.

However, as the control pressure rises above the residual pressure in the memory chamber, the valve in the base of this chamber opens, suddenly adding its volume to that of the control chamber, with the result that the rate of pressure rise equally suddenly falls, remaining low for the remainder of the brake reapplication cycle. This reduces the tendency for the brake re-application pressure to overshoot its target value. At the same time, the point at which the fall in rate of pressure rise starts – dependent on the residual pressure in the memory chamber – is such that the target pressure for brake reapplication is related to the maximum pressure in the preceding cycle and, after the initial cycle, quickly adjusts to a value appropriate to the coefficient of friction in the tyre/road contact patch. Because of the reduced rate of rise to the target pressure and the absence of overshoot, the braking force remains longer at the optimum level, as indicated in Fig. 33.43, the stopping distance is reduced, and the whole sequence of operations is much smoother, thus avoiding the excitation of vibrations of either the cab or the whole vehicle.

In Fig. 33.44, the lower sine curve represents a typical characteristic of a simple commercial vehicle anti-lock system, for comparison with that of Skidchek GX, as depicted in Fig. 33.43. The upper sine curve in Fig. 33.44 is an indication of what can be achieved with an anti-skid system for a car. The difference between the two is attributable to the faster reaction time of a small

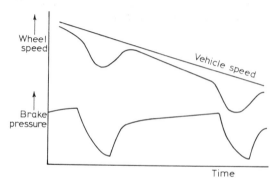

Fig. 33.43 GX anti-lock cycles

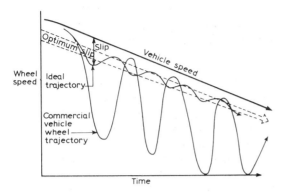

Fig. 33.44 Wheel-speed trajectory during anti-lock stop

hydraulic brake control system of the car, as compared with that of a much larger pneumatic system, and the relatively small rotational inertia of car wheels. A heavy commercial vehicle wheel takes a much longer time to spin up to normal speed after it has locked, especially on a very slippery surface. Obviously, under maximum braking conditions, a small degree of wheelslip is inevitable, the optimum being the range between the two dotted lines in Fig. 33.44.

A continuously-operating electronic monitoring unit, working in conjunction with the electronic control, causes the system to revert to normal braking, without anti-lock, and a warning lamp to be illuminated, in the following circumstances—

(1) Energisation of the solenoid for longer than 3 s.
(2) Solenoid open circuit.
(3) Inadequate output from the sensor.
(4) Regularly missing sensor pulses – due, perhaps, to damage to the sensor ring.
(5) Low supply voltage.

Should the fault clear during the journey, anti-lock is reinstated and the warning lamp extinguished. Similarly, a spurious wheel-lock signal caused by one wheel ceasing to spin following the application of excessive drive torque is cancelled as soon as rotation of both wheels is synchronised.

Chapter 34

Front axle and steering mechanism

Generally, the function of the steering system is thought of simply as that of providing a means whereby the driver can place his vehicle as accurately as practicable where he wants it to be on the road, for selection of the course he wants to steer round corners, and so that he can avoid other road users and obstructions. It must also, however, keep the vehicle stably on course regardless of irregularities in the surface over which the vehicle is travelling.

For the achievement of these basic aims, the first requirement is that, when the vehicle is travelling very slowly, all the wheels shall roll truly, that is, without any lateral slip. In Fig. 34.1, motion of the wheel along YY is pure rolling, along XX it is wholly slip. Motion along any other axis, ZZ for example, will have both rolling and slip components.

Since for all the wheels on a vehicle to roll truly they must all move in the same direction perpendicular to their axes XX, these axes must all intersect at a common point. If the vehicle is on a straight course, this point will be at infinity, in other words the axes will be parallel. On the other hand, if the vehicle is turning a corner, this point will always be the centre about which the vehicle as a whole is turning, and the tighter the turn the closer it will be to the vehicle.

Unless both the front and rear wheels are to be steered – impracticable on grounds of complexity, except in special circumstances, such as on vehicles having more than eight wheels, in which it may be virtually inescapable – the common centre must lie somewhere along the lines of the axis produced of the fixed rear axle. As can be seen from Fig. 34.2, this means that, when the front wheels are steered, their axes must be turned through different angles so that the point O of their intersection is always on that axis produced. With a beam axle this can be done by pivoting the whole axle assembly about a vertical axis midway between its ends. However, such an arrangement is impracticable for any but very slow vehicles.

Generally, the wheels are carried on stub axles, A and B in Fig. 34.2. Except with independent suspension, these stub axles are pivoted on the ends of the axle beam C which, since it is connected by the road springs to the chassis frame, remains in effect parallel to the rear axle, as viewed in plan. With independent suspension, the principle remains the same, even though the

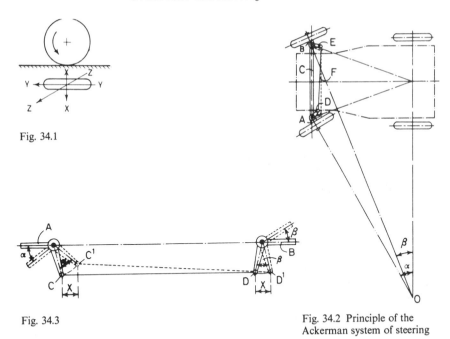

Fig. 34.1

Fig. 34.3

Fig. 34.2 Principle of the
Ackerman system of steering

mechanism is different in detail. The arms D and E together with their associated stub axles form what amounts to bell-crank levers pivoted on the kingpins and are used for coupling the two wheels so that they move together when they are steered. These arms are termed the *track arms* and are interconnected by the *track rod*. The actual steering is usually effected by a connecting link, called a *drag link*, between the steering gear and either what is termed the *steering arm* on the adjacent stub axle assembly or, in some instances, part of the track rod system.

34.1 Ackerman linkage

From the illustration it can be seen that there is a difference between the angles α and β through which the wheels on the inside and outside respectively of the curve have to be turned. In practice, this difference is obtained by setting the arms D and E at angles such that, in the straight-ahead position, shown dotted, lines drawn through the centres of the two pivots on each intersect near the centre of the rear axle. The exact position of this intersection point depends on the relationship between the wheelbase and track, and other factors.

From Fig. 34.3 can be seen how the stub axles are steered differentially by this linkage, the full lines depicting the straight-ahead and the dotted lines a steered condition. In the latter, the stub axle B has turned through an angle β and the end D of its track arm has moved to D′, a distance x parallel to the axle beam. Neglecting the slight angle of inclination of the track rod, it follows that the end C of the other track arm must move the same distance x parallel to the axle beam. This, however, entails movement of the arm C through a greater angle than D, because the latter is swinging across bottom dead centre, as viewed in plan, while the former is moving further from its corresponding lowest point. Although, for practical reasons, these arms may have to be curved, perhaps to clear some other part of the wheel or brake assembly, the effective arm remains that of a straight line joining the centres of the kingpin and the pivot at the opposite end.

The illustrations show the track rod behind the axle, but sometimes it is in front, again with suitably inclined arms. An advantage of placing it to the rear is the protection afforded to it by the axle beam, but it is then loaded in compression and therefore must be of stiffer construction. On the other hand, when it is in front, difficulty is generally experienced in providing clearance between its ball joints and the wheels.

With Ackerman steering, the wheels roll truly in only three positions – straight ahead, or when turned through a specifically chosen angle to the right and left. Even in the last two positions, true rolling occurs only at low speeds. At all other angles, the axes of the front wheels do not intersect on that of the rear wheels, while at the higher speeds the slip angles of the front and rear tyres usually differ and certainly those of the tyres on the outside will always differ from those on the inside of the curve. In all instances, the slip angle on both the front and rear wheels has the effect of turning their effective axes forwards.

Linkages giving virtually perfect static steering geometry on all locks have been devised, they have been complex and in practice have not proved satisfactory because they cannot take into account the variations in slip angle. The Ackerman principle, based on the best practicable compromise – usually slip angles are assumed to be equal on all four wheels – is satisfactory in practice, probably because flexing of the tyres accommodates the errors.

34.2 Multi-wheel vehicles

With six-wheel vehicles, perfect steering geometry is unobtainable because the two rear axles always remain parallel. An assumption is made, therefore, that the vehicle turns about a centre on an axis midway between and parallel to those of the two rear axles, and the steering linkage for the front axle is laid out on this basis. There will of course be some tyre scrub on both rear axles, but this is accommodated by the lateral flexibility of these tyres. Even so, the two back axles are always placed as close together as practicable within the limitations imposed by legal requirements and the size of tyres needed to support the rear end of the vehicle in the fully laden condition.

On eight-wheel vehicles, all four front wheels must be steered. A similar requirement arises on certain special-purpose six-wheel vehicles having a preponderance of weight at the front, and which therefore may have only a single axle at the rear. Both cases are represented in Fig. 34.4. On the six-wheel vehicle, each front axle has an Ackerman linkage, that for the leading axle designed for wheelbase W_2, and that for the second axle for wheelbase W_1. For the eight-wheel vehicle, the two wheelbases would be extended to midway between the two rear axles. All four front wheels have to be steered through different angles α, β, γ and δ. These differences are taken care of partly by the Ackerman linkage, there being several ways of catering also for those between the first and second axles. One is illustrated in the scrap view, in Fig. 34.4. Here, the two steering arms, V and U, on the foremost and second axles respectively are interconnected by a link. The arm U on the first is slightly longer than that at V, on the second axle, as measured between the axes of their kingpins and

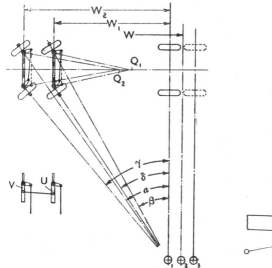

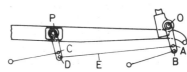

Fig. 34.4   Fig. 34.5

the pins connecting them to the link, so the leading axle is steered through the greater angle. One of these two arms is actuated in the usual way by a drag link from the drop arm on the steering box. Comments on the layout of the drag link are given in Section 35.12.

A more common method of interlinking two steered axles is illustrated in Fig. 34.5. Two drop arms OB and PD are used, one for each axle, the former being splined on to the spindle of the steering box and the latter pivoted on a bracket on the chassis frame. The arm PD is actuated by a link E connecting its end D to A on OB. Links are taken from B and C to the steering arms on the first and second axles respectively. Since PC is shorter than OB and, furthermore, OA is shorter than PD, the second axle is always moved through a smaller angle than the leading one.

34.3 Steering linkages for independent suspension

The Ackerman linkage already described is occasionally used with independent suspension, especially the sliding, or pillar, type. A system suitable for the single or double leading or trailing link type suspension is illustrated in Fig. 34.6. Its Ackerman linkage comprises two bell-cranks AX and BY interconnected by the track rod C and pivoted on brackets on the vehicle structure. Two drag links G and H connect the ends of the bell-cranks to the steering arms E and F, on the stub axles.

Ball joints must be used at E, F, X and Y. When the wheels are steered straight ahead, the centres X and Y must lie on the axes about which the stub axle assemblies oscillate as the suspension deflects. Then the suspension motion will not have any steering effect.

With single or double transverse link suspension systems, a divided track rod is necessary, the inner end of each part of which lies on the axis about

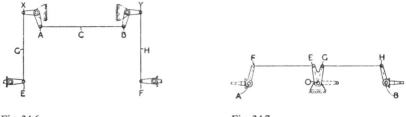

Fig. 34.6   Fig. 34.7

which the adjacent stub axle assembly oscillates as the suspension rises and falls. Such a system is illustrated in Fig. 34.7, where A and B are the stub axle assemblies and OEG is a triangular link, or an acute angle bell-crank lever, pivoted at O on the vehicle structure, or frame. Links EF and GH are the two parts of the divided track rod.

Clearly the links OE, EF, FA constitute an Ackerman system for the offside wheel, and OG, GH, HB form a similar system for the nearside wheel. The arm shown dotted on the triangular link is connected by a rod to the steering gear drop arm, which is arranged so that it oscillates in a horizontal plane. Obviously, if the triangular lever cannot be accommodated as illustrated, its pivot O can be moved to the other side of the track rod, and the two pieces of the track rod do not necessarily have to be in line. Where a rack-and-pinion steering gear is used, the triangular link can be omitted and the joints E and G are either on the ends of the rack, as in Fig. 34.8, or linked separately to only one end of it.

34.4   Centre-point steering

In Fig. 34.9 it will be noticed that the axis about which the stub axle (with its road wheel) turns when steering is effected intersects the ground at the point O,

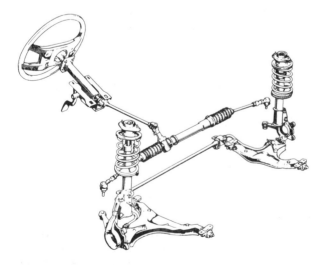

Fig. 34.8 Lancia *Delta* front suspension and steering system

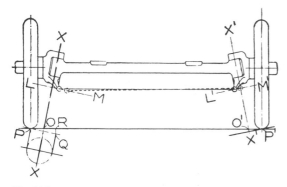

Fig. 34.9

while the centre area of contact between the tyre and the road is P. Now, when a car is travelling along a road there is a force acting in the plane of contact between the front tyres and the road in a direction opposite to that of the motion. On a good road this force may be small, but on a bad road, and when the front wheels are braked hard, it may be considerable. This force acts perpendicular to the OP (in plan) and hence has a moment, about the axis XX, which tends to turn the stub axle about its pivot pin. This tendency has, of course, to be resisted. Now, the stub axle on the other side tends to turn in the opposite direction, and since the two stub axles are connected together by the track rod the two tendencies towards rotation will, if they are equal, neutralise each other, and the only result will be a stress in the track arms and the track rod. If, however, the two tendencies are not equal, the difference between them has to be resisted by the friction in the steering mechanism or by the driver.

In order to reduce to the minimum the moment of the force tending to turn the stub axles the point O is brought very much closer to P than might be inferred from the diagram, and in some instances these two points may coincide; alternatively, point O may even be outboard of P. When O and P coincide, the geometry is said to have *zero-scrub* or to be of the *centre-point steering* layout; if O is inboard of P, it is said to have a *positive scrub radius*; while if outboard it has a *negative scrub radius*. Incidentally, the radius is not OP, but is the distance measured perpendicularly from P to the axis of the swivel pin.

There are three ways of changing the relative positions of O and P, as follows—

(1) Keeping the swivel pin axes vertical the wheels can be inclined so as to bring the point P nearer the point O.
(2) Both the wheels and the swivel pins can be inclined.
(3) The arrangement can be such that part or all of the axis of the swivel pin is closer to, or even coincident with, the central plane of the tyre by, for instance, dishing the wheel discs to bring the rims further inboard.

Positive scrub is rarely used. Zero scrub is applicable in certain circumstances, to lighten the loads on the steering controls and is especially suitable where tyres of very large section are used, for example in specialist cross-country vehicles. Negative scrub, with toe-in, is the most widely favoured, for the reasons outlined in Section 34.20.

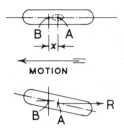

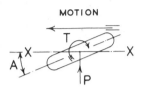

Fig. 34.10 Diagram of castor action   Fig. 34.11 Diagram of slip-angle and
                                          self-righting torque

With inclined swivel pin axes, two points should be noted. First, the connections between the track arms and track rod must be of the ball-and-socket type since, as can be seen from Fig. 34.9, when the stub axles are rotated about these inclined axes, the ends of the track arms not only rotate round them but one also swings up and the other down, and of course a simple bush-type bearing could not accommodate all these movements. Secondly, except with centre-point steering, turning the wheel in either direction away from its straight ahead alignment lifts slightly the front end of the vehicle. This is because the point P rotates in a circle in the plane PQ, which is perpendicular to the axis of the swivel pin. It follows that, since the weight of the vehicle is at all times tending to cause it to sink to the lowest possible position, it has a self-centring effect on the steering, tending to return it the straight-ahead position.

34.5   Castoring or trailing action

The swivel axis about which the wheel is turned for steering purposes is generally inclined in a fore-and-aft direction by a few degrees so that its intersection with the ground (B in Fig. 34.10) lies slightly in front of the point of contact A of the tyre with the ground or rather, since this contact is over a small area, in front of the centre of this contact patch. The distance, x, between the two points is called the *trail.*

The object sought is to stabilise the wheel under running conditions so that the wheel tends to remain with its plane parallel to the direction of motion. Fig. 34.10 shows the wheel at a slight angle to the direction of motion and the force R, acting at the road surface, which will be parallel to the direction of motion, is seen to have a moment about the point B (that is about the swivel axis) that tends to bring the wheel back into the plane of motion just as the castor wheel on a chair leg does; hence the term *castor action.* If however the force acted in the opposite direction, the trail would be disadvantageous since it would tend to make the wheel turn through 180°.

Since in front-wheel-driven cars the driving force is in the forwards direction it would seem that in such cars the trail provided should be negative and this is sometimes the case. However a compromise has to be made in any case between the requirements during driving and braking conditions.

In practice the steering of a wheeled vehicle is not quite such a simple matter as these considerations would imply, because it has been found by experiment that a wheel which is rolling along a road cannot sustain any side force unless it is held so that its plane makes an angle with the direction of motion. Thus in

Fig. 34.11 if the wheel is required to travel in the direction XX while a side force *P* is applied to it the wheel must be held so that its plane makes an angle A (called the *slip angle*) with XX, as shown. It has been found that the side force that can be sustained is proportional to the magnitude of the slip angle for values of the latter up to about 6°. The ratio *side force sustained/slip angle* has been taken as a measure of the cornering ability of a tyre and has been called the *cornering power.*

34.6   Cornering power

The cornering power of tyres has been found to depend on many factors such as the construction of the tyre itself, the value of the vertical force between the tyre and the road (referred to in what follows as the load on the tyre), the inflation pressure, the size of the tyre and the extent of any tilting of the wheel. Thus the cornering power of a tyre falls off as the load on the tyre departs from the normal load for the tyre, but the extent of this falling off is small provided the variation in the load does not exceed plus or minus about 50% of the normal load. The cornering power increases as the inflation pressure is increased, but is smaller for large tyres than for small ones of the same type of construction. As regards camber the cornering power falls off as the top of the wheel in Fig. 34.11 is moved in the direction of the force *P* (this being called *positive camber*) and increases with the amount of negative camber. The cornering power has been found to be independent of the speed.

34.7   Self-righting torque

When a wheel is travelling along the line XX as in Fig. 34.11 there will be a torque *T* acting between it and the road which will tend to turn the wheel so that its plane becomes parallel to the direction of motion, and in order to keep the wheel travelling as shown an equal and opposite torque must be applied to the axle carrying the wheel. The torque *T* may be called the *self-righting torque* and it has been found to increase in direct proportion to the load on the wheel and to be greater for wheels with positive camber than for vertical wheels.

34.8   Steering characteristics – oversteer and understeer

The result of the above actions is that when a vehicle moves in a circular path the centre of that path does not coincide with the point of intersection of the wheel axes. This is indicated in Fig. 34.12, where O is the intersection of the wheel axes and O_1 is the actual centre of rotation. The slip angles for the wheels (A_1, A_2, A_3 and A_4) will in general all have different magnitudes.

If a side force *P* acts on a car that is travelling in a straight line XX, Fig. 34.13, that force must be balanced by side forces acting between the road and the wheels and the wheels must be set at the appropriate slip angles to the direction of motion. If the slip angles for all the wheels were equal then the car would continue to move in a straight line but inclined at that slip angle to the original path, but if the slip angles of the wheels are not equal then the car will generally move in some curve. If, for example, the slip angles of the back wheels are a little greater than those for the front wheels, the car will begin to move about some centre O as shown. This would introduce a centrifugal force which

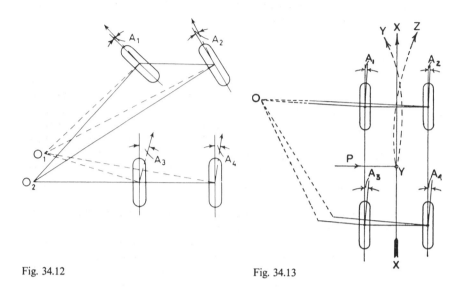

Fig. 34.12   Fig. 34.13

in effect would increase the magnitude of the side force *P* and would thus accentuate the action. The car will tend to move in a curve such as YY, veering towards the side force *P*. This action has been called *oversteer* and tends to increase with any increase in the speed, because the centrifugal force increases as the square of the speed, while the side forces between the wheels and the road do not increase with the speed but only as a result of an increase in the slip angles. A car with this steering characteristic when going round a corner may require the steering wheel to be turned back towards the straight-ahead position in order to keep it from turning too sharply and is generally less stable and more difficult to handle than a car with an *understeer* characteristic. The latter would be obtained if the slip angles of the front wheels were greater than those of the back wheels, for then the centre of rotation would lie on the opposite side of the line XX and the car would tend to move in a curve such as YZ. The centrifugal force due to this motion opposes the side force *P* and the action will tend to decrease with any increase in the speed. A car with an understeer characteristic when turning a corner will tend to straighten out and the steering wheel will generally have to be turned a little more to counteract this tendency. Such a car is more stable and easier to handle. An oversteer characteristic can sometimes be corrected by decreasing the slip angles of the rear tyres and increasing those of the front tyres by increasing the inflation pressure of the rear tyres and decreasing that of the front tyres or by altering the weight distribution so that a smaller proportion of both the load and the side force comes on to the rear wheels.

Since both the slip angle and the self-righting torque are affected by alteration of camber the steering characteristic can be controlled to some extent by controlling the change of camber of the wheels due to a tilt of the body of the car relative to the ground. This can be done with independent suspensions of the double-arm type (see Sections 35.12 and 36.3). Most drivers prefer oversteer, though understeer is quite acceptable to those accustomed to it. In jest, it has been said that the preference depends on whether one prefers to slide off the road forwards or backwards!

34.9   Axle beam

When the front wheels are not braked the axle beam is usually a simple forging of I section with suitable seats for the attachment of the springs and with the ends suitably shaped to carry the stub axles. The I section is adopted because it is the best adapted to withstand the bending action to which the beam is subjected. This action arises, as in the back axle, because the axle beam is supported at its ends, while the loads are transmitted to it at the spring seats which are situated nearer to its centre. The action tends to bend the beam in a vertical plane. Simple rectangular and tubular sections are also sometimes used.

The axle beam is also subjected to a bending action in the horizontal plane, but usually this action is small compared with that in the vertical plane. But when the front wheels are braked the horizontal bending action becomes considerable. In this case a rectangular or a circular section may be more suitable than an I section. Again, when the front wheels are braked the portions of the axle beam between the stub axles and the spring seats (and possibly the rest of the axle also) are subjected to a twisting action, the torque reaction of the braking torque applied to the brake drums. A circular section is best adapted to withstand this twisting action. Hence, when front wheel brakes are fitted, circular sections may be used.

34.10   Stub-axle construction

The principal methods of pivoting the stub axles on the ends of the axle beam are shown in Fig. 34.14. The first goes by the name of the *Elliot stub axle*, the second is called the *reversed Elliot*, the third is sometimes called the *Lemoine*. The third and fourth methods are sometimes used upside down, that is, with the wheel axis below the axle beam instead of above it.

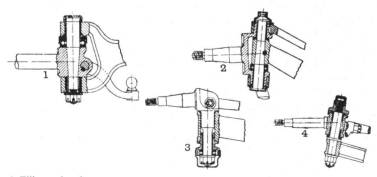

1 Elliot stub axle
2 Reversed Elliot
3 Lemoine
4 Lemoine (inverted)

Fig. 34.14

In the Elliot construction the swivel pin is usually fixed in the stub-axle forging and its ends therefore turn in the forked end of the axle beam. In the reversed Elliot construction the swivel pin is usually fixed in the axle beam, so that in each construction the bearing surfaces are situated as far apart as possible and when wear occurs on those surfaces the resulting angular shake or play is as small as possible.

Although the use of a front axle is still the commonest construction for commercial vehicles, they are rarely now used in cars. The above discussion of the principles involved is, however, relevant to the independent suspensions used in cars, but a description of these suspensions is deferred to Chapter 36.

34.11   Wheel bearings

On cars, wheel bearings are not severely loaded. The static load is generally in the range 250 to 400 kg and maximum speeds are of the order of 700 rev/min. In addition to the dynamic loading accompanying suspension deflections, the following loads also have to be reacted: brake drag, traction in the case of driving wheels, and centrifugal forces due to cornering. Obviously the cornering force, applied at the rolling radius of the tyre, produces a couple about a wheel centre, which has to be reacted by the two bearings and therefore adds to the vertical load – including any extra due to weight transfer – on one and subtracts from that on the other. However, on average, a car spends only about 5% of its running time cornering, divided equally between left- and right-hand turns.

Ball or roller bearings are generally used, and the impact loading on them is not normally high. After all, there are few other applications in which such bearings are cushioned on both sides – in the case under discussion, by the tyre on one hand and suspension spring on the other. The choice between ball and roller bearings is based on two main factors: first, whether radial or axial compactness is required and, secondly, the life specified.

Where the roller type exclusively are used, taper-roller bearings are virtually essential for taking the combined axial and radial loading. Because they make line contact with their races, they tend to have long lives. They certainly are more compact radially and, for any given design loading and outside diameter, can accommodate a shaft of larger diameter than a ball bearing. Assembly is easy because a ring of caged rollers on an inner race can be placed in position on the axle independently of the outer race in the hub. Adjustment of bearing clearance, using shims between the inner races, is also easy. The principles are outlined in an article in *Automotive Engineer*, December 1979.

A typical taper roller bearing assembly for a driven wheel is illustrated in Fig. 34.15, which is that of the Ford *Fiesta*, while for a wheel that is not driven, Fig. 34.16 illustrates a good design, also Ford *Fiesta*, and Fig. 34.17 an even more compact layout. All have Timken bearings and are greased and sealed for life. Interesting features in two of these designs include the abutment of the inner races so that only the shoulder separating the two outer bearings has to be machined to close tolerances – all the other tight tolerances are in the bearings themselves which, in any case, are precision components.

In Fig. 34.17, a major factor in keeping the axial length of the assembly to a minimum is the housing of the seals so that they bear on the inner races. With this design, the stub axle is so short that it is loaded more in shear

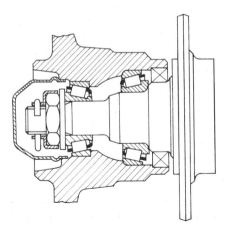

Fig. 34.16

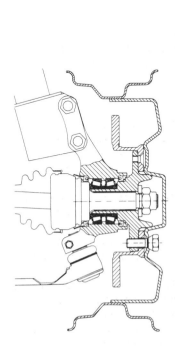

Fig. 34.15 The Ford *Fiesta* front
drive wheel bearing arrangement using
set-right adjusted bearings

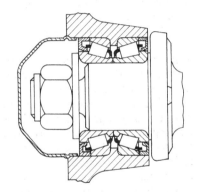

Fig. 34.17

than in bending, which makes it inherently more resistant to failure. In the
design in Fig. 34.16, on the other hand, advantage has been taken of the
bending moment diagram characteristic of a cantilever to reduce the diameter of the hub and bearing at the outer end. In Fig. 34.15, full benefit has
been derived from the capability of taper roller bearings to accept the large
diameter shaft needed for transmitting the drive as well as serving as a stub
axle.

The most widely used ball bearing arrangement is that illustrated in Fig.
34.18, because it is the most economical of all, but double-row angular contact
ball bearings, as in Fig. 34.19 and 34.20 are being looked on with increasing
favour. These are all RHP bearings and are of the angular-contact type
designed to take both axial and radial loading. The illustrations are from an
article on hub design published in *Automotive Engineer*, April/May 1980.

Although each ball has, in theory, only point contacts with the races, this has
two advantages: first, it keeps friction to a minimum; secondly, because of the
relative ease with which this tiny area of contact can be increased elastically
under load without excessive generation of heat, such a bearing is more

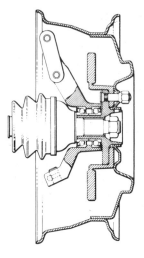 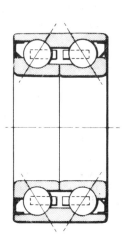 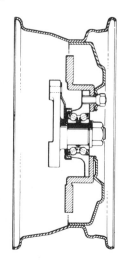

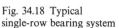

Fig. 34.18 Typical
single-row bearing system

Fig. 34.19 Double-row
ball bearing, first generation

Fig. 34.20 Double-row ball
bearing, second generation

tolerant of pre-loads than are taper rollers. On the other hand, shock loading or severe overloading can quickly cause fatigue failure.

Owing to the absence of play in pre-loaded bearings, precise control is obtainable over both steering and the concentricity of brakes, so they are favoured for racing and sports cars. Another advantage of ball bearings is that, if they fail, the symptoms are simply excessive noise, whereas a taper-roller bearing is liable to seize and therefore might be dangerous.

In all types, a most important requirement is good sealing, to keep lubricant in and dirt and water out. With a double row angular contact bearing, generally a double-lip-type seal at each end bears on the inner race and is a tight interference fit in the outer race, the complete assembly being pre-greased and sealed for life. Lip-type seals with garter springs are usually fitted externally with single row bearings. Shrouds, thrower rings and baffles are widely used, in conjunction with the seals at the inner ends of road wheel hubs, as in Fig. 34.15, to keep water out. The outer ends are generally well protected by hub caps and nave plates. These last mentioned terms, incidentally, are often confused: a hub cap is a small cup shape screwed on or snapped over or into the end of the hub to retain the grease and keep the bearing clean; a nave plate is a large diameter circular disc, usually chromium-plated, snapped on or sprung into the outer face of the wheel to hide its retaining nuts and improve its tidiness and general appearance.

34.12 Steering column

Rigid steering columns are now ruled out by legal requirements in most countries, because of the risk of their penetrating or otherwise severely damaging the chests of drivers in the event of collisions. Either the steering shaft is in three parts set at angles relative to each other and connected by universal joints, as in Fig. 34.8, so that an impact will cause them to fold, or a sliding coupling is interposed in the column, or one that is made of a convoluted section that will concertina if an axial load is applied.

At some point between the wheel at the top and the connection at the bottom to the steering arm on the stub axle – usually at the bottom of the column – a system of levers or gearing is incorporated to reduce the effort that has to be applied to the wheel at the top to steer the vehicle. It follows that the steering wheel has to be turned through a larger angle than the stub axle assemblies. The actual ratio may range up to well over 25:1, according to the weight of the vehicle, size and type of tyre and other factors such as whether ladies are likely to drive, and the top speed. For cars, the average ratio ranges from about 15 to 22:1, giving between three and $4\frac{1}{2}$ turns from lock to lock. Where power assistance is provided, the ratios are smaller. A disadvantage of a high ratio is that it may make it difficult for the driver to actuate the steering rapidly enough to correct a skid or, on heavy commercial vehicles, even to get round a sharp corner.

34.13 Reversible and irreversible steering

With most steering gears, if the road wheels are gripped, they can be swung about their kingpins, causing the steering wheel inside the car to rotate. If, however, the friction forces within the steering gear are high enough to prevent this, the gear is said to be irreversible. In general, the higher the steering ratio, the greater is the tendency to irreversibility, since a high ratio helps the driver to overcome the frictional resistance when he uses the steering wheel but multiplies that resistance if he applies his effort at the road wheels.

Early steering gears were of the screw-and-nut type. These were succeeded by the worm-and-sector, which was followed by the cam-, or screw-and-lever with first a fixed stud, or cam follower, and later a bearing-mounted stud or follower. A still later development was the recirculating-ball type, in which a train of balls was interposed between the threads of a screw and nut, with a return channel for recirculating them. Now, however, the most popular type, at least for cars, is the rack-and-pinion.

There have been many variants of these steering gears. Examples of two of the cam-and-lever type are illustrated in Figs 34.24 and 34.25. A third is the Ross cam-and-twin lever type which is similar to that in Fig. 34.24, but the lever is forked to carry two studs, one registering at each end of the cam, or screw. Its advantages are a halving of the wear in the straight-ahead position and an increasing mechanical advantage towards the extremes of wheel movement, because the relevant lever is then nearest to its top dead centre position. The worm-and-sector type is simply a variant of the worm-and-worm-wheel alternative, but only a sector of the worm-wheel is used in it. Various other forms of gearing have been employed, including straight and helical spur gear pairs and epicyclic trains.

34.14 Rack-and-pinion steering mechanism

This is especially suitable for cars with double transverse wishbone or MacPherson strut-type suspension. Its principal advantages, in addition to accommodating relatively easily with the suspension geometry, are the positive feel of such a system, high efficiency, simplicity, the relative ease with which it can be rigidly mounted on the vehicle structure and the consequent precision of the system. Moreover, by increasing the spiral angle of the teeth, a high ratio is obtainable with a pinion that is small yet has teeth of adequate section for taking the loads to which they are subjected.

In Fig. 34.21, the rack is housed in a tubular casing, which has rubber bushes around its ends by means of which it is secured to the structure of the vehicle. These bushes help to absorb high-frequency vibrations and reduce the severity of shocks transmitted back to the steering wheel.

A spring-loaded pad presses the rack into contact with the pinion, which is integral with a short spindle splined to the steering wheel shaft. This spindle is carried by plain bearings in the cast housing on one end of the tubular housing for the rack. Ball joints connect the ends of the rack to the track rods. The centres of these ball joints are approximately in line with the axes about which the stub axle assemblies oscillate, so suspension deflection does not have any significant steering effect.

34.15   Screw-and-nut mechanism

An example of this type of steering gear is shown in Fig. 34.22. A multiple-threaded screw B is free to rotate in bearings in a cast iron casing C. Axial motion is prevented by thrust bearings DD and the screw is connected by the shaft A to the steering wheel. A nut H fits on the screw and is prevented from

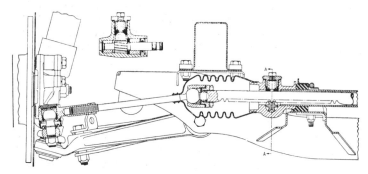

Fig. 34.21

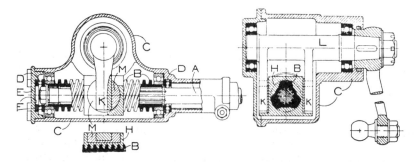

Fig. 34.22 Screw-and-nut steering mechanism

rotating. Hence, if the screw is turned the nut must move axially up or down the screw. This motion is caused to rotate the drop arm through the medium of a two-armed spindle L, which is carried in ball bearings in the casing and which carries the drop arm at its outer end. The two arms KK of the spindle L straddle the nut, to which they are connected by the bronze pads MM. These pads are free to turn in cylindrical recesses formed in the sides of the nut, and they are provided with parallel grooves to receive the arms of the spindle L. The pads are necessary because the nut moves in a straight line while the arms of the spindle L move in circular paths.

In this design the nut is prevented from rotating by the arms K of the spindle L. In some designs this is done by guiding the nut in guides in the casing. In some designs also short connecting rods are used, instead of the pads, to connect the nut with the arms of the spindle L.

Backlash in the above mechanism can arise from end play of the screw and from wear of the threads of the screw and nut, etc. End play of the screw is eliminated in assembly by means of the selected washer F. Wear of the threads cannot usually be remedied except by replacement, but in view of the large area of contact between the nut and the screw the wear should be small provided that lubrication is attended to.

This type of mechanism has the advantage that the leverage provided increases as the steering approaches full lock.

A slightly different form of screw-and-nut mechanism is shown in Fig. 34.23. The screw is formed on the end of the shaft A to which the steering wheel is fixed. The nut is formed with integral trunnions BB which pivot in holes in the ends of arms CC of the fork member, which is splined to the drop arm shaft D. At its upper end the shaft A is supported in the steering column in a ball and socket joint E which secures it against axial motion but allows it to turn about its axis and also to swing slightly about an axis perpendicular to the axis of the shaft itself. This last freedom is required because the trunnions BB must move in the arc XX as the nut moves to and fro along the screw, and so the lower end of the shaft A is moved slightly in the direction YY. Actually, instead of a plain ball-and-socket joint at E, a self-aligning ball bearing is used. Alternatively an ordinary journal bearing supported in a rubber bush may be used, the rubber bush accommodating the rocking of the shaft A. This arrangement eliminates the pads M of the mechanism of Fig. 34.22, reduces the number of bearings required and so cheapens the construction. The fixing of the nut against rotation is also more easily done by the arms CC than by the arms KK of Fig. 34.22.

34.16   Cam steering mechanisms

Many cam steering mechanisms have been invented but few have been commercially successful. One example which is used by many makers is shown in Fig. 34.24. It is the Marles steering gear. Carried by the drop arm spindle A on the ball bearings shown is a V-shaped roller B which engages a groove cut in the member C. The latter is keyed to the steering wheel shaft but is fixed axially. When the steering wheel is turned the spiral groove in the member C constrains the roller B to move to the right or left from the position shown, thus turning the drop arm. End play of the member C can be eliminated by screwing the casing E farther into the steering box, a clamping screw F being

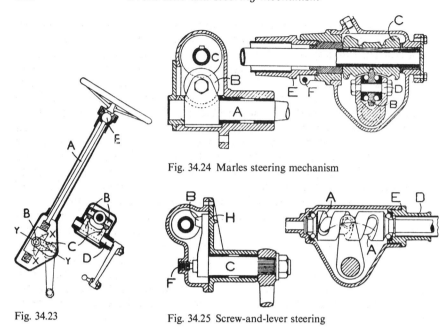

Fig. 34.24 Marles steering mechanism

Fig. 34.23

Fig. 34.25 Screw-and-lever steering

provided for locking purposes. Proper meshing of the roller with the groove in the member C may be obtained by turning the pin D. The centre portion of this pin, which carries the races that support the roller, is eccentric to the end portions, hence when the pin is turned the roller is moved into closer mesh with the member C.

34.17   Screw-and-lever mechanism

This steering unit is a form of cam mechanism and an example is shown in Fig. 34.25. At the bottom end of the steering wheel shaft a helical groove A is formed, and this engages the projection B of the drop arm spindle lever. When the steering wheel shaft is turned the drop arm is rocked to and fro. End play of the steering wheel shaft is eliminated by putting in a suitable washer at E. The drop arm spindle lever is supported by bearing on the cover plate at H, and a screw F prevents it from meshing too deeply with the groove A. The pin B is sometimes made in the form of a roller, being carried in bearings in the drop arm spindle lever. A recent modification is to provide the lever with two pins which, in the central position of the gear, engage the screw near its ends. This enables increased leverage to be obtained when the steering is locked hard over.

In the Marles Weller steering gear, whose general arrangement is similar to that of Fig. 34.25, the pin B is formed with two hemispherical recesses one on each side and half-balls fit in these recesses. The approximately flat faces of the half-balls engage the sides of the screw thread.

34.18   Steering connections

The drop arm, or rack, of the steering mechanism is connected to the steering arm of the stub axle and the connections must be such as will allow angular motion in two planes. Ball-and-socket joints are now used almost always but at one time universal joints were used. An example of modern ball-and-socket joint construction appears in Fig. 34.26, and this is a design by Automotive Products. The bush A is split on a vertical diameter and can thus be made to contact the ball member both above and below the central horizontal plane and so this bush carries all the actual steering loads, the lower member B being spring loaded to eliminate backlash and rattle. The bushes are moulded in a composition which contains a lubricant and the joint requires no other lubrication throughout its life. The ball and bush assembly is retained in the solid eye of the link C by rolling the lip of the eye over, as seen at D. Dust and water are excluded by the rubber boot E. The surface of the ball is plated to give protection against corrosion and to provide a fine surface finish. Joints of this kind can provide considerable angular freedom in all directions.

In the joint of Fig. 34.27 the upper, load bearing, surface is formed on a bush A which is free to rotate on the conical end of the pin C. This enables a large spherical surface to be provided while at the same time reducing the relative motion between that surface and the eye of the rod. This is because motions about the axis XX will occur at the surface C. A spring-loaded cup B holds the parts in contact and limits any separation to the amount of the clearance between the bottom of the cup and the cap D. When the angular motions about XX are not large the member A is usually made integral with the pin.

34.19 Alignment of the front wheels

When a vehicle is moving in a straight line all its wheels should be parallel to the direction of motion. Unless the back axle has been badly damaged the back wheels will be so, but proper adjustment is necessary if the front wheels are to be so. In order that the front wheels shall be parallel when the vehicle is moving they must usually be set slightly out of parallel when the vehicle is stationary, the distance between the wheels in the plan view being made slightly less at the front than at the back. Then, when the vehicle is moving along the road the forces acting on the wheels will cause small deflections in the steering connections, which will bring the wheels into parallelism. This setting of the wheels is known as *toe-in* and the difference between the 'track' of the wheel rims at the front and back varies with different makers from almost nothing up

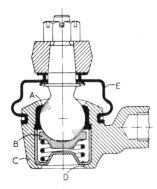

Fig. 34.26

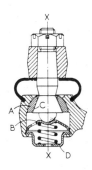

Fig. 34.27

to as much as 6 mm, though it is generally restricted to about 3 mm, to avoid significantly increasing tyre wear.

34.20 Effect of toe-in on steering

A tyre steered straight ahead is in an inherently unstable condition because it cannot exert any side force. Consequently, if it is disturbed by, for example, rotational imbalance or hitting a bump, and there is any play or elasticity in the bearings of the wheel, suspension or in the steering system, it will shimmy. One reason for toe-in therefore is to prevent shimmy by ensuring that there is a constant side force large enough to overcome the small disturbing forces to which the tyre is constantly subjected while rotating. Another, and the one that determines that it should be toe-in rather than toe-out, is the unavoidable deflections in the steering system when braking or acceleration loads transmitted through the wheels and tyres are counteracted by it, as will be seen below. Others are related to the interactions between the suspension and steering mechanisms.

In Section 34.4 it was noted that the point O is in most instances not coincident with P. Its position, governed by the previously mentioned interactions, has hitherto been set on the basis of experience and further improved upon by road tests. Recently, however, computer programs have been developed for getting it as nearly right as possible before the design has progressed beyond the drawing board stage.

The aim is at avoiding self-steering effects regardless of whether the vehicle is being run at constant speed or braked or accelerated, with either identical or icy or other surfaces under the wheels on either side of the vehicle. It is also required as a measure for the avoidance of variations in steering characteristics with either changes in speed or radius of turn. This latter mode of stability is necessary to prevent the driver from being taken by surprise by a sudden change in response to the effort he is applying to the steering wheel. Most important among the local dependent variables are: toe-in, camber angle, scrub radius and deflections of the suspension bushes. Deflections and other characteristics of tyres and rubber bushes are obtainable from the manufacturers of those components, while motions dependent on suspension and steering geometry can of course be represented graphically. A particularly good example of design to take into account all these factors is the General Motors ACT suspension, described in the August/September issue of *Automotive Engineer*, 1987.

34.21 Power steering

It has been seen that when the size and weight of a vehicle become large it may be necessary to have servo-assisted or power-operated brakes and, for the same reasons, it may also become necessary to employ power steering. Several systems have been developed by various makers and three will be described. They are all similar in principle and are all operated by oil under pressure. This oil is supplied by a pump driven off the engine, or sometimes off the rear end of the gearbox. An accumulator or reservoir is usually provided. The pressures used are fairly high, reaching $7000 \, \text{kN/m}^2$. Systems using compressed air have also been developed but the principle of operation is the same as that of the systems about to be described. This is that the first slight movement of the steering wheel operates a valve so as to open up a passage for pressure oil from the pump and accumulator to the appropriate end of a *slave cylinder* whose piston is connected to the steering linkage. A pressure difference is thus built up across the sides of the piston which therefore moves and actuates the steering. The movement of the piston is arranged to 're-set' the valve, that is, to bring it back to the equilibrium position and so the movement of the piston and steering linkage is made proportional to the movement of the steering wheel. The systems are always arranged so that the steering linkage can be actuated direct by the steering wheel if the oil pressure should not be available. They are also sometimes arranged so that some appreciable effort must be applied to the steering wheel before the valve of the system is operated; this effort is transmitted direct to the steering linkage and, if the forces and moments opposing motions of the road wheels are small, this direct effort may be sufficient to overcome them and the steering is then done direct without any power assistance. This is claimed to give a good 'feel' to the steering. When the resistances are larger then the valve is moved and the power effort is added to the direct manual effort.

34.22   Vickers system

The principal features of the system developed by Vickers Incorporated of Detroit are shown in Figs 34.28 and 34.29. The slave cylinder A, Fig. 34.28, takes the place of the drag link of the steering linkage being connected by a ball-and-socket joint B to the steering arm of the stub axle while the piston rod is coupled by the ball-and-socket joint C to the frame. The drop arm D of the steering box is arranged to actuate the valve of the system, as will be seen from Fig. 34.29. Supposing the steering wheel to be turned so that the drop arm turns clockwise then the valve E will be moved to the left, thus opening the port x to the pressure supply and the port y to the exhaust. The pressure in the end F of the slave cylinder will therefore rise while that in the end G will fall. The cylinder will therefore move bodily to the left, thus actuating the steering. The movement of the cylinder brings the valve E back to the equilibrium position and the desired movement of the steering will have been effected, chiefly by the oil pressure but partly by the direct pressure of the drop arm on the left-hand spring of the valve. The movement of the valve is limited to a very small amount either way and if the oil pressure should not be available then the drop arm can actuate the steering direct as soon as the valve comes up against its stop. There is therefore some backlash, the amount depending on the free movement allowed to the valve, when the steering is being operated manually.

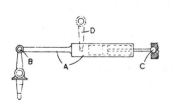

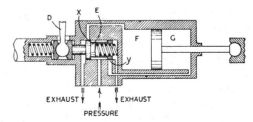

Fig. 34.28   Fig. 34.29

In order that the slave cylinder shall not exert any drag on the system when it is being operated manually a valve is provided which opens a direct passage between the ends of the cylinder if the oil pressure fails and the movement of the control valve exceeds the normal amount.

34.23   Ross system

This is shown in Figs 34.30 to 34.32. The steering wheel is fixed to the shaft A which at its lower end is provided with a screw or cam B, the groove of which is engaged by the pin C. The latter is carried in taper roller bearings in an arm secured to the drop arm shaft D, the drop arm E is also fixed to this shaft and is actuated by the piston of the slave cylinder through a yoke member that engages a sliding block carried on the pin G of the arm F. The latter arm is also fixed to the drop arm shaft. The shaft A and cam B are, in effect, integral with the valve of the system which is situated at H and is shown in more detail in Figs 34.31 and 34.32. Supposing the steering wheel to be turned clockwise then for the first slight movement the pin C will remain fixed and the cam B, shaft A and valve will move to the left. The detailed action of the valve will be described later but when it moves to the left the pressure in the pipe *j* will rise and that in *k* will fall and so the piston of the slave cylinder will be urged to the left thereby rotating the drop arm anti-clockwise and actuating the steering. The anti-clockwise movement of the lever and pin C will move the valve back to its equilibrium position. The action of the valve is as follows. Referring to Fig. 34.31, the valve position, when the system is in equilibrium, will be such

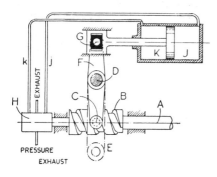

Fig. 34.30

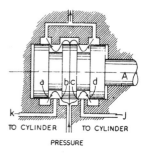

Fig. 34.31

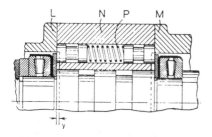

Fig. 34.32

that the areas of the passages *a*, *b*, *c* and *d* will be equal (neglecting for the moment the effect of the piston rod of the slave cylinder), the pressure drops across the passages *b* and *c* will be equal and the pressure drops across the passages *a* and *d* will also be equal. The pressures in the recesses of the valve will therefore also be equal and these pressures are transmitted by the pipes *j* and *k* to the ends of the slave cylinder. If now the valve is moved to the left the passages *b* and *d* will be reduced in size while the passages *a* and *c* will be increased. The pressure drop across *b* will increase while that across *c* will decrease, the pressure drop across *a* will decrease while that across *d* will increase. Hence the pressure in the left-hand recess and pipe *k* will fall and the pressure in the right-hand recess and pipe *j* will rise as described above. The effect of the piston rod of the slave cylinder is to reduce the effective area of the end K of the cylinder and as the forces acting on the piston must be balanced in the equilibrium position the pressure in the end K must be a little higher than in the end J. This is obtained automatically, the valve setting itself so that the area *c* is slightly larger than *b* and the area *d* slightly smaller than *a*. The valve is provided with a number of centring springs P, Fig. 34.32; in the equilibrium position these springs exert no force on the valve because the plungers against which the springs act abut against the end covers L and M. Before the valve can move from the equilibrium position a torque must be exerted on the steering wheel which is sufficient to produce a force equal to the spring force. The reaction of this force acts on the pin C, Fig. 34.30, and is transmitted to the steering linkage. If the forces opposing the motion of the road wheels are small steering may be effected without any movement of the valve and purely by the manual effort of the driver. Under these conditions there will be no backlash in the steering. If the forces opposing the motion of the road wheels are large then the valve will be moved and the steering will be done mainly by the slave cylinder. If no oil pressure is available the motion of the valve is limited to the amount *y* (shown exaggerated in Fig. 34.32), the depth of the recess in the end of the body N of the valve housing, and the backlash is limited accordingly. The space between the plungers acted on by the centring springs is connected to the pressure side of the system and consequently when oil pressure is available the force of the centring springs is supplemented by the force exerted on the plungers by the oil.

34.24 Marles-Bendix Varamatic system

This is shown in Figs 34.33 and 34.34. The mechanical part of the system is based on the Marles mechanism which is described in Section 34.16 while the servo part is basically the same as that of the Ross system described in Section 34.22. There are, however, some important differences. The cam A, Fig. 34.33, now has a variable pitch so that when the roller follower B is in the central position shown, the ratio *angular motion of cam: angular motion of drop arm* (C), for a very small motion, is about 21:1 but when the drop arm and follower have moved about 12° away from the central position that ratio has fallen to 13:1 and thereafter it remains constant at that value. The valve which controls the servo action now operates by the rotational displacement of its two main components instead of by their axial displacements as in the Ross system.

The cam is carried in the casing on two angular contact ball bearings and is coupled at the left-hand side to the torsion bar D by a cross-pin. At the

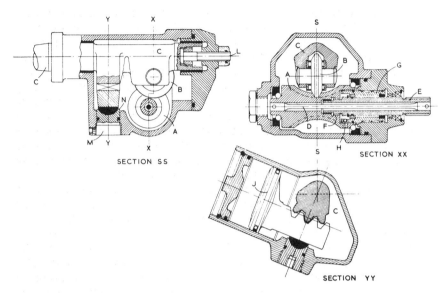

SECTION SS

SECTION XX

SECTION YY

Fig. 34.33

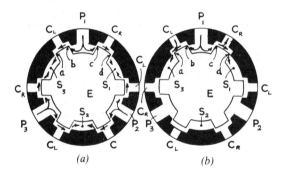

(a)   (b)

Fig. 34.34

right-hand end the torsion bar is coupled, also by a cross-pin, to the sleeve E which is splined to the steering-wheel shaft and which forms the inner member of the valve. The left-hand end of the sleeve E is formed with splines F which engage splines formed in the right-hand end of the cam A but these splines are machined so as to allow 7° of freedom of rotation and are only to provide a safeguard against over-stressing of the torsion bar when the steering is operated without servo assistance. Pipes connect the pump, which is driven by the engine, to the valve and the latter to the outer end of the servo cylinder; the inner end of the servo cylinder forms the casing which houses the cam A and follower assembly and the valve is directly connected to that space. The outer member G of the valve is coupled by the ball-ended pin H to the right-hand end of the cam A. The servo piston J is integral with a stem on which rack teeth are formed and these teeth engage teeth machined on the end of the drop arm forging C.

The principle of operation of the valve is shown by the diagrams Figs 34.34(*a*) and (*b*). It is really three valves in parallel, parts relative to which are denoted by suffixes 1, 2 and 3 – following the letters P and S in the diagram – but the action will be described in relation to one of them. The ports P are connected to the delivery of the pump and when the valve sleeve E is centrally placed, as shown at (*a*), fluid flows equally to each of the pockets S_1 and S_2 which are connected at their ends to the return pipe to the pump. The pressure drops across the apertures *b* and *a* are equal to those across the apertures *c* and *d* and so there is no pressure difference between the spaces C_L and C_R which are connected to the ends of the servo cylinder. Hence there is no net hydraulic force acting on the servo piston. But when a torque is applied to the steering wheel to overcome a resistance to a steering motion of the road wheels the torsion bar is twisted and relative motion occurs between the inner valve sleeve E and the outer one G, as is shown at (*b*). The passage *c* is thereby increased while the passage *d* is decreased and so the pressure in the space C_R is raised. Conversely, the passage *b* is decreased and *a* is increased so that the pressure in the space C_L is lowered. A pressure difference is thus established across the servo piston and the drop arm is rotated. As this occurs the cam and outer sleeve of the valve rotate so as to follow up the inner valve sleeve and bring the valve to a central position. The drop arm having thus been rotated the required amount the servo action ceases and the system remains in equilibrium. The use of three sets of ports provides a valve in which the radial hydraulic pressures are balanced and the required port areas are obtained with a valve only one-third the length that would be needed if only one set was provided.

Provision is made for adjusting the mesh of the roller B with the cam A; this is done by means of the screw L which, when turned, will move the drop arm shaft C in or out and thus bring the roller into closer or looser mesh. Similarly, the mesh of the rack teeth on the servo piston stem can be adjusted by means of the screwed plug M which bears on the underside of the stem through the spherically seated pad N.

Chapter 35

Suspension principles

Obviously, if the loads applied to the rolling wheels of a vehicle were transmitted directly to the chassis, not only would its occupants suffer severely but also its structure would be subjected to an excessive degree of fatigue loading. The primary function of the suspension system, therefore, is to isolate the structure, so far as is practicable, from shock loading and vibration due to irregularities of the road surface. Secondly, it must do this without impairing the stability, steering or general handling qualities of the vehicle. The primary requirement is met by the use of flexible elements and dampers, while the second is achieved by controlling, by the use of mechanical linkages, the relative motions between the unsprung masses – wheel-and-axle assemblies – and the sprung mass. These linkages may be either as simple as a semi-elliptic spring and shackle or as complex as a double transverse link and anti-roll bar or some other such combination of mechanisms.

35.1 Road irregularities and human susceptibility

Some indication of the magnitudes of the disturbances caused by road irregularities can be gained from *Surface Irregularity of Roads*, DSIR Road Research Board Report, 1936–7. From this report it appears that surface undulations on medium-quality roads have amplitudes in general of 0.013 m or less, while amplitudes of 0.005 m are characteristic of very good roads. The average pitch of these undulations is under 4 m while most road vehicle wheels roll forwards at about 2 m/rev. In addition to the conventional tarmac roads, there are pavé and washboard surfaces, the latter occurring largely on un-surfaced roads and tracks. Representative replicas of these two types of surface are described in *The MIRA Proving Ground*, by A. Fogg, *Proc. A.D. Inst. Mech. Engrs.* 1955–56.

Obviously the diameter of the tyre, size of contact patch between tyre and road, the rate of the tyre acting as a spring, and weight of wheel and axle assembly affect the magnitude of the shock transmitted to the axle, while the amplitude of wheel motion is influenced by all these factors plus the rate of the suspension springs, damping effect of the shock absorbers, and the weights of the unsprung and sprung masses. The unsprung mass can be loosely defined as that between the road and the main suspension springs, while the sprung mass is that supported on these suspension springs, though both may also include the weights of parts of the springs and linkages.

Two entirely different types of shock are applied to the wheel: that due to the wheel's striking a bump, and that caused by the wheel's falling into a pot-hole. The former will be influenced to a major extent by the geometry of the bump and the speed of the vehicle, while the major influence on the latter, apart from the geometry of the hole, is the unsprung masses and spring rates, speed being an incidental influencing factor.

Human sensitivity to these disturbances is very complex, and a more detailed discussion can be found in *Car Suspension and Handling* by Donald Bastow, Pentech Press, London, 1980. It is widely held that vertical frequencies associated with walking speeds between 2.5 and 4 mph – that is, 1.5 to 2.3 Hz – are comfortable, and that fore-and-aft or lateral frequencies of the head should be less than 1.5 Hz. Dizziness and sickness are liable to be experienced if the inner ear is subjected to frequencies between 0.5 and about 0.75 Hz. Serious discomfort may be felt in other important organs at frequencies between 5 and 7 Hz.

35.2   Suspension system

A suspension system can be represented, in simplified form, as illustrated in Fig. 35.1. The natural frequency of the sprung mass – that at which it would bounce up and down if momentarily disturbed and then left to bounce freely on its springs – is determined by the combined rate of the tyres and the suspension springs in series, which is—

$$\frac{1}{R} = \frac{1}{R_s} + \frac{1}{R_t}$$

where R is the overall suspension rate
R_s is the suspension spring rate
R_t is the tyre rate

In Fig. 35.1, the shock absorber is the hydraulic damper at D. Any friction in the suspension system will be additional to the hydraulic damping. However,

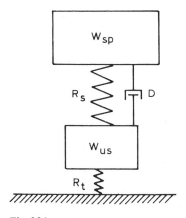

Fig. 35.1

whereas the hydraulic damping force of the shock absorber can be taken as proportional to the square of the vertical velocity of the sprung mass relative to that of the unsprung mass, the dynamic friction damping force is, in effect, constant regardless of velocity. It follows that while small amplitude, small velocity movements of the suspension are virtually unaffected by the hydraulic damping, the force applied by the friction damping is the same for these small movements as it is for large ones.

With, for example, a new multi-leaf semi-elliptic spring, there is only a small difference between the static and dynamic interleaf friction forces – sometimes differentiated by calling them respectively *stiction* and *friction*. When, however, the same spring becomes rusty and dirty, this difference can become considerable, with the result that mean value of the work done by the friction damping during high frequency, small amplitude motions becomes excessive. Indeed, in extreme cases, the spring may become so stiff that, for small amplitude disturbances, it in effect does not deflect at all. This can of course lead to a harsh uncomfortable ride. It must be borne in mind, too, that even the hydraulic forces are in any case transmitted directly – that is, uncushioned by the suspension springs – to the sprung mass.

Dampers have a two-fold function. First, they are for reducing the tendency for the carriage unit to continue to bounce up and down on its springs after the disturbance that caused the initial motion has ceased. Secondly, they prevent excessive build-up of amplitude of bounce as a result of periodic excitation at a frequency identical to the natural frequency of vibration of the spring–mass system. This natural frequency is a function of the weight of the sprung mass and the spring rate, and in fact can be shown to be directly proportional to $1/\sqrt{\delta}$, where δ is the static deflection of the spring.

Each of two forms of disturbances can cause either of two entirely different resonances. One such form is the passage of the wheel over a series of equidistant bumps at a speed such that the frequency of the disturbance that they generate coincides with the natural frequency of the suspension system. The second is imbalance of the wheel, the out-of-balance force of which will increase as the square of the speed of rotation.

Of the two different frequencies: one is that of the sprung mass on the suspension spring system, and the second is that of the unsprung mass – the wheel and axle assembly – on the tyre. Obviously the latter is affected by the suspension spring rate, but only marginally. The former will be experienced as a relatively low frequency – perhaps about 1 to 1.5 Hz – bouncing of the carriage unit, while the latter is that of wheel hop, at a higher frequency – generally 10 to 15 Hz – and is generated almost totally independently of the motions of the carriage unit. For minimising the amplitudes of wheel hop – not only resonant but also isolated hops – the unsprung weight must be kept as small as possible. Resonances of either the sprung or unsprung masses can affect adversely, and indeed to a dangerous extent, the handling characteristics of the vehicle. Obviously, therefore, it is important to maintain the dampers, or shock absorbers, in good working condition.

These same disturbances can also cause pitching or rolling vibrations of the vehicle. In these instances, the natural frequencies are functions of the spring rates in the rolling and pitching modes and the moment of inertia of the sprung masses about the lateral and longitudinal axes respectively. Rolling oscillations of an axle alone, that is about an axis parallel to the longitudinal axis of the vehicle, are generally termed *tramp* because the effect is something like the tramping motion of a man – advancing one step at a time.

Supplementary roll stiffness can be obtained, without affecting two-wheel bounce stiffness – resistance to vertical motion of complete axle or, with independent suspension, of both wheels simultaneously – by the use of a torsion-bar spring. This spring is generally called an *anti-roll bar*, and it is mounted transversely beneath the vehicle in two bearings – usually rubber bushes – its ends being connected by levers and, sometimes, shackles to the axles. If simple pivots, instead of shackles, are used to connect the levers to the axles, these levers can be utilised as radius rods, for guidance of the axles as they move up and down, as in the MacPherson suspension illustrated in Fig. 36.10. Since the use of an anti-roll bar inevitably implies the imposition of extra vertical loading on the tyres during roll, it has an effect on steering and handling. This is because the extra vertical deflection of the tyre of the outer wheel, in the region of its contact with the ground, renders it more susceptible to lateral deflection – increases the slip angle – when the vehicle is cornering. This effect has to be considered by the designer when he decides whether to instal an anti-roll bar at the front or rear, or both.

Clearly there must be an interaction – a vibration coupling effect – between the motions of the front and rear suspensions, and this must affect the tendency to pitch. The magnitude of the interaction will depend on the frequency of the disturbances, or bumps over which the car rolls, and the natural frequencies of the front and rear suspensions. Obviously, the forcing frequencies depend on the spacing of the bumps, the speed of the vehicle, and its mass moment of inertia about the axis of pitch, while the magnitude of the response in pitching of the vehicle will depend not only on these two factors but also on its wheelbase.

It can be shown, if the rear suspension has a lower natural frequency than the front, the pitching motion tends to persist longer than if the front has the lower frequency. Moreover, if the rear suspension has the higher natural frequency, the initial pitching motion is less severe. Consequently, the natural frequency of the rear suspension is normally made higher than that of the front. The higher the speed of the vehicle, the less severe is the initial pitching motion. This effect arises because, as the speed increases, the time between the front and rear wheels striking the bump becomes a smaller proportion of the periodic time of the pitching motion of the vehicle at its natural frequency, and in theory could ultimately become zero. The principle is analogous to that of vibration isolation – for instance of an engine vibrating freely at a relatively high frequency on a rubber mounting system having a low natural frequency.

35.3   Damping

As mentioned previously, the dampers, or shock absorbers as they are sometimes called, are required to cause a rapid die-away of any vibrations forced either randomly or periodically at the natural frequency of the suspension system and thus introducing a state of resonance. To do this, they apply a force in a direction opposite to that of the instantaneous motion of the suspension. Early cars had friction dampers, which were generally packs of friction material interleaved between blades, or arms, which were attached alternately to the sprung and unsprung masses. Semi-rotary vane type hydraulic dampers have also been used. However, these were abandoned because the ratio of sealing length around their vanes to volume displaced was so high that these units were rapidly adversely affected by wear.

Modern dampers are almost invariably either telescopic hydraulic struts interposed between the sprung and unsprung masses – carriage unit and axle – or, less frequently, of the lever type, which also are hydraulic units. The body, incorporating the hydraulic cylinder, of the lever-type damper is usually mounted on the carriage unit, with its actuating lever connected to the axle. If the body were mounted on the axle, the high-frequency, high-velocity motions to which it would be subjected might cause aeration of the hydraulic fluid and hence adversely affect the damping capacity of the unit. The minimum vertical accelerations of an axle under dynamic conditions can be of the order of 20 to 30g.

Damping is effected by the damper piston, or pistons, forcing the hydraulic fluid at high velocities through small holes. Thus energy is absorbed by the fluid, converted into heat, and then dissipated partly by conduction into the surrounding structure of the vehicle, but ultimately all passing into the air stream flowing past these components. The amount of energy thus absorbed and dissipated, for any given rate of energy input, is a function of the volume and viscosity of the fluid and the numbers, sizes and geometry of the holes through which it is forced. A major advantage of hydraulic damping is that, as previously mentioned, the resistance to deflection of the damper is a function of the square of its velocity. Therefore, slow movements of the wheels can occur with relative freedom, but the resistance increases rapidly with the velocity of motion.

Ideally, the aim in damper design is to obtain maximum possible potential for energy absorption for any given size, and this would imply equal damping on the bump and rebound strokes. However, since the bump stroke is often a violently forced motion, and it is undesirable to transmit such forces directly through the damper to the sprung mass, the damping on the bump stroke is often arranged to be less than that on the rebound stroke, which of course is a gentler motion effected by the weight of the axle and the force exerted by the suspension spring. To relieve the bump stroke of all damping, and thus the carriage unit of all directly transmitted shocks, is impracticable for two reasons: first, it would halve the energy-absorbing ability of the damping system; secondly, damping only on the rebound would tend to jack the carriage unit down to a mean level below that of its static deflection on the springs. The transmission of high frequency, small amplitude vibrations through the dampers directly to the carriage unit is avoided by the interposition of rubber bushes or blocks between their end-fittings and their anchorages on the axle and carriage unit – that is, on the sprung and unsprung masses.

35.4 Dampers in practice

A characteristic common to all telescopic dampers is that, as the piston moves into the cylinder, its whole area is effective in transmitting the load, and thus pressurising, the fluid but, as it moves outwards, the effective area of the piston is reduced to that of the annulus between its periphery and that of the piston rod. If equal damping is required in both directions, therefore, some compensation is obtained by adjusting, to a different value for each direction of flow, the pressures at which the valves open the small holes in the piston, through which the fluid is forced to provide the damping. Also, the total cross-sectional areas of the holes for the flows in the two directions must differ too. The latter effect can of course be obtained by use of a simple plate valve to close some of the holes during motion in one direction only.

Another effect of the intrusion of the piston rod into the cylinder is that the volumes available for accommodation of the fluid on the two sides of the piston differ. Compensation for this can be provided by the incorporation of a flexible element in the cylinder, so that the total volume within it can adjust automatically, as required. This flexible element can be an elastic sphere containing an inert gas, or a free piston with an inert gas between it and the closed end of the cylinder. An alternative is the use of the double-tube design.

35.5 Double-tube damper

A double-tube design is illustrated in Fig. 35.2. It consists of a cylinder A, to which is welded a head B. The latter is screwed into the outer tube C to which is welded a pressed steel cap and eye D by means of which the cylinder A is secured to the axle or wheel assembly. A piston E, in the cylinder A, is secured to the piston rod F which at its upper end has an eye welded to it by which it is attached to the frame of the vehicle. The part of the piston rod that emerges from the cylinder is protected by a cover welded to the fixing eye. A gland G prevents leakage where the piston road passes through the head B; any fluid scraped off by the gland packing passes down a drain hole to the reservoir space between the cylinder A and the outer tube C. At the bottom of the cylinder A is a *foot-valve* assembly L.

Piston E has two concentric rings of holes drilled through it: the outer ring is covered by a disc valve H held down by a star-shaped disc spring I, while the inner ring is covered by the disc valve J held up by the coil spring K. The valve assembly in the foot is similar to that in the piston except that the lower disc valve, covering the inner rings of the holes is held up by a disc spring instead of a coil spring. This is to reduce the dead length of the shock absorber – that is, the length not available for the working stroke.

Both ends of cylinder A are completely filled with fluid, but the space between A and C is only partly filled. If the eye D is moved upwards, fluid must be displaced from beneath to above the piston E. This fluid will pass through the outer ring of holes, by lifting the valve H against the spring I. But since the increase in the volume of the upper end of the cylinder is less – by the volume of the portion of piston rod that enters the cylinder – than the decrease in volume of the lower end, fluid will also be displaced through the inner ring of holes in the foot valve and the level of the fluid in the reservoir space will rise.

The pressure set up will depend on the sizes of the passage opened by the valves in the piston and foot valve, and, of course, on the square of the speed at which the cylinder is moved upwards. For downward motion of the cylinder the valve J will be forced off its seating, and fluid will be displaced from the upper end of the cylinder, through the inner ring of holes in the piston, to the lower end, but the removal of the volume of the piston rod from the cylinder will cause fluid to be drawn also from the reservoir space, through the outer ring of holes in the foot valve into the lower end of the cylinder.

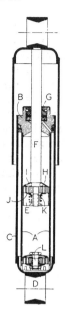

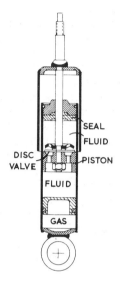

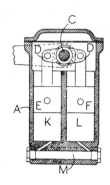

Fig. 35.2 Double-tube
telescopic damper

Fig. 35.3 Girling single-
tube telescopic damper

Fig. 35.4 Lever-
arm-type damper

An advantage of the double-tube damper is that the oil displaced from the main tube into the outer one carries heat with it, which is then readily conducted away. This tends to keep all the fluid at a moderate temperature. Obviously, the higher the level of fluid in the outer tube the greater is the heat transfer effect. Another advantage is that damage – a dent for example – in the outer tube will not interfere with the working of the piston.

35.6 Single-tube damper

A single-tube damper is shown in Fig. 35.3. The spaces immediately above and below the piston are filled with oil, and the damping action arises from the viscous losses that occur in the orifices, as in the double-tube type. However the effect of the volume of the piston rod is allowed for by the use of a volume of gas under compression at the bottom of the tube, where it is separated from the oil by a floating piston. When the tube of the damper moves upwards, the gas will be further compressed and the floating piston will be moved downwards relative to it by the amount required to accommodate the changes in the volume of the two oil spaces. This compression of the gas results in a progressive change in the characteristic of the damping, such that the force required to move the damper tube upward at a constant speed will rise at an increasing rate. In contrast, in the double-tube type, the rate of increase is constant.

In a single-tube damper, the pressure of the gas – usually nitrogen – must be higher than the maximum operating pressure in the fluid below the main piston, and may be of the order of 2.5 MN/m^2. This of course adds to the total spring-rate of the suspension by an amount equal to the gas pressure multiplied by the effective cross-sectional area of the piston rod.

The Woodhead Manufacturing Company has produced a single-tube damper without a floating piston. In it the inert gas is free in the cylinder and therefore tends to emulsify the fluid. Although the gas and oil separate while the vehicle is stationary, re-emulsification occurs rapidly owing to the large flow rates inherent in the design, the performance of the unit therefore becomes normal equally rapidly. An advantage of this type of damper is the impossibility of jacking up and subsequent bending of the piston rod, in the event of leakage of fluid past a free piston. Also, again because of the absence of the free piston, the dead length of the damper is small. Moreover, the performance of these emulsion-type dampers is affected less than that of the fluid type by variations in working temperature. Single-tube dampers in general have the advantage that, for a given overall diameter, the piston area is greater, but they have the disadvantage of a higher built-in pressure.

Telescopic dampers are also used in the MacPherson, or strut-type, suspension, Fig. 36.10. Here, a major problem arises because of the imposition of side loads at the top of the strut, and therefore bending moments on the piston rod. Because such a piston rod, taking these loads and performing guidance function, has to be of large diameter, the problems arising owing to its intrusion into the damper are aggravated.

35.7 Lever-arm-type damper

In Fig. 35.4, two pistons E and F are actuated by the ends of a double lever DD carried on a shaft C, which is coupled externally to the axle. The bores K and L are interconnected by the passage M, which houses some form of orifice assembly. Damping again arises from the viscous losses that occur in the orifices. The lever arm of such a damper can be made to serve as one of the links of an independent suspension system.

35.8 Springs

The function of the suspension spring can be understood from the following. When the wheel hits a bump, it rises extremely rapidly, so if a spring were not interposed between it and the carriage unit, the shock transmitted would be considerable. With the traditional suspension system, the only force transmitted to the carriage unit is that required to compress the spring far enough for the wheel to ride over the bump. This force causes the body to accelerate upwards, but at much smaller rate than would otherwise be experienced.

When the wheel falls into a depression, the force in the spring, acting on the relatively light unsprung mass – wheel and axle – forces it down at a rapid rate so that it generally reaches the base of the depression almost before the relatively much larger mass of the carriage unit has, owing to its inertia, had time to begin to descend. Since the variations in the spring force over such deflections are relatively small, the downward acceleration of the carriage unit supported by the spring is correspondingly moderate as compared with that which, under the influence of gravity, it would be if there were no spring.

When the disturbance has passed, whether it is a bump or a depression, the subsequent motion of the carriage unit is its free vibration on the springs, the acceleration being small. This vibration is rapidly reduced to zero by the dampers.

The following types of spring are currently in use for the suspension of cars and commercial vehicles: laminated, or leaf; taper-leaf; coil; torsion bar; rubber; air; and gas. The principal characteristics governing the choice of type of spring are: overall cost of the installation, relative capacity for storing energy, total weight of suspension system, fatigue life, and location or guidance linkages required.

Leaf springs in general have approximately only a quarter of the energy-storage capacity, for a given stress level, of either coil or torsion-bar springs. Calculated on the basis of weight for a given energy storage capacity, the ratio is about 3.9 : 1 in favour of coil or torsion-bar springs. For rubber springs this ratio varies according to the form such springs take, but it would probably be of the order of between 7.4 and 15 : 1. If the stress in a leaf spring has to be limited to $700 \, \text{MN/m}^2$, its energy storage capacity is about $50.4 \, \text{Nm/kg}$. By using the ratios just quoted, it is easy to calculate the approximate specific energy storage capacities of the other springs.

Gas springs of course inherently have an extremely high energy storage capacity per kg, but this is considerably reduced if the weight of the ancillary equipment required is taken into consideration. This equipment may include a compressor, reservoir, gas drier, de-icer, pipework, valves, dampers and filters. On commercial vehicles, however, some or all of these components may be carried in any case for the air brake system.

35.9 Types of leaf spring

In its most commonly used form on road vehicles, the leaf spring is a beam simply supported at each end, with a point load at the centre. Consequently, its bending moment diagram is a triangle, with its apex above the centre and its base a straight line extending between its end supports. If the spring were of lozenge shape, therefore, as in the top diagram in Fig. 34.5, and appropriately proportioned, the bending stresses would be uniform along its length. It follows that if we were then to cut it into strips, as indicated by the dotted lines in the diagram, and assemble them together in pairs of equal length, as in the plan and elevation below the top diagram, we should have a laminated, or leaf, spring of compact form in which optimum use was made of the material because the stresses would be fairly well distributed throughout the volume of the material.

In other words, if a leaf spring is deflected until the stresses in the outer layers of all the leaves rise to the elastic limit, the energy stored greatly exceeds that which could be stored in, for example, a solid steel spring of the same shape, as viewed in plan and elevation, similarly stressed to its elastic limit. The leaf spring is of course much more flexible because the stiffness of a beam is proportional to the cube of its depth multiplied by its width. In the case of the leaf spring illustrated, the depth is in effect that of only one leaf, though the width is the mean of that of the original lozenge shape.

Obviously, as such a laminated spring deflects, there is relative sliding between the surfaces of the leaves, and friction between them tends to oppose the motion. In an effort to reduce and control this friction, the spring leaves are often either interleaved with plastics materials having a low coefficient, or pads or buttons of similar materials are interposed between the ends of their adjacent leaves. Also, the springs themselves are wrapped and sealed to

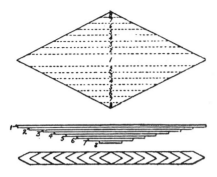

Fig. 35.5

prevent the entry of dirt and water. Even so, to obviate this problem altogether, coil springs or torsion bars have now been widely substituted for leaf springs in cars. Another disadvantage of the leaf spring is that it is heavier than the coil or torsion-bar spring. Moreover, whereas two-thirds of the weight of a semi-elliptic leaf spring is carried by the axle – that is, has to be considered as unsprung weight – only half of that of a coil spring and virtually none of that of a torsion bar is unsprung weight.

35.10   Laminated spring details

On road vehicles, semi-elliptic springs are usually slung beneath the axles, but on cross-country vehicles, to give more ground clearance beneath the body and chassis, they may seat on top of them. In either case, the spring seating pad is generally either welded to the axle tube or is integral with it, and the spring is clamped between it and a saddle plate by either four bolts, or by two U-bolts either embracing grooves around the saddle or seating in separate stirrup pieces around the tube. The bolts used are generally of special steel and should not be replaced by any other type. They have been known to stretch, leaving the spring loose, so exceptionally heavy shock loading can shift the axle back along the springs. These are points that should be looked for in service.

The ends of the longest, or master, leaf of a laminated spring are usually coiled to form the eyes for receiving the pivot or shackle pins. Sometimes, however, these eyes are forged on the ends or other special arrangements made. Rubber bushes are widely used in the eyes of springs for light vehicles, but metal bearing-bushes, with provision for lubrication, are necessary for very heavily loaded springs. Sometimes bushes of threaded form are used in conjunction with similarly threaded pins, to increase their effective bearing areas and better to retain the lubricant. Although rubber bushes need no lubrication and help to prevent the transmission of high-frequency, low-amplitude vibrations to the carriage unit, if they work loose they can squeak and deteriorate rapidly. Nylon or other plastics and composite self-lubricating bearings have been used to obviate both the squeaking and the need for lubrication.

Generally, the ends of the second leaf are extended in order partly to embrace the eyes, for safety in the event of breakage of the master leaf. A single bolt, usually in a vertical hole through the centre of the spring, holds all the leaves together. The head of this bolt is generally accommodated in a hole in the spring seating pad.

Various methods have been used to locate the outer ends of the leaves laterally within the pack. For example pips, or projections, on their upper surfaces can register in corresponding recesses in the underside of the leaves above, as shown in Fig. 35.6. Alternatively, the upper and lower surfaces respectively can have ribs and grooves rolled along their whole lengths. Sometimes the clips alone, also illustrated in Fig. 35.6, are used. Their main function, however, is to distribute the load over the whole length of the spring in the rebound condition, when without the clips, the ends would drop away from the leaves above them. This separation of the leaves, due to the reversal of the load on the spring, would throw all the rebound load on to the master leaf and possibly cause it to break. Sometimes short *rebound leaves* are fitted on top of the master leaf to help to take this reverse loading.

In a patented form of spring, the leaves are symmetrically arranged above and below the master leaf. With clips, such a spring is equally effective in both directions, but such characteristics are rarely required.

The function of the shackle is of course to accommodate the variations in length of the spring, as it deflects upwards through the straight condition to an inverted semi-elliptic shape. Because of this change of length, the rate of the spring changes, being lowest in the straight condition and increasing as its effective length decreases during deflection in either direction from that condition.

Alternatives to the swinging link type of shackle include the slider-block, and rubber types, illustrated in Figs 35.7 and 35.8. In the former, the ends of the two top leaves project into a slot in the cylindrical member A, in which they are free to slide. The cylindrical member is free to rotate in its housing B, which is secured to the chassis frame. A disadvantage of this scheme is that it requires lubrication, but modern commercial vehicles generally have the simpler arrangement that can be seen in Fig. 36.18, where the spring ends seat on a curved pad in a tunnel-shaped member bolted to the chassis frame. There is plenty of vertical clearance between the spring and the tunnel, but the lateral clearance is restricted to an amount needed to allow a limited degree of articulation as the spring is twisted by one end of the axle rising to ride over a single-wheel bump.

With the Metalastik rubber shackle, Fig. 35.8, lateral location of the end of the spring is afforded by the wedge-shaped base of the mounting, leaving it free to deflect the rubber in shear for fore-and-aft movements – with any shackle of

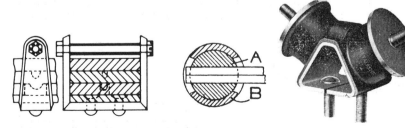

Fig. 35.6   Fig. 35.7   Fig. 35.8

course, fore-and-aft location of the spring as a whole is effected by the eye at its other end. Advantages of the rubber shackle include the absence of any maintenance requirement, silence of operation, and absorption of high frequency vertical vibrations, which are taken in compression in the rubber. Tension in the rebound condition is avoided by a rebound stop, usually a strap fitted loosely between the axle and frame. An example of the use of this system is the 1980 *Ailsa* Mk III double-deck bus chassis.

A semi-elliptic spring arrangement is illustrated at (*a*) in Fig. 35.9. Since the axle is secured by U-bolts beneath the centre of this spring, the shorter leaves are the lower ones. In suspension arrangements such as some of those illustrated in Chapter 36, where the sprung mass, instead of the unsprung mass, is secured to the centre of the spring and the axles to its outer ends, the shorter leaves are uppermost.

Another form of semi-elliptic spring layout is termed the *full-cantilever spring* – rarely used today except possibly on special mechanisms for construction equipment. This type of spring is pivoted at its centre and at one end, these pivots being on the frame; its other end is then either pivoted or shackled to the axle, according to whether the spring is to take thrust loads and guide the axle or whether other links are to be used to perform these functions.

The quarter-elliptic spring is illustrated at (*b*) in Fig. 35.9. This is a cantilever multi-leaf spring the thick end of which is bolted rigidly to the frame, or sprung mass, and the other end is pivoted or shackled to the axle. It has the advantage of extreme simplicity; for example, its thick end can be clamped in the end of a chassis frame, which therefore can be much shorter than if a semi-elliptic spring were used.

Semi-elliptic springs have also been installed with their axes arranged transversely, Figs 36.6 and 36.7, one of the last examples in a quantity-produced car being for the independent front suspension of the Fiat 600, introduced in 1955. An earlier example, used on a beam axle, is illustrated at (*c*) in Fig. 35.9. This arrangement has the advantage of economy – one spring and fixings instead of two – but the disadvantage of small vertical deflections unless thin leaves are used or high stresses tolerated, or both. Moreover, such springs can be unstable if called upon to react brake torque and drag simultaneously. With the Fiat 600 and some other independent front-suspension systems, the ends of the transverse leaf spring form the lower links of a double transverse link suspension.

35.11   Taper-leaf springs

Despite the measures taken to improve the stress distribution throughout the depth of a semi-elliptic spring, by dividing it into a multiplicity of separate leaves, the maximum stresses in each leaf still occur only adjacent to where it is

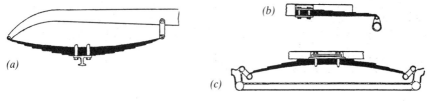

Fig. 35.9

supported between the adjacent leaf or leaves and its seating pad on the axle, Fig. 35.10(a). Therefore, because so much of the material remains relatively lightly stressed, such a spring is heavier than it otherwise might be.

In recent years, therefore, the taper-leaf spring, sometimes known as the *parabolic-leaf* or *minimum-leaf spring*, has become increasingly popular. It was pioneered in the early nineteen-sixties by the Rockwell Corporation in the USA and, later, by what used to be Rockwell's part-owned company in the UK, Bramber Engineering Ltd.

For any given maximum permissible stress level and width, the stiffness of a semi-elliptic spring is proportional to nt^3, where n is the number of leaves and t is the thickness of each. Consequently if n is halved, to maintain the same stiffness, it is necessary to increase t by a factor of only $\sqrt[3]{2}$, which is about 1.26. If, for example, a spring has six leaves, each 0.5 cm thick, its overall thickness is 3 cm, so its stiffness is a constant $K \times 6 \times 0.5^3 = K \times 0.75$. So, if we halve n, t must be increased to $0.5 \times 1.26 = 0.63$ so that the stiffness again becomes $K \times 3 \times 0.63^3 = 0.75$. This reduction in the number of leaves and overall thickness, together with the fact that the leaves can be tapered from the centre to each end fairly accurately for efficient material utilisation, means that such a spring can be about 30% lighter than its conventional multi-leaf spring equivalent.

For cars, a single leaf is in many instances adequate but, for heavy commercial vehicles, two or three leaves are common, often used in conjunction with a helper spring. The upper main, or master, leaf has eyes or slider pads at its ends, as in a conventional multi-leaf spring, while the second and third leaves are shaped so that they bear on each other, or on the main leaf, only at their ends, as in Fig. 35.12, and of course they may have spacers between them at the centre. Consequently, inter-leaf friction is extremely small and remains, for all practicable purposes, constant throughout the life of the spring. The Bedford two-leaf front suspension is illustrated in Fig. 35.10(b). This is of interest too because the lower end of the damper is spring mounted.

As always, the cross-sectional area of the outer ends of the master leaf, and therefore its depth, must be adequate to take the shear plus any torsional

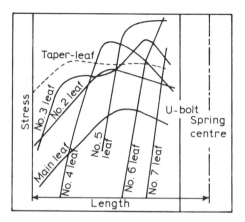

Fig. 35.10(a) Stresses in a single taper-leaf spring compared with those in a plain multi-leaf spring

Fig. 35.10(*b*) The Bedford BL truck has two taper leaves in each front spring. One of the lower leaf is wrapped around the eye at the front end of the master leaf and the other end is simply dipped beneath the rear one. Of special interest is the resilient mounting of the lower end of the telescopic damper for cushioning exceptionally severe shocks

loading that might be applied when the vehicle rolls or one wheel only deflects. Inwards from this point, while the bending load increases linearly towards the spring seating in the centre, the stress is a function of the square of the thickness of the leaf. Consequently, the taper is profiled according to an inverse square law – hence the use of the term *parabolic leaf spring*. Sometimes, such a spring is also tapered as viewed in plan, to obtain more clearance between its ends and the tyre. In this case, a straight taper is of course employed, but the ends must be wide enough to avoid instability, or twisting under combined vertical and lateral or torsional loading.

For multi-leaf springs a silico-manganese steel – 0.53 to 0.62% carbon, 1.70 to 2.10% silicon and 0.70 to 1.00% manganese – is adequate because the leaves are so thin that there is no difficulty in obtaining the necessary 80% martensitic structure at the centre of the section. Where the leaves may be up to 28.5 mm thick, however, a carbon 0.60 to 0.90% chromium steel is needed, while for even thicker sections a chromium-molybdenum steel is required. Since, owing to their through-hardening properties, the high-quality steels are employed for necessity for taper-leaf springs, and their yield points are of the order of 620 N/mm², as compared with 448 N/mm² for a silico-manganese steel, even further weight reduction is obtained. Moreover with these better steels, reduction in strength at the surface as a result of decarburisation is not a problem. Finally, with taper-leaf springs, no allowance has to be made for loss of fatigue strength owing to interleaf fretting and corrosion in service.

In practice, fatigue strength is further increased by 'scragging' – loading beyond the yield point, as a manufacturing operation – to leave residual compressive stresses in the surfaces that are in tension during deflection. Strain-peening – shot-peening while the spring is under load – is also used for this purpose and to round off the edges of any scratch marks left in the surface by the manufacturing operations, again improving fatigue resistance.

Despite the inherently light weight of the taper-leaf spring, the multi-leaf type can still be the better choice in certain circumstances. First, the tooling costs for the manufacture of taper-leaf springs are high, so such springs can be too costly for small quantity production. Secondly, because of the thickness of leaf, the taper-leaf spring must be longer than the equivalent multi-leaf type – stiffness is proportional to t^3 – and the chassis layout may be such as to limit the length of spring that can be accommodated. Thirdly, in some applications, the hysteresis of a multi-leaf spring may be an advantage. In some cases, the attraction of the multi-leaf spring is that it can be relatively easily made up from any number of leaves of virtually any thickness, or of various thicknesses, to meet a given design requirement.

35.12 Steering effects of leaf springs

Although leaf springs have the advantage of performing the two functions of springing and axle guidance, they have steering effects on their axles. These effects arise from two sources: the first is simply by the motion they induce in the axle, and the second is attributable to the effect of this motion on the steering linkage. The motion of the axle, if it is not regulated by any other linkage, is controlled by the motion of the spring acting as if it were a solid link pivoting about the end secured by its eye directly to the sprung mass – the other end is of course so connected by a shackle. The effective length of this link, however, is modified by the initial straightening and subsequent inversion of the curved shape, or camber, of the spring as it moves from the rebound to the full bump condition. This tends to pull the axle towards the eye as the spring deflects either up or down from the straight condition, and thus reduces the effective radius about which the axle swings. On the other hand, the effect of the shackle in tending to keep the centre of the spring horizontal, is to increase that effective radius.

A rule of thumb for calculating the effective radius of the spring acting as an axle guidance link is as follows: assume the spring to be horizontal and deflected to zero camber. If the centre of the eye is then a height h above the centre of the master leaf, which is the leaf on the end of which the eye is formed, the other end of the effective link will be $\frac{1}{2}h$ below the centre of the master leaf and on a vertical line three-quarters the effective half-length of the spring from the eye, Fig. 35.11(a). If the axle is clamped rigidly to the spring, the effective half-length is the distance between the centre of the eye and the point at which the spring protrudes from the clamp. Sometimes, however, a resilient pad is interposed between the spring and its seating, for absorption of high frequency, small amplitude vibrations; in this case the half-length will be a short distance inside the clamp and is most easily determined experimentally.

The effect of the spring, acting as a radius rod about its eye, is therefore to pull the axle forward as it moves towards the full bump and rebound positions – or rearwards if the shackle is at the front end the eye at the rear end of the spring. Consequently, when only one wheel rises over a bump, there is a steering effect on the axle, one end only being pulled forward. This phenom-

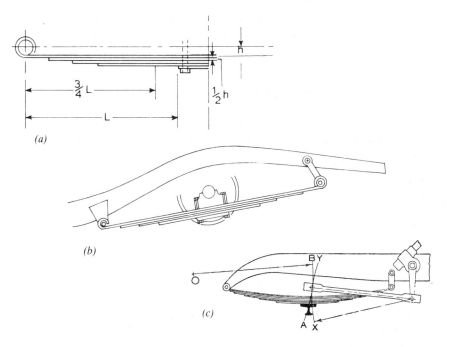

Fig. 35.11

enon can be employed to provide what is termed *rear-end roll-steer*, and is sometimes utilised by designers to compensate for excessive oversteer or understeer arising from other factors such as unequal front–rear weight distribution or the extra deflection of the tyres on one axle owing to the effect of an anti-roll bar on that axle. A steering effect is also, and more usually, obtained by inclining the spring, front to rear as in Fig. 35.11(b), so that it pulls the axle further forward when deflecting in one direction than in the other. It can be modified further by setting the axle off-centre along the length of the spring, taking advantage of the effect of the swinging of the shackle. This measure can also be applied to reduce squatting of the rear end during braking and rearing up during acceleration.

Nowadays only commercial vehicles have semi-elliptic springs at the front. Here, the axle-guidance effect of the spring can interact with the action of the steering rod, as shown in Fig. 35.11(c), unless the centre of the eye at the lower end of the steering drop arm is in line with the axis about which the front eye of the spring pivots. Similar principles apply to the positioning of steering-rod joints with independent front suspension, but these will be dealt with later.

35.13   Coil and torsion springs

Both these types of spring were originally used mainly with independent suspension systems, details of which are given in the next chapter. Now, however, coil springs are widely employed with both live and dead axles. Various linkage systems, as described in the latter part of Chapter 28, are used to control the motions of these axles and to react tractive and brake torques.

Coil springs have the advantage of fitting into a compact space, though the accommodation of a torsion bar parallel to the longitudinal axis of the vehicle is not necessarily difficult. Both of course are stressed in torsion, though there is some bending in a coil spring, and their lives are increased by shot-peening their surfaces to induce compressive stresses in them, and to reduce the effects of scratches in initiating fatigue cracks. Immediately after shot-peening, such springs may be given an anti-corrosion treatment, again to increase fatigue life.

Torsion bars are *scragged* – that is, overloaded in torsion during manufacture – to stretch their outer layers beyond the elastic limit. Because this leaves a residual stress in the outer layers, the maximum stress under service loading occurs beneath the surface, where it is less likely to initiate cracks. This again helps to increase fatigue life. The main reason why torsion bars are not as popular as coil springs is that their end fixings are more costly and provision has to be made for adjustment of the ride height on the vehicle assembly line.

35.14   Variable-rate springs

The rate of a spring is the increment of static load it will carry per unit of deflection – that is, in kg/cm – and with nearly all simple spring arrangements it is constant throughout the normal range of deflection. A rising rate, therefore, is one that increases, instead of remaining constant, as the spring progressively deflects. It has two advantages. First, when the vehicle is only lightly laden – either statically or dynamically – a high degree of isolation of the carriage unit from loading due to small amplitude deflections of the wheels is obtained, without the penalty of excessive deflections under heavy loading. Secondly, since the natural frequency of a mass suspended on a spring is a function of its rate, resonance will not – which it otherwise would – cause a dramatic increase in amplitude of vibration of that mass since such an increase would imply a change in natural frequency, taking the system off resonance.

A rising rate is often obtained by installing what is termed a *helper spring*, which comes into operation only towards the end of the upward deflection of the main spring. With coil-spring suspension, such a helper may take the form of either a rubber spring, or a second coil spring shorter than the main spring, which is mounted beneath the chassis frame so that the axle comes up against it only in the latter stages of its upward deflection. Alternatively, with leaf-spring suspension, a second semi-elliptic spring, having a smaller radius of curvature than the main one, can be centrally clamped on top of the latter in a manner such that its ends come up against stops on the chassis frame only on the latter stages of the upward deflection of the main spring, as in Fig. 35.12. With such an arrangement, the rate of the helper spring increases with deflection too since, as it flattens, the slender sections of its ends progressively slide outside the span between its two stops, or supports. A progressive rate is even more easily obtained with rubber springs, as described in the next section, and it is an inherent characteristic of a gas spring compressed adiabatically, as described in Section 35.17.

35.15   Composite leaf springs

In the early nineteen-seventies GKN began its development of leaf springs made of composite materials and, in late 1986, was the first to go into large

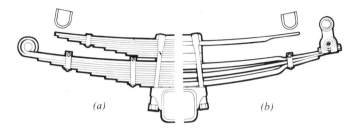

Fig. 35.12 Alternative multi-leaf (*a*) and taper-leaf (*b*) spring assemblies with helper springs designed for mounting on common fixing on a Scania range of chassis

quantity production, with a design illustrated in Fig. 35.13. Very high strength Kevlar fibres could have been laminated in longitudinally to reinforce the polyurethane resin of the leaf, but glass fibres are employed because they cost much less and are perfectly adequate.

A major advantage of such composite springs is that they weigh about 50% less than a multi-leaf and 30% less than a taper leaf steel spring. On an articulated vehicle this saving could amount to as much as 500 kg. Another significant feature is that, whereas failure of a steel spring generally means a transverse fracture and loss of axle location, if a composite spring fails it splits longitudinally along its laminations, the fibres of which are continuous from end to end, and it therefore continues to locate the axle. Other advantages include good resistance to fatigue, wear and corrosion, good noise and vibration damping properties and by virtue of light weight, a significant contribution to the quality of ride.

Optimisation of the spring design depends upon consistent distribution of the reinforcing layers of fibre throughout the thickness of the spring, right through to its ends. Although the spring illustrated has eyes at its ends, slipper fittings can be embodied.

GKN extends the leaf beneath the steel-sleeved rubber-bushed bearing in each eye to cradle it. The outer sleeve of the bearing is projection-welded in the fabricated pressed steel eye, the two ends of which are clamped by a steel band over the end of the leaf. This arrangement is much simpler to produce than to form a complete circular eye in the laminated material. Moreover, it is simpler than moulding or otherwise inserting a tongue projecting from a steel eye into each end of the laminated leaf and securing it by bolts through the spring. Additionally, it obviates points of potentially high stress concentration. At the centre of the leaf, polyurethane pads form a resilient platform for the axle and

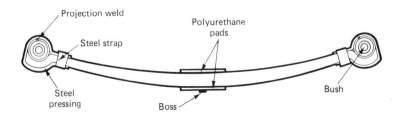

Fig. 35.13 Fabricated steel eyes are cradled on the ends of the GKN composite leaf spring

spread the loading uniformly over the contact area with the spring. For locating the axle a boss can be moulded into the polyurethane pad or a steel pin inserted into a hole in it.

35.16 Rubber springs

Various forms of rubber springs are used, mainly for commercial vehicles. They have been employed in passenger-carrying vehicles too, but the precise control of ride essential for comfort is difficult to achieve with such springs. A reasonably successful system was that used on the first *Minis* introduced by BMC. This had rubber springs similar to those in the Hydrolastic system described in Section 35.21, but each actuated directly by a rod attached to its centre, and of course without the hydraulic actuation, damping and interconnection system. Separate conventional telescopic hydraulic dampers were used.

An advantage of rubber springs is that by designing their fixing brackets for loading them initially in shear and then progressively changing to compression they can easily be given a rising rate – because rubber is much stiffer in compression than in shear. The arrangements for the progressive change from shear to compression loading can be seen from Fig. 35.14, where a wedge-shaped seating is employed.

Another advantage of this sort of arrangement is that the fatigue strength of rubber is best in compression. It is worst in tension because of the tendency for cracks to be opened out into which impurities are drawn and trapped. Consequently, some form of check-strap or tie is generally used to prevent tension loads from being applied to the rubber, in rebound conditions, see Figs 35.16 and 35.17. From the operator's viewpoint, the main advantage of rubber springs is a complete absence of any maintenance requirement. Leaf springs are especially liable to break in rough conditions such as are encountered in quarrying or construction sites. The damping, due to hysteresis of the rubber, is much more predictable than interleaf friction and helps to improve the ride in the unladen condition. Furthermore with an inherently high roll centre, the roll stiffness of a rubber suspension system is good.

The suspension in Fig. 35.14 is for a Norde trailer bogie, made by North Derbyshire Engineering Co. Ltd. On each side of the bogie are longitudinal beams A pivoted, on rubber bushes B at their centres, on the ends of a transverse beam. These longitudinal beams are rubber-bushed at their ends too, where they are pinned to the lower ends of vertical brackets on dead axles. The upper ends of these brackets are connected to rubber-bushed radius rods

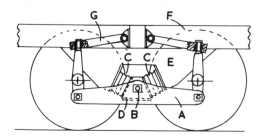

Fig. 35.14

Fig. 35.15 Norde Mk II suspension for Dodge GR 1600 tipper (16 tonnes gvw)

Fig. 35.16 Norde 20-tonne bogie

G, pivoted in front of and behind the transverse member of the chassis frame. Under each end of this transverse member are two inverted V-shaped brackets, D and E, one beneath the frame side member and the other mounted on the end of the transverse beam. The rubber springs C are interposed, chevron fashion, between these inverted V-shaped brackets, so that they are loaded partly in shear and partly in compression.

Radius rods G are usually A-shaped links as viewed in plan, to locate the axles laterally. This can be seen in Fig. 35.15, in which the outer ends – apices –

of these A-shaped links are attached to the tops of the differential housings of the axles, thus obviating the need for brackets. Fig. 35.15 illustrates a dual-drive live axle bogie, which is fitted to a Dodge GR1600, 16-tonne tipper chassis.

Another live axle arrangement, the Norde 20-tonne bogie, is shown in Fig. 35.16, and a Foden 40-tonne arrangement in Fig. 35.17. In the latter, two pairs of rubber springs are used, to take the heavier load, and the rebound stop is a cable, instead of the twin vertical tie-rods of Fig. 35.16, which are perhaps noisier. A further difference is the use in the Foden version of metal bump stops and, in the Norde bogie, rubber ones – the latter in effect are helper springs acting solely in compression. In each case, these stops are in the apex of the inverted V-bracket. A Norde single-axle system, with rubber-bushed trailing links and a Panhard rod for axle location, is shown in Fig. 35.18, again with a rubber bump stop.

In each instance, the rubber springs are interleaved with bonded-in steel plates. The primary function of these plates is to stabilise the rubbers, preventing them from bowing and ultimately being squeezed out bodily from between their abutments. They also help to conduct heat, generated by hysteresis, out from the centre of the rubber blocks.

Two other forms of rubber springs have been used to a limited extent, mainly on trailers and for special-purpose vehicles. The first is simply a steel tube within which is carried coaxially a spindle, either solid or hollow, both components being bonded to a sleeve of rubber in the annulus between them. The tube is usually secured, with its axis horizontal, to the frame. Then the spindle is connected by means of an arm to the axle, thus loading the rubber sleeve in torsion.

The second type of springing is the Neidhart system. This is similar to the first, except that both the tube and spindle are of square section, the sides of the squares of one being set 45° to those of the other, and rubber rollers are interposed between the two, as shown at A in Fig. 35.19. A variant is shown at B, and there are others, for example with hexagonal section spindles and tubes. Significant features of all these arrangements are that, because the rubber elements roll as well as being compressed, the deflection for any given value of torque is large and, by virtue of the shapes of the members between which the rubber rollers are contained, a rising rate is obtained.

35.17   Air springs

A volume of air, enclosed either in a cylinder fitted with a piston or in a flexible bellows, can be used as a spring, as in Fig. 35.20. Under the static load, the air is compressed to a predetermined pressure, and subsequent motion of the piston either increases or decreases the pressure and consequently increases or decreases the force acting on the piston. If this force is plotted against the piston travel, a curve similar to the compression curve of an engine indicator card will be obtained, so obviously the rate at which the force varies with the piston travel becomes greater as the air pressure increases. It follows that, whereas with a metal spring, equal increments of force result in equal increments of deflection, the rate of an air spring is not constant. This varying rate is an advantage in that a low rate can be obtained for small deflections from the mean riding position while keeping the total rise and fall of the axles within reasonable limits.

Fig. 35.17 Foden 20-tonne suspension

Fig. 35.18 Single-axle 10-tonne suspension with rubber spring, rubber bump-stop, and trailing link rotation for axle

Air springs are fairly widely employed on vehicles whose laden and unladen weights differ greatly. This includes principally tractors for semi-trailers, the semi-trailers themselves, and large drawbar trailers. They are also used to some extent on coaches, more especially in Continental Europe and the USA, because of the very high quality ride obtainable with them, particularly if used with independent suspension. The disadvantages are high cost, complexity of compressed air ancillary system, and therefore risk of breakdown, more

TYPE A

TYPE B

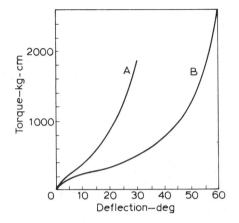

Fig. 35.19 Cross-sections of the type A and B spring units, and a typical torque-deflection curve for each

maintenance than other types of springing, and freezing of moisture in the air in cold weather, which can cause malfunction of valves. Air suspension systems of this sort are, in general, too bulky and too complex for cars, though Citroën cars for instance have their hydro-pneumatic system, Fig. 35.23.

In double-wishbone-type suspensions a rubber bellows, circular in section and having two convolutions, is generally used and simply replaces the coiled spring of the conventional design. Rubber bellows-type springs are used also in the Dunlop Stabilair suspension, Fig. 35.21. Alternatively, a metal air-container in the form of an inverted drum is fixed to the frame and a piston, or plunger, is attached to the lower wishbone. Since the piston is considerably smaller than the drum, sealing is effected by a flexible diaphragm secured to its periphery and the lip of the drum. This construction enables the load-deflection characteristic of an air spring to be varied considerably by using profiled guides, such as E and F in Fig. 35.20, to control the form assumed by the diaphragm, and thus its effective area, as the inner member moves relative to the outer one.

Elongated convoluted bellows such as are indicated in Fig. 35.22 have been used in trucks and coaches, with radius rods to deal with the driving and braking torques and thrusts, and a Panhard rod for lateral location.

35.18   Adjustable and self-adjusting suspensions

When steel torsion-bar springs are used, some method of adjusting the standing height of the suspension is needed. This is because, owing to the multiplying effect of the lever arm connected to the active end of the torsion bar, even a small tolerance on the angular relationship between the fittings at its ends can make a significant difference to the attitude of the vehicle. Moreover, it is generally difficult to maintain tight tolerances on the angular relationship between the ends, especially when the bar has been overstressed, or scragged, to increase its fatigue resistance. The adjustment device is generally a screw stop against which a short lever on the static end of the bar bears.

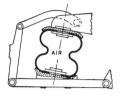
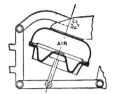

Fig. 35.20

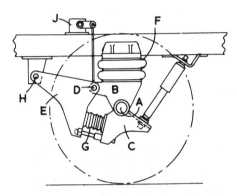

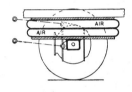

Fig. 35.21 Dunlop Stabilair suspension

Fig. 35.22

There are also variants of this principle, in which a worm-and-wheel drive is used, the wheel being on the static end of the torsion bar and the worm on a spindle that can be rotated by the driver while seated in the vehicle. Whereas the screw-type adjustment is for the initial setting on the production line and only rarely used when the vehicle is being serviced, the worm-and-wheel or other mechanism – sometimes actuated by a small electric motor – is employed also for adjusting the fore-and-aft trim of the vehicle to cater for variations in the load distribution – for example, when heavy luggage is carried in the boot. While provision for such manual adjustment systems is uncommon, automatic adjustment is the norm for air suspension.

There are two distinctly different types of automatic adjustment system for air suspension. One is the Citroën arrangement, Fig. 35.22, in which an engine-driven hydraulic pump supplies fluid under pressure to an accumulator and thence through levelling valves to combined air spring and strut-damper units. This is the *constant mass* system, in which the mass of the air, or an inert gas, enclosed in the spring is constant. The principle is illustrated diagrammatically, but greatly simplified, in Fig. 35.24 where the hydraulic accumulator is omitted and a floating piston P is depicted instead of the flexible diaphragm of the Citroën system, and the hydraulic damping system is omitted from the chamber O. The constant mass of gas A is compressed above the floating piston.

Space O, between the floating piston P and the piston *b* attached to the axle, is filled with oil O, which moves up and down with the pistons *b* and P, the air being correspondingly compressed or expanded. If the load is increased so that the assembly C, which is fixed to the body B, moves downwards, the valve V

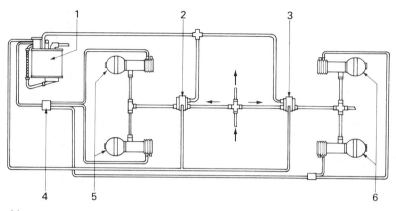

(a)
1 Reservoir
2 Front height corrector
3 Rear height corrector

4 Return reservoir
5 Front suspension cylinders
6 Rear suspension cylinders

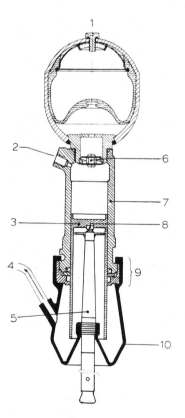

(b)

1 Filler screw
2 High-pressure fluid inlet
3 Piston
4 Overflow return
5 Piston rod
7 Cylinder
8 Thrust pad
9 Sealing system
10 Dust shield

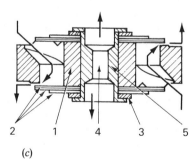

(c)

1 Body
2 Disc valves
3 Distance spacer
4 Calibrated by-pass hole
5 Spindle

Fig. 35.23 (a) Citroën GS suspension, (b) a spring unit, (c) damper valve

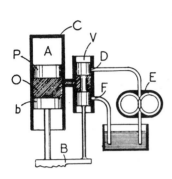

Fig. 35.24

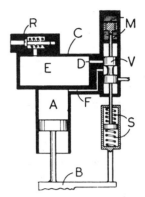

Fig. 35.25

opens port D, so oil from the pump E passes into the space O. The piston P and assembly C, together with the body, therefore move upwards, and this continues until the port D closes again. Similarly, if the load decreases, the port F is opened and oil escapes from the space O until the port F closes again. Thus the basic ride height of the suspension can be kept constant. This self-adjusting action is damped so that the motions between the body and the axle due to irregularities of the road do not influence the basic setting of the ride height.

The second type of automatic adjustment system is the *constant volume* one, Fig. 35.25. In this, for ride height adjustment, the volume of space A must be kept constant, so air has to be pumped into and released from that space, thus varying the mass of air in it.

If the load carried by the vehicle is increased, valve spool V opens port D, so that air can pass from reservoir E through passage F into space A, until the appropriate ride height is regained. The reservoir is kept charged with air, from a compressor, through non-return valve R. If the load carried is reduced, the valve spool V drops, until passage F is connected through the waisted portion of the spool to the outlet to atmosphere. Since such systems call for a compressor, they are used almost exclusively on commercial vehicles with air suspension – for these vehicles, a compressor is needed in any case for their brake systems.

To avoid excessive air consumption, a dashpot is introduced at M to prevent rapid movement of the spool V by suspension deflections as the wheels roll over the road, but allowing the longer term adjustments in ride height to be made to cater for changes in load. The introduction of this dashpot of course necessitates the interposition of a spring connection of some kind – such as that at S or, perhaps, resembling a hairpin spring – so that, if the axle lifts or falls rapidly, the spring deflects leaving the spool valve to move only at the speed it is allowed to by the dashpot.

35.19 Interconnected suspension systems

In a four-wheeled vehicle the suspensions of the front and back axles or wheels are usually quite independent of each other, but sometimes they are connected together and then the suspension system is said to be *interconnected, com-*

pensated or *equalised*. The basic principle involved is shown in Fig. 35.26, where a lever C connects the front and back springs. If the body D is assumed to move up and down so that it remains parallel to the ground, the lever has no effect, and the natural frequency of a vibration of this nature (which is commonly referred to as *bounce*) would be the same as if no lever were used and would depend chiefly on the stiffness of the springs A. But if the body is assumed to vibrate about the axis O, the frequency of this type of vibration (known as *pitch*) will depend chiefly on the stiffness of the springs B and, by making these relatively soft in comparison with the springs A, the natural frequency in pitch can be made very low while keeping the natural frequency in bounce reasonably high. This is the chief claim made for interconnected suspensions.

An equivalent arrangement which is more easily carried out in practice and which is used by Citroën in some of their vehicles is shown in Fig. 35.27.

An alternative arrangement is shown in Fig. 35.28, where the springs B are placed between the body and the axles, and only the springs A are coupled by the lever C. This produces the same kind of effect as the previous arrangement, but the bounce frequency now depends chiefly on the stiffness of the springs B, and the pitch frequency on that of the springs A. A suspension system using this principle has been used, with torsion bar springs, by the Packard company.

An article on the theory of interconnected suspension systems will be found in the *Automobile Engineer*, Vol. 47, No. 1.

35.20   Interconnected air and liquid suspensions

The interconnection of air springs is a simple matter provided that the actual spring units are of a suitable type. It is essential that they should give an increase in spring force with upward motion of the road wheel even though the air pressure in the unit remains constant. This will be seen from the diagrams in Fig. 35.29(*a*) and (*b*). In the first the air spring units are shown as simple cylinders such that the effective area of the piston or diaphragm on which the pressure acts remains constant as the wheel rises or falls relative to the body of the vehicle. Thus, assuming that both air spring units are of the same size, it follows that the pressures in them when they are interconnected will be equal as also will be the forces P exerted by them. The resultant vertical load will thus be $2P = W$ acting at the mid-point between the wheels. The centre of gravity (CG) of the body must therefore always lie at that mid-point in order to obtain equilibrium and this is obviously impossible. Any slight shift of the CG would

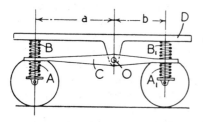

Fig. 35.26

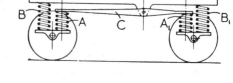

Fig. 35.27

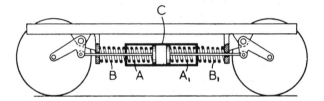

Fig. 35.28

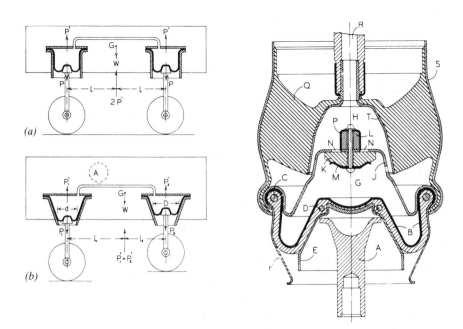

Fig. 35.29   Fig. 35.30

result in one air unit moving up to its extreme upward position and the other to its lowest.

In diaphragm (*b*), however, the air units are shown such that the effective area of the diaphragm increases as the road wheels move upward. Thus the effective area of the left-hand unit in the position shown is approximately $\pi d^2/4$ and that of the right-hand unit is $\pi D^2/4$ and so since the air pressure in the units is the same the force P_2 will be greater than P_1. The system will now be stable since any shift of the CG will cause one air unit to move up, thus increasing the force it exerts, while the other unit will move down, thus reducing its force, and equilibrium will be reached when the resultant of these unequal forces acts along the same line as the weight of the body.

The argument is equally valid if the areas of the units are unequal, the CG being no longer at the mid-point.

A system as shown at (*b*) would be effective if it were filled with liquid instead of air but would give a very stiff, or hard, suspension because liquids are only slightly compressible. The effective stiffness of the system can, however, be adjusted to a suitable value by making some part of the system elastic. This is indicated in the diagram where A represents a flexible container or reservoir whose volume can change considerably with changes of internal pressure. Alternatively, the reservoir can be rigid but fitted with a spring-loaded piston or diaphragm or could contain compressible material.

By putting suitable valves in the piping connecting the units of an interconnected fluid system sufficient damping can be obtained to eliminate the necessity for separate dampers.

35.21   BL Hydrolastic suspension systems

In these the elastic element is rubber acting in compression and shear and the interconnection is by fluid, with which the systems are filled. The general form of the construction is shown in Fig. 35.30, which shows a front-wheel unit for one of the BL *Mini* vehicles. It is used in an approximately vertical position but in the BL 1800 a similar unit is used in a horizontal position with its axis transverse to the centre line of the vehicle and for the rear-wheel units the position is again approximately horizontal but parallel to the centre line. The stem A is attached to a rod which bears at its lower end on the upper link of a double wishbone suspension at a point fairly close to the pivot of the wishbone so that the stem motion is only about one-fifth of the wheel movement. The wheel load is supported by the pressure acting on the upper surface of the diaphragm B. This is made of two rubber materials, one providing the required strength and the other the necessary seal for the liquid; it is reinforced by steel beads C and D. A piston member E in conjunction with a skirt F provides support for the diaphragm and helps to give the required spring characteristic. The fluid in the chamber G can at all times pass into the chamber H through a bleed hole provided in the member J which separates the chambers. A damper valve assembly K, L also provides additional passages. Thus the rubber flap valve K which is loaded by the spring M will open downwards when the pressure in H rises sufficiently above that in G, thus allowing fluid to pass through the holes N. The similar valve L, which is at right angles to K, will let fluid pass from G to H. These valves are kept from rotating out of position by fingers integral with the springs M and P which are bent down (as is seen in the case of P) so as to engage the holes N. The fluid in the chamber H acts on the underside of the rubber element Q and through the hose R is transmitted to the other wheel unit on the same side of the car. The rubber Q is bonded on the outside to the canister S which is fixed to the body structure and at its inside to the pot member T. The skirt F, diaphragm B, member J and the canister S are all secured by rolling over the edges of the metal components as shown.

35.22   Moulton Hydragas suspension

The Moulton Hydragas suspension, Fig. 35.21, superseded the Hydrolastic system on the BL car models. Moreover, since September 1977, its use has not been restricted, as it was by the original agreement, solely to this vehicle

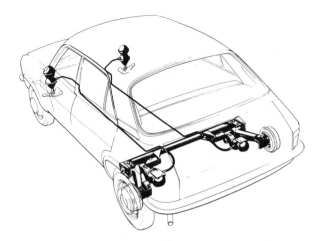

Fig. 35.31 Hydragas installation in Austin *Allegro*

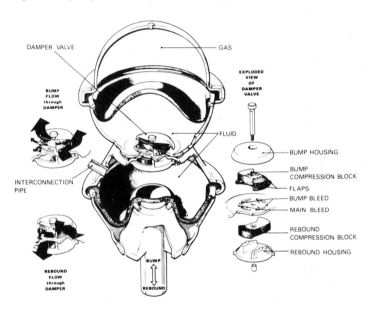

Fig. 35.32 Exploded view of Hydragas unit

manufacturer. Now, a constant volume of nitrogen gas has replaced the rubber springs of the Hydrolastic system.

The spring unit, Fig. 35.32, is manufactured for Moulton Developments Ltd by Dunlop. It comprises a fabricated pressed-steel container, divided into three chambers, one above the other. The uppermost chamber, beneath the domed top of the unit, contains the nitrogen gas and is separated by a rubber diaphragm from the intermediate chamber below it, which is full of hydraulic fluid – water with anti-freeze agents and a corrosion inhibitor. Below this is the third chamber, again full of hydraulic fluid, and its lower end is also closed by a diaphragm. Beneath this diaphragm bears the tapered piston that is moved up and down by the suspension. The interconnection passage between the two lowest chambers contains the damper valve.

From the illustration it can be seen that, as the piston rises, it forces fluid from the lower chamber, past the elastic rebound valve block which remains seated on the valve plate, through the holes each side of it in that plate, thus lifting the elastic upper, or bump, valve block and flowing into the central chamber where, acting on the upper flexible diaphragm, it compresses the gas in the top chamber. On the rebound stroke, the flow reverses, allowing the upper damper valve block to close and forcing the lower one to open. The shapes and sizes of these two valve blocks determine which holes they cover and uncover, the arrangement being such as to give more damping on the rebound than the bump stroke. Such a spring unit in the *Allegro* rear suspension is illustrated in Fig. 35.33.

With both Hydrolastic and Hydragas suspension, the spring rate increases with deflection. This rising rate is obtained in several ways: the first, in the Hydragas system only, by a progressive increase in the area of the piston contacting the diaphragm; secondly, in the Hydrolastic system only, by the progressive transition from shear to compression loading in the rubber spring; thirdly, by the geometry of the lever system – that is, the suspension – actuating the piston; and finally, the torsional characteristics of the rubber bushes in the pivots of that system. With Hydragas, the front–rear interconnection is made between the lower chambers of each unit.

35.23 Austin *Mini Metro* Suspension

The *Metro* suspension, Fig. 35.34, is a modified Hydragas system. All four wheels have gas springs but, because of the short wheelbase and potentially large variations in front–rear weight distribution, the front and rear springs are not interconnected. This avoids both a high frequency of pitching when lightly laden and a tail-down attitude with a heavy load of luggage at the back. As in the *Mini*, a twin unequal-length traverse link system is used at the front, and a trailing link actuates each rear gas spring through a push rod, the leverage ratio being 5 : 1.

Fig. 35.33 Hydragas spring unit in the Austin *Allegro* rear suspension

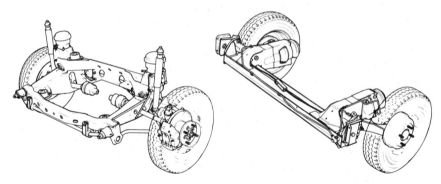

Fig. 35.34 Austin *Metro* suspension

With only the driver in the car, a high proportion of the total load is that of the engine and suspension assembly on the front sub-frame. To provide adequate damping control, therefore, between the structure of the vehicle and both the front suspension and the engine, telescopic hydraulic dampers are installed, and there are no damping valves in the front Hydragas units.

The rear Hydragas springs, on the other hand, are internally damped, as described in Section 35.22. This avoids the need for separate shock absorbers and consequent vertical intrusion into the load space above. The spring units are installed horizontally within the confines of a U-shaped sub-frame. To reduce the suspension deflection from the unladen to fully laden condition, the gas springs at the rear are pre-loaded by coil springs coaxial with the push-rods. This restricts the range of operation of the gas spring to the top part of its load-deflection curve, where the slope of that curve steepens. The front–rear attitude of the vehicle is set by charging the rear springs appropriately with fluid. These springs are interconnected laterally so a single charging connection is provided. On the other hand, since there is no front–rear interconnection, the front springs have to be charged individually – they are not laterally interconnected.

Because the front springs act independently and an anti-roll bar is installed there too, all tendency to roll is reacted at the front, while the rear end, on its interconnected rear suspension, is allowed to float freely so far as roll is concerned. This arrangement is analogous to that of a three-legged stool, which has the advantage that all the legs are at all times firmly planted on the ground regardless of whether it is horizontal, flat, or uneven. Consequently, the adhesion between all four wheels is constantly in balance and therefore road-holding is good.

35.24 Chassis lubrication

Most of the more heavily-loaded pivots and control joints in cars either have greased and sealed-for-life bearings or rubber or other types of bushes that do not need lubrication. There may still be, however, one or two points where periodic lubrication is needed, either under pressure through nipples or from an oil can through holes provided for this purpose. On commercial vehicles, where the loading is much heavier, more gun lubrication points may be provided. Oil is preferred to grease on the grounds that it will flow more readily to all parts of the bearing. To reduce the labour involved, nipples are in some instances grouped, with short pipes connecting them to the points to be lubricated.

Alternatively, all the points may be fed automatically by a pump, from a reservoir, through pipelines and restrictors to the bearings. Various types of restrictor are employed, from simple calibrated orifices to orifices in which are tapered pins for individual adjustment. The pump may be of the plunger type actuated electrically or by air pressure or induction manifold or other source of depression, by a spring which, after having been compressed by the driver extends progressively, or even by the oscillation of a mass suspended on a spring. Other systems have pumps driven mechanically from the engine or some part of the transmission system. In all instances the rate of delivery from the pump is of course very slow.

Some systems have individual pipes from the pump to each bearing, while others have a single pipe delivering to distributor blocks, close to groups of bearings, each of which are then served by short pipes from their distributor. The advantages of the latter arrangement are simplicity, a great reduction in the length and complexity of piping, and therefore light weight and low cost. Its disadvantage, however, is that should a breakage or blockage occur in the primary pipe run the whole system will be starved of lubricant, instead of only one bearing.

35.25 Some autolubrication systems

An example of a central automatic lubrication system is the Clayton Dewandre Milomatic. It is shown in Fig. 35.35 and consists essentially of three units: the distributor valve, the intensifier and the delivery valve block. The distributor valve, seen in the lower view, comprises a gearbox giving a reduction of 955 : 1 between the input shaft A (which is driven off any convenient part of the transmission, the speedometer drive being frequently used) and the gear wheel B which carries the eccentric C. The latter is thus caused to make one revolution per vehicle mile and it operates the valve plunger F through the bell crank D and spring E. As the plunger is pushed to the left it first makes contact with the disc valve G and thus cuts off the passage from the pipe J to atmosphere which has hitherto been open through the hole in the plunger. The disc valve is then pushed off its seat so that air pressure can pass from the inlet H, which is connected through a pressure regulating valve to the reservoir of the brake system, to the outlet J which is connected to the inlet J in the upper view. This view shows the intensifier; in the position shown the space N is open, through the valve M, to the oil tank and so is full of oil. When the distributor valve passes air to the inlet J the piston K is forced to the right and immediately releases the valve M which then seals off the inlet from the oil tank. The movement of the piston then raises the pressure in the space N and consequently in the spaces P of the delivery valve block which is shown on the right. This consists of a housing Q which is shown bolted up to the face of the intensifier unit but can be mounted as a separate unit and connected to the intensifier by a pipe. In the housing Q injector units S are mounted radially and each one is coupled by a pipe to a chassis point requiring to be lubricated. The spaces T of the delivery valve block are connected to the oil tank via the

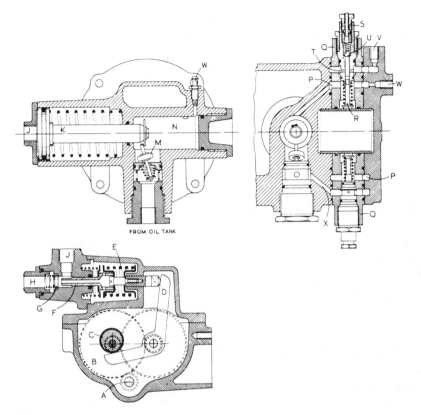

FROM OIL TANK

Fig. 35.35

passage X or by a separate pipe if the valve block is a separate unit. When the pressure in the spaces P is raised by the action of the intensifier it acts on the area of the plungers R and forces the plungers outward. As soon as the plungers seal off the inlet ports T oil will be forced out through the non-return valves U to the lubrication point. By varying the position of the inlet port the amount of oil delivered per cycle can be varied and three standard positions are available corresponding to deliveries of 0.025, 0.05 and 0.075 cm^3 per cycle. Initially, the system must be filled with oil and so bleeding points are provided as shown at W. Any excess oil can return to the oil tank through the pipe connected to the outlet V. Each valve unit contains 12 injector units and up to six valve units can be operated by one injector unit. The system will continue to operate satisfactorily for all the other lubrication points if one point should become blocked or if the pipe to it should be fractured.

Other manufacturers who supply such systems include Interlube Systems Ltd, which offers both mechanical and electrical systems, some with electronic controllers. Lincoln make the Ecolube, multi-line, and Quicklub, single-line, systems. Both are electrically actuated, but Quicklub is available also with either manual or automatic control. Manufactured in France are the range of Mechafluid single-line systems and the Compact single-line, air-actuated system for light commercial vehicles. From Holland, though originally from the USA, come the IMS-Groenewald air-actuated, single-line systems. Romatic produces, in addition to manual and automatic systems, some that are electrically or pneumatically actuated, while Telehoist, a Rockwell company, offers the Telelube and Greaselub single-line air-actuated systems.

Chapter 36

Suspension systems

Independent suspension is the term used to describe any arrangement by which the wheels are connected to the carriage unit in a manner such that the rise and fall of one wheel has no direct effect on the others. Although widely used only since the Second World War, its first employment was at least as early as 1878. Some of the reasons for the majority of manufacturers abandoning – progressively for about a decade from 1935 – the beam axle and semi-elliptic spring front suspension on cars are given at the end of Section 35.9. The benefits derived from the adoption of independent front suspension are as follows—

(1) Because of the close approximation to vertical travel of the wheels, gyroscopic kicks in the steering system are obviated – with a beam axle, these reactions, which tend to initiate wheel shimmy, are inevitable if only one wheel rises over a bump.
(2) Steering effects due to lateral movements of the tyre/road contact patch, as the wheel rises and falls, are obviated.
(3) Axle tramp, and particularly that associated with alternating wind-up and release of leaf springs subjected to brake torque, is obviated.
(4) Variations in castor angle – defined in Section 34.5 – due to this wind-up are obviated.
(5) A coil spring can be more easily accommodated close to a wheel that has to be steered than can be a semi-elliptic spring.
(6) The roll centre can be lower and the spring base wider, so the roll resistance is higher. A secondary effect of the inherently high roll resistance is that softer springing can be used which, as explained in Section 35.2, reduces the tendency to pitch.
(7) The engine can be positioned further forward, since it does not have to clear a beam axle – this leaves more space for the occupants of the car.
(8) Unsprung mass is lighter, and therefore the ride quality improved.

There are many independent suspension layouts, and they can be classified as in Table 36.1.

Most of these can be adapted for use with any of several different types of spring. It would be impossible in the space available to describe all the systems, but the most important types are dealt with approximately in order of merit in this chapter.

Table 36.1—INDEPENDENT SUSPENSION LAYOUTS

Types	*Examples*
Single transverse link (Section 36.5)	A Mercedes-Benz 300d swing axle (rear)
Double transverse link (Section 36.3)	Majority of modern front suspensions
Single leading or trailing link (Section 36.6)	Citroën 2 CV front and rear, *Mini* and Renault *Frégate* rear
Double leading or trailing link (Section 36.7)	VW *Beetle*, rear
Broulhiet (Section 36.8)	Broulhiet
Girling (Section 36.9)	Daimler *Regency*, 1951
Single link with pivot axis not parallel to either longitudinal or transverse axes of car (Section 36.19)	Triumph 2000 (1963) and Fiat 600 (1965) rear suspension
Double link with pivot axis not parallel to either longitudinal or transverse axes of car (Section 36.19)	Humber *Super Snipe* (1952)
MacPherson strut type (Section 36.4)	Many modern cars, including Ford range
Dubonnet (Section 36.10)	Vauxhall ranges, just prior to and after the Second World War
Slider, or pillar, type (Section 36.11)	Morgan

36.1 Camber angle

The overwhelming majority of cars now have a double transverse-link front suspension, though the MacPherson strut-and-link type, to be described in Section 36.4, is gaining ground. In both instances the aim is at a vertical motion of the wheel approximating to what it would be if it were mounted on an arm whose pivot was a point of infinity on the opposite side of the vehicle. This implies no changes in camber angle, which is the angle of the plane of the wheel relative to the vertical.

Camber angle is positive when the lower edge of the wheel and tyre is tucked under the vehicle, and negative when the upper part slopes in towards the centre of the vehicle – easy to remember by association of negative with knock-kneed. Negative camber, by setting the wheel at a favourable inclination for reacting the centrifugal force on the vehicle, increases its cornering ability. Positive camber, on the other hand, is sometimes adopted to help in obtaining centre point steering, as described in Section 34.4.

36.2 Roll centre

When a force acts on a car from the side, making the body tilt or roll, the motion will be about some line lying in the vertical plane containing the longitudinal axis of the car. This line is called the *roll axis*. Its position depends on the type of suspensions at the front and rear. In Fig. 36.1 is a front view of a car having an axle and laminated springs – the latter are indicated simply by zigzag lines. When no side force acts, these springs will be equally compressed and each will exert a force P equal to half the weight carried by the axle. The vertical forces between each wheel and the ground will also be equal to P plus half the weight W of the axle and wheels. When a side force F acts on the body it sets up forces f_1 and f_2 at the points of connection of the body and springs. The relative magnitudes of these forces and the exact positions at which they

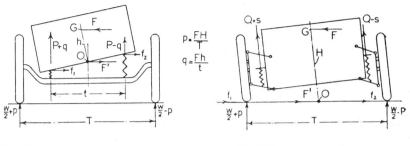

Fig. 36.1   Fig. 36.2

act will always be somewhat uncertain. However, assuming them to be equal and to act as shown, their resultant is a force F' equal and opposite to F. These two forces F and F' thus constitute a couple of magnitude Fh, h being the perpendicular distance between them which, unless the tilt is very large, may be assumed to be equal to OG, where O is the roll centre and, if F is the centrifugal force, G is the centre of gravity. For equilibrium, there must be an equal and opposite couple to balance the couple Fh. This balancing couple is supplied by an increase q in the left-hand vertical reaction and a decrease of the same magnitude in the right-hand one. Consequently, the left-hand spring will sink a little lower than before while the right-hand one will rise by the same amount. The body will thus tilt about the point O, on the roll axis. Similarly there will be a corresponding centre O′ at the rear, so the line OO′ is the roll axis. The change q in the spring forces on each side will be equal to Fh/t, t being the spring base.

The vertical forces between the wheels and the ground will change by amounts p, where $p = FH/T$, T being the wheel track and H the height of the line of action of the force F above the ground. If, under the action of the side forces f_1 and f_2 or, more accurately, under the reactions to those forces, the springs deflect slightly sideways, the point O would move sideways and therefore the centre of tilt would be slightly lower.

Considering now the car with independent suspension as shown in Fig. 36.2, the side force F again sets up a tilting couple and this has to be balanced by an increase p in the force exerted by the left-hand spring and an equal decrease in that exerted by the right-hand spring. If the tilt is not excessive, these changes in spring force will be equal, one increasing from Q to $Q + s$ and the other decreasing to $Q - s$. The compression of the left-hand spring will be increased by some amount and that of the right-hand spring will be decreased by the same amount. Therefore the suspension will assume the position shown and the body will have tilted about the point O. Thus O is now a point on the roll axis and the line joining it to the similar centre of tilt O′ at the rear will be the roll axis.

For a car having rigid axles at front and back, the roll axis will be some distance above ground level while for a car having independent suspension at front and back, the roll axis will be at or near ground level. The roll axis of a car having independent suspension at the front and a rigid axle at the rear will be inclined, from approximately ground level at the front rising to about axle level at the rear.

With the MacPherson-type independent front suspension, a roll centre height just above ground level is usually chosen because it offers minimum lateral scrub, coupled with relatively little change in roll centre height with deflection of the suspension. On the other hand, a higher roll centre is obtained with the angled single link now popular for rear suspension, Section 36.19, but because the axis along which the pivots are set is fixed, Fig. 36.24, its height does not vary much with deflection.

If the roll axis lies at ground level, the overturning couple FH will be greater than Fh obtaining when the roll axis is above ground level. Although this will tend to make the tilt of the car with independent suspension greater than that of the car with rigid axles front and rear, this tendency is offset because the effective spring base of the independent suspension is the wheel track T, which is considerably wider than the spring base t. It can be shown that, for a vehicle with rigid axles, the angle of tilt is approximately proportional to $2q/t^2 = 2Fh/t^2$ while, for a car with independent suspension, it is proportional to $2p/T^2 = 2FH/T^2$. In most cases, the latter will be the smaller because of the greater effective spring base, which will outweigh the effect of the increase in the arm H of the overturning couple.

The greater the roll stiffness – resistance to roll – at one end of the car, the larger will be the proportion of the tilting couple that will be reacted at that end. Indeed, if the roll stiffness were infinite at the front, the whole of the tilting couple would be reacted there, unless the vehicle were flexible enough to deflect torsionally and thus throw some of the load on to the rear suspension. Cars are very rigid torsionally relative to the loads they carry. This is because the whole body can be regarded as a torque tube. Heavy commercial vehicles, however, because of the great weights that they must carry – relative to which their chassis frames are shallow – are flexible torsionally, which is one reason why independent suspension is not normally used on them.

In practice, however, 'roll centre' is not such a simple concept as it might at first appear. Although it is generally assumed that the vehicle rolls about an axis represented by a line passing through the roll centres of the front and rear suspension, this could be true only if the wheel and tyre assemblies were rigid and did not move sideways on the road. The roll centre, as determined from the kinematics of the suspension links, moves as the suspension deflects and does so increasingly towards the extremes of this deflection. Obviously the actual motions of the carriage unit, taking into account not only the movements of the roll centre due to variations of the suspension geometry with wheel deflection but also both vertical and lateral deflections of the tyre calls for the use of a computer. Even so, a first approximation accurate enough for practical purposes can be obtained from consideration of the suspension link geometry.

The method of finding the roll centre is illustrated in Fig. 36.3. It involves extending the axes of the suspension links until they meet at O, and then joining O and the centre of contact T between the tyre and the road. Point O is then the instantaneous centre about which rotate all parts of the suspension and its pivots on the carriage unit on that side of the vehicle. If we then do the same for the suspension on the other side, we find that the two lines OT intersect at C, which is the instantaneous centre about which the two points T rotate and is therefore the roll centre of the vehicle as a whole. By re-drawing this diagram with one suspension deflected up and the other down – the situation when the body rolls – it can be shown that the points O move down

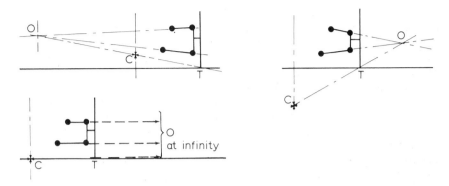

Fig. 36.3 Note: If the suspension on the other side of the vehicle were drawn, the diagram for obtaining the instantaneous centre would be a mirror image of that shown here, which is why C is always on the vertical centre line of the vehicle, except when the wheels on each side deflect in opposite directions

and up respectively on the two sides which of course move the roll centre C not only up or down, according to the geometry, but also laterally, off the vertical centre line of the vehicle. Some typical linkage systems and their roll centres, with the suspension in its static position, are shown in Fig. 36.4.

36.3   Double transverse-link suspension

A double transverse-link suspension, with a torsion-bar spring, is illustrated diagrammatically in Fig. 36.5. To eliminate wheel tilt with deflection, that is change in the camber angle ϕ, the two arms would have to be of equal lengths, which would place the roll centre at ground level. This, however, would have the effect of varying the track with roll, having undesirable steering effects and could also adversely affect tyre wear. By shortening the upper link it is possible to keep the track almost constant without introducing too much variation in camber angle. Moreover, the slight change in the camber angle is negative on the outer wheel when the vehicle is turning, which increases the cornering power. In the early independent suspension systems, these links were generally both parallel to the ground. After the Second World War, however, the practice of inclining them, to move the roll centre, became widespread. Exactly what these effects are can be seen easily by drawing sketches on the lines indicated in Fig. 36.3, but with the arms at various angles. It will be seen that sloping the upper link down towards the wheel raises the roll centre, and *vice versa*.

Where a coil spring is to be used, instead of a torsion bar, it is usually installed coaxially with the telescopic damper S in Fig. 36.5. Incidentally, it is more common for the upper ends of the spring and shock absorber to be on the vehicle structure instead of on the upper transverse arm. On some vehicles, the springs and dampers have been mounted separately, so that access could be gained more easily for servicing the shock absorber. An alternative is to arrange for removal of the shock absorber through a hole in the spring seating pan. With front-wheel-drive cars, the coil spring and shock absorber are in most instances interposed between the upper transverse link and the vehicle structure, to leave space for the driveshaft to the wheel.

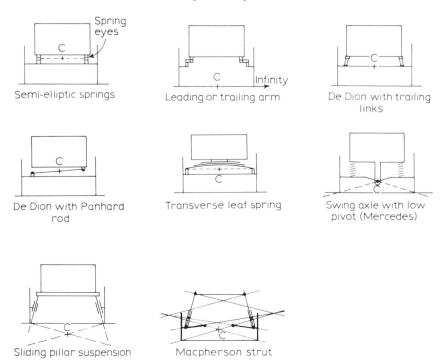

Fig. 36.4

With unequal-length wishbones, the end B of the steering arm, Fig. 36.5, moves in a curve that is not an arc of a circle, so vertical deflection of the suspension inevitably has some steering effect if the centre of the other end – that of the joint at its connection to the remainder of the steering linkage – is in line with the axis X_1X_1 of the pivot for the lower transverse link. This problem is generally overcome in one of two ways: either a position for the centre of the connection to the remainder of the steering linkage can be chosen so that it coincides with the centre of an arc approximating to the curve through which B moves, in which case the undesirable steering effect may become negligible; or the centre of B is placed in line with the axis x_1 of the lower arm A_1, in which case, with the other connection in line with the axis X_1, the motion would be truly circular and there would be no steering effect. Similarly, the two ends could be in line with the axes X_2 and x_2 of the upper arm A_2, though this is more difficult to arrange.

Two variants of this type of suspension linkage are shown in Figs 36.6 and 36.7. In the first, a laminated spring serves as the upper arm, while in the second, two pairs of laminated springs, one at the top and the other at the bottom, replace both transverse arms. The first can of course be inverted, as in the case of the Fiat 600, which had a wishbone-type upper link and no drag link. When the lever arm type shock absorber was widely used, its arm was sometimes made to serve as the upper transverse link, a notable example having been in the Morris *Minor*.

Another variant is one in which the lower arm is replaced by the driveshaft. One example is the Triumph *Herald*, in which the upper transverse link is

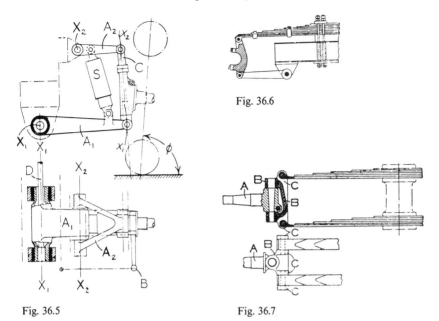

Fig. 36.6

Fig. 36.5

Fig. 36.7

formed by the ends of a transverse leaf spring, and the lower by what is sometimes termed a *swinging halfshaft*. This shaft swings up and down about a universal joint at its inner end, where it is connected to the final drive differential gear. Provision for the necessary articulation at its outer end is made in an unusual manner. The wheel is keyed directly on to the tapered outer end of the halfshaft, and the wheel bearings are immediately inboard of it. These bearings, with the outer end of the halfshaft rotating in them, are carried in a housing that is pivot-mounted on the lower end of the vertical link to the upper end of which the eye of the transverse leaf spring is connected. The axes of these two pivot connections are of course parallel, so that the two links can articulate together, carrying the vertical link and thus the wheel assembly up and down with them. To provide fore-and-aft location, a drag link, pivoted at both ends, serves as a radius rod between the wheel bearing housing and a transverse member further forward on the chassis frame.

A similar independent rear-suspension layout, but with coil springs, Fig. 36.8, is used on several Jaguar cars. Here the swinging halfshaft forms the upper transverse link, but it has at its outer end a universal joint instead of the pivoted bearing housing arrangement of the Triumph *Herald*, to give it the required freedom to articulate. The lower link comprises a single transverse arm and a drag link. The disc brakes are mounted inboard, on the final-drive assembly, thus considerably reducing the unsprung weight and relieving the suspension linkage of the brake torque except in so far as it is transmitted back through the swinging halfshaft to the brakes. Since the stub axle carrier pivots about the outer end of the lower link, the vertical loads applied at the wheel put the halfshaft in compression while, during cornering, the side load on the outer wheel places it in tension, the two tending to cancel each other out. On the inner wheel, although the two loads are additive, they are in any case lighter and, during extreme cornering, fall to zero.

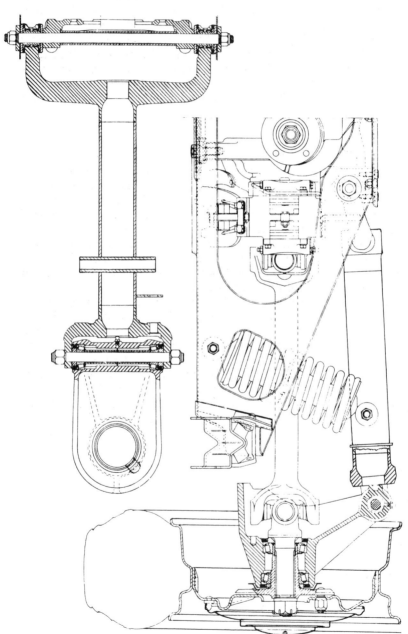

Fig. 36.8 Jaguar Mk X independent rear suspension assembly

There are some of the variants of the double transverse link system, in which each of the two links can be a single arm used in conjunction with a drag link, as in the Jaguar system, or it may be triangulated to form what is termed a *wishbone link*. In the latter, the apex of the link is adjacent to the stub-axle carrier, and the base is secured by a pivot bearing at each corner to the vehicle structure. The more widely spaced are these bearings, the lighter becomes their loading due to brake drag and torque. There are examples in which the axis of the pivot bearings are not parallel to the longitudinal axis of the vehicle. This arrangement is usually adopted, as in the Humber *Super Snipe* chassis, simply to enable the front transverse member of the chassis frame to clear the engine sump or, in other words, to enable the engine to be installed as far forward as possible. The frame of this vehicle is illustrated in Fig. 31.2, and the wishbone links (not shown) trail at the same angle as the outer ends of the front transverse member.

36.4 MacPherson strut type

The principle of the strut-type suspension invented in the nineteen-forties by Earle S. MacPherson, a Ford engineer in the USA, is illustrated in Fig. 36.9 and details in Figs 36.10 and 34.8. This type is common because, with its widely spaced attachments to the carriage unit, it fits in well with the basic concepts of chassisless construction. Moreover, with transverse engines there may be no room for upper transverse links and, even if there is, the strut type leaves more space around the engine, so access for maintenance is easier. The only significant disadvantage is the radial loading on the piston, due to lateral forces during cornering and to brake torque.

Two links, one taking the lateral and the other the drag loading, and one nearly vertical telescopic strut, make up the complete mechanism, Fig. 36.9. A single transverse arm is pivoted on the structure at B and connected by a ball-and-socket joint D to the base of the strut C. Sliding in C is the member E, the upper end of which is secured to the body by the equivalent of a ball-and-socket joint – in practice, since the articulation at F is small, a rubber joint is generally used. The spring is compressed between two flanges, one on the member C and the other on E. This telescopic strut also serves as the hydraulic damper. For steering, member C rotates about the axis DF. In the

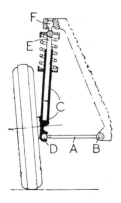

Fig. 36.9

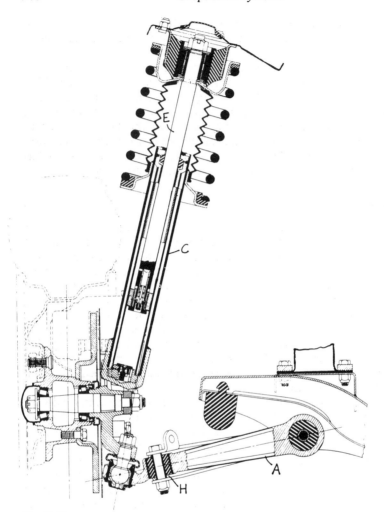

Fig. 36.10

mid-nineteen-eighties Ford introduced, for their *Transit* vans and *Escort* models, a variant with its coil springs interposed between the transverse links and the frame, so that it would not intrude into the space where the clutch and brake pedals had to be placed.

The fore-and-aft forces acting on the road wheel are taken by a tie or drag link, the rear end of which is pivoted at H, near the outer end of the link A; at its other end, this drag link is pivoted to the body structure. Both pivots are usually rubber bushes. In plan view, it is usually set approximately 45° relative to the longitudinal axis of the car and, since it is in front of the link A, it is in tension. In many instances, it is formed simply by bending the ends of a transversely-installed anti-roll bar, so that they will perform the dual functions of drag links and lever arms for actuation of the anti-roll bar, which is usually carried in rubber bushes adjacent to its cranked ends. To clear the engine sump, the anti-roll bar is usually mounted at the front, but if it can be mounted behind the engine, the application of brake torque to its forward extending lever arms tends to lift their pivot bearings and thus to counteract brake drive.

In Fig. 36.10, the strut and stub-axle forging are made as separate components, for ease of replacement of a damper. The top end of the piston rod E is free to rotate in a bush in the centre of a large rubber mounting by means of which the unit is secured to the body. This bush serves as a swivel bearing for the steering, while the rubber mounting insulates the body from noise, vibration and strains due to deflection of the suspension unit. The lower transverse link A is connected by a rubber bush H to the drag link, which has another rubber bush for its connection to the vehicle structure.

36.5   Single transverse link

What might be described as a single transverse-link system was used for the front suspension of the Allard vehicles in the nineteen-fifties. In effect, the front axle beam was divided in the centre, where each half was pivoted to the chassis frame. Coil springs were interposed between the outer ends of these half-axles and the frame. Drag loads were taken by two radius rods, both pivoted beneath the centre of the front transverse member of the frame and extending rearwards and outwards to a point beneath the spring seats, where they were rigidly attached to the axle. The axis of the pivots of the radius rods were in line with those of the divided axle. Thus, each half-axle and drag rod together formed a single transverse link. This type of system has been more commonly used in rear suspensions, as the swing axle arrangement described in Section 36.18, and it suffers the same disadvantages.

Fig. 36.11 illustrates diagrammatically a front axle arrangement, the dotted line showing how a wishbone-type link can be used, instead of a drag link, to react brake drag. Both the track T and the camber angle ϕ change with suspension deflection. The steering rods have to pivot on ball joints the centres of which are in line with the axes X, so that the rods rise and fall parallel with the transverse link. If the forward arm of the wishbone is shorter than the rear one, and thus sweeps the link forward, the application of brake torque tends to lift the pivot bearings and therefore to counteract brake dive, though the magnitude of this effect depends on the angle the axis XX makes with the

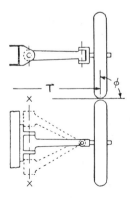

Fig. 36.11

longitudinal axis of the vehicle, and therefore is not normally of great significance.

36.6   Single leading or trailing link

This type, shown diagrammatically, in Fig. 36.12, is the simplest but has several disadvantages, which include variations in angle of inclination θ of the kingpin axis, the difficulty of obtaining adequate stiffness to resist satisfactorily the couple due to lateral loading at ground level during cornering, and the fact that the wheels tilt to the same angle as the sprung mass when the vehicle rolls. Such a system has been used also with torsion-bar springing, including with laminated torsion bars. Normally the two torsion bars, one for the left-hand and the other for the right-hand arms, are attached one at each of the pivots X of the arms A, and their other ends are anchored to the frame or vehicle structure, usually on the side remote from the pivots, so that they overlap instead of being coaxial, and therefore can be longer.

When the wheel is steered straight ahead, the joint B on the end of the steering arm moves through arc x_1 while the stub axle moves through arc x as the suspension deflects. Therefore, if unwanted steering effects are to be avoided, the other end of the steering rod should pivot somewhere along axis XX, which can be difficult with torsion-bar springing. Because of this and the variation of castor angle, or angle of trail, this type of suspension has been used more for rear than for front suspension systems.

36.7   Double leading or trailing link

With two parallel links, A_1 and A_2, of equal lengths and pivoting about axes X_1 and X_2, Fig. 36.13, the vertical member C, which carries the stub axle and

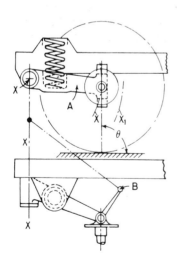

Fig. 36.12

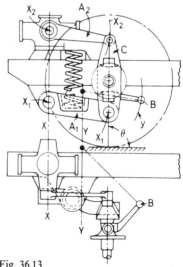

Fig. 36.13

kingpin, maintains the angle θ constant. The end B of the steering arm moves through arc y, the radius of which is equal to that of arcs x_1 and x_2. Therefore, if the steering rod end is pivoted anywhere on the axis YY, vertical deflection of the suspension will not have any steering effect. If torsion-bar springing is used, one bar can be connected to the upper link on one side and the other bar to the lower link on the other side.

36.8 Broulhiet suspension

The Broulhiet arrangement, Fig. 36.14, is a variant of the double transverse-link system. A single link is used at the top and a wishbone- or V-shaped arm at the bottom. The bottom link is made in two parts, B and C, pinned together. This joint is not, however, a working joint so B and C could be made integral. The axes XX of all the pivots are parallel to each other but inclined – converging towards the front – relative to the longitudinal axis of the car. This arrangement places the arm C in a good position for taking the fore-and-aft forces; it is also longer than in the more conventional double transverse arm type, and therefore can more easily react brake torque. Since the pivots E and F are a considerable distance apart, it is difficult to keep their axes coincident. To allow for any slight misalignment, one or both of the pivots can be rubber bushed. Except for misalignment errors, the movements of the linkage do not involve any but ordinary pivot action at any of the joints.

36.9 Girling suspension

This is a variant of the Broulhiet system, and is illustrated diagrammatically in Fig. 36.15. The stub-axle and swivel-pin assembly is in effect carried by the arm C, and if the length of that arm is appropriately chosen, the brake torque can apply at its pivot an upward force appropriate for counteracting the nose-down effect due to weight transfer from rear to front axle. Because of the conflicting motions of the links, rubber bushes have to be used at all the pivot points, so this type of suspension is not favoured for modern cars where precise steering is required.

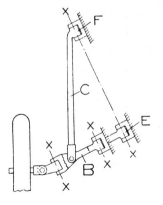

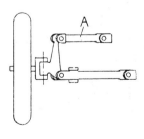

Fig. 36.14

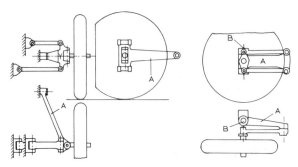

Fig. 36.15 Girling suspension

Fig. 36.16 Dubonnet suspension. Member A pivots about swivel pin B on the end of a chassis transverse member. The wheel is carried by the parallel arms which pivot about the other end of A

36.10 Dubonnet suspension

With the Dubonnet system, Fig. 36.16, the kingpins are mounted on the ends of a transverse beam rigidly secured to the vehicle structure, and each stub axle is on the end of either a leading- or a trailing-arm assembly deflecting vertically about the end of a torsion-bar spring and, for steering the vehicle, swivelling about the kingpin. In other words, the kingpin bearings and steering arm are on the sprung mass, and the mechanism allowing vertical deflection of the suspension is interposed between the kingpin and wheel. Several American cars had such a system in the nineteen-thirties but the most recent example was that used on Vauxhall cars in the late nineteen-forties. Fig. 36.16 is a diagrammatic representation of the 1936 Vauxhall DX Light Six, which had a torque-reaction rod parallel to the leading arm.

The Vauxhall system, Fig. 36.17, the 1949 *Velox* and *Wyvern* L-type, was in fact a modified Dubonnet arrangement, in which a coil spring supplemented the torsion-bar spring. A rigid hollow trailing arm extended rearwards from the kingpin bearing. It contained the coil spring and damper and carried at its rear end both the outer end of the transverse torsion-bar spring and, just outboard of it, the bearings for the forward extending arm carrying the stub axle. Because of the restricted space for the torsion-bar spring, owing to the proximity of the engine, it comprised two coaxial halves – a hollow and a solid bar. The outer end of the solid bar was secured to the pivot end of the suspension arm and the inner end splined into the inner end of the hollow torsion bar, the outer end of which was flanged and bolted to the rear end of the rigid arm housing the coil spring and damper.

The functioning of the forward extending arm and torsion-bar spring is simple, so no explanation is necessary. Rather more complicated, however, is the action of the powerful supplementary spring and the simultaneous operation of the damper. The latter is a twin side-by-side cylinder type, hydraulic damping unit actuated by a pair of arms diametrically opposed on the pivot pin extension of the torsion bar, and therefore rocking with it as it oscillates in each direction. In principle, it is similar to the lever-type damper illustrated in Fig. 35.4.

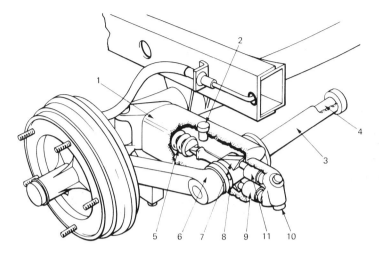

1 Suspension housing	5 Compensating ring	9 Shock damper piston
2 Filler plug and	6 Wheel carrier arm	10 Shock damper relief
bump-stop	7 Control lever	valve plug
3 Torsion tube	8 Oil seal assembly	11 Shock damper return
4 Torsion shaft		piston spring

Fig. 36.17 Vauxhall front suspension, sectioned view

The damper is housed behind, and the coil spring in front of, the torsion bar, both of course being in the rigid hollow arm. A toggle arm extends forwards from the pivot pin extension of the torsion bar, and the coil spring is compressed between its end and the swivel-pin bearing housing assembly. When the suspension arm is in its normal static position, the axes of the spring and the toggle arm are in line, so the spring has no effect on the overall spring rate. As the toggle rotates further, allowing the coil spring to extend, and thus unloading it, the overall rate increases again, ultimately approaching, at both extremes of deflection, that of the fairly short, and therefore stiff, torsion bar. The advantages of this system are light unsprung mass, variable-rate suspension, accurate steering geometry, and an anti-dive effect because the arm carrying the stub axle tends to lift the front of the car when brake torque is applied. Its disadvantages, however, are its complexity, and shortness of the torsion-bar spring.

36.11   Slider, or pillar, type

Lancia were probably the first to introduce this type of suspension, which has been used for many years too on Morgan cars. The principle is illustrated in Fig. 36.18, in which, for steering, the stub axle A, on pillar B, swivels in plain bearings in the housing C for the coil spring D. Broulhiet produced a similar suspension but with the lower end of the pillar carried in bearings, for accommodating steering motions, in the end of a rigid transverse member secured to the frame. The slider portion of the Broulhiet pillar had splines and linear ball bearings to reduce the friction in the system. This arrangement also had the advantage of placing the steering linkage on the sprung mass.

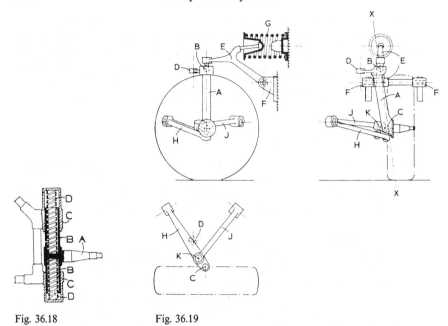

Fig. 36.18   Fig. 36.19

In general, the disadvantages of the pillar type of suspension, other than the Broulhiet arrangement, include a steering effect owing to the vertical motion of the wheel in relation to the arc through which the steering rod must move. With all types, there is difficulty in accommodating a pillar and coil spring of adequate length without seriously reducing the ground clearance beneath them. Moreover, the resultant need to limit the vertical deflection can cause severe understeer when, under extreme conditions during cornering, the roll has taken up all the suspension movement and roll reaction is consequently taken mainly through the front wheels. The principal advantage is the absence of variations in camber and castor angles or track, with vertical deflection of the suspension.

36.12   Rover 2000 front suspension

This type of suspension, Fig. 36.19, does not fall neatly into any of the 11 categories listed at the beginning of this chapter. The road wheel is carried on a stub axle which is integral with the tubular member A. This member is held top and bottom by ball-and-socket joints B and C and so can turn about the axis XX for steering – the steering arm is at D. The ball of the joint B is part of a lever E, which is pivoted in rubber bushes, on the axis FF, and carried by brackets fixed to the body structure. The lever E is also used to transfer the vertical load of the wheel to the spring G and thus to the body. At the lower end of the member A, the ball of the joint is fixed in the composite wishbone member H and J, whose two parts are pivoted in rubber bushes on the vehicle structure at their inner ends and are pivoted to each other at their outer ends, at K, by a ball-and-socket joint at an angle to each other. A telescopic damper (not shown) is coupled at its upper end to the lever E and at its lower end to the body structure.

The advantages include a reduction in weight of the vehicle, because the main load-bearing structure extending forward from the dash is limited to that needed to carry the engine mountings and lower link only of the suspension. Additionally, camber and castor angles and track vary very little with vertical deflection. A major disadvantage is that, because the suspension loads are taken mainly directly by the dash, it is difficult to keep the noise out of the passenger compartment.

36.13 Driven-wheel suspension

Many of the systems previously described have been adapted for driven wheels in both front- and rear-wheel-drive cars, and others will be described in the sections on rear suspension that follow. Some early examples had transverse leaf springs serving also as either one or both of the transverse links on each side, as in Figs 36.6 and 36.7. The telescopic dampers then had to be offset to the front or rear of the driveshafts.

Some of the Renault models have a coil spring and double transverse-wishbone system, but the coil spring and damper are placed above the upper link, thus leaving the space between the links clear for the driveshaft. In the Ford *Fiesta*, the MacPherson strut-type suspension is adapted simply by cranking the bottom transverse link, as viewed in plan, so that the pivot at its inboard end is positioned far enough forward to clear the driveshaft as the suspension drops to the rebound position.

36.14 Rear suspension – live axle

Some comments on control systems for live axles have already been given in Chapter 28. With such a rear axle, the unsprung weight comprises that of the axle assembly – differential, its casting, the halfshafts and their tubular housings, the hub and wheel assemblies, about two-thirds of the weight of the springs, if semi-elliptic, plus that of the propeller shaft. If coil springs are employed, only half their weight is unsprung, and in any case, they are much lighter than leaf springs. On the other hand, some additional links must then be provided to control the motion of the axle and, generally, half the weight of these is unsprung. The greater the unsprung mass the larger will be the amplitudes of the hop and tramp motions of the axle and the more difficult it becomes to obtain good ride, roadholding and stability.

Another disadvantage of the live rear axle is the tendency for the propeller shaft torque to press the wheel on one side down and to lift that on the other side. In the early days, engines were designed to be started with a cranked handle, using the right hand, so they traditionally rotate clockwise as viewed from the front. When they are running, their torque, from the viewpoint of a man in the driving seat, tends to press the left-hand rear wheel down and to lift the right-hand wheel. Consequently during acceleration in low gear from a standing start – when the torque can be very high – the right-hand wheel may spin and traction be lost. Additionally, excessive roll may occur if the car is turning to the right under these conditions.

36.15 Torque reaction and axle guidance

Methods that have been used to obviate torque-reaction effects and for axle guidance include the provision of a torque-reaction tube coaxial with the propeller shaft and attached at its front end to the vehicle structure or the gearbox casing and at its rear end to the nose of the differential gear carrier. Alternatively, if a torque-reaction rod as described in Section 28.8 is used, the reaction point, at its forward end, can be offset to the right of the vehicle so that the lift due to the reaction to the torque in the halfshafts counteracts the downward pressure due to the reaction to the torque in the propeller shaft. Otherwise, the only remedy apart from the use of complex torque-reaction rods is a de Dion or independent suspension system – in other words, either a dead axle or an axleless suspension.

To obviate the problems of interleaf friction, the unsprung weight of semi-elliptic springs, and the imprecision of motion of axles when guided only by such springs, coil springs and axle guidance and torque-reaction links are now fairly widely used at the rear. An example is the Vauxhall *Viva*, Fig. 36.20, where each extremity of the live back axle is pivoted on the trailing end of an arm, the forward end of which pivots beneath the vehicle structure. A coil spring is interposed between the rear end of the arm and the structure, or sprung mass. Lateral location is effected by two semi-trailing links, arranged V-fashion with their rear ends pivoted on top of the differential casing and their fronts ends in brackets outboard, again on the vehicle structure. All the pivot points are rubber-bushed, to introduce some compliance, to prevent the transmission of some of the vibration from the road to the vehicle, and to accommodate the motions due to single-wheel deflections and the skewing of the axes of the upper links relative to those of the lower ones. Brake torque reaction on the links tends to pull the rear end down, thus at least partially counteracting the front end dive effect, while drive torque tends to offset the tendency to rear end squat.

Several manufacturers use twin parallel trailing links on each side, for axle location and to react brake torque, in which case a Panhard rod is required for lateral location. To reduce lateral movement of the axle to a minimum, the Panhard rod is generally made as long as possible by attaching it to the vehicle structure on one side and, to the axle, as far away as practicable on the other side.

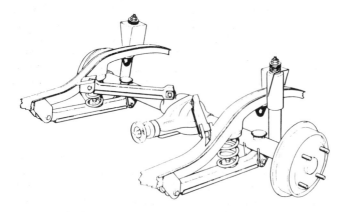

Fig. 36.20 Vauxhall *Viva* HC rear axle

In some instances, the rear end of the lower trailing link on each side is pivoted to a bracket below the outer end of the axle, while a single central upper rod, triangulated in the form of an A-link or wishbone link, is pivoted at its apex to the top of the differential casing and at its base to a transverse member on the vehicle structure. Such a system, shown diagrammatically in Fig. 27.10, and used by Renault for example, has the advantage that the single triangulated link replaces not only the two upper ones, but also the Panhard rod, since it can locate the axle laterally. This practice is fairly common on heavy commercial trailer axles used with air or rubber springs, Fig. 33.14.

The single trailing link system has been used with live axles, but only rarely. A good example is the London Transport *Routemaster* bus, on which the single trailing links, one each side, project behind the axle to carry at their ends a transverse beam that extends outwards behind the rear wheels. Coil springs seat on the ends of this beam, so the spring base is as wide as possible. Air springs have also been used with this layout.

An essential with the single trailing-link arrangement is a flexible connection between the links on each side and the axle, to relieve the latter of the severe torsional loading that would otherwise be applied to it during single wheel deflection. On the *Routemaster*, this flexibility is provided by interposing on each side two Metalastik rubber sandwich-type mountings, chevron fashion, between the axle and the trailing link.

36.16   Watt's linkage

Watt's linkages have been employed for both lateral and fore-and-aft location of axles, so that they will rise and fall in a straight line. However, since such an arrangement is both costly and difficult to accommodate, it tends to be used mainly on sports and racing cars, where extreme precision is important. Typically, the Watt's linkage comprises a leading and trailing link, the adjacent ends of which are inter-connected by a vertical link, as in Fig. 36.21, or for the lateral location of a de Dion axle, Fig. 36.28. To the centre of the vertical link is rigidly attached the axle. There are pivot joints at the ends of all the links so, as the axle rises and falls, it does so in a straight line. If the two horizontal links are of equal length, the axle has to be attached to the centre of the vertical one; if not, it has to be set off-centre, towards the end of the longer horizontal link by an amount such that the distance between these two centres is $LA/(A + B)$, where L is the overall centre-to-centre length of the vertical link, and A and B are respectively the centre-to-centre lengths of the longer and shorter horizontal links. The centre-to-centre distance from the axle and the attachment of the vertical link to the shorter horizontal link therefore is $LB/(A + B)$.

36.17   Rear suspension – dead axles

The considerations applying to a dead axle at the front also apply at the rear, but without the complication inherent in the need to steer the wheels. However, except on some commercial front-wheel-drive special-purpose vehicles, and of course trailers, dead rear axles are rarely used. For cars, on the other hand, a particular form – the de Dion axle – is not uncommon.

The de Dion axle (see Section 20.3) is most suitable for cars with a low centre of gravity, hard suspension and a long wheelbase. This is because the axle location links are usually approximately at the height of the wheel centres, which makes it impossible for the roll centre to be near the ground, and therefore extra roll stiffness is desirable. Moreover, with a high roll centre, the contact point between both wheels and the ground moves sideways as one wheel is deflected upwards, and this has a steering effect. With independent suspension, even though there may be a similar degree of sideways movement on the rising wheel, it is resisted by the other wheel. A long wheelbase makes up for the absence of any anti-dive effect during braking and anti-squat effect during acceleration. Lateral location of a de Dion axle is generally effected by a Panhard rod, but a Watt's linkage is used in some instances, as in Fig. 36.28.

An especially interesting, but complex, de Dion system – that which was used on the Rover 2000 – is illustrated diagrammatically in Fig. 36.21. The stub axles of the wheels are carried in bearings B in the ends of the two parts A_1 and A_2 of the dead axle. These two parts of the dead axle are free both to slide and rotate relative to each other on the axis XX. The outer ends of the driveshafts G_1 and G_2 are coupled by universal joints to the short flanged shafts I in the wheel hubs, while their inner ends are similarly connected to the differential gear F. There are no variations in track, since their swinging motions are accommodated by the sliding joint on XX in the dead axle – the freedom of rotation in this joint allows for the differential rise and fall of the wheels on each side.

Lateral location of the wheels is effected by the shafts G_1 and G_2, and lateral location of the differential and final drive unit F by the radius rod K. This radius rod is needed because the final drive unit F is flexibly mounted to prevent the transmission of noise and vibration to the vehicle structure.

The links C and D, together with the vertical bracket carrying the bearings B, form a Watt's linkage to guide the axle vertically in a straight line and to locate it fore-and-aft. These links also react both drive and brake thrusts and torques. Disc brakes are installed inboard on the differential gear casing, the discs being mounted on the driveshafts, so the brake and drive torques are both ultimately reacted through the final drive mountings. To reduce to a minimum their vertical loading due to these torque reactions, the mountings are spaced very widely – on the ends of a transverse yoke at the rear, and under an exceptionally long extension of the nose of the final drive casing at the front.

36.18   Rear suspension – independent

At one time, the most common form of independent rear suspension was the swing axle, in which the axle tube and driveshaft on each side formed a single transverse link, in some instances with a radius rod extending forwards to take the reaction to brake and tractive thrust. This arrangement was used for many years by Porsche, VW and Daimler-Benz. In principle, it differed from the Rover system just described in that there was a universal joint only at the inner end of the halfshaft, as shown diagrammatically in Fig. 36.22, so the camber angle changed as the wheel rose and fell.

In Fig. 36.22, the wheel hubs are carried on the outer ends – not shown – of the axle tubes A and the halfshafts rotate in bearings in their inner ends. At (*a*) the axle tubes swing about trunnions, while at (*b*) they have semi-cylindrical flanges B on which they pivot in semi-cylindrical housings in the final drive

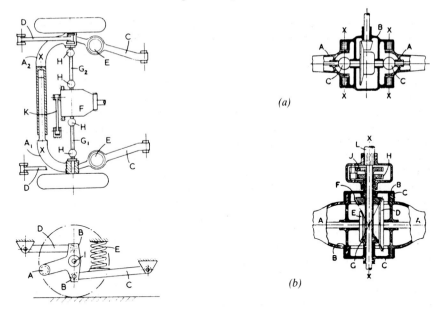

Fig. 36.21   Fig. 36.22

casing C. The axes XX of the trunnions at (*a*) are in line with the centres O of the universal joints. At (*b*), rotation of the centres of the semi-cylindrical flanges is accommodated by rolling of the bevel gears D and E about the pinions F and G. These pinions are mounted on sleeves driven by the wheels H and J of a spur-type differential.

There have been many variants of the swing-axle layout, but most commonly a fabricated link was substituted for the axle tube. As with an axle tube, however, the wheel hub was attached to its outer end. The inner end was usually forked to carry the two bearings about which it swung, and the pivots for these bearings were generally in the differential casing, though in some instances the whole final drive unit was integral with the inner end of one link and the other link swung about a pivot pin common to both and mounted on the chassis frame.

On the Mercedes-Benz 220, this pin was below the final drive unit, which gave a roll centre fairly close to the ground. The object was, by making the shafts as long as possible, to reduce not only the wheel camber variations but also, by lowering the roll centre, what are termed the *tuck-under* and *jacking-up* effects. Tuck-under occurs if either the whole rear end, or one side only, is lifted so that the wheels, or wheel, drop to the rebound position. When the car falls again, the forces on the wheels are such that it will roll along for a certain distance with the wheels tucked under before they roll out to their normal positions again. Jacking-up occurs when the couple due to the lateral loading as a result of the centrifugal force on the vehicle times the height of the roll centre above the ground is greater than the righting couple due to the weight of the vehicle acting vertically downwards multiplied by the horizontal distance from the centre of the outer wheel contact patch on the ground to the projected centre of gravity of the vehicle. The latter effect leads to instability because, as the jacking-up progresses, the roll centre moves upwards, progressively increasing the overturning couple. Both effects tend ultimately to cause rear-end breakaway during cornering, which is why this system is rarely used nowadays.

36.19   Single link with angled pivot axis

Of the types of suspension listed at the beginning of this chapter, only two remain to be discussed. These are single and double link with pivot axes not parallel to either the longitudinal or the transverse axes of a car. The single-arm systems of this type are commonly used in rear suspensions. Brief mention of a double-arm system is made at the end of Section 35.3. Its characteristics are a combination of those of the other twin-link systems – with longitudinal or with transverse pivot axes – and the principles of the single arm apply so far as roll axis, etc., are concerned. Consequently, there is no need to go into it in any more detail.

The single-arm system is attractive as a rear suspension principally because of the simplicity of a single link and the relative ease with which it can be accommodated. It would be difficult to fit into the front end because the engine would be in the way.

The basis of the evolution of this system has been as follows. Swing axle systems suffer the disadvantages outlined at the end of Section 36.18, and the single trailing arm has the disadvantages set out at the beginning of Section 36.6. These include too much camber angle change with the former and too little with the latter, so a combination of the two could be just right. Again, the swing-axle system lacks fore-and-aft rigidity unless a drag link is added and the trailing arm system lacks lateral rigidity, so again a combination of both could give adequate rigidity in all directions.

We can reject the swing-axle system totally on grounds of instability when cornering, but the trailing-arm system is worthy of more serious consideration and, indeed, is used on a number of modern cars. How, then, can we improve it and combine with it some of the desirable features of the swing-axle system? The answer is to stiffen the arm by triangulating it and set its pivot axis somewhere between the lateral and longitudinal orientations.

Consider first a simple conventional trailing arm. An obvious step is to add to it an extra bracing member pivoted inboard, on an axis in line with that of the main bearing. The whole link will then be in the form of a right-angle triangle with the wheel hub at its apex. This places the right angle at the outboard end of its base, where the pivot bearing for the trailing arm is situated; and the brace forms the hypotenuse and, where it joins the base, carries a second pivot bearing. Both bearings are carried in brackets on a transverse member of the vehicle structure. A disadvantage of this system is that either the hypotenuse becomes too long, since it is both loaded in compression and subject to fairly severe vertical inertia loading (see Section 35.3) or the two bearings are not spaced apart widely enough.

A wide bearing-spacing is important for reducing to a minimum the loading imposed on them by the couple applied by centrifugal force in a lateral direction when the vehicle is cornering. Rubber bushes are usually used for the pivot bearings and, if they are heavily loaded, they have to be made too stiff. This is an especial disadvantage when modern radial-ply tyres are used, since these are much less resilient, so far as absorption of fore-and-aft shock loading is concerned, than cross-ply tyres. With radial ply tyres, therefore, fore-and-aft resilience has to be provided by using a soft rubber in these bushes, otherwise unpleasant vibrations are felt by the occupants of the car.

A good example of a semi-trailing link rear suspension can be seen in Fig. 36.23, which illustrates that of the Triumph 2000. A similar system is used on the BMW 1500. In the Triumph suspension, the trailing arm is swept from inboard of the wheel to outboard towards its pivot. This is to widen the spread between the pivot and the bearing for the bracing arm that forms the hypotenuse of the triangulation system. That this lengthens the trailing arm does not matter, since it is subjected only to tensile loads when the vehicle is moving forwards. Moreover, the fact that the curvature thus introduced reduces the rigidity of the arm – since it will cause it to be subjected to bending loads – does not matter either, because some flexibility has to be introduced in any case if radial-ply tyres are to be fitted.

By setting the axis of the two pivot bearings at an angle intermediate between those required for a true trailing link and a swing axle, the bracing arm, or hypotenuse of the triangle, is shortened so that it can better take the compression loads which, because of weight transfer during cornering, are heavier than any tension loads to which it might be subjected. The driveshafts have universal joints at both ends and sliding joints at their inner ends to cater for the variations in their lengths as the suspension arms rise and fall.

An additional advantage of this angular setting of the pivot bearings is that the transverse member on which they are pivoted, being of necessity of V shape, can be used to carry the nose of the final drive unit. The rear end of this unit is carried by another transverse member, or yoke. Both these transverse members have circular rubber sandwich mountings at their ends, by means of which they are secured to the structure of the vehicle. On the

Fig. 36.23

BMW, conical mountings, which offer greater resistance to horizontal than to vertical loading, are used and, while there are two for its V-shaped front member, there is only one for supporting the rear of the final drive unit. The absence of the rear transverse member and of one mounting no doubt compensates, at least partly, for the employment of the more costly conical-type mountings.

On both vehicles, the innermost of the two pivot bearings on each side is lower than the outer one. The effect of this is to lower the roll centre, which reduces the variation in track as the wheels rise and fall. With a trailing link suspension, the roll centre is initially at ground level, Fig. 36.4, and, with a swing axle it is either above and half way between the two pivots or, if there is only one pivot, the roll centre will coincide with its axis.

The method of finding the roll centre with a semi-trailing link is shown in Fig. 36.24. Here it can be seen that increasing the degree of trail of the suspension arm moves the point O outboard of the wheel on the opposite side and, in effect, increases the length of the imaginary swing axle and therefore reduces the variation in camber as the wheel rises and falls. The point O in Fig. 36.24 can be arrived at by drawing in the axes of the pivots on the other side, but this is not really necessary because it can be deduced from the symmetry of the overall system.

An example of an early suspension of this type, with a transverse leaf spring and a swing axle, is shown in Fig. 36.25, and similar arrangements have been used, with coil springs seating on the outer ends of the swing axles A, or on the ends of the drag links D. In the illustration, the final drive unit is carried on a sub-frame C, and a spherical bearing is employed at E. With these arrangements, the axis about which the suspension arms swing is set at an angle approximating too closely to that of a swing axle systems, so the roll axis is too high and the jacking-up and tuck-under effects too pronounced.

36.20   Influence of angle of pivot axis on camber and toe-in

It is easiest to visualise the effect of a compound angle of inclination of the pivot axis of a semi-trailing link, as in Fig. 36.24, by viewing independently the two components of its angularity. Consider first an end-elevation of the

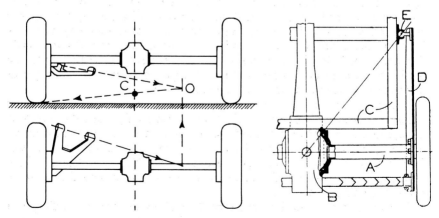

Fig. 36.24   Fig. 36.25

vehicle. If the pivot axis is horizontal and in a transverse plane perpendicular to the longitudinal axis of the vehicle, rotation of the suspension arm about its pivot axis will produce zero change in both camber angle and toe-in. If, at the other extreme, the pivot axis is vertical, there will again be no change in camber angle, but the angle of toe-in or toe-out will swing through the same angle as the suspension arm, as it rotates about its pivot. If follows that the variation of toe-in with deflection of the suspension arm is a function of the angle between the pivot axis and the horizontal, as viewed in front or rear elevation, while camber angle is totally unaffected.

Now consider the plan view. If the pivot axis is horizontal and, in the previously mentioned transverse plane, rotation of the arm about the pivot produces zero change in both toe-in and camber angle. Again, at the other extreme, if the pivot axis has been swung round until it is parallel to the longitudinal axis of the vehicle there will still be no change in toe-in, but the camber angle will at all times be equal to that of the deflection of the suspension arm from the horizontal. It follows that the variation of camber angle with vertical deflection of the suspension is a function of the angle between the pivot axis and the longitudinal axis of the vehicle, as viewed in plan. If, owing to a sudden increase in vertical loading, the camber angles of both wheels change simultaneously in the same sense, either positive or negative, there will be a change in track so tyre scrub will tend to occur. However, the faster the wheel is rolling along the road, the smaller will be the scrub in real terms, because a combination of the rolling action of the tyre simultaneously with the lateral elastic deflection of the tread as the vertical motion of the suspension progresses will easily accommodate it. In practice, the scrub is rarely of significance unless it is excessive or in the extremely unlikely circumstance of the vehicle's being habitually operated in tight turns at very low speeds.

36.21   Vehicle handling considerations

Now consider the effects of the geometry on handling. During a turn, the vehicle will tend to roll on to the outer wheel which, owing to weight transfer, will then have a greater influence than the inner wheel on the behaviour of the vehicle. The larger the angle between the pivot axis and the transverse plane, the greater will be the movement of the outer wheel towards negative camber as the roll progresses. Since negative camber tends to cause the tyre to steer inwards, the result will be a tendency to understeer (see Section 34.8). Furthermore, the lower the outer pivot relative to the inner one, the greater the toe-in effect on the outer wheel as the roll angle increases. Again, this has an understeer tendency, as indicated in Section 36.20.

Another factor that the designer has to take into account is the steering effect of deflections in rubber-bushed pivots, especially when the vehicle is being braked or, if of rear-wheel-drive layout, accelerated. Bush deflections, however, may have a greater influence in the front suspension than the rear. This is because they can change the relative position and attitude of the steering links and levers relative to the suspension arms, and thus actually rotate the axle about its kingpin.

Apart from the suspension geometry, the major influence on handling is of course the front–rear weight distribution: the higher the proportion of the load on the rear, the greater is the oversteer effect. Weight distribution in a car depends to a major extent on whether it is of the front- or rear-wheel-drive layout and, if front-wheel-drive, whether the engine is installed transversely or longitudinally. Since it is usually easier to manipulate the suspension geometry than the weight distribution, the designer will generally fine-tune the handling characteristics by making adjustments to the former. Indeed, the geometrical factors may be arranged so that all give a tendency in the same direction, or some may be designed to counteract the steering tendencies of others over either the whole or only part of the vertical deflection range. Correction over part of the range can be particularly beneficial in counteracting a sudden and unexpected change in steering characteristic during a turn, since such changes could take drivers by surprise and even cause accidents.

36.22   MacPherson strut rear suspension

A particularly good example of a MacPherson strut-type rear suspension is that of the Lancia *Delta*, Fig. 36.26. With this arrangement, the springs occupy space above the wheels, which in any case is normally wasted so far as capacity for luggage in the boot is concerned. An alternative arrangement is that of the Ford *Escort*, Fig. 36.27.

Among the interesting features of the Lancia system are the use of twin transverse links pivoted close to the longitudinal axis of the car, one in front of and the other behind a substantial transverse member of the structure. A single trailing link on each side reacts the brake drag, while the torque is taken equally between this link and the top end attachment of the strut.

The axis of the spring is offset slightly relative to that of the strut, to apply a couple tending to counteract that due to the offset of the roadwheel on the strut. This relieves the piston and rod assembly of some side load and therefore the friction between it and its cylinder. The piston rods are Teflon coated to reduce the sliding friction between them and the glands at the top ends of the cylinders.

As can be seen from the illustration, the ends of the anti-roll bar pivot in rubber bushes on the stub-axle carrier assemblies. The straight central portion of the anti-roll bar is carried in two rubber mountings, one at each end. These mountings are on the lower ends of pendant links, the upper ends of which are pivoted on the vehicle structure. This uncommon arrangement has been adopted so that the roll bar cannot exert any constraint on the movement of the suspension strut and linkage.

In the front suspension of this vehicle, Fig. 34.8, the method of avoiding such constraint has been simply to place the rubber mountings in line with the axes of the pivots of the wishbone links. Yet another method – the use of shackles on the ends of the anti-roll bar – has been adopted by Alfa Romeo for the rear suspension of the *Alfa 6*, Fig. 36.28. The *Alfa 6* rear suspension layout is virtually identical to those of the *Alfetta* and *Giulietta* models. Its driveshafts have plunge-type constant-velocity joints, as described at the end of Chapter 26, to accommodate variations in their lengths as they swing up and down. Mounted on the de Dion axle, adjacent to the centrally pivoted arm of the Watt's linkage, is a tuned vibration damper to counteract axle hop. Fore-and-aft location of the axle is effected by a trailing V-shaped link the apex of which is pivoted on a ball joint beneath a transverse member. The ends of the

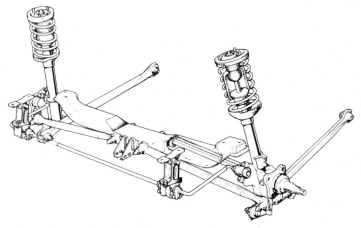

Fig. 36.26 Lancia *Delta* rear suspension

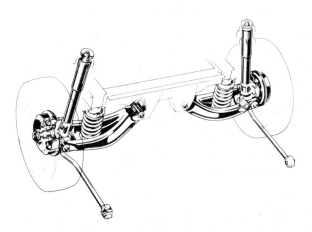

Fig. 36.27 Ford *Escort* rear suspension

anti-roll bar are connected by shackles to the arms of the V, so that they can articulate freely. Another feature of note is the installation of a combined bump stop and hollow rubber spring coaxially inside each coil spring, to give a progressively rising rate.

36.23 Active suspension

Conventional suspension systems, comprising spring, damper and resilient tyre, Fig. 35.1, can be termed *passive*: they are self-contained and react automatically to the loads applied to them at the road surface. However, there is absolutely no active control over these reactions. An *active* suspension system, on the other hand, is one in which the reactions to the applied loads are positively supplied by automatically controlled powered actuators, usually hydraulic with electronic controls. Power consumption by such active systems

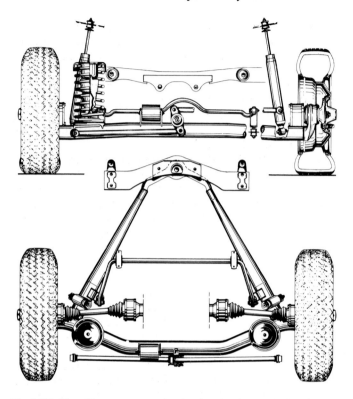

Fig. 36.28 The *Alfa* 6 rear suspension in elevation (*top*) and plan (*bottom*)

can, in severe conditions, be a significant proportion of the output from the engine. They can be costly, therefore, not only because of their complexity but also in terms of fuel consumption and, perhaps, maintenance. Even though, in some instances, some of this cost can be offset by reduced complexity of the actual suspension linkage, it nevertheless seems likely to remain a major obstacle except for very expensive and special-purpose vehicles.

A fully active suspension is one in which there is no spring at all and the wheel deflections are controlled entirely by the actuators. Semi-active suspensions are those in which a conventional spring and damper system is used but its reactions are supplemented by a control system. The control may be effected only over the damper, in which case it is generally termed an *active ride control* system, or it can vary the rate of the suspension by supplementing the reaction forces afforded by the spring, or, of course, it can perform both functions, in which case it is truly a semi-active suspension.

Semi-active suspensions can be designed so that, in the event of failure of the control, the vehicle will still ride on its springs when it is, for instance, driven in for repair. On the other hand, such a system, because of the presence of its passive element, can be ideal for only one set of operating conditions such as speed, road surface, etc., and may be far from ideal for at least some of the others. However, in the conditions of operation for which it is designed, including speed range, it can be almost as good as a fully active system.

Fully active suspension, at least in theory, can be designed to adapt to any operating conditions, provided the potential for vertical deflection of the wheels is adequate. In this connection it must be accepted that even though there may be no actual spring in the suspension system, the tyre remains as a resilient element in the overall system. Consequently, the vertical deflection of the wheel can be limited only at the expense of increasing the loading on the tyre. Given adequate vertical deflection, the advantages of a fully active over a passive system can be considerable. However, as the vertical deflection for the wheel is reduced so also is this advantage.

Active and semi-active suspension systems could be of particular value for applications where the loads carried vary widely, such as buses, delivery vans and trucks, vehicles that are required to travel fast across very rough terrain, and ambulances which may have to transport seriously ill patients smoothly at high speeds. They might be of use also for racing cars and vehicles for which very small or constant ground clearances are advantageous.

36.24   Suspension control systems

Suspension control entails more than just regulation of the vertical movement of the wheels. The many factors that have to be taken into account include comfort for the occupants, roll, both longitudinal and lateral weight transfer, and the maintenance of contact pressure between the wheels and the ground consistent with good stability and handling.

In a fully active system there is a pump and hydraulic fluid reservoir and, generally, one hydraulic actuator and one control valve for each wheel or pair of wheels. There may also be one or more hydraulic accumulators, to supplement the rate of flow from the pump to cater for sudden deflections of the suspension. The control valves may be all in one unit. They execute commands from an on-board computer which is served with information by sensors. The computer may be programmed to make instantaneous responses to changes in road surface or equilibrium (speed, roll, brake dive, or acceleration squat) as indicated by the sensors or, in a simpler control arrangement, it may issue its commands periodically, to adjust the system to suit average conditions over periods of seconds or even minutes, depending on the precision of control required. Correction can be made also to under- or over-steer, by adjustment of the front : rear roll stiffness ratio, even automatically while the car is being driven.

Ideally, perhaps, sensors would detect the rise and fall of the ground in front of each wheel, so that the suspension could be made to deflect a precisely equal amount and thus keep the vehicle riding at a constant height above the ground. A speed sensor would be necessary, too, so that the computer could synchronise the movements of the suspension with the passage of the measured surface profile under the wheels. A further refinement might be a transducer to measure the actual response of the wheel so that, by taking into account hardness or softness of the surface, the dynamic loading on the tyres could be limited.

Such a system, however, is impracticable at the current state of the art, and it is simpler to use a pressure transducer in each hydraulic actuator to signal to the computer any increase or decrease in the load applied by the ground to the wheel. Responding virtually instantaneously to this signal, the control system

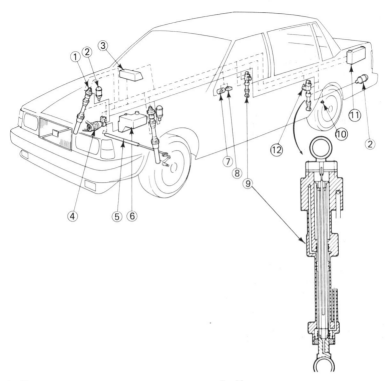

1	Front actuator	7	Yaw gyro
2	Accumulator	8	Longitudinal and lateral accelerometer
3	Control panel	9	Rear actuator
4	Hydraulic pump	10	Hub accelerometer
5	Steer angle transducer	11	Computer
6	Oil reservoir	12	Servo vale

———— Hydraulic pipeline   – – – – –Transducer signal   ·······Control signal

Fig. 36.29 Schematic diagram of the Volvo Computer Control Suspension (CCS). A major advantage of such a system is that changes to the performance characteristics of its different components, which of course may affect those of others, can be made simply by changing the program on the computer, and the overall effect assessed immediately without having to modify a vehicle and take it on to the test track

can direct fluid to flow either into or out of the actuator, as necessary.

The sensors actually used may include one transducer on each axle to measure the variations in height of the sprung mass above it under varying loads, a speed sensor, a yaw detection gyroscope or steering motion transducer, or a lateral accelerometer for measuring tendency to roll, a longitudinal accelerometer to detect braking and acceleration forces, and an accelerometer or strain guage on each hub for assessing the quality of the road surface. In some instances a degree of simplification has been obtained by using for the rear axle only one height sensor and one actuator, the motions of the two rear wheels being made interdependent through a hydraulic interconnection. Alternatives to some of the above-mentioned sensors might be transducers

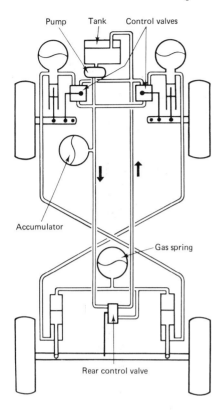

Pump   Tank   Control valves

Accumulator

Gas spring

Rear control valve

Fig. 36.30 With gas springs, the AP suspension system is of the semi-active type. As with a fully active suspension system, it has the advantage of a soft ride coupled with high roll stiffness. Unlike such a system, though, it does not allow the car to sink on its suspension when the engine is stopped. There are three gas springs, one for the rear axle and one each for the two front wheels

sensing the displacements of the throttle, brake pedal, steering gear and axle.

A typical fully active suspension system is illustrated in Fig. 36.29 and a semi-active one in Fig. 36.30. Neither, however, is incorporated even as a standard option in a quantity-produced vehicle. Since, if studied in relation to the basic principles already outlined, Figs 36.29 and 36.30 are self-explanatory, it is not proposed to describe them in detail here. However, some comments are necessary regarding Figs 36.31 and 36.32.

Fig. 36.31 shows a single suspension unit in the AP system. An increase in the static load on the vehicle causes the lever actuating the spool-type control valve, on the right in the illustration, to be deflected upwards about its pivot. This of course is provided the relative movement between suspension arm and body is slow, so that the coil spring and damper in the linkage between the suspension arm and the lever are not compressed. The consequent upward deflection of the lever pulls the spool valve to the left, causing it to direct hydraulic fluid into the hydraulically damped gas spring, thus extending it until the ride height returns to what it was before.

Rapid upward movements of the lever, on the other hand, are opposed by

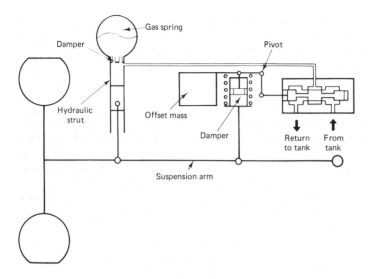

Fig. 36.31 A single AP suspension module. The pendulus offset mass is on one end of a bell-crank lever, the other end of which actuates the spool valve. The connection between the first-mentioned arm and the suspension arm is a link in which is incorporated a small coil spring and damper unit

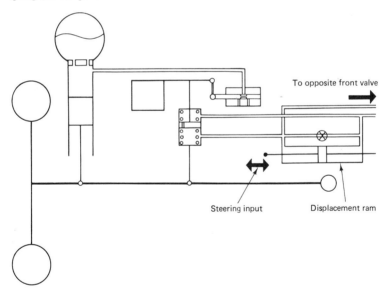

Fig. 36.32 The hydraulic ram is actuated by the steering gear, which moves it to displace fluid in either direction through the damper unit, according to which way the vehicle is steered

the inertia of the offset mass, so the coil spring and damper are compressed and there is little or no movement of the spool valve. In these circumstances the gas spring performs as in a conventional suspension system and, since little or no fluid motion is involved, the energy consumption by the engine-driven pump is correspondingly small. To obtain this effect, the coil spring and damper have to be tuned so that the force exerted on the offset pendulous mass gives it the same vertical acceleration as that imparted to the body of the vehicle by its gas spring and integral damper. This system has two major advantages. First, the body does not sink down on its suspension when it is switched off. Secondly, it does not need an electronic control.

The AP system can be tuned to take into account not only the lateral weight transfer but also the vertical deflection of the tyres during cornering. In some instances, however, especially for heavy vehicles, it may react too slowly. This can be overcome by arranging for a steering input, as shown in Fig. 36.32. A double-acting ram, actuated by the steering mechanism, transfers hydraulic fluid from one side to the other of the vehicle through the dampers in the links between the suspension arms and the levers actuating the spool valve. It does this in a direction such as to lift the lever on the outer and lower that on the inner side of the turn.

Further information on active suspension in general, including design calculation, can be obtained from two papers, by Sharp and Hassan, *Proc. I.Mech.E.,* Vol. 200 D3, 1986, and Vol. 201 D2, 1987.

Chapter 37

Six-wheel vehicles

In most countries a limit is set by law to the weight that may be carried on any one axle of a road vehicle; this limit is regulated, presumably, according to the capacity of the roads upon which the vehicles will operate, but factors other than the road also set a limit, for example, the tyres. With unshod steel wheels that were used on traction engines the permissible load could be increased readily by increasing the width of the wheel, but with rubber-tyred wheels this is not so practicable. The use of twin pneumatic tyres enables the load to be approximately doubled, but this solution of the problem has many drawbacks. The great width of the wheels necessitates the use of a comparatively narrow frame and somewhat hampers the coach builder; stones get wedged between the tyres and cause damage; it is difficult to change the inner tyre and the load is not always equally distributed between the tyres. These drawbacks to the twin-tyred wheel have led to the adoption of the alternative method of carrying heavier loads, which is by using more than two axles.

Since the use of three or more axles is primarily to enable heavier loads to be carried, while keeping the load per axle within definite limits, the arrangement used ought to be such that the load carried will be properly distributed between the axles, however uneven the road may be. This requirement should, strictly, rule out the adding of axles having independent springing since, if one of these axles stood on a hump, it would obviously carry more than its proper proportion of the total load. However, this compromise is sometimes adopted and when the road surfaces are good it may be satisfactory.

There are two quite different solutions of the problem, and six-wheelers can be divided into two classes accordingly—

(1) 'Flexible' or 'articulated' vehicles.
(2) 'Rigid' vehicles.

Vehicles of the first class consist of a two- or three-axle tractor unit with, mostly, a short wheelbase, and a one- or two-axle trailer, which may be permanently or detachably connected to it. The connection between the two units to provide for road inequalities must consist of a ball and socket joint or its equivalent. The trailer wheels are almost always merely weight carriers and not driving wheels, owing to the difficulty of arranging a satisfactory drive to them. They are usually carried on a simple axle, no provision being made for steering them. The trailer then 'cuts-in' when the vehicle is cornering, and manoeuvring in reverse requires some skill. Sometimes, however, the trailer wheels are carried by an axle fixed to a turntable and arranged to steer so that the trailer wheels follow in the tracks of the tractor wheels, thus overcoming the cutting-in difficulty, which, however, is not a serious disadvantage so that steering trailers are not much used.

This type of six-wheeler cannot operate on very uneven ground: first, because it is not practicable to provide for sufficient relative movement between the two units to ensure constant weight distribution; secondly, because such vehicles are somewhat unstable when turning on sloping ground and thirdly, because of insufficient adhesion between the ground and the driving wheels, which can be especially serious owing to the drag of the trailer wheels.

37.1   The rigid six-wheeler

This type of vehicle has been intensively developed during the last 20 years and many different arrangements have been evolved both as regards their suspension and transmission systems. The earliest vehicles consisted essentially of a four-wheeled lorry with the addition of a third axle placed behind and as close as possible to the existing rear axle. The third axle was sometimes merely a weight-carrying dead axle and sometimes a live driving axle. This type of vehicle is still widely used, but other types have been developed.

37.2   Suspensions for rigid six-wheelers

Considering first the suspension systems, Fig. 37.1 shows four simple arrangements. In (a) the loads carried by the axles are equalised by connecting the adjacent ends of the springs to a balance beam A pivoted at its centre to the frame. As shown, the springs are pivoted at their other ends to the frame and are used to take the torque reactions and driving thrust, but when torque-thrust members are provided for the axles then shackles would be fitted. The balance beam cannot be made very long without making the distance between the axle centres rather large, which is undesirable because of steering considerations, and this results in either excessive angular movement of the lever or undue limitation of axle movement. The difficulty is avoided in (b) where the lever is connected to the rear ends of the springs, and the arrangement at (c) overcomes the difficulty as regards the centre distance of the

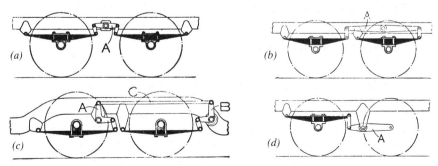

Fig. 37.1

axles. In this latter arrangement the springs are connected by bell-crank levers A and B and the rod C. A variation of (b) is to connect the lever A to the outer ends of the springs and a variation of (c) is to connect the adjacent ends of the springs to the bell-cranks. The centre distance between the axle can be reduced by using the arrangement (a), but with the rear spring connected to the front end of the lever and the front spring connected to the rear of the lever; as the ends of the springs then overlap one has to be placed above the other, and one is usually placed on the top of the axle casing, the other being underslung. In (d) a single spring is used and the rear wheels are carried on the ends of levers A, no rear axle being used. The rear wheels are then not driven but are merely weight carriers. For equality of loading the lever arms must be unequal. In Fig. 37.2(a) a single spring is again used and rigid lever A is pivoted to that spring at its centre and to the axles at its ends. The single spring has now to carry the total weight supported by the two axles. The same principle is used in the Scammell design described later. At (b) is shown a commonly used arrangement, a single laminated spring being pivoted at its centre to a bracket fixed to the frame and at its ends to the axles. When the spring is required to take the driving and braking torque reactions it becomes somewhat difficult to obtain a satisfactory connection between the spring and the axle casings. This is achieved in one design by providing the spring with an extra leaf (as shown dotted), which is separately pivoted to the axle casings. This difficulty is avoided in the arrangement shown in Fig. 37.3 by using two springs secured rigidly to a trunnion block A at their centres and pivoted to the axle casings at their ends. The trunnion block is free to pivot on the end of a cross member of the frame. Fig. 37.4 illustrates a design used by Thornycroft, in which, again, two springs were used, but where each was separately pivoted at its centre on a pin carried by a bracket fixed to the frame.

All the above arrangements will equalise the loads carried by the middle and rear wheels when the vehicle to which they are fitted is motionless, but this equality may be destroyed as soon as the wheels are driven or braked, and a driving or braking torque reaction is set up. To maintain equality during driving or braking the system adopted for dealing with the torque reactions must satisfy certain conditions which will be considered later. However, the resulting inequality when it does occur is seldom of any great importance in vehicles operating on good or moderately good roads, but in vehicles operating on bad ground any serious inequality of loading may render the vehicle useless, and for such vehicles special attention must be given to the arrangements for dealing with the driving and braking torque reactions.

With any of the above systems the rise of the frame when one axle goes over a bump is only half the height of that bump, as shown in Fig. 37.5. The shocks transmitted to the frame and to the road are consequently less than in a

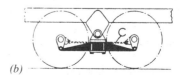

(a)   (b)

Fig. 37.2

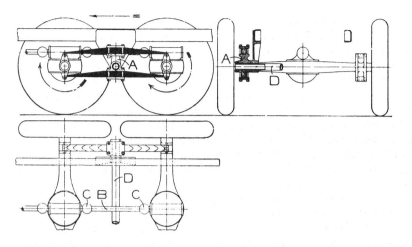

Fig. 37.3

Fig. 37.4   Fig. 37.5

four-wheeled vehicle, but two shocks are experienced for each bump instead of only one.

37.3   Transmissions of six-wheelers

The most important of these are shown in Fig. 37.6. That at (*a*) is probably the simplest and most widely used, the employment of worm-driven axles enabling the worm shaft of the middle axle to be extended through the back of the axle casing and coupled by an intermediate shaft to the rear axle. The intermediate shaft must be provided with two universal joints (indicated by circles) and a sliding joint S as shown. An overhead worm gives large road clearances and is used for cross-country vehicles, whereas an underhung worm gives a low body position and is used for buses and coaches. The use of bevel-driven axles is not so simple, as the diagrams (*b*) and (*c*) will show. In (*b*) a second pinion is mounted at the back of the middle axle in order to drive the intermediate shaft, and it will be seen that the crown wheel of the rear axle has to be mounted on the offside of the pinion in order to make the wheels of the two axles rotate in the same directions. The driving torque for both middle and rear axles is transmitted through the pinion and crown wheel of the middle axle, and the design of those gears is made more difficult. For a given distance between the axles this drive results in the shortest distance between the centres of the universal joints on the intermediate shaft, and thus for a given relative movement of the axles in the greatest angularity of those joints. In (*c*) the drive-shaft is mounted above the

(a)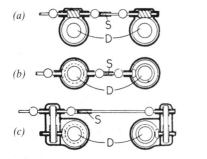

(d)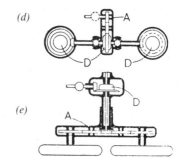

(b)

(c)

(e)

Fig. 37.6

axles, the drive being carried down to the bevel pinion shafts by chains or gears. By using similar axles turned back to back as shown, the length of the intermediate shaft is increased and the angularity of the universal joints is decreased. This arrangement was evolved and is used by the FWD Company. At (d) is shown an American design. The casing A is fixed to the frame and the drive is taken through a train of gears to the bottom shaft, the ends of which are coupled by universal joints to the bevel pinion shafts of the axles. The latter are provided with torque thrust tubes which are anchored to the casing A by ball and socket connections surrounding the universal joints; sliding joints are consequently unnecessary. The length of the torque thrust tubes cannot be made very great without making the centre distance between the axles large, and consequently the angles at which the universal joints may have to work may be rather large. Constant-velocity-type joints are used but even so it is questionable whether this transmission could be used successfully on a cross-country vehicle. For vehicles operating on good roads it should be very satisfactory, and for such vehicles the use of a third differential to divide the torque equally between the two axles is desirable and is very easily provided for as shown. A third differential can be arranged if required when transmission (a) is used and an example, the design of the Association Equipment Company (AEC), is shown in Fig. 37.7. The shaft A is coupled by a universal joint to the propeller shaft coming from the gearbox and carries the four-armed spider B on splines at its right-hand end. On the arms of this spider the differential pinions C are free to turn and are kept in place by the ring D, the inside of which is made spherical to fit the spherical ends of the pinions. The ring D is itself kept in place

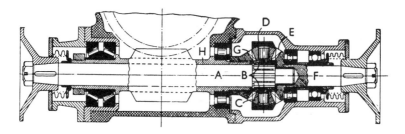

Fig. 37.7 AEC third differential

by the overlapping portions of the teeth of the differential wheel E. The latter is made integral with its shaft F, which is coupled by a universal joint to the intermediate shaft going to the third axle. The other differential wheel G is splined to the hollow worm shaft H. A transmission for an early Scammell six-wheeler is shown in Fig. 37.6(*e*), a single reduction bevel-driven axle being shown, but the double-reduction axle shown in Fig. 30.4 can be used equally well. The drive to the road wheels is through a train of gears housed in the lever casing A. It is very difficult to arrange a third differential to divide the drive equally in this transmission, but since it is intended for cross-country work this is not of importance. A suspension similar to the Scammell has been used by the Saurer company, who used a shaft drive in place of the train of gears used in the Scammell, thus a bevel pinion was mounted on the end of the axle shaft and meshed with two bevel wheels mounted on longitudinal shafts housed in the lever casing; at their outer ends these shafts carried other bevel wheels that meshed with bevel wheels fixed to the road wheel shafts. Fig. 37.8 shows a transmission that has been used in America and on the Continent. Two separate propeller shafts are used, one driving the middle axle and the other the rear axle. The drive coming from the gearbox is divided between the two propeller shafts in a 'transfer' case A fixed to the frame and a third differential (the principle of whose construction is similar to that of the AEC design shown in Fig. 37.7) is sometimes provided. To enable the rear axle shaft to clear the middle axle casing it is provided with an intermediate bearing B fixed to the frame. An alternative to this is to carry this intermediate bearing on the casing of the middle axle. Offset final drive axles must be used and it is difficult to provide for any great relative movement of the axles. The system is consequently used only for vehicles operating on roads. The use of worm-driven axles facilitates the design somewhat. A German Büssing chassis which used this transmission employed torque thrust tubes to take the driving and braking reactions.

37.4   A Scammell design

The general arrangement of the rear bogie transmission system of the Scammell *Constructor* chassis, which is designed for off-the-road use, is shown in Fig. 37.9. The axles A and B are each provided with a torque-thrust member, C and D respectively. These members are connected by ball joints to cross members of the frame and transmit the driving and braking efforts and torque reactions. The drive to the axle A is through a conventional propeller shaft with universal joints, H and J, at the ends. The shaft is provided with axial

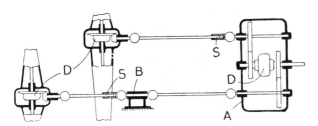

Fig. 37.8

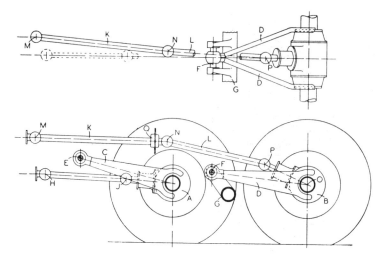

Fig. 37.9

sliding freedom. The centre E, about which the axle pivots, is placed relatively to the joints H and J, so that the angles between the propeller shaft and the transfer gearbox shaft and between the propeller shaft and the bevel pinion shaft of the axle are always maintained approximately equal in magnitude so that a constant velocity drive is obtained on the lines of those used in front-wheel drives and described in Section 26.4. The axle B is driven by a propeller shaft L through an intermediate propeller shaft K, which is coupled to a shaft of the transfer gearbox and which is supported in a flexibly mounted bearing Q at its right-hand end. The centre F is again placed in relation to the joints N and P so that a constant-velocity drive is obtained.

The axles are positioned sideways by Panhard rods. Flat leaf springs are employed, and these are pivoted to the frame at their centres. As the springs play no part in the positioning of the axles, the connections between them and the axles are designed to avoid any constraint of the axles and any undue distortion of the springs. They are shown in Fig. 37.10. The master leaf of the spring has a cup member B bolted to its end, and the spherical part of a pedestal member C fits into this cup, being retained by a collar D, made in two parts, and a cover E. The member C is free to slide on a hardened steel plate F, which is mounted on the axle casing G. The member C is prevented from rotating about a vertical axis by a pin J, whose ends engage slots formed in the collar D. A flexible wire stirrup H prevents separation of the assembly during rebounds of the axle.

37.5 Torque reaction in rigid six-wheelers

Referring to Fig. 37.3, when torque reactions are exerted on the axle casings as indicated by the arrows they will tend to turn the whole unit consisting of the two axles, the springs and the trunnion blocks about the transverse member on which these blocks are free to pivot; this tendency has to be balanced by an increase of load between the rear wheels and the ground and an equal decrease

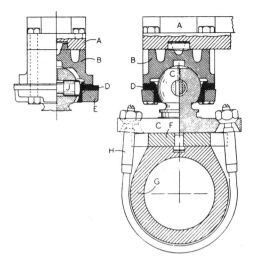

Fig. 37.10

of load between the middle wheels and the ground. Calling the torque reactions t_1 and t_2 and denoting the distance between the axles by l then the magnitude of this alteration in load is given by—

$$Q = \frac{t_1 + t_2}{l}$$

In cross-country vehicles fitted with auxiliary gearboxes giving very low ratios for emergency purposes, it is possible for this alteration Q to become equal to the normal load carried by each axle so that the middle axle would be carrying zero load and would be about to lift off the ground, while the rear axle would be carrying twice the normal load. Under these conditions the rear wheels will probably sink into the ground if the latter is at all soft or sandy and the vehicle will be stalled. The same result will occur with any suspension that deals with the torque reactions in the manner indicated. During braking the reverse of the above action will occur, the middle axle load increasing and the rear axle load decreasing. The arrangement shown in Fig. 37.4 avoids the trouble, the torque reactions being transferred to the frame of the vehicle by means of horizontal forces acting in the springs as shown and which have no effect on the balance of the vertical forces, that is, the loads, acting at the axles. When this arrangement, which was evolved by Thornycroft, is used, the total torque reaction $t_1 + t_2$ is balanced by a decrease q in the load carried by the front axle and an equal increase in the load transmitted through the spring trunnions at the rear. This latter increase however will be shared equally by the middle and rear axles, whose loads will remain equal. The alteration q is given by—

$$q = \frac{t_1 + t_2}{L}$$

where L is the mean wheelbase of the whole vehicle (that is, the distance between the spring trunnions and the front axle), and since L is much greater than l the alteration q will be much smaller than Q. The same result is achieved in the War Department (WD) design shown in Fig. 37.11. The springs are bolted rigidly to the trunnion blocks D, which are pivoted on the ends of the frame cross member E, but at their outer ends the springs are pivoted to boxes A which are mounted on spherical members C which are free to slide outwards along the axle tubes for a reason that will appear later. Torque rods B, connected at one end to the axle casings and at the other end to the frame, are now used to transmit the torque and brake reactions to the frame. Comparing this arrangement with the Thornycroft it will be seen that the torque rods B are acting as the upper spring in Fig. 37.4, and the two springs in Fig. 37.11 are acting as the lower spring in Fig. 37.4. As the torque reactions are transferred to the frame by horizontal forces, the equilibrium of the vertical forces is unaltered, and this design gives the same results as the Thornycroft. Torque rods, arranged as in the WD suspension, are used by some makers in conjunction with the single spring suspension shown in Fig. 37.2(*b*). When the suspensions in Fig. 37.1 are used with the springs secured rigidly to the axle casing at the centre and pivoted to the frame at one end so that they take both the torque reactions and thrusts, it can easily be shown that the equality of wheel loading is upset by the torque reactions and also by the couple due to the driving thrust being exerted at axle level and resisted at the level of the spring eyes, the alteration being however that the load on the middle axle increases and that on the rear axle decreases when the vehicle is driven forwards, the opposite action occurring during braking. When each axle is provided with a torque thrust tube then the equality of loading will depend on whether the torque reactions of the two axles are equal or unequal. In the Scammell design, Fig. 37.6(*e*), equality of loading can only be obtained if the gear ratio between the driveshafts and the road wheels is unity and if those members revolve in the same direction. Actually the gear ratio is a little greater than $2:1$, so that torque reaction does produce some alteration in wheel loading. Brake torques

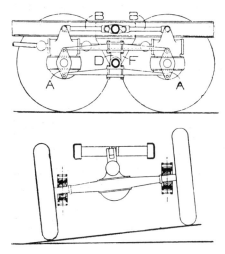

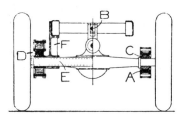

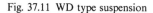

Fig. 37.11 WD type suspension

will also produce an alteration which may be much greater than that due to the driving torque. Experience has shown, however, that these alterations do not affect the performance of the Scammell design, which is outstanding among cross-country vehicles.

37.6 Spring stresses in rigid six-wheelers

The suspension systems of six-wheelers must be considered from another point of view than that of weight distribution, namely, the effect of axle or wheel movements on the springs themselves. When the springs are rigidly bolted or are pivoted by pins to the axle and are connected to the frame by pin joints, then clearly when the axles assume angular positions relatively to the frame, as indicated in Fig. 37.12, the springs will be twisted. While this twisting is unimportant in vehicles operating on roads where the axle movements will be comparatively small, it is very serious in cross-country vehicles, and if permitted would lead to fracture of the springs. It is avoided in the WD design by the use of the ball members C, Fig. 37.11, which allow the axles to tilt without producing any corresponding tilt of the boxes A. Twisting of the springs is avoided in the suspension shown in Fig. 37.2(b) by using the connection shown in Fig. 37.13, which is self-explanatory. A design, used by the Kirkstall Forge Company, which relieves the springs of longitudinal twisting and of side forces, is shown in Fig. 37.14. It comprises a ball-and-socket joint, the ball of which is capable of sliding along the pin on which it is mounted. Hardened steel pads prevent damage to the joint when unduly large movements occur.

In the Thornycroft design the same result is achieved by using gimbals A, Fig. 37.16, pivoted to the axle casings on trunnions whose axes are fore-and-aft. The spring ends are pin-jointed to the gimbals. This gives the same freedom as in the WD design while permitting the torque reactions to be transmitted through the springs to the frame. There is another action which must be avoided however. When the axle tilts, the horizontal distance between the points of connection of the springs becomes smaller, while the horizontal distance between the points of connection of the springs and frame remains unaltered. There is consequently a side bending action which in cross-country vehicles would be serious. It is avoided in the WD design by leaving the balls C free to slide outwards along the axle tubes and in the Thornycroft design by

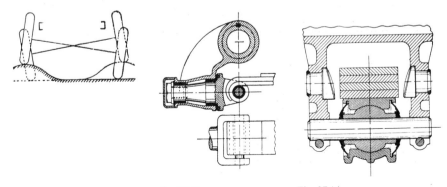

Fig. 37.12   Fig. 37.13   Fig. 37.14

leaving the springs free to slide inwards along their central pivots and outwards along the pins connecting them to the gimbals. To avoid bending and twisting the torque rods themselves they are sometimes connected to the axle and frame by ball-and-socket joints, but sometimes universal joints are used. In the latter case the rods must be made in two portions free to turn relatively about their longitudinal axis, the joint being made capable of transmitting tension or compression. Some freedom from twisting of the springs is obtained by using swivelling boxes between the spring eyes and the frame brackets or shackles, but while these are satisfactory for vehicles operating on roads it is doubtful whether they can cope with the large movements experienced in cross-country work.

37.7   Scammell articulated trailer

It has been seen that in the Scammell *Constructor* chassis the springs have been relieved of all stresses, except those due to the weight of the body. In the four-wheeled trailer portion of one of their articulated vehicles, Scammell have used the springs to take the braking forces and, in conjunction with torque rods, the brake torque reactions, but the springs are relieved of the sideways forces by transverse radius rods. Undesirable stresses are minimised by employing the connections shown in Fig. 37.16 between the axles and the springs. The end of the spring is provided with a ball A fitting in a socket B which is made in two parts for assembly and which is free to slide sideways, that is parallel to the axle, in the housing C. The latter is bolted to the bracket D of the axle. Lubricating nipples are provided at E. This arrangement enables the springs to take longitudinal and vertical forces, but relieves them of other actions.

The general arrangement of the suspension is shown in Fig. 37.17. The springs B are bolted to seats, which are pivoted on pins C carried by brackets. These brackets are secured to the frame members of the trailer and are braced by the cross member D. The braking forces and torque reactions are transferred to the frame by longitudinal forces in the springs B and torque rods E, and side forces are taken by the transverse radius rods F, which are pivoted at one end to the axles and at the other end to the spring-pivot brackets. Both the rods E and F are provided with rubber bushes at the pivots, and these

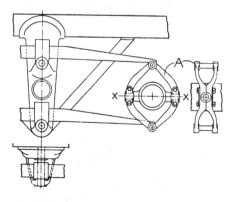

Fig. 37.15

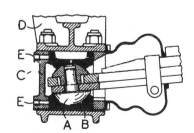

Fig. 37.16

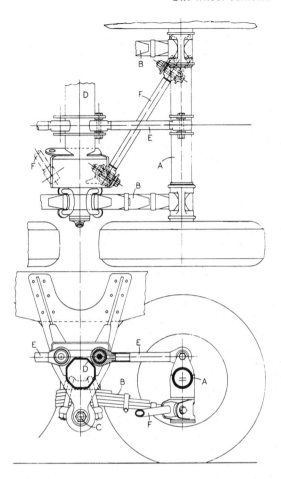

Fig. 37.17

accommodate the very small difference in the motion of the axle due to the flexing of the springs and that due to the constraint of the radius rod; they also relieve the rods E of undesirable bending stresses.

37.8   Scammell *Routeman*

Many of the principles described in this book are exemplified in the Scammell *Routeman* and *Constructor 8* tandem drive bogie, Fig. 37.18. The drive is taken by the propeller shaft to a four-pinion, lockable, inter-axle differential unit in the top of the final-drive casing of the front axle, and then split two ways: one is through a pair of spur gears down to the final drive pinion of the front axle, and the other straight through to the rear, through a short propeller shaft, to the trailing axle. Thus, the drive is normally shared – by the inter-axle differential – between the two axles. However, to stop the wheels on one axle from spinning, in icy conditions or on soft ground, the inter-axle differential can be locked by means of an air-actuated dog clutch on the leading axle. This clutch of course

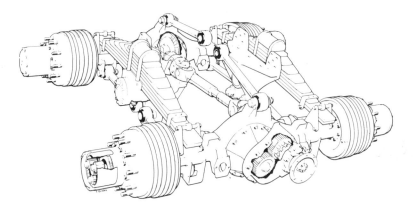

Fig. 37.18 Scammell *Routeman 3* tandem-drive bogie

must be disengaged as soon as the vehicle regains hard ground, to avoid transmission wind-up in the short propeller shaft interconnecting the two axles.

These axles are of the double-reduction type. The first reduction is effected by the crown-wheel-and-pinion and the second by a planetary gearset at the outer end of each halfshaft. Fore-and-aft location of each axle is effected by two radius rods at the bottom and one at the top. Those on top are pivoted at one end on top of each final-drive casing and their other ends are pivoted on a transverse member of the frame, above the pivot mountings for the two road springs. Each of the four lower rods, beneath the springs, is pivoted at one end under the axle tube and at the other on the bracket that carries the pivot mounting for the spring above it. Thus all the connections to the frame are grouped compactly around the transverse member forming the main support for the bogie. These rods of course also react drive and brake torques.

Lateral location is effected by the road springs. The slipper ends of these springs seat between lugs on top of the axle tubes, and their centres, of course, are clamped by large U-bolts to the seating pads on top of their pivot, or trunnion, bearings. Relatively narrow pads bear on the tops of the ends of the springs, to hold them down on their seats between the lugs on the axle tubes under rebound conditions. These look like rollers but, in fact, are not free to roll. Because they are narrow, they allow the springs freedom to articulate sufficiently to accommodate vertical motions of the wheels on one side relative to those on the other.

BELFAST
INSTITUTE
MILLFIELD
LIBRARY

Index

97/HH9/3
£25·00